CHECK FOR A DISK IN BACK
OF BOOK

TA 169 .E52 1996

Elsayed, Elsayed A.

Reliability engineering

WITHDRAWN

Reliability Engineering

12/2007

Reliability Engineering

Elsayed A. Elsayed

ADDISON WESLEY LONGMAN, INC.

Reading, Massachusetts · Menlo Park, California · New York · Don Mills, Ontario
Wokingham, England · Amsterdam · Bonn · Sydney · Singapore
Tokyo · Madrid · San Juan · Seoul · Milan · Mexico City · Taipei

Many of the designations used by manufacturers and sellers to distinguish their products are claimed as trademarks. Where those designations appear in this book and Addison Wesley Longman, Inc. was aware of a trademark claim, the designations have been printed with initial capital letters.

The programs and applications presented in this book have been included for their instructional value. They have been tested with care but are not guaranteed for any particular purpose. The publisher does not offer any warranties or representations, nor does it accept any liabilities with respect to the programs or applications.

The publisher offers discounts on this book when ordered in quantity for special sales.

For more information, please contact:

Corporate & Professional Publishing Group
Addison Wesley Longman, Inc.
One Jacob Way
Reading, Massachusetts 01867

Library of Congress Cataloging-in-Publication Data

Elsayed, Elsayed A.
Reliability engineering / Elsayed A. Elsayed.
p. cm.
Includes bibliographical references and index.
ISBN 0-201-63481-3 (alk. paper)
1. Reliability (Engineering) I. Title.
TA169.E52 1996
620'.00452—dc20

96-12121
CIP

TA 169 .E52 1996
Elsayed, Elsayed A.

Reliability engineering

Copyright © 1996 By Addison Wesley Longman, Inc.

All rights reserved. No part of this publication may be reproduced, stored in a retrieval system, or transmitted, in any form or by any means, electronic, mechanical, photocopying, recording, or otherwise, without the prior written permission of the publisher. Printed in the United States of America. Published simultaneously in Canada.

Text design by Wilson Graphics & Design (Kenneth J. Wilson)
Text printed on recycled and acid-free paper

ISBN 0-201-63481-3
1 2 3 4 5 6 7 8 9-MA-99989796
First printing, June 1996

Contents

Preface xv
Prelude xix

PART I Reliability and System Design 1

1 Reliability and Hazard Functions 3
1.1. Introduction 3
1.2. Reliability Definition and Estimation 4
1.3. Hazard Functions 14
1.3.1. Constant Hazard 15
1.3.2. Linearly Increasing Hazard 18
1.3.3. Linearly Decreasing Hazard 20
1.3.4. Weibull Model 20
1.3.5. Mixed Weibull Model 24
1.3.6. Exponential Model (the Extreme Value Distribution) 26
1.3.7. Normal Model 27
1.3.8. Lognormal Model 31
1.3.9. Gamma Model 35
1.3.10. Log-Logistic Model 40
1.3.11. Beta Model 40
1.3.12. Other Forms 43
1.4. Multivariate Hazard Rates 45
1.5. Mixture of Failure Rates 49
1.6. Mean Time to Failure (MTTF) 52
1.7. Mean Residual Life (MRL) 55
Problems 57
References 67

2 System Reliability Evaluation 69
2.1. Introduction 69
2.2. Reliability Block Diagrams 70

2.3. Series Systems 73
2.4. Parallel Systems 75
2.5. Parallel-Series, Series-Parallel, and Mixed-Parallel Systems 77
2.5.1. Parallel-Series 77
2.5.2. Series-Parallel 78
2.5.3. Mixed-Parallel 79
2.5.4. Optimal Assignments of Units 80
2.6. Consecutive-k-out-of-n:F System 83
2.6.1. Consecutive-2-out-of-n:F System 84
2.6.2. Generalization of the Consecutive-k-out-of-n:F Systems 86
2.6.3. Reliability Estimation of the Consecutive-k-out-of-n:F Systems 88
2.6.4. Optimal Arrangement of Components in Consecutive-2-out-of-n:F Systems 91
2.7. Reliability of k-out-of-n Systems 93
2.8. Complex Reliability Systems 96
2.8.1. Decomposition Method 96
2.8.2. Tie-Set and Cut-Set Methods 100
2.8.3. Event-Space Method 103
2.8.4. Boolean Truth Table Method 104
2.8.5. Reduction Method 106
2.8.6. Path-Tracing Method 107
2.8.7. Factoring Algorithm 108
2.9. Multistate Models 112
2.9.1. Series Systems 112
2.9.2. Parallel Systems 114
2.9.3. Parallel-Series and Series-Parallel 116
2.10. Redundancy 118
2.10.1. Redundancy Allocation for a Series System 120
2.11. Importance Measures of Components 122
2.11.1. Birnbaum's Importance Measure 123
2.11.2. Criticality Importance 126
2.11.3. Fussell-Vesely Importance 129
2.11.4. Barlow-Proschan Importance 131
2.11.5. Upgrading Function 131
Problems 135
References 147

3 Time- and Failure-Dependent Reliability 151

3.1. Nonrepairable Systems 151
3.1.1. Series Systems 152
3.1.2. Parallel Systems 153
3.1.3. *k*-out-of-*n* Systems 157
3.2. Mean Time to Failure (MTTF) 159
3.2.1. MTTF for Series Systems 160
3.2.2. MTTF for Parallel Systems 162
3.2.3. *k*-out-of-*n* Systems 165
3.2.4. Other Systems 167
3.3. Repairable Systems 169
3.3.1. Alternating Renewal Process 170
3.3.2. Markov Models 175
3.4. Availability 180
3.4.1. Instantaneous (Point) Availability, $A(t)$ 181
3.4.2. Average Uptime Availability, $A(T)$ 181
3.4.3. Steady-State Availability, $A(\infty)$ 183
3.4.4. Inherent Availability, A_i 184
3.4.5. Achieved Availability, A_a 184
3.4.6. Operational Availability, A_o 184
3.4.7. Other Availabilities 185
3.5. Dependent Failures 189
3.5.1. Markov Model for Dependent Failures 189
3.5.2. The Joint Density Function Approach 191
3.6. Redundancy and Standby 195
3.6.1. Nonrepairable Simple Standby Systems 197
3.6.2. Nonrepairable Multiunit Standby Systems 198
3.6.3. Repairable Standby Systems 201
Problems 206
References 216

PART II Parameter Estimation and Reliability Testing 219

4 Estimation Methods of the Parameters of Failure-Time Distributions 221

4.1. Method of Moments 222
4.1.1. Confidence Intervals 228

4.2. The Likelihood Function 230
4.2.1. The Method of Maximum Likelihood 236
4.2.2. Exponential Distribution 237
4.2.3. The Rayleigh Distribution 239
4.2.4. The Normal Distribution 240
4.2.5. Information Matrix and the Variance-Covariance Matrix 244
4.3. Method of Least Squares 247
Problems 251
References 256

5 Parametric Reliability Models 259

5.1. Approach 1: Historical Data 259
5.2. Approach 2: Operational Life Testing 260
5.3. Approach 3: Burn-In Testing 261
5.4. Approach 4: Accelerated Life Testing 261
5.5. Types of Censoring 263
5.5.1. Type 1 Censoring 263
5.5.2. Type 2 Censoring 263
5.5.3. Random Censoring 263
5.5.4. Hazard Rate Calculations Under Censoring 264
5.6. The Exponential Distribution 265
5.6.1. Testing for Abnormally Short Failure Times 268
5.6.2. Testing for Abnormally Long Failure Times 271
5.6.3. Data with Type 1 Censoring 275
5.6.4. Data with Type 2 Censoring 278
5.7. The Rayleigh Distribution 280
5.7.1. Estimation of Rayleigh's Parameter for Data Without Censored Observations 280
5.7.2. Estimation of Rayleigh's Parameter for Data with Censored Observations 282
5.7.3. Best Linear Unbiased Estimate for Rayleigh's Parameter for Data with and Without Censored Observations 283
5.8. The Weibull Distribution 289
5.8.1. Failure Data Without Censoring 290
5.8.2. Failure Data with Censoring 295
5.8.3. Unbiased Estimates of the Weibull Parameters 296
5.8.4. Variance of the Estimates 296

5.8.5. Unbiased Estimates of $\hat{\gamma}$ 297
5.8.6. Confidence Interval for $\hat{\gamma}$ 300
5.8.7. Inferences on $\hat{\theta}_1$ 301
5.9. Lognormal Distribution 303
5.9.1. Failure Data Without Censoring 305
5.9.2. Failure Data with Censoring 309
5.10. The Gamma Distribution 312
5.10.1. Failure Data Without Censoring 312
5.10.2. Failure Data with Censoring 315
5.10.3. Variance of $\hat{\gamma}$ and $\hat{\theta}$ 317
5.10.4. Confidence Intervals for γ 319
5.11. The Extreme Value Distribution 320
5.12. The Half-Logistic Distribution 323
5.13. Linear Models 331
5.14. Multicensored Data 333
5.14.1. Product-Limit Estimator (PLE) 333
5.14.2. Cumulative-Hazard Estimator (CHE) 334
Problems 337
References 349

6 Models for Accelerated Life Testing 353

6.1. Accelerated Testing 353
6.2. Accelerated Failure Data Models 354
6.3. Statistics-Based Models: Parametric 355
6.3.1. Exponential Distribution Acceleration Model 356
6.3.2. Weibull Distribution Acceleration Model 359
6.3.3. Rayleigh Distribution Acceleration Model 361
6.3.4. Lognormal Distribution Acceleration Model 363
6.4. Statistics-Based Models: Nonparametric 368
6.4.1. The Linear Model 368
6.4.2. Proportional Hazards Model 371
6.5. Physics-Statistics-Based Models 378
6.5.1. The Arrhenius Model 378
6.5.2. The Eyring Model 380
6.5.3. The Inverse Power Rule Model 384
6.5.4. Combination Model 387

6.6. Physics-Experimental-Based Models 388
6.6.1. Electromigration Model 389
6.6.2. Humidity Dependence Failures 389
6.6.3. Fatigue Failures 391
6.7. Degradation Models 392
6.7.1. Resistor Degradation Model 392
6.7.2. Laser Degradation 394
6.7.3. Hot-Carrier Degradation 395
6.8. Accelerated Life Testing Plans 396
Problems 398
References 409

PART III **Reliability Improvement: Warranty and Preventive Maintenance** 413

7 Renewal Processes and Expected Number of Failures 415

7.1. Parametric Renewal Function Estimation 416
7.1.1. Continuous Time 416
7.1.2. Discrete Time 428
7.2. Nonparametric Renewal Function Estimation 432
7.2.1. Continuous Time 432
7.2.2. Discrete Time 438
7.3. Alternating Renewal Process 443
7.3.1. Expected Number of Failures in an Alternating Renewal Process 444
7.3.2. Probability That Type j Component Is in Use at Time t 445
7.4. Approximations of $M(t)$ 446
7.5. Other Types of Renewal Processes 448
7.6. The Variance of the Number of Renewals 449
7.7. Confidence Intervals for the Renewal Function 457
7.8. Remaining Life at Time t 459
7.9. Poisson Processes 462
7.9.1. Homogeneous Poisson Process (HPP) 462
7.9.2. Nonhomogeneous Poisson Process (NHPP) 463
Problems 466
References 474

8 Warranty Models 475

8.1. Introduction 475
8.2. Warranty Models for Nonrepairable Products 477
8.2.1. Warranty Costs for Nonrepairable Products 477
8.2.2. Warranty Reserve Fund: Lump-Sum Rebate 481
8.2.3. Mixed Warranty Policies 486
8.2.4. Optimal Replacements for Items Under Warranty 492
8.3. Warranty Models for Repairable Products 497
8.3.1. Warranty Cost for Repairable Products 497
8.3.2. Warranty Models for a Fixed Lot Size: Arbitrary Failure Time Distribution 501
8.3.3. Warranty Models for a Fixed Lot Size: Minimal Repair Policy 503
8.3.4. Warranty Models for a Fixed Lot Size: Good-as-New Repair Policy 504
8.3.5. Warranty Models for a Fixed Lot Size: Mixed Repair Policy 507
8.4. Warranty Claims 513
8.4.1. Warranty Claims with Lag Times 514
8.4.2. Warranty Claims for Grouped Data 519
Problems 520
References 525

9 Preventive Maintenance and Inspection 527

9.1. Preventive Maintenance and Replacement Models: Cost Minimization 528
9.1.1. The Constant Interval Replacement Policy (CIRP) 529
9.1.2. Replacement at Predetermined Age 533
9.2. Preventive Maintenance and Replacement Models: Downtime Minimization 537
9.2.1. The Constant Interval Replacement Policy (CIRP) 537
9.2.2. Preventive Replacement at Predetermined Age 538
9.3. Minimal Repair Models 540
9.3.1. Optimal Replacement Under Minimal Repair 541
9.4. Optimum Replacement Intervals for Systems Subject to Shocks 545
9.4.1. Periodic Replacement Policy: Time-Independent Cost 546
9.4.2. Periodic Replacement Policy: Time-Dependent Cost 548

9.5. Preventive Maintenance and Number of Spares 549
9.5.1. Number of Spares and Availability 554
9.6. Group Maintenance 557
9.7. Periodic Inspection 561
9.7.1. An Optimum Inspection Policy 561
9.7.2. Periodic Inspection and Maintenance 565
9.8. On-Line Surveillance and Monitoring 570
9.8.1. Vibration Analysis 570
9.8.2. Acoustic Emission and Sound Recognition 571
9.8.3. Temperature Monitoring 572
9.8.4. Fluid Monitoring 572
9.8.5. Corrosion Monitoring 573
9.8.6. Other Diagnostic Methods 573
Problems 574
References 581

10 Case Studies 585

Case 1: A Crane Spreader Subsystem 585
Case 2: Design of a Production Line 592
Case 3: An Explosive Detection System 601
Case 4: Reliability of Furnace Tubes 609
Case 5: Reliability Modeling of Telecommunication Networks for the Air Traffic Control System 615
Case 6: System Design Using Reliability Objectives 625
References 636

Appendixes 639

Appendix A Gamma Table 641
Appendix B Coefficients of b_i's for $i = 1, \ldots, n^*$ 647
Appendix C Variance of θ_2^*'s in Terms of θ_2^2/n and K_3/K_2^* 661
Appendix D Coefficients (a_i and b_i) of the Best Estimates of the Mean (μ) and Standard Deviation (σ) in Censored Samples Up to $n = 20$ from a Normal Population 665
Appendix E Standard Normal Distribution 679
Appendix F Computer Program to Calculate the Reliability of a Consecutive-k-out-of-n:F System 685

Appendix G Optimum Arrangement of Components in Consecutive-2-out-of-n:F Systems 687

Appendix H Computer Program for Solving the Time-Dependent Equations Using Runge-Kutta's Method 695

Appendix I The Newton-Raphson Method 697

Appendix J Computer Listing of the Newton-Raphson Method 703

Appendix K Baker's Algorithm 705

Appendix L Critical Values of χ^2 709

Appendix M Solutions of Selected Problems 713

Index 725

Preface

Reliability is one of the most important quality characteristics of components, products, and large and complex systems. The role of reliability is observed daily by all of us—when we turn the ignition key of a vehicle, attempt to place a phone call, or try to use a copier, computer, or fax machine. In all these instances, the user expects that the machine will provide the function it is designed for when the function is requested. As you probably have experienced, machines do not always function or deliver the desired quality of service when needed.

Engineers spend a significant amount of time and resources during the design and production phases of the product life cycle to ensure that the product or system will provide the desired service level. In doing so, engineers start with a conceptual design, select its components, test its functionality, and estimate its reliability. Then they usually make modifications and design changes and repeat these steps until the product (or service) satisfies its requirements.

Designing the product may require redundancy of components (or subsystems) or introduction of newly developed components or changes in the design configuration. These will have a major impact on the product reliability.

This book is an *engineering* reliability book. It is organized according to the sequence followed when engineers design a product or service. The book consists of three parts. Part I focuses on system reliability estimation for time-independent and time-dependent models. Chapter 1 focuses on the basic definitions of reliability, its measures, and methods for its calculation. Extensive coverage of different hazard functions is provided. Chapter 2 describes, in greater detail, methods for estimating reliabilities of a variety of engineering systems configurations starting with series systems, parallel systems, series-parallel, parallel-series, consecutive k-out-of-n:F, and complex network systems. It also addresses systems with multi-state devices and concludes by estimating reliabilities of redundant systems and the optimal allocation of components in a redundant system. The next step in product design is to study the effect of time on the system reliability. Hence, Chapter 3 discusses, in detail, time- and failure-dependent reliability and the calculation of mean time to failure (MTTF) of a variety of system configurations. The chapter also introduces availability as a measure of system reliability.

Once the design is "firm," the engineer assembles the components and configures them to achieve the desired reliability objectives. This may require conducting reliability tests on components or using field data from similar components. Part II of this book, starting with Chapter 4, presents the concept of constructing the likelihood function and its use in estimating the parameters of failure-time distributions. Chapter 5 provides a comprehensive coverage of parametric and nonparametric reliability models for failure data. The extensive examples and methodologies presented in this chapter will aid the engineer in appropriately modeling the test data. Confidence intervals for the parameters of the models are also discussed. More important, the book devotes a full chapter, Chapter 6, to accelerated life testing. The main objective of this chapter is to provide varieties of statistical-based models, physics-statistics-based models, and physics-experimental-based models to relate the failure time and data at accelerated conditions to the normal operating conditions at which the product is expected to operate. The computer software *Reliability Analysis Software*™ that accompanies this book provides useful tools for reliability estimation, failure-time distributions, and a wide range of accelerated life models as described in Chapters 5 and 6.

Finally, once a product is produced and sold, the manufacturer must ensure its reliability objectives by providing preventive and scheduled maintenance and warranty policies. Part III of the book focuses on these topics. Beginning with Chapter 7, different methods (exact and approximate) for estimating the expected number of system failures during a specified time interval are presented. These estimates are used in Chapter 8 to determine different warranty policies for the product, including the length of warranty and its reserve fund. Finally, Chapter 9 discusses optimal preventive maintenance schedules, optimum inspection policies, and methods for estimating the inventory levels of spares that are required to ensure predetermined reliability values.

Chapter 10 concludes the book. Its case studies use the approaches and methodologies discussed throughout the book to demonstrate how to solve real-life cases. The role of reliability during the design phase of a product or a system is particularly emphasized.

Two features contribute to the usefulness of this book: every theoretical development discussed in this book is followed by an engineering example that illustrates its application, and several problems are included at the end of each chapter. These features increase the usefulness of the book as both a comprehensive reference for practitioners and professionals in the quality and reliability engineering area and also as a text for a one- or two-semester course on reliability engineering for senior undergraduates or graduate students in industrial and systems, mechanical, and electrical engineering programs. In addition, it can be adapted for use in a life data analysis course offered in many graduate programs in statistics.

The book presumes a background in statistics and probability theory and differential calculus.

Acknowledgments

This book represents the work of many authors who are referenced throughout. I have tried to give adequate credit to all whose work has influenced this book. Particular acknowledgment is given to the Institute of Electrical and Electronic Engineers, CRC Press, Institute of Mathematical Statistics, American Society of Mechanical Engineers, Siemens AG, Electronic Products, and Elsevier Applied Science Publishers for the use of figures, tables in the appendixes, and permission to include previously published material in this book.

Special thanks to Jai-Hyun Byun of Gyeongsang National University, Korea, for his tireless effort in reading several drafts of this manuscript. I also wish to acknowledge the feedback from Hoang Pham of Rutgers University; Mike Tortorella of AT&T Bell Laboratories; Melike Gursoy of Rutgers University; Yesim Erke of Lehigh University; Jose L. Ribeiro of the Universidade Federal do Rio Grande do Sul, Brazil, who provided extensive help for Chapters 5 and 6 and developed the software (*Reliability Analysis Software*™) that accompanies this book; Khatab Hassanein of the University of Kansas Medical Center, who developed an approach to estimate the parameters of Rayleigh distribution and also developed Appendixes B and C; N. Balakrishnan of McMaster University, who provided input about the log logistic distribution; William Mayo of Rutgers University for helping with Case Study 3; four anonymous reviewers; L. Lamberson of Western Michigan University; Christos Alexopoulas and David Goldsman of Georgia Institute of Technology; Nick Zaino and Skip Creveling of Eastman Kodak; and my colleague K. Kapur of the University of Washington, who provided me with excellent comments that shaped this book into its present form.

I would like to thank the students of the Department of Industrial Engineering at Rutgers University who used earlier and much shorter versions of this book over the last ten years and provided me with valuable input. Thanks are extended to Suzanne Tobias, who tirelessly worked many hours in typing and retyping this manuscript and attending to details, and to Cindy Ielmini and Doris Clark for their help in typing revisions of this book. I am also indebted to Joe Lippencott and Aladdin Elsayed, who provided great help in computer programming and creating some of the figures.

Thanks to my enthusiastic and helpful editor, Jennifer Joss, and to many others at Addison Wesley Longman, Inc. whose efforts have greatly benefited this book, including Tara Herries and Rosa Aimée González.

Special thanks are reserved for my wife, Linda, who spent many late hours carefully editing this manuscript, and my children, for their patience and understanding.

E. A. Elsayed
Piscataway, New Jersey

Prelude

THE DEACON'S MASTERPIECE,

or The Wonderful One-Hoss Shay[1]

A *Design for Reliability* Logical Story

"The Deacon's Masterpiece, or the Wonderful One-Hoss Shay" is a perfectly logical story that demonstrates the concept of designing a product for reliability. It starts by defining the objective of the product or service to be provided. The reliability structure of the system is then developed and its components and subsystem are selected. A prototype is constructed and tested. The failure data of the components are collected and analyzed. The system is then redesigned and retested until its reliability objectives are achieved. These logical steps are elegantly described below.

I. System's Objective and Structure

Have you heard of the wonderful one-hoss shay,
It ran a hundred years to a day,
And then, of a sudden, it—ah, but stay,
I'll tell you what happened without delay,
Scaring the parson into fits,
Frightening people out of their wits,—
Have you ever heard of that, I say?

Seventeen hundred and fifty-five.
Georgius Secundus was then alive,—
Snuffy old drone from the German hive.
That was the year when Lisbon-town
Saw the earth open and gulp her down,
And Braddock's army was done so brown,
Left without a scalp to its crown.
It was on the terrible Earthquake-day
That the Deacon finished the one-hoss shay.

[1] Oliver Wendell Holmes, "The Deacon's Masterpiece," in *The Complete Poetical World of Oliver Wendell Holmes*, Fourth Printing, 1968, by Houghton Mifflin Company.

II. System Prototyping and Failures of Components

Holmes' preface to the poem: Observation shows us in what point any particular mechanism is most likely to give way. In a wagon, for instance, the weak point is where the axle enters the hub or nave. When the wagon breaks down, three times out of four, I think, it is at this point that the accident occurs. The workman should see to it that this part should never give way; then find the next vulnerable place, and so on, until he arrives logically at the perfect result attained by the deacon.

Now in building of chaises, I tell you what,
There is always *somewhere* a weakest spot,—
In hub, tire, felloe, in spring or thill,
In panel, or crossbar, or floor, or sill,
In screw, bolt, thoroughbrace,—lurking still,
Find it somewhere you must and will,—
Above or below, or within or without,—
And that's the reason, beyond a doubt,
That a chaise *breaks down*, but does n't *wear out*.

But the Deacon swore (as Deacons do,
With an "I dew vum," or an "I tell *yeou*")
He would build one shay to beat the taown
'N' the keounty 'n' all the kentry raoun';
It should be so built that it *could n'* break daown:
"Fur," said the Deacon, "'t's mighty plain
Thut the weakes' place mus' stan' the strain;
'N' the way t' fix it, uz I maintain, Is only jest
T' make that place us strong uz the rest."

III. System Redesign

So the Deacon inquired of the village folk
Where he could find the strongest oak,
That could n't be split nor bent nor broke,—
That was for spokes and floor and sills;
He sent for lancewood to make the thills;
The crossbars were ash, from the straightest trees,
The panels of white-wood, that cuts like cheese,
But last like iron for things like these;
The hubs of logs from the "Settler's ellum,"—
Last of its timber,—they could n't sell 'em,
Never an axe had seen their chips,
And the wedges flew from between their lips,

Their blunt ends frizzled like celery-tips;
Step and prop-iron, bolt and screw,
Spring, tire, axle, and linchpin too,
Steel of the finest, bright and blue;
Thoroughbrace bison-skin, thick and wide;
Boot, top, dasher, from tough old hide
Found in the pit when the tanner died.
That was the way he "put her through."
"There!" said the Deacon, "naow she'll dew!"

Do! I tell you, I rather guess
She was a wonder, and nothing less!
Colts grew horses, beards turned gray,
Deacon and deaconess dropped away,
Children and grandchildren—where were they?
But there stood the stout old one-hoss shay
As fresh as on Lisbon-earthquake-day!

IV. Observation of the System During Its Operation

EIGHTEEN HUNDRED;—it came and found
The Deacon's masterpiece strong and sound.
Eighteen hundred increased by ten;—
"Hahnsum kerridge" they called it then.
Eighteen hundred and twenty came;—
Running as usual; much the same.
Thirty and forty at last arrive,
And then come fifty, and FIFTY-FIVE.

Little of all we value here
Wakes on the morn of its hundredth year
Without both feeling and looking queer.
In fact, there's nothing that keeps its youth,
So far as I know, but a tree and truth.
(This is a moral that runs at large;
Take it.—You're welcome.—No extra charge.)

FIRST OF NOVEMBER,—the Earthquake-day,—
There are traces of age in the one-hoss shay,
A general flavor of mild decay,
But nothing local, as one may say.
There could n't be,—for the Deacon's art
Had made it so like in every part

That there was n't a chance for one to start.
For the wheels were just as strong as the thills,
And the floor was just as strong as the sills,
And the panels just as strong as the floor,
And the whipple-tree neither less nor more,
And the back crossbar as strong as the fore,
And spring and axle and hub *encore*.
And yet, *as a whole*, it is past a doubt
In another hour it will be *worn out*!

V. System Reaches Its Expected Life

First of November, 'Fifty-five!
This morning the parson takes a drive.
Now, small boys, get out of the way!
Here comes the wonderful one-hoss shay,
Drawn by a rat-tailed, ewe-necked bay.
"Huddup!" said the parson.—Off went they.
The parson was working his Sunday's text,—
Had got to *fifthly*, and stopped perplexed
At what the—Moses—was coming next.

All at once the horse stood still,
Close by the meet'n'-house on the hill.
First a shiver, and then a thrill,
Then something decidedly like a spill,—
And the parson was sitting upon a rock,
At half past nine by the meet'n'-house clock,—
Just the hour of the Earthquake shock!
What do you think the parson found,
When he got up and stared around?
The poor old chaise in a heap or mound,
As if it had been to the mill and ground!
You see, of course, if you're not a dunce,
How it went to pieces all at once,—
All at once, and nothing first,—
Just as bubbles do when they burst.

End of the wonderful one-hoss shay.
Logic is logic. That's all I say.

PART I

Reliability and System Design

CHAPTER 1

Reliability and Hazard Functions

1.1. Introduction

One of the quality characteristics that consumers require from the manufacturer of products is reliability. Unfortunately, when consumers are asked what reliability means, the response is usually unclear. Some consumers may respond by stating that the product should always work properly without failure or by stating that the product should always function properly when required for use, while others will completely fail to explain what they mean by reliability.

What is reliability from your viewpoint? Take, for instance, the example of starting your car. Would you consider your car reliable if it starts immediately? Would you still consider your car reliable if it takes you two times to turn on the ignition key for the car to start? How about three times? As you can see, without quantification, it becomes more difficult to define or measure reliability. I define reliability later in this chapter, but for now, to further illustrate the importance of reliability as a field of study and research, consider the following cases.

On April 9, 1963, the *U.S.S. Thresher*, a nuclear submarine, slipped beneath the surface of the Atlantic and began a run for deep waters (1,000 feet below surface). *Thresher* exceeded its maximum test depth and imploded. Its hull collapsed, causing the death of 129 crew members and civilians. It should be noted that the *Thresher* was the most advanced submarine of its day, with a destructive power beyond that of the Navy's entire submarine force in World War II.

In 1979, a DC-10 commercial aircraft crashed killing all passengers aboard. The cause of its failure was poor maintenance procedure. Engineers had specified that engines should have been taken off before engine mounting assemblies because of the excessive weight of the engines. Apparently, those guidelines were not followed during maintenance, causing excessive stresses that cracked the engine mounts.

On December 2, 1982, a team of doctors and engineers at Salt Lake City, Utah, performed an operation to replace a human heart with a mechanical one—

the Jarvik heart. Two days later, the patient underwent further operations due to a malfunction of the valve of the mechanical heart. Here, the failure of a system directly affected a human life. In January 1990, the Food and Drug Administration stunned the medical community by recalling the world's first artificial heart because of deficiencies in manufacturing quality, training, and other areas. This heart affected the lives of 157 patients over an eight-year period. Now, consider the following case, where failures of systems have a much wider effect.

On April 26, 1986, two explosions occurred at the newest of the four operating nuclear reactors at the Chernobyl site in the former USSR. It was the worst commercial disaster in the history of the nuclear industry. A total of thirty-one site workers and members of the emergency crew died as a result of the accident and about 200 people were treated for symptoms of acute radiation syndrome. Economic losses were estimated at $3 billion, and the full extent of the long-term damage has yet to be determined.

The explosion of the space shuttle *Challenger* in 1986 and the loss of the two external fuel tanks of the space shuttle *Columbia* (at a cost of $25 million each) are other examples of the importance of reliability in the design, operation, and maintenance of a system.

Reliability plays an important role in the service industry. For example, to provide virtually uninterrupted communications for its customers, American Telephone and Telegraph Company (AT&T) installed the first transatlantic cable with a reliability goal of a maximum of one failure in twenty years of service. The cable surpassed the reliability goal and is being replaced by new fiber optic cables for economic reasons. The reliability goal of the new cables is one failure in 80 years of service!

A final example of the reliability role in structural design is illustrated by the Point Pleasant Bridge (West Virginia/Ohio border), which collapsed on December 15, 1967, causing the death of forty-six persons and the injuries of several dozen persons. The failure was attributed to the metal fatigue of a crucial eyebar, which started a chain reaction of one structural member falling after another.

Reliability also has a great effect on the consumers' perception of a manufacturer. For example, consumers' experiences with car recalls, repairs, and warranties will affect the future sales of that manufacturer.

1.2. Reliability Definition and Estimation

A formal definition of reliability is given as:

Reliability Reliability is the probability that a product or service will operate properly for a specified period of time (design life) under the design operating conditions (such as temperature or volt) without failure.

In other words, reliability may be used as a measure of the system's success in providing its function properly. Consider the following illustration.

Suppose n_o identical components are subjected to a design operating conditions test. During the interval of time $(t - \Delta t, t)$, we observed $n_f(t)$ failed components and $n_s(t)$ surviving components $[n_f(t) + n_s(t) = n_o]$. Since reliability is defined as the cumulative probability function of success, then at time t, the reliability $R(t)$ is

$$R(t) = \frac{n_s(t)}{n_s(t) + n_f(t)} = \frac{n_s(t)}{n_o}. \tag{1.1}$$

In other words, if $\mathbf{t}$ is a random variable denoting the time to failure, then the reliability function at time t can be expressed as

$$R(t) = P(\mathbf{t} > t). \tag{1.2}$$

The cumulative distribution function of failure $F(t)$ is the complement of $R(t)$—that is,

$$R(t) + F(t) = 1. \tag{1.3}$$

If the time to failure $\mathbf{t}$ has a probability density function (p.d.f.) $f(t)$, then Eq. (1.3) can be rewritten as

$$R(t) = 1 - F(t) = 1 - \int_0^t f(\zeta)\, d\zeta. \tag{1.4}$$

Taking the derivative of Eq. (1.4) with respect to t, we obtain

$$\frac{d\,R(t)}{dt} = -f(t). \tag{1.5}$$

For example, if the time to failure distribution is exponential with parameter λ, then

$$f(t) = \lambda e^{-\lambda t}, \tag{1.6}$$

and the reliability function is

$$R(t) = 1 - \int_0^t \lambda e^{-\lambda\zeta}\, d\zeta = e^{-\lambda t}. \tag{1.7}$$

From Eq. (1.7), we express the probability of failure of a component in a given interval of time $[t_1, t_2]$ in terms of its reliability function as

$$\int_{t_1}^{t_2} f(t)\,dt = R(t_1) - R(t_2). \tag{1.8}$$

We define the failure rate in a time interval $[t_1, t_2]$ as the probability that a failure per unit time occurs in the interval given that no failure has occurred prior to t_1, the beginning of the interval. Thus the failure rate is expressed as

$$\frac{R(t_1) - R(t_2)}{(t_2 - t_1)R(t_1)}. \tag{1.9}$$

If we replace t_1 by t and t_2 by $t + \Delta t$, then we rewrite Eq. (1.9) as

$$\frac{R(t) - R(t + \Delta t)}{\Delta t\, R(t)}. \tag{1.10}$$

The hazard function is defined as the limit of the failure rate as Δt approaches zero. In other words, the hazard function or the instantaneous failure rate is obtained from Eq. (1.10) as

$$h(t) = \lim_{\Delta t \to 0} \frac{R(t) - R(t + \Delta t)}{\Delta t\, R(t)} = \frac{1}{R(t)}\left[-\frac{d}{dt}R(t)\right]$$

or

$$h(t) = \frac{f(t)}{R(t)}. \tag{1.11}$$

From Eqs. (1.5) and (1.11), we obtain

$$R(t) = e^{\left[-\int_0^t h(\zeta)\,d\zeta\right]}, \tag{1.12}$$

$$R(t) = 1 - \int_0^t f(\zeta)\,d\zeta, \tag{1.13}$$

and

$$h(t) = \frac{f(t)}{R(t)}. \tag{1.14}$$

Equations (1.5), (1.12), (1.13), and (1.14) are the key equations that relate $f(t)$, $F(t)$, $R(t)$, and $h(t)$.

The following example illustrates how the hazard rate and reliability are estimated from failure data.

EXAMPLE 1.1

A manufacturer of light bulbs is interested in estimating the mean life of the bulbs. Two hundred bulbs are subjected to a reliability test. The bulbs are observed, and the failures in 1,000-hour intervals are recorded as shown in Table 1.1.

Table 1.1. Number of Failures in the Time Intervals

Time Interval (Hours)	*Failures in the Interval*
0–1,000	100
1,001–2,000	40
2,001–3,000	20
3,001–4,000	15
4,001–5,000	10
5,001–6,000	8
6,001–7,000	7
Total	200

Plot the failure density function estimated from data $f_e(t)$, the hazard rate function estimated from data $h_e(t)$, the cumulative probability function estimated from data $F_e(t)$, and the reliability function estimated from data $R_e(t)$. The subscript e refers to *estimated*. Comment on the hazard-rate function.

SOLUTION

We estimate $f_e(t)$, $h_e(t)$, $R_e(t)$, and $F_e(t)$ by using the following equations:

$$f_e(t) = \frac{n_f(t)}{n_o\,\Delta t}, \tag{1.15}$$

$$h_e(t) = \frac{n_f(t)}{n_s(t)\,\Delta t}, \tag{1.16}$$

$$R_e(t) = \frac{f_e(t)}{h_e(t)}, \tag{1.17}$$

and

$$F_e(t) = 1 - R_e(t). \tag{1.18}$$

Note that $n_s(t)$ is the number of surviving units at the beginning of the period Δt. Summaries of the calculations are shown in Tables 1.2 and 1.3. The plots are shown in Figures 1.1 and 1.2.

Table 1.2. Calculations of $f_e(t)$ and $h_e(t)$

Time Interval (Hours)	*Failure Density* $f_e(t) \times 10^{-4}$	*Hazard Rate* $h_e(t) \times 10^{-4}$
0–1,000	$\frac{100}{200 \times 10^3} = 5.0$	$\frac{100}{200 \times 10^3} = 5.0$
1,001–2,000	$\frac{40}{200 \times 10^3} = 2.0$	$\frac{40}{100 \times 10^3} = 4.0$
2,001–3,000	$\frac{20}{200 \times 10^3} = 1.0$	$\frac{20}{60 \times 10^3} = 3.33$
3,001–4,000	$\frac{15}{200 \times 10^3} = 0.75$	$\frac{15}{40 \times 10^3} = 3.75$
4,001–5,000	$\frac{10}{200 \times 10^3} = 0.5$	$\frac{10}{25 \times 10^3} = 4.0$
5,001–6,000	$\frac{8}{200 \times 10^3} = 0.4$	$\frac{8}{15 \times 10^3} = 5.3$
6,001–7,000	$\frac{7}{200 \times 10^3} = 0.35$	$\frac{7}{7 \times 10^3} = 10.0$

As shown in Figure 1.1, the hazard rate is constant until time $t = 6{,}000$ hours and then increases with t. Thus $h_e(t)$ can be expressed more or less as

$$h_e(t) = \begin{cases} \lambda_0 & 0 \leq t \leq 6{,}000 \\ \lambda_1 t & t > 6{,}000 \end{cases},$$

where λ_0 and λ_1 are constants.

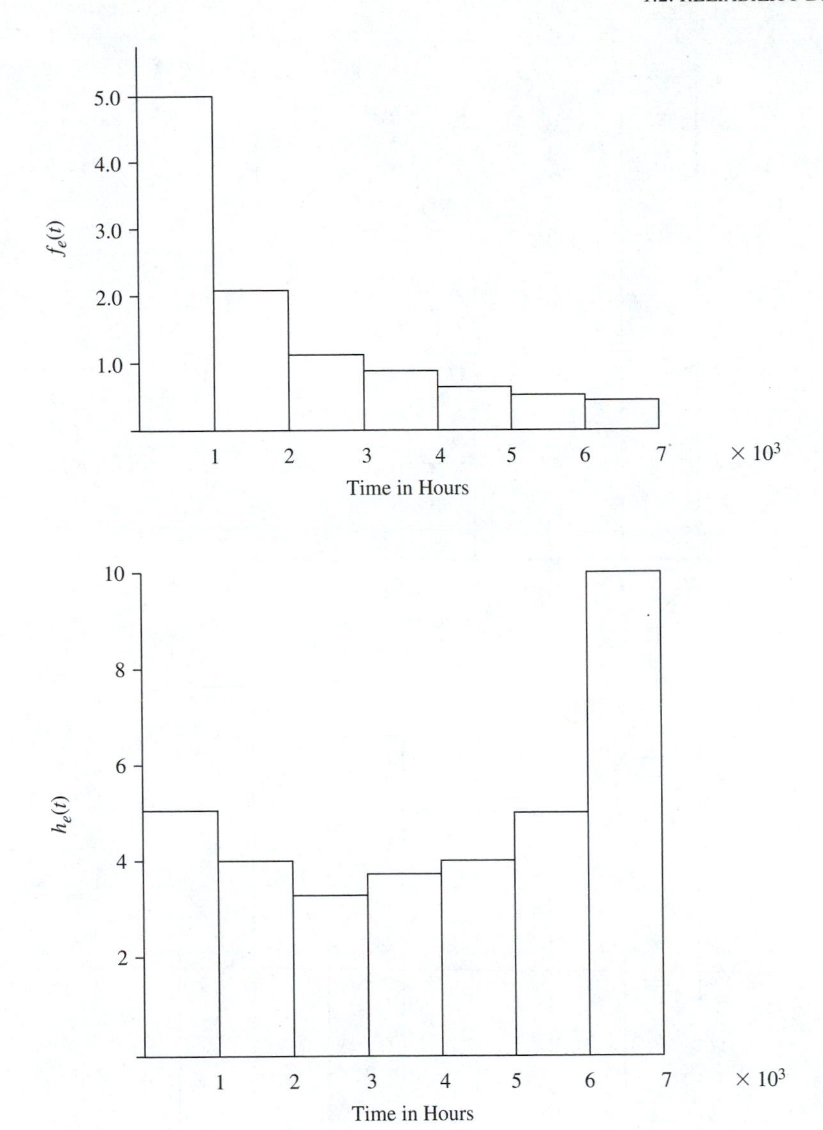

Figure 1.1. Plots of $f_e(t)$ and $h_e(t)$

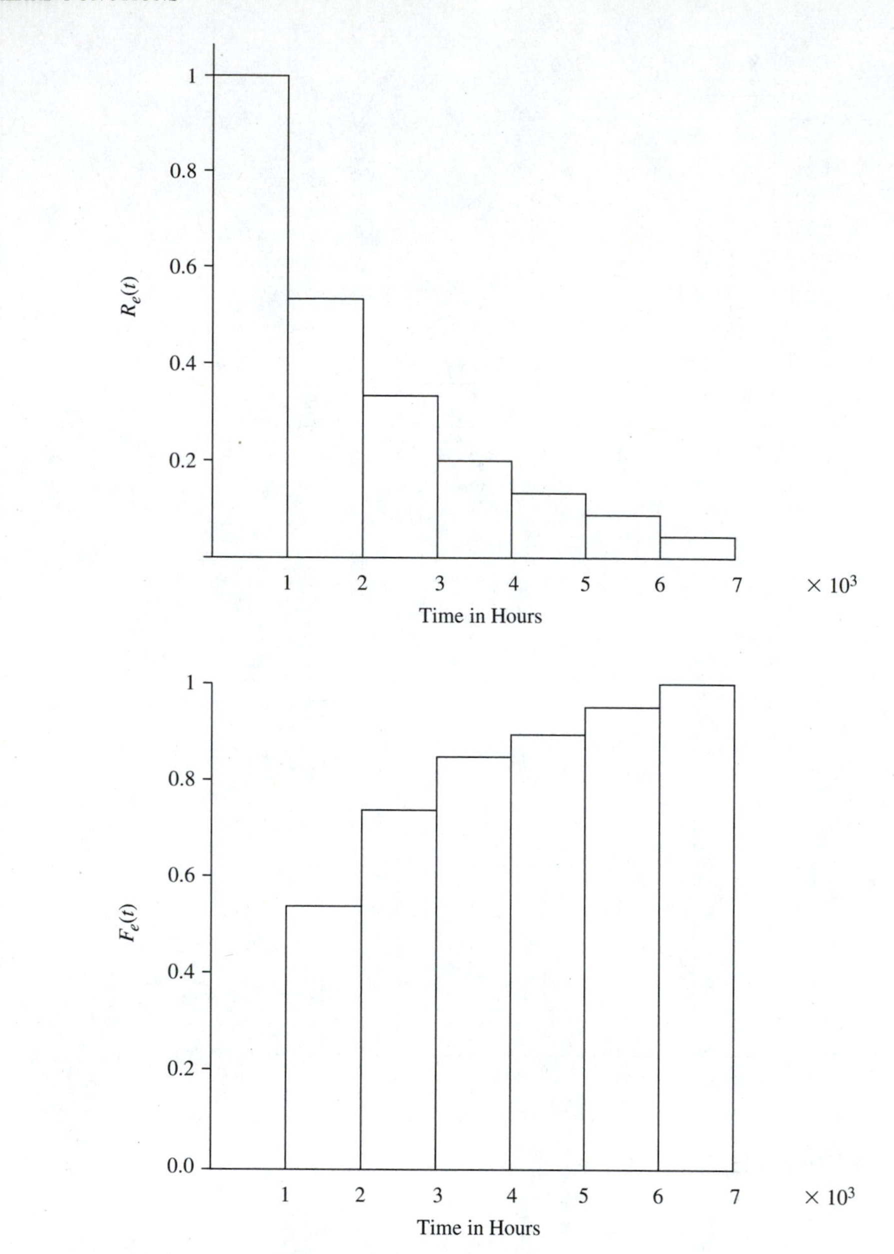

Figure 1.2. Plots of $R_e(t)$ and $F_e(t)$

Table 1.3. Calculations of $R_e(t)$ and $F_e(t)$

Time Interval	*Reliability* $R_e(t) = f_e(t)/h_e(t)$	*Unreliability* $F_e(t) = 1 - R_e(t)$
0–1,000	$\frac{5.0}{5.0} = 1.000$	0.000
1,001–2,000	$\frac{2.0}{4.0} = 0.500$	0.500
2,001–3,000	$\frac{1.0}{3.33} = 0.300$	0.700
3,001–4,000	$\frac{0.75}{3.75} = 0.200$	0.800
4,001–5,000	$\frac{0.5}{4.0} = 0.125$	0.875
5,001–6,000	$\frac{0.4}{5.3} = 0.075$	0.925
6,001–7,000	$\frac{0.35}{10.0} = 0.035$	0.965

The above example shows the hazard-rate function is constant for a period of time and then linearly increases with time. In other situations the hazard-rate function may be decreasing, constant, or increasing and the rate at which the function decreases or increases may be constant, linear, polynomial, or exponential with time. The following example is an illustration of an exponentially increasing hazard rate.

EXAMPLE 1.2

Facsimile (fax) machines are designed to transmit documents, figures, and drawings between locations via telephone lines. The principle of a fax machine is shown in Figure 1.3. The document on the sending unit drum is scanned in both the horizontal and rotating directions. The document is divided into graphic elements, which are converted into electrical signals by a photoelectric reading head. The signals are transmitted via telephone lines to the receiving end where they are demodulated and reproduced by a recording head.

The quality of the received document is affected by the reliability of the photoelectric reading head in converting the graphic elements of the document being sent into proper electrical signals. A manufacturer of fax machines performs a

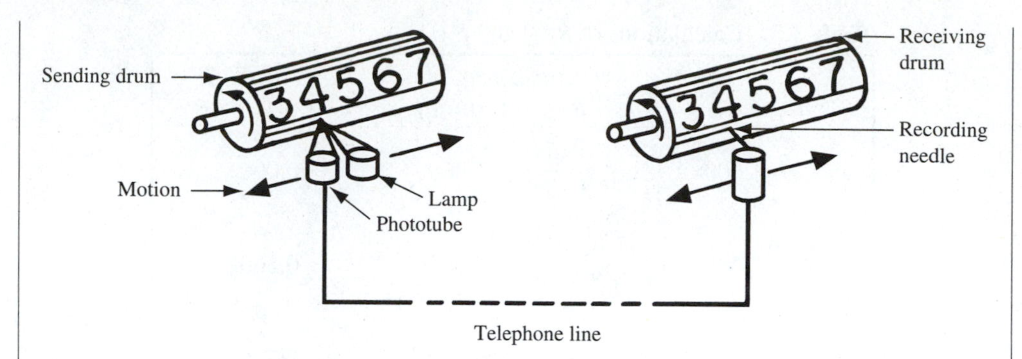

Figure 1.3. The Principle of a Fax Machine

Table 1.4. Failure Data of the Facsimile Machines

Time interval (hours)	*0–150*	*151–300*	*301–450*	*451–600*	*601–750*	*751–900*
Number of failures	20	28	27	32	33	40

reliability test to estimate the mean life of the reading head by subjecting 180 heads to repeated cycles of readings. The threshold times, at which the quality of the received document is unacceptable, are recorded in Table 1.4.

Estimate the hazard rate and reliability function of the machines.

SOLUTION

Using Eqs. (1.15), (1.16), and (1.17), we calculate $f_e(t)$, $h_e(t)$, and $R_e(t)$ as shown in Table 1.5. Plots of the hazard rate and the reliability function are shown in Figures 1.4 and 1.5, respectively.

Table 1.5. Calculations for $f_e(t)$, $h_e(t)$, and $R_e(t)$

t	$f_e(t) \times 10^{-4}$	$h_e(t) \times 10^{-4}$	$R_e(t)$
0–150	7.407	7.407	1.000
151–300	10.370	11.666	0.889
301–450	10.000	13.636	0.733
451–600	11.852	20.317	0.583
601–750	12.222	30.137	0.406
751–900	14.815	66.667	0.222

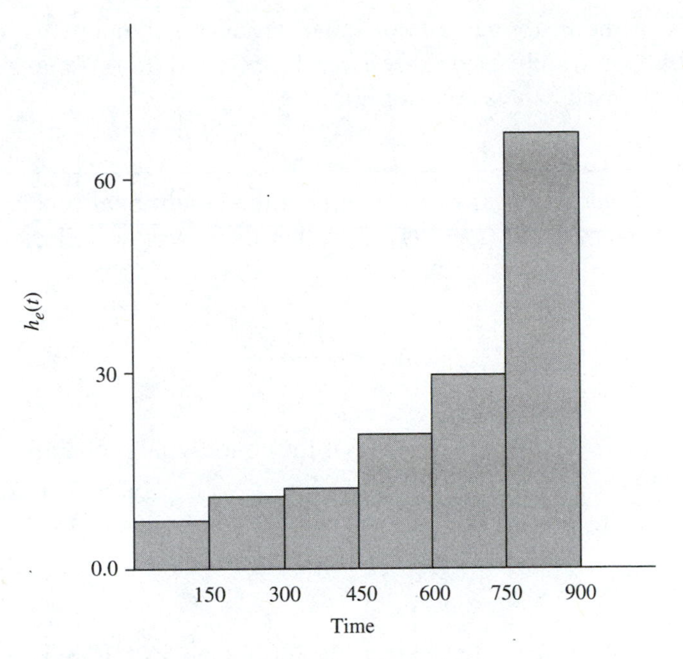

Figure 1.4. Plot of the Hazard-Rate Function Versus Time

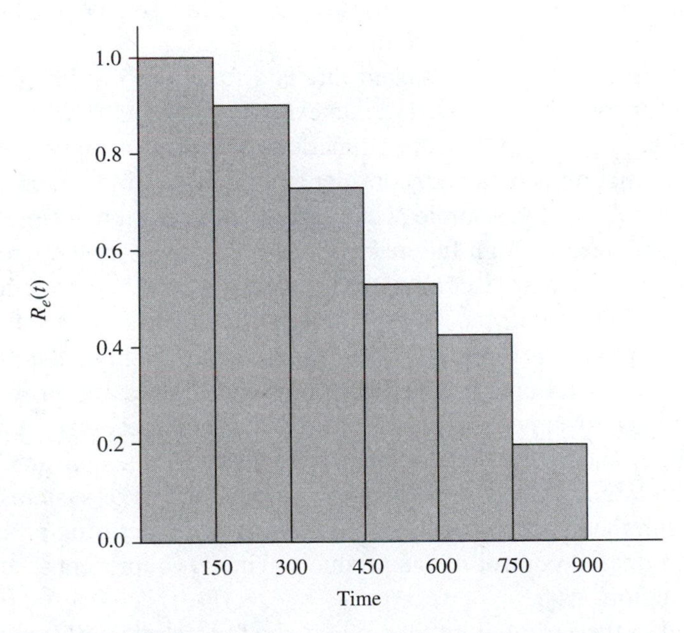

Figure 1.5. Plot of the Reliability Function Versus Time

Analysis of the historical data of failed products, components, devices, and systems resulted in widely used expressions for $h(t)$ and $R(t)$. We now consider the most commonly used expressions for $h(t)$.

1.3. Hazard Functions

The *hazard function* or *hazard rate* $h(t)$ is the conditional probability of failure in the interval t to $(t + dt)$, given that there was no failure at t. It is expressed as

$$h(t) = \frac{f(t)}{R(t)}. \tag{1.19}$$

The *cumulative hazard function* $H(t)$ is the conditional probability of failure in the interval 0 to t:

$$H(t) = \int_0^t h(\zeta)\, d\zeta. \tag{1.20}$$

The hazard rate is also referred to as the instantaneous failure rate. The hazard rate expression is of the greatest importance for system designers, engineers, and repair and maintenance groups. The expression is useful in estimating the time to failure (or time between failures), repair crew size for a given repair policy, the availability of the system, and the warranty cost. It can also be used to study the behavior of the system's failure with time.

As shown in Eq. (1.19), the hazard rate is a function of time. One may ask, What type of function does the hazard rate exhibit with time? The *general* answer to this question is the bathtub-shaped function as shown in Figure 1.6. To illustrate how this function is obtained, consider a population of identical components from which we take a large sample N and place it in operation at time $T = 0$. The sample will experience a high failure rate at the beginning of the operation time due to weak or substandard components, manufacturing imperfections, design errors, and installation defects. This period of decreasing failure rate is referred to as the "infant mortality region," the "shake-down" region, the "debugging" region, or the "early failure" region. This is an undesirable region from both the manufacturer and consumer viewpoints as it causes an unnecessary repair cost for the manufacturer and an interruption of product usage for the consumer. The early failures can be minimized by increasing the burn-in period of systems or components before shipments are made, by improving the manufacturing process, and by improving the quality control of the products. Time T_1 represents the end of the early failure region.

At the end of the early failure-rate region, the failure rate will eventually reach a constant value. During the constant failure-rate region (between T_1 and T_2), the failures do not follow a predictable pattern but occur at random due to the changes

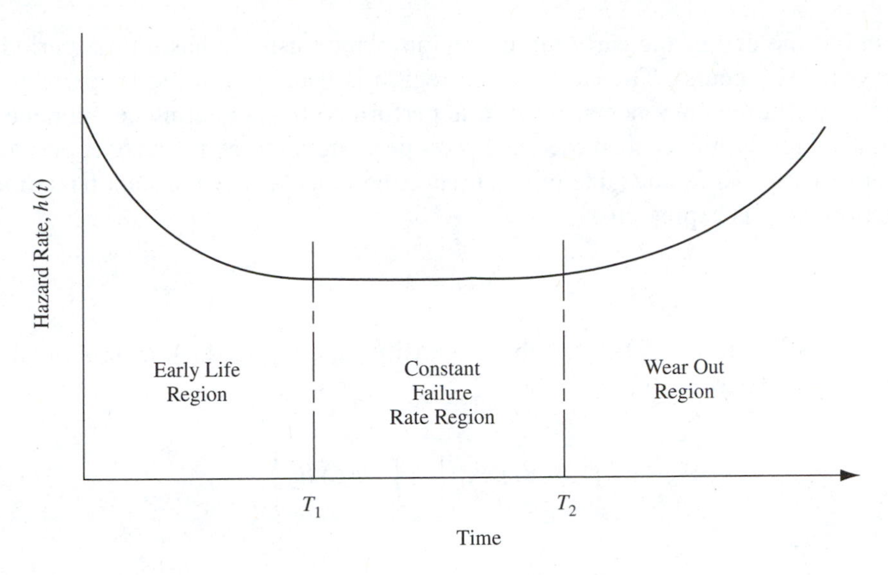

Figure 1.6. The General Failure Curve

in the applied load (the load may be higher or lower than the designed load). A higher load may cause overstressing of the component, while a lower load may cause derating (application of a load in the reverse direction of what the component experiences under normal operating conditions) of the component, and both will lead to failures. The randomness of the material flaws or manufacturing flaws will also lead to failures during the constant failure-rate region.

The third and final region of the failure-rate curve is the wear-out region, which starts at T_2. The beginning of the wear-out region is noticed when the failure rate starts to increase significantly more than the constant failure-rate value, and the failures are no longer attributed to randomness but are due to the age and wear of the components. Within this region, the failure rate increases rapidly as the product reaches its useful (designed) life. To minimize the effect of the wear-out region, one must use periodic preventive maintenance or consider replacement of the product.

Obviously, not all components exhibit the bathtub-shaped failure-rate curve. Most electronic and electrical components do not exhibit a wear-out region. Some mechanical components may not show a constant failure rate region but may exhibit a gradual transition between the early failure rate and wear-out regions. The length of each region may also vary from one component (or product) to another. Other forms of the bathtub curve are presented later in this chapter.

1.3.1. CONSTANT HAZARD

Many electronic components—such as transistors, resistors, integrated circuits, and capacitors—exhibit constant failure rate during their lifetimes. Of course, this occurs at the end of the early failure region, which usually has a time period of one year (10^4 hours). The early failure region is usually reduced by performing burn-in of these components. Burn-in is performed by subjecting components to stresses slightly higher than the expected operating stresses for a short period in order to weed out failures due to manufacturing defects. The constant hazard rate function, $h(t)$, is expressed as

$$h(t) = \lambda, \tag{1.21}$$

where λ is a constant. The probability density function (p.d.f.), $f(t)$, is obtained from Eq. (1.19) as

$$f(t) = h(t) \exp\left[-\int_o^t h(\zeta)\, d\zeta\right] \tag{1.22}$$

or

$$f(t) = \lambda e^{-\lambda t} \tag{1.23}$$

and

$$F(t) = \int_0^t \lambda e^{-\lambda\zeta}\, d\zeta = 1 - e^{-\lambda t}. \tag{1.24}$$

The reliability function, $R(t)$, is

$$R(t) = 1 - F(t) = e^{-\lambda t}. \tag{1.25}$$

Plots of $h(t)$, $f(t)$, $F(t)$, and $R(t)$ are shown in Figures 1.7 and 1.8. At $t = 1/\lambda$, $f(1/\lambda) = \lambda/e$, $F(1/\lambda) = 1 - 1/e = 0.632$, and $R(1/\lambda) = 1/e = 0.368$. This is an

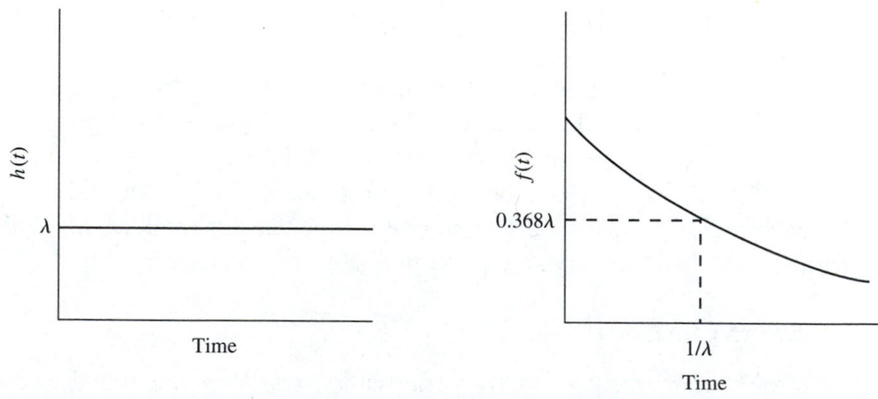

Figure 1.7. Plots of $h(t)$ and $f(t)$

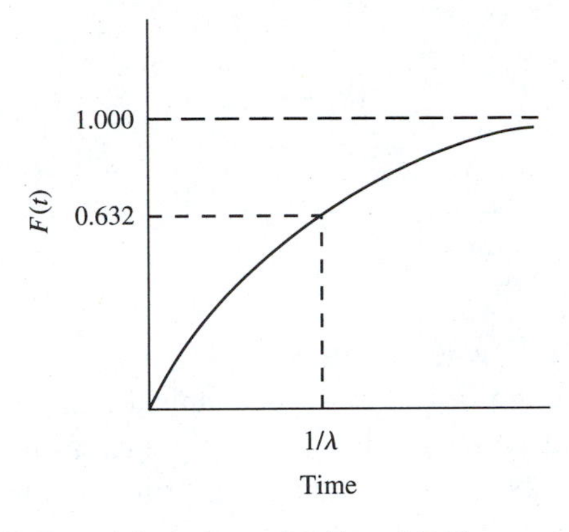

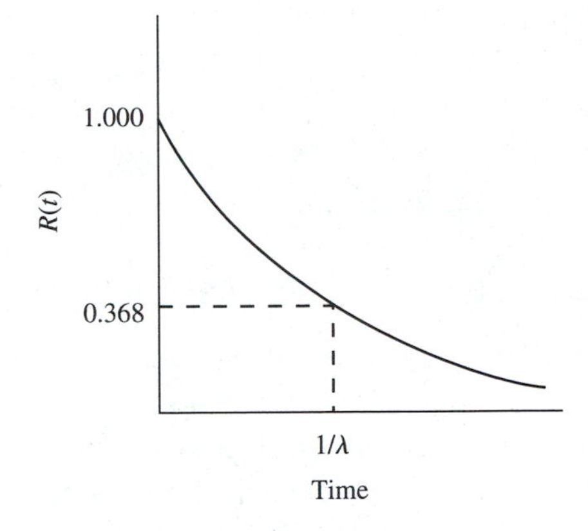

Figure 1.8. Plots of $F(t)$ and $R(t)$

important result since it states that the probability of failure of a product by its estimated mean time to failure ($1/\lambda$) is 0.632. Also, note that the failure time for the constant-hazard model is exponentially distributed.

EXAMPLE 1.3

A manufacturer performs an Operational Life Test (OLT) on ceramic capacitors and finds that they exhibit constant failure rate (used interchangeably with hazard rate) with a value of 3×10^{-8} failures per hour. What is the reliability of a capacitor after one year (10^4 hours)? In order to accept a large shipment of these capacitors, the user decides to run a test for 5,000 hours on a sample of 2,000 capacitors. How many capacitors are expected to fail during the test?

SOLUTION

Using Eqs. (1.21) and (1.25), we obtain

$$h(t) = 3 \times 10^{-8} \text{ failures per hour,}$$

and

$$R(t) = e^{-\int_0^t 3\times 10^{-8}\,dt} = e^{-3\times 10^{-8}t}, \text{ and}$$

$$R(10^4) = e^{-3\times 10^{-4}} = 0.99970.$$

To determine the expected number of failed capacitors during the test, we define the following:

- n_o number of capacitors under test,
- n_s expected number of surviving capacitors at the end of the test, and
- n_f expected number of failed capacitors during the test.

Thus,

$$n_s = e^{-3\times 10^{-8}\times 5000} \times 2000 = 1999 \text{ capacitors and}$$

$$n_f = 2000 - 1999 = 1 \text{ capacitor.}$$

1.3.2. LINEARLY INCREASING HAZARD

A component exhibits an increasing hazard rate when it either experiences wear-out or when it is subjected to deteriorating conditions. Most mechanical components—such as rotating shafts, valves, and cams—exhibit linearly increasing hazard rate. Few electrical components such as relays exhibit linearly increasing hazard rate. The hazard-rate function is expressed as

$$h(t) = \lambda t, \tag{1.26}$$

where λ is constant. The probability density function, $f(t)$, is a Rayleigh distribution and is obtained as

$$f(t) = \lambda t e^{-\frac{\lambda t^2}{2}} \tag{1.27}$$

and

$$F(t) = 1 - e^{-\frac{\lambda t^2}{2}}. \tag{1.28}$$

The reliability function, $R(t)$, is

$$R(t) = e^{\frac{-\lambda t^2}{2}}. \tag{1.29}$$

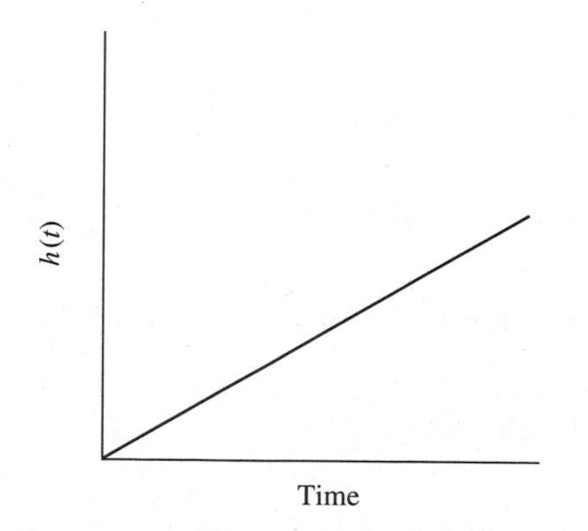

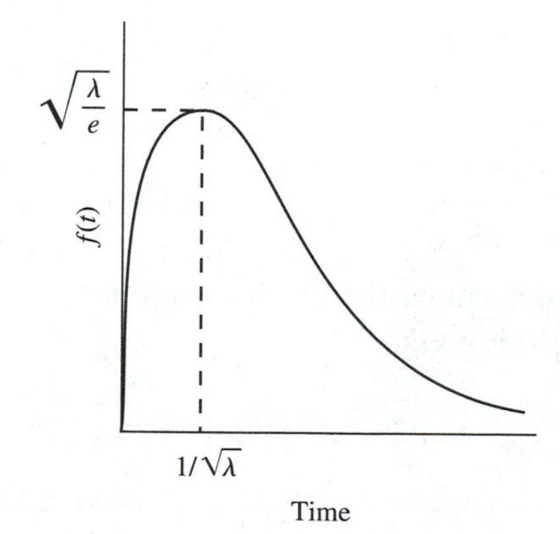

Figure 1.9. Plots of $h(t)$ and $f(t)$

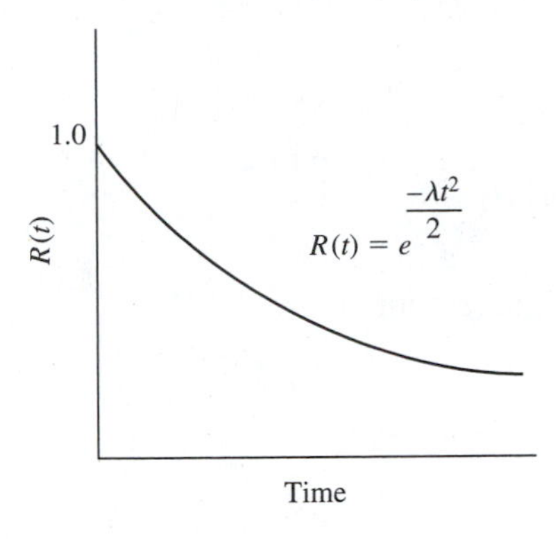

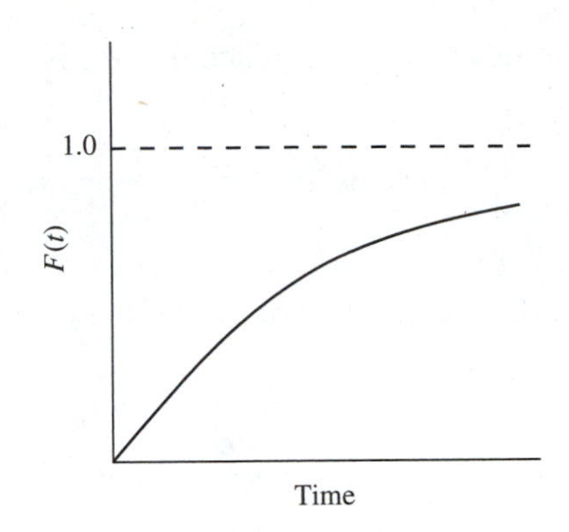

Figure 1.10. Plots of $R(t)$ and $F(t)$

Plots of $h(t), f(t), R(t)$, and $F(t)$ are shown in Figures 1.9 and 1.10. It should be noted that the failure-time distribution of the linearly increasing hazard is a Rayleigh distribution. The mean (expected value) and the variance of the distribution are

$$\sqrt{\frac{\pi}{2\lambda}} \quad \text{and} \quad \frac{2}{\lambda}\left(1 - \frac{\pi}{4}\right), \text{ respectively.}$$

EXAMPLE 1.4

Rolling resistance is a measure of the energy lost by a tire under load when it resists the force opposing its direction of travel. In a typical car, traveling at sixty miles per hour, about 20 percent of the engine power is used to overcome the rolling resistance of the tires. A tire manufacturer introduces a new material that, when added to the tire rubber compound, significantly improves the tire rolling resistance but increases the wear rate of the tire tread. Analysis of a laboratory test of 150 tires shows that the failure rate of the new tire is linearly increasing with time (in hours). It is expressed as

$$h(t) = 0.50 \times 10^{-8} t.$$

Determine the reliability of the tire after one year of use. What is the mean time to replace the tire?

SOLUTION

Using Eq. (1.29) we obtain the reliability after one year as

$$R(10^4) = e^{-\frac{0.5}{2} \times 10^{-8} \times 10^8} = 0.7788.$$

The mean time to replace the tire is

$$\text{Mean time} = \sqrt{\frac{\pi}{2\lambda}} = \sqrt{\frac{\pi}{2 \times 0.5 \times 10^{-8}}} = 17{,}724 \text{ hours},$$

and the standard deviation of the time to tire replacement is

$$\sigma = \sqrt{\frac{2}{\lambda}\left(1 - \frac{\pi}{4}\right)} = 9{,}265 \text{ hours}.$$

1.3.3. LINEARLY DECREASING HAZARD

Most components (both mechanical and electrical) show decreasing hazard rates during their early lives. The hazard rate decreases linearly or nonlinearly with time. In this section, we shall consider linear hazard functions while nonlinear functions will be considered in the next section. The linearly decreasing hazard-rate function is expressed as

$$h(t) = a - bt \tag{1.30}$$

and

$$a \geq bt,$$

where a and b are constants. Similar to the linearly increasing hazard-rate function, we can obtain expressions for $f(t)$, $R(t)$, and $F(t)$. The failure model and the reliability of a component exhibiting such hazard function depend on the values of a and b.

1.3.4. WEIBULL MODEL

A nonlinear expression for the hazard-rate function is used when it clearly cannot be represented linearly with time. A typical expression for the hazard function (decreasing or increasing) under this condition is

$$h(t) = \frac{\gamma}{\theta} t^{\gamma - 1}. \tag{1.31}$$

This model is referred to as the Weibull Model, and its $f(t)$ is given as

$$f(t) = \frac{\gamma}{\theta} t^{\gamma - 1} e^{-\frac{t^{\gamma}}{\theta}} \qquad t > 0, \tag{1.32}$$

where θ and γ are positive and are referred to as the characteristic life and the shape parameter of the distribution respectively. For $\gamma = 1$ this $f(t)$ becomes an exponential density. When $\gamma = 2$, the density function becomes a Rayleigh distribution. It is also well known that the Weibull p.d.f. is almost identical to a normal distribution if a suitable value for the shape parameter γ is chosen. Makino (1984) approximated the normal distribution to a Weibull distribution using the mean hazard rate and found that the shape parameter that approximates the two distributions is $\gamma = 3.43927$. This value of γ is near the value $\gamma = 3.43938$, which is the value of the shape parameter of the Weibull distribution at which the mean is equal to the median. The p.d.f.'s of the Weibull distribution for different γ's are shown in Figure 1.11. The distribution and reliability functions of the Weibull distribution $F(t)$ and $R(t)$ are given below:

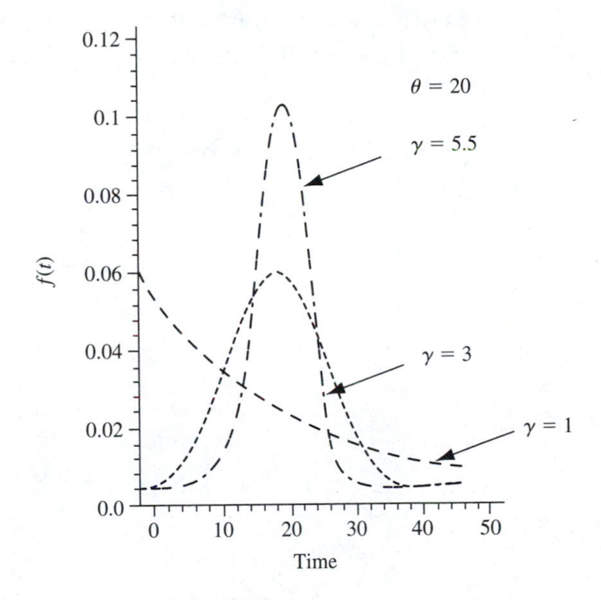

Figure 1.11. The Weibull p.d.f. for Different γ

$$F(t) = \int_0^t \frac{\gamma}{\theta} \zeta^{\gamma-1} e^{-\zeta^{\gamma}/\theta} \, d\zeta \tag{1.33}$$

or

$$F(t) = 1 - e^{\frac{-t^{\gamma}}{\theta}} \qquad t > 0 \quad \text{and} \tag{1.34}$$

$$R(t) = e^{\frac{-t^{\gamma}}{\theta}} \qquad t > 0. \tag{1.35}$$

The Weibull distribution is widely used in reliability modeling since other distributions such as exponential, Rayleigh, and normal are special cases of the Weibull distribution. Again, the hazard-rate function follows the Weibull model:

$$h(t) = \frac{f(t)}{1 - F(t)} = \frac{\gamma}{\theta} t^{\gamma - 1}. \tag{1.36}$$

When $\gamma > 1$, the hazard rate is a monotonically increasing function with no upper bound that describes the wear-out region of the bathtub curve. When $\gamma = 1$, the hazard rate becomes constant (constant failure-rate region), and when $\gamma < 1$, the hazard-rate function decreases with time (the early failure-rate region). This enables the Weibull model to describe the failure rate of many failure data in practice. The mean and variance of the Weibull distribution are as follows:

$$E[T(\text{time to failure})] = \theta^{\frac{1}{\gamma}} \Gamma\left(1 + \frac{1}{\gamma}\right) \tag{1.37}$$

$$\text{Var}[T] = \theta^{\frac{2}{\gamma}} \left\{ \Gamma\left(1 + \frac{2}{\gamma}\right) - \left[\Gamma\left(1 + \frac{1}{\gamma}\right)\right]^2 \right\}, \tag{1.38}$$

where $\Gamma(n)$ is the gamma function

$$\Gamma(n) = \int_0^{\infty} x^{n-1} e^{-x}\, dx \quad \text{and}$$

$$\int_0^{\infty} x^{n-1}\, e^{-x/\theta}\, dx = \Gamma(n)\, \theta^n.$$

EXAMPLE 1.5

To determine the fatigue limit of specially treated steel bars, the Prot method (Collins, 1981) for performing fatigue test is utilized. The test involves the application of a steadily increasing stress level with applied cycles until the specimen under test fails. The number of cycles to failure is observed to follow a Weibull distribution with $\theta = 250$ (measurements are in 10^5 cycles) and $\gamma = 2$.

1. What is the reliability of a bar at 10^6 cycles? What is the corresponding hazard rate?
2. What is the expected life (in cycles) for a bar of this type?

SOLUTION

Since the shape parameter γ equals 2, the Weibull distribution becomes a Rayleigh distribution, and we have a linearly increasing hazard function. Its probability density function is given by Eq. (1.32).

1. The reliability expression for the Weibull model is given by Eq. (1.35),

$$R(10^6) = e^{-(10)^2/250}$$

$$= e^{-0.4} = 0.6703.$$

The hazard rate at 10^6 cycles is

$$h(t) = \frac{\gamma}{\theta} t^{\gamma-1} = \frac{2}{250} \times 10$$

or

$$h(10^6) = 0.08 \text{ failures}/10^5 \text{ cycles}.$$

2. The expected life of a bar is

$$E[T(\text{cycles to failure})] = \theta^{\frac{1}{\gamma}} \Gamma\left(1 + \frac{1}{\gamma}\right)$$

$$= (250)^{1/2} \Gamma\left(\frac{3}{2}\right)$$

$$= (15.81138)\left(\frac{1}{2}\right) \Gamma\left(\frac{1}{2}\right)$$

$$= (15.81138)\left(\frac{1}{2}\right) \sqrt{\pi} = 14.01247.$$

The expected life of a bar from this steel is 14.01247×10^5 cycles.

In the above example, the Weibull model became a Rayleigh model since the failure rate is linearly increasing with time. In the following example, we consider the situation when the failure rate is nonlinearly increasing with time.

EXAMPLE 1.6 A manufacturing engineer observes the wear-out rate of a milling machine tool insert and fits a Weibull hazard model to the tool wear data. The parameters of the model are $\gamma = 2.25$ and $\theta = 300$. Determine the reliability of the tool insert after ten hours, the expected life of the insert, and the standard deviation of the life.

SOLUTION The reliability after ten hours of operation is

$$R(10) = e^{\frac{-10^{2.25}}{300}} = 0.553.$$

The mean life of the insert is

$$\text{Mean life} = \theta^{\frac{1}{\gamma}}\,\Gamma\left(1 + \frac{1}{\gamma}\right)$$

$$= (300)^{\frac{1}{2.25}}\,\Gamma\left(1 + \frac{1}{2.25}\right),$$

or

$$\text{Mean life} = 12.6166\,\Gamma(1.444) = 11.176 \text{ hours.}$$

The value of $\Gamma(1.444)$ is obtained from the tables of the gamma function given in Appendix A.

Using Eq. (1.38), we obtain the variance of the mean life as

$$\text{Variance} = \theta^{\frac{2}{\gamma}}\left\{\Gamma\left(1 + \frac{2}{\gamma}\right) - \left[\Gamma\left(1 + \frac{1}{\gamma}\right)\right]^2\right\}$$

$$= 300^{\frac{2}{2.25}}\left\{\Gamma(1.888) - \left[\Gamma(1.444)\right]^2\right\}$$

or

Variance = 27.133, and the standard deviation of the life is 5.21 hours.

1.3.5. MIXED WEIBULL MODEL

This model is applicable when components or products experience two or more failure modes. For example, a mechanical component, such as a load-carrying bearing or a cutting tool, may fail due to wear-out or when the applied stress exceeds the design strength of component material (catastrophic failure is a failure that destroys the system, such as a missile failure). Each type of these failures may be modeled by a separate simple Weibull model. Since the component or the tool can fail in either of the failure modes, it is then appropriate to describe the hazard rate by a mixed Weibull model. It is expressed as

$$f(t) = p\frac{\gamma_1}{\theta_1}t^{\gamma_1-1}\, e^{\frac{-t^{\gamma_1}}{\theta_1}} + (1-p)\frac{\gamma_2}{\theta_2}t^{\gamma_2-1}\, e^{\frac{-t^{\gamma_2}}{\theta_2}} \tag{1.39}$$

for $\theta_1, \theta_2 > 0$, and $0 < \gamma_1 < \gamma_2$.

The quantity p $(0 \leq p \leq 1)$ is the probability that the component or the tool fails in the first failure mode, and $1 - p$ is the probability that it fails in the second failure mode. Clearly, if a product experiences more than two failure modes, the model given by Eq. (1.39) can be expanded to include all failure modes and associated probabilities such that $\Sigma_{i=1}^{n} p_i = 1$ where p_i is the probability that the product fails in the ith failure mode, and n is the total number of failure modes.

Following Kao (1959), the time t_e at which the proportion of the catastrophic failure is equal to that of wear-out failure is obtained as

$$1 - e^{\frac{-t_e^{\gamma_1}}{\theta_1}} = 1 - e^{\frac{-t_e^{\gamma_2}}{\theta_2}}$$

or

$$t_e = \left(\frac{\theta_2}{\theta_1}\right)^{\frac{1}{\gamma_2-\gamma_1}} = \exp\left(\frac{\ln\theta_2 - \ln\theta_1}{\gamma_2 - \gamma_1}\right). \tag{1.40}$$

The reliability expression of the mixed Weibull model is

$$R(t) = 1 - p\left[1 - e^{\frac{-t^{\gamma_1}}{\theta_1}}\right] - (1-p)\left[1 - e^{\frac{-t^{\gamma_2}}{\theta_2}}\right]. \tag{1.41}$$

Clearly, if the second failure mode occurs after a delay time δ, from the first failure mode, we rewrite Eqs. (1.39) and (1.41) as follows:

$$f_d(t) = p\,\frac{\gamma_1}{\theta_1}\,t^{\gamma_1-1}\, e^{\frac{-t^{\gamma_1}}{\theta_1}} + (1-p)\,\frac{\gamma_2}{\theta_2}\,(t-\delta)^{\gamma_2-1}\, e^{\frac{-(t-\delta)^{\gamma_2}}{\theta_2}} \tag{1.42}$$

and

$$R_d(t) = 1 - p\left[1 - e^{\frac{-t^{\gamma_1}}{\theta_1}}\right] - (1-p)\left[1 - e^{\frac{-(t-\delta)^{\gamma_2}}{\theta_2}}\right], \tag{1.43}$$

where the subscript d denotes delay.

1.3.6. EXPONENTIAL MODEL (THE EXTREME VALUE DISTRIBUTION)

The extreme value distribution is closely related to the Weibull distribution. It is useful in modeling cases when the hazard function is initially constant and then begins to increase rapidly with time. This is illustrated by the Weibull hazard rate when $\gamma = 5.5$ as shown in Figure 1.11.

The distribution is used to describe the failure time of products (or components) that will operate properly at normal operating conditions and will fail owing to a secondary cause of failure (such as overheating or fracture) when subjected to extreme conditions. In other words, the interest is in the tails of the failure distribution. Here, the hazard-rate function, the failure-time density function, and the reliability function are expressed as

$$h(t) = be^{\alpha t} \tag{1.44}$$

$$f(t) = be^{\alpha t}\, e^{-\int_0^t h(\zeta)\, d\zeta} \tag{1.45}$$

$$f(t) = be^{\alpha t}\, e^{-\frac{b}{\alpha}(e^{\alpha t}-1)} \tag{1.46}$$

$$R(t) = e^{-\frac{b}{\alpha}(e^{\alpha t}-1)}, \tag{1.47}$$

where b is a constant and e^{α} represents the increase in failure rate per unit time. For example, if it is found that the failure rate of a component increases about 10 percent each year, then $h(t) = b(1.1)^t$ where $\alpha = \ln(1.1) = 0.0953$. The function $f(t)$ as given by Eq. (1.46) is also known as the *Gompertz distribution*.

Plots of the hazard rate and the reliability functions of the *extreme value distribution* for different values of α and b are shown in Figure 1.12. Some electronic components show such a hazard function. There are mechanical assemblies that exhibit extreme value hazard functions when subjected to high stresses. An example of such assemblies is a gear box that operates properly at the recommended speeds. Excessive speeds may cause wearout of bearings that results in misalignments of shafts and an eventual failure of the assembly.

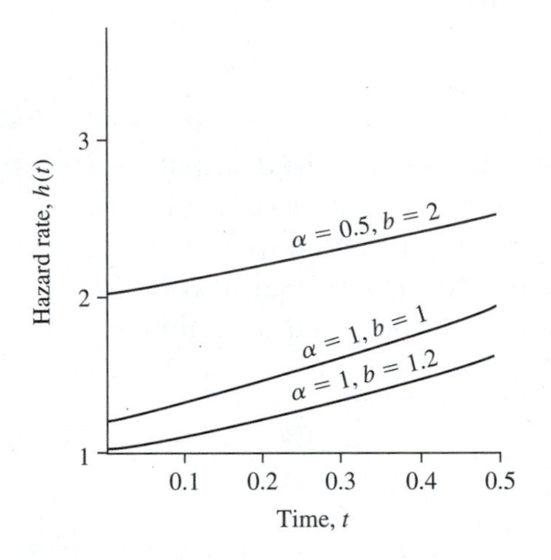

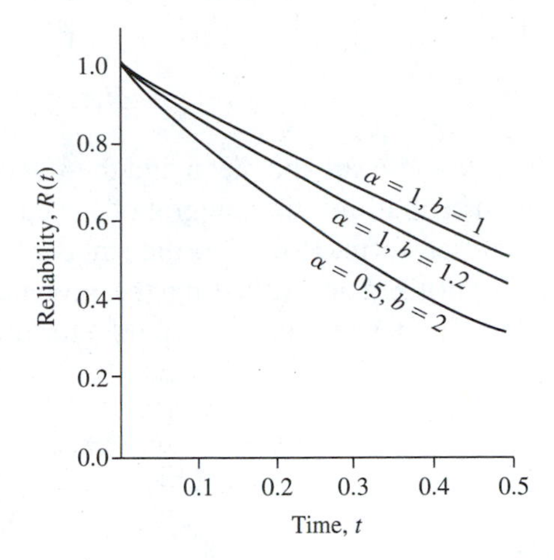

Figure 1.12. Plots of $h(t)$ and $R(t)$

EXAMPLE 1.7

Excessive vibrations due to high-speed cutting on a CNC (computer numerical control) machine may lead to the failure of the cutting tool. The failure time of the tool follows an extreme value distribution. The failure rate increases about 15 percent per hour. Assuming that the constant $b = 0.01$, calculate the reliability of the tool at $t = 10$ hours.

SOLUTION

Since the failure rate increases by 15 percent per hour, then $\alpha = \ln(1.15) = 0.1397$. Substituting the parameters α and b into Eq. (1.47), we obtain

$$R(10) = e^{-\frac{0.01}{0.1397}(e^{0.1397\times 10} - 1)}$$

$$R(10) = 0.8042.$$

1.3.7. NORMAL MODEL

There are many practical situations where the failure time of components (or parts) can be described by a normal distribution. For example, most of the mechanical components that are subjected to repeated cyclic loads, such as a fatigue test, exhibit normal hazard rates. Unlike other continuous probability distributions, there are no closed-form expressions for the reliability or hazard-rate functions. The cumulative distribution function of the life of a component is given by

$$F(t) = P[\mathbf{t} \leq t] = \int_{-\infty}^{t} \frac{1}{\sigma\sqrt{2\pi}} \exp\left[-\frac{1}{2}\left(\frac{\tau - \mu}{\sigma}\right)^2\right] d\tau, \qquad (1.48)$$

and

$$R(t) = 1 - F(t),$$

where μ and σ are the mean and the standard deviation of the distribution. Unlike other distributions, the integral of the cumulative distribution cannot be evaluated in a closed form. However, the standard normal distribution ($\sigma = 1$ and $\mu = 0$) can be utilized in evaluating the probabilities for any normal distribution. The probability density function (p.d.f.) for the standard normal distribution is

$$\phi(z) = \frac{1}{\sqrt{2\pi}} \exp\left(-\frac{z^2}{2}\right) \qquad -\infty < z < \infty, \tag{1.49}$$

where

$$z = \frac{\tau - \mu}{\sigma}.$$

The cumulative distribution function is

$$\Phi(\tau) = \int_{-\infty}^{\tau} \frac{1}{\sqrt{2\pi}} \exp\left(-\frac{z^2}{2}\right) dz. \tag{1.50}$$

Therefore, when the failure time of a component is expressed as a normally distributed random variable **t**, with mean μ and standard deviation σ, one can easily determine the probability that the component will fail at time t (that is, the unreliability of the component) by using the following equation:

$$P(\mathbf{t} \leq t) = P\left(\mathbf{t} \leq \frac{t - \mu}{\sigma}\right) = \Phi\left(\frac{t - \mu}{\sigma}\right). \tag{1.51}$$

The right side of Eq. (1.51) can be evaluated using the standard normal tables. The hazard function, $h(t)$, of the normal distribution is

$$h(t) = \frac{f(t)}{R(t)} = \frac{\phi\left(\frac{t - \mu}{\sigma}\right) \Big/ \sigma}{R(t)}. \tag{1.52}$$

It can be shown that the hazard function for a normal distribution is a monotonically increasing function of t:

$$h(t) = \frac{f(t)}{1 - F(t)}$$

$$h'(t) = \frac{(1-F)f' + f^2}{(1-F)^2}. \tag{1.53}$$

The denominator is nonnegative for all t. Hence, it is sufficient to show that the numerator of Eq. (1.53) is ≥ 0:

$$(1-F)f' + f^2 \geq 0. \tag{1.54}$$

The p.d.f. of the normal distribution is

$$f(t) = \frac{1}{\sqrt{2\pi\sigma^2}} e^{-(t-\mu)^2/2\sigma^2}, \qquad -\infty < t < \infty,$$

and Eq. (1.54) can be rewritten as

$$R(t)\frac{d}{dt}f(t) + f^2(t) \geq 0.$$

Now, the derivative term is

$$\frac{d}{dt}f(t) = \frac{1}{\sqrt{2\pi\sigma^2}}\frac{d}{dt}e^{-(t-\mu)^2/2\sigma^2} = \frac{1}{\sqrt{2\pi\sigma^2}}\frac{-(t-\mu)}{\sigma^2}e^{-(t-\mu)^2/2\sigma^2}$$

$$= \frac{-(t-\mu)}{\sigma^2}f(t),$$

so now the condition that must be satisfied is

$$f(t)\left(\frac{-(t-\mu)}{\sigma^2}R(t) + f(t)\right) \geq 0.$$

Since $f(t) \geq 0$ by definition and $R(t) = \int_t^\infty f(x)\,dx$, we may use the condition

$$\frac{(t-\mu)}{\sigma^2}\int_t^\infty f(x)\,dx \leq \int_t^\infty \frac{(x-\mu)}{\sigma^2}f(x)\,dx = \int_t^\infty -df(x) = f(t)$$

to obtain

$$f(t) \geq \frac{t-\mu}{\sigma^2}\int_t^\infty f(x)\,dx$$

so

$$f(t)\left(f(x) - \frac{(t-\mu)}{\sigma^2}\int_t^{\infty} f(x)\,dx\right) \geq 0,$$

and therefore the Gaussian hazard function is a monotonically increasing function of time. The plots of $f(t)$, $F(t)$, $R(t)$, and $h(t)$ for $u = 20$ are shown in Figure 1.13.

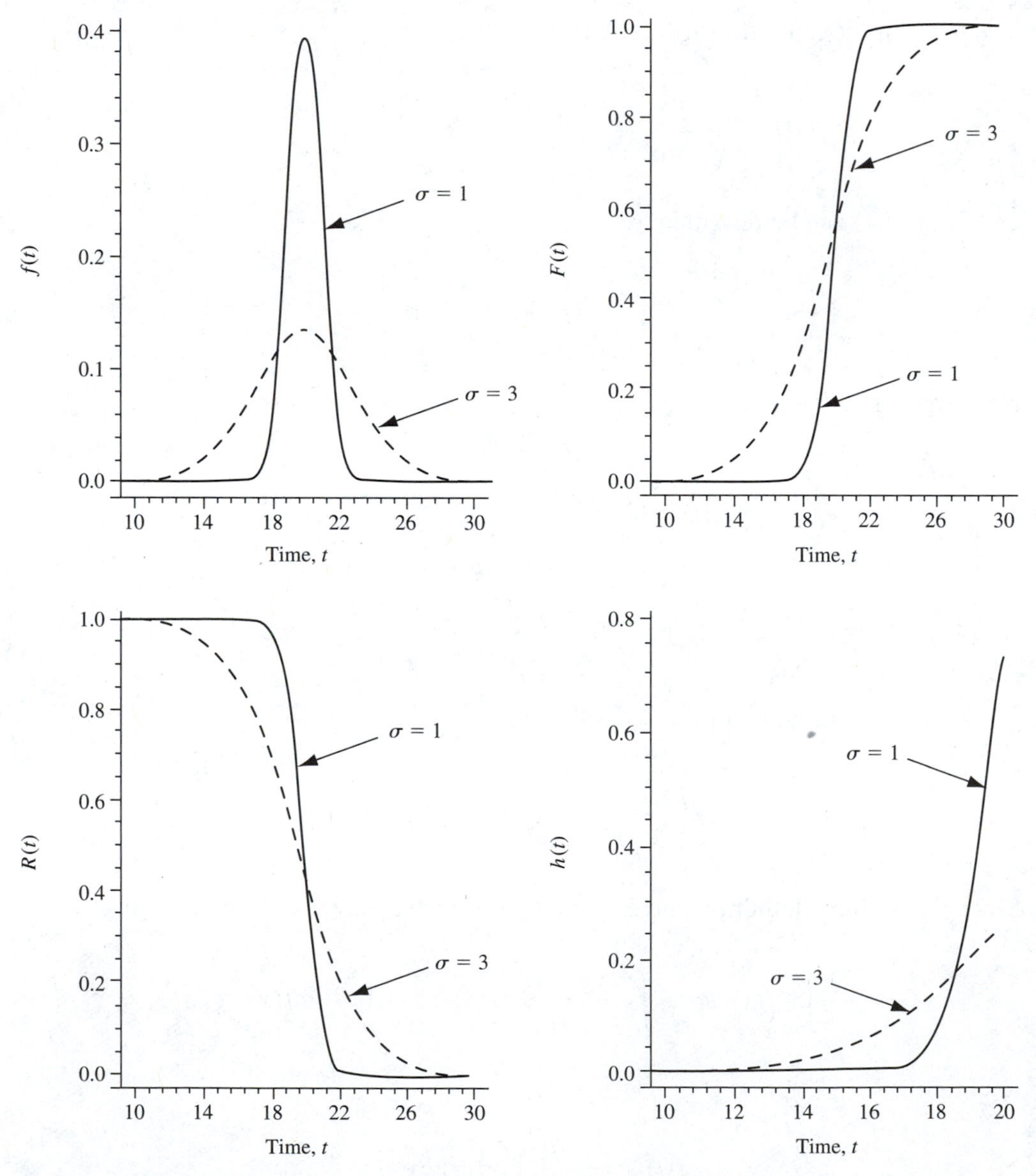

Figure 1.13. $f(t)$, $F(t)$, $R(t)$, and $h(t)$ for the Normal Model

EXAMPLE 1.8

A component has a normal distribution of failure times with $\mu = 40{,}000$ cycles and $\sigma = 2{,}000$ cycles. Find the reliability and hazard function at 38,000 cycles.

SOLUTION

The reliability function is

$$R(t) = P\left(z > \frac{t-\mu}{\sigma}\right)$$

$$R(38{,}000) = P\left(z > \frac{38{,}000 - 40{,}000}{2{,}000}\right)$$

$$= P[z > -1.0] = \Phi(1.0)$$

$$= 0.8413.$$

The value of $h(38{,}000)$ is

$$h(38{,}000) = \frac{f(38{,}000)}{R(38{,}000)} = \frac{\phi\left(z = \frac{38{,}000 - 40{,}000}{2{,}000}\right) \Big/ 2{,}000}{R(38{,}000)}$$

$$= \frac{\phi(-1.0)}{2000 \times 0.8413} = \frac{0.2420}{2000 \times 0.8413}$$

$$= 0.0001438 \text{ failures per cycle.}$$

1.3.8. LOGNORMAL MODEL

One of the most widely used probability distributions in describing the life data resulting from a single semiconductor failure mechanism or a closely related group of failure mechanisms is the lognormal distribution. It is also used in predicting reliability from accelerated life test data. The probability density function of the lognormal distribution is

$$f(t) = \frac{1}{\sigma t\sqrt{2\pi}} \exp\left[-\frac{1}{2}\left(\frac{\ln t - \mu}{\sigma}\right)^2\right] \qquad -\infty < \mu < \infty, \qquad \sigma > 0, t > 0. \tag{1.55}$$

Figure 1.14 shows the p.d.f. of the lognormal distribution for different μ and σ.

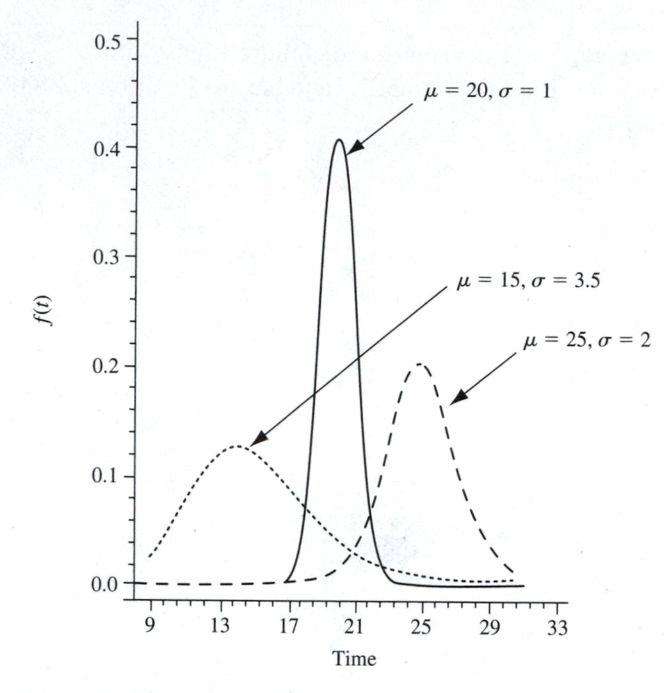

Figure 1.14. $f(t)$ of the Lognormal Distribution for Different μ and σ

If a random variable X is defined as $X = \ln T$, where T is lognormal, then X is normally distributed with mean μ and standard deviation σ:

$$E[X] = E[\ln(t)] = \mu$$

$$\text{Var}[X] = \text{Var}[\ln(t)] = \sigma^2.$$

Since $T = e^X$, then the mean of the lognormal can be found by using the normal distribution.

$$E(T) = E(e^X) = \int_{-\infty}^{\infty} \frac{1}{\sigma\sqrt{2\pi}} \exp\left[x - \frac{1}{2}\left(\frac{x-\mu}{\sigma}\right)^2\right] dx$$

$$E(T) = \exp\left(\mu + \frac{\sigma^2}{2}\right) \int_{-\infty}^{\infty} \frac{1}{\sigma\sqrt{2\pi}} \exp\left[-\frac{1}{2\sigma^2}(x - (\mu + \sigma^2))^2\right] dx.$$

The mean of the lognormal is

$$E(T) = \exp\left[\mu + \frac{\sigma^2}{2}\right].$$

The second moment is obtained as

$$E(T^2) = E[e^{2X}] = \exp[2(\mu + \sigma^2)],$$

and the variance of the lognormal is

$$\mathrm{Var}(T) = [e^{2\mu+\sigma^2}]\,[e^{\sigma^2} - 1].$$

The distribution function of the lognormal is

$$F(t) = \int_0^t \frac{1}{\tau\sigma\sqrt{2\pi}} \exp\left[-\frac{1}{2}\left(\frac{\ln \tau - \mu}{\sigma}\right)^2\right] d\tau$$

or

$$F(t) = P(T \leq t) = P\left[z \leq \frac{\ln t - \mu}{\sigma}\right].$$

The reliability is

$$R(t) = P[T > t] = P\left[z > \frac{\ln t - \mu}{\sigma}\right]. \tag{1.56}$$

Thus, the hazard function is

$$h(t) = \frac{f(t)}{R(t)} = \frac{\phi\left(\frac{\ln t - \mu}{\sigma}\right)}{t\sigma\, R(t)}. \tag{1.57}$$

Figure 1.15 shows the reliability and the hazard-rate functions of the lognormal distribution for different values of μ and σ.

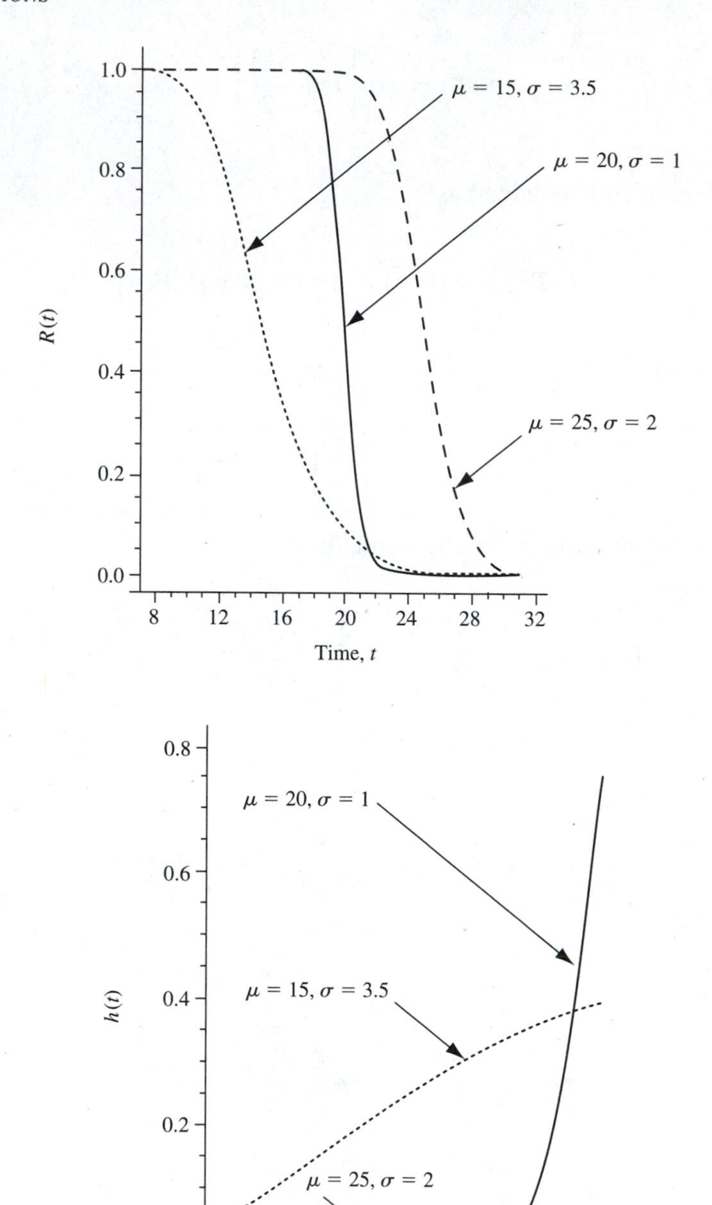

Figure 1.15. $R(t)$ and $h(t)$ for the Lognormal Model

EXAMPLE 1.9

The failure time of a component is lognormally distributed with $\mu = 6$ and $\sigma = 2$. Find the reliability of the component and the hazard rate for a life of 200 time units.

SOLUTION

$$R(200) = P\left[z > \frac{\ln 200 - 6}{2}\right] = P[z > -0.350] = 0.6386.$$

The hazard function is

$$h(200) = \frac{\phi\left(\frac{\ln 200 - 6}{2}\right)}{200 \times 2 \times 0.6386}$$

$$= \frac{\phi(-0.350)}{200 \times 0.6386} = \frac{0.3752}{200 \times 2 \times 0.6386}$$

$$= 0.001472 \text{ failures per unit time.}$$

1.3.9. GAMMA MODEL

Like the Weibull model, the gamma model covers a wide range of the hazard-rate functions: decreasing, constant, or increasing hazard rates. The gamma distribution is suitable for describing the failure time of a component whose failure takes place in n stages or the failure time of a system that fails when n independent sub-failures have occurred.

The gamma distribution is characterized by two parameters: shape parameter γ and scale parameter θ. When $0 < \gamma < 1$, the failure rate monotonically decreases from infinity to $1/\theta$ as time increases from 0 to infinity. When $\gamma > 1$, the failure rate monotonically increases from $1/\theta$ to infinity. When $\gamma = 1$ the failure rate is constant and equals $1/\theta$.

The probability density function of a gamma distribution is

$$f(t) = \frac{t^{\gamma-1}}{\theta^{\gamma}\Gamma(\gamma)} e^{\frac{-t}{\theta}}. \tag{1.58}$$

When $\gamma > 1$, there is a single peak of the density function at time $t = \theta(\gamma - 1)$. The cumulative distribution function, $F(t)$, is

$$F(t) = \int_0^t \frac{\tau^{\gamma-1}}{\theta^{\gamma}\Gamma(\gamma)} e^{\frac{-\tau}{\theta}} \, d\tau.$$

Substituting $\tau/\theta = u$, we obtain

$$F(t) = \frac{1}{\Gamma(\gamma)} \int_0^{t/\theta} u^{\gamma-1} e^{-u} \, du$$

or

$$F(t) = I\left(\frac{t}{\theta}, \gamma\right),$$

where $I(t/\theta, \gamma)$ is known as the incomplete gamma function and is tabulated in Pearson (1957).

The reliability function $R(t)$ is

$$R(t) = \int_t^{\infty} \frac{1}{\theta\Gamma(\gamma)} \left(\frac{\tau}{\theta}\right)^{\gamma-1} e^{\frac{-\tau}{\theta}} \, d\tau. \tag{1.59}$$

When the shape parameter γ is an integer n, the gamma distribution becomes the well-known Erlangian distribution. In this case, the cumulative distribution function is written as

$$F(t) = 1 - e^{\frac{-t}{\theta}} \sum_{k=0}^{n-1} \frac{\left(\frac{t}{\theta}\right)^k}{k!} \tag{1.60}$$

and the reliability function is

$$R(t) = e^{\frac{-t}{\theta}} \sum_{k=0}^{n-1} \frac{\left(\frac{t}{\theta}\right)^k}{k!}. \tag{1.61}$$

The hazard rate of the gamma model, when γ is an integer n, is obtained by dividing Eq. (1.58) by Eq. (1.61):

$$h(t) = \frac{\dfrac{1}{\theta}\left(\dfrac{t}{\theta}\right)^{n-1}}{(n-1)!\displaystyle\sum_{k=0}^{n-1}\frac{\left(\dfrac{t}{\theta}\right)^{k}}{k!}}. \tag{1.62}$$

Figures 1.16, 1.17, and 1.18 show the gamma density function, the reliability function, and the hazard rate for different γ values and a constant $\theta = 20$.

The mean and variance of the gamma distribution are obtained as

$$\begin{aligned}
\text{Mean life} &= \int_{-\infty}^{\infty} t\, f(t)\, dt \\
&= \int_{0}^{\infty} t \frac{1}{\Gamma(\gamma)\theta^{\gamma}}\, e^{\frac{-t}{\theta}}\, dt \\
&= \frac{1}{\Gamma(\gamma)\theta^{\gamma}} \int_{0}^{\infty} t^{\gamma}\, e^{\frac{-t}{\theta}}\, dt
\end{aligned}$$

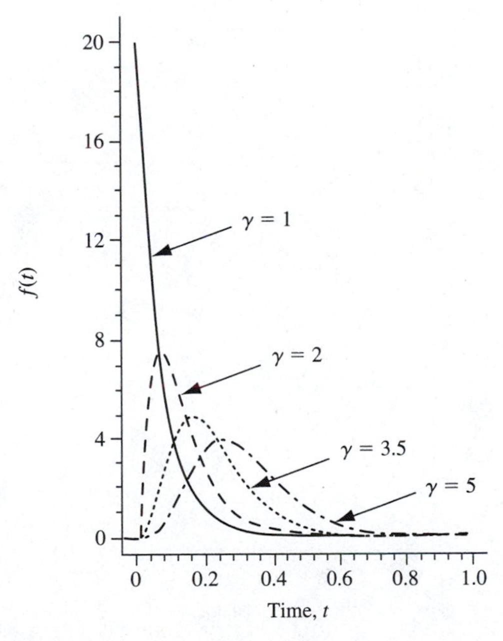

Figure 1.16. Gamma Density Function with Different γ Values, $\theta = 20$

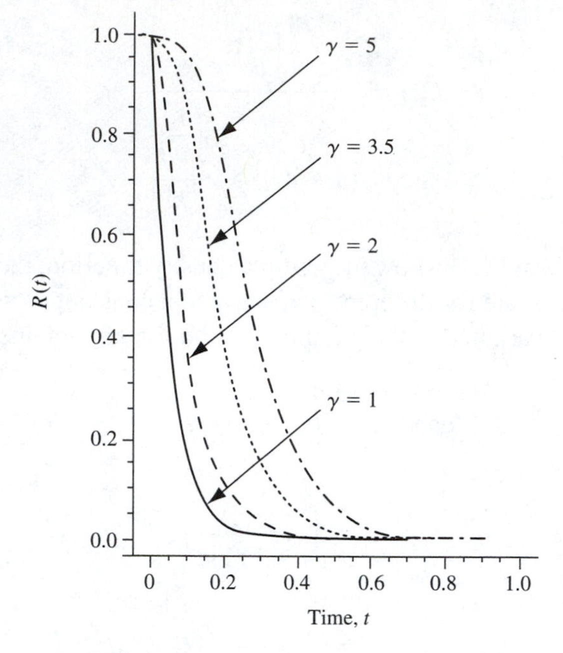

Figure 1.17. Gamma Reliability Function for Different γ Values, $\theta = 20$

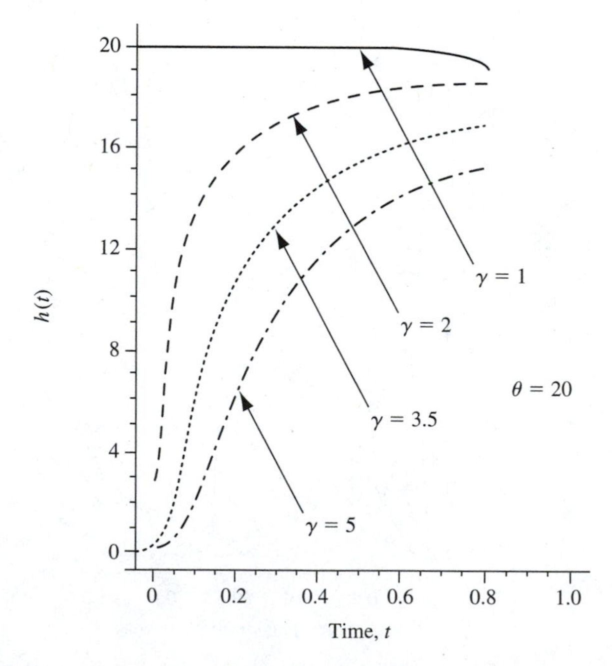

Figure 1.18. Gamma Hazard Rate for Different γ Values

or

$$\text{Mean life} = \frac{1}{\Gamma(\gamma)\theta^{\gamma}}\Gamma(\gamma + 1)\theta^{\gamma+1} = \gamma\theta.$$

Similar manipulations yield $E[T^2] = \gamma(\gamma + 1)\theta^2$ and the variance of the life is

$$\text{Var}(T) = \gamma(\gamma + 1)\theta^2 - \gamma^2\theta^2 = \gamma\theta^2.$$

EXAMPLE 1.10

A mechanical system requires a constant supply of electric current, which is provided by a main battery having life length T_1 with an exponential distribution of mean 120 hours. The main battery is supported by two identical backup batteries with mean lives of T_2 and T_3. When the main unit fails, the first backup battery provides the necessary current to the system. The second backup battery provides the current when the first backup unit fails. In other words, the batteries provide the current independently but sequentially.

Determine the reliability and the hazard rate of the mechanical system at $t =$ 280 hours. What is the mean life of the system?

SOLUTION

Since the life lengths of the batteries are independent exponential random variables with means T_1, T_2, and T_3, then the total life length of the mechanical system is $T = T_1 + T_2 + T_3$. The distribution of T is a gamma distribution with $\gamma = n = 3$ and $\theta = 120$. Using Eq. (1.61) we obtain

$$R(280) = e^{\frac{-280}{120}} \sum_{k=0}^{2} \frac{\left(\frac{280}{120}\right)^k}{k!} = 0.85119.$$

The hazard rate at 280 hours is obtained by substituting into Eq. (1.62):

$$h(280) = \frac{\frac{1}{120}\left(\frac{280}{120}\right)^2}{2!(8.777)} = 0.00258 \text{ failures per hour.}$$

The mean life of the mechanical system is given by

$$\text{Mean life} = \gamma\theta = 3(120) = 360 \text{ hours.}$$

1.3.10. LOG-LOGISTIC MODEL

If $T > 0$ is a random variable representing the failure time of a system and t represents a typical time instant in its range, we use $Y \equiv \log T$ to represent the log failure time (Kalbfleisch and Prentice, 1980). The log-logistic distribution for T is obtained if we express $Y = \alpha + \sigma W$ and W has the logistic density

$$f(w) = \frac{e^w}{(1 + e^w)^2}. \tag{1.63}$$

The logistic density is symmetric with mean $= 0$ and variance $= \pi^2/3$ with slightly heavier tails than the normal density function (Kalbfleisch and Prentice, 1980). The probability density function of the failure time t is

$$f(t) = \lambda p(\lambda t)^{p-1}[1 + (\lambda t)^p]^{-2}, \tag{1.64}$$

where $\lambda = e^{-\alpha}$ and $p = 1/\sigma$.

The reliability and hazard functions of the log-logistic model are

$$R(t) = \frac{1}{1 + (\lambda t)^p} \tag{1.65}$$

and

$$h(t) = \frac{\lambda p(\lambda t)^{p-1}}{1 + (\lambda t)^p}. \tag{1.66}$$

This model has the same advantage as both the Weibull and exponential models; it has simple expressions for $R(t)$ and $h(t)$.

Examination of Eq. (1.66) reveals that the hazard function is monotonically decreasing when $p = 1$. If $p > 1$, the hazard rate increases from 0 to a peak at $t = (p - 1)^{1/p}/\lambda$ and then decreases with time thereafter. The hazard rate is monotonically decreasing if $p < 1$. Figures 1.19 and 1.20 show the reliability function and the hazard rate for different values of p and a constant $\lambda = 20$.

1.3.11. BETA MODEL

The hazard function models discussed thus far are defined as nonzero functions over the time range of zero to infinity. However, the life of some products or

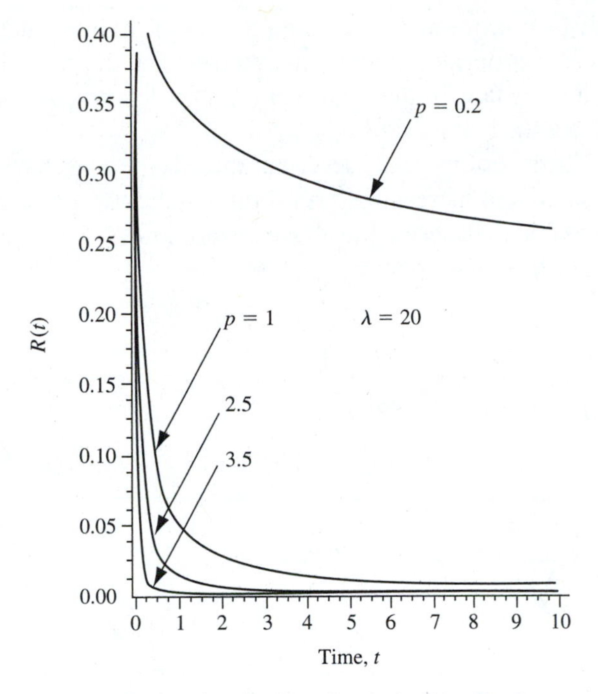

Figure 1.19. Reliability Function for the Log-Logistic Distribution

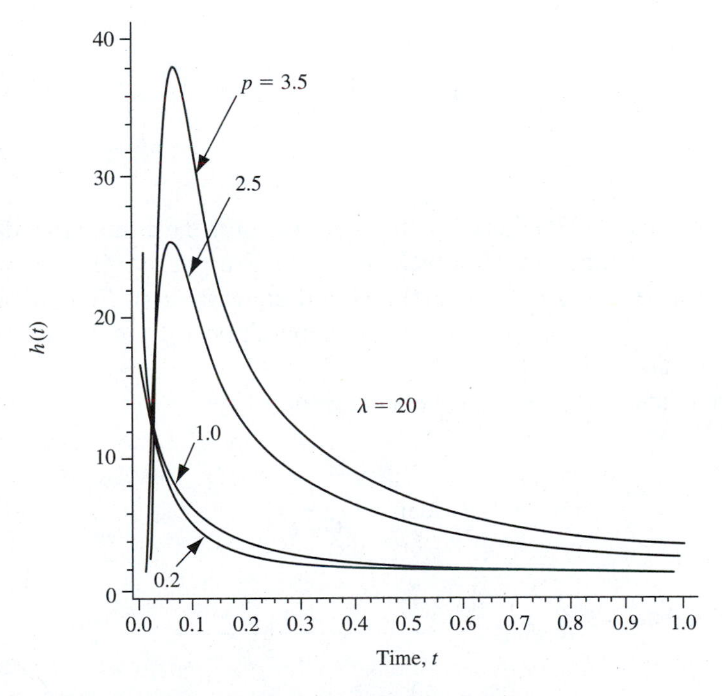

Figure 1.20. Hazard Rate for the Log-Logistic Distribution

components may be constrained to a finite interval of time. In such cases, the beta model is the most appropriate model that can describe the reliability behavior of the product during the constrained interval (0, 1). Clearly, any finite interval can be transformed to a (0, 1) interval.

Like other distributions that describe three types of hazard functions—decreasing, constant, and increasing hazard rates—the two parameters of the beta model make it flexible to describe the above hazard rates. The standard form of the density function of the beta model is

$$f(t) = \begin{cases} \dfrac{\Gamma(\alpha+\beta)}{\Gamma(\alpha)\Gamma(\beta)} t^{\alpha-1}(1-t)^{\beta-1} & 0 < t < 1 \\ 0 & \text{otherwise.} \end{cases} \tag{1.67}$$

The parameters α and β are positive. Since

$$\int_0^1 f(t)\, dt = 1,$$

then

$$\int_0^1 t^{\alpha-1}(1-t)^{\beta-1}\, dt = \frac{\Gamma(\alpha)\Gamma(\beta)}{\Gamma(\alpha+\beta)} \tag{1.68}$$

for positive α and β.

In general, there is no closed-form expression for the cumulative distribution or the hazard rate function. However, if α or β is a positive integer, a binomial expansion can be used to obtain $F(t)$ and consequently $h(t)$. $F(t)$ will be a polynomial in t, and the powers of t will be, in general, positive real numbers ranging from 0 through $\alpha + \beta - 1$.

The mean and variance of the beta distribution are

$$\text{Mean} = \frac{\alpha}{\alpha+\beta}$$

$$\text{variance} = \frac{\alpha\beta}{(\alpha+\beta)^2(\alpha+\beta+1)}.$$

1.3.12. OTHER FORMS

1.3.12.1. *The Generalized Pareto Model*

When the hazard rate is either monotonically increasing or monotonically decreasing, it can be described by a three-parameter distribution with a hazard-rate function of the form

$$h(t) = \alpha + \frac{\beta}{t + \gamma}, \tag{1.69}$$

where α, β, and γ are the parameters of the model.

1.3.12.2. *The Gompertz-Makeham Model*

This is a generalized model of the Gompertz hazard model with hazard rate

$$h(t) = \rho_0 + \rho_1 \, e^{\rho_2 t}, \tag{1.70}$$

where ρ_0, ρ_1, and ρ_2 are the parameters of the model.

1.3.12.3. *The Power Series Model*

There are many practical situations where none of the above-mentioned models is suitable to accurately fit the hazard-rate values. In such a case, a general power series model can be used to fit the hazard-rate values. Clearly, the number of terms in the power-series model relates to the desired level of fitness of the model to the empirical data. A good measure for the appropriateness of fitting the model to the data is the mean squared error between the hazard values obtained from the model and the actual data. The hazard-rate function of the power series model is

$$h(t) = a_0 + a_1 t + a_2 t^2 + \ldots + a_n t^n. \tag{1.71}$$

The reliability function, $R(t)$, is

$$R(t) = \exp\left[-\left(a_0 t + \frac{a_1 t^2}{2} + \frac{a_2 t^3}{3} + \ldots + \frac{a_n t^{n+1}}{n+1}\right)\right]. \tag{1.72}$$

EXAMPLE 1.11

Electromigration is a common failure mechanism in semiconductor devices. It is a phenomenon whereby a metal line in a device "grows" a link to another line or creates an open condition, due to movement (migration) of metal ions toward the anode at high temperatures or current densities (Comeford, 1989). Two hundred integrated circuits (ICs) are subjected to an elevated temperature of 250°C to accelerate their failures. The failures observed due to electromigration during the test intervals are given in Table 1.6.

Table 1.6. Failure Data for the ICs

Time Interval (Hours)	*Failures in the Interval*
0–100	10
101–200	20
201–300	35
301–400	40
401–500	45
501–600	50
Total	200

Assume that the hazard-rate function is expressed as a power-series function. Determine the hazard rate and the reliability after ten hours of operation at the same elevated temperature.

SOLUTION We calculate the hazard rate from the data as shown in Table 1.7.

Table 1.7. Hazard Rate Calculation for Example 1.11

Time Interval (Hours)	*Failures in the Interval*	*Hazard Rate* $\times 10^{-3}$
0–100	10	$10/(200 \times 100) = 0.50$
101–200	20	$20/(190 \times 100) = 1.05$
201–300	35	$35/(170 \times 100) = 2.05$
301–400	40	$40/(135 \times 100) = 2.92$
401–500	45	$45/(95 \times 100) = 4.73$
501–600	50	$50/(50 \times 100) = 10.00$

We use the hazard-rate data in Table 1.7 to fit the model given by Eq. (1.71) using the least squares method to obtain

$$h(t) = 3.653 \times 10^{-3} - 0.171 \times 10^{-4} t + 4.86 \times 10^{-8} t^2$$

$$h(10 \text{ hours}) = 3.484 \times 10^{-3}.$$

The reliability is obtained using Eq. (1.72) as

$$R(10) = \exp\left[-\left(3.653 \times 10^{-2} - \frac{0.171}{2} \times 10^{-2} + \frac{4.86}{3} \times 10^{-5}\right)\right]$$

$$= 0.9649.$$

1.4. Multivariate Hazard Rate

When a system is composed of two or more components, the joint life lengths are described by a multivariate distribution whose nature depends on the individual component life length. For example, consider a two-component system connected in parallel with each component having an exponential distribution life length. The system fails when the two components fail. When the effect of the operating conditions is accounted for, the joint life lengths of the components are shown to have a bivariate distribution whose marginals are univariate Paretos.

Assume that λ_i is the parameter of component i ($i = 1, 2$). If the lives of the two components are assumed to be independent, then the reliability of the system is

$$R(t) = e^{-\lambda_1 t} + e^{-\lambda_2 t} - e^{-(\lambda_1+\lambda_2)t}.$$

Suppose that the operating conditions affect the parameter λ_i by a common positive factor η. Then the system reliability is expressed as

$$R(t) = e^{-\eta\lambda_1 t} + e^{-\eta\lambda_2 t} - e^{-\eta(\lambda_1+\lambda_2)t}.$$

Following Lindley and Singpurwalla (1986), if η is an unknown quantity whose uncertainty is described by the distribution function $G(\eta)$, then the system reliability becomes

$$R(t) = G^*(\lambda_1 t) + G^*(\lambda_2 t) - G^*[(\lambda_1 + \lambda_2)t],$$

where

$$G^*(y) = \int \exp(-\eta y)\, dG(\eta)$$

is the Laplace transform of G.

When $G(\eta)$ is a gamma distribution with density,

$$g(\eta) = \beta^{\alpha+1} \frac{\eta^\alpha}{\alpha!} e^{-\eta\beta}, \tag{1.73}$$

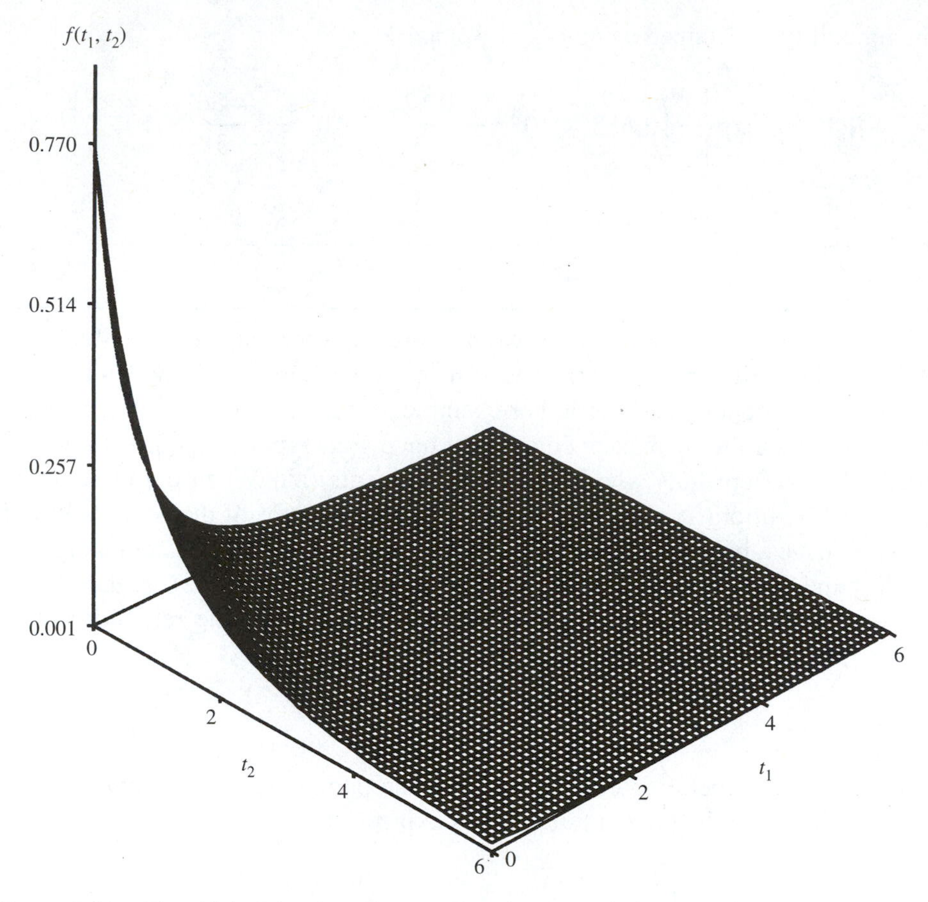

Figure 1.21. Plot of the Bivariate Gamma Density ($\lambda_1 = 0.5, \lambda_2 = 0.3, \alpha = 0.6, \beta = 0.9$)

where $\alpha > -1$ and $\beta > 0$, then

$$R(t) = \left(\frac{\beta}{\lambda_1 t + \beta}\right)^{\alpha+1} + \left(\frac{\beta}{\lambda_2 t + \beta}\right)^{\alpha+1} - \left(\frac{\beta}{(\lambda_1 + \lambda_2)t + \beta}\right)^{\alpha+1}. \quad (1.74)$$

The joint density of T_1 and T_2, the times to failure of the two components at t_1 and t_2, respectively, is

$$f(t_1, t_2, \lambda_1, \lambda_2, \alpha, \beta) = \frac{\lambda_1 \lambda_2 (\alpha + 1)(\alpha + 2)\beta^{\alpha+1}}{(\lambda_1 t_1 + \lambda_2 t_2 + \beta)^{\alpha+3}}. \quad (1.75)$$

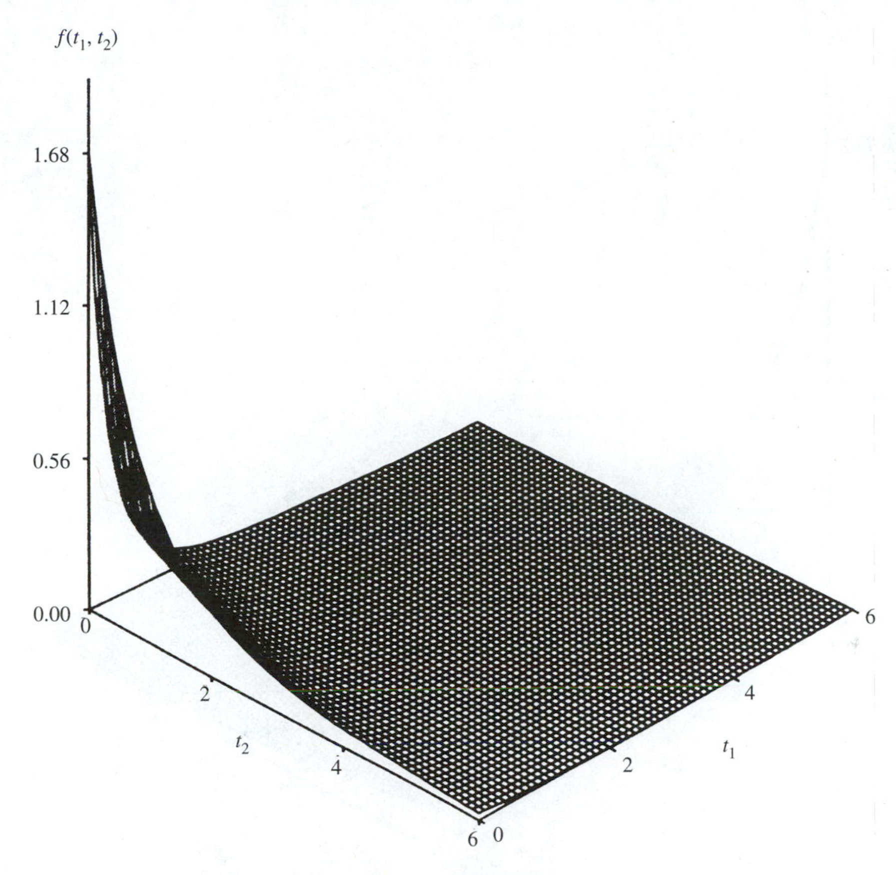

Figure 1.22. Plot of the Bivariate Gamma Density ($\lambda_1 = 0.9, \lambda_2 = 0.3, \alpha = 0.8, \beta = 0.9$)

Plots of Eq. (1.75) for different values of λ_1, λ_2, α, and β are shown in Figures 1.21 and 1.22.

The bivariate hazard rate of the system is

$$h(t_1, t_2, \lambda_1, \lambda_2, \alpha, \beta) = \frac{(\alpha + 1)(\alpha + 2)\lambda_1\lambda_2}{(\beta + \lambda_1 t_1 + \lambda_2 t_2)^2}. \tag{1.76}$$

The plots of the bivariate hazard rates for different λ_1, λ_2, α, and β are shown in Figures 1.23 and 1.24. Like univariate hazard rates, the bivariate hazard exhibits similar shapes—decreasing, constant, and increasing hazard rate.

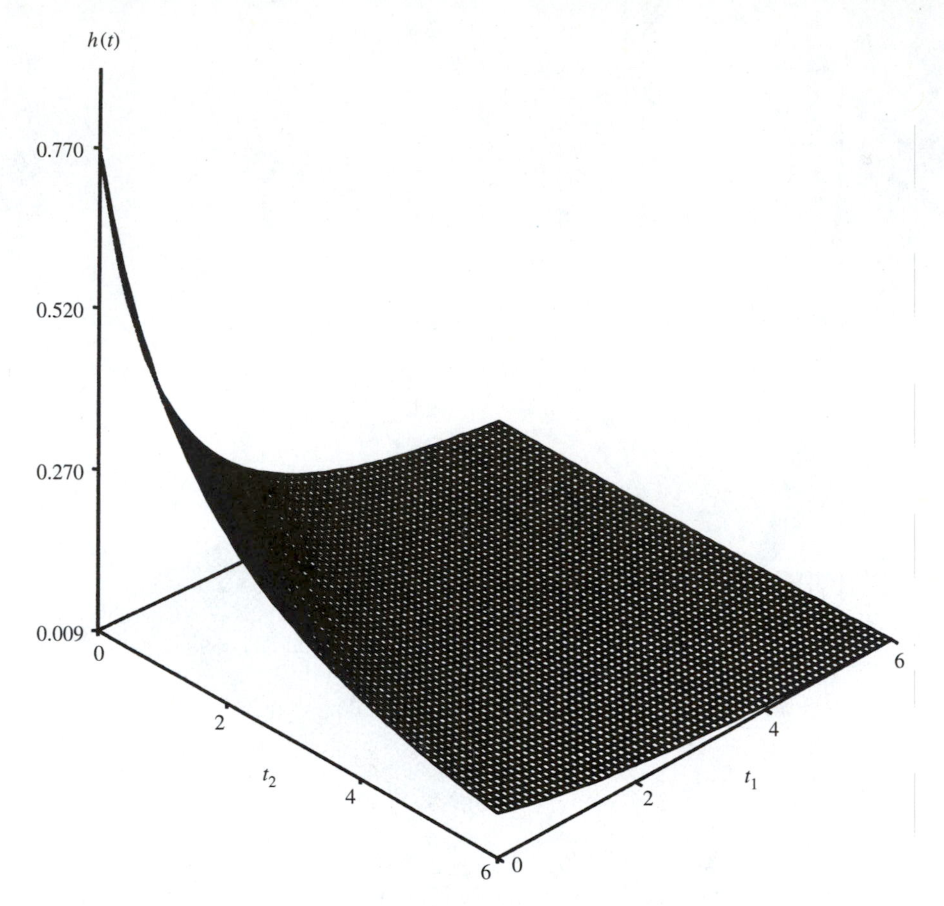

Figure 1.23. Plot of the Bivariate Hazard Rate ($\lambda_1 = 0.5$, $\lambda_2 = 0.3$, $\alpha = 0.6$, $\beta = 0.9$)

The marginal density function of t_1 is obtained by integrating Eq. (1.75) with respect to t_2, which yields

$$f(t_1, \lambda_1, \alpha, \beta) + \frac{\lambda_1(\alpha + 1)\beta^{\alpha+1}}{(\lambda_1 t_1 + \beta)^{\alpha+2}}. \tag{1.77}$$

The density function given by Eq. (1.77) is a Pearson Type VI whose mean and variance exist only for certain values of the shape parameter α. This distribution is also referred to as the "Pareto distribution of the second kind" (Lindley and Singpurwalla, 1986). Johnson and Kotz (1972) refer to Eq. (1.77) as the *Lomax distribution*.

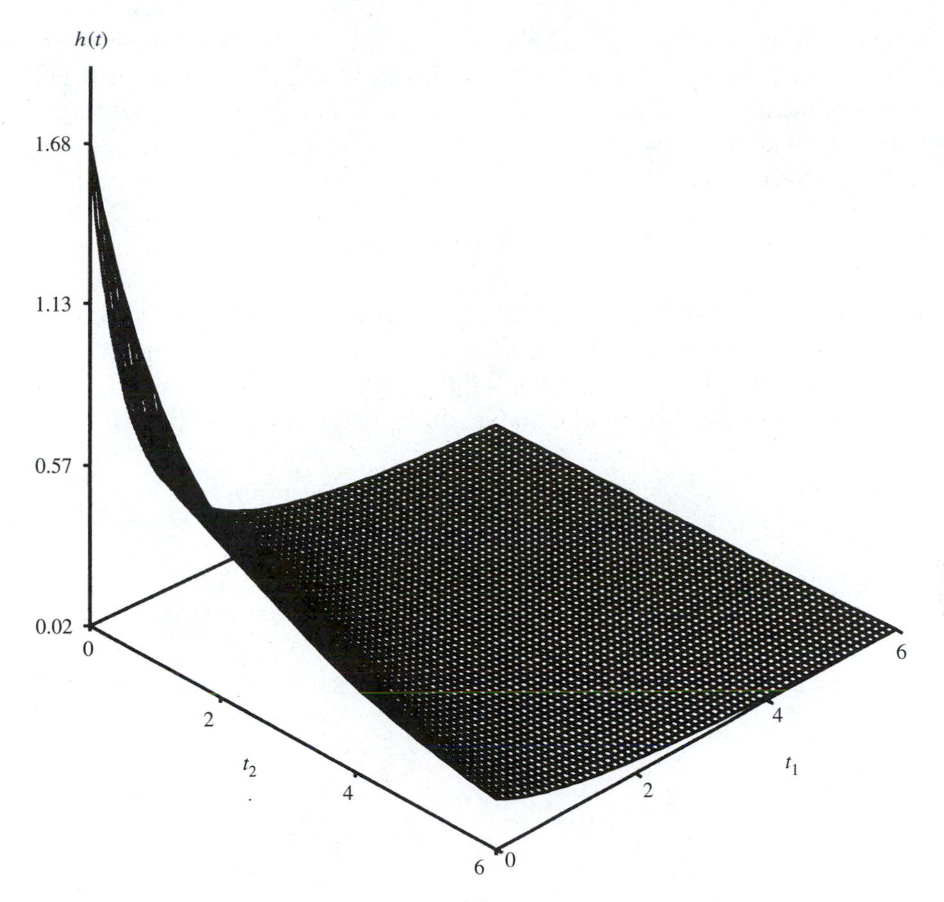

Figure 1.24. Plot of the Bivariate Hazard Rate ($\lambda_1 = 0.9, \lambda_2 = 0.3, \alpha = 0.8, \beta = 0.9$)

1.5. Mixture of Failure Rates

It is obvious that mixtures of distributions with decreasing failure rates (DFR) are always DFR. On the other hand, it may be intuitive to assume that the mixtures of distributions with increasing failure rates (IFR) are also IFR. Unfortunately, some mixtures of distributions with IFR may exhibit DFR. In this section we discuss the conditions that guarantee that mixtures of IFR distributions will exhibit a DFR.

This is very important since, in practice, different IFR distributions are usually pooled in order to enlarge the sample size. In doing so, the analysis of data may actually reverse the IFR property of the individual samples to a DFR property for the mixture. Proschan (1963) shows that the mixture of two exponential distributions (each has a constant failure rate) exhibits the DFR property.

Based on the work of Gurland and Sethuraman (1993), we consider mixtures of two arbitrary IFR distribution functions $F_i(t)$, $i = 1, 2$. The pooled distribution function of the mixture of the two distributions is $F_p(t) = p_1F_1(t) + p_2F_2(t)$ where $\mathbf{p} = (p_1, p_2)$ with $0 \le p_1, p_2 \le 1$, and $p_1 + p_2 = 1$ is a mixing vector.

We use the notation

$$h'_i(t) = H''_i(t) \quad \text{and} \quad \Re_i(t) = p_iR_i(t), \qquad i = 1, 2,$$

where $h_i(t)$, $H_i(t)$, and $R_i(t)$ are the hazard-rate function, the cumulative-hazard function, and the reliability function of component i at time t. From Section 1.2, $R_i(t) = 1 - F_i(t)$, $H_i(t) = -\log R_i(t)$, and $h_i(t) = H'_i(t)$.

The reliability function of the mixture of the two IFR distributions is

$$R_p(t) = p_1R_1(t) + p_2R_2(t).$$

But

$$H_p(t) = -\ln R_p(t)$$

$$H_p(t) = -\ln[p_1R_1(t) + p_2R_2(t)]$$

and

$$h_p(t) = H'_p(t) = \frac{p_1R_1(t)h_1(t) + p_2R_2(t)h_2(t)}{p_1R_1(t) + p_2R_2(t)}$$

$$= \frac{\Re_1(t)h_1(t) + \Re_2(t)h_2(t)}{\Re_1(t) + \Re_2(t)}. \tag{1.78}$$

A hazard-rate function $h_p(t)$ is a DFR if $h'_p(t) \le 0$. Therefore, we take the derivative of Eq. (1.78) with respect to t to obtain

$$\begin{aligned}(\Re_1(t) + \Re_2(t))^2h'_p(t) &= [\Re_1(t) + \Re_2(t)]\,[\Re_1(t)h'_1(t) \\ &\quad + \Re_2(t)h'_2(t) - \Re_1(t)h_1^2(t) - \Re_2(t)h_2^2(t)] \\ &\quad + [\Re_1(t)h_1(t) + \Re_2(t)h_2(t)]^2 \\ &= (\Re_1(t) + \Re_2(t))(\Re_1(t)h'_1(t) + \Re_2(t)h'_2(t)) \\ &\quad - \Re_1(t)\,\Re_2(t)(h_1(t) - h_2(t))^2.\end{aligned} \tag{1.79}$$

Using the fact that $\Re'_i(t) = -\Re_i(t)h_i(t)$ in the above equation, we show that the necessary and sufficient condition for $h'_p(t) \le 0$ and thus, for the mixture $F_p(t)$ to be DFR is

$$[\Re_1(t) + \Re_2(t)][\Re_1(t)h'_1(t) + \Re_2(t)h'_2(t)] \le \Re_1(t)\Re_2(t)[h_1(t) - h_2(t)]^2. \tag{1.80}$$

EXAMPLE 1.12

The failure-time distribution of a failure mode of a system is described by a truncated extreme distribution whose failure rate is $h_1(t) = \theta e^t$. Another mode of the system's failure exhibits a constant failure rate $h_2(t) = \lambda$. Although one failure mode of the system exhibits IFR while the other is a constant failure rate if treated separately, the analyst pools the data from both failure modes to obtain a pooled hazard rate function. Prove that the pooled hazard rate is a DFR.

SOLUTION

The reliability functions of the failure modes of the system are

$$R_1(t) = e^{-\theta(e^t-1)} \quad \text{and} \quad R_2(t) = e^{-\lambda t}.$$

The corresponding hazard rates are

$$h_1(t) = \theta \mathrm{e}^t \quad \text{and} \quad h_2(t) = \lambda.$$

Let $F_p(t) = (1 - p)\, F_1(t) + pF_2(t)$. Then the failure rate of the pooled data is

$$h_p(t) + \frac{(1-p)R_1(t)h_1(t) + pR_2(t)h_2(t)}{(1-p)R_1(t) + pR_2(t)}.$$

The necessary and sufficient condition that makes $h_p(t)$ a DFR function is given by Eq. (1.80). Substituting the parameters of the individual distributions and $\Re_1(t) = (1-p)\, e^{-\theta(et\, -\, 1)}$ and $\Re_2(t) = pe^{-\lambda t}$ it is easy to check that there is a $t_0(p)$ such that the derivative of the pooled hazard rate with respect to t is negative for $t \geq t_0(p)$ for each value of p. Thus the mixture is DFR.

The class of IFR distributions that, when mixed with an exponential, become DFR is large. It includes, for example, the Weibull, truncated extreme, gamma, truncated normal, and truncated logistic distributions. This phenomenon of the reversal of increasing failure rates could be troublesome in practice when much of the data conform to an IFR distribution and the remainder (perhaps a small amount) of the data conform to an exponential distribution, and yet the overall pooled data would conform to a DFR distribution (Gurland and Sethuraman, 1994).

Before concluding the presentation of the hazard functions, it is important to mention that some recent work argues that the bathtub curve is not a general failure-rate function that describes the failure rate of most, if not all, components. For example, Wong (1989) claims that the "roller-coaster" hazard-rate curve is more appropriate to describe the hazard rate of electronic systems than the bathtub curve. It is shown that semiconducting devices exhibit a generally decreasing hazard-rate curve with one or more humps on the curve. Data from a burn-in test of some electronic board assemblies demonstrate the trimodal (hump) characteristic on the cumulative failure rate. The wear-out (increasing failure rate) region starts immediately at the end of the decreasing failure-rate region without experiencing the constant failure-rate region, a main characteristic of the bathtub curve.

1.6. Mean Time to Failure (MTTF)

One of the measures of the system's reliability is the mean time to failure (MTTF). It should not be confused with the mean time between failure (MTBF). We refer to the expected time between two successive failures as the MTTF when the system is nonrepairable. Meanwhile, when the system is repairable we refer to it as the MTBF.

Now, let us consider n identical nonrepairable systems and observe the time to failure for them. Assume that the observed times to failure are $t_1, t_2, \ldots, t_n$. The estimated mean time to failure, $\widehat{\text{MTTF}}$, is

$$\widehat{\text{MTTF}} = \frac{1}{n}\sum_{i=1}^{n} t_{\text{i}}. \tag{1.81}$$

Since t_i is a random variable, then its expected value can be determined by

$$\text{MTTF} = \int_0^\infty t f(t)\, dt. \tag{1.82}$$

But $R(t) = 1 - F(t)$ and $f(t) = d\,F(t)/dt = -d\,R(t)/dt$. Substituting in Eq. (1.82), we obtain

$$\begin{aligned}\text{MTTF} &= -\int_0^\infty t \frac{d\,R(t)}{dt}\, dt \\ &= -\int_0^\infty t \; d\,R(t) \\ &= t\,R(t)\,\big|_0^\infty + \int_0^\infty R(t)\, dt.\end{aligned}$$

Since $R(\infty) = 0$ and $R(0) = 1$, then the first part of the above equation is 0 and the MTTF is

$$\text{MTTF} = \int_0^\infty R(t)\, dt. \tag{1.83}$$

The mean time to failure for a constant hazard-rate model is

$$\text{MTTF} = \int_0^\infty e^{-\lambda t}\,dt = \frac{1}{\lambda}. \tag{1.84}$$

The mean time to failure of a linearly increased hazard-rate model is:

$$\text{MTTF} = \int_0^\infty e^{\frac{-\lambda t^2}{2}}\,dt = \frac{\Gamma\left(\frac{1}{2}\right)}{2\sqrt{\frac{\lambda}{2}}} = \sqrt{\frac{\pi}{2\lambda}}. \tag{1.85}$$

Similarly, the MTTF for the Weibull model is

$$\text{MTTF} = \int_0^\infty e^{-\frac{t^\gamma}{\theta}}\,dt.$$

Substituting $x = t^\gamma/\theta$, the above equation becomes

$$\begin{aligned}
\text{MTTF} &= \frac{\theta}{\gamma}\int_0^\infty e^{-x}\frac{1}{\theta^{1-\frac{1}{\gamma}}}x^{\frac{1}{\gamma}-1}\,dx \\
&= \frac{\theta^{\frac{1}{\gamma}}}{\gamma}\int_0^\infty e^{-x}x^{\frac{1}{\gamma}-1}\,dx \\
&= \theta^{\frac{1}{\gamma}}\frac{1}{\gamma}\Gamma\left(\frac{1}{\gamma}\right) \\
&= \theta^{1/\gamma}\Gamma\left(1+\frac{1}{\gamma}\right).
\end{aligned} \tag{1.86}$$

EXAMPLE 1.13

The MTTF for a robot controller that will be operating in different stress conditions is specified to be warranted for 20,000 hours. The hazard-rate function of a typical controller is found to fit a Weibull model with $\theta = 100$ and $\gamma = 1.5$. Does the controller meet the warranty requirement? If not, what should the value of the characteristic life be to meet the requirement (measurements are in 10^3 hours)?

SOLUTION Substituting $\theta = 100$ and $\gamma = 1.5$ in Eq. (1.86), we obtain the MTTF as

$$\begin{aligned} \text{MTTF} &= 100^{\frac{1}{1.5}}\,\Gamma\left(1+\frac{1}{1.5}\right) \\ &= 19.383. \end{aligned}$$

Thus the MTTF is $19.383 \times 10^3 = 19{,}383$ hours. The MTTF does not meet the warranty requirement. The characteristic life that meets the requirement is calculated as

$$20{,}000 = \theta^{\frac{1}{1.5}}\,\Gamma(1.666).$$

θ should equal 104.46.

EXAMPLE 1.14 The failure time of an electronic device is described by a Pearson type V distribution. The density function of the failure time is

$$f(t) = \begin{cases} \dfrac{t^{-(\alpha+1)}e^{-\beta/t}}{\beta^{-\alpha}\,\Gamma(\alpha)} & \textit{if } t > 0 \\ 0 & \text{otherwise.} \end{cases}$$

The shape parameter $\alpha = 3$ and the scale parameter $\beta = 4{,}000$ hours. Determine the MTTF of the device.

SOLUTION Using Eq. (1.82), we obtain

$$\begin{aligned} \text{MTTF} &= \int_0^\infty \frac{t^{-\alpha}e^{-\beta/t}}{\beta^{-\alpha}\,\Gamma(\alpha)}\,dt \\ &= \frac{1}{\beta^{-\alpha}\,\Gamma(\alpha)} \int_0^\infty t^{-\alpha}e^{-\beta/t}\,dt \end{aligned}$$

or

$$\text{MTTF} = \frac{\beta}{\alpha - 1} = \frac{4{,}000}{3 - 1} = 2{,}000 \text{ hours.}$$

1.7. Mean Residual Life (MRL)

A measure of the reliability characteristic of a product, component, or a system is the *mean residual life* function $L(t)$. It is defined as

$$L(t) = E[T - t \mid T \geq t], \qquad t \geq 0. \tag{1.87}$$

In other words, the mean residual function is the expected remaining life, $T - t$, given that the product, component, or a system has survived to time t (Leemis, 1995).

The conditional probability density function for any time $\tau \geq t$ is

$$f_{T|T \geq t}(\tau) = \frac{f(\tau)}{R(t)}, \qquad \tau \geq t. \tag{1.88}$$

The conditional expectation of the function given in Eq. (1.88) is

$$E[T \mid T \geq t] = \int_t^\infty \tau f_{T|T \geq t}(\tau)\, d\tau = \int_t^\infty \tau \frac{f(\tau)}{R(t)}\, d\tau. \tag{1.89}$$

Since the component, product, or system has survived up to time t, the mean residual life is obtained by subtracting t from Eq. (1.89), thus

$$\begin{aligned} L(t) &= E[T - t \mid T \geq t] \\ &= \int_t^\infty (\tau - t)\frac{f(\tau)}{R(t)}\, d\tau = \int_t^\infty \tau \frac{f(\tau)}{R(t)}\, d\tau - t \end{aligned}$$

or

$$L(t) = \frac{1}{R(t)} \int_t^\infty \tau f(\tau)\, d\tau - t. \tag{1.90}$$

EXAMPLE 1.15

A manufacturer uses rotary compressors to provide cooling liquid for a power generating unit. Experimental data show that the failure times (between 0 and 1 year) of the compressors follow a beta distribution with $\alpha = 4$ and $\beta = 2$.

What is the mean residual life of a compressor given that the compressor has survived five months?

SOLUTION The p.d.f. of the failure time is

$$f(t) = \begin{cases} \dfrac{\Gamma(\alpha + \beta)}{\Gamma(\alpha)\Gamma(\beta)} t^{\alpha-1} (1 - t)^{\beta-1} & 0 < t < 1 \\ 0 & \text{otherwise,} \end{cases}$$

or

$$f(t) = \frac{\Gamma(6)}{\Gamma(4)\,\Gamma(2)} t^3(1 - t)$$

$$= 20(t^3 - t^4).$$

But

$$R(t) = 1 - F(t) = 1 - \int_0^t 20(\tau^3 - \tau^4)\, d\tau.$$

The value of t corresponding to five months is $5/12 = 0.416$, thus

$$R(0.416) = 1 - 20 \int_0^{0.416} (t^3 - t^4)\, dt = 0.900.$$

Using Eq. (1.90), we obtain the mean residual life of a compressor that survived five months as

$$L(0.416) = \frac{20}{0.900} \int_{0.416}^{1} t(t^3 - t^4)\, dt - 0.416$$

$$= 0.288$$

or the mean residual life is 3.46 months.

Table 1.8 summarizes the characteristics of the hazard functions discussed in this chapter.

Table 1.8. Characteristics of the Hazard Functions

Hazard Function	*h(t)*	*f(t)*	*R(t)*	*Parameters*
Constant	λ	$\lambda e^{-\lambda t}$	$e^{-\lambda t}$	λ
Linearly increasing	λt	$\lambda t\, e^{\frac{-\lambda t^2}{2}}$	$e^{\frac{-\lambda t^2}{2}}$	λ
Weibull	$\frac{\gamma}{\theta} t^{\gamma-1}$	$\frac{\gamma}{\theta} t^{\gamma-1}\, e^{\frac{-t^\gamma}{\theta}}$	$e^{\frac{-t^\gamma}{\theta}}$	γ, θ
Exponential	$b\, e^{\alpha t}$	$b\, e^{\alpha t}\, e^{\frac{-b}{\alpha}\left(e^{\alpha t}-1\right)}$	$e^{\frac{-b}{\alpha}\left(e^{\alpha t}-1\right)}$	α, b
Normal	$\dfrac{\phi\left(\frac{t-\mu}{\sigma}\right)}{\sigma\, R(t)}$	$\dfrac{1}{\sqrt{2\pi\sigma^2}}\, e^{\frac{-1}{2}\left(\frac{t-\mu}{\sigma}\right)^2}$	$1-\int_{-\infty}^{t} \dfrac{1}{\sigma\sqrt{2\pi}}\, e^{\frac{-1}{2}\frac{(\tau-\mu)}{\sigma^2}}\, d\tau$	μ, σ
Lognormal	$\dfrac{\phi\left(\frac{\ln t-\mu}{\sigma}\right)}{t\sigma\, R(t)}$	$\dfrac{1}{\sigma t\sqrt{2\pi}}\, e^{\frac{-1}{2}\left(\frac{\ln t-\mu}{\sigma}\right)^2}$	$1-\int_{0}^{t} \dfrac{1}{\tau\sigma\sqrt{2\pi}}\, e^{\frac{-1}{2}\left(\frac{\ln \tau-\mu}{\sigma}\right)^2}\, d\tau$	μ, σ
Gamma	$\dfrac{f(t)}{R(t)}$	$\dfrac{t^{\gamma-1}}{\theta^\gamma\, \Gamma(\gamma)}\, e^{\frac{-t}{\theta}}$	$\int_{t}^{\infty} \dfrac{1}{\theta\, \Gamma(\gamma)} \left(\dfrac{\tau}{\theta}\right)^{\gamma-1} e^{-\frac{\tau}{\theta}}\, d\tau$	θ, γ
Log-logistic	$\dfrac{\lambda p(\lambda t)^{p-1}}{1+(\lambda t)^p}$	$\dfrac{\lambda p(\lambda t)^{p-1}}{[1+(\lambda t)^p]^2}$	$\dfrac{1}{1+(\lambda t)^p}$	λ, p

PROBLEMS

It is appropriate to use *Reliability Analysis Software*™ (the software included with this book) to plot $R(t), f(t)$, $h(t)$, and $F(t)$ when required.

1-1. Determine the mean and the variance of a uniform random variable X whose p.d.f. is

$$f(x) = \frac{1}{b-a} \qquad a < x < b$$

$$= 0 \qquad \text{otherwise.}$$

1-2. Determine the first and second moments for a normal distribution with parameters μ and σ^2.

1-3. The p.d.f. of the lognormal distribution is given by

$$f(t) = \frac{1}{\sigma t \sqrt{2\pi}} e^{-\frac{1}{2}\left\{\frac{\ln(t)-\mu}{\sigma}\right\}^2}.$$

Determine the variance and the median. (Hint: Median is defined as $\int_{med}^{\infty} f(x)\,dx = 1/2$).

1-4. A mechanical fatigue test is conducted on 100 specimens of a new polymer. The applied stress is identical for all specimens. The number of cycles observed and the corresponding number of failed specimens are given below:

Number of Cycles $\times$ *10*^{5}	*Cumulative Number of Failed Specimens*
10	35
20	59
30	72
40	84
50	93
60	100

a. Plot graphs for $f_e(t)$, $R_e(t)$, $h_e(t)$, and $F_e(t)$.

b. Comment on the above results.

c. Derive an analytical expression for $h_e(t)$ and estimate the MTTF of a bar made of the same material and subjected to the same loading conditions.

1-5. The reliability of disk drives can be predicted by increasing the operational machine hours accumulated in the field or in the laboratory as part of the initial design process. The following failures have been accumulated:

Hour of Operation $\times$ *10*^{3}	*Number of Failed Disks*
0–10.0	0
10.1–14.0	10
14.1–18.0	15
18.1–22.0	18
22.1–26.0	20
26.1–30.0	16
30.1–34.0	22
34.1–38.0	20

a. Plot graphs for $f_e(t)$, $R_e(t)$, $h_e(t)$, and $F_e(t)$.

b. Comment on the above results.

c. Derive an analytical expression for $h_e(t)$ and estimate the MTTF of a bar made of the same material and subjected to the same loading conditions.

d. Would you buy a disk produced by the above manufacturer? Why?

1-6. One of the modern methods for stress screening is called highly accelerated stress screening (HASS), which uses the highest possible stresses (well beyond the normal operating level) to attain time compression on the screens. The HASS exhibits an exponential acceleration of screen strength with stress level. A manufacturer employs a HASS test on newly designed leaf springs for light trucks. A cyclic load was applied on a number of springs and the failure times are recorded below:

Time Interval (Minutes)	*Number of Failed Units*
0–1.999	10
2–3.999	15
4–5.999	22
6–7.999	34
8–9.999	49
10–11.999	63
12–14	70

a. Fit a nonlinear polynomial hazard function to describe the hazard rate of the springs.

b. What is the reliability at $t = 8$?

c. Assume that we obtained 500 springs that require testing under the same conditions. What is the expected time to failure? What is the least time needed to ensure that all units fail under test?

1-7. Show that the variance of a component whose hazard rate can be described by $h(t) = (\gamma/\theta)t^{\gamma-1}$ is

$$\mathrm{Var}[T] = \theta^{\frac{2}{\gamma}}\left\{\Gamma\left(1+\frac{2}{\gamma}\right) - \left[\Gamma\left(1+\frac{1}{\gamma}\right)\right]^2\right\},$$

where

$$\Gamma(n) = \int_0^{\infty} \tau^{n-1} e^{-\tau} \, d\tau$$

and

$$\int_0^{\infty} \tau^{n-1} \, e^{-\tau/\theta} \, d\tau = \Gamma(n) \, \theta^n.$$

1-8. Use Weibull graph paper to estimate the parameters of a Weibull distribution that fits the data given in Problem 1-6.

1-9. Plot the $h(t)$ and $R(t)$, for $t = 0$ to 1,000, for different shape parameters of 0.5 to 3.5 with an increment of 0.5, and for different characteristic lives of 200 to 300 with an increment of 25. What is the effect of the characteristic life on the hazard-rate function? What is the best combination of shape parameter and characteristic life that results in the highest reliability at $t = 1{,}000$? (Weibull distribution).

1-10. Dhillon (1979) proposes a hazard-rate model given by

$$h(t) = k\lambda c t^{c-1} + (1 - k) b t^{b-1} \beta e^{\beta t^b}$$

for

$$b, c, \beta, \lambda > 0 \qquad 0 \leq k \leq 1 \qquad t \geq 0,$$

where

b, c	=	shape parameters,
β, λ	=	scale parameters, and
t	=	time.

Derive the reliability function and determine the conditions that make the hazard rate increasing, decreasing, or constant.

1-11. A rolling bearing rotating under load may ultimately suffer from material fatigue. Typically, fatigue damage is characterized by a small piece of material breaking away from the raceway leaving a cavity. This cavity may then propagate into a crack and the bearing will fail. If a large batch of identical bearings is run under the same conditions until 10 percent of the batch has failed from the material fatigue damage, then the batch is said to have attained its L_{10} life. In other words, the remaining 90 percent of the bearings in the batch will survive for periods longer than the L_{10} life. Consider a rolling bearing that has a hazard rate function in the form

$$h(t) = \frac{\dfrac{1}{\theta}\left(\dfrac{t}{\theta}\right)^{n-1}}{(n-1)! \displaystyle\sum_{k=0}^{n-1} \dfrac{(t/\theta)^k}{k!}},$$

where $n = 3$ and $\theta = 290$ hours. Determine the reliability of the bearing at $t = 100$ hours. Assuming $L_{10} = 100$ hours, determine the mean residual life of the bearing.

1-12. Find $f(t)$, $h(t)$, $R(t)$ and MTTF, assuming

$$F(t) = 1 - \frac{8}{7}e^{-t} + \frac{1}{7}e^{-8t}.$$

1-13. Find $f(t)$, $F(t)$, $R(t)$ and MTTF, assuming

$$h(t) = \frac{1}{25}t^{-1/4}.$$

If 200 units are placed in operation at the same time, how many failures are expected during one year of operation?

1-14. The failure rate of a brake system is found to be

$$h(t) = 0.006(1.5 + 2t - 3t^2) \text{ failures per year.}$$

a. What is the reliability at $t = 10^4$ hours?

b. If twenty systems are subjected to a test at the same time, how many would have survived at time $t = 10^3$ hours? What is the expected number of failures in one year of operation?

1-15. The failure rate of a hydraulic system is found to be

$$h(t) = 0.003\,(1 + 2.5\mathrm{e}^{-3t} + \mathrm{e}^{-t/50}) \text{ failures per year.}$$

a. What is the reliability at $t = 10^5$ hours?

b. What is the mean time to failure?

c. If ten systems are subjected to a test at the same time, how many would have survived at time $t = 10^3$ hours? What is the expected number of failures in one year of operation?

1-16. Consider the general hazard failure rate (Hjorth, 1980) that is given by

$$h(t) = \delta t + \frac{\theta}{1 + \beta t}.$$

Special cases are:

$\theta = 0$	The Rayleigh distribution,
$\delta = \beta = 0$	The exponential distribution,
$\delta = 0$	Decreasing failure rate,
$\delta \geq \theta\beta$	Increasing failure rate, and
$0 < \delta < \theta\beta$	Bathtub curve.

The reliability function corresponding to this general hazard rate is

$$R(t) = \frac{e^{-\delta t^2/2}}{(1 + \beta t)^{\theta/\beta}}, \qquad t \geq 0.$$

Let T have the above reliability function, and define

$$I(a, b) = \int_0^\infty \frac{e^{-a t^2/2}}{(1 + t)^b} \, dt.$$

Find the mean and the variance of T. Plot the hazard rate for different parameters.

1-17. The probability density function of the failure time of the PCBs (printed circuit boards) to be used in a plug-compatible video display terminal is found to follow a Cauchy distribution, which is given by

$$f(t) = \frac{1}{\pi\beta\{1 + [(t - \alpha)/\beta]^2\}} \qquad -\infty < \alpha < \infty,\ \beta > 0,\ -\infty < t < \infty,$$

$$f(t) = 0 \qquad \text{otherwise.}$$

Find the reliability function, the hazard rate, and the MTTF for the special case when $\beta = 1$ and $\alpha = 0$. Is the hazard rate increasing, decreasing, or constant?

1-18. The probability density function of the early failure times of the circuit boards used in 9600 baud rate modems are found to follow a Pearson type V distribution which is given by

$$f(t) = \frac{t^{-(\alpha+1)}e^{-\beta/t}}{\beta^{-\alpha}\Gamma(\alpha)} \qquad \text{if } t > 0$$

$$f(t) = 0 \qquad \text{otherwise,}$$

where α and β are the shape and scale parameters, respectively. Find the reliability function, the hazard rate, and the MTTF for the special case when $\beta = 1$ and $\alpha = 3$. Is the hazard rate increasing, decreasing, or constant?

1-19. Let t denote the time to failure of a component whose p.d.f. is given by

$$f(t) = \frac{1}{\ln 2}\frac{1}{t}, \qquad 25{,}000 < t < 50{,}000 \text{ hours.}$$

a. Verify that f is a density for a continuous random variable.

b. What is the hazard function of this component?

c. What is the expected life of the component?

1-20. In most electronic manufacturing operations, the role of process control has traditionally fallen to automated board-test systems. These systems are typically placed at the end of the manufacturing line in order to monitor fault trends and thus help control the process. The failure data collected at board-test system show that the failure time follows a triangular distribution with the following p.d.f.:

$$f(t) = \begin{cases} \dfrac{2(t-a)}{(b-a)(c-a)} & \text{if } a \leqq t \leqq c \\[2ex] \dfrac{2(b-t)}{(b-a)(c-a)} & \text{if } c < t \leqq b \\[2ex] 0 & \text{otherwise,} \end{cases}$$

where a, b, and c are real numbers with $a < c < b$. a is a location parameter, $b - a$ is a scale parameter, c is a shape parameter. Assume that $a = 2$, $b = 4$, and $c = 3$. What is the expected mean time to failure? What is the variance of time to failure?

1-21. A manufacturer intends to introduce a new product. Five products are subjected to a reliability test. The mean of the failure times is 300 hours and the variance is 90,000 hours^2. Since the number of failure data is limited, it is difficult to determine with an acceptable confidence level the type of the failure time distribution.

a. What is the expected number of failures at 500 hours?

b. The similarity between this product and another product that has been already in the market for the last ten years indicates that the failure-time distribution is likely to follow a Gamma distribution. What is the expected number of failures under these conditions at 500 hours? Compare the results with (a) above. What do you conclude?

1-22. The failure time of a new brake drum design is observed to follow a Gamma distribution with a p.d.f. of

$$f(t) = \frac{\lambda(\lambda t)^{\gamma-1}e^{-\lambda t}}{\Gamma(\gamma)}.$$

For $\gamma = 2$ and $\lambda = 0.0002$, determine

a. The expected number of failures in one year of operation,

b. The mean time to failure, and

c. The reliability at $t = 1{,}000$ hours.

1-23. Solve the above problem when $\gamma = 3$ and $\lambda = 0.0002$. Compare the results. Which brake system is better? Why?

1-24. Most fractional horsepower motor controllers use a silicon-controlled rectifier (SCR) to vary the power applied to the motor and thereby control armature voltage and thus the motor's speed. The SCR is made of different layers of semiconductor materials. The heat dissipation from the motor increases the failure rate of the SCR. Failure data from the field show that the failure time follows a beta distribution with the following p.d.f.:

$$f(t) = \begin{cases} \dfrac{\Gamma(\alpha+\beta+2)}{\Gamma(\alpha+1)\Gamma(\beta+1)} t^{\alpha}(1-t)^{\beta} & 0 < t < 1,\ a > -1,\ \beta > -1 \\ 0 & \text{otherwise.} \end{cases}$$

Assuming that $\alpha = 1.8$ and $\beta = 4.7$, what is the expected mean time to failure? What is the variance? What is the expected number of failures at $t = 2.5$?

1-25. Consider the case where the failure time of components follows a logistic distribution with a p.d.f. of

$$f(t) = \frac{(1/\beta)e^{-(t-\alpha)/\beta}}{(1 + e^{-(t-\alpha)/\beta})^2}, \qquad -\infty < \alpha < \infty,\ \beta > 0,\ -\infty < t < \infty.$$

Determine the expected number of failures in the interval $[t_1, t_2]$.

1-26. In order for a manufacturer to determine the length of the warranty period for newly developed ICs (integrated circuits), 100 units are placed under test for 5,000 hours. The hazard rate function of the units is

$$h(t) = 5 \times 10^{-9}\, t^{0.9}.$$

What is the expected number of failures by the end of the test? Should the manufacturer make the warranty period longer or shorter if the ICs were redesigned and its new hazard-rate function became $h(t) = 6 \times 10^{-8}\, t^{0.75}$?

1-27. The manufacturer of diodes subjects 100 diodes to an elevated temperature testing for a two-year period. The failed units are found to follow a Weibull distribution with parameters $\theta = 50$ and $\gamma = 2$ (in thousands of hours). What is the expected life of the diodes? What is the expected number of failures in a two-year period?

1-28. In Problem 1-27, if a diode survives one year of operation, what is its mean residual life?

1-29. The hazard rate function of a manufacturer's jet engines is a function of the amount of silver and iron deposits in the engine oil. If the metal deposit readings are "high," the engine is removed from the aircraft and overhauled. The hazard-rate function (Jardine and Buzacott, 1985) is

$$h(t; z(t)) = \frac{5.335}{3255.19}\left(\frac{t}{3255.19}\right)^{4.335} \exp[0.506\, z_1(t) + 1.25\, z_2(t)],$$

where

t = flight hours,

$z_1(t)$ = iron deposits in parts per million at time t, and

$z_2(t)$ = silver deposits in parts per million at time t.

Analysis of the deposits over time shows that

$$z_1(t) = 0.0005 + 0.00006t$$

$$z_2(t) = 0.00008t + 8 \times 10^{-8}t^2.$$

Plot the reliability of the engine against flying hours. What is the mean time to failure?

1-30. Determine the mean life and the variance of a component whose failure time is expressed by

$$f(t) = \sum_{i=1}^{n} p_i \frac{\gamma_i}{\theta_i} t^{\gamma_i - 1} e^{\frac{-t^{\gamma_i}}{\theta_i}},$$

where

$$\sum_{i=1}^{n} p_i = 1.$$

1-31. Assume that the mean hazard rate is given by

$$E[h(T)] = \int_0^\infty h(t) f(t)\, dt$$

and the mean time to failure $E[T]$ is

$$E[T] = \int_0^\infty R(t)\, dt.$$

Prove that $\{E[h(T)] \cdot E[T]\}$ is an increasing function of the shape parameter of the Weibull model.

1-32. Consider a Weibull distribution with a reliability function $R(t) = \exp(-\theta\lambda t^{\gamma})$ for $t \geq 0$. For $\gamma > 1$, $\theta > 0$, and $\lambda > 0$, the Weibull density becomes an IFR distribution (the wear-out region of the bathtub curve). Suppose that the values of λ follow a gamma distribution with p.d.f. $f(\lambda)$ given by

$$f(\lambda) = \frac{\alpha^{\beta}}{\Gamma(\beta)} e^{-\alpha\lambda}\lambda^{\beta-1} \qquad \alpha > 0,\ \beta > 0,\ \lambda > 0.$$

The reliability function of the mixture is given by

$$R_{\text{mixture}}(t) = \int_0^{\infty} R(t) f(\lambda)\, d\lambda.$$

a. Show that the failure-rate function of the mixture is given by (Gurland and Sethuraman, 1994):

$$h_{\text{mixture}}(t) = \beta \frac{\theta\gamma t^{\gamma-1}}{\alpha + \theta\, t^{\gamma}}.$$

b. Plot $h_{mixture}(t)$ for large values of t. What do you conclude?

c. Plot the hazard rate for different values of α, β, θ, and γ. What are the conditions at which $h_{mixture}(t)$ is an IFR function? A DFR function? A constant failure-rate function?

1-33. Data from a linearly increasing failure-rate distribution is mixed with some data from a constant failure-rate distribution. Assume that the linearly increasing failure rate is a Rayleigh distribution with $R_R(t) = e^{-\lambda t^2/2}$, where λ is a constant, and the reliability function of the constant failure rate is $R_c(t) = e^{-\theta t}$. Investigate $h(t)$ for the mixture of the distributions.

1-34. The failure time of a component follows a Pareto distribution with a p.d.f. of

$$f(t) = \frac{\gamma\lambda^{\gamma}}{t^{\gamma+1}}, \qquad \lambda > 0,\ \gamma > 0,\ \lambda < t < \infty.$$

Determine the MTTF of the component and its mean residual life function.

REFERENCES

Collins, J. A. (1981). *Failure of Materials in Mechanical Design.* New York: Wiley.

Comeford, R. (1989). "Reliability Testing Surges Forward." *Electronics Test* (December), 52–53.

Cox, D. R. (1962). *Renewal Theory*. London: Methuen.

Dhillon, B. S. (1979). "A Hazard Rate Model." *IEEE Transactions on Reliability* R28, 150.

Gurland, J., and Sethuraman, J. (1993). "How Pooling Failure Data May Reverse Increasing Rates." Technical Report No. 907, Department of Statistics, University of Wisconsin.

Gurland, J., and Sethuraman, J. (1994). "Reversal of Increasing Failure Rates When Pooling Failure Data." *Technometrics* 36(4), 416–418.

Hjorth, U. (1980). "A Reliability Distribution with Increasing, Decreasing, Constant, and Bathtub-Shaped Failure Rates." *Technometrics* 22(1), 99–107.

Jardine, A. K. S. (1973). *Maintenance, Replacement, and Reliability*. New York: Wiley.

Jardine, A. K. S., and Buzacott, J. A. (1985). "Equipment Reliability and Maintenance." *European Journal of Operational Research* 19, 285–296.

Johnson, N. L., and Kotz, S. (1972). *Distributions in Statistics: Continuous Univariate Distributions*. New York: Wiley.

Kalbfleisch, J. D., and Prentice, R. L. (1980). *The Statistical Analysis of Failure Time Data*. New York: Wiley.

Kao, J. H. K. (1959). "A Graphical Estimation of Mixed Weibull Parameters in Life-Testing of Electron Tubes." *Technometrics* 1(4), 389–407.

Leemis, L. M. (1995). *Reliability: Probabilistic Models and Statistical Methods*. Englewood Cliffs, NJ: Prentice-Hall.

Lindley, D. V., and Singpurwalla, N. D. (1986). "Multivariate Distributions for the Life Lengths of Components of a System Sharing a Common Environment." *Journal of Applied Probability* 23, 418–431.

Makino, T. (1984). "Mean Hazard Rate and Its Application to the Normal Approximation of the Weibull Distribution." *Naval Research Logistics Quarterly* 31, 1–8.

Pearson, K. (1957). *Tables of Incomplete Γ-Function*. Cambridge: Cambridge University Press.

Proschan, F. (1963). "Theoretical Explanation of Observed Decreasing Failure Rate." *Technometrics* 5, 373–383.

Wong, K. L. (1989). "The Roller-Coaster Curve Is In." *Quality and Reliability Engineering International* 5, 29–36.

CHAPTER 2

System Reliability Evaluation

2.1. Introduction

In Chapter 1, we presented definitions of reliability, hazard functions, and other measures of reliability, such as MTTF and the expected number of failures in a given time interval. These definitions and measures are applicable to both components and systems. A system (or a product) is a collection of components arranged according to a specific design in order to achieve desired functions with acceptable performance and reliability measures.

Clearly, the type of components used, their qualities, and the design configuration in which they are arranged have a direct effect on the system performance and its reliability. For example, a designer may use a smaller number of high-quality components (made of prime material) and configure them in such a way to result in a highly reliable system, or a designer may use a larger number of *lower*-quality components and configure them differently in order to achieve the same level of reliability. A system configuration may be as simple as a series system where all components are connected in series; a parallel system where all components are connected in parallel; a series-parallel; or a parallel-series, where some components are connected in series and others in parallel and a complex configuration such as networks. Once the system is configured, its reliability must be evaluated and compared with an acceptable reliability level. If it does not meet the required level, the system should be redesigned and its reliability should be reevaluated. The design process continues until the system meets the desired performance measures and reliability level.

As seen above, system reliability needs to be evaluated as many times as the design changes. This chapter presents methods for evaluating reliability of systems with different configurations and methods for assessing the importance of a component in a complex structure. The presentation is limited to those systems that exhibit constant probability of failure. In the next chapter, time-dependent reliability systems are discussed.

2.2. Reliability Block Diagrams

The first step in evaluating a system's reliability is to construct a reliability block diagram, which is a graphical representation of the components of the system and how they are connected. A block (rectangle) does not show any details of the component or the subsystem it represents. The second step is to create a reliability graph that corresponds to the block diagram. The reliability graph is a line representation of the blocks that indicates the path on the graph. The following examples illustrate the construction of both the reliability block diagram and the reliability graph.

EXAMPLE 2.1

A computer tomography system is used as a nondestructive method to inspect welds from outside when the inner surfaces are inaccessible. It consists of a source for illuminating the rotating welded part with a fan-shaped beam of X rays or gamma rays as shown in Figure 2.1. Detectors in a circular array on the opposite

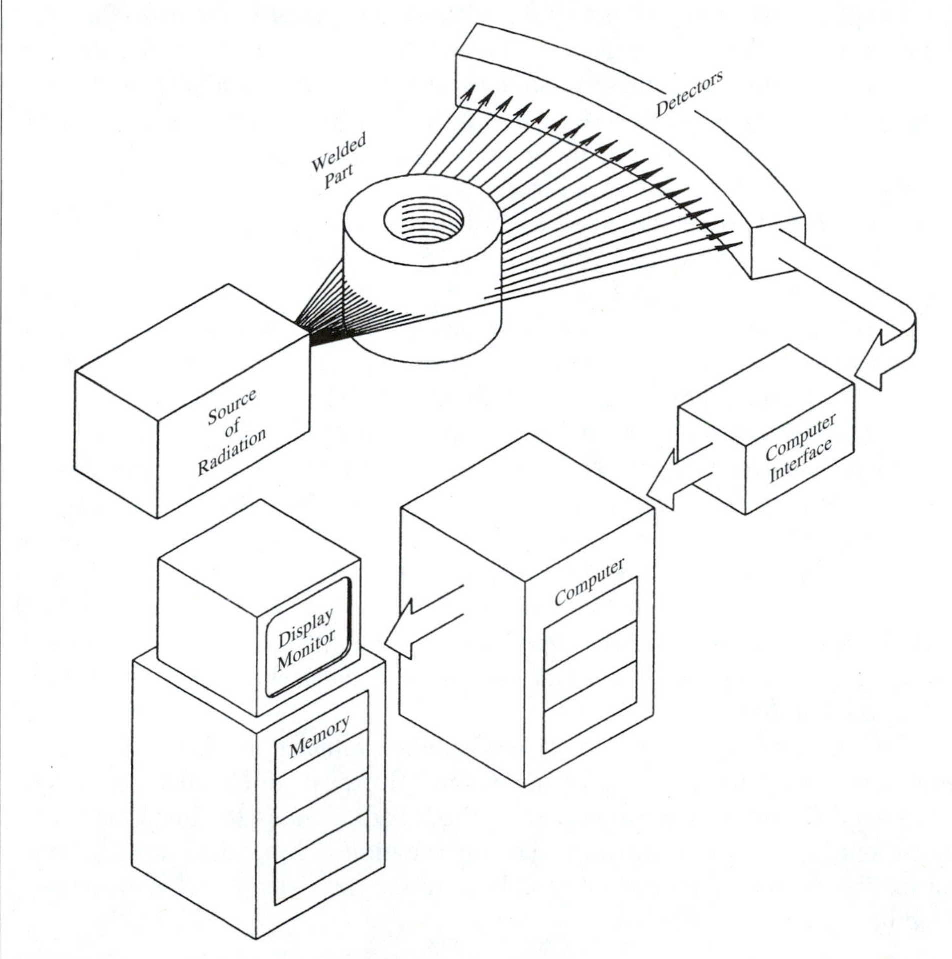

Figure 2.1. A Computer Tomography System

side of the part intercept the beam and convert it into electrical signals. A computer processes the signals into an image of a cross-section of the weld, which is displayed on a video monitor (NASA, 1990). Draw the reliability block diagram and the reliability graph of the system.

SOLUTION

The reliability block diagram and the reliability graph are shown in Figure 2.2.

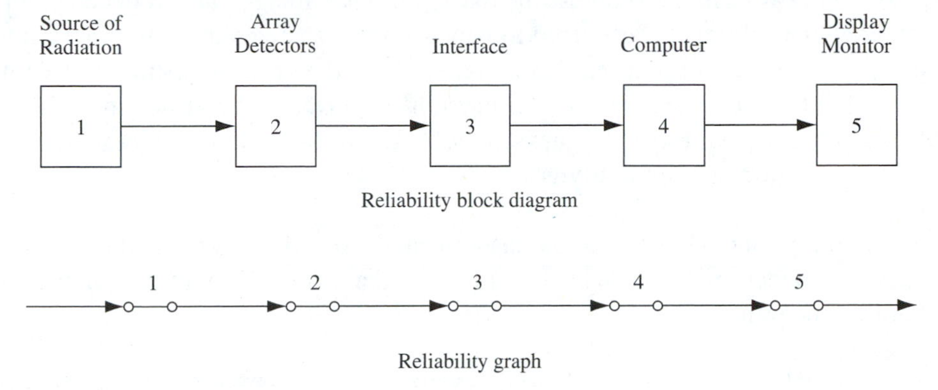

Figure 2.2. Reliability Block Diagram and Reliability Graph

EXAMPLE 2.2

The operating principle of a typical laser printer is explained as follows (refer to Figure 2.3). The main component of the laser printer is a photoconductor drum

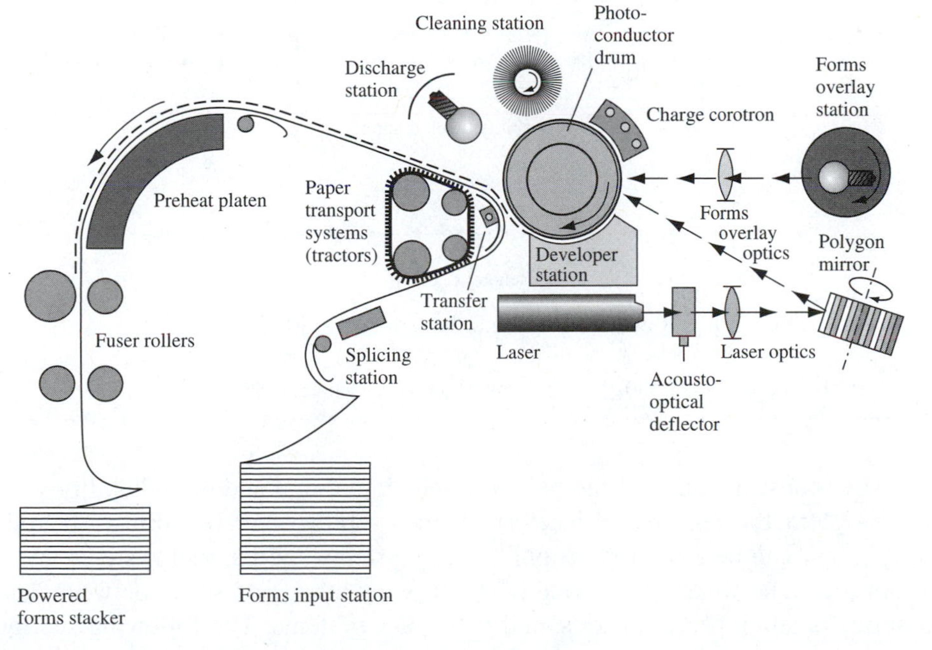

Figure 2.3. Operating Principle of the Laser Printer

that rotates at a constant speed with a semiconductor layer. During each revolution, this layer is electrostatically charged by a charge corotron. A laser beam, deflected vertically by an acousto-optical deflector and horizontally by a rotating polygon mirror, writes the print information onto the semiconductor layer by partially discharging this layer. Subsequently, as the drum passes through a "toner bath" in the developer station, the locations on the drum that have been discharged by the laser beam will then capture the toner. The print image thus produced on the photoconductor drum is transferred to paper in the transfer station and fused into the paper surface to prevent image smudging. After the printing operation is complete, a light source discharges the semiconductor layer on the drum and a brush removes any residual toner (Siemens, 1983)[1]. Draw the reliability block diagram and the corresponding reliability graph.

SOLUTION The reliability block diagram of the laser printer is a series system and the failure of any component will result in the failure of the system. Figure 2.4 shows the reliability diagram.

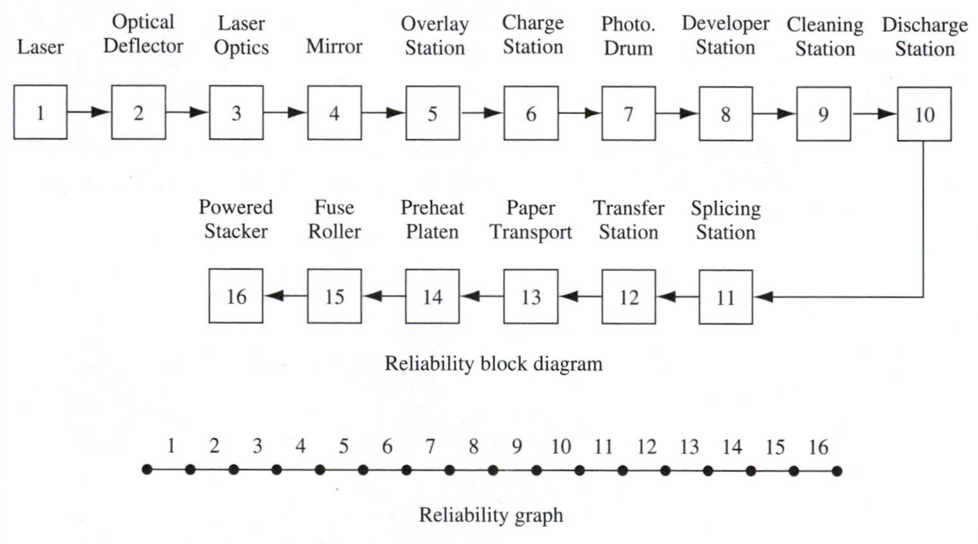

Figure 2.4. Reliability Block Diagram and the Corresponding Graph

[1] Source: "The Laser Printer, bt020e," Siemens Aktiengesellschaft, 1983.

After constructing both the reliability block diagram and the reliability graph of the system, the next step is to determine the overall system reliability. The reliability graph can be as simple as pure series systems and parallel systems and as complex as a network with a wide range of many other systems in between, such as series-parallel, parallel-series, and k-out-of-n systems. The following sections present different approaches for determining the reliability of systems, starting with the simplest system.

2.3. Series Systems

A series system is composed of n components (or subsystems) connected in series. A failure of any component results in the failure of the entire system. A car, for example, has several subsystems connected in series, such as the ignition subsystem, the steering subsystem, and the braking subsystem. The failure of any of these subsystems causes the car not to perform its function, and thus a failure of the system is considered to occur. In the car example, each subsystem may consist of more than one component connected in any of the configurations mentioned earlier. However, in estimating the reliability of the car the subsystems are treated as components connected in series. In all situations, a system or a subsystem can be analyzed at different levels down to the component level. Another example of a simple series system is a flashlight system, which consists of three components connected in series—bulb, battery, and switch. The three must function properly for the flashlight to operate properly.

In order to determine the reliability of a series system, assume that the probability of success of every unit in the system is known at the time of the system's evaluation. Assume also the following notations:

x_i = the ith unit is operational,

$\bar{x}_i$ = failure of the ith unit,

$P(x_i)$ = probability that unit i is operational,

$P(\bar{x}_i)$ = probability that unit i is not operational (failed),

R = reliability of the system, and

P_f = unreliability of the system ($P_f = 1 - R$).

Since the successful operation of the system consisting of n components requires that all units must be operational, then the reliability of the system can be expressed as

$$R = P(x_1 x_2 \ldots x_n)$$

or

$$R = P(x_1)P(x_2 / x_1)P(x_3 / x_1 x_2) \ldots P(x_n / x_1 x_2 x_3 \ldots x_{n-1}). \quad (2.1)$$

The conditional probabilities in Eq. (2.1) reflect the case when the failure mechanism of a component affects other components' failure rates. A typical example of such a case is the heat dissipation from a failing component, which causes the failure rate of adjacent components to increase. When the components' failures are independent, then Eq. (2.1) can be written as

$$R = P(x_1)P(x_2)\ldots P(x_n)$$

or

$$R = \prod_{i=1}^{n} P(x_i). \tag{2.2}$$

Alternatively, the reliability of the system can be determined by computing the probability of system failure and subtracting it from unity. The system fails if any of the components fails. Thus,

$$P_f = P(\bar{x}_1 + \bar{x}_2 + \ldots + \bar{x}_n), \tag{2.3}$$

where "+" means the union of events.

From the basic laws of probability, the probability of either event A or B occurring is

$$P(A + B) = P(A) + P(B) - P(AB). \tag{2.4}$$

Following Eq. (2.4), we rewrite Eq. (2.3) as follows

$$P_f = [P(\bar{x}_1) + P(\bar{x}_2) + \ldots + P(\bar{x}_n)] - [P(\bar{x}_1\,\bar{x}_2) + P(\bar{x}_1\,\bar{x}_3) + \ldots]$$
$$+ \ldots + [-1]^{n-1} P(\bar{x}_1\,\bar{x}_2\,, \ldots, \bar{x}_n). \tag{2.5}$$

The reliability of the system is

$$R = 1 - P_{f.}$$

It should be noted that the reliability of a series system is always less than or equal to that of the component with the lowest reliability.

EXAMPLE 2.3

Consider a series system that consists of three components and the probabilities that components 1, 2, and 3 being operational are 0.9, 0.8, and 0.75, respectively. Estimate the reliability of the system.

SOLUTION

Assuming independent failures, we use Eq. (2.2) to obtain the reliability of the system

$$R = 0.9 \times 0.8 \times 0.75 = 0.54.$$

Alternatively, we use Eq. (2.5) to get

$$P_f = [P(\bar{x}_1) + P(\bar{x}_2) + P(\bar{x}_3)] - [P(\bar{x}_1)P(\bar{x}_2) + P(\bar{x}_1)P(\bar{x}_3) + P(\bar{x}_2)P(\bar{x}_3)]$$

$$+ [P(\bar{x}_1)P(\bar{x}_2)P(\bar{x}_3)]$$

$$= 0.55 - 0.095 + 0.005 = 0.46$$

and

$$R = 1 - P_f = 1 - 0.46 = 0.54.$$

As shown above, the reliability of the system, 0.54, is less than the reliability of the worst component, 0.75.

2.4. Parallel Systems

In a parallel system, components or units are connected in parallel such that the failure of one or more paths still allows the remaining path(s) to perform properly. In other words, the reliability of a parallel system is the probability that any one path is operational. The block diagram and reliability graph of a parallel system consisting of n components (units) connected in parallel are shown in Figure 2.5.

Similar to the series systems, the reliability of parallel systems can be determined by estimating the probability that any one path is operational or by estimating the unreliability of the system then subtracting it from unity. In other words,

$$R = P(x_1 + x_2 + \ldots + x_n)$$

or

$$R = [P(x_1) + P(x_2) + \ldots + P(x_n)]$$

$$- [P(x_1 x_2) + P(x_1 x_3) + \ldots + P_{i \neq j}(x_i x_j)] \qquad (2.6)$$

$$+ \ldots + (-1)^{n-1} P(x_1 x_2 \ldots x_n).$$

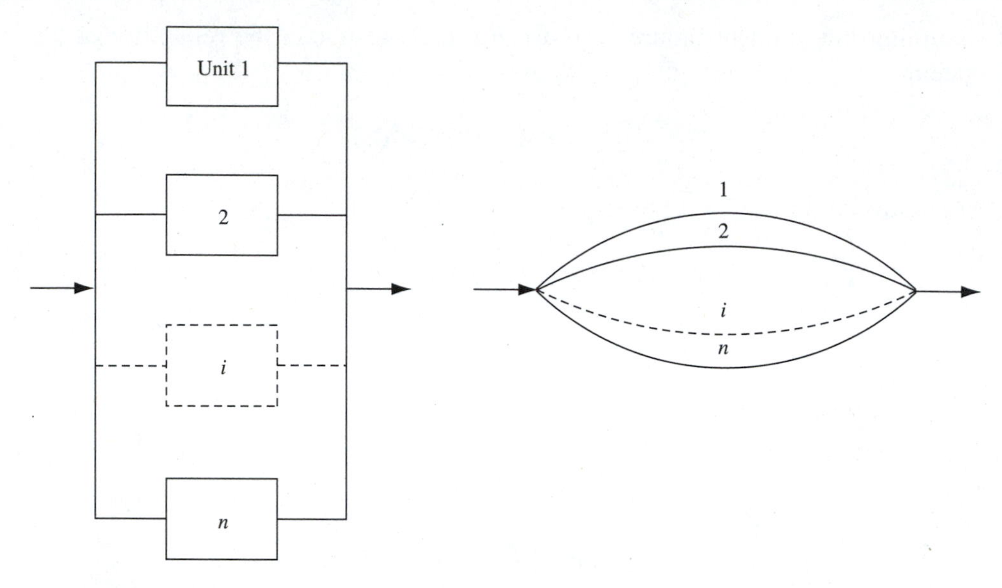

Figure 2.5. Block Diagram and Reliability Graph of a Parallel System

Alternatively,

$$P_f = P(\bar{x}_1\bar{x}_2 \ldots \bar{x}_n)$$

$$R = 1 - P_f$$

or

$$R = 1 - P(\bar{x}_1)P(\bar{x}_2 / \bar{x}_1)P(\bar{x}_3 / \bar{x}_1\bar{x}_2) \ldots \tag{2.7}$$

Again, if the components are independent, then Eq. (2.7) can be rewritten as

$$R = 1 - P(\bar{x}_1)P(\bar{x}_2) \ldots P(\bar{x}_n)$$

or

$$R = 1 - \prod_{i=1}^{n} P(\bar{x}_i). \tag{2.8}$$

If the components are identical, then the reliability of the system is

$$R = 1 - (1 - p)^n,$$

where p is the probability that a component is operational.

EXAMPLE 2.4

Consider a system that consists of three components in parallel. The probabilities of the three components being operational are 0.9, 0.8, and 0.75. Determine the reliability of the system.

SOLUTION

The reliability of a parallel system is obtained by using Eq. (2.6) as follows:

$$R = P(x_1 + x_2 + x_3)$$

$$= P(x_1) + P(x_2) + P(x_3) - [P(x_1)P(x_2) + P(x_1)P(x_3) + P(x_2)P(x_3)]$$

$$+ P(x_1)P(x_2)P(x_3)$$

or

$$R = 2.450 - 1.995 + 0.540 = 0.995.$$

One can also obtain the reliability of the system by using Eq. (2.8):

$$R = 1 - \prod_{i=1}^{n} P(\bar{x}_i)$$

$$R = 1 - (1 - 0.90)\,(1 - 0.80)\,(1 - 0.75) = 0.995.$$

The reliability of a parallel system is greater than the reliability of the most reliable unit (or component) in the system. This may imply that the more units we have in parallel the more reliable the system. This statement is only valid for systems whose components exist only in two states, either operational or failure. As we shall show later, there is an optimal number of multistate components (units) that can be connected in parallel, and adding more units in parallel results in lower values of reliability.

2.5. Parallel-Series, Series-Parallel, and Mixed-Parallel Systems

The systems discussed in Sections 2.3 and 2.4 are referred to as pure series and pure parallel systems, respectively. There are many situations where a system is composed of a combination of series and parallel subsystems. This section considers three systems: parallel-series, series-parallel, and mixed-parallel.

2.5.1. PARALLEL-SERIES

A parallel-series system consists of m parallel paths. Each path has n units connected in series as shown in Figure 2.6. Let $P(x_{ij})$ be the reliability of component j

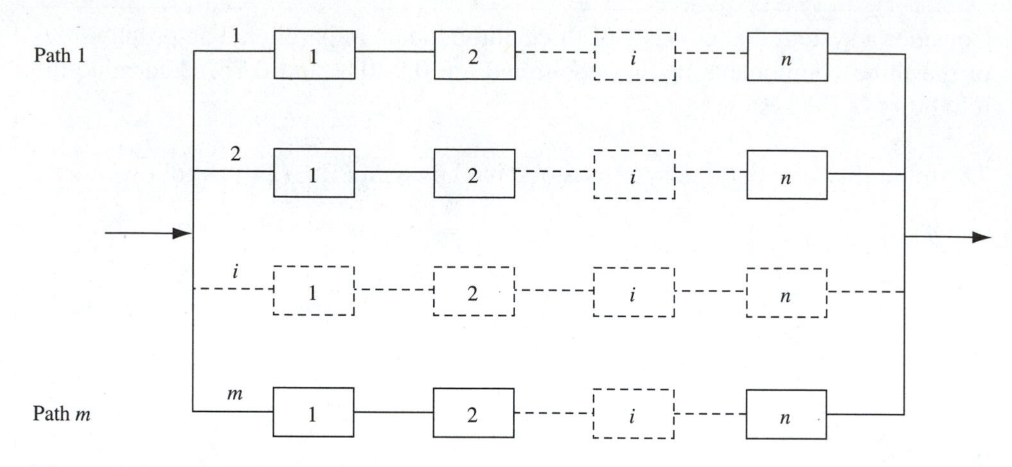

Figure 2.6. A Parallel-Series System

($j = 1, 2, \ldots, n$) in path i ($i = 1, 2, \ldots, m$) and x_{ij} be an indicator that component j in path i is operational. The reliability of path i is

$$P_i = \prod_{j=1}^{n} P(x_{ij}) \qquad i = 1, 2, \ldots, m \text{ and } j = 1, 2, \ldots, n.$$

The unreliability of path i is $\bar{P}_i$ and the reliability of the system is

$$R = 1 - \prod_{i=1}^{m} \bar{P}_i$$

or

$$R = 1 - \prod_{i=1}^{m} \left[1 - \prod_{j=1}^{n} P(x_{ij}) \right].$$

If all units are identical and the reliability of a single unit is p, then the reliability of the system becomes

$$R = 1 - (1 - p^n)^m. \tag{2.9}$$

2.5.2. SERIES-PARALLEL

A general series-parallel system consists of n subsystems in series with m units in parallel in each subsystem as shown in Figure 2.7. Following the parallel-series systems, we derive the reliability expression of the system as

$$R = \prod_{i=1}^{n} \left[1 - \prod_{j=1}^{m} (1 - P(x_{ij})) \right],$$

where $i = 1, 2, \ldots, n$; $j = 1, 2, \ldots, m$; and $P(x_{ij})$ is the probability that component j in subsystem i is operational. When all units are identical and the reliability of a single unit is p, then the reliability of the series-parallel system becomes

$$R = [1 - (1 - p)^m]^n. \tag{2.10}$$

In general, series-parallel systems have higher reliabilities than parallel-series systems when both have an equal number of units and each unit has the same probability of operation.

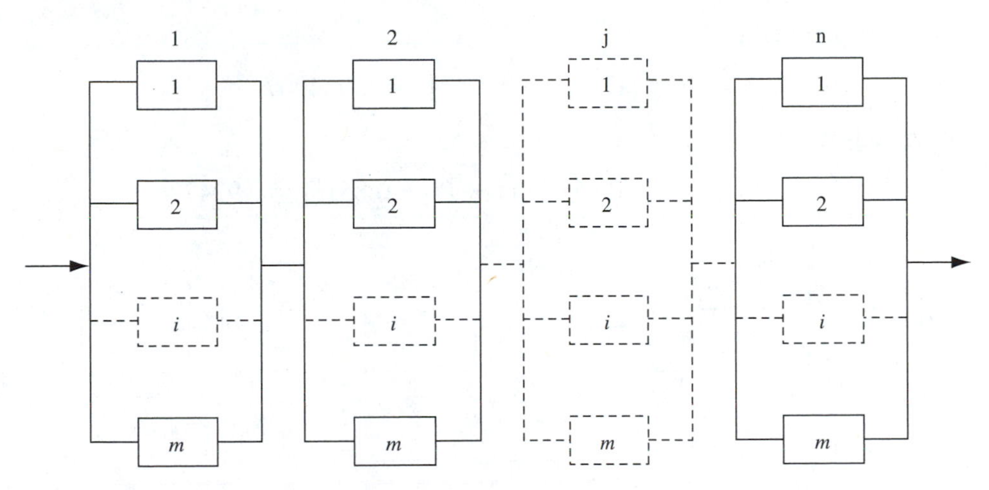

Figure 2.7. A Series-Parallel System

2.5.3. MIXED-PARALLEL

A mixed-parallel system has no specific arrangement of units other than the fact that they are connected in parallel and series configurations. Figure 2.8 illustrates two possible mixed-parallel systems for eight units. The reliability of a mixed-parallel system can be estimated using Eqs. (2.2) and (2.8).

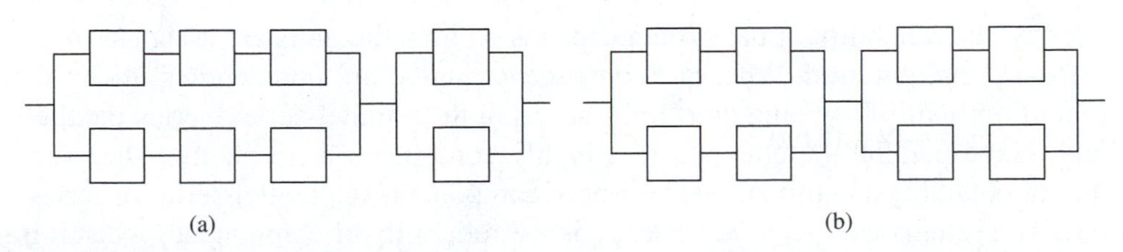

Figure 2.8. Mixed-Parallel Systems

EXAMPLE 2.5 Given six identical units each having a reliability of 0.85, determine the reliability of three systems resulting from the arrangements of the units in parallel-series, series-parallel, and mixed-parallel configurations.

SOLUTION Assume that the six units can be arranged in three series and parallel configurations as shown in Figure 2.9. The reliabilities of the systems are as follows:

1. Parallel-series:

$$R = 1 - (1 - p^n)^m,$$

when $m = 2$, $n = 3$.

$$R = 1 - (1 - 0.85^3)^2 = 0.85110.$$

2. Series-parallel:

$$R = [1 - (1 - 0.85)^2]^3 = 0.934007.$$

3. Mixed-parallel:

$$R = 1 - (1 - (0.85)^2)^2\,[1 - (1 - 0.85)^2] = 0.924726.$$

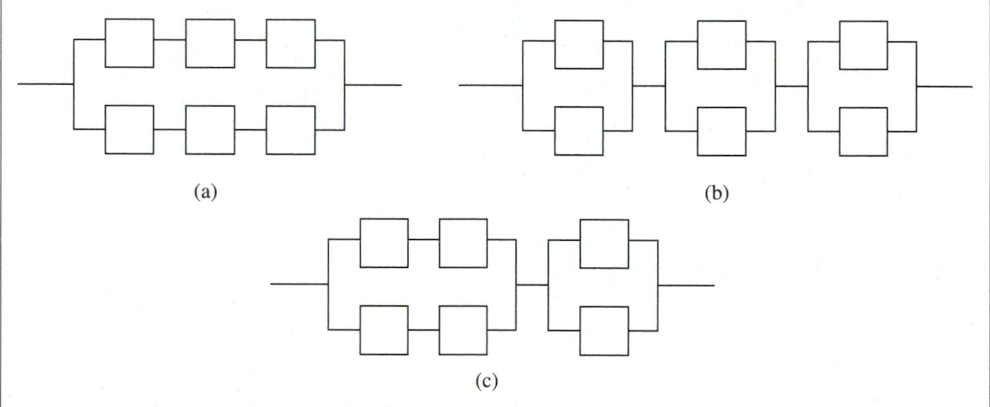

Figure 2.9. (a) Parallel-Series, (b) Series-Parallel, (c) Mixed-Parallel

2.5.4. OPTIMAL ASSIGNMENTS OF UNITS

Clearly, the reliability of the system depends on how the units are assigned in the system's configuration. When the components (units) are nonidentical, the problem of optimally assigning units to locations in the parallel-series, series-parallel, and mixed-parallel systems becomes highly combinatorial in nature. The problem of obtaining an optimal assignment of components in parallel-series or series-parallel systems with $p_{ij} = p_j$ where p_{ij} is the reliability of component j when it is assigned to position i has been analytically solved by El-Neweihi, Proschan, and Sethuraman (1986). More recently, Prasad, Nair, and Aneja (1991) present three algorithms to determine the optimal assignments of components in such systems with the assumption that $p_{ij} = r_i p_j$ where r_i is a probability whose value is dependent on the position i.

In this section one approach, though not optimal, illustrates how units can be assigned to a series-parallel configuration in order to maximize the overall system reliability.

As shown in Section 2.5.2 a series-parallel system consists of n subsystems connected in series with m units in parallel in each system. Let $K_1, K_2, \ldots, K_n$ represent the subsystems 1, 2, . . . , and n. Consider the case when all subsystems have an equal number of units m. Thus, the total number of units in the entire system is $u = n \times m$. The problem of assigning units to the series-parallel system can simply be stated as: given a system of u units with reliabilities given by the vector $\boldsymbol{p} = (p_1, p_2, \ldots, p_u)$ where p_i is the probability that unit C_i is operating, $(i = 1, 2, \ldots, u)$. We wish to allocate these units, which are assumed to be interchangeable, in such a manner that the reliability function $\mathfrak{R}$ of the system is maximized (Baxter and Harche, 1992).

The reliability function $\mathfrak{R}$ is

$$\mathfrak{R} = \prod_{i=1}^{n} R_i, \tag{2.11}$$

where

$$R_i = 1 - \prod_{j \in K_i} q_j \tag{2.12}$$

and

$$q_j = 1 - p_j, \qquad j = 1, 2, \ldots, u.$$

Revisiting the series-parallel configuration discussed in Section 2.5.2, we observe that the reliability of the system is maximum when the R_i's (reliability of individual subsystems) are as equal as possible. Based on this, we utilize the *top down heuristic* (TDH) proposed by Baxter and Harche (1992) to assign units to positions in the system configuration. The steps of the heuristic are

Step 1 Rank and label the units such that $p_1 \geq p_2 \geq \ldots \geq p_u$.

Step 2 Allocate units C_j to subsystem K_j, $\quad j = 1, 2, \ldots, n$.

Step 3 Allocate units C_j to subsystem K_{2n+1-j}, $\quad j = n + 1, \ldots, 2n$.

Step 4 Set $\nu := 2$.

Step 5 Evaluate $R_i^{(\nu)} = 1 - \Pi_{j \in K_i} q_j$ for $i = 1, 2, \ldots, n$. Allocate unit $C_{\nu n+i}$ to subsystem K_i for which $R_i^{(\nu)}$ is jth smallest, $j = 1, 2, \ldots, n$.

Step 6 If $\nu < m$, set $\nu := \nu + 1$ and repeat Step 5. If $\nu = m$, stop.

It should be noted that there exists a bottom-up heuristic (BUH) corresponding to the TDH, which begins by allocating the n least reliable units, one to each subsystem, and then allocating the n next least reliable units in reverse order, and so on. Clearly, this heuristic is inferior to the TDH in practice, since (as is shown later) the most important unit of a parallel system is the most reliable component. It should also be noted that the TDH results in optimal allocations when only two units are allocated to each subsystem.

EXAMPLE 2.6

An engineer wishes to design a redundant system that includes six resistors, all having the same resistance value. Their reliabilities are 0.95, 0.75, 0.85, 0.65, 0.40, and 0.55. The resistors are interchangeable within the system. Space within the enclosure where the resistors are connected limits the designer to allocate the resistor in a series-parallel arrangement of the form (3, 2)—that is, two subsystems connected in series, and each subsystem consists of three resistors connected in parallel. Use the TDH to allocate units to the subsystems. Compare the reliability of the resultant system with that obtained from the application of the BUH.

SOLUTION

We follow the steps of the TDH to allocate units to the subsystems K_1 and K_2 as follows:

Step 1 Rank the resistors in a decreasing order of their reliabilities: 0.95, 0.85, 0.75, 0.65, 0.55, and 0.40.

Step 2 Allocate resistors C_1 to K_1 and C_2 to K_2.

Step 3 Allocate resistors C_3 to K_2 and C_4 to K_1.

Step 4 $\nu := 2$.

Step 5 Calculate the reliabilities of subsystems K_1 and K_2, respectively, as

$$R_1^{(2)} = 1 - (1 - 0.95)(1 - 0.65) = 0.9825, \text{ and}$$

$$R_2^{(2)} = 1 - (1 - 0.85)(1 - 0.75) = 0.9625.$$

Since $R_2^{(2)} < R_1^{(2)}$, we allocate unit C_5 to K_2.

Step 6 $\nu := 3 = k$, allocate C_6 to K_1 and stop.

The resultant allocation of the resistors is as follows:

Subsystem K_1	*Subsystem* K_2
C_1	C_2
C_4	C_3
C_6	C_5

The reliability of the system is

$$\Re_s = [1 - (1 - 0.95)(1 - 0.65)(1 - 0.40)][1 - (1 - 0.85)(1 - 0.75)(1 - 0.55)]$$

or

$$\Re_s = 0.972802.$$

Application of the BUH results in the following allocation:

Subsystem K_1	*Subsystem* K_2
C_6	C_5
C_3	C_4
C_2	C_1

The reliability of the system is

$$\Re_s = [1 - (1 - 0.40)(1 - 0.75)(1 - 0.85)][1 - (1 - 0.55)(1 - 0.65)(1 - 0.95)]$$

or

$$\Re_s = 0.9698.$$

In general, allocation of units to subsystems using the TDH results in a higher reliability of the system than the BUH.

2.6. Consecutive-k-out-of-n:F System

In Section 2.3 we presented a series system consisting of n components. In such a system, the failure of one or more components results in system failure. However, there exist systems that are not considered failed until at least k components have failed. Moreover, those k components must be consecutively ordered within the system. Such systems are known as *consecutive*-k-*out-of*-n:F *systems*. An example of a consecutive-k-out-of-n:F system is presented in Chiang and Niu (1981), which considers a telecommunications system with n relay stations (either satellites or ground stations). The stations are named consecutively 1 to n. Suppose a signal emitted from Station 1 can be received by both Stations 2 and 3, and a signal relayed from Station 2 can be received by both Stations 3 and 4 and so on. Thus, when Station 2 fails, the telecommunications system is still able to transmit a signal from Station 1 to Station n. However, if both Stations 2 and 3 fail, a signal cannot be transmitted from Station 1 directly to Station 4; therefore, the system fails. Similarly, if any two consecutive stations in the system fail, the system fails. This is considered a consecutive-2-out-of-n:F system.

Determining the reliability of the consecutive-2-out-of-n:F system is simple. First we define the following notations:

n = the number of components in a system,

k = the minimum number of consecutive failed components that cause system failure,

p = the probability that a component is functioning properly (all components have identical and independent life distributions),

$R(p, k, n)$ = the reliability of a consecutive-k-out-of-n:F system whose components are identical and each component has a probability p of functioning properly,

x_i = the state of component i,

X = the vector of component states,

Y = a random variable indicating the index of first 0 in X,

M = a random variable indicating the index of first 1 after the position Y in X,

$\lfloor a \rfloor$ = the largest integer less than or equal to a, and

$\cup_k p = 1 - (1 - p)^k$.

2.6.1. CONSECUTIVE-2-OUT-OF-n:F SYSTEM

We follow the work of Chiang and Niu (1981). The reliability of a consecutive-2-out-of-n:F system is

$$R(p, 2, n) = P\,[\text{the system is functioning}]$$

$$= \sum_{j=0}^{\lfloor (n+1)/2 \rfloor} P\,[\text{system is functioning and } j \text{ components failed}]. \tag{2.13}$$

If the number of failed components is greater than $\lfloor (n+1)/2 \rfloor$, then there exist two consecutive failed components in the system—that is, the system fails. Hence, the above expression of system reliability does not include the terms for $j > \lfloor (n+1)/2 \rfloor$.

If j components have failed, $j \leq \lfloor (n+1)/2 \rfloor$, the system functions if there is at least one functioning component between every two failed components. The number of such combinations between functioning and failed components is

$$\binom{(j+1)+(n-2j+1)-1}{n-2j+1} = \binom{n-j+1}{j}, \tag{2.14}$$

which follows directly from Feller (1968) and Pease (1975). Substituting Eq. (2.14) into Eq. (2.13), we obtain

$$R(p, 2, n) = \sum_{j=0}^{\lfloor (n+1)/2 \rfloor} \binom{n-j+1}{j} (1-p)^j p^{n-j} \tag{2.15}$$

EXAMPLE 2.7

Consider four components that are connected in series. Each component has a reliability p. The system fails if two consecutive components fail. This system is referred to as consecutive-2-out-of-4:F system. Determine the reliability of the system when $p = 0.95$.

SOLUTION

Using Eq. (2.15), we obtain

$$R(p, 2, 4) = \sum_{j=0}^{2} \binom{4-j+1}{j} (1-p)^j p^{4-j}$$

$$= \binom{5}{0}(1-p)^0 p^4 + \binom{4}{1}(1-p)p^3 + \binom{3}{2}(1-p)^2 p^2$$

$$= 3p^2 - 2p^3.$$

When $p = 0.95$, then

$$R(0.95, 2, 4) = 3(0.95)^2 - 2(0.95)^3$$

$$= 0.992750.$$

2.6.2. GENERALIZATION OF THE CONSECUTIVE-k-OUT-OF-n:F SYSTEMS

We define X as an n-vector with element i having a value of 0 or 1 depending on whether component i is failing or not. The procedure for determining system reliability is based on observing the first sequence of consecutive 0's in the X vector. The system is considered to be failed if at least k consecutive 0's are observed in X. Since the reliability of a consecutive-k-out-of-n:F system for all $n < k$ is 1 by definition, we can recursively compute the reliability of consecutive-k-out-of-n:F system for $n \geq k$. The reliability (Chiang and Niu, 1981) is

$$
\begin{aligned}
R(p, k, n) &= P\,[\text{the system is functioning}] \\
&= \sum_{y} \sum_{m} P\,[\text{the system is functioning} \,/\, Y{=}y, M{=}m]\; P[Y{=}y, M{=}m] \\
&= \sum_{y=1}^{n-k+1} \sum_{m=y+1}^{y+k-1} P\,[\text{the system is functioning} \,/\, Y{=}y, M{=}m] p^{y} \\
&\quad \times (1-p)^{m-y} + p^{n-k+1}.
\end{aligned}
$$

Since the system has less than k failed components for $Y > n - k + 1$, then P [system is functioning / $Y = n - k + 1$] = 1 and $P\,[Y > n - k + 1] = p^{n-k+1}$. When $m \geq y + k$, the system already has k failed components and is considered failed.

For $y + 1 \leq m \leq y + k - 1$, the first sequence of 0's does not constitute a cut-set. Furthermore, since $x_m = 1$, the event that the consecutive-k-out-of-n:F system is functioning now is equivalent to the event that a consecutive-k-out-of-$(n - m)$:F system is functioning. Thus, the recursive formula for determining $R(p, k, n)$ is

$$
R(p, k, n) = \sum_{y=1}^{n-k+1} \sum_{m=y+1}^{y+k-1} R(p, k, n-m) p^{y} (1-p)^{m-y} + p^{n-k+1} \tag{2.16}
$$

$$
R(p, k, j) = \begin{cases} 1, & 0 \leq j < k \\ 0, & j < 0 \end{cases}.
$$

As discussed earlier, the failure of k consecutive components results in the system failure. Therefore, k consecutive components are the only minimum cut-sets and there are $n - k + 1$ such sets in the consecutive-k-out-of-n:F system. If the system is functioning, then there is at least one functioning component in every cut-set. Hence the lower bound for system reliability is

$$
R_L(p, k, n) \geq (\cup_k\, p)^{n-k+1}. \tag{2.17}
$$

Similarly, an upper bound for system reliability can be obtained as

$$R_U(p, k, n) \leq (\cup_k p)^{\lfloor n/k \rfloor}. \quad (2.18)$$

EXAMPLE 2.8

Derive an expression for the reliability of a consecutive-2-out-of-7:*F* system. Each component has a reliability p. Calculate the reliability and its lower and upper bounds when $p = 0.90$.

SOLUTION

Using Eq. (2.15) we obtain

$$R(p, 2, 2) = 2p - p^2$$

$$R(p, 2, 3) = p + p^2 - p^3$$

$$R(p, 2, 4) = 3p^2 - 2p^3 \text{ (see Example 2.7)}$$

$$R(p, 2, 5) = p^2 + 3p^3 - 4p^4 + p^5$$

$$R(p, 2, 6) = 4p^3 - 2p^4 - 2p^5 + p^6.$$

We use Eq. (2.16) and the above expressions to obtain $R(p, 2, 7)$ as follows:

$$R(p, 2, 7) = \sum_{y=1}^{6} R(p, 2, 6 - y)p^y(1 - p) + p^6$$

$$R(p, 2, 7) = p^3 + 6p^4 - 9p^5 + 3p^6$$

$$R(0.9, 2, 7) = 0.945513.$$

The lower and upper bounds of the reliability are obtained using Eqs. (2.17) and (2.18), respectively:

$$R_L(0.9, 2, 7) \geq [1 - (1 - 0.9)^2]^6$$

or

$$R_L(0.9, 2, 7) \geq 0.941480$$

and

$$R_U(0.9, 2, 7) \leq [1 - (1 - 0.9)^2]^3$$

or

$$R_U(0.9, 2, 7) \leq 0.970299.$$

The effect of the component reliability on the system reliability is shown in Figure 2.10.

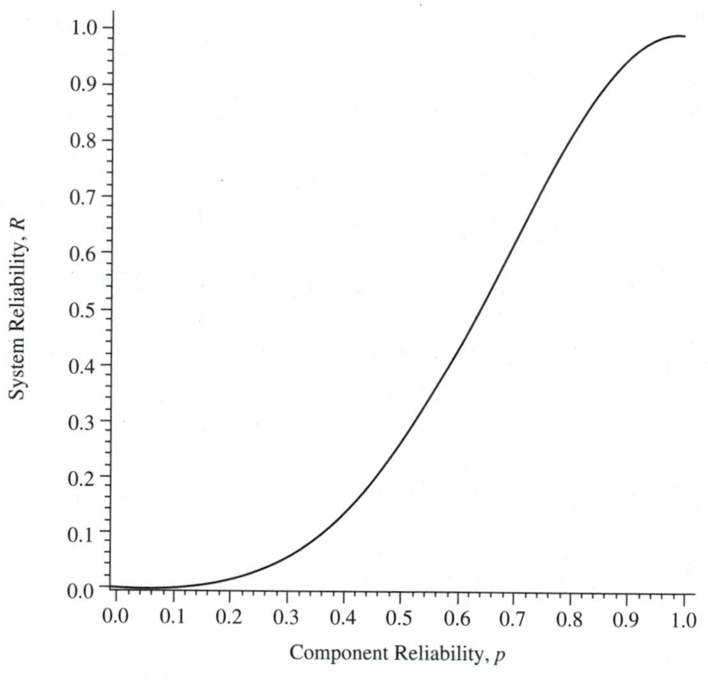

Figure 2.10. Effect of Change of p on R

2.6.3. RELIABILITY ESTIMATION OF THE CONSECUTIVE-k-OUT-OF-n:F SYSTEMS

In the previous sections, we discussed methods for estimating the reliability of 2-out-of-n:F systems for both cases when all units are identical and when the reliabilities of the units are not equal. This section presents an algorithm for reliability of estimation of the consecutive-k-out-of-n:F systems when the p_i's are not equal. After an approach for reliability estimation of such systems is introduced, we present a general and more efficient algorithm.

The first approach is based on finding the reliability by determining the failure states of the system—that is, determining all the combinations of component failures that result in a system failure. For example, consider a system that consists of five components connected in series. The system fails when two consecutive components fail. The minimum possible states that cause the system failure are

$$\bar{x}_1\bar{x}_2,\ \bar{x}_2\bar{x}_3,\ \bar{x}_3\bar{x}_4,\quad \text{and}\quad \bar{x}_4\bar{x}_5.$$

From these states, the reliability of the system can be estimated as

$$R = 1 - P[\bar{x}_1\,\bar{x}_2 + \bar{x}_2\,\bar{x}_3 + \bar{x}_3\,\bar{x}_4 + \bar{x}_4\,\bar{x}_5]$$

$$\begin{aligned} R = 1 - [&P(\bar{x}_1\,\bar{x}_2) + P(\bar{x}_2\,\bar{x}_3) + P(\bar{x}_3\,\bar{x}_4) + P(\bar{x}_4\,\bar{x}_5) \\ &- P(\bar{x}_1\,\bar{x}_2\,\bar{x}_3) - P(\bar{x}_1\,\bar{x}_2\,\bar{x}_3\,\bar{x}_4) - P(\bar{x}_1\,\bar{x}_2\,\bar{x}_4\,\bar{x}_5) \\ &- P(\bar{x}_2\,\bar{x}_3\,\bar{x}_4) - P(\bar{x}_2\,\bar{x}_3\,\bar{x}_4\,\bar{x}_5) - P(\bar{x}_3\,\bar{x}_4\,\bar{x}_5) \\ &+ P(\bar{x}_1\,\bar{x}_2\,\bar{x}_3\,\bar{x}_4) + P(\bar{x}_1\,\bar{x}_2\,\bar{x}_3\,\bar{x}_4\,\bar{x}_5) \\ &+ P(\bar{x}_1\,\bar{x}_2\,\bar{x}_3\,\bar{x}_4\,\bar{x}_5) + P(\bar{x}_2\,\bar{x}_3\,\bar{x}_4\,\bar{x}_5) - P(\bar{x}_1\,\bar{x}_2\,\bar{x}_3\,\bar{x}_4\,\bar{x}_5)] \end{aligned}$$

or

$$\begin{aligned} R = 1 - [&P(\bar{x}_1\,\bar{x}_2) + P(\bar{x}_2\,\bar{x}_3) + P(\bar{x}_3\,\bar{x}_4) + P(\bar{x}_4\,\bar{x}_5) \\ &- P(\bar{x}_1\,\bar{x}_2\,\bar{x}_3) - P(\bar{x}_1\,\bar{x}_2\,\bar{x}_4\,\bar{x}_5) - P(\bar{x}_2\,\bar{x}_3\,\bar{x}_4) \\ &- P(\bar{x}_3\,\bar{x}_4\,\bar{x}_5) + P(\bar{x}_1\,\bar{x}_2\,\bar{x}_3\,\bar{x}_4\,\bar{x}_5)]. \end{aligned} \tag{2.19}$$

EXAMPLE 2.9

Repeaters are devices that amplify signals and send them to other network segments. They play a vital role in building large networks. A repeater amplifies signals such that they can reach two repeaters away without loss or distortion. The repeaters are connected in series, and the signals are considered lost when two consecutive repeaters fail.

Assume that five repeaters are connected in series and the probabilities of failure of repeaters 1 through 5 are $q_1 = 0.62$, $q_2 = 0.079$, $q_3 = 0.25$, $q_4 = 0.22$, and $q_5 = 0.42$. Determine the reliability of the system.

SOLUTION

The reliability of the system can be obtained by substituting the reliability values of the components in Eq. (2.19):

$$\begin{aligned} R = 1 - [&0.62 \times 0.079 + 0.079 \times 0.25 + 0.25 \times 0.22 + 0.22 \times 0.42 \\ &- 0.62 \times 0.079 \times 0.25 - 0.62 \times 0.079 \times 0.22 \times 0.42 \\ &- 0.079 \times 0.25 \times 0.22 - 0.25 \times 0.22 \times 0.42 \\ &+ 0.62 \times 0.079 \times 0.25 \times 0.22 \times 0.42] \end{aligned}$$

or

$$R = 0.826.$$

Clearly, when n is large and $1 \le k \le n$, the above approach becomes quite complex and difficult to apply. This has prompted researchers to investigate different algorithms that can efficiently estimate system reliability. Among them are Bollinger (1982), Lambiris and Papastavridis (1985), Pham (1988), Shanthikumar (1982), and Zuo and Kuo (1990). We summarize Shanthikumar's algorithm as follows:

Step 1 Choose k, n. Test that $1 \le k \le n$.
$(i = 1, 2, \ldots, n)$. $q_i = 1 - p_i$.

Step 2 Set $F(r; k) = 0$ for $r = 0, 1$.
Set $Q \leftarrow \Pi_{i=1}^{k} q_i$.
Set $F(k; k) = Q$.

Step 3 Do for $r = k + 1$ to n.
$Q \leftarrow Q \times q_r / q_{r-k}$.
$F(r; k) = F(r - 1; k) + [\, 1 - F(r - k - 1; k)\,] \times P_{r-k} \times Q$.

Step 4 $R(n; k) = 1 - F(n; k)$.
End.

The above algorithm is coded in a computer program and is listed in Appendix F.

EXAMPLE 2.10 Use the above algorithm to estimate the reliability of the system described in Example 2.9.

SOLUTION We apply the steps of the algorithm as follows:

Step 1 $k = 2, n = 5$,
$p_1 = 0.38, p_2 = 0.921, p_3 = 0.75, p_4 = 0.78, p_5 = 0.58$.

Step 2 $F(0; 2) = 0$,
$F(1; 2) = 0$,
$Q = q_1 q_2 = 0.62 \times 0.079 = 0.0489$,
$F(2; 2) = 0.0489$.

Step 3 $r = 3$,
$Q = 0.0489 \times q_3 / q_1 = 0.0197$,
$F(3; 2) = F(2; 2) + [\, 1 - F(0; 2)\,] \times 0.38 \times 0.0197 = 0.0563$.

Step 3 $r = 4$,
$Q = 0.0197 \times q_4/q_2 = 0.0548$,
$F(4; 2) = F(3; 2) + [\, 1 - F(1; 2)\,] \times 0.921 \times 0.0548 = 0.1068$,

Step 3 $r = 5$,
$Q = 0.0548 \times q_5/q_3 = 0.0920$,
$F(5; 2) = F(4; 2) + [\, 1 - F(2; 2)\,] \times 0.75 \times 0.0920 = 0.1724$.

Step 4 $R = 1 - F(5; 2) = 0.827$.

The reliability of the system obtained by this algorithm is identical to that obtained by Eq. (2.19).

2.6.4. OPTIMAL ARRANGEMENT OF COMPONENTS IN CONSECUTIVE-2-OUT-OF-*n*:*F* SYSTEMS

As shown earlier, the reliability of a consecutive-2-out-of-n system, when the components have different failure probabilities, depends on the arrangement of the components in the system. The designer of such a system may wish to assign n components simultaneously to n positions within the system such that the reliability of the system is maximum. The components are ranked such that $p_1 < p_2 < \ldots < p_n$. Derman, Lieberman, and Ross (1982) have conjectured that the optimal arrangement is

$$(1, n, 3, n-2, \ldots, n-3, 4, n-1, 2),$$

obtained by placing the least reliable pair of components outermost, followed by the most reliable pair, and so on in an alternating fashion. We shall now prove this conjecture. Define $\psi = (\psi(1), \ldots, \psi(n))$ as a policy where component $\psi(1)$ is assigned to position one in the system, $\psi(2)$ to the second position, ..., $\psi(n)$ to position n. Let $r(\psi)$ be the reliability of the system when policy ψ is used.

Consider the case where $n = 2$. The optimum arrangement is either (1, 2) or (2, 1).

When $n = 3$, the reliability $\psi(r)$ is

$$r(\psi) = 1 - q_{\psi(1)}q_{\psi(2)} - q_{\psi(2)}q_{\psi(3)} + \prod_{i=1}^{3} q_{\psi(i)},$$

where $q_{\psi(i)}$ is the unreliability of $\psi(i)$. For $r(\psi)$ to be maximum, the value of $[q_{\psi(1)}q_{\psi(2)} + q_{\psi(2)}q_{\psi(3)}]$ should be minimum. It can be verified that the arrangement $\psi = (1, 3, 2)$ yields maximum reliability. Similarly, when $n = 4$, the reliability of the system is

$$r(\psi) = 1 - [q_{\psi(1)}q_{\psi(2)} + q_{\psi(2)}q_{\psi(3)} + q_{\psi(3)}q_{\psi(4)} - q_{\psi(1)}q_{\psi(2)}q_{\psi(3)}$$

$$- q_{\psi(1)}q_{\psi(2)}q_{\psi(3)}q_{\psi(4)} - q_{\psi(2)}q_{\psi(3)}q_{\psi(4)} + q_{\psi(1)}q_{\psi(2)}q_{\psi(3)}q_{\psi(4)}]$$

$$r(\psi) = 1 - [q_{\psi(1)}q_{\psi(2)} + q_{\psi(3)}q_{\psi(4)} + q_{\psi(2)}q_{\psi(3)}(1 - q_{\psi(1)} - q_{\psi(4)})].$$

The arrangement $\psi^* = (1, 4, 3, 2)$ maximizes $r(\psi)$. This follows from the observation that ψ^* simultaneously minimizes the sum of the first two terms and the last term within the bracket (Derman, Lieberman, and Ross, 1982).

Generalization of the above for any $n \geq 1$ yields

$$\psi^* = (1, n, 3, n - 2, \ldots, n - 3, 4, n - 1, 2).$$

Appendix G is a listing of a computer program that finds the optimal arrangement of components in a consecutive-2-out-of-n:F system and estimates its reliability.

EXAMPLE 2.11

A collision-avoidance system for articulated robot manipulators uses infrared proximity sensors grouped together in an array of sensor modules. The modules are distributed processing board-level products for acquiring data from proximity sensors mounted on robot manipulators. Each module consists of eight sensing elements, discrete electronics, a microcontroller, and communications components. The sensor system detects objects made of various materials at a distance of up to 50 cm. The module fails to detect the object if consecutive-2-out-of-8 sensing elements fail. The unreliabilities of the eight sensing elements are 0.01, 0.02, 0.03, 0.04, 0.05, 0.06, 0.07, and 0.08. Determine the optimal arrangement of these elements such that reliability of the module is maximized.

SOLUTION

We rank the sensing elements in decreasing order of the q_is:

Element	1	2	3	4	5	6	7	8
q_i	0.08	0.07	0.06	0.05	0.04	0.03	0.02	0.01

Applying Derman, Lieberman, and Ross's (1982) conjecture yields the following optimal arrangement of the sensing elements:

$$\psi^* = [1, 8, 3, 6, 5, 4, 7, 2].$$

The maximum reliability of the system is 0.9915.

Wei, Hwang, and Sös (1983) partially support the conjecture developed by Derman, Lieberman, and Ross (1982). Malon (1985) characterizes all other values of k and n for which an optimal configuration can be determined without knowledge of the component failure probabilities. As we mentioned earlier, the reliability of the system depends on the particular failure probabilities and the positions of the components in the system. However, for certain values of k and n, there is an arrangement that is optimal regardless of the failure probabilities. We refer to such an arrangement as an invariant optimal arrangement. We now characterize all values of k and n for such arrangements.

Rank the components such that $q_1 \geq q_2 \geq \ldots \geq q_n$. Malon (1985) states that the consecutive-k-out-of-n:F system admits an invariant arrangement if and only if $k \in \{1, 2, n-2, n-1, n\}$. The optimal arrangements are given in Table 2.1.

Table 2.1. Optimal Arrangements of Components

k	*Invariant Optimal Arrangement*
1	Any arrangement
2	$[1, n, 3, n-2, \ldots, n-3, 4, n-1, 2]$
$n-2$	[1, 4, (any arrangement), 3, 2]
$n-1$	[1, (any arrangement), 2]
n	Any arrangement

EXAMPLE 2.12

Solve Example 2.11 for a consecutive-6-out-of-8:F system when the sensing elements have the following unreliability values:

Element	1	2	3	4	5	6	7	8
q_i	0.4	0.35	0.32	0.28	0.25	0.21	0.18	0.15

SOLUTION

Using Table 2.1, any of the following arrangements will result in a maximum reliability of 0.999651:

Arrangements [1, 4, any arrangement of elements (5, 6, 7, 8), 3, 2].

2.7. Reliability of k-out-of-n Systems

In Section 2.6, we presented a consecutive-k-out-of-n:F system where a system fails if at least k *consecutive* components fail. In many cases, the k failures need not be consecutive and the system fails if any k or more components fail. For example, large airplanes usually have three or four engines, but two engines may be the minimum number required to provide a safe journey. Similarly, in many power-generating systems that have two or three generators, one generator may be sufficient to provide the power requirements. Also, in a typical wire cable for cranes and bridges, the cable may contain thousands of wires, but only a fraction of them may be required to carry the desired load. Assuming that all units have identical and independent life distributions and the probability that a unit is functioning is p, then the probability of having exactly k functioning units out of n is

$$P(k; n, p) = \binom{n}{k} p^k (1-p)^{n-k} \qquad k = 0, 1, \ldots, n. \tag{2.20}$$

The system is considered to be functioning properly if k or $k+1$ or ... or $n-1$ or n units are functioning. Therefore, the reliability of the system is

$$R(k; n, p) = \sum_{r=k}^{n} \binom{n}{r} p^r (1-p)^{n-r}. \tag{2.21}$$

If the units are all different, then in order to determine the reliability of the system, all possible operational combinations should be evaluated as shown in Example 2.13.

EXAMPLE 2.13 Consider a telecommunication system that consists of four different parallel channels. A system is considered operational if any three channels are operational. Determine the reliability of the system.

SOLUTION This is a three-out-of-four system. Let x_1, x_2, x_3, and x_4 be indicators when channels 1 through 4 are functioning properly and $\bar{x}_1$, $\bar{x}_2$, $\bar{x}_3$, and $\bar{x}_4$ be the indicators when the channels fail. The reliability of the system is

$$R = P(x_1 x_2 x_3 + x_1 x_2 x_4 + x_1 x_3 x_4 + x_2 x_3 x_4). \tag{2.22}$$

Let

$$A_1 = x_1 x_2 x_3$$

$$A_2 = x_1 x_2 x_4$$

$$A_3 = x_1 x_3 x_4$$

$$A_4 = x_2 x_3 x_4.$$

In estimating the reliability one may include $A_5 = x_1 x_2 x_3 x_4$ in Eq. (2.22). However, the interaction terms will result in cancellations of some probabilities, which in turn causes Eq. (2.22) to be valid without the inclusion of A_5.

We rewrite Eq. (2.22) as

$$\begin{aligned}R &= P(A_1+A_2+A_3+A_4)\\ &= P(A_1)+P(A_2)+P(A_3)+P(A_4)-P(A_1A_2)-P(A_1A_3)-P(A_1A_4)-P(A_2A_3)\\ &\quad -P(A_2A_4)-P(A_3A_4)+P(A_1A_2A_3)+P(A_1A_2A_4)+P(A_1A_3A_4)\\ &\quad +P(A_2A_3A_4)-P(A_1A_2A_3A_4). \end{aligned} \tag{2.23}$$

But

$$A_1A_2 = x_1x_2x_3x_4$$

$$A_1A_3 = x_1x_2x_3x_4$$

$$A_1A_4 = x_1x_2x_3x_4$$

$$A_2A_3 = x_1x_2x_3x_4$$

$$A_2A_4 = x_1x_2x_3x_4$$

$$A_3A_4 = x_1x_2x_3x_4$$

$$A_1A_2A_3 = A_1A_2A_4 = A_1A_3A_4 = A_1A_2A_3A_4 = x_1x_2x_3x_4.$$

Substitution in Eq. (2.23) yields

$$\begin{aligned}R &= P(x_1x_2x_3) + P(x_1x_2x_4) + P(x_1x_3x_4) + P(x_2x_3x_4)\\ &\quad - 6P(x_1x_2x_3x_4) + 4P(x_1x_2x_3x_4) - P(x_1x_2x_3x_4)\\ &= P(x_1x_2x_3) + P(x_1x_2x_4) + P(x_1x_3x_4) + P(x_2x_3x_4) - 3P(x_1x_2x_3x_4).\end{aligned}$$

If the units are independent and identical, then

$$R = 4p^3 - 3p^4.$$

The above expression can also be obtained using Eq. (2.21):

$$R = \sum_{r=3}^{4} \binom{4}{r} p^r (1-p)^{4-r}$$

$$= \binom{4}{3} p^3(1-p) + \binom{4}{4} p^4(1-p)^0$$

or

$$R = 4p^3 - 3p^4.$$

2.8. Complex Reliability Systems

Telecommunication systems, computer networks, electric power utility systems, and water utility distribution systems are typical examples of complex networks. Some of the networks are referred to as *directed* networks when the flow from one node to another is unidirectional. When the flow is bidirectional we refer to the network as *undirected*. The examples, procedures, and problems discussed in this chapter are applicable to both the undirected and directed networks.

Consider the network shown in Figure 2.11. This network is a more complex system than those presented earlier in this chapter since it cannot be modeled (or is difficult to model) as series, parallel, parallel-series, series-parallel, or k-out-of-n systems. The reliability of such systems can be determined using any of the following methods (Shooman, 1968).

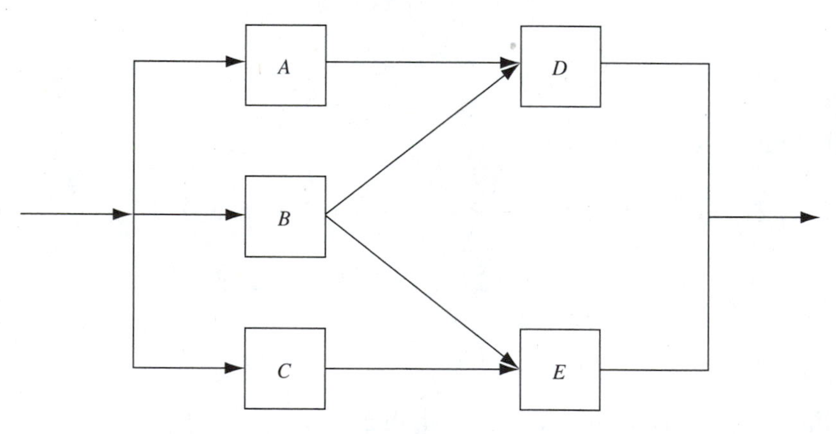

Figure 2.11. A Complex Reliability System

2.8.1. DECOMPOSITION METHOD

This method begins by selecting a *keystone* component, x, which appears to link (bind) together the reliability structure of the system. The reliability may then be expressed in terms of the keystone component based on the theorem of total probability as

$$R = P(\text{system good}/x)P(x) + P(\text{system good}/\bar{x})P(\bar{x}), \qquad (2.24)$$

where $P(\text{system good}/x)$ is the probability that the system is functioning given that x is functioning and $P(\text{system good}/\bar{x})$ is the probability that the system is functioning given that x is not functioning. Obviously, the choice of the keystone component has a direct effect on the necessary calculations for $P(\text{system good}/x)$ and $P(\text{system good}/\bar{x})$. An experienced engineer should be able to identify the keystone components. Nevertheless, if a component is selected as a keystone component, when in fact it is not, we still can determine the reliability of the system with little or no difficulty.

EXAMPLE 2.14

Determine the reliability of the network shown in Figure 2.11 when component B is selected as the keystone component.

SOLUTION

In this case, B is the keystone component and the reliability of the network can be determined as

$$R = P(\text{system good}/B)P(B) + P(\text{system good}/\bar{B})P(\bar{B}). \qquad (2.25)$$

Now, we estimate $P(\text{system good}/B)$ by determining the working paths in the network when B is functioning as shown in Figure 2.12. Similarly, the $P(\text{system good}/\bar{B})$ is obtained using the block diagram shown in Figure 2.11.

$$P(\text{system good}/B) = P(D) + P(E) - P(D)P(E) \qquad (2.26)$$

$$P(\text{system good}/\bar{B}) = P(A)P(D) + P(C)P(E) - P(A)P(D)P(C)P(E). \qquad (2.27)$$

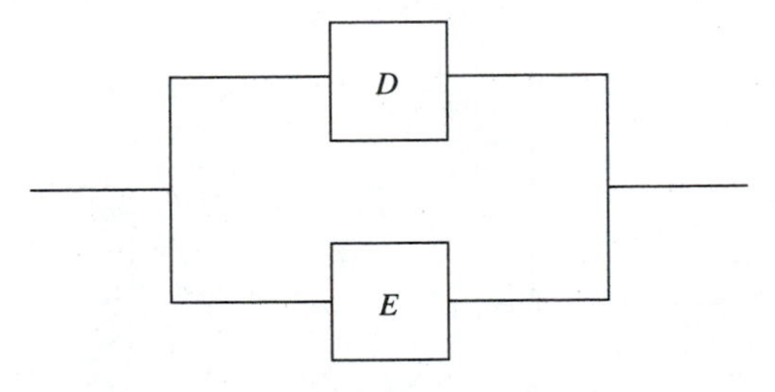

Figure 2.12. Block Diagram when B Is Working

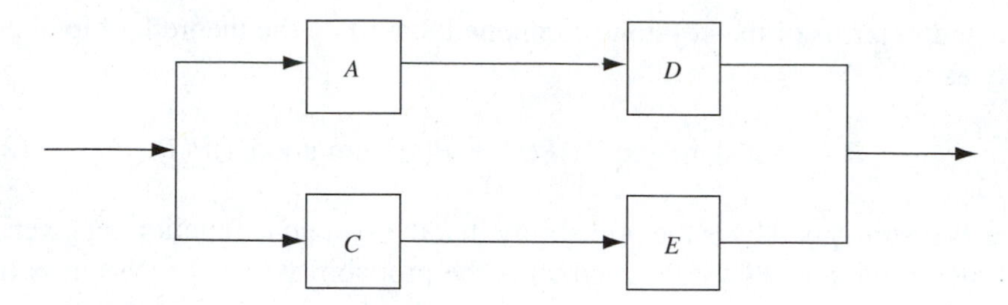

Figure 2.13. Block Diagram When B Fails

Substituting Eqs. (2.26) and (2.27) into Eq. (2.25) we obtain

$$\begin{aligned} R = {} & [P(D) + P(E) - P(D)P(E)]P(B) \\ & + [P(A)P(D) + P(C)P(E) - P(A)P(D)P(C)P(E)]\,[1 - P(B)]. \end{aligned} \tag{2.28}$$

If all components have equal probabilities (p) of functioning properly, then

$$R = 4p^2 - 3p^3 - p^4 + p^5. \tag{2.29}$$

EXAMPLE 2.15

Solve Example 2.14 using A as the keystone component.

SOLUTION

$$R = P(\text{system good}/A)P(A) + P(\text{system good}/\bar{A})P(\bar{A}). \tag{2.30}$$

Following Example 2.14, we estimate P(system good/A) using the diagrams shown in Figure 2.14. We start with diagram (a) which is reduced to diagram (b)

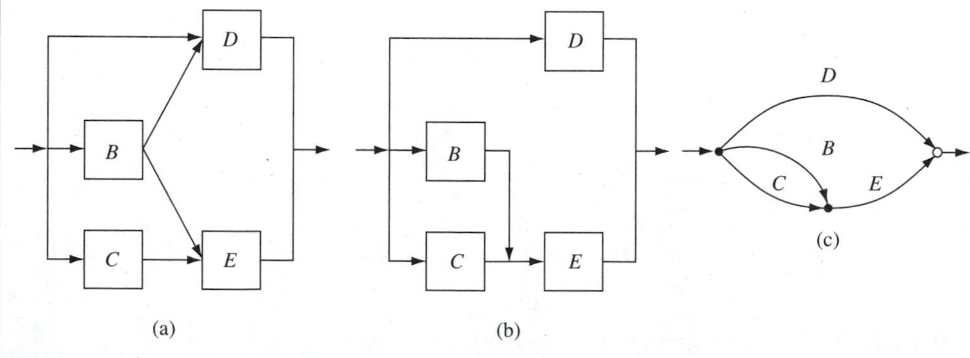

Figure 2.14. System Diagram When A Is Working

and finally to graph (c). Assume that all components are independent and identical. Then

$$P(\text{system good}/A) = 1 - (1-p)[1 - p(1-(1-p)^2)]$$
$$= p + 2p^2 - 3p^3 + p^4. \quad (2.31)$$

Similarly, we consider the system reliability when Component A fails. The corresponding block diagram is shown in Figure 2.15. The diagram in Figure 2.15 is still complex and does not decompose into series/parallel arrangements. Therefore, we choose another keystone Component C and the block diagram in Figure 2.15 can be redrawn as shown in Figure 2.16 to represent a subsystem of the main network:

$$P(\text{subsystem good}/C) = 1 - (1-p)(1-p^2). \quad (2.32)$$

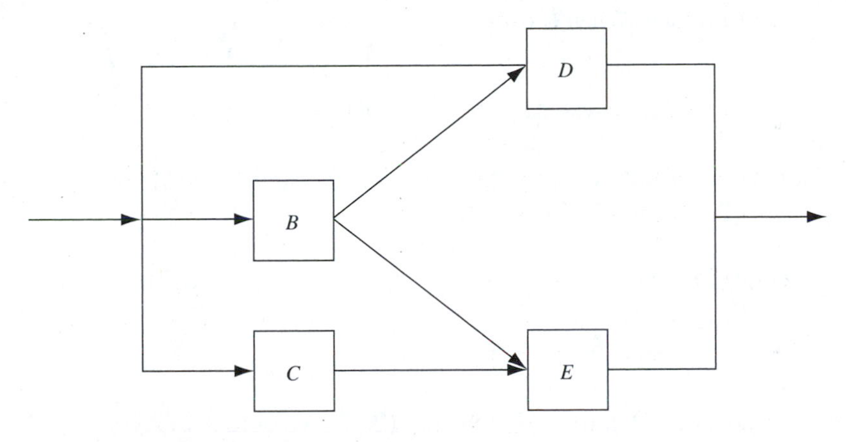

Figure 2.15. Block Diagram When A Fails

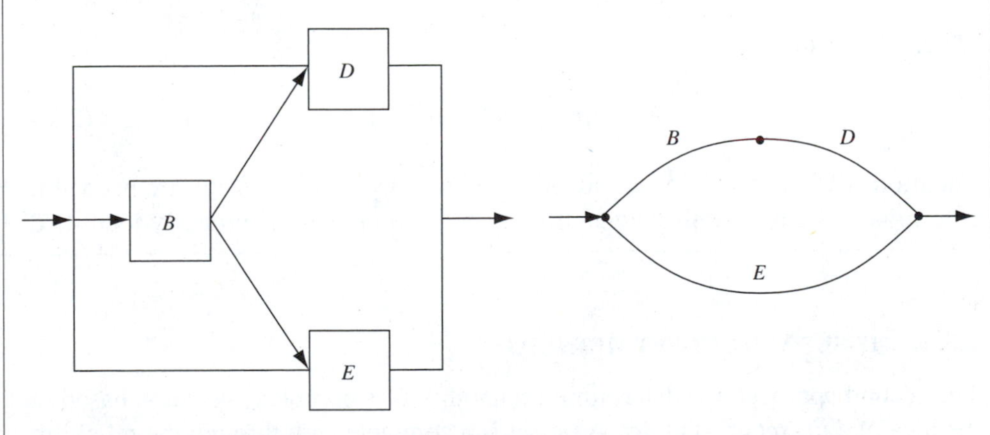

Figure 2.16. Block Diagram and Reliability Graph When C Is Working

$P(\text{subsystem good}/\bar{C})$ is estimated using Figure 2.17:

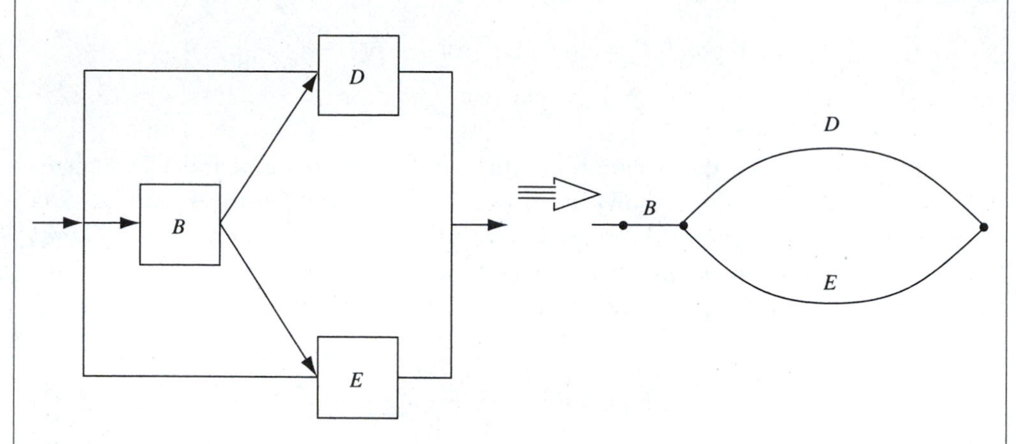

Figure 2.17. Subsystem When C Fails

$$P(\text{subsystem good}/\bar{C}) = p(1 - (1 - p)^2). \tag{2.33}$$

The reliability of the subsystem is the same as $P(\text{system good}/\bar{A})$ and is obtained by using Eqs. (2.32) and (2.33) as follows:

$$\begin{aligned} P(\text{system good}/\bar{A}) &= [1 - (1 - p)(1 - p^2)]p + p(1 - (1 - p)^2)(1 - p) \\ &= 3p^2 - 2p^3. \end{aligned} \tag{2.34}$$

Substituting Eqs. (2.31) and (2.34) into Eq. (2.30) we obtain

$$R = (p + 2p^2 - 3p^3 + p^4)p + (3p^2 - 2p^3)\,(1 - p),$$

which results in

$$R = 4p^2 - 3p^3 - p^4 + p^5. \tag{2.35}$$

Equations (2.29) and (2.35) are identical. However, far fewer steps are needed to determine system reliability when the keystone component is properly identified.

2.8.2. TIE-SET AND CUT-SET METHODS

The second approach for determining reliability of a complex system is based on the idea of a *tie-set* or a *cut-set*. A tie-set is a complete path through the reliability block diagram. It is not sufficient to determine all tie-sets since some of the tie-sets are contained within others. Therefore, it is important to define the *minimum* tie-set as the tie-set that contains no other tie-sets within it. The reliability of the system is given by the union of all minimum tie-sets.

A cut-set is a set of blocks (components) that interrupts all connections between the input and the output ends when removed from the reliability block diagram. A *minimum* cut-set is the one that contains no other cut-sets within it. The unreliability of the system is given by the probability that at least one minimal cut-set fails.

The following examples illustrate the use of tie-set and cut-set methods for estimating reliability of a complex system.

EXAMPLE 2.16

Consider the system shown in Figure 2.18. Use the tie-set and cut-set methods to estimate the system reliability.

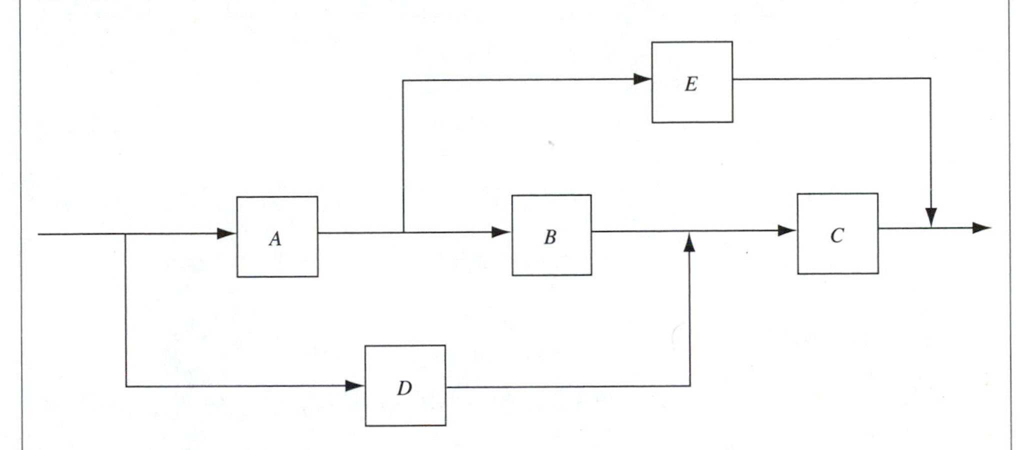

Figure 2.18. A Complex System

SOLUTION

The minimum tie-sets of the system are

$$T_1 = AE$$

$$T_2 = DC$$

$$T_3 = ABC.$$

The reliability of the system is the union of the tie-sets:

$$\begin{aligned} R &= P(AE \cup DC \cup ABC) \\ &= P(AE) + P(DC) + P(ABC) \\ &\quad - P(AEDC) - P(AEBC) - P(DCAB) + P(AEDCB). \end{aligned} \tag{2.36}$$

Assuming independence of probabilities, then Eq. (2.36) can be written as

$$R = P(A)P(E) + P(D)P(C) + P(A)P(B)P(C)$$

$$- P(A)P(E)P(D)P(C) - P(A)P(E)P(B)P(C). \qquad (2.37)$$

$$- P(D)P(C)P(A)P(B) + P(A)P(E)P(D)P(C)P(B).$$

If all units are identical and each has a probability p of functioning properly, then Eq. (2.37) becomes

$$R = 2p^2 + p^3 - 3p^4 + p^5. \qquad (2.38)$$

We can also apply the cut-set method to determine R. The minimum cut-sets are

$$C_1 = \bar{A}\bar{D}$$

$$C_2 = \bar{E}\bar{C}$$

$$C_3 = \bar{A}\bar{C}$$

$$C_4 = \bar{B}\bar{E}\bar{D}.$$

The reliability of the system is

$$R = 1 - P(\bar{A}\bar{D} \cup \bar{E}\bar{C} \cup \bar{A}\bar{C} \cup \bar{B}\bar{E}\bar{D}). \qquad (2.39)$$

Again, assuming independence of probabilities, Eq. (2.39) becomes

$$R = 1 - [P(\bar{A}\bar{D}) + P(\bar{E}\bar{C}) + P(\bar{A}\bar{C}) + P(\bar{B}\bar{E}\bar{D}) - P(\bar{A}\bar{D}\bar{E}\bar{C})$$

$$- P(\bar{A}\bar{D}\bar{C}) - P(\bar{A}\bar{D}\bar{B}\bar{E}) - P(\bar{E}\bar{C}\bar{A}) - P(\bar{E}\bar{C}\bar{B}\bar{D})$$

$$- P(\bar{A}\bar{C}\bar{B}\bar{E}\bar{D}) + P(\bar{A}\bar{D}\bar{E}\bar{C}) + P(\bar{A}\bar{D}\bar{C}\bar{B}\bar{E}) + P(\bar{A}\bar{D}\bar{C}\bar{B}\bar{E})$$

$$+ P(\bar{E}\bar{C}\bar{A}\bar{B}\bar{D}) - P(\bar{A}\bar{B}\bar{C}\bar{D}\bar{E})].$$

Substituting $P(\bar{A}) = [1 - P(A)]$ and $P(A) = P(B) = P(C) = P(D) = P(E) = p$ in the above equation, we obtain

$$R = 2p^2 + p^3 - 3p^4 + p^5. \qquad (2.40)$$

Equations (2.38) and (2.40) are identical.

2.8.3. EVENT-SPACE METHOD

The event-space method is based on listing all possible logical occurrences of the system. In other words, all components are considered functioning initially, and then they are allowed to fail individually, two at a time, three at a time, and so on. The reliability of the system is then determined by the union of all successful occurrences. Clearly, the number of occurrences depends on the number of components in the system. For example, a system with five components, where each component can either be working or failing, will have $2^5 = 32$ occurrences. There is only one occurrence with no failure $\{\binom{5}{0} = 1\}$, and five occurrences containing one failure $\{\binom{5}{1} = 5\}$ and so on. The following example illustrates the use of the event-space method in estimating the reliability of complex systems.

EXAMPLE 2.17

Determine the reliability of the network given in Figure 2.18 using the event-space method.

SOLUTION

Since there are five blocks in the network, the number of system occurrences is $2^5 = 32$. These occurrences are shown in Table 2.2 (Shooman, 1968). The reliability of the system is the probability of the union of operational occurrences. Thus,

$$R = P(X_1 + X_2 + \ldots + X_7 + X_{10} + \ldots + X_{14} + X_{16} + X_{20} + X_{24}). \quad (2.41)$$

Assuming that all components are independent, then Eq. (2.41) can be rewritten as

$$\begin{aligned} R = {} & P(X_1) + P(X_2) + \ldots + P(X_7) + P(X_{10}) + P(X_{11}) \\ & + P(X_{12}) + P(X_{13}) + P(X_{14}) + P(X_{16}) + P(X_{20}) + P(X_{24}). \end{aligned} \quad (2.42)$$

If all components are identical and each has a probability p of functioning properly, then

$$P(X_1) = P(ABCDE) = p^5$$

$$P(X_2) = P(X_3) = \ldots = P(X_6) = (1 - p)p^4$$

$$P(X_7) = P(X_{10}) = \ldots = P(X_{14}) = P(X_{16}) = (1 - p)^2 p^3$$

$$P(X_{20}) = P(X_{24}) = (1 - p)^3 p^2$$

and

$$R = p^5 + 5(1 - p)p^4 + 7(1 - p)^2 p^3 + 2(1 - p)^3 p^2$$

or

$$R = 2p^2 + p^3 - 3p^4 + p^5.$$

This is the same result obtained by both tie-set and cut-set methods.

Table 2.2. All Possible Logical Occurrences for Figure 2.18

Group 0 (No Failures)	$X_1 = ABCDE$
Group 1 (one failure)	$X_2 = \bar{A}BCDE, X_3 = A\bar{B}CDE,$ $X_4 = AB\bar{C}DE, X_5 = ABC\bar{D}E, X_6 = ABCD\bar{E}$
Group 2 (two failures)	$X_7 = \bar{A}\bar{B}CDE, X_8 = \underline{\bar{A}B\bar{C}DE},$ $X_9 = \underline{\bar{A}BC\bar{D}E}, X_{10} = \bar{A}BCD\bar{E},$ $X_{11} = A\bar{B}\bar{C}DE, X_{12} = A\bar{B}C\bar{D}E,$ $X_{13} = A\bar{B}CD\bar{E}, X_{14} = AB\bar{C}\bar{D}E,$ $X_{15} = \underline{AB\bar{C}D\bar{E}}, X_{16} = ABC\bar{D}\bar{E}$
Group 3 (three failures)	$X_{17} = \underline{AB\bar{C}\bar{D}\bar{E}}, X_{18} = \underline{A\bar{B}C\bar{D}\bar{E}},$ $X_{19} = \underline{A\bar{B}\bar{C}D\bar{E}}, X_{20} = A\bar{B}\bar{C}\bar{D}E, X_{21} = \underline{\bar{A}BC\bar{D}\bar{E}},$ $X_{22} = \underline{\bar{A}B\bar{C}D\bar{E}}, X_{23} = \underline{\bar{A}B\bar{C}\bar{D}E}, X_{24} = \bar{A}\bar{B}CD\bar{E},$ $X_{25} = \underline{\bar{A}\bar{B}C\bar{D}E}, X_{26} = \underline{\bar{A}\bar{B}\bar{C}DE}$
Group 4 (four failures)	$X_{27} = \underline{A\bar{B}\bar{C}\bar{D}\bar{E}}, X_{28} = \underline{\bar{A}B\bar{C}\bar{D}\bar{E}}, X_{29} = \underline{\bar{A}\bar{B}C\bar{D}\bar{E}},$ $X_{30} = \underline{\bar{A}\bar{B}\bar{C}D\bar{E}}, X_{31} = \underline{\bar{A}\bar{B}\bar{C}\bar{D}E}$
Group 5 (five failures)	$X_{32} = \underline{\bar{A}\bar{B}\bar{C}\bar{D}\bar{E}}$

Note: Underlined occurrence implies failure of the system.

2.8.4. BOOLEAN TRUTH TABLE METHOD

This method is based on the construction of a Boolean truth table for the system. This method is tedious if done manually, but recent developments in computer software and hardware have made it possible to construct large truth tables in a relatively small amount of time. A truth table is similar to the event-space method where every possible state of the system is listed. A state refers to the condition of a component as functioning or not. A column is created in the table for each component, and a value of 1 or 0 is assigned to the column to indicate that the component is functioning or not, respectively. Each row in the table then represents a state of the system. Each row is examined to determine the state of the system as functioning or not. This is indicated by assigning 1 or 0 to the system state column. The state probability for every functioning row is computed and the reliability of the system is obtained by adding all functioning state probabilities.

EXAMPLE 2.18

Use the Boolean truth table method to obtain the reliability of the system given in Figure 2.18.

SOLUTION

We construct the Boolean truth table for the system as shown in Table 2.3. The reliability of the system is obtained by adding the probabilities of functioning

Table 2.3. Boolean Truth Table for Example 2.18

A	*B*	*C*	*D*	*E*	*System State*	*State Probability*
1	1	1	1	1	1	$P(A)P(B)P(C)P(D)P(E)$
1	1	1	1	0	1	$P(A)P(B)P(C)P(D)P(\bar{E})$
1	1	1	0	1	1	$P(A)P(B)P(C)P(\bar{D})P(E)$
1	1	1	0	0	1	$P(A)P(B)P(C)P(\bar{D})P(\bar{E})$
1	1	0	1	1	1	$P(A)P(B)P(\bar{C})P(D)P(E)$
1	1	0	1	0	0	
1	1	0	0	1	1	$P(A)P(B)P(\bar{C})P(\bar{D})P(E)$
1	1	0	0	0	0	
1	0	1	1	1	1	$P(A)P(\bar{B})P(C)P(D)P(E)$
1	0	1	1	0	1	$P(A)P(\bar{B})P(C)P(D)P(\bar{E})$
1	0	1	0	1	1	$P(A)P(\bar{B})P(C)P(\bar{D})P(E)$
1	0	1	0	0	0	
1	0	0	1	1	1	$P(A)P(\bar{B})P(\bar{C})P(D)P(E)$
1	0	0	1	0	0	
1	0	0	0	1	1	$P(A)P(\bar{B})P(\bar{C})P(\bar{D})P(E)$
1	0	0	0	0	0	
0	1	1	1	1	1	$P(\bar{A})P(B)P(C)P(D)P(E)$
0	1	1	1	0	1	$P(\bar{A})P(B)P(C)P(D)P(\bar{E})$
0	1	1	0	1	0	
0	1	1	0	0	0	
0	1	0	1	1	0	
0	1	0	1	0	0	
0	1	0	0	1	0	
0	1	0	0	0	0	
0	0	1	1	1	1	$P(\bar{A})P(\bar{B})P(C)P(D)P(E)$
0	0	1	1	0	1	$P(\bar{A})P(\bar{B})P(C)P(D)P(\bar{E})$
0	0	1	0	1	0	
0	0	1	0	0	0	
0	0	0	1	1	0	
0	0	0	1	0	0	
0	0	0	0	1	0	
0	0	0	0	0	0	

states. Assume that the components are independent, are identical, and have the same probability p of operating properly. The reliability is

$$R = p^5 + 5p^4(1-p) + 7p^3(1-p)^2 + 2p^2(1-p)^3$$

or

$$R = 2p^2 + p^3 - 3p^4 + p^5. \tag{2.43}$$

As shown above, the reliability obtained using the Boolean truth table is the same as that obtained using the tie-set and cut-set methods.

2.8.5. REDUCTION METHOD

The reduction method is based on the standard Boolean truth table method and then applying the resulting mutually exclusive sum-of-products (s-o-p) terms (Case, 1977). The procedure starts by constructing a 2^n *truth table* (n is the number of components in the system). Each row in the table is then examined, and rows resulting in a system success (functioning properly) are indicated. A reduction table is then constructed by listing all success rows in column 1. By a comparative process, product terms are formed for those terms in column 1 that differ by a letter inverse. Once a term is used in a comparison, it is eliminated from all further comparisons ensuring that all remaining terms are still mutually exclusive. This procedure is repeated until no further comparisons are possible. The reliability of the system is the union of all terms that cannot be further compared.

It is important to note that the order of terms selected for the comparison process has no effect on the estimation of system reliability.

EXAMPLE 2.19

Use the reduction method to determine the reliability of the system given in Figure 2.18.

SOLUTION

We utilize the results obtained in Table 2.3. The functioning states of the system are listed under column 1 in Table 2.4. The probability of the system is obtained by the union of all the states that cannot be further combined:

$$R = P(A\bar{B}C\bar{D}E + A\bar{B}CD + ABC + A\bar{C}E + \bar{A}CD)$$

$$= p^3(1-p)^2 + p^3(1-p) + p^3 + p^2(1-p) + p^2(1-p)$$

or

$$R = 2p^2 + p^3 - 3p^4 + p^5. \tag{2.44}$$

The reliability obtained by Eq. (2.44) is the same as that obtained by other methods.

Table 2.4. Reduction Table for Figure 2.18

Column 1 *Functional States*	*Column 2*	*Column 3*
$ABCDE$	$ABCD$	
$ABCD\bar{E}$		ABC
$ABC\bar{D}E$	$ABC\bar{D}$	
$ABC\bar{D}\bar{E}$		
$AB\bar{C}DE$	$AB\bar{C}E$	$A\bar{C}E$
$AB\bar{C}\bar{D}E$		
$A\bar{B}CDE$	$A\bar{B}CD$	
$A\bar{B}CD\bar{E}$		
$A\bar{B}C\bar{D}E$		
$A\bar{B}\bar{C}DE$	$A\bar{B}\bar{C}E$	
$A\bar{B}\bar{C}\bar{D}E$		
$\bar{A}BCDE$	$\bar{A}BCD$	
$\bar{A}BCD\bar{E}$		$\bar{A}CD$
$\bar{A}\bar{B}CDE$	$\bar{A}\bar{B}CD$	
$\bar{A}\bar{B}CD\bar{E}$		

2.8.6. PATH-TRACING METHOD

The path-tracing method is simple and efficient in estimating the reliability of complex structures. The method starts by assuming that all blocks in the reliability diagram are missing initially, and the components are replaced singly, in pairs, in triplets, and so on. The successful paths found by using the least number of components are then used in calculating system reliability as shown below.

EXAMPLE 2.20

Use the path-tracing method to determine the reliability of the system given in Figure 2.18.

SOLUTION

As shown in the block diagram, no single component forms a successful path by itself, but the pairs AE and DC and the triplet components ABC form successful paths. The reliability of the system is then obtained by the probability of the union of these paths:

$$R = P(AE) + P(DC) + P(ABC) - P(AEDC) - P(AEBC) - P(DCAB) + P(ABCDE). \tag{2.45}$$

If all components are independent and identical and each has a probability p of functioning properly, then Eq. (2.45) becomes

$$R = 2p^2 + p^3 - 3p^4 + p^5,$$

which is the same as the reliability estimated by other methods.

2.8.7. FACTORING ALGORITHM

The complex structures presented in this chapter are simple and limited when compared with large-scale structures such as computer and telephone communication networks and electric power utility networks. Reliability estimation of such networks is, in a sense, more difficult than many standard combinatorial optimization problems. However, researchers have developed algorithms that can efficiently estimate the reliability of networks with specified characteristics. In this section, we present one of these algorithms—the factoring algorithm.

Consider a complex structure that is represented by a reliability network or graph (a *graph* is a pictorial representation of the network). A typical graph consists of nodes and arcs, where a node represents a location (or a point) that communicates with other nodes via arcs. An arc can represent a means of communication between the nodes, such as components, cables, and pipes.

The factoring algorithm is based on the decomposition method discussed in Section 2.8.1. Following Eq. (2.30), if $R(G/e)$ is the reliability of graph G under the condition that component e (arc or edge e of the graph) is working and $R(G/\bar{e})$ the reliability of G under the condition that component e is not working, then the reliability of the graph G is

$$R(G) = p_e\, R(G/e) + (1 - p_e)\, R(G/\bar{e}), \tag{2.46}$$

where p_e is the reliability of component (edge) e.

The reliability of any graph G can be computed by repeated application of Eq. (2.46). Undirected graphs have some special properties that can be used to simplify this method. If the vertices (nodes) are assumed to be working, then $R(G/e)$ coincides with $R(G_e)$, where G_e is the graph obtained from G by deleting edge e and merging its end points. Similarly, $R(G/\bar{e})$ equals $R(G - e)$, where $G - e$ is the graph with e deleted and no vertex is deleted. It is important to note that this factoring algorithm can be employed using graph representation but without knowing the minimal path sets. Moreover, unless some kind of probability reductions are performed (such as parallel and series reductions) after the deletion of an edge, the factoring algorithm will be equivalent to state space enumeration (Agrawal and Barlow, 1984). We now illustrate the use of the factoring algorithm in estimating network reliability.

EXAMPLE 2.21

In large cities, gas is produced by vaporizing LNG (liquefied natural gas). The quality of the gas is checked for calorific values, combustibility, and other characteristics. The gas is then sent out through transmission pipelines to distribution centers where gas pressure is decreased before delivery to customers. Figure 2.19 shows a simplified network of a gas distribution system. Node A represents the location at which gas is vaporized, nodes B, C, D, E, and F are major distribution centers where gas is received from A (directly or indirectly). Gas must reach the distribution center F since it provides gas to critical services of the city. Thus, the reliability of the network is the probability that gas sent from node A reaches the distribution center F. Assume that the reliability of every transmission pipe is p. Use the factoring algorithm to determine the reliability of the network.

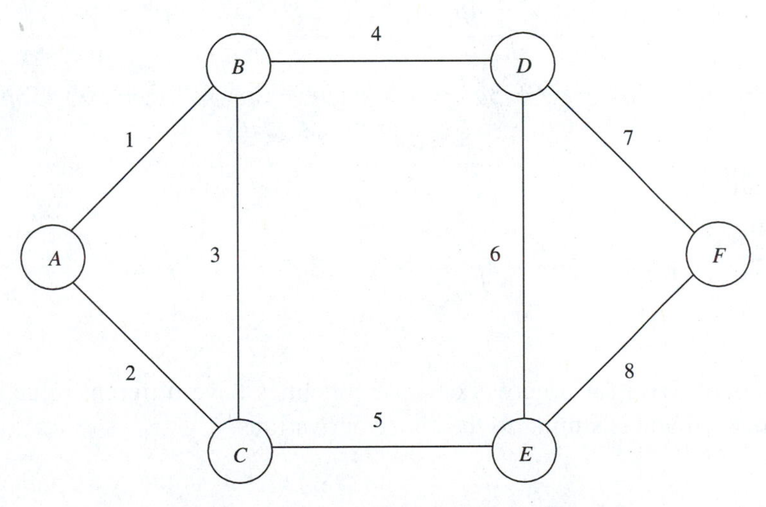

Figure 2.19. Simplified Network of the Gas Distribution System

SOLUTION We use the operator $\oplus$ to correspond to calculating the reliability of parallel pipelines as

$$p_i \oplus p_j = p_i + p_j - p_i p_j.$$

The initial step in applying the algorithm is to select an arc—say, arc 1—and form two subgraphs: G_1 corresponding to arc 1 working, and $G - 1$ corresponding to arc 1 failed. We now apply a parallel probability reduction by replacing the two arcs 2 and 3 in graph G_1 by a single arc with associated reliability $p_2 + p_3 - p_2 p_3$. Likewise in G_1, this new arc and arc 5 form a series system, and these two arcs can now be replaced by a single arc having reliability $(p_2 + p_3 - p_2 p_3)p_5$. The factoring algorithm now proceeds by considering arc 4, which results in two additional subgraphs, each of which can be reduced to a single arc by series and parallel probability reduction. The algorithm continues until no further reductions can be made. Figure 2.20 shows the steps of the algorithm as described above (Agrawal and Barlow, 1984).

Using Eq. (2.46) and the four subgraphs at the bottom of Figure 2.20 we obtain the reliability of the network as

$$\begin{aligned}
R(G) = p^2 (((((p \oplus p)p) \oplus p)p) \oplus p) + p(1-p)(((p \oplus p)p)(p^2 \oplus p)) \\
+ p(1-p)((p(p \oplus p)) \oplus p^2)p + (1-p)^2 ((p^3 \oplus p)p^2) \\
= (p^3 + p^4 + p^5 - 5p^6 + 4p^7 - p^8) + (2p^4 - p^5 - 4p^6 + 4p^7 - p^8) \\
+ (3p^4 - 4p^5 - p^6 + 3p^7 - p^8) + (p^3 - 2p^4 + 2p^5 - 3p^6 + 3p^7 - p^8)
\end{aligned}$$

or

$$R(G) = 2p^3 + 4p^4 - 2p^5 - 13p^6 + 14p^7 - 4p^8. \tag{2.47}$$

Derivations of $R(G)$ for a network whose pipelines have different reliabilities is straightforward and is similar to the above derivations.

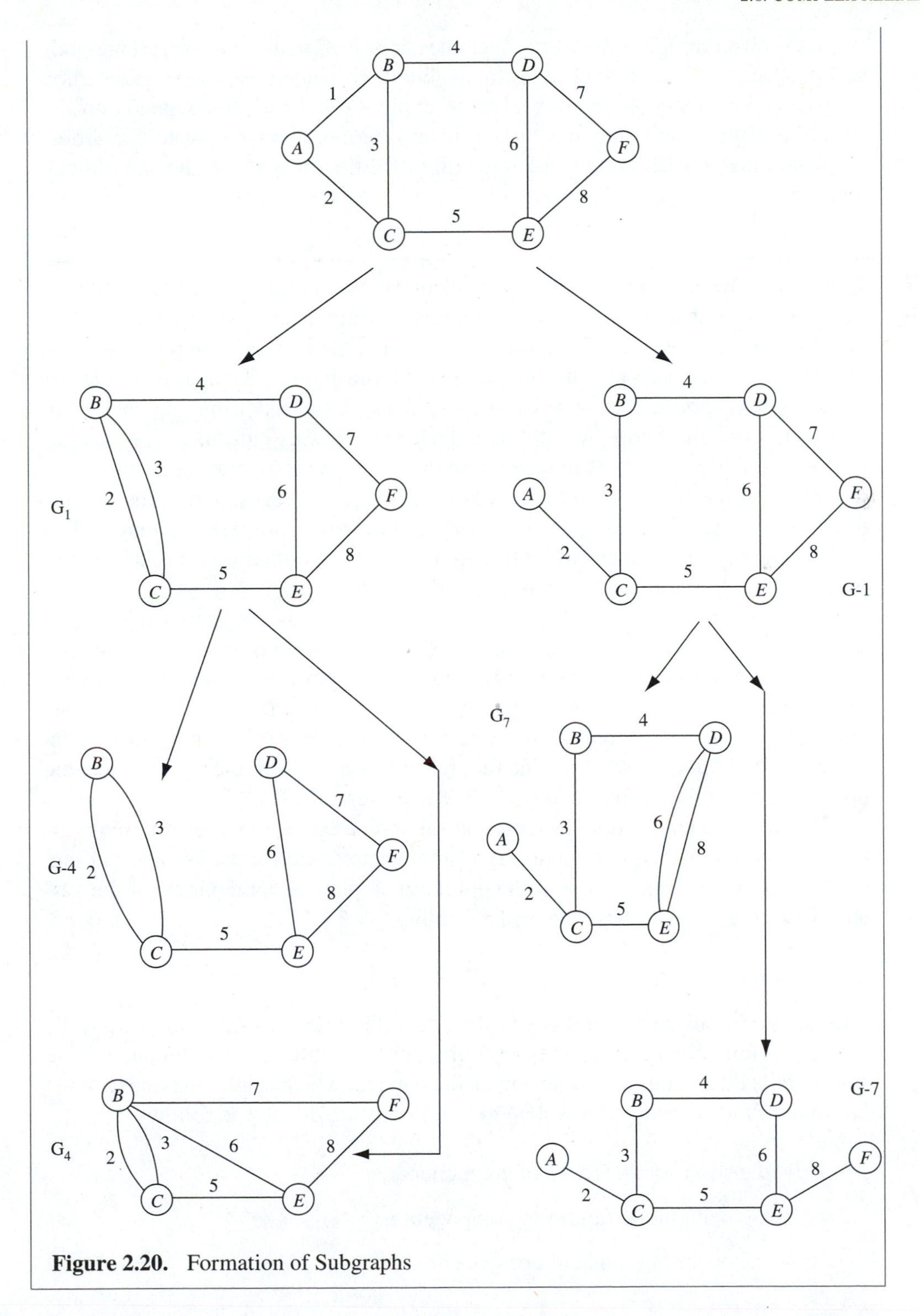

Figure 2.20. Formation of Subgraphs

As shown through many of the examples presented in this chapter, all methods will produce the same reliability estimate, but in any individual case, one method may be considerably more convenient to apply. Of course, this depends on the structure of the block diagram. In fact, it may be more convenient to use different techniques for estimating the reliability of different parts of the same block diagram.

2.9. Multistate Models

So far, we have assumed that a component can be in either one of two states—operational or failure. In many situations, a component may experience more than two states—for example, a three-state component may operate properly in its normal mode but may fail in either of two failure modes. Typical examples of three-state components are transistors and diodes. A transistor may operate properly or fail open or short. A *diode* is a device that passes current in the forward direction and blocks current in the reverse direction. When operating properly, the resistance in the forward direction is zero, whereas the resistance in the reverse direction is essentially infinite. The diode may operate properly or may fail in either state: (1) it may open circuit (that is, resistance in both directions is infinite), or (2) it may short circuit (resistance in both directions is zero).

As mentioned earlier, redundancy is one of the means of increasing system reliability. Increasing the number of redundant components in a system whose components have only two states (operational or failure) increases the reliability of the system. Unlike the two-state components, adding multistate components may either increase or decrease the system reliability. This, of course, depends on the dominant mode of component failure, configuration of the system and the number of redundant components (Dhillon and Singh, 1981).

In the following sections, we present reliability expressions for different system configurations composed entirely of multistate components. We also present methods for the determination of the optimum number of components of the system that achieve the highest levels of reliability.

2.9.1. SERIES SYSTEMS

This section considers components that have three states—x (good), $\bar{x}_s$ (fails short), $\bar{x}_o$ (fails open). In a series configuration of n three-state components, the system fails if any component fails in an open mode, whereas all components must fail in the short mode for the system to fail. Terms are defined as follows:

$\bar{x}_{si}$ = the short-mode failure of component i,

$\bar{x}_{oi}$ = the open-mode failure of component i,

x_i = the operating mode of component i,

n = the number of nonidentical but independent three-state components,

q_{si} = the probability of short-mode failure of component i, and

q_{oi} = the probability of open-mode failure of component i.

The reliability of a system composed of one three-state component is

$$R = P(x_1) = 1 - P(\bar{x}_{o1}) - P(\bar{x}_{s1})$$

or

$$R = (1 - q_{o1}) - q_{s1}. \tag{2.48}$$

Consider now a system composed of two three-state components in series. Its reliability is obtained as

$$\begin{aligned} R &= 1 - P(\text{system failure}) \\ &= 1 - P(\bar{x}_{o1} + \bar{x}_{o2} + \bar{x}_{s1}\bar{x}_{s2}) \\ &= 1 - [P(\bar{x}_{o1}) + P(\bar{x}_{o2}) - P(\bar{x}_{o1}\bar{x}_{o2}) + P(\bar{x}_{s1}\bar{x}_{s2})] \end{aligned}$$

or

$$R = 1 - [(q_{o1} + q_{o2} - q_{o1}q_{o2}) + q_{s1}q_{s2}]. \tag{2.49}$$

Rewriting Eq. (2.49), we obtain

$$R = \prod_{i=1}^{2} (1 - q_{oi}) - q_{s1}q_{s2}. \tag{2.50}$$

By induction from Eqs. (2.48) and (2.50), the reliability of an n components system is

$$R = \prod_{i=1}^{n} (1 - q_{oi}) - \prod_{i=1}^{n} q_{si}. \tag{2.51}$$

If all components are independent and identical, then Eq. (2.51) becomes

$$R = (1 - q_o)^n - q_s^n. \tag{2.52}$$

Unlike the standard series system with identical two-state components, the reliability of series systems with identical three-state components will reach its maximum by connecting an optimum number of components. Any number of components less or greater than the optimum will result in lower reliability values. To obtain the optimum number of three-state components in series that maximizes the reliability of the system, take the derivative of Eq. (2.52) with respect to n and equate it to zero. Then solve the resultant equation to determine the optimum number (n^*) of components. Thus,

$$\frac{\partial R}{\partial n} = (1 - q_o)^n \ln(1 - q_o) - q_s^n \ln q_s = 0$$

or

$$n^* = \frac{\ln[\ln q_s / \ln(1 - q_o)]}{\ln[(1 - q_o) / q_s]}. \tag{2.53}$$

If n^* is not an integer, then $\lfloor n^* \rfloor$ and $\lfloor n^* \rfloor + 1$ are also optimum solutions. Note that $\lfloor n^* \rfloor$ is the largest integer less than or equal to n^*.

2.9.2. PARALLEL SYSTEMS

We now consider a parallel system that is composed of two components connected in parallel. Using the same notations given in Section 2.9.1, we derive the reliability of the system as

$$\begin{aligned} R &= 1 - P(\bar{x}_{o1}\bar{x}_{o2} + \bar{x}_{s1} + \bar{x}_{s2}) \\ &= 1 - [P(\bar{x}_{o1})P(\bar{x}_{o2}) + P(\bar{x}_{s1}) + P(\bar{x}_{s2}) - P(\bar{x}_{s1})P(\bar{x}_{s2})] \\ &= 1 - [q_{o1}q_{o2} + q_{s1} + q_{s2} - q_{s1}q_{s2}] \end{aligned}$$

or

$$R = \prod_{i=1}^{2} (1 - q_{si}) - \prod_{i=1}^{2} q_{oi}. \tag{2.54}$$

Equation (2.54) can be generalized for systems with n components in parallel as

$$R = \prod_{i=1}^{n} (1 - q_{si}) - \prod_{i=1}^{n} q_{oi}. \tag{2.55}$$

If all components are identical, the reliability of the system becomes

$$R = (1 - q_s)^n - q_o^n. \tag{2.56}$$

For any range of q_o and q_s, the optimum number of parallel components that maximizes system reliability is one if $q_s > q_o$. For most practical values of q_o and q_s, the optimum number is two (Von Alven, 1964). In general, for a given q_s and q_o, the reliability function in terms of n would have the form shown in Figure 2.21. Therefore, we take the derivative of Eq. (2.56) with respect to n and equate the result to zero to find the optimum number of components:

$$\frac{\partial R}{\partial n} = \frac{\partial[(1 - q_s)^n - q_o^n]}{\partial n}$$

$$0 = (1 - q_s)^n \ln(1 - q_s) - q_o^n \ln q_o$$

and

$$n^* = \frac{\ln\left[\dfrac{\ln q_o}{\ln(1 - q_s)}\right]}{\ln\left[(1 - q_s)/q_o\right]}. \tag{2.57}$$

Again, if n^* is not an integer, then $\lfloor n^* \rfloor$ and $\lfloor n^* + 1 \rfloor$ are also optimum solutions.

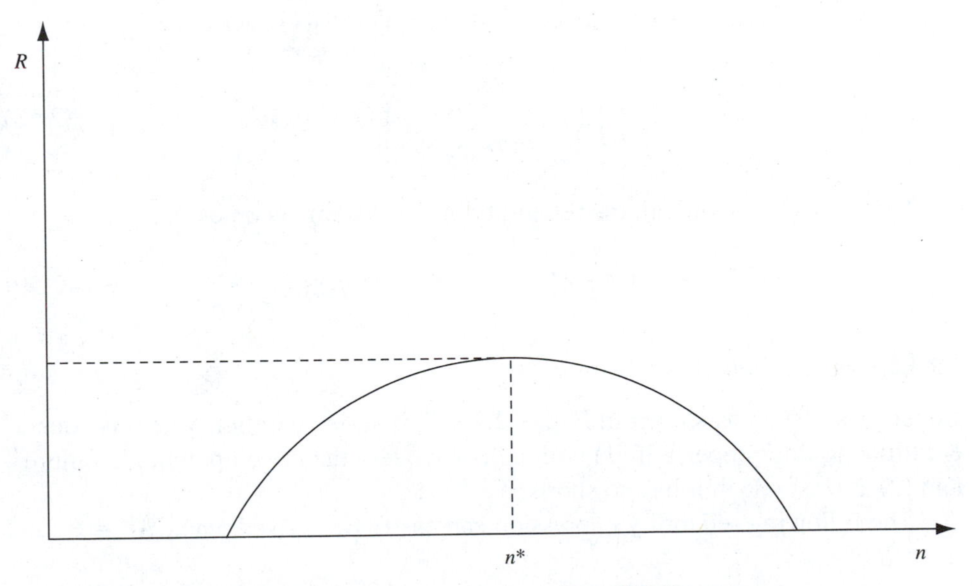

Figure 2.21. System Reliability Versus Number of Parallel Components

2.9.3. PARALLEL-SERIES AND SERIES-PARALLEL

2.9.3.1. *Parallel-Series*

Consider a parallel-series system that consists of four components as shown in Figure 2.22. The components in the same path are identical. The system is considered to be properly functioning if (1) at least one path has no open mode failures, and (2) each path has less than two shorts.

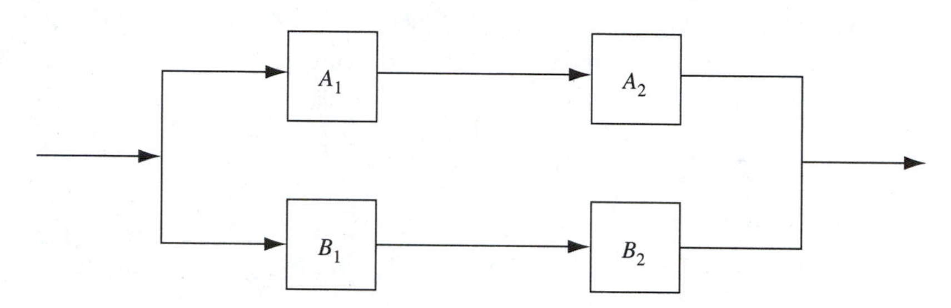

Figure 2.22. A Parallel-Series System

The reliability of the system when $A_1 = B_1$ and $A_2 = B_2$ can then be obtained as

$$R = [1 - q_{sa}q_{sb}]^2 - [1 - (1 - q_{oa})(1 - q_{ob})]^2.$$

For m identical parallel paths each containing n elements in series,

$$R = [1 - \prod_{i=1}^{n} q_{si}]^m - [1 - \prod_{i=1}^{n} (1 - q_{oi})]^m. \tag{2.58}$$

If all elements are identical, the reliability of the system becomes

$$R = [1 - q_s^n]^m - [1 - (1 - q_o)^n]^m. \tag{2.59}$$

2.9.3.2. *Series-Parallel*

Consider a system as shown in Figure 2.23. This series-parallel system is considered functioning properly if (1) both units have less than two open mode failures, and (2) at least one unit has no shorts.

The following reliability expression can easily be derived when $A_1 = B_1$ and $A_2 = B_2$:

$$R = [1 - q_{oa}q_{ob}]^2 - [1 - (1 - q_{sa})(1 - q_{sb})]^2. \tag{2.60}$$

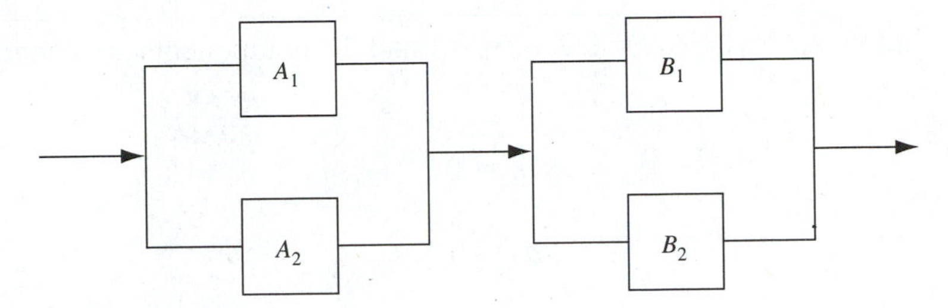

Figure 2.23. A Series-Parallel System

For n identical subsystems each containing m components in parallel, the reliability is

$$R = [1 - \prod_{i=1}^{m} q_{oi}]^n - [1 - \prod_{i=1}^{m} (1 - q_{si})]^n. \tag{2.61}$$

If all components are identical, the reliability of the system becomes

$$R = [1 - q_o^m]^n - [1 - (1 - q_s)^m]^n. \tag{2.62}$$

To find the optimum configurations for either the parallel-series or series-parallel, set the partial derivatives of R with respect to m and n equal to zero, and then iteratively solve the resulting equations simultaneously for n^* and m^*.

EXAMPLE 2.22

A series system consists of six identical three-state components. The probabilities that a component fails in an open mode and a short mode are 0.1 and 0.2, respectively. What is the reliability of the system? What is the optimum number of components that maximizes the system reliability?

SOLUTION

$$q_o = 0.1$$

$$q_s = 0.2.$$

Using Eq. (2.52), we obtain the reliability of the system as

$$R = (1 - 0.1)^6 - 0.2^6 = 0.53137.$$

The optimum number of components is obtained using Eq. (2.53):

$$n^* = \frac{\ln [\ln 0.2 / \ln 0.9]}{\ln [0.9 / 0.2]} = 1.8 \cong 2 \text{ units}.$$

The reliability corresponding to this system is 0.77.

EXAMPLE 2.23 Solve Example 2.22 when $q_o = 0.2$, $q_s = 0.1$ and the components are connected in parallel.

SOLUTION

$$q_s = 0.1$$

$$q_o = 0.2.$$

From Eq. (2.56),

$$R = (1 - 0.1)^6 - 0.2^6 = 0.53137.$$

The optimum number of components in parallel is obtained using Eq. (2.57):

$$n^* = \frac{\ln(\ln 0.2 / \ln 0.9)}{\ln(0.9 / 0.2)} \cong 2.$$

The reliability corresponding to this system is 0.77. This is identical to the result obtained in Example 2.22.

In other words, a series system is equivalent to a parallel system if the same number of components is used in both systems and if the values of q_s and q_o are reversed from one system to the other.

2.10. Redundancy

Redundancy is defined as the use of additional components or units beyond the number actually required for satisfactory operation of a system for the purpose of improving its reliability. A series system has no redundancy since a failure of any component causes failure of the entire system, whereas a parallel system has redundancy since the failure of a component (or possibly more) does not result in a system failure. Similarly, consecutive-k-out-of n:F systems, k-out-of-n, parallel-series, and series-parallel systems have redundancy.

In a pure parallel system, redundancy is a function of the number and type of components connected in parallel. As stated earlier in this chapter, if only two-state components are used, then increasing the number of parallel components will increase the reliability of the system. However, if the components have more than two states, then there is an optimum number of components, which maximizes the system reliability. In other words, improving the system reliability through redundancy is not as simple as doubling, tripling, or adding more components in parallel.

There are two types of redundancy: active and inactive. In *active redundancy*, all redundant components are in operation and are sharing the load with the main unit. Under *nonactive standby*, the redundant components do not share any amount of the load with the main components, and they start operating only when one or more operating components fail. When the failure rate of the standby component is the same as the main unit, we refer to this arrangement as *hot* standby. When the failure rate of the standby unit is less than that of the main unit, we then have a *warm* standby, and when the failure rate of the standby unit when it is not operating is zero, then we have a *cold* standby. Clearly, the application of the type of redundancy depends on the criticality of the system and the consequences of a major failure. For example, an airplane that requires two out of three engines for successful operation usually has all its engines in active redundancy, whereas a computer system uses an uninterrupted power supply (UPS) in an inactive redundancy to provide the needed power when a failure occurs in the main power source. There is no difference between operating a system under active or inactive redundancy if the switching system (which connects the inactive components to the system) is perfect—that is, does not fail—and if the failure rate of the redundant component is the same whether it is operating or not. The following example illustrates the difference between active and inactive redundancy.

EXAMPLE 2.24

A two-component system may be configured as active or inactive redundancy as shown in Figure 2.24. Assume that the switch S is perfect. What are the reliabilities of both systems?

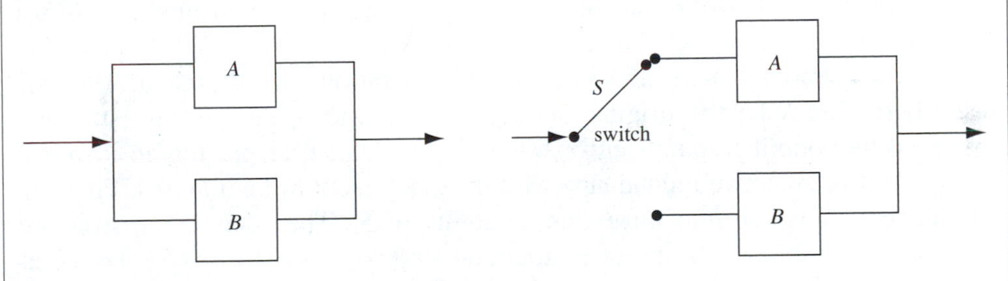

Figure 2.24. Active and Inactive Redundancy

SOLUTION

The two components active redundancy system fails only if both components A and B fail. Thus, the reliability of the system is

$$R_{\text{active}} = 1 - P(\bar{A}\bar{B}) = 1 - P(\bar{A})P(\bar{B}/\bar{A}). \tag{2.63}$$

If the components A and B are identical and independent, each having reliability p, then

$$R_{\text{active}} = 2p - p^2. \tag{2.64}$$

In inactive redundancy, the system fails if Component A fails, the perfect switch switches to B, and then B fails. The reliability of the system is

$$R_{\text{inactive}} = 1 - P(\bar{A}\bar{B}) = 1 - P(\bar{A})P(\bar{B}/\bar{A}). \tag{2.65}$$

It appears that the active and inactive redundancies result in the same value of reliability. This is not true, since the interpretations of the conditional probabilities in Eqs. (2.63) and (2.65) are distinctly different. In Eq. (2.63) $P(\bar{B}/\bar{A})$ may simply be $P(\bar{B})$ if events $\bar{A}$ and $\bar{B}$ are independent, or it may be slightly different if there is a small dependency. In the active redundancy case, Component B is assumed to have operated since time $t = 0$. In inactive redundancy $P(\bar{B}/\bar{A})$ is always a dependent probability since Component B does not start to operate until A fails. Clearly, this conditional probability is a function of time. Further discussion of redundancy is presented in Chapter 3.

2.10.1. REDUNDANCY ALLOCATION FOR A SERIES SYSTEM

As shown earlier, the reliability of a system composed entirely of two-state components increases by adding components in parallel with the main components of the system. An engineer may be interested in increasing the reliability of an n components series system. In order to do so, the engineer must allocate components in parallel with the main components of the system. We intend to determine the minimum number of redundant components that can be allocated to a series structure so that a given reliability level is achieved.

The sequential search method proposed by Barlow and Proschan (1965) is used here. Let S be the original series structure and S_i be the new structure obtained by doubling component x_i. *Doubling* is defined as placing an identical component in an active redundancy with the component to be doubled. First use component x_i, which maximizes the reliability of S_i. Then denote the structure obtained by doubling component x_j (after doubling component x_i) as S_{ij}. The component x_j is chosen so that the reliability of S_{ij} is maximal. The process is continued until the desired reliability level is achieved. Choosing a component to be doubled depends on the reliability of the individual components, as shown below.

Suppose that the original system S is composed of n components $x_1, x_2, \ldots, x_n$ connected in series and their respective reliabilities are $p_1, p_2, \ldots, p_n$. The reliability of the system is

$$R = p_1 p_2 \ldots p_n. \tag{2.66}$$

If component x_i is doubled, the reliability of the new system is

$$R_i = p_1 p_2 \ldots [1 - (1 - p_i)^2] \ldots p_n$$

$$= p_1 p_2 \ldots p_i (2 - p_i) \ldots p_n$$

$$= (2 - p_i) p_1 p_2 \ldots p_n$$

or

$$R_i = (2 - p_i)R. \tag{2.67}$$

Thus, the reliability R_i is maximum when p_i is minimum. Therefore doubling the least reliable component results in the largest gain in the reliability of the system. Repeating this reasoning, we either add another component in parallel with x_i or double the least reliable component other than x_i, and so on (Kaufmann, Grouchko, and Croun, 1977).

EXAMPLE 2.25

A series system consists of three components x_1, x_2, and x_3, and their reliabilities are 0.70, 0.75, and 0.85, respectively. Determine the minimal number of components that can be added in parallel (active redundancy) to the initial components such that the reliability becomes at least 0.82. Note: components used in active redundancy are identical to the components of the original system.

SOLUTION

The original system is shown in Figure 2.25. The reliability of the original system is

$$R = 0.70 \times 0.75 \times 0.85 = 0.4462.$$

Now apply the procedure given above by doubling component x_1 using an identical component as shown in Figure 2.26. Then

$$R_1 = [1 - (1 - 0.70)^2] \times 0.75 \times 0.85 = 0.5801.$$

The least reliable component in the structure shown in Figure 2.26 is component x_2. Therefore, we choose to double x_2, and the resulting structure is shown in Figure 2.27. The reliability becomes

$$R_{12} = 0.91 \times [1 - (1 - 0.75)^2] \times 0.85 = 0.7251.$$

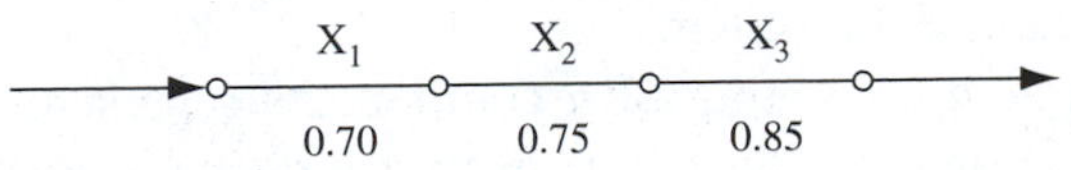

X1 X2 X3

0.70 0.75 0.85

Figure 2.25. Original Series System

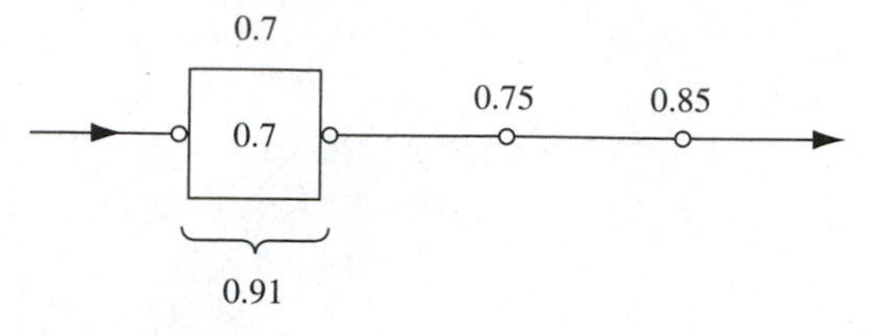

Figure 2.26. Doubling Component x_1

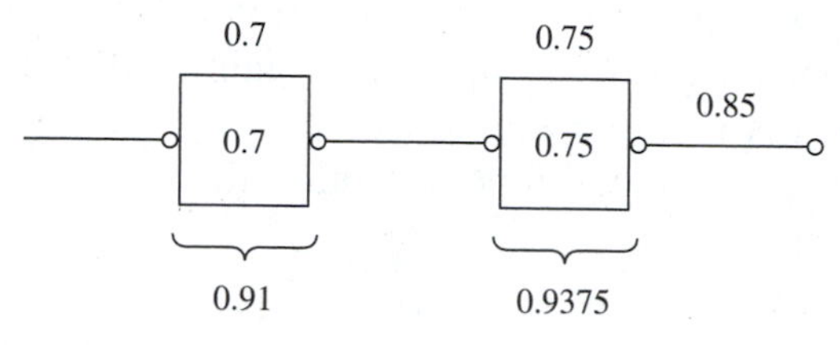

Figure 2.27. Doubling Component x_2

Now double the least reliable component, x_3, to get

$$R_{123} = 0.91 \times 0.9375 \times [1 - (1 - 0.85)^2] = 0.8339.$$

The optimal solution that results in the desired reliability of 0.82 is obtained by using the following active redundancies: x_1 doubled, x_2 doubled, and x_3 doubled.

2.11. Importance Measures of Components

After the reliability and design engineers configure a system that is usually composed of many components, they often face the problem of identifying design weaknesses and component failures that are crucial to the proper functioning of the system. By doing so, the designers may allocate additional resources or redundancy to these components in order to improve the overall system reliability. This section presents methods for measuring the importance of the components in terms of the system reliability. In assessing the importance of a component, most of the methods are based on observing the reliabilities (or unreliabilities) of the system when the component is functioning properly and when it is not. These reliabilities, in conjunction with the component reliability, are then algebraically manipulated to obtain different importance measures. To simplify the calculations necessary for these measures, we show how the reliabilities can be observed using the *structure function* of the system.

Consider a system consisting of n components represented by the set $N = (1, 2, \ldots, n)$. The system and the components can be in either of two states, working or not working as denoted by 0 or 1, respectively. The state of the system depends only on the state of its components. Let $X = (X_1, X_2, \ldots, X_n)$ be the random vector representing the state of the components at a given instant of time, where X_i is the random variable denoting the state of component i at the given instant of time and $X_i = 0$ or 1 representing that component i is working or not for $i = 1, 2, \ldots, n$. Let $\phi(X)$ be the structure function of the system. Then the random variable $\phi(X)$ denotes the state of the system with $\phi(X) = 0$ meaning the system is working and $\phi(X) = 1$ that the system is not working (Seth and Ramamurthy, 1991). Obviously, $P[\phi(X) = 1] = E[\phi(X)]$. We shall assume that $X_1, X_2, \ldots, X_n$ are independently distributed binary random variables with $P[X_i = 1] = q_i$. In this case, $E[\phi(X)]$ is a function of $\boldsymbol{q} = (q_1, q_2, \ldots, q_n)$. Let $G(\boldsymbol{q}) = E[\phi(X)]$, then $G(\boldsymbol{q})$ is called the unreliability (or unavailability) function of the system. This function will now be used in evaluating the structural importance measures discussed below.

2.11.1. BIRNBAUM'S IMPORTANCE MEASURE

The Birnbaum reliability importance $I_B^i(t)$ of component i_i is defined to be the probability that the ith component is critical to the functioning of the system at time t. It can be expressed as

$$I_B^i(t) = \frac{\partial G(\boldsymbol{q}(t))}{\partial q_i(t)} = G(1_i, \boldsymbol{q}(t)) - G(0_i, \boldsymbol{q}(t)) \tag{2.68}$$

or

$$I_B^i(t) \equiv \Delta G_i(t), \tag{2.69}$$

where $G(1_i, \boldsymbol{q}(t))$ is the unavailability of the system when component i is not working and $G(0_i, \boldsymbol{q}(t))$ is the unavailability when component i is working. $I_B^i(t)$ can also be interpreted as (Henley and Kumamoto, 1981):

$$\begin{aligned} I_B^i(t) &= E[\phi(1_i, X(t)) - \phi(0_i, X(t))] \\ &= 1 \times P[\phi(1_i, X(t)) - \phi(0_i, X(t)) = 1] \\ &\quad + 0 \times P[\phi(1_i, X(t)) - \phi(0_i, X(t)) = 0] \end{aligned}$$

or

$$I_B^i(t) = P[\phi(1_i, X(t)) - \phi(0_i, X(t)) = 1]. \tag{2.70}$$

EXAMPLE 2.26

The measurement of electrical resistance has many important applications, such as the determination of continuity in an electrical circuit and the measurement of changes in resistance on the order of 10^{-6} ohms. One of the simplest methods of measuring resistance is accomplished by imposing a voltage across the unknown resistance and measuring the resulting current flow using a galvanometer. A manufacturer of such galvanometers requires standard cells (batteries) to provide the necessary voltage. The manufacturer has the following options for placing four batteries with constant failure rates of $\lambda_1 = 0.005$, $\lambda_2 = 0.009$, $\lambda_3 = 0.003$, and $\lambda_4 = 0.05$ failures per hour in any of the following configurations:

1. All batteries are connected in series;
2. Batteries 1 and 2 are connected in series with batteries 3 and 4 connected in parallel;
3. The four batteries are connected in parallel; or
4. Two out of four batteries are needed for the galvanometer to function properly.

Find Birnbaum's importance measure of every battery in the above configuration at $t = 40$ hours.

SOLUTION

We estimate the unreliability of the batteries at $t = 40$ hours as

$$q_1 = 1 - R_1(t) = 1 - e^{-\lambda_1 t} = 0.181$$

$$q_2 = 0.302$$

$$q_3 = 0.113$$

$$q_4 = 0.864.$$

1. Batteries are connected in series. The structure function is obtained as

$$\phi(X) = 1 - (1 - X_1)(1 - X_2)(1 - X_3)(1 - X_4)$$

and

$$G(\boldsymbol{q}) = 1 - (1 - q_1)(1 - q_2)(1 - q_3)(1 - q_4). \tag{2.71}$$

Birnbaum's importance measures for batteries 1 through 4 are obtained using Eq. (2.68) at $t = 40$ hours:

$$I_B^1(40) = (1 - q_2)(1 - q_3)(1 - q_4) = 0.084$$

$$I_B^2(40) = (1 - q_1)(1 - q_3)(1 - q_4) = 0.098$$

$$I_B^3(40) = (1 - q_1)(1 - q_2)(1 - q_4) = 0.077$$

$$I_B^4(40) = (1 - q_1)(1 - q_2)(1 - q_3) = 0.507.$$

Battery 4 has the highest importance measure. Accordingly, it has the most impact on the overall system reliability. Therefore, in order to improve the system reliability, the designer may wish to replace this battery with one having a smaller failure rate or may add a redundant battery.

2. Batteries 1 and 2 are connected in series with batteries 3 and 4 connected in parallel. The structure function of this system is

$$\phi(X) = (X_1 \bigvee X_2) \bigvee (X_3 \bigwedge X_4),$$

where $\bigvee$ is the OR Boolean operator and $\bigwedge$ is the AND Boolean operator, respectively. Thus,

$$\phi(X) = [1 - (1 - X_1)(1 - X_2)] \bigvee (X_3 X_4)$$

or

$$\phi(X) = X_1 + X_2 - X_1 X_2 + X_3 X_4 - X_1 X_3 X_4 - X_2 X_3 X_4 + X_1 X_2 X_3 X_4$$

and

$$G(\boldsymbol{q}) = q_1 + q_2 - q_1 q_2 + q_3 q_4 - q_1 q_3 q_4 - q_2 q_3 q_4 + q_1 q_2 q_3 q_4. \tag{2.72}$$

Birnbaum's importance measures are

$$I_B^1(40) = 1 - q_2 - q_3 q_4 + q_2 q_3 q_4 = 0.629$$

$$I_B^2(40) = 1 - q_1 - q_3 q_4 + q_1 q_3 q_4 = 0.739$$

$$I_B^3(40) = q_4 - q_1 q_4 - q_2 q_4 + q_1 q_2 q_4 = 0.493$$

$$I_B^4(40) = q_3 - q_1 q_3 - q_2 q_3 + q_1 q_2 q_3 = 0.065.$$

In this case, the importance measure places more emphasis on battery 2 since it is the most likely battery to fail.

3. Batteries are connected in parallel. The structure function is obtained as

$$\phi(X) = X_1 \bigwedge X_2 \bigwedge X_3 \bigwedge X_4$$

and

$$G(\boldsymbol{q}) = q_1 q_2 q_3 q_4. \tag{2.73}$$

The importance measures are

$$I_B^1(40) = q_2 q_3 q_4 = 0.029$$

$$I_B^2(40) = q_1 q_3 q_4 = 0.018$$

$$I_B^3(40) = q_1 q_2 q_4 = 0.047$$
$$I_B^4(40) = q_1 q_2 q_3 = 0.006.$$

In the parallel configuration, battery 3 is considered the most critical for the overall system reliability. This is rather unexpected since battery 4 is the least reliable unit. This is because Birnbaum's importance measure is related to the probability that the system is in a state at time t in which the functioning of a battery is critical. Since battery 3 fails last among other batteries, then it is considered, according to Birnbaum's measure, to be most critical. This anomalous result in parallel systems exists in other importance measures as well. Note that Birnbaum's importance measure for one-event cut-sets is always, and usually incorrectly, numerically equal to one (Henley and Kumamoto, 1981).

4. Batteries are connected in a 2-out-of-4 configuration. The structure function of the system is derived as

$$\phi(X) = 1 - [1 - X_1 X_2][1 - X_1 X_3][1 - X_1 X_4] \times [1 - X_2 X_3][1 - X_2 X_4][1 - X_3 X_4].$$

The unavailability function $G(\boldsymbol{q})$ is obtained as

$$G(\boldsymbol{q}) = q_1 q_2 + q_1 q_3 + q_1 q_4 + q_2 q_3 + q_2 q_4 + q_3 q_4 - 2q_1 q_2 q_3 - 2q_1 q_2 q_4 - 2q_1 q_3 q_4 - 2q_2 q_3 q_4 + q_1 q_2 q_3 q_4. \quad (2.74)$$

Birnbaum's importance measures are

$$I_B^1(40) = q_2 + q_3 + q_4 - 2q_2 q_3 - 2q_2 q_4 - 2q_3 q_4 + q_2 q_3 q_4 = 0.522$$
$$I_B^2(40) = q_1 + q_3 + q_4 - 2q_1 q_3 - 2q_1 q_4 - 2q_3 q_4 + q_1 q_3 q_4 = 0.626$$
$$I_B^3(40) = q_1 + q_2 + q_4 - 2q_1 q_2 - 2q_1 q_4 - 2q_2 q_4 + q_1 q_2 q_4 = 0.449$$
$$I_B^4(40) = q_1 + q_2 + q_3 - 2q_1 q_2 - 2q_1 q_3 - 2q_2 q_3 + q_1 q_2 q_3 = 0.383.$$

In this case, battery 2 has the most critical effect on the system reliability. The result of this configuration can be explained in a similar way as that of the parallel system. This measure is not a useful importance criterion except for a simple series system whose results are obvious (Henley and Kumamoto, 1981).

2.11.2. CRITICALITY IMPORTANCE

Criticality importance corresponds to the conditional probability that the system is in a state at time t such that component i is critical and has failed, given that the system has failed by this same time (Gandini, 1990). This importance measure is based on the fact that it is more difficult to improve the more reliable components than to improve the less reliable components. The criticality importance measure is expressed as

$$I^i_{CR}(t) = \frac{\partial G(\boldsymbol{q}(t))}{\partial q_i(t)} \times \frac{q_i(t)}{G(\boldsymbol{q}(t))}.$$

Equation (2.75) can be rewritten as

$$I^i_{CR}(t) = \frac{[G(1_i, \boldsymbol{q}(t)) - G(0_i, \boldsymbol{q}(t))]\, q_i(t)}{G(\boldsymbol{q}(t))}. \tag{2.75}$$

We now illustrate the application of this measure.

EXAMPLE 2.27

Calculate the criticality importance measure for the four system configurations given in Example 2.26.

SOLUTION

1. Batteries are connected in series. We use the unavailability expression of the series system given by Eq. (2.71):

$$I^1_{CR}(40) = \frac{(1 - q_2)(1 - q_3)(1 - q_4)q_1}{1 - (1 - q_1)(1 - q_2)(1 - q_3)(1 - q_4)} = \frac{0.084 \times 0.181}{0.931} = 0.016$$

$$I^2_{CR}(40) = \frac{0.098 \times 0.302}{0.931} = 0.032$$

$$I^3_{CR}(40) = \frac{0.77 \times 0.113}{0.931} = 0.009$$

$$I^4_{CR}(40) = \frac{0.507 \times 0.864}{0.931} = 0.471.$$

This importance measure results in the same ranking of the batteries' importance Birnbaum's measures when components are connected in a series configuration.

2. Batteries 1 and 2 are connected in series with batteries 3 and 4 connected in parallel. We use the unavailability of this configuration as given by Eq. (2.72) to obtain the criticality importance measures of the batteries:

$$I^1_{CR}(40) = \frac{(1 - q_2 - q_3 q_4 + q_2 q_3 q_4)\, q_1}{q_1 + q_2 - q_1 q_2 + q_3 q_4 - q_1 q_3 q_4} = \frac{0.629 \times 0.181}{0.484} = 0.235$$

$$I^2_{CR}(40) = \frac{0.739 \times 0.302}{0.484} = 0.461$$

$$I_{CR}^{3}(40) = \frac{0.493 \times 0.113}{0.484} = 0.115$$

$$I_{CR}^{4}(40) = \frac{0.065 \times 0.864}{0.484} \approx 0.116.$$

3. Batteries are connected in parallel. This measure places equal importance on all batteries as shown below:

$$I_{CR}^{1}(40) = \frac{q_2\, q_3\, q_4\, q_1}{q_1\, q_2\, q_3\, q_4} = 1$$

$$I_{CR}^{2}(40) = \frac{q_1\, q_3\, q_4\, q_2}{q_1\, q_2\, q_3\, q_4} = 1$$

$$I_{CR}^{3}(40) = \frac{q_1\, q_2\, q_4\, q_3}{q_1\, q_2\, q_3\, q_4} = 1$$

$$I_{CR}^{4}(40) = \frac{q_1\, q_2\, q_3\, q_4}{q_1\, q_2\, q_3\, q_4} = 1.$$

4. Batteries are connected in a 2-out-of-4 configuration. We use Eq. (2.74) to obtain the criticality importance measures as

$$I_{CR}^{1}(40) = \frac{[q_2 + q_3 + q_4 - 2(q_2 q_3 + q_2 q_4 + q_3 q_4) + q_2 q_3 q_4]q_1}{G(\boldsymbol{q})}$$

$$= \frac{0.522 \times 0.181}{0.429} = 0.220$$

$$I_{CR}^{2}(40) = \frac{0.626 \times 0.302}{0.429} = 0.441$$

$$I_{CR}^{3}(40) = \frac{0.449 \times 0.113}{0.429} = 0.118$$

$$I_{CR}^{4}(40) = \frac{0.383 \times 0.864}{0.429} = 0.771.$$

The above measures show that battery 4 has the most impact on the overall system unavailability.

2.11.3. FUSSELL-VESELY IMPORTANCE

The Fussell-Vesely importance measure of component i, I_{FV}^{i}, suggests consideration of the probability that the system life coincides with the failure of a cut-set containing component i (Boland and El-Neweihi, 1995). The importance measure is given by

$$I_{FV}^{i}(t) = \frac{G_i(\boldsymbol{q}(t))}{G(\boldsymbol{q}(t))}, \tag{2.76}$$

where $G_i(\boldsymbol{q}(t))$ is the probability of component i contributing to cut set failure.

EXAMPLE 2.28

Determine the Fussell-Vesely importance measures for the battery configurations given in Example 2.26.

SOLUTION

1. Batteries are connected in series. The probability of cut-sets containing battery i in a series configuration is $G_i(\boldsymbol{q}(t)) = q_i(t) = q_i$, $i = 1, 2, 3$, and 4. The importance measures are

$$I_{FV}^{1}(40) = \frac{q_1}{G(\boldsymbol{q}(40))} = \frac{0.181}{0.931} = 0.194$$

$$I_{FV}^{2}(40) = \frac{q_2}{G(\boldsymbol{q}(40))} = \frac{0.302}{0.931} = 0.324$$

$$I_{FV}^{3}(40) = \frac{q_3}{G(\boldsymbol{q}(40))} = \frac{0.113}{0.931} = 0.121$$

$$I_{FV}^{4}(40) = \frac{q_4}{G(\boldsymbol{q}(40))} = \frac{0.864}{0.931} = 0.928.$$

   The importance rankings of the batteries are identical to those obtained by Birnbaum's importance measures. Again, battery 4 has the most impact on the overall system reliability.

2. Batteries 1 and 2 are connected in series with batteries 3 and 4 connected in parallel. The probability of cut-sets containing battery i in this configuration is

$$G_1(\boldsymbol{q}(t)) = q_1$$
$$G_2(\boldsymbol{q}(t)) = q_2$$
$$G_3(\boldsymbol{q}(t)) = q_3 q_4$$
$$G_4(\boldsymbol{q}(t)) = q_3 q_4.$$

The importance measures of the batteries are

$$I_{FV}^{1}(40) = \frac{q_1}{G(\boldsymbol{q}(t))} = \frac{0.181}{0.484} = 0.373$$

$$I_{FV}^{2}(40) = \frac{q_2}{G(\boldsymbol{q}(t))} = 0.623$$

$$I_{FV}^{3}(40) = \frac{0.046}{0.484} = 0.201$$

$$I_{FV}^{4}(40) = \frac{0.046}{0.484} = 0.201.$$

3. Batteries are connected in parallel:

$$G_i(\boldsymbol{q}(t)) = q_1 q_2 q_3 q_4 \qquad i = 1, 2, 3, \text{ and } 4$$

$$G(\boldsymbol{q}(t)) = q_1 q_2 q_3 q_3.$$

Thus $I_{FV}^{i} = 1$ for $i = 1, 2, 3$, and 4. In other words, the Fussell-Vesely importance measures rank all the batteries equally in terms of their impact on the overall reliability of the system. This is a shortcoming of the measure since in a parallel system the most reliable component has the most impact on the system reliability.

4. Batteries are connected in a 2-out-of-4 configuration:

$$G_1(\boldsymbol{q}(t)) = q_1q_2 + q_1q_3 + q_1q_4 - q_1q_2q_3 - q_1q_2q_4 - q_1q_3q_4 + q_1q_2q_3q_4 = 0.165$$

$$G_2(\boldsymbol{q}(t)) = q_1q_2 + q_2q_3 + q_2q_4 - q_1q_2q_3 - q_1q_2q_4 - q_2q_3q_4 + q_1q_2q_3q_4 = 0.272$$

$$G_3(\boldsymbol{q}(t)) = q_1q_3 + q_2q_3 + q_3q_4 - q_1q_2q_3 - q_1q_3q_4 - q_2q_3q_4 + q_1q_2q_3q_4 = 0.104$$

$$G_4(\boldsymbol{q}(t)) = q_1q_4 + q_2q_4 + q_3q_4 - q_1q_2q_4 - q_1q_3q_4 - q_2q_3q_4 + q_1q_2q_3q_4 = 0.425.$$

The importance measures are

$$I_{FV}^{1}(40) = \frac{G_1(\boldsymbol{q}(t))}{G(\boldsymbol{q}(t))} = \frac{0.165}{0.429} = 0.384$$

$$I_{FV}^{2}(40) = \frac{0.272}{0.429} = 0.634$$

$$I_{FV}^{3}(40) = \frac{0.104}{0.429} = 0.242$$

$$I_{FV}^{4}(40) = \frac{0.425}{0.429} = 0.990.$$

2.11.4. BARLOW-PROSCHAN IMPORTANCE

This measure corresponds to the conditional probability that component i causes the system to fail in the time interval (t_0, t_F), given that the system has failed in the same period (Barlow and Proschan, 1974). It is expressed as

$$I_{BP}^{i} \equiv \frac{\displaystyle\int_{t_0}^{t_F} \frac{\partial G(\boldsymbol{q}(t))}{\partial q_i} \frac{dq_i}{dt} dt}{\displaystyle\sum_{k=1}^{N} \int_{t_0}^{t_F} \frac{\partial G(\boldsymbol{q}(t))}{\partial q_k} \frac{dq_k}{dt} dt}, \tag{2.77}$$

where N is the total number of components in the system.

2.11.5. UPGRADING FUNCTION

This function is developed by Lambert (1975). It is defined as the fractional reduction in the probability of the system failure when component failure rate λ_i is reduced fractionally. It is given by

$$I_{UF}^{i}(t) = \frac{\lambda_i}{G(\boldsymbol{q}(t))} \frac{\partial G(\boldsymbol{q}(t))}{\partial \lambda_i}. \tag{2.78}$$

Lambert and Yadigaroglu (1977) have applied this measure of importance to the problem of determining the optimal choice of system upgrade. This function is limited to measuring the importance of components in nonrepairable systems.

EXAMPLE 2.29

An O ring is a rubber doughnut squeezed into a groove between parts that are to be sealed. Pressure from the sealed gas pushes the O ring ahead of it into the gap between the body parts so that the O ring obstructs passage of the gas. This is called a self-energizing seal. The gas must exert pressure on the entire left side of the O ring, or instead of pushing it forward and upward to block the escape route, the gas will push it down, out of the way of the escape route, and the gas will escape. Therefore, the O ring groove must be wider than the compressed O ring; otherwise, the O ring will touch all four sides of its enclosure and will not seal as shown in Figure 2.28 (Kamm, 1991).

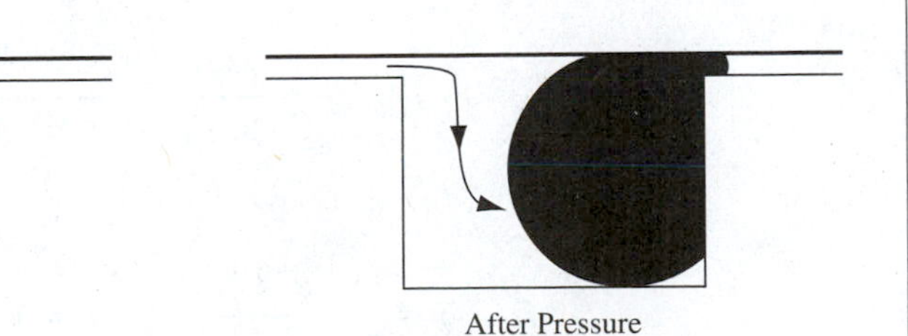

Figure 2.28. O-Ring Before and After Gas Pressure Is Applied

A manufacturer of satellite booster rockets uses three O rings located two inches away from each other to prevent leakage of gases. The manufacturer considers two designs, both of which meet the reliability requirements:

- All of the three O rings must not leak under the maximum pressure; and
- Two of three O rings must not fail under the maximum allowable pressure.

The O rings exhibit constant failure rates of $\lambda_1 = 0.004$, $\lambda_2 = 0.009$, and $\lambda_3 = 0.025$ failures per hour. Determine the upgrading functions for each O ring in both designs. Plot the functions against time. What are the most critical O rings? Why?

SOLUTION The unreliabilities of the O rings are

$$q_1(t) = 1 - e^{-\lambda_1 t} = 1 - e^{-0.004t},$$

$$q_2(t) = 1 - e^{-0.009t}, \text{ and}$$

$$q_3(t) = 1 - e^{-0.025t}.$$

We now consider the two designs.

1. All of the O rings must operate. This is a series system, and its $G(\boldsymbol{q}(t))$ is

$$G(\boldsymbol{q}(t)) = 1 - e^{-\lambda_1 t} e^{-\lambda_2 t} e^{-\lambda_3 t}$$

$$\frac{\partial G(\boldsymbol{q}(t))}{\partial \lambda_1} = te^{-(\lambda_1 + \lambda_2 + \lambda_3)t}$$

$$\frac{\partial G(\boldsymbol{q}(t))}{\partial \lambda_2} = te^{-(\lambda_1 + \lambda_2 + \lambda_3)t}$$

$$\frac{\partial G(\boldsymbol{q}(t))}{\partial \lambda_3} = te^{-(\lambda_1 + \lambda_2 + \lambda_3)t}.$$

Thus

$$I_{UF}^{i}(t) = \frac{\lambda_i te^{-(\lambda_1 + \lambda_2 + \lambda_3)t}}{1 - e^{-(\lambda_1 + \lambda_2 + \lambda_3)t}}$$

or

$$I_{UF}^{i}(t) = \frac{\lambda_i te^{-0.038t}}{1 - e^{-0.038t}} \qquad i = 1, 2, \text{ and } 3. \tag{2.79}$$

2. For the 2-out-of-3 O ring system,

$$G(\boldsymbol{q}(t)) = q_1 q_2 + q_2 q_3 + q_3 q_1 - 2q_1 q_2 q_3$$

or

$$G(\boldsymbol{q}(t)) = 1 - e^{-(\lambda_1 + \lambda_2)t} - e^{-(\lambda_1 + \lambda_3)t} - e^{-(\lambda_2 + \lambda_3)t} + 2\,e^{-(\lambda_1 + \lambda_2 + \lambda_3)t}$$

$$\frac{\partial G(\boldsymbol{q}(t))}{\partial \lambda_1} = te^{-(\lambda_1 + \lambda_2)t} + te^{-(\lambda_1 + \lambda_3)t} - 2te^{-(\lambda_1 + \lambda_2 + \lambda_3)t}$$

$$\frac{\partial G(\boldsymbol{q}(t))}{\partial \lambda_2} = te^{-(\lambda_1 + \lambda_2)t} + te^{-(\lambda_2 + \lambda_3)t} - 2te^{-(\lambda_1 + \lambda_2 + \lambda_3)t}$$

$$\frac{\partial G(\boldsymbol{q}(t))}{\partial \lambda_3} = te^{-(\lambda_1 + \lambda_3)t} + te^{-(\lambda_2 + \lambda_3)t} - 2te^{-(\lambda_1 + \lambda_2 + \lambda_3)t}.$$

We now use Eq. (2.78) to obtain

$$I_{UF}^{i}(t) = \frac{\lambda_i}{G(\boldsymbol{q}(t))} \frac{\partial G(\boldsymbol{q}(t))}{\partial \lambda_i} \qquad i = 1, 2, \text{and } 3. \tag{2.80}$$

Graphs of Eqs. (2.79) and (2.80) are shown in Figures 2.29 and 2.30, respectively. As shown in Figures 2.29 and 2.30, the values of the importance measures for both systems decrease with time. Moreover, the differences between the importance measures within the same system decrease rapidly with time. This is a very important observation since allocation of resources for the

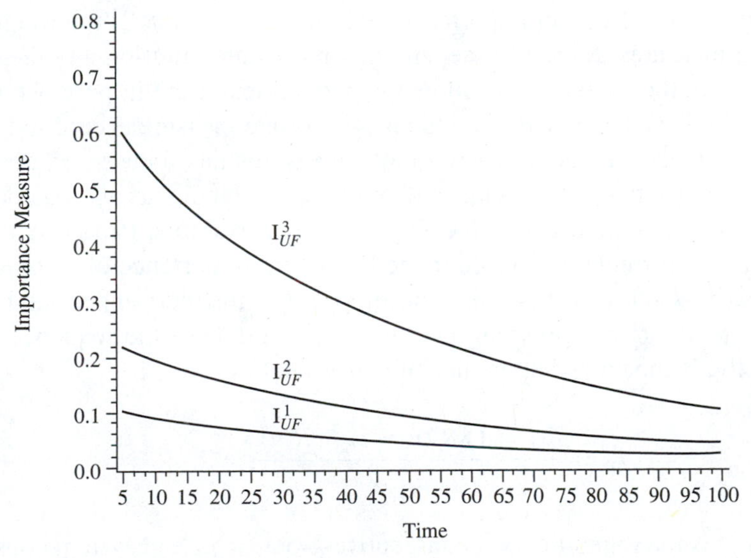

Figure 2.29. $I_{UF}^{i}(t)$ for the Series System

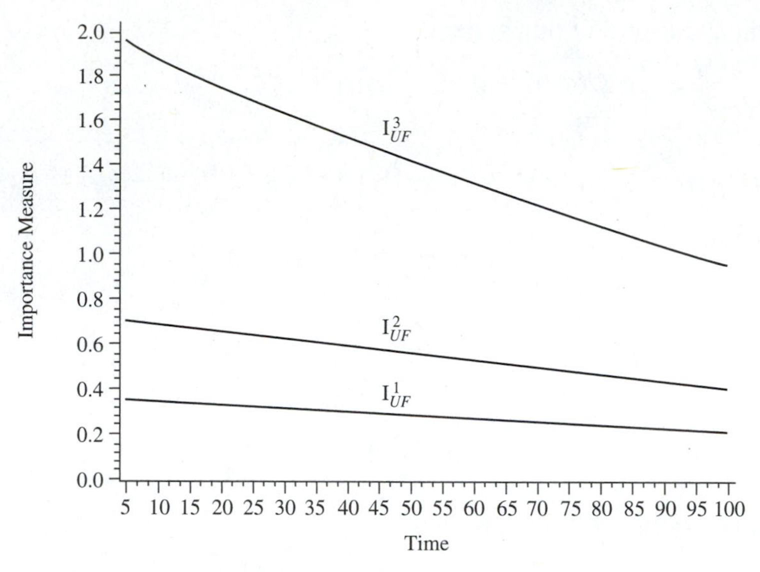

Figure 2.30. $I^i_{UF}\,(t)$ for the 2-out-of-3 System

improvement of the component reliability should ensure that the critical component's reliability is achieved at the desired time.

As shown in the previous examples, none of the importance measures is valid for all configurations—that is, the rankings of the system's components in terms of their importance are not consistently valid or intuitive. Therefore, it may be more appropriate to use an importance measure whose value is the weighted sum of several measures. Moreover, depending on the configuration and the objective of the system, the analyst may follow the derivations of the importance measures presented in this chapter and develop an appropriate measure accordingly.

Finally, recent research in importance measures has focused on determining component importance in systems with multistate components and component importance of consecutive-k-out-of-n:F systems. For example, Papastavridis (1987) gives a simple formula to determine the Birnbaum importance of a component in a consecutive-k-out-of-n:F system and proves that for independent and identical components, the most important ones are in the middle of the sequence. The formula for the Birnbaum's importance of component i is

$$I_B^i = \frac{R(i-1)\,R'(n-i) - R(n)}{q_i}, \tag{2.81}$$

where $R(j)$ is the reliability of a consecutive-k-out-of-j:F subsystem consisting of components $1, 2, \ldots, j$ and $R'(j)$ is the reliability of a consecutive-k-out-of-j:F subsystem, consisting of components $(n-j+1), (n-j+2), \ldots, (n-1), n$.

PROBLEMS

2-1. Figure 2.31 shows the block diagram of a closed-loop servo accelerometer. The accelerometer functions as follows: a pendulous mass reacts to an acceleration input and begins to move. A position sensor detects this minute motion and develops an output signal. This signal is demodulated, amplified, and applied as negative feedback to an electrical torque generator (torquer) coupled to the mass. The torquer develops a torque proportional to the current applied to it. The magnitude and direction of this torque just balance out the torque attempting to move the pendulous mass as a result of the acceleration input, preventing further movement of the mass.

Since both torques are equal and the torque generator output is proportional to its input current, the input current is, therefore, proportional to the torque attempting to move the pendulous mass. In fact, this torque is proportional to the product of the moment of inertia and acceleration. Therefore, the torque generator current is proportional to applied acceleration. If this current is passed through a stable resistor, the voltage developed is proportional to applied acceleration.

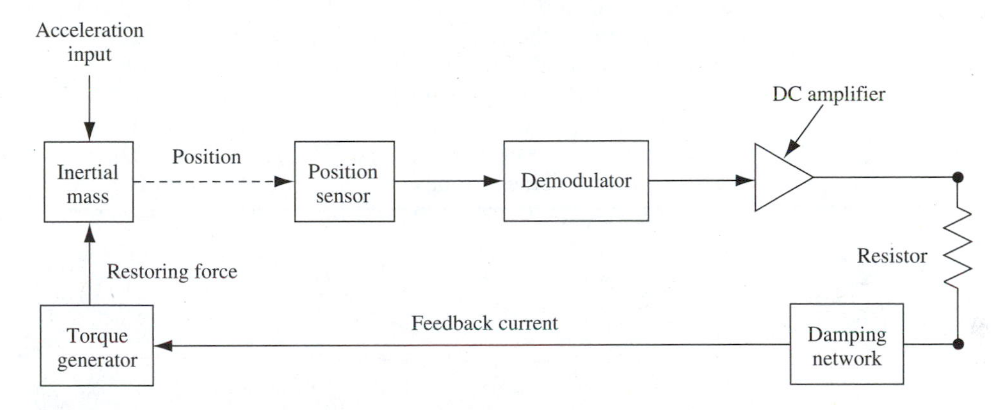

Figure 2.31. A Block Diagram for Problem 2-1

a. Draw a reliability graph of the feedback system.

b. Assuming that the probability of a component functioning properly is 0.9, what is the total system reliability when all components have the same probability of success?

2-2. Consider the head of a computer disk drive that must quickly transverse the radius of a rotating disk drive. The head moves from a known position to another known position on the disk. The head rests at the end of an arm. A large magnet surrounds the other end of the arm. The part of the arm that is next to the magnet is an electromagnet. Current is supplied to it as needed to produce a force to move the arm. A sensor detects the position of the head in relation to the target position and decreases the current (force) proportionally. Moreover, a damper is connected to the arm to ensure that the head will not oscillate around the desired position. Construct a block diagram and a reliability graph to represent the operation of the disk drive.

2-3. The cassette tape recorder is a commonly used device in modern life. It consists of seven components necessary for the proper function of the tape. These components are shown in Figure 2.32 and listed in Table 2.5. A cassette tape recorder has two units for recording from one recorder to the other. Construct a block diagram for the cassette tapes. Assuming that the components exhibit constant failure rates as given in Table 2.5, estimate the reliability of a successful recording from one tape to another.

Table 2.5. Components and Failure Rates of the Tape

Component	*Function*	*Failure Rate*
1	Feed-spool, advances the tape	0.0003
2	Take-up spool, guides the tape	0.0002
3	Erase head, erases the contents of the tape	0.0005
4	Record/Replay head, transforms magnetized particles on tape to electric signals	0.0008
5	Pressure pad, supports tape	0.0001
6	Pinch wheel, provides tension in tape	0.00025
7	Capstan, ensures flatness of tape	0.0002

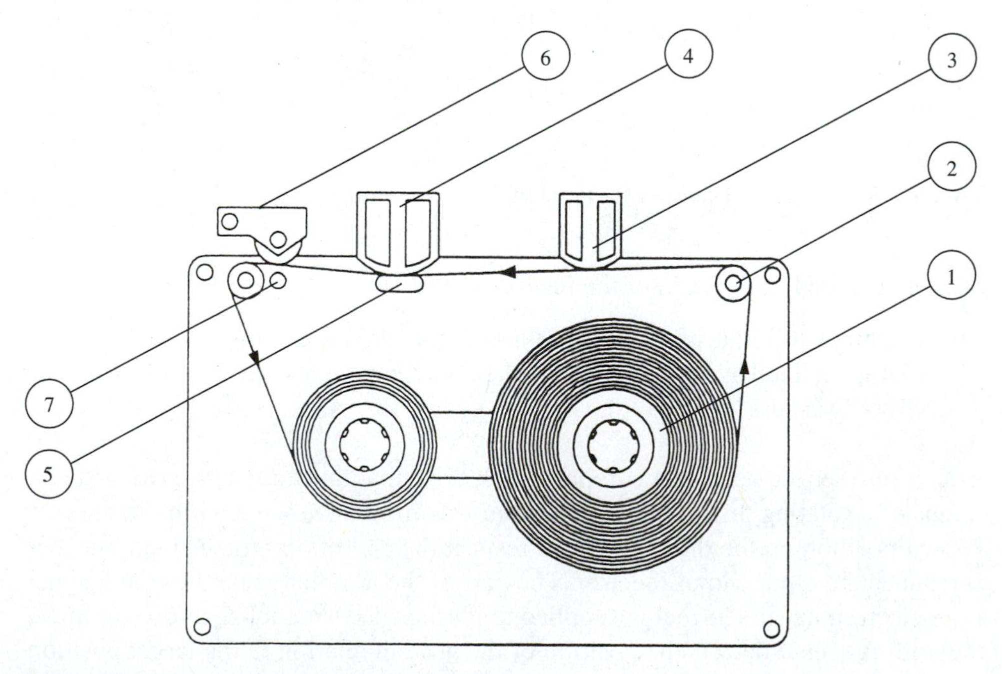

Figure 2.32. Components of a Tape Cassette

2-4. Determine the reliability of the system shown in Figure 2.33 when (a) two out of four units (9 through 12) or (b) three out of four units (9 through 12) are needed for successful operation. What is the reliability of the system if one unit out of Units 3, 4, and 5 and one unit out of 6, 7, and 8 must function in addition to at least three out of four of the remaining units (9 through 12)?

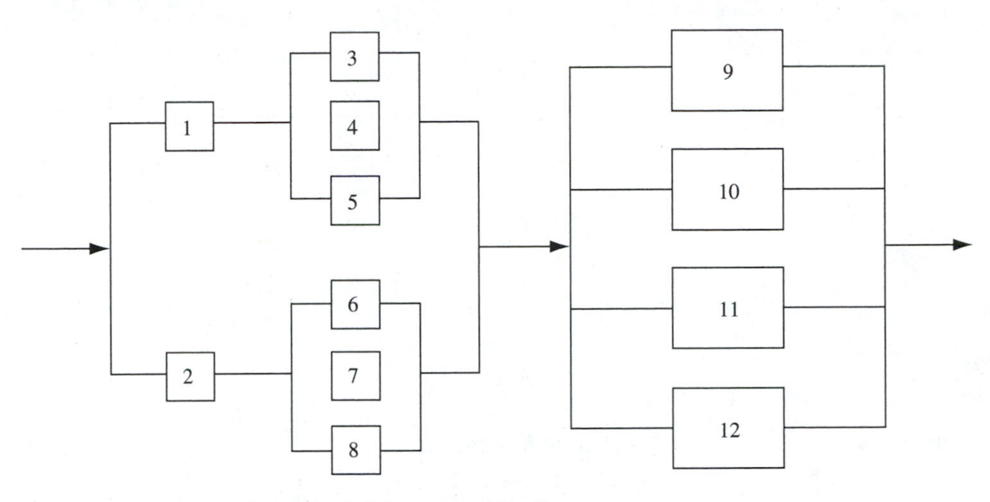

Figure 2.33. Reliability Block Diagram of the System

2-5. Walski and Pelliccia (1982) developed break-rate equations (hazard-rate equations) for the Binghamton, New York, water system. These equations are

$$\text{Pit cast iron (PCI)} \qquad N(t) = 0.02577\, e^{0.027t}$$

$$\text{Sandspun cast iron (SCI)} \qquad N(t) = 0.0627\, e^{0.0137t},$$

where

$N(t)$ = the break rate in breaks per mile per year and

t = the age of the pipe in years.

An engineer wishes to design a new water distribution system as shown in Figure 2.34.

a. What is the reliability of the system after two years of service? (Reliability is measured as the probability of successful water delivery from node 1 to node 6)

b. What is the mean time to failure?

c. What do you suggest to ensure a reliability of 0.98 after two years of service?

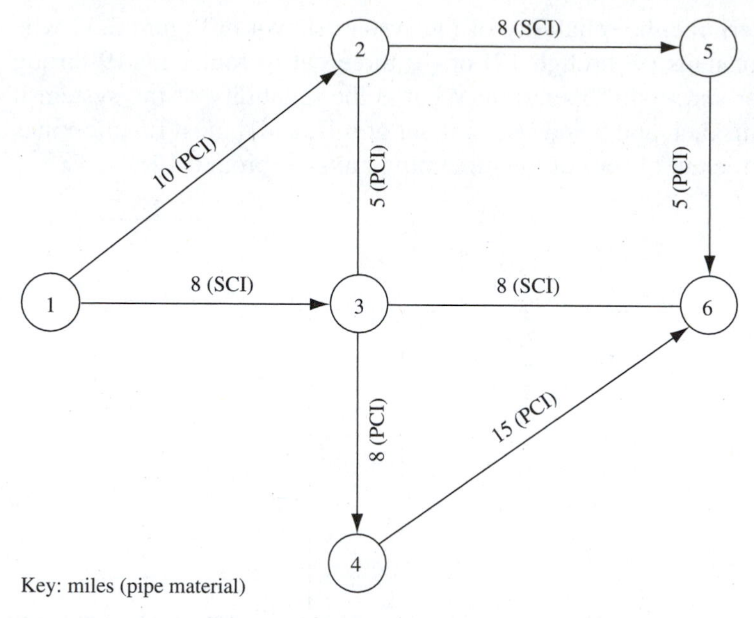

Figure 2.34. A New Water Distribution System

2-6. Consider a diode that can function properly but can malfunction by either short circuiting or by open circuiting. Let the probabilities of these be as follows:

p = probability of proper operation

p_s = probability of short circuit

p_o = probability of open circuit.

If we consider four identical diodes for improving the total system reliability, these diodes can be arranged in two possible configurations as shown in Figure 2.35. What is the ratio of the reliability improvement for both configurations?

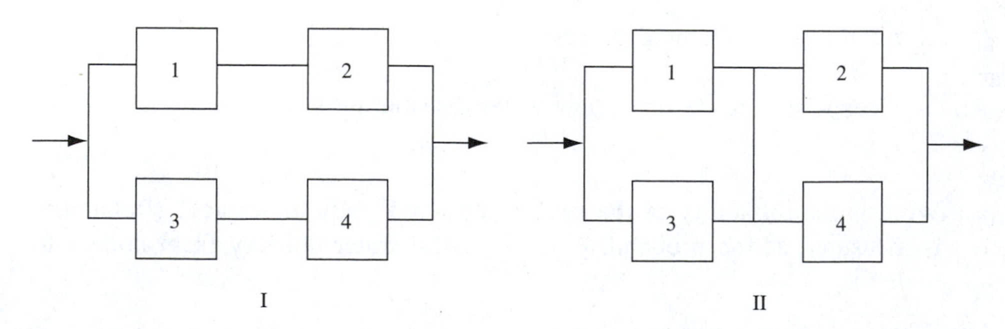

Figure 2.35. Configurations of Diodes

2-7. Consider the case of three elements, as shown in Figure 2.36. At least two of the three elements must function properly for the system to function properly. All three elements are different. Elements a and b have three states each. An element being in State 1 implies that it is working properly, whereas State 2 represents a short failure mode and State 3 represents an open failure mode of the unit. Let p_{ij} represent the probability that element i ($i = a, b$) is in state j ($j = 1, 2, 3$). Element c has two states only (working or not). Determine the reliability of the system.

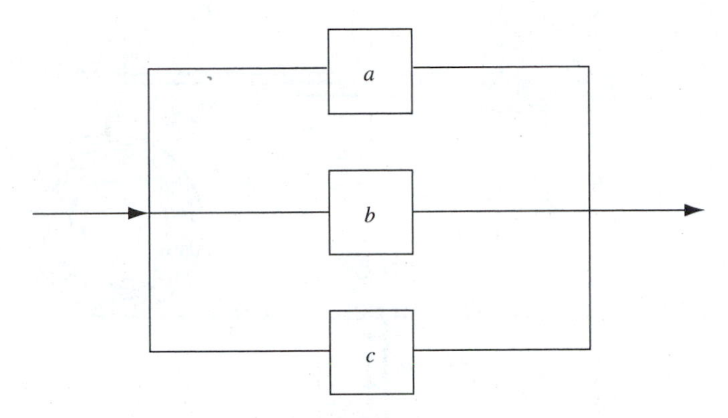

Figure 2.36. Reliability Block Diagram of the System

2-8. When a fax machine receives a document, it converts electrical signals into a copy of the original document. There are two types of recording systems: (a) the thermal recording system, which uses a set of fine wires positioned across the recording paper that produce hot spots as current passes through them, burning the image into the paper, or (b) the electrostatic recording system, which is shown in Figure 2.37. In this system, a charge is applied on the recording paper where a mark is needed. Black powder toner adheres to the charged areas, which are either fused to the paper with heat or pressed onto it by rollers. The paper is then cut to length. The main components of the fax machine and their constant failure rates are given in Table 2.6. All components exhibit constant failure rates per year. A fax machine uses the "group faxing" concept by sending the same document to a group of machines simultaneously. Assume that a fax machine uses the group faxing concept to send a document to six other machines. The reliability of the system is measured as the probability that all machines receive the document.

a. Draw a block diagram and reliability graph of the system.

b. Assuming that the communication links between machines do not fail, determine the reliability of the system at $t = 200$ hours.

c. Calculate I_{FV}^{i} for all components of a fax machine at $t = 400$ hours. What are the components that should be improved to increase the overall reliability of the fax machine?

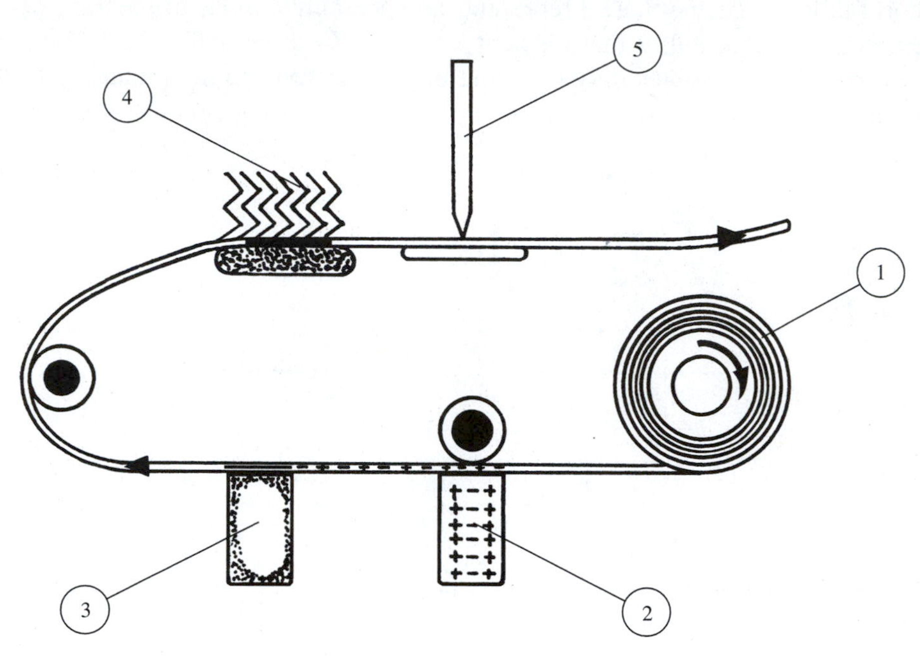

Figure 2.37. Schematic of an Electrostatic Fax Machine

Table 2.6. Components of the Fax Machine

Component	*Description and Function*	*Failure Rate*
1	Paper feeder	0.001
2	Printer head, applies charge	0.009
3	Toner, contains powder	0.0005
4	Heater, fuses powder onto paper	0.018
5	Cutter, cuts paper to length	0.0085

2-9. Consider the reliability block diagram of the system shown in Figure 2.38, which is composed of four subsystems: Sub1, Sub2, Sub3, and Sub4. These subsystems and the reliabilities of their components are listed below:

Sub1 is a series subsystem with $p_1 = 0.95, p_2 = 0.98, p_3 = 0.999$;

Sub2 is a redundant subsystem with $p_4 = 0.90, p_5 = 0.95, p_6 = 0.90$;

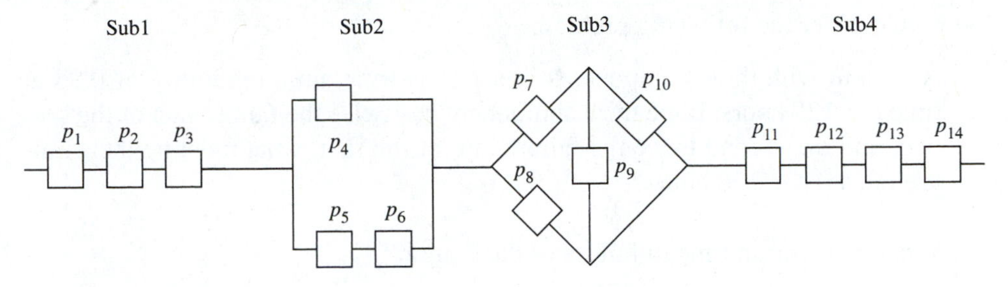

Figure 2.38. Reliability Block Diagram for Problem 2-9

Sub3 is a network subsystem with $p_7 = 0.98$, $p_8 = 0.85$, $p_9 = 0.93$, $p_{10} = 0.99$;

Sub4 is a consecutive 2-out-of-4:F series system with $p_{11} = p_{12} = p_{13} = p_{14} = 0.96$.

a. Determine the reliability of the system.

b. Use Birnbaum's importance measure to determine whether component 10 is more critical than component 13.

2-10. Four elements are configured as shown in Figure 2.39. At least two of the four elements must function properly for the system to function properly. All four elements are different, having reliabilities of p_a, p_b, p_c, and p_d, respectively. Find the reliability of the system.

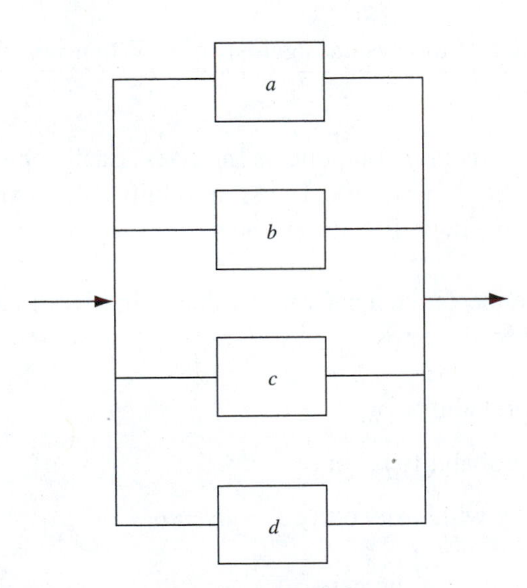

Figure 2.39. Reliability Block Diagram of the System

2-11. Consider the following problem:

a. A system with three components in series is to attain a reliability of 0.95 at time $t = 120$ hours. If the third component has twice the failure rate of the second, and the second twice the failure rate of the first, what must be their failure rates if $t = 120$ hours?

b. What is the mean time to failure of the system?

c. What is the probability of having 0, 1, and 2 failures in 100 hours?

d. What failure rates of components 1, 2, and 3 would you require if you demanded 0.95 reliability for the duration of the MTTF (use the same ratio of failure rates)?

e. If redundancy is allowed for all three components and if the cost of each component is the same, how much and where would you impose redundancy if you require a reliability of 0.98 for a duration of 1,000 hours?

2-12.

a. In a 3-out-of-n system with components having a linearly increasing hazard rate $h(t) = 0.5 \times 10^{-8}\ t$ failures per hour, determine the number of components for the system such that a reliability of 0.98 is achieved at $t = 10^3$ hours. What is the MTTF?

b. Solve (a) when the components are connected in parallel. Are the results identical? Explain why.

c. Plot the reliability of the system against time. When will the reliability reach 0.96?

2-13. A system consists of n components in series. Each component is subject to failure and its reliability is $p = 0.98$. The system fails if any two consecutive components fail. Determine the reliability of the system when $n = 6$.

2-14. Diodes are connected in a network as shown in Figure 2.40. Each diode can be in any of the following states:

- Fail open with probability p_o,
- Fail short with probability p_s, or
- Function properly with probability $1 - p_o - p_s$.

a. What is the reliability of the network?

b. Assuming all components are identical, what is the reliability of the network?

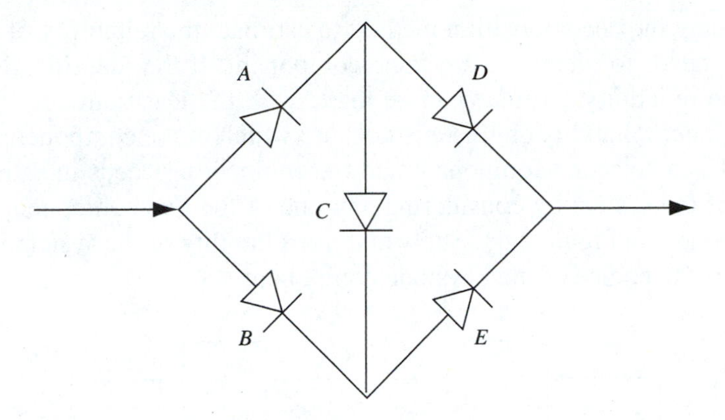

Figure 2.40. Network for Problem 2-14

2-15. Consider a 3-out-of-4 system at time $t = 50$ hours. The failure rates of the components are

$$h_1(t) = 0.005$$

$$h_2(t) = 0.005t$$

$$h_3(t) = 0.006t^{1.1}$$

$$h_4(t) = 0.009t^{1.05}.$$

Calculate the following importance measures:

a. Birnbaum's importance $\Delta G_i(t)$.

b. $I_{\mathrm{B}}^{i}(t)$.

c. Solve (a) and (b) if the system is a consecutive-2-out-of-4:F system.

2-16. What are the reliability expressions for the following systems whose components are identical, independent, and exhibit a constant failure rate λ?

a. Five components in series.

b. A 2-out-of-5 system.

c. A 3-out-of-5 system.

d. A parallel system of five components.

2-17. What is the mean time to failure of a system composed of two components having hazard rates of $k_1 \, t^m$ and $k_2 \, t^m$? What is the reliability of the system at $t = k_1/k_2$?

2-18. In using the decomposition method to estimate the reliability of a complex system one needs to identify a *keystone* component. If the identification is done properly the reliability estimate can be made with the least amount of computation. A beginner reliability engineer is not sure which of the components shown in Figure 2.41 is a keystone component and the engineer proceeds in estimating the reliability of the system by considering any one of the five components as a keystone component in Figure 2.41. Show that the reliability of the system is the same regardless of the choice of the keystone component.

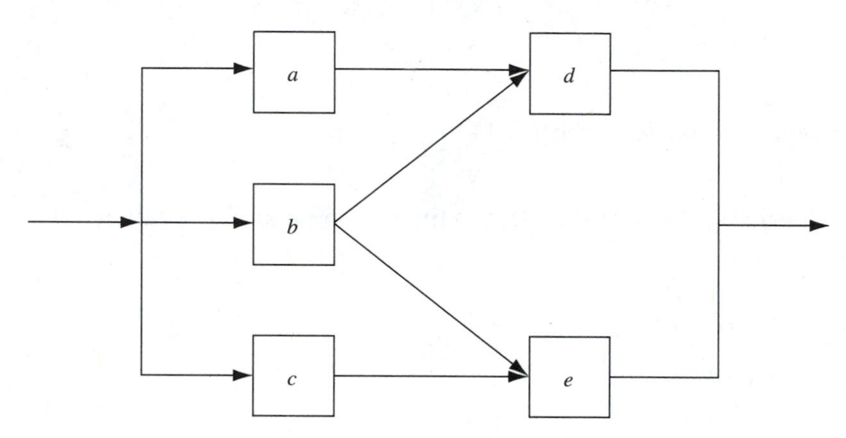

Figure 2.41. Reliability Block Diagram for Problem 2-18

2-19. Repeat the above problem for the system shown in Figure 2.42.

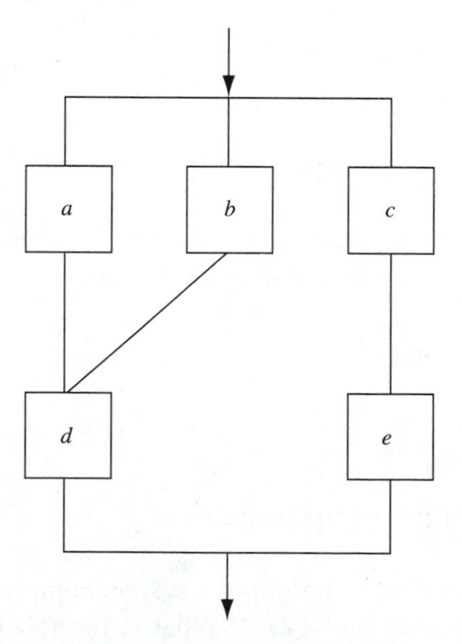

Figure 2.42. Reliability Block Diagram for Problem 2-19

2-20. Figure 2.43 represents a four-node communications network. The four nodes *a*, *b*, *c*, and *d* represent the four stations. The six branches represent two-way communication links between each pair of stations.

a. Find the minimum cut-sets and tie-sets between *a* and *b*.

b. Approximate the system reliability when all links are independent and identical with probability of success *p*.

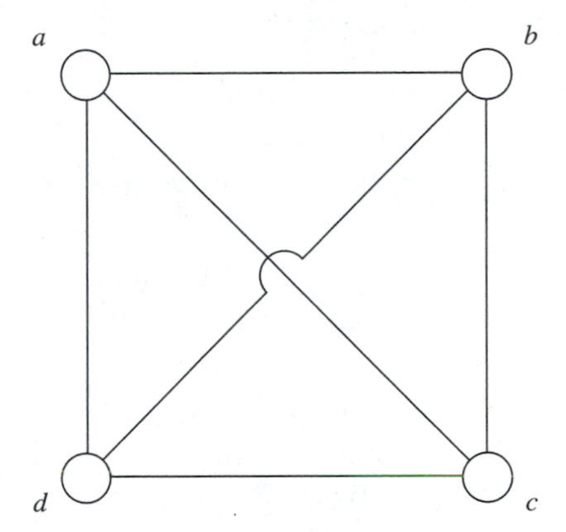

Figure 2.43. Reliability Block Diagram for Problem 2-20

2-21. The reliability graph of a parallel-series system is shown in Figure 2.44. Assume that each component has a linearly increasing hazard function of the type $h(t) = a + bt$. The number of components connected in series is n whereas the number of parallel paths is m. What is the mean time to failure of this system? What is the effect of m and n on the MTTF?

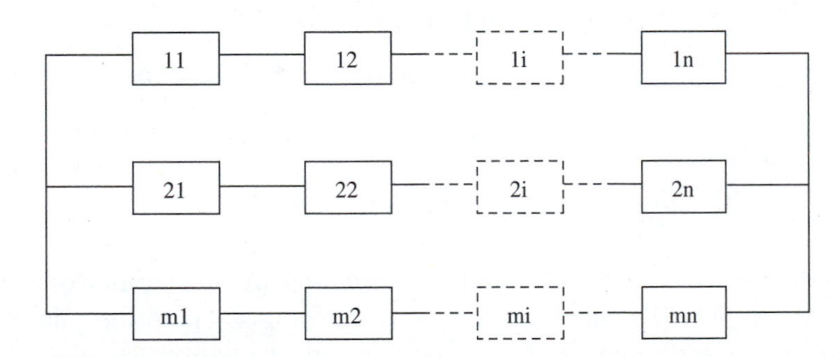

Figure 2.44. Figure for Problem 2-21

2-22. A computer chip has 160,000 transistors connected in parallel, and at least k transistors are required to operate properly for the chip to perform its function.

Assuming that each transistor has a constant hazard model $h(t) = 5 \times 10^{-6}$ failures per hour, what value of k ensures a chip reliability of 0.95 at $t = 10{,}000$ hours?

2-23. A system consists of three components with hazard rates $h_1(t)$, $h_2(t)$, and $h_3(t)$. Assuming that the three components are connected in series, determine the reliability and MTTF under the following conditions:

a. $h_1(t) = \lambda_1$, $h_2(t) = \lambda_2$, and $h_3(t) = \lambda_3$.

b. $h_1(t) = \lambda_1 t$, $h_2(t) = \lambda_2 t$, and $h_3(t) = \lambda_3 t$.

c. $h_1(t) = \lambda_1$, $h_2(t) = \lambda_2 t$, and $h_3(t) = \lambda_3 t^m$.

2-24. Solve Problem 2-23 when the three components are connected in parallel.

2-25. Solve Problem 2-23 when components 1 and 2 are connected in parallel while component 3 is connected in series with them.

2-26. Using the numerical values given below, compare the MTTF for the three systems in Problems 2-23, 2-24, and 2-25. Sketch the reliability functions for all conditions when $\lambda_1 = 0.001$, $\lambda_2 = 0.003$, $\lambda_3 = 0.009$, and $m = 1.5$.

2-27. Consider components 1, 2, 3, and 4. Their failure rates at $t = 40$ hours are $\lambda_1 = 0.006$, $\lambda_2 = 0.008$, $\lambda_3 = 0.0002$, and $\lambda_4 = 0.07$. The following four configurations are to be made:

a. Four components are connected in series;

b. Components 1 and 2 are connected in series with components 3 and 4 connected in parallel.

c. The four components are connected in parallel.

d. Two out of four components are needed for system functions.

e. Three out of four components are needed for the system to operate properly.

Determine the reliability of each configuration.

2-28. Most color laser printers use a combination of four colors of cyan (C), magenta (M), yellow (Y), and black (K) to create colors. The printer has a heated print head to transfer pigment from a thin plastic ribbon onto paper or transparency film. The ribbon contains successive panels of pigment in C, M, Y, and K. After the printer has applied one color's dots, the drive mechanism pulls the media back for the next pass. After the application of the colors, the paper or transparency is transferred to the fuser, which ensures the permanency of the colors. A diagram representing the elements and operation of the color laser printers is shown in Figure 2.45.

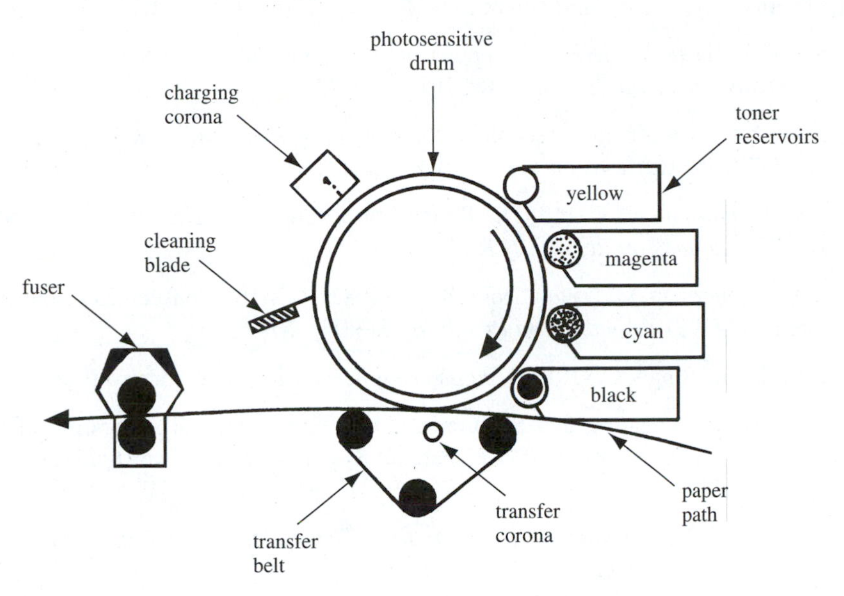

Figure 2.45. Figure for Problem 2-28

a. Draw both the block reliability diagram and the reliability graph.

b. Assume that all toner reservoirs have the same constant failure rate with parameter λ_r and the drum's failure rate $\lambda_d = 2\lambda_r$. The failure rates of the fuser, transfer belt, transfer corona, cleaning blade, and the charging corona are $h_f(t) = at$, $h_{belt}(t) = bt$, $h_{corona}(t) = ct$, $h_{blade}(t) = dt$, and $h_{cc}(t) = et^2$, respectively. Determine the reliability of the printer.

c. What is the most critical component of the printer?

d. Recommend an alternative design for the most critical component.

REFERENCES

Agrawal, A., and Barlow, R. E. (1984). "A Survey of Network Reliability and Domination Theory." *Operations Research* 32(3), 478–492.

Barlow, R. E., and Proschan, F. (1965). *Mathematical Theory of Reliability*. New York: Wiley.

Barlow, R. E., and Proschan, F. (1974). "Importance of System Components and Fault Tree Analysis." ORC-74-3, Operations Research Center, University of California, Berkeley.

Baxter, L. A., and Harche, F. (1992). "On the Optimal Assembly of Series-Parallel System." *Operations Research Letters* 11, 153–157.

Boland, P. J., and El-Neweihi, E. (1995). "Measures of Component Importance in Reliability Theory." *Computers and Operations Research* 22(4), 455–463.

Bollinger, R. C. (1982). "Direct Computation for Consecutive-k-out-of-n:F Systems." *IEEE Transactions on Reliability* R-31(5), 444–446.

Case, T. (1977). "A Reduction Technique for Obtaining a Simplified Reliability Expression." *IEEE Transactions on Reliability* R-26(4), 248–249.

Chiang, D. T., and Niu, S. C. (1981). "Reliability of Consecutive-k-out-of-n:F System." *IEEE Transactions on Reliability* R-30, 87–89.

Derman, C., Lieberman, G. L., and Ross, S. M. (1982). "On the Consecutive-k-out-of-n:F System." *IEEE Transactions on Reliability* R-31(1), 57–63.

Dhillon, B. S., and Singh, C. (1981). *Engineering Reliability*. New York: Wiley.

El-Neweihi, E., Proschan, F., and Sethuraman, J. (1986). "Optimal Allocation of Components in Parallel-Series and Series-Parallel Systems." *Journal of Applied Probability* 23, 770–777.

Feller, W. (1968). *An Introduction to Probability Theory and Its Applications* (3rd ed.) (vol. 1). New York: Wiley.

Gandini, A. (1990). "Importance and Sensitivity Analysis in Assessing System Reliability." *IEEE Transactions on Reliability* 39(1), 61–69.

Henley, E. J., and Kumamoto, H. (1981). *Reliability Engineering and Risk Assessment*. Englewood Cliffs, NJ: Prentice-Hall.

Kamm, L. J. (1991). *Real-World Engineering*. New York: IEEE Press.

Kaufmann, A., Grouchko, D., and Croun, R. (1977). *Mathematical Models for the Study of the Reliability of Systems*. New York: Academic Press.

Lambert, H. E. (1975). "Fault Trees for Decision Making in Systems Analysis." Ph.D. Thesis, UCRL-51829, University of California, Livermore.

Lambert, H. E., and Yadigaroglu, G. (1977). "Fault Trees for Diagnosis of System Fault Conditions." *Nuclear Science Engineering* 62, 20–34.

Lambiris, M., and Papastavridis, S. (1985). "Exact Reliability Formulas for Linear and Circular Consecutive-k-out-of-n:F Systems." *IEEE Transactions on Reliability* R-34(2), 124–126.

Malon, D. M. (1985). "Optimal Consecutive-k-out-of-n:F Component Sequencing." *IEEE Transactions on Reliability* R-34(1), 46–49.

NASA. (1990). *Nasa Tech Briefs*, p. 79, July.

Papastavridis, S. (1987). "The Most Important Component in a Consecutive-k-out-of-n:F System." *IEEE Transactions on Reliability* R-36(2), 266–267.

Pease, R. W. (1975). "General Solution to the Occupancy Problem with Variable Sized Runs of Adjacent Cells Occupied by Single Balls." *Mathematics Magazine* 48, 131–134.

Pham, H. (1988). "The Efficiency of Computing the Reliability of *k*-out-of-*n* Systems." *IEEE Transactions on Reliability* 37(5), 521–523.

Prasad, V. R., Nair, K. P. K., and Aneja, Y. (1991). "Optimal Assignment of Components to Parallel-Series and Series-Parallel Systems." *Operations Research* 39, 407–414.

Seth, A., and Ramamurthy, K. G. (1991). "Structural Importance of Components." *OPSEARCH* 28(2), 88–101.

Shanthikumar, J. G. (1982). "Recursive Algorithm to Evaluate the Reliability of a Consecutive-*k*-out-of-*n*:*F* System." *IEEE Transactions on Reliability* R-31(5), 442–443.

Shooman, M. L. (1968). *Probabilistic Reliability: An Engineering Approach*. New York: McGraw-Hill.

Siemens Aktiengesellschaft. (1983). "The Laser Printer." bt020e, Siemens.

Von Alven, W. H. (Ed.). (1964). *Reliability Engineering*. Englewood Cliffs, NJ: Prentice-Hall.

Walski, T. M., and Pelliccia, A. (1982). "Economic Analysis of Water Main Breaks." *Journal of the American Water Works Association* (March), 140–147.

Wei, V. K., Hwang, F. K., and Sös, V. T. (1983). "Optimal Sequencing of Items in a Consecutive-2-out-of-*n* System." *IEEE Transactions on Reliability* R-32(1), 30–33.

Zuo, M., and Kuo, W. (1990). "Design and Performance Analysis of Consecutive-k-out-of-n Structure." *Naval Research Logistics* 37, 203–230.

CHAPTER 3

Time- and Failure-Dependent Reliability

In Chapter 2, we presented different system configurations and the appropriate methods for estimating their reliabilities. The reliability values were not time dependent since the reliabilities of the components were considered constant and the failure-time distributions have not been incorporated in estimating the reliability of the system. In other words, we had a snapshot of the system at a specified instant and did not observe the reliability of the system over time (or over the life of the system). Moreover, we have not fully considered the dependence between component failures—that is, the effect of the failure of a component on the failure rates of other components in the system. Likewise, we have not considered the effect of repairs on the system performance in terms of its reliability, availability, mean time to failure (MTTF), and mean time between failures (MTBF).

In this chapter, we develop time-dependent reliability expressions for both nonrepairable and repairable systems. We also present different approaches for estimating the reliability of failure-dependent systems—for example, when the failure of a component affects the failure rate of other components in the system. Finally, we estimate different performance measures of the system such as MTTF, MTBF, and availability. We begin by presenting time-dependent reliability estimates of nonrepairable systems and progress gradually to the repairable systems.

3.1. Nonrepairable Systems

The number of nonrepairable systems and products is on the rise due to the increasing cost of labor and the high rate of technological obsolescence of many products. For example, the rate of technological advances in the developments of computer chips renders the repair of a two-year-old personal computer unnecessary since the advances in these two years may result in significantly less expensive but faster (clock speed) computers. Other nonrepairable systems include, until recently, satellites, single-mission products such as rockets, and inexpensive radios and telephone sets.

3.1.1. SERIES SYSTEMS

Assume n independent components arranged in series with a reliability of one for each component at time $t = 0$—that is, $R_i(0) = 1$, $(i = 1, 2, \ldots, n)$. The reliability of the system at time t is the probability that all components survive to time t, thus

$$R_s(t) = R_1(t)\, R_2(t) \ldots R_n(t) = \prod_{i=1}^{n} R_i(t). \tag{3.1}$$

When each component has a constant hazard, the reliability of component i at time t is expressed as

$$R_i(t) = e^{-\lambda_i t}, \tag{3.2}$$

where $R_i(t)$ is the reliability of component i at time t and λ_i is a constant failure rate of component i. Substituting Eq. (3.2) into Eq. (3.1), we obtain

$$R_s(t) = \prod_{i=1}^{n} e^{-\lambda_i t} = e^{-\sum_{i=1}^{n} \lambda_i t}. \tag{3.3}$$

Thus, the effective failure rate of a series system composed of n components is the sum of the failure rates of the individual components.

Equation (3.3) is valid only under the assumptions that all components are independent and that each one of them exhibits a constant hazard. If the hazard rate of component i is $h_i(t)$ and the cumulative hazard is $H_i(t) = \int_0^t h_i(\zeta) d\zeta$, then we can generalize Eq. (3.3) for a series system as

$$R_s(t) = \prod_{i=1}^{n} e^{-H_i(t)} = e^{-\sum_{i=1}^{n} H_i(t)}. \tag{3.4}$$

We now illustrate the use of Eq. (3.4) to estimate the reliability of a series system when the components have different hazard rates:

- For components with linearly increasing hazard rate—$h_i(t) = k_i t$—the reliability of the system is obtained as

$$R_s(t) = \prod_{i=1}^{n} e^{-k_i t^2/2} = e^{-\sum_{i=1}^{n} \frac{k_i t^2}{2}}. \tag{3.5}$$

- For components with Weibull hazard—$h_i(t) = (\gamma_i/\theta_i)t^{\gamma_i - 1}$—

$$R_s(t) = \exp\left[-\sum_{i=1}^{n} \frac{t^{\gamma_i}}{\theta_i}\right]. \tag{3.6}$$

- r components have constant hazard rates and $n - r$ components have Weibull hazard rates:

$$R_s(t) = \prod_{i=1}^{r} e^{-\lambda_i t} \prod_{i=r+1}^{n} e^{\frac{-t^{\gamma_i}}{\theta_i}}$$

or

$$R_s(t) = \exp\left[-\sum_{i=1}^{r} \lambda_i t - \sum_{i=r+1}^{n} \frac{t^{\gamma_i}}{\theta_i}\right]. \tag{3.7}$$

EXAMPLE 3.1

A series system consists of five components, three of which have constant failure rates $\lambda_1 = 5 \times 10^{-6}$, $\lambda_2 = 3 \times 10^{-6}$, and $\lambda_3 = 9 \times 10^{-6}$. The remaining two components exhibit Weibull hazards that have the following parameters: $\theta_1 = 3.5 \times 10^8$, $\gamma_1 = 2.2$, $\theta_2 = 5.5 \times 10^8$, and $\gamma_2 = 2.1$. Determine the reliability of the system at $t =$ 1,000 hours.

SOLUTION

The exponent of Eq. (3.7) at $t =$ 1,000 is

$$= -\sum_{i=1}^{3} \lambda_i t - \sum_{i=1}^{2} \frac{t^{\gamma_i}}{\theta_i}$$

$$= -(17 \times 10^{-6})1{,}000 - \frac{1{,}000^{2.2}}{3.5 \times 10^8} - \frac{1{,}000^{2.1}}{5.5 \times 10^8}$$

$$= -0.032$$

The reliability of the system is

$$R_s(1{,}000) = e^{-0.032} = 0.9685.$$

3.1.2. PARALLEL SYSTEMS

As shown in Chapter 2, a parallel system fails if and only if all parallel components fail. The reliability of an n components parallel system is expressed as

$$R_s(t) = P(x_1 + x_2 + \ldots + x_n) = 1 - P(\bar{x}_1 \bar{x}_2 \ldots \bar{x}_n). \tag{3.8}$$

In the case of constant-hazard independent components, the unreliability of component i is $1 - e^{-\lambda_i t}$ and the reliability of the system is obtained by using Eq. (3.8) as follows:

$$R_s(t) = 1 - \prod_{i=1}^{n} (1 - e^{-\lambda_i t}). \tag{3.9}$$

The effective hazard rate of a two-component parallel system is obtained as follows. Using Eq. (3.9) with $n = 2$, the reliability of the system, $R_s(t)$, is

$$\begin{aligned} R_s(t) &= 1 - (1 - e^{-\lambda_1 t})(1 - e^{-\lambda_2 t}) \\ &= e^{-\lambda_1 t} + e^{-\lambda_2 t} - e^{-(\lambda_1 + \lambda_2)t}. \end{aligned} \tag{3.10}$$

Since $h(t) = f(t)/R(t)$ and $f(t) = -dR(t)/dt$, then using Eq. (3.10),

$$f(t) = \lambda_1 e^{-\lambda_1 t} + \lambda_2 e^{-\lambda_2 t} - (\lambda_1 + \lambda_2) e^{-(\lambda_1 + \lambda_2)t}, \tag{3.11}$$

and the effective hazard rate (failure rate) of the system is

$$h(t) = \frac{\lambda_1 e^{-\lambda_1 t} + \lambda_2 e^{-\lambda_2 t} - (\lambda_1 + \lambda_2) e^{-(\lambda_1 + \lambda_2)t}}{e^{-\lambda_1 t} + e^{-\lambda_2 t} - e^{-(\lambda_1 + \lambda_2)t}}. \tag{3.12}$$

EXAMPLE 3.2

Consider a parallel system with two components having constant hazard rates of $\lambda_1 = 0.5 \times 10^{-6}$ and $\lambda_2 = 0.3 \times 10^{-6}$ failures per hour. What is the reliability of the system at $t = 1{,}000$ hours? What is the effective hazard rate of the system? What is the effect of λ_1 and λ_2 on $h(t)$ at $t = 800$ hours?

SOLUTION

From Eq. (3.10), we obtain

$$R_s(1{,}000) = e^{-0.5\times10^{-6}\times10^3} + e^{-0.3\times10^{-6}\times10^3} - e^{-0.8\times10^{-6}\times10^3}$$

or

$$R_s(1{,}000) = 0.99999.$$

The effective hazard rate at 1,000 hours is obtained by substituting the hazard rate parameters in Eq. (3.12) as follows:

$$\begin{aligned} h(1{,}000) = \Big[&0.5 \times 10^{-6} e^{-0.5\times10^{-6}\times1000} + 0.3 \times 10^{-6} e^{-0.3\times10^{-6}\times1000} \\ &-0.8 \times 10^{-6} e^{-0.8\times10^{-6}\times1000}\Big]/R_s(1{,}000) \end{aligned}$$

or

$$h(1{,}000) = \frac{2.99820 \times 10^{-10}}{0.99999} = 2.99823 \times 10^{-10} \text{ failures per hour.}$$

The effect of λ_1 and λ_2 on the hazard rate $h(t)$ is shown in Figure 3.1. In general, the reliability of a parallel system with n components each having a hazard rate $h_i(t)$ is expressed as

$$R_s(t) = 1 - \prod_{i=1}^{n} (1 - e^{-H_i(t)}), \tag{3.13}$$

where

$$H_i(t) = \int_0^t h_i(\zeta)\, d\zeta.$$

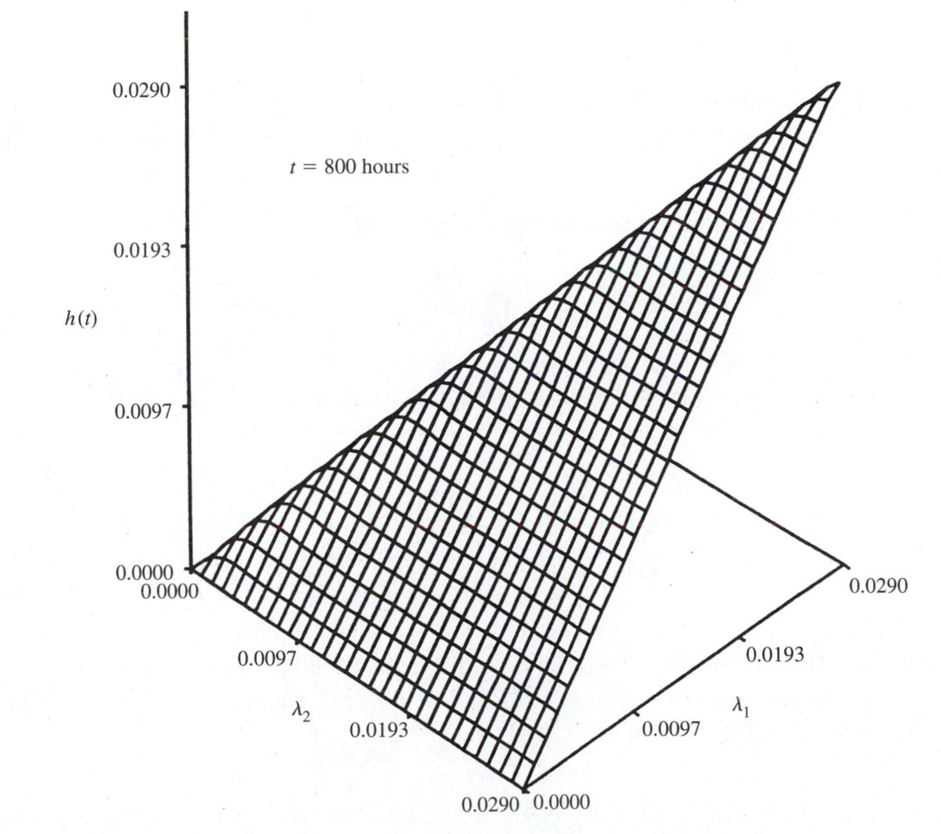

Figure 3.1. Effect of λ_1 and λ_2 on the Effective Hazard Rate of a Parallel System

The expansion of Eq. (3.13) can be easily obtained as follows (Shooman, 1968):

$$(1-y_1)(1-y_2)\dots(1-y_n) = 1 - \sum_{i=1}^{n} y_i + \sum_{i=1}^{n}\sum_{j=i+1}^{n} y_i y_j$$

$$-\sum_{i=1}^{n}\sum_{j=i+1}^{n}\sum_{k=j+1}^{n} y_i y_j y_k + \dots + (-1)^n \prod_{i=1}^{n} y_i. \tag{3.14}$$

Using the above expansion, we now simplify Eq. (3.13) as

$$R_s(t) = \left[\sum_{i=1}^{n} e^{-H_i(t)}\right] - \left[\sum_{i=1}^{n}\sum_{j=i+1}^{n} e^{-(H_i(t)+H_j(t))}\right]$$

$$+\left[\sum_{i=1}^{n}\sum_{j=i+1}^{n}\sum_{k=j+1}^{n} e^{-(H_i(t)+H_j(t)+H_k(t))}\dots\right] - \dots \tag{3.15}$$

The rth parentheses in Eq. (3.15) contains $n!/[r!(n-r)!]$ terms.

The reliability of a parallel system for different hazard rates of the components is given when

- The hazard rates are linearly increasing with time,

$$R_s(t) = 1 - \prod_{i=1}^{n}(1 - e^{-k_i t^2/2}), \text{ and} \tag{3.16}$$

- The hazard rates follow the Weibull model as follows:

$$R_s(t) = 1 - \prod_{i=1}^{n}(1 - e^{-\frac{t^{\gamma_i}}{\theta_i}}). \tag{3.17}$$

EXAMPLE 3.3

Determine the reliability of a three-component parallel system at time $t =$ 100 hours when the components exhibit linearly increasing hazard rates. The coefficients of the hazard rates are

$$k_1 = 2.5 \times 10^{-6}, k_2 = 4 \times 10^{-6}, \text{and } k_3 = 3.5 \times 10^{-6}.$$

SOLUTION

Using Eq. (3.15), we obtain

$$R_s(1{,}000) = e^{\frac{-2.5}{2}\times 10^{-6}\times 10^4} + e^{\frac{-4}{2}\times 10^{-6}\times 10^4} + e^{\frac{-3.5}{2}\times 10^{-6}\times 10^4}$$

$$-\left[e^{\frac{-6.5}{2}\times 10^{-6}\times 10^4} + e^{\frac{-6}{2}\times 10^{-6}\times 10^4} + e^{\frac{-7.5}{2}\times 10^{-6}\times 10^4}\right] + \left[e^{\frac{-10}{2}\times 10^{-6}\times 10^4}\right]$$

$$R(1{,}000) = 0.9999957.$$

3.1.3. k-OUT-OF-n SYSTEMS

In these types of systems, any combination of k operating components out of n independent components will guarantee successful operation of the system. If the components are not identical, we should investigate every possible successful path of the reliability structure in order to accurately estimate the reliability of the system. Fortunately, most k-out-of-n systems have independent and identical components, and the reliability of the system is much simpler to estimate by using the binomial distribution. In a typical k-out-of-n system with components having constant failure rates, the reliability of the system is

$$R_s(t) = \sum_{r=k}^{n} \binom{n}{r} (e^{-\lambda t})^r (1 - e^{-\lambda t})^{n-r}$$

or

$$R_s(t) = \sum_{r=k}^{n} \binom{n}{r} e^{-r\lambda t}(1 - e^{-\lambda t})^{n-r} = 1 - \sum_{r=0}^{k-1} \binom{n}{r} e^{-r\lambda t}(1 - e^{-\lambda t})^{n-r}. \quad (3.18)$$

Similarly, the reliabilities of a k-out-of-n system when the components exhibit linear or Weibull hazard rates are given by Eqs. (3.19) and (3.20), respectively. (In order to avoid confusion, for the moment, we replace the constant k of the linear hazard model by another constant λ).

$$R_s(t) = \sum_{r=k}^{n} \binom{n}{r} e^{-r\lambda t^2/2}(1 - e^{-\lambda t^2/2})^{n-r} \quad (3.19)$$

and

$$R_s(t) = \sum_{r=k}^{n} \binom{n}{r} e^{\frac{-rt^\gamma}{\theta}} \left(1 - e^{\frac{-t^\gamma}{\theta}}\right)^{n-r}. \quad (3.20)$$

EXAMPLE 3.4

Consider a 2-out-of-3 system with components that exhibit constant failure rates with parameter λ. What is the reliability of the system? If $\lambda = 3.0 \times 10^{-5}$ failures per hour, determine the reliability at time $t = 1{,}000$ hours.

SOLUTION

Using Eq. (3.18), we obtain

$$R_s(t) = \sum_{r=2}^{3} \binom{n}{r} e^{-r\lambda t}\left[1 - e^{-\lambda t}\right]^{n-r}$$

$$= \binom{3}{2} e^{-2\lambda t}\left[1 - e^{-\lambda t}\right] + \binom{3}{3} e^{-3\lambda t}$$

$$= 3e^{-2\lambda t} - 3e^{-3\lambda t} + e^{-3\lambda t} = 3e^{-2\lambda t} - 2e^{-3\lambda t}.$$

Substitute $\lambda = 3.0 \times 10^{-5}$ and $t = 1{,}000$.

$$R_s(1{,}000) = 3e^{-6\times 10^{-2}} - 2e^{-9\times 10^{-2}} = 0.9974.$$

EXAMPLE 3.5

In a 2-out-of-n system with components having a constant hazard rate 0.4×10^{-4} failures per hour, determine the number of components for the system such that a reliability of 0.966 is achieved at $t = 1{,}000$ hours.

SOLUTION

The reliability of any component in the system is $\exp(-0.4 \times 10^{-4} \times 1{,}000) = 0.96078$. Thus,

$$0.966 = 1 - \sum_{r=0}^{1} \binom{n}{r} (0.96078)^r (0.03922)^{n-r}$$

$$= 1 - \left[\binom{n}{0} (0.03922)^n + \binom{n}{1} (0.96078)(0.03922)^{n-1}\right]$$

$$= 1 - 0.03922^n - n(0.96078)(0.03922)^{n-1}$$

or

$$0.034 \leq 0.03922^n + n(0.96078)(0.03922)^{n-1}.$$

From the above equation, $n = 2$ components. In other words, at most, two components should be used to achieve the desired reliability over a time period of 1,000 hours.

3.2. Mean Time to Failure (MTTF)

MTTF is one of the most widely used measures of reliability. It is simply defined as the expected or mean value $E[T]$ of the failure time T. Hence

$$MTTF = \int_0^\infty t f(t)\, dt. \tag{3.21}$$

The $MTTF$ may be expressed directly in terms of the system reliability by substituting the following relationship into Eq. (3.21):

$$f(t) = -\frac{dR(t)}{dt}$$

$$MTTF = -\int_0^\infty t \frac{dR(t)}{dt}\, dt$$

or

$$MTTF = -tR(t) \Big|_0^\infty + \int_0^\infty R(t)\, dt. \tag{3.22}$$

Since $tR(t) \rightarrow 0$ as $t \rightarrow 0$ and $tR(t) \rightarrow 0$ as $t \rightarrow \infty$, then Eq. (3.22) can be written as

$$MTTF = \int_0^\infty R(t)\, dt. \tag{3.23}$$

The MTTF in itself is the mean of the failure time; it does not provide additional information about the distribution of the TTF (time to failure). In order to do so, we need to determine the standard deviation of the TTF.

By definition, the standard deviation of the TTF is given as

$$\sigma_{TTF} = \sqrt{\int_0^\infty t^2 f(t)\, dt - MTTF^2}. \tag{3.24}$$

The following sections show how the MTTF is calculated for different systems.

3.2.1. MTTF FOR SERIES SYSTEMS

The MTTF for series systems with n components each having constant, linearly increasing, and Weibull hazard rates is given below.

3.2.1.1. *Constant Hazard*

The reliability expression for a series system with constant hazard rates is given by Eq. (3.3). The MTTF of such a system is

$$MTTF = \int_0^\infty e^{-\sum_{i=1}^{n} \lambda_i t} \, dt$$

or

$$MTTF = \frac{1}{\sum_{i=1}^{n} \lambda_i}.$$

3.2.1.2. *Linearly Increasing Hazard*

The reliability of a component with linearly increasing hazard is

$$R(t) = e^{-kt^2/2},$$

and the MTTF is

$$MTTF = \int_0^\infty e^{-kt^2/2} \, dt = \frac{\Gamma(1/2)}{2\sqrt{k/2}} = \sqrt{\frac{\pi}{2k}}.$$

For a system with n components in series and each having a linearly increasing hazard, the MTTF is

$$MTTF = \sqrt{\frac{\pi}{2 \sum_{i=1}^{n} k_i}}. \tag{3.25}$$

3.2.1.3. *Weibull Hazard*

For a system composed of one component having a Weibull hazard rate, the MTTF is obtained as follows

$$MTTF = \int_0^\infty R(t)\, dt$$

or

$$MTTF = \int_0^\infty e^{\frac{-t^\gamma}{\theta}}\, dt. \tag{3.26}$$

Let

$$x = \frac{1}{\theta} t^\gamma$$

and then

$$dt = \frac{\theta^{\frac{1}{\gamma}}}{\gamma} x^{\frac{1}{\gamma}-1}\, dx.$$

Substituting in Eq. (3.26), we obtain

$$MTTF = \frac{\theta^{\frac{1}{\gamma}}}{\gamma} \int_0^\infty e^{-x} x^{\frac{1}{\gamma}-1}\, dx$$

or

$$MTTF = \theta^{\frac{1}{\gamma}} \frac{1}{\gamma} \Gamma\left(\frac{1}{\gamma}\right) = \theta^{\frac{1}{\gamma}} \Gamma\left(1 + \frac{1}{\gamma}\right). \tag{3.27}$$

The values of $\Gamma(x)$ for different x are given in Appendix A.

If n components form a series configuration and all components exhibit Weibull hazards with the same value of γ, then Eq. (3.27) can be rewritten as

$$MTTF = \left(\sum_{i=1}^{n} \theta_i\right)^{\frac{1}{\gamma}} \Gamma\left(1 + \frac{1}{\gamma}\right). \tag{3.28}$$

EXAMPLE 3.6 A series system consists of six components that exhibit the same shape parameter of a Weibull distribution. The shape parameter is 1.75 and the scale parameters of the components are 7.0×10^5, 8.2×10^5, 4.6×10^5, 6.5×10^5, 6.8×10^5, and 5×10^5. Determine the MTTF of the system.

SOLUTION Using Eq. (3.28), we obtain

$$MTTF = (38.1 \times 10^5)^{\frac{1}{1.75}} \Gamma\left(1 + \frac{1}{1.75}\right) = 18{,}551 \text{ hours.}$$

3.2.2. MTTF FOR PARALLEL SYSTEMS

The calculations of the MTTF for parallel systems are similar to those of the series systems. Again, the MTTF's for different hazard functions are obtained as shown below.

3.2.2.1. *Constant Hazard*

Consider a parallel system consisting of n independent components and that the failure rate, λ_i, of component i is constant. The MTTF of the system is

$$MTTF = \int_0^\infty R(t)\, dt$$

$$= \int_0^\infty \left[\sum_{i=1}^{n} e^{-\lambda_i t} - \sum_{i=1}^{n} \sum_{j=i+1}^{n} e^{-(\lambda_i+\lambda_j)t} + \ldots\right] dt$$

or

$$MTTF = \sum_{i=1}^{n} \frac{1}{\lambda_i} - \sum_{i=1}^{n-1} \sum_{j=i+1}^{n} \frac{1}{\lambda_i + \lambda_j} + \sum_{i=1}^{n-2} \sum_{j=i+1}^{n-1} \sum_{k=j+1}^{n} \frac{1}{\lambda_i + \lambda_j + \lambda_k} - \ldots + (-1)^{n+1} \frac{1}{\sum_{i=1}^{n} \lambda_i}. \tag{3.29}$$

If all components are identical and each component has a failure rate λ, then

$$R_s(t) = 1 - (1 - e^{-\lambda t})^n$$

and

$$MTTF = \frac{1}{\lambda}\left[1 + \frac{1}{2} + \ldots + \frac{1}{n}\right]. \tag{3.30}$$

Equation (3.30) implies that in active redundancy where each component exhibits one type of failure mode, the MTTF of the system exceeds the MTTF of the individual component and the contribution of the second component and other additional components would have a diminishing return on the system's MTTF as n increases. In other words, there is an optimum n at which the cost of adding a component in parallel far exceeds the gained benefit in the MTTF.

3.2.2.2. *Linearly Increasing Hazard*

The components are assumed to have linearly increasing hazard rates. In other words, each component i has a linearly increasing hazard, $k_i t$. The MTTF of such a system is

$$MTTF = \int_0^\infty \left[\sum_{i=1}^{n} e^{-1/2k_i t^2} - \sum_{i=1}^{n-1}\sum_{j=i+1}^{n} e^{-1/2(k_i+k_j)t^2} + \sum_{i=1}^{n-2}\sum_{j=i+1}^{n-1}\sum_{k=j+1}^{n} e^{-1/2(k_i+k_j+k_k)t^2} + \ldots\right] dt$$

or

$$MTTF = \sum_{i=1}^{n}\sqrt{\frac{\pi}{2k_i}} - \sum_{i=1}^{n-1}\sum_{j=i+1}^{n}\sqrt{\frac{\pi}{2(k_i+k_j)}}$$

$$+ \sum_{i=1}^{n-2}\sum_{j=i+1}^{n-1}\sum_{k=j+1}^{n}\sqrt{\frac{\pi}{2(k_i+k_j+k_k)}} - \ldots \tag{3.31}$$

If all components are identical with a hazard rate kt, then

$$MTTF = \sqrt{\frac{\pi}{2k}}\left[n - \binom{n}{2}\sqrt{\frac{1}{2}} + \binom{n}{3}\sqrt{\frac{1}{3}} - \binom{n}{4}\sqrt{\frac{1}{4}} + \ldots\right]. \tag{3.32}$$

EXAMPLE 3.7

An active redundant system consists of four identical parallel components each having a linearly increasing hazard rate, kt, with $k = 3.5 \times 10^{-6}$ failures per hour. Determine the MTTF of the system.

SOLUTION Using Eq. (3.32) with $n = 4$ and $k = 3.5 \times 10^{-6}$, we obtain the MTTF of the system as

$$MTTF = \sqrt{\frac{\pi}{7 \times 10^{-6}}}\left[4 - 2\sqrt{\frac{1}{2}} + \frac{4}{3}\sqrt{\frac{1}{3}} - \sqrt{\frac{1}{4}}\right]$$

or

$$MTTF = 1{,}913 \text{ hours.}$$

3.2.2.3. *Weibull Hazard*

The MTTF of an active redundancy system that consists of n components in parallel and each component exhibits a Weibull hazard of the form $(\gamma/\theta_i)t^{\gamma-1}$, where θ_i is a constant for component i and γ is the same shape parameter for all the components, can be obtained as

$$MTTF = \Gamma\left(1 + \frac{1}{\gamma}\right)\left[\sum_{i=1}^{n} \theta_i^{\frac{1}{\gamma}} - \sum_{i=1}^{n-1} \sum_{j=i+1}^{n} (\theta_i + \theta_j)^{\frac{1}{\gamma}} + \ldots + (-1)^{n+1}\left[\sum_{i=1}^{n} \theta_i\right]^{\frac{1}{\gamma}}\right]. \tag{3.33}$$

EXAMPLE 3.8 Solve Example 3.7 when the system consists of three components in parallel and their hazard rates are

$$h_1(t) = \frac{2.5}{4 \times 10^6} t^{1.5}$$

$$h_2(t) = \frac{2.5}{4.9 \times 10^6} t^{1.5}$$

$$h_3(t) = \frac{2.5}{4.1 \times 10^6} t^{1.5}.$$

SOLUTION

Using Eq. (3.33), we obtain

$$MTTF = \Gamma\left(1 + \frac{1}{2.5}\right)\left[(4 \times 10^6)^{\frac{1}{2.5}} + (4.9 \times 10^6)^{\frac{1}{2.5}} + (4.1 \times 10^6)^{\frac{1}{2.5}} - (8.9 \times 10^6)^{\frac{1}{2.5}} - (8.1 \times 10^6)^{\frac{1}{2.5}} - (9 \times 10^6)^{\frac{1}{2.5}} + (13 \times 10^6)^{\frac{1}{2.5}}\right]$$

or

$$MTTF = 267.036 \times 0.8873 = 236.94 \text{ hours.}$$

This implies that the failure rate is increasing rapidly with time and the system should be either redesigned or the components should be replaced by others with a much reduced failure rate.

3.2.3. *k*-OUT-OF-*n* SYSTEMS

The reliability expression for a *k*-out-of-*n* system whose components are independent and identical is

$$R_s(t) = \sum_{r=k}^{n} \binom{n}{r} [p(t)]^r [1 - p(t)]^{n-r}, \tag{3.34}$$

where $p(t)$ is the reliability of the component at time t. There is no general expression for the MTTF of a *k*-out-of-*n* system since it depends on the values of k and n. Therefore, we illustrate the procedure for obtaining the MTTF for a *k*-out-of-*n* system for different hazard rates through the following examples.

3.2.3.1. *Constant Hazard*

Consider a *k*-out-of-*n* system whose components are independent and identical. Each component exhibits a constant hazard rate λ. The MTTF of this system is obtained by substituting $p(t) = e^{-\lambda t}$ into Eq. (3.34):

$$MTTF = \int_0^\infty \sum_{r=k}^{n} \binom{n}{r} (e^{-\lambda t})^r (1 - e^{-\lambda t})^{n-r}\, dt. \tag{3.35}$$

EXAMPLE 3.9

Determine the MTTF of a 2-out-of-4 system with independent components each having a constant hazard of 8.5×10^{-6} failures per hour.

SOLUTION

First derive a reliability expression for the system, and then estimate its MTTF as $\int_0^\infty R_s(t)\, dt$.

$$R_s(t) = \sum_{r=2}^{4} \binom{4}{r} (e^{-\lambda t})^r (1 - e^{-\lambda t})^{4-r}$$

$$= \binom{4}{2} e^{-2\lambda t} (1 - 2e^{-\lambda t} + e^{-2\lambda t}) + \binom{4}{3} e^{-3\lambda t} (1 - e^{-\lambda t}) + \binom{4}{4} e^{-4\lambda t}$$

$$= 6e^{-2\lambda t} - 12e^{-3\lambda t} + 6e^{-4\lambda t} + 4e^{-3\lambda t} - 4e^{-4\lambda t} + e^{-4\lambda t}$$

or

$$R_s(t) = 6e^{-2\lambda t} - 8e^{-3\lambda t} + 3e^{-4\lambda t}$$

$$MTTF = \int_0^\infty R_s(t)\, dt = \frac{13}{12\lambda} = 1.2745 \times 10^5 \text{ hours.}$$

3.2.3.2. *Linearly Increasing Hazard*

The MTTF of a k-out-of-n system, when all components are independent, identical, and exhibit linearly increasing hazards, is determined by substituting $p(t) = e^{-kt^2/2}$ in Eq. (3.34) to obtain a reliability expression of the system. Then the resulting expression is integrated with respect to t from 0 to ∞ as shown in the following example.

EXAMPLE 3.10 Determine the MTTF for the system given in Example 3.9 when the failure rates of the components are linearly increasing with parameter $k = 2.7 \times 10^{-4}$.

SOLUTION The reliability of the system is

$$R_s(t) = \binom{4}{2} (e^{-kt^2/2})^2 (1 - e^{-kt^2/2})^2 + \binom{4}{3} (e^{-kt^2/2})^3 (1 - e^{-kt^2/2}) + \binom{4}{4} (e^{-kt^2/2})^4$$

$$= 6e^{-kt^2} (1 - 2e^{-kt^2/2} + e^{-kt^2}) + 4e^{-3kt^2/2} (1 - e^{-kt^2/2}) + e^{-2kt^2}$$

or

$$R_s(t) = 6e^{-kt^2} - 8e^{-3kt^2/2} + 3e^{-2kt^2}.$$

The mean time to failure of the system is

$$MTTF = \int_0^\infty R_s(t)\, dt = 6\sqrt{\frac{\pi}{4k}} - 8\sqrt{\frac{\pi}{6k}} + 3\sqrt{\frac{\pi}{8k}} = 85.7 \times 10^3 \text{ hours.}$$

3.2.3.3. *Weibull Hazard*

Similar to the linearly increasing hazard, we calculate the MTTF of a k-out-of-n system composed of independent and identical components that exhibit Weibull hazard by first deriving an expression for the system reliability as shown in the following example.

EXAMPLE 3.11

Determine the MTTF of the system given in Example 3.9 if the components are independent, identical, and exhibit a Weibull hazard with parameters $\theta = 4 \times 10^5$ and $\gamma = 2.1$.

SOLUTION

The reliability of the system is

$$R_s(t) = \binom{4}{2}\left(e^{\frac{-t^\gamma}{\theta}}\right)^2\left(1 - e^{\frac{-t^\gamma}{\theta}}\right)^2 + \binom{4}{3}\left(e^{\frac{-t^\gamma}{\theta}}\right)^3\left(1 - e^{\frac{-t^\gamma}{\theta}}\right) + \binom{4}{4}\left(e^{\frac{-t^\gamma}{\theta}}\right)^4$$

or

$$R_s(t) = 6e^{\frac{-2t^\gamma}{\theta}} - 8e^{\frac{-3t^\gamma}{\theta}} + 3e^{\frac{-4t^\gamma}{\theta}}.$$

The mean time to failure is obtained as

$$MTTF = \int_0^\infty R_s(t)\, dt$$

$$= \frac{1}{\gamma}\left[6\left(\frac{\theta}{2}\right)^{\frac{1}{\gamma}}\Gamma\left(\frac{1}{\gamma}\right) - 8\left(\frac{\theta}{3}\right)^{\frac{1}{\gamma}}\Gamma\left(\frac{1}{\gamma}\right) + 3\left(\frac{\theta}{4}\right)^{\frac{1}{\gamma}}\Gamma\left(\frac{1}{\gamma}\right)\right]$$

or

$$MTTF = 0.88\,[2006.55 - 2205.65 + 721.23] = 462.45 \text{ hours.}$$

3.2.4. OTHER SYSTEMS

The estimation of the MTTF of any system requires the derivation of an expression for the reliability of the system. This expression is then integrated over time from 0 to ∞. When the system structure is not a standard structure such as series, parallel, or a k-out-of-n, we follow the same procedures described in Chapter 2 for the reliability estimation of complex structures to obtain an expression for $R_s(t)$ as shown in Example 3.12.

EXAMPLE 3.12

Determine the MTTF of the complex reliability structure system shown in Figure 3.2. Assume that the components are independent, identical, and exhibit a constant failure rate $\lambda = 3.5 \times 10^{-5}$ failures per hour.

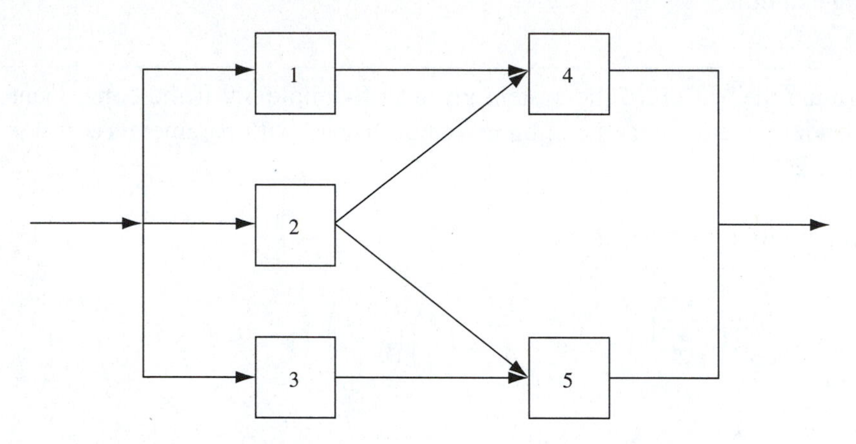

Figure 3.2. A Complex Reliability Structure

SOLUTION

The reliability of this structure can be estimated using any of the methods discussed in Chapter 2 such as cut-set or tie-set approaches. The reliability of the system is

$$R_s(t) = e^{-5\lambda t} - e^{-4\lambda t} - 3e^{-3\lambda t} + 4e^{-2\lambda t},$$

and the mean time to failure is

$$MTTF = \int_0^{\infty} (e^{-5\lambda t} - e^{-4\lambda t} - 3e^{-3\lambda t} + 4e^{-2\lambda t})\, dt$$

$$= \frac{1}{5\lambda} - \frac{1}{4\lambda} - \frac{3}{3\lambda} + \frac{4}{2\lambda} = \frac{57}{60\lambda} = 2.7142 \times 10^4 \text{ hours}$$

or

$$MTTF \cong 2.7142 \text{ years.}$$

A summary of the MTTF expressions for different configurations and hazard rates is given in Table 3.1.

Table 3.1. MTTF for Different Configurations

Configuration	*Hazard Rate*	*MTTF*
Series	λ_i	$\dfrac{1}{\sum_{i=1}^{n} \lambda_i}$
(n units in series)	$k_i t$	$\sqrt{\dfrac{\pi}{2\sum_{i=1}^{n} k_i}}$
	$\dfrac{1}{\theta_i} t^{\gamma-1}$	$\left[\sum_{i=1}^{n} \theta_i\right]^{\frac{1}{\gamma}} \Gamma\left(1+\dfrac{1}{\gamma}\right)$
Parallel	λ_i	$\sum_{i=1}^{n} \dfrac{1}{\lambda_i} - \sum_{i=1}^{n-1} \sum_{j=i+1}^{n} \dfrac{1}{\lambda_i+\lambda_j} + \sum_{i=1}^{n-2} \sum_{j=i+1}^{n-1} \sum_{k=j+1}^{n} \dfrac{1}{\lambda_i+\lambda_j+\lambda_k} - \ldots$
(n units in parallel)	$k_i t$	$\sum_{i=1}^{n} \sqrt{\dfrac{\pi}{2k_i}} - \sum_{i=1}^{n-1} \sum_{j=i+1}^{n} \sqrt{\dfrac{\pi}{2(k_i+k_j)}} + \sum_{i=1}^{n-2} \sum_{j=i+1}^{n-1} \sum_{k=j+1}^{n} \sqrt{\dfrac{\pi}{2(k_i+k_j+k_k)}} - \ldots$
	$\dfrac{1}{\theta_i} t^{\gamma-1}$	$\Gamma\left(1+\dfrac{1}{\gamma}\right)\left[\sum_{i=1}^{n} \theta_i^{\frac{1}{\gamma}} - \sum_{i=1}^{n-1} \sum_{j=i+1}^{n} (\theta_i+\theta_j)^{\frac{1}{\gamma}} + \ldots + (-1)^{n+1}\left[\sum_{i=1}^{n} \theta_i\right]^{\frac{1}{\gamma}}\right]$
k-out-of-n	λ	$\int_0^{\infty} \sum_{r=k}^{n} \binom{n}{r} (e^{-\lambda t})^r (1-e^{-\lambda t})^{n-r}\, dt$

3.3. Repairable Systems

Repairable systems are those systems that are repaired upon failure. Repairable systems include large and complex systems, automobiles, airplanes, HVAC (heating, ventilation, and air conditioning), mainframe computers, telephone networks, and many others. In Chapter 2, we illustrated the use of redundant components (or systems) to improve the overall system reliability. Other methods of improving system reliability include the use of "highly" reliable components (prime material free of manufacturing defects) and the use of efficient repair and maintenance systems. Two of the most important performance criteria of repairable systems are availability (there are several measures of availability that will be discussed later in this chapter) and the MTBF (mean time between failures). Conventionally, we use MTBF for repairable systems and MTTF for nonrepairable systems. For now, we define availability as the probability that the system is operating properly when it is requested for use.

Since repair and maintenance have a major impact on the system availability, we devote Chapter 9 to discuss in detail, different repairs, replacements, and preventive maintenance policies.

In this section, we present two approaches for estimating the time dependent reliability and availability of repairable systems. The first is the alternating renewal process and the second is the Markov process.

3.3.1. ALTERNATING RENEWAL PROCESS

Consider a repairable system that has a failure-time distribution with a probability density function $w(t)$, and a repair-time distribution with a probability density function $g(t)$. When the system fails, it is repaired and restored to its initial working condition. The process of failure and repair is repeated. We refer to this process as an alternating renewal process. The underlying density function $f(t)$ of the renewal process is the convolution of w and g. In other words,

$$f^*(s) = w^*(s)g^*(s), \tag{3.36}$$

where $f^*(s)$, $w^*(s)$, and $g^*(s)$ are the Laplace transforms of the corresponding density functions.

As shown later in Chapter 7, the Laplace transform of the renewal density equation is

$$m^*(s) = \frac{f^*(s)}{1 - f^*(s)}. \tag{3.37}$$

Substituting Eq. (3.37) into Eq. (3.36) results in

$$m^*(s) = \frac{w^*(s)g^*(s)}{1 - w^*(s)g^*(s)}. \tag{3.38}$$

The component (or system) may be functioning at time t if either it had not failed during the time interval $(0, t]$ with probability $R(t)$ or the last repair occurred at time x, $0 < x < t$, and the component (or system) continued to function properly since that time with probability $\int_0^t R(t - x)m(x)\,dx$. Thus, the availability of the component (or system) at time t is the sum of the two probabilities, or

$$A(t) = R(t) + \int_0^t R(t - x)\, m(x)\, dx. \tag{3.39}$$

The Laplace transform of Eq. (3.39) is

$$A^*(s) = R^*(s)\,[1 + m^*(s)]. \tag{3.40}$$

Substituting Eq. (3.38) into Eq. (3.40), we get

$$A^*(s) = R^*(s)\left[1 + \frac{w^*(s)g^*(s)}{1 - w^*(s)g^*(s)}\right]$$

or

$$A^*(s) = \frac{R^*(s)}{1 - w^*(s)g^*(s)}.$$

Since $R(t) = 1 - W(t)$, then $R^*(s)$ is

$$R^*(s) = \frac{1 - w^*(s)}{s}$$

and

$$A^*(s) = \frac{1 - w^*(s)}{s\,[1 - w^*(s)g^*(s)]}. \tag{3.41}$$

The Laplace inverse of $A^*(s)$ results in obtaining the point availability $A(t)$. Often, a closed-form expression of the inverse of $A^*(s)$ is difficult to obtain, and numerical solutions or approximations become the only alternatives for obtaining $A(t)$. The steady-state availability, A, is

$$A = \lim_{t\to\infty} A(t) = \lim_{s\to 0} s\,A^*(s).$$

When s is small, then $e^{-st} \cong 1 - st$ and

$$w^*(s) = \int_0^\infty e^{-st}\,w(t)\,dt \cong \int_0^\infty w(t)\,dt - s\int_0^\infty t\,w(t)\,dt$$

$$w^*(s) = 1 - \frac{s}{\alpha},$$

where $1/\alpha$ is the MTBF (mean time between failures). Similarly, $g^*(s) = 1 - s/\beta$, where $1/\beta$ is the MTTR (mean time to repair). Therefore, the steady-state availability is obtained by taking the limit of Eq. (3.41) as $s \to 0$:

$$A = \lim_{s \to 0} \frac{1 - \left(1 - \dfrac{s}{\alpha}\right)}{1 - \left(1 - \dfrac{s}{\alpha}\right)\left(1 - \dfrac{s}{\beta}\right)} = \frac{\dfrac{1}{\alpha}}{\dfrac{1}{\alpha} + \dfrac{1}{\beta}}$$

or

$$A = \frac{MTBF}{MTBF + MTTR}. \tag{3.42}$$

EXAMPLE 3.13

The failure time of the system follows a Weibull distribution with a p.d.f. of the form

$$w(t) = \frac{\gamma t^{\gamma-1}}{\theta^{\gamma}} \exp\left[-\left(\frac{t}{\theta}\right)^{\gamma}\right]$$

and its repair time follows an exponential distribution with a p.d.f. of

$$g(t) = \mu e^{-\mu t}.$$

Determine the point availability of the system $A(t)$ and its steady-state value.

SOLUTION

We first obtain the Laplace transforms of $w(t)$ and $g(t)$ as

$$w^*(s) = \int_0^{\infty} e^{-st}\, w(t)\, dt = \sum_{j=0}^{\infty} (-1)^j \frac{(\theta s)^j}{j!} \Gamma\left(\frac{j+\gamma}{\gamma}\right)$$

and

$$g^*(s) = \int_0^{\infty} e^{-st}\, g(t)\, dt = \frac{\mu}{1 + s\mu}.$$

Substituting $w^*(s)$ and $g^*(s)$ into Eq. (3.41) results in

$$A^*(s) = \frac{1 - \sum_{j=0}^{\infty} (-1)^j \frac{(\theta s)^j}{j!} \Gamma\left(\frac{j+\gamma}{\gamma}\right)}{s\left[1 - \frac{\mu}{(1+s\mu)} \sum_{j=0}^{\infty} (-1)^j \frac{(\theta s)^j}{j!} \Gamma\left(\frac{j+\gamma}{\gamma}\right)\right]}.$$

A closed form expression of $A(t)$ cannot be obtained from $A^*(s)$. Therefore, $A(t)$ can only be estimated numerically or by approximation. Though $A(t)$ is difficult to obtain, its steady-state value can be easily estimated by using Eq. (3.42):

$$MTBF = \theta\, \Gamma\left(\frac{1+\gamma}{\gamma}\right)$$

and

$$MTTR = \mu.$$

If $\theta = 5 \times 10^6$, $\gamma = 2.15$, and $\mu = 10{,}000$, then

$$A = \frac{4.428 \times 10^6}{4.428 \times 10^6 + 10^4} = 0.997746.$$

An alternative to the use of Laplace transform in obtaining the availability of the system is now described. The pointwise availability of a system at time t is defined as the probability of the system being in a working state (operating properly) at t. As shown in Eq. (3.42), the limiting availability of a system that has constant failure rate λ and repair rate μ is

$$A = \frac{\mu}{\lambda + \mu}.$$

The unavailability of the system $\bar{A}(t)$ is

$$\bar{A}(t) = 1 - A(t),$$

and the limiting unavailability is

$$\bar{A} = \frac{\lambda}{\lambda + \mu} = \frac{\beta\lambda}{1 + \beta\lambda}, \tag{3.43}$$

where $\beta = 1/\mu$ (the MTTR).

Equation (3.43) can be rewritten as the power series (Holcomb, 1981)

$$\bar{A} = \sum_{i=1}^{\infty} (-1)^{i+1} (\beta\lambda)^i$$

$$= \beta\lambda - (\beta\lambda)^2 + (\beta\lambda)^3 - \ldots \tag{3.44}$$

Since $\lambda << 1/\beta$, the above series falls off very quickly. Thus, when $\beta\lambda$ is small, the first-order approximation of $\bar{A}$ is $\beta\lambda$. The same reasoning can be used to estimate the unavailability $\bar{A}(t)$ when the failure rate is time dependent. In this case, the expected number of failures, E, during the interval $(t - \beta, t)$ is

$$E = \int_{t-\beta}^{t} \lambda(x)\, dx \cong \bar{A}(t). \tag{3.45}$$

This is approximately equal to the probability of being nonworking at t as long as $E << 1$. Therefore, the integral of Eq. (3.45) can be approximated as

$$\bar{A}(t) \approx \beta\lambda(t). \tag{3.46}$$

Equation (3.43) can now be rewritten as

$$\bar{A}(t) \cong \frac{\lambda(t)}{\lambda(t) + \mu}. \tag{3.47}$$

This approximation is reasonable if the rate of failure changes relatively little near t—that is, over a time range on the order of β. For example, the failure rate of the Weibull model is

$$\lambda(t) = h(t) = \frac{\gamma}{\theta} t^{\gamma - 1}.$$

As t increases, the relative change of $\lambda(t)$ decreases. In other words, for a large enough time, the failure rate changes slowly enough to satisfy the conditions for Eqs. (3.46) and (3.47). Details of the derivation of the unavailability and how large t should be to ensure the validity of the approximations are given in Holcomb (1981).

EXAMPLE 3.14

Use the approximation given by Eq. (3.47) to estimate the availability of the system described in Example 3.13 for different values of γ. Determine the times at which the availability obtained from the approximation is equal to those of the steady state.

SOLUTION

The steady-state values of the availabilities for systems with $\theta = 5 \times 10^6$, $\mu = 10{,}000$, and different γ's are shown in Table 3.2. Figure 3.3 shows the effect of γ on $A(t)$. The times to reach steady-state availability values increase as γ decreases.

Table 3.2. Availability Values for Different γ

γ	*Time to Reach Steady State*	*Steady-State Availability*
2.15	5.17×10^6	0.997746
2.0	5.6930×10^7	0.997748
1.9	4.30×10^9	0.997766

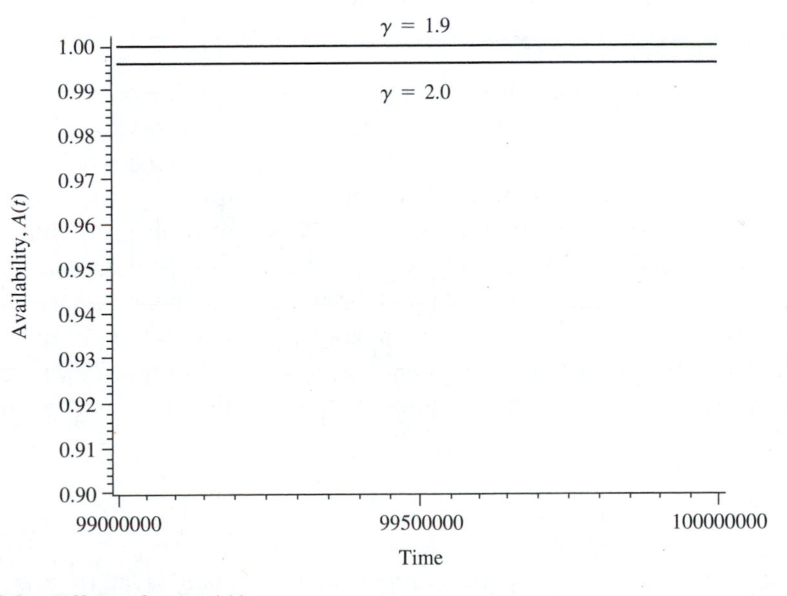

Figure 3.3. Effect of γ on $A(t)$

3.3.2. MARKOV MODELS

This is the second approach that can be used to estimate the time-dependent reliability of the system. This approach is valid when both the failure and repair rates are constant. When these rates are time dependent, the Markov process breaks down, except in some special cases. In this section the presentation of the Markov models is limited to constant failure and repair rates.

The first step in formulating a Markov model requires the definition of all the mutually exclusive states of the system. For example, a nonrepairable system may have two states: state $s_0 = x$, the system is working properly (that is, good), and state $s_1 = \bar{x}$, the system is not working properly (that is, failed), where x is the indicator that the system is good and $\bar{x}$ is the indicator that the system is not working.

The second step is to define the initial and final conditions of the system. For example, it is reasonable to assume that initially the system is working properly at $t = 0$ with reliability $R_s(0) = 1$. It is also reasonable to assume that the system will eventually fail as time approaches infinity—that is, $R_s(\infty) = 0$.

The third step involves the development of the Markov state equations, which describe the probabilistic transitions of the system. In doing so, the probability of transition from one state to another in a time interval Δt is $h(t)\,\Delta t$ where $h(t)$ is the rate associated with the two states ($h(t)$ can be a failure rate or a repair rate as shown later in this section). Moreover, the probabilities of more than one transition in Δt can be neglected.

3.3.2.1. *Nonrepairable Component*

We now consider a nonrepairable component that has a failure rate λ. We are interested in the development of a time-dependent reliability expression of the component. Define $P_0(t)$ and $P_1(t)$ as the probabilities of the component being in state s_0 (working properly) and in state s_1 (not working) at time t, respectively. Let us examine the states of the component at time $t + \Delta t$. The probability that the component is in state s_0 at $t + \Delta t$ is given by the probability of the component being in state s_0 at time t, $P_0(t)$, times the probability that the component does not fail in Δt, $1 - \lambda\,\Delta t$, plus the probability of the component being in state s_1 at time t, $P_1(t)$, times the probability that the component is repaired during Δt (this probability equals zero for nonrepairable components or systems). We write the state-transition equations as

$$P_0(t + \Delta t) = [1 - \lambda\,\Delta t]\,P_0(t) + 0\,P_1(t). \tag{3.48}$$

Likewise, the probability that the component is in state s_1 at time $t + \Delta t$ is expressed as

$$P_1(t + \Delta t) = \lambda\,\Delta t\,P_0(t) + 1P_1(t). \tag{3.49}$$

Note that the transition probability $\lambda\,\Delta t$ is the probability of failure in Δt (change from state s_0 to state s_1) and the probability of remaining in state s_1 is unity (Shooman, 1968).

Rearranging Eqs. (3.48) and (3.49) and dividing by Δt, we obtain

$$\frac{P_0\,(t+\Delta t) - P_0(t)}{\Delta t} = -\lambda P_0(t)$$

and

$$\frac{P_1(t+\Delta t) - P_1(t)}{\Delta t} = \lambda P_0(t).$$

Taking the limit of the above equations as $\Delta t \to 0$, then

$$\frac{dP_0(t)}{dt} + \lambda P_0(t) = 0 \tag{3.50}$$

$$\frac{dP_1(t)}{dt} - \lambda P_0(t) = 0. \tag{3.51}$$

Using the initial conditions $P_0(t=0) = 1$ and $P_1(t=0) = 0$, we solve Eq. (3.50) as

$$\frac{dP_0(t)}{dt} = -\lambda P_0(t)$$

$$\ln P_0(t) = -\int_0^t \lambda\, d\xi + c$$

and

$$P_0(t) = c_1 \exp\left[-\int_0^t \lambda\, d\xi\right].$$

Since $P_0(t=0) = 1$, then $c_1 = 1$ and

$$P_0(t) = e^{-\int_0^t \lambda\, d\xi}. \tag{3.52}$$

When λ is constant, Eq. (3.52) becomes

$$P_0(t) = e^{-\lambda t}.$$

In other words, the reliability of the component at time t is

$$R(t) = P_0(t) = e^{-\lambda t}. \tag{3.53}$$

The solution of $P_1(t)$ is obtained from the condition $P_0(t) + P_1(t) = 1$,

$$P_1(t) = 1 - e^{-\lambda t}. \tag{3.54}$$

3.3.2.2. *Repairable Component*

We now illustrate the development of a Markov model for a repairable component. Consider a component that exhibits a constant failure rate λ. When the component fails, it is repaired with a repair rate μ. Similar to the nonrepairable component, we define two mutually exclusive states for the repairable component: state s_0 represents a working state of the component, and state s_1 represents the nonworking state of the component. The state-transition equations of the component are

$$P_0(t + \Delta t) = [1 - \lambda\,\Delta t]\,P_0(t) + \mu\,\Delta t\,P_1(t) \tag{3.55}$$

$$P_1(t + \Delta t) = [1 - \mu\,\Delta t]\,P_1(t) + \lambda\,\Delta t\,P_0(t). \tag{3.56}$$

Rewriting Eqs. (3.55) and (3.56) as

$$\frac{\mathrm{d}P_0(t)}{\mathrm{d}t} = \dot{P}_0(t) = -\lambda P_0(t) + \mu P_1(t) \tag{3.57}$$

$$\frac{\mathrm{d}P_1(t)}{\mathrm{d}t} = \dot{P}_1(t) = -\mu P_1(t) + \lambda P_0(t). \tag{3.58}$$

Solutions of these equations can be obtained using Laplace transform and the initial conditions $P_0(0) = 1$. Thus,

$$sP_0(s) - P_0(0) = -\lambda P_0(s) + \mu P_1(s) \tag{3.59}$$

$$sP_1(s) = -\mu P_1(s) + \lambda P_0(s). \tag{3.60}$$

From Eq. (3.60) we obtain

$$P_1(s) = \frac{\lambda}{s + \mu} P_0(s).$$

Substituting into Eq. (3.59) to get $P_0(s)$

$$P_0(s) = \frac{s + \mu}{s(s + \lambda + \mu)}. \tag{3.61}$$

Using the partial-fraction method, we write Eq. (3.61) as

$$P_0(s) = \frac{\dfrac{\mu}{\lambda+\mu}}{s} + \frac{\dfrac{\lambda}{\lambda+\mu}}{(s+\lambda+\mu)} \tag{3.62}$$

and the inverse of Eq. (3.62) is

$$P_0(t) = \frac{\mu}{\lambda+\mu} + \frac{\lambda}{\lambda+\mu} e^{-(\lambda+\mu)t}. \tag{3.63}$$

Of course, $P_0(t)$ is the availability of the component at time t. The unavailability $\bar{A}(t)$ is

$$\bar{A}(t) = 1 - P_0(t) = P_1(t) = \frac{\lambda}{\lambda+\mu} - \frac{\lambda}{\lambda+\mu} e^{-(\lambda+\mu)t}. \tag{3.64}$$

When the number of the mutually exclusive states is large, the solution of state-transition equations by using the Laplace transform becomes difficult, if not impossible, to obtain. In such a case, the equations may be solved numerically for different values of time. The following example illustrates the numerical solution of the transition equations.

EXAMPLE 3.15

An electric circuit that provides constant current for direct current (DC) motors includes one diode that may be in any of the following states: (1) s_0 represents the diode operating properly; (2) s_1 represents the short failure mode of the diode—that is, the diode allows the current to return in the reverse direction; (3) s_2 represents the open failure mode of the diode—that is, the diode prevents the passage of current in either direction; and (4) s_3 represents the assembly failure mode of the diode—that is, when the diode is not properly assembled on the circuit board, it generates hot spots that result in not providing the current for the motor to function properly. Let the failure rates from state s_0 to state s_i be constant with parameters λ_i $(i = 1, 2, 3)$. The repair rate from any of the failure states to s_0 is constant with parameter μ. Transitions occur only between state s_0 and other states and vice versa. Graph the availability of the circuit against time for different values of failure and repair rates.

SOLUTION

Let $P_i(t)$ be the probability that the diode is in state i $(i = 0, 1, 2, 3)$ at time t. The state-transition equations are

$$\dot{P}_0(t) = -[\lambda_1 + \lambda_2 + \lambda_3]\, P_0(t) + \mu P_1(t) + \mu P_2(t) + \mu P_3(t) \tag{3.65}$$

$$\dot{P}_1(t) = -\mu P_1(t) + \lambda_1 P_0(t) \tag{3.66}$$

$$\dot{P}_2(t) = -\mu P_2(t) + \lambda_2 P_0(t) \tag{3.67}$$

$$\dot{P}_3(t) = -\mu P_3(t) + \lambda_3 P_0(t). \tag{3.68}$$

The initial conditions of the diode are $P_0(0) = 1$, $P_i(0) = 0$ ($i = 1, 2, 3$). The solution of the above equations can be obtained by using the matrix-geometric approach and the analytical perturbations method (Baruh and Altiok, 1991; Schendel, 1989) or by using the Runge-Kutta method for solving differential equations. We utilize the computer program given in Appendix H and graph the availability, $P_0(t)$, of the circuit due the diode failure as shown in Figure 3.4. We choose $\lambda_1/\lambda_2 = \lambda_2/\lambda_3 = 0.5$ and $\mu/\lambda_1 = 10$ with $\lambda_1 = 0.0001, 0.0005, 0.0008$, and 0.0018. The availability decreases rapidly as t increases, and then it reaches an asymptotic value for large values of t.

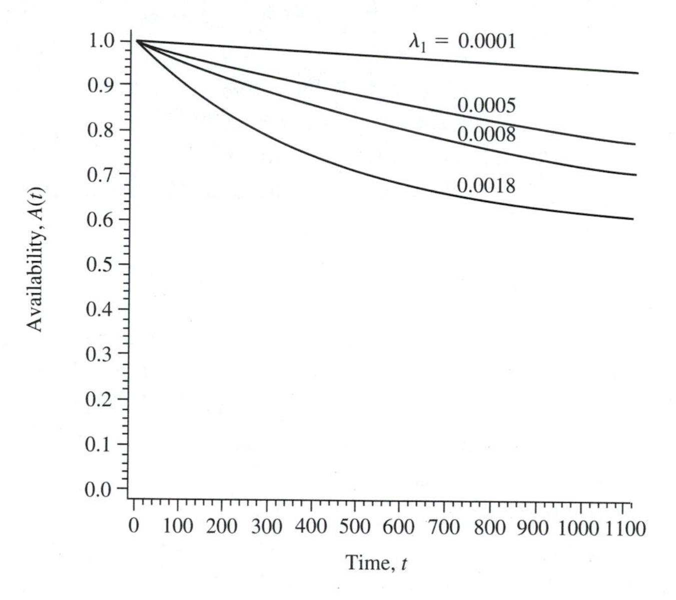

Figure 3.4. Effect of the Failure Rate on the Circuit's Availability

3.4. Availability

Availability is considered to be one of the most important reliability measures of maintained systems since it includes both reliability and maintainability. Indeed, the importance of availability has prompted manufacturers and users of critical systems to state the availability values in the systems' specifications. For example, manufacturers of mainframe computers that are used in large financial institutions and banks provide guaranteed availability values for their systems. In this section, we present different classifications of availability and methods for its estimation.

Availability can be classified either according to (1) the time interval considered or (2) the type of down time (repair and maintenance). The time-interval availability includes instantaneous (or point availability), average up time, and steady-state availabilities. The availability classification according to down time includes inherent, achieved, and operational availabilities (Lie, Hwang, and Tillman, 1977). Other classifications include mission-oriented availabilities.

3.4.1. INSTANTANEOUS (POINT) AVAILABILITY, $A(t)$

Instantaneous point availability is the probability that the system is operational at any random time t. The instantaneous availability can be estimated for a system whose states are characterized by an alternating renewal process by using Eq. (3.41):

$$A^*(s) = \frac{1 - w^*(s)}{s\,[1 - w^*(s)g^*(s)]},$$

where $w^*(s)$ and $g^*(s)$ are the Laplace transforms of the failure-time and repair-time distributions, respectively.

When the failure and repair rates are constant, the availability can be obtained using the state-transition equations as described in Section 3.3. For the case when either the failure rate or the repair rate is time-dependent, $A(t)$ can be estimated by using semi-Markov state-transition equations or by using an appropriate approximation method as given by Eq. (3.47).

3.4.2. AVERAGE UP-TIME AVAILABILITY, $A(T)$

In many applications it is important to specify availability requirements in terms of the proportion of time in specified intervals $(0, T)$ that the system is available for use. We refer to this availability requirement as the average up-time availability. It is expressed as

$$A(T) = \frac{1}{T}\int_0^T A(t)\,dt. \tag{3.69}$$

$A(T)$ can be estimated by obtaining an expression for $A(t)$ as a function of time, if possible, and substituting in Eq. (3.69), or by numerically solving the state-transition equations and summing the probabilities of the "up" states over the desired time interval T or by fitting a function to the probabilities of the "up" state and substituting this function in Eq. (3.69).

EXAMPLE 3.16 Estimate the average up-time availability of the circuit described in Example 3.15 during the interval 0 to 1,135 hours under the following conditions: $P_0(0) = 1$, $P_i(0) = 0$ for $i = 1, 2, 3$, $\lambda_1/\lambda_2 = \lambda_2/\lambda_3 = 0.5$, $\mu/\lambda_1 = 10$ and $\lambda_1 = 0.0018$ failures per hour.

SOLUTION We solve the state-transition Eqs. (3.65) through (3.68) using the Runge-Kutta method to obtain the probability of the "up" state, $P_0(t)$. A partial listing of $P_0(t)$ is shown in Table 3.3. The average up-time availability $A(T)$ is obtained by two methods:

Table 3.3. Point Availability $P_0(t)$

Time, t	$P_0(t)$
1.135	.998867
2.270	.997738
3.405	.996611
4.540	.995488
5.675	.994368
6.810	.993250
7.945	.992136
...	...
...	...
993.111	.625207
994.246	.625105
995.381	.625004
996.516	.624903
997.651	.624802
998.786	.624701
999.921	.624601
1001.056	.624501

- Adding $P_0(t)$ over the interval 0 to 1,135 hours and dividing by 1,135:

$$A(1135) = \frac{\sum_{i=1}^{1135} P_0(t_i)}{1135} = 0.6414196.$$

- Fitting a function of the form $P_0(t) = Ae^{Bt}$ to the data obtained from the solution of the state-transition equation, we obtain

$$A(t) = P_0(t) \cong 0.90661264\, e^{-0.0004t} \qquad (3.70)$$

and

$$A(1135) \cong \frac{1}{1135} \int_0^{1135} A(t)\, dt$$

$$A(1135) \cong 0.7287.$$

Obviously the average up-time availability obtained by the second method is more accurate than that obtained by the first method since the availability is integrated over a continuous time interval.

The average up-time availability may be the most satisfactory measure for systems whose usage is defined by a duty cycle such as a tracking radar system, which is called upon only when an object is detected and is expected to track the system continuously during a given time period (Lie, Hwang, and Tillman, 1977).

3.4.3. STEADY-STATE AVAILABILITY, $A(\infty)$

The steady-state availability is the availability of the system when the time interval considered is very large. It is given by

$$A(\infty) = \lim_{T \to \infty} A(T).$$

The steady-state availability can be easily obtained from the state-transition equations of the system by setting $\dot{P}_i(t) = 0$, $i = 0, 1, \ldots$

EXAMPLE 3.17

Determine the steady-state availability of the system given in Example 3.16.

SOLUTION

Since we are seeking $A(\infty)$, we set $\dot{P}_i(t) = 0$ and let $P_i(t) = P_i$, $i = 0, 1, 2, 3$ in Eqs. (3.65) through (3.68). This results in

$$-(\lambda_1 + \lambda_2 + \lambda_3)P_0 + \mu P_1 + \mu P_2 + \mu P_3 = 0 \tag{3.71}$$

$$-\mu P_1 + \lambda_1 P_0 = 0 \tag{3.72}$$

$$-\mu P_2 + \lambda_2 P_0 = 0 \tag{3.73}$$

$$-\mu P_3 + \lambda_3 P_0 = 0. \tag{3.74}$$

Using the condition $P_0 + P_1 + P_2 + P_3 = 1$ and solving Eqs. (3.71) through (3.74), we obtain

$$A(\infty) = P_0 = \frac{\mu}{\lambda_1 + \lambda_2 + \lambda_3 + \mu} = 0.5882.$$

The steady-state availability may be a satisfactory measure for systems that operate continuously, such as a detection radar system and an undersea communication cable.

3.4.4. INHERENT AVAILABILITY, A_i

Inherent availability includes only the corrective maintenance of the system (the time to repair or replace the failed components) and excludes ready time, preventive maintenance down time, logistics (supply) time, and waiting or administrative time. It is expressed as

$$A_i = \frac{MTBF}{MTBF + MTTR}. \tag{3.75}$$

The inherent availability is identical to the steady-state availability when the only repair time considered in the steady-state calculation is the corrective maintenance time.

3.4.5. ACHIEVED AVAILABILITY, A_a

Achieved availability, A_a, includes corrective and preventive maintenance down time. It is expressed as a function of the frequency of maintenance and the mean maintenance time as

$$A_a = \frac{MTBM}{MTBM + M}, \tag{3.76}$$

where *MTBM* is the mean time between maintenance and *M* is the mean maintenance down time resulting from both corrective and preventive maintenance actions (Lie, Hwang, and Tillman, 1977).

3.4.6. OPERATIONAL AVAILABILITY, A_o

Operational availability is a more appropriate measure of availability since the repair time includes many elements: the direct time of maintenance and repair and the indirect time, which includes ready time, logistics time, and waiting or administrative down time.

$$A_o = \frac{MTBM + \text{ready time}}{(MTBM + \text{ready time}) + MDT}, \tag{3.77}$$

where ready time = operational cycle − $(MTBM + MDT)$ and the mean delay time, MDT, equals M + delay time.

3.4.7. OTHER AVAILABILITIES

Other availability definitions include the *mission-availability*, $A_m(T_o, t_f)$, which is defined as

$$A_m(T_o, t_f) = \text{Probability of each individual failure that occurs in a mission of a total operating time } T_o \text{ is repaired in a time} \leq t_f. \tag{3.78}$$

This availability is used for specifying the availabilities of military equipment. Clearly, the repair time in Eq. (3.78) includes all direct and indirect elements.

We follow Birolini (1985) and consider that the end of the mission falls within an operating period. The mission availability is obtained by summing over all the possibilities of having n failures ($n = 1, 2, 3, \ldots$) during the total operating time T_o. Each failure can be repaired in a time shorter than (or equal to) t_f. In other words,

$$A_m(T_o, t_f) = 1 - F(T_o) + \sum_{n=1}^{\infty} [F_n(T_o) - F_{n+1}(T_o)]\,(G(t_f))^n, \tag{3.79}$$

where

$F(t)$ = the distribution function of the failure time,

$F_n(T_o) - F_{n+1}(T_o)$ = the probability of n failures in T_o (see Chapter 7 for further details),

$(G(t_f))^n$ = the probability that the time of each of the n repairs is shorter than t_f, and

$G(t)$ = the distribution function of the repair time.

Assume that the system has a constant failure rate λ—that is, $f(t) = \lambda e^{-\lambda t}$. Then

$$A_m(T_o, t_f) = e^{-\lambda T_o} + \sum_{n=1}^{\infty} \frac{(\lambda T_o)^n}{n!} e^{-\lambda T_o} (G(t_f))^n = e^{-\lambda T_o (1 - G(t_f))}. \tag{3.80}$$

EXAMPLE 3.18

In a musical play the heroine is lowered to the stage on a 37,000 lb. set of the inside of a mansion. The mansion set, along with the other scenery, is powered by an integrated hydraulic motor pump designed to orchestrate the operations of the stage reliably until the final act of the first part of the play.

A winch system consisting of steel cables controls the movements of the set. The cables are connected to a hydraulic brake, which is digitally regulated by proportional control valves. The hydraulic system is powered by an integrated motor pump that generates 30 horsepower and is capable of flow rates of up to 33 gallons per minute and 1,000 pounds per square inch (psi). To provide hydraulic power during the musical, the pump operates at flow rates of up to 28 gallons per minute and pressures of up to 1,450 psi. The hydraulic system is regulated by a microprocessor-based controller (O'Connor, 1995). The heat generated by the electric motor and hydraulic pump raises the temperature inside the room where the equipment is installed and, in turn, affects the life of the controller.

The failure rate of the controller is constant with $\lambda = 0.006$ failures per hour. The repair follows a gamma distribution with a p.d.f. of

$$g(t) = \frac{t^{\beta - 1}}{\alpha^{\beta}\Gamma(\beta)} \exp\left(\frac{-t}{\alpha}\right), \tag{3.81}$$

where $\beta = 3$, and $\alpha = 1{,}000$.

The mansion and the other hydraulic equipment are used in the play for 60 minutes. Determine the mission availability of the system if $t_f = 2$ minutes.

SOLUTION

$$T_o = 1 \text{ hour}$$

$$t_f = 0.0333 \text{ hours}$$

$$1 - G(t_f) = \exp\left(\frac{-t_f}{\alpha}\right) \sum_{j=0}^{\beta-1} \left(\frac{t_f}{\alpha}\right)^j \frac{1}{\Gamma(j+1)}$$

or

$$1 - G(t_f) = 0.99996667\,[1 + 0.00000333 + 0.0]$$

$$1 - G(t_f) = 0.999970.$$

Substituting $[1 - G(t_f)]$ into Eq. (3.80), we obtain

$$A_m(T_o, t_f) = e^{-0.006 \times 0.999970} = 0.994.$$

The *work-mission availability*, $A_{wm}(T_o, t_d)$, is a variant of mission availability. It is defined as

$$A_{wm}(T_o, t_d) = \text{Probability of the sum of all repair times for failures occurring in a mission with total operating time } T_o \text{ is} \leq t_d. \quad (3.82)$$

Using Eq. (3.79) we rewrite Eq. (3.82) as

$$A_{wm}(T_o, t_d) = 1 - F(T_o) + \sum_{n=1}^{\infty} [F_n(T_o) - F_{n+1}(T_o)]\, G_n(t_d), \quad (3.83)$$

where $G_n(t_d)$ is the probability that the sum of n repair times, which are distributed according to $G(t)$, is shorter than t_d (Birolini, 1985).

EXAMPLE 3.19

Membrane keyboards are widely used in the personal computer industry. A membrane keyswitch has a rubber dome-shaped actuator at the bottom of the keyswitch plunger as shown in Figure 3.5. When the key is depressed, the rubber dome compresses, and a small rubber nib or bump inside the dome is pushed down onto the membranes, bringing them together and closing the contacts (Johns, 1995).

The membrane consists of metallic pads that are screen-printed onto two membrane sheets. A third spacer membrane with holes is placed between the two

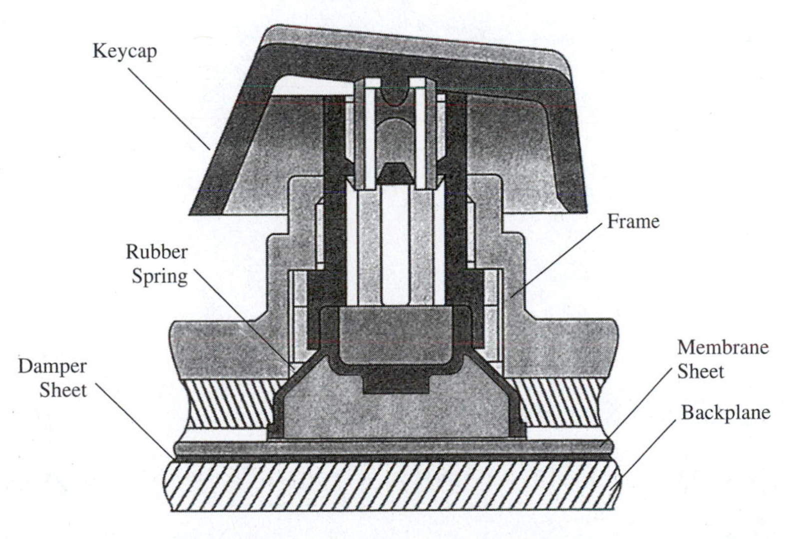

Source: Reprinted with permission from "Membrane Versus Mechanical Keyboards," Garden City: Electronic Products, June 1995, Don Johns.

Figure 3.5. A Sketch of a Key in a Membrane Keyboard

printed membrane sheets. When the keyswitch is depressed, the top and bottom printed membranes are squeezed together, allowing the pad on one to touch its corresponding pad on the other through the hole on the third sheet, thus forming the contact. The expected life of a membrane keyboard is about 25 million keystrokes (approximately three years). The failure rate of the keyboard is constant with $\lambda = 0.000114$ failures per hour. The repair rate is also constant with $\mu = 0.002$. Assume that the keyboard is attached to a computer that is used for a 120-hour task. What is the work-mission availability of the keyboard if the time required to repair all failures does not exceed one hour?

SOLUTION

$t_d = 1$ hour,

$T_o = 120$ hours,

$\lambda = 0.000114$ failures per hour, and

$\mu = 0.002$ repairs per hour.

In addition,

$$1 - F(T_o) = e^{-\lambda T_o} = e^{-0.000114 \times 120} = 0.986413.$$

Let x_1 and x_2 be the time to repair the first and second failures, respectively. Then

$$G_1(t_d) = (1 - e^{-\mu t_d}) = 0.001998$$

$$G_2(x_1 + x_2 \leq t_d) = \int_0^{t_d} \int_0^{t_d - x_1} \mu e^{-\mu x_1} \mu e^{-\mu x_2}\, dx_2\, dx_1$$

$$G_2(t_d) = \int_0^{t_d} [\mu e^{-\mu x_1} - \mu e^{-\mu t_d}]\, dx_1$$

$$G_2(t_d) = 1 - e^{-\mu t_d} - \mu t_d\, e^{-\mu t_d}$$

$$G_2(1) = 2 \times 10^{-6}.$$

Substituting in Eq. (3.83), we obtain

$$A_{wm}(120, 1) = 0.986413 + 0.000114 \times 120\, e^{-0.000114 \times 120} \times 0.001998$$

$$+ \frac{1}{2}(0.000114 \times 120)^2\, e^{-0.000114 \times 120} \times 2 \times 10^{-6}$$

$$A_{wm}(120, 1) = 0.98643996.$$

3.5. Dependent Failures

In Chapters 2 and 3, we estimate the performance measures of the system reliability under the assumption that the failure-time distributions of the components are identical and independent. In other words, we consider only the situations where the failure of a component has no effect on the failure rate of other components in the system. This assumption, though valid in many situations, needs to be relaxed when the failure of a component or a group of components may change the failure rate of the remaining components. For example, consider a twin engine airplane. The engines have identical failure-time distributions, and they operate in parallel—that is, both engines share the load. When either one of the engines fails, the other engine will provide the additional power requirement for safe operation of the airplane. This, in turn, causes the failure rate of the surviving engine to increase, and the reliability analysis of the system should reflect such change.

Similarly, the advances in computer technology have resulted in an increase in the number of components placed on a computer chip, which causes significant heat dissipation from the chip to the adjacent components. Insufficient cooling of the computer board results in an elevated operating temperature of the components, which, in turn, increases their failure rates.

Reliability analysis of systems whose components experience dependent failures can be performed using the Markov model. The model performs well when the number of state-transition equations is small and when the failure-time and repair-time distributions are exponential. When these conditions are not satisfied, alternative approaches, such as the *joint density function* and the *compound events*, can be used. Although both approaches are applicable for situations when the failure rates are time dependent, they rapidly break down as the joint density function of the failure times is too complex to solve analytically. This section briefly presents approaches for reliability analysis of systems with dependent failures.

3.5.1. MARKOV MODEL FOR DEPENDENT FAILURES

The Markov model for dependent failures is similar to the models discussed in Section 3.3.2 with the exception that the failure and repair rates are dependent on the state of the system (or component). The following example illustrates the development of such a Markov model.

EXAMPLE 3.20

Time-dependent dielectric breakdown (TDDB) of gate oxides of MOS transistors and of other thin oxide structures has been, and continues to be, one of the principal mechanisms of failure of MOS integrated circuits (Hawkins and Soden, 1986). Extensive studies of dielectric breakdown of MOS device structures show distribution of breakdown voltage and the effects of device processing, voltage, and temperature on the rate of failure of gate oxides, which are subject to an electric field (Crook, 1979; Domangue, Rivera, and Shepard, 1984; Dugan, 1986; Edwards, 1982; Swartz, 1986). These studies show that the use of higher voltages is far more effective than the use of higher temperatures in screening to eliminate devices with defective oxide sites that would be susceptible to time-dependent dielectric breakdown. This prompts the designers of integrated circuits (ICs) to improve the reliability of the circuits by using redundant devices.

Utilizing this information, a designer of an integrated circuit connects three oxide structures, such as transistors, in parallel in order to improve the reliability of the device. The device functions properly when no more than one transistor fails. Failure times of the transistors are exponentially distributed with the following parameters:

$\lambda_0 = 9 \times 10^{-5}$ failures per hour when all transistors are working properly,

$\lambda_1 = 16 \times 10^{-5}$ failures per hour when one unit fails, and

$\lambda_2 = 21 \times 10^{-5}$ failures per hour when two units fail.

Graph the reliability of the device over the period of 0 to 9,500 hours. Also, graph the reliability when $\lambda_2 = 2\lambda_1 = 2\lambda_0$.

SOLUTION Let $P_{si}(t)$ be the probability that the three transistors are in state i ($i = 0, 1, 2, \ldots, 7$), where

$$s0 = x_1 x_2 x_3 \text{ (no failures of the transistors)},$$

$$s1 = \bar{x}_1 x_2 x_3 \text{ (one unit fails)},$$

$$s2 = x_1 \bar{x}_2 x_3 \text{ (one unit fails)},$$

$$s3 = x_1 x_2 \bar{x}_3 \text{ (one unit fails)},$$

$$s4 = \bar{x}_1 \bar{x}_2 x_3 \text{ (two units fail)},$$

$$s5 = \bar{x}_1 x_2 \bar{x}_3 \text{ (two units fail)},$$

$$s6 = x_1 \bar{x}_2 \bar{x}_3 \text{ (two units fail), and}$$

$$s7 = \bar{x}_1 \bar{x}_2 \bar{x}_3 \text{ (all units fail)}.$$

The state-transition equations are

$$\dot{P}_{s0}(t) = (1 - 3\lambda_0)P_{s0}(t)$$

$$\dot{P}_{s1}(t) = (1 - 2\lambda_1)P_{s1}(t) + \lambda_0 P_{s0}(t)$$

$$\dot{P}_{s2}(t) = (1 - 2\lambda_1)P_{s2}(t) + \lambda_0 P_{s0}(t)$$

$$\dot{P}_{s3}(t) = (1 - 2\lambda_1)P_{s3}(t) + \lambda_0 P_{s0}(t)$$

$$\dot{P}_{s4}(t) = (1 - \lambda_2)P_{s4}(t) + \lambda_1 P_{s1}(t) + \lambda_1 P_{s2}(t)$$

$$\dot{P}_{s5}(t) = (1 - \lambda_2)P_{s5}(t) + \lambda_1 P_{s1}(t) + \lambda_1 P_{s3}(t)$$

$$\dot{P}_{s6}(t) = (1 - \lambda_2)P_{s6}(t) + \lambda_1 P_{s2}(t) + \lambda_1 P_{s3}(t)$$

$$\dot{P}_{s7}(t) = \lambda_2 P_{s4}(t) + \lambda_2 P_{s5}(t) + \lambda_2 P_{s6}(t).$$

Solutions of the above equations under the conditions $P_{s0}(0) = 1$ and $P_{si}(0) = 0$ for $i = 1, 2, \ldots, 7$ can be obtained numerically using Appendix H. The reliability of the device over the time interval of 0 to 9,500 hours, $R(t) = \Sigma_{i=0}^{6} P_{si}(t)$, for dependent failures is shown in Figure 3.6. The reliability of the device decreases rapidly when λ_2 is significantly greater than λ_1 and λ_1 is significantly greater than λ_0.

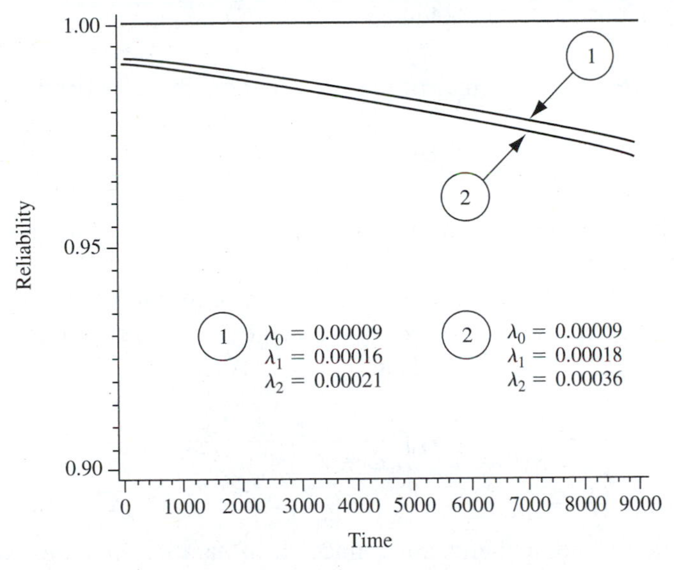

Figure 3.6. Reliability of the Device for Dependent Failure Rates

3.5.2. THE JOINT DENSITY FUNCTION APPROACH

This approach requires that the probability density function of the failure-time distribution of each component in the system as well as the joint density functions (j.d.f.) of all components be known. Development of the j.d.f. is not an easy task since field or test failure data are usually collected under the assumption that the failure times of the components are independent. Careful testing of components under all possible configurations and failure conditions leads to the development of a j.d.f. that reflects the actual dependence of components. Once the configuration of the components in the system is known—that is, series, parallel,

k-out-of-n, and so on—we proceed by developing the j.d.f. and then follow the reliability analysis using an appropriate method as described in Chapters 2 and 3.

We follow Shooman (1968) by considering two identical components having constant failure rates of λ_s when they operate singularly and λ_b when both operate simultaneously. Let τ be the time of the first failure and $g_1(t)$ be the density function for the first failure $(0 < \tau < t)$. The time of the second failure is t and its dependent density function, $g_2(t/\tau)$, holds for $\tau < t$. In other words,

$$g_1(\tau) = 2\lambda_b\, e^{-2\lambda_b \tau} \qquad 0 < \tau < t$$

$$g_2(t/\tau) = \begin{cases} \lambda_s e^{-\lambda_s(t-\tau)} & 0 < \tau < t \\ 0 & \tau > t \end{cases}.$$

The density function, $g_1(\tau)$, is obtained as

$$g_1(\tau) = P\,[\text{if either of the two components fails first}]$$

$$g_1(\tau) = 2\lambda_b e^{-2\lambda_b \tau}.$$

The j.d.f. of the components can only be estimated once the configuration of the components in the system is known. For example, if the two components are connected in series, the system fails when either of the components fails and the j.d.f. is $\phi(\tau, t) = g_1(t)$. The system failure is governed by the marginal density function, $f(t) = \int_0^\infty \phi(\tau, \zeta)\, d\zeta = 2\lambda_b\, e^{-2\lambda_b t}$. Thus,

$$R(t) = 1 - \int_0^t \phi(\tau, \zeta)\, d\zeta = e^{-2\lambda_b t}.$$

Similarly, if the two components are connected in parallel, then the j.d.f., $\phi(\tau, t)$, is defined as

$$\phi(\tau, t) = g_1(\tau) g_2(t\,/\,\tau) \qquad 0 < \tau < t. \tag{3.84}$$

The marginal density function, $f(t)$, is

$$f(t) = \int_0^t \phi(\tau, t)\, d\tau$$

$$= \int_0^t (2\lambda_b e^{-2\lambda_b \tau}) \lambda_s e^{-\lambda_s(t-\tau)}\, d\tau$$

$$= 2\lambda_b \lambda_s e^{-\lambda_s t} \int_0^t e^{-(2\lambda_b - \lambda_s)\tau}\, d\tau$$

or

$$f(t) = \frac{2\lambda_b \lambda_s}{2\lambda_b - \lambda_s}\left(e^{-\lambda_s t} - e^{-2\lambda_b t}\right)$$

and

$$R(t) = 1 - \int_0^t f(\zeta)\, d\zeta$$

$$R(t) = \frac{2\lambda_b}{2\lambda_b - \lambda_s} e^{-\lambda_s t} - \frac{\lambda_s}{2\lambda_b - \lambda_s} e^{-2\lambda_b t}.$$

The MTTF is obtained as

$$MTTF = \int_0^\infty R(t)\, dt = \frac{4\lambda_b^2 - 2\lambda_s^2}{2\lambda_b \lambda_s \left[2\lambda_b - \lambda_s\right]}. \tag{3.85}$$

EXAMPLE 3.21

The main function of the generator regulator in a car is to set the battery voltage to a nominal value and to keep this value within a tight tolerance over the range of the operating conditions of the car (Fostner, 1994). The regulator monitors the generator (alternator) voltage and controls the alternator's current. Because of the critical role of the regulator, some manufacturers install two regulators in parallel. When the two regulators are functioning properly, their failure rates are identical and constant with parameter $\lambda_b = 6 \times 10^{-6}$ failures per hour. When either one of the regulators fails, the remaining unit operates at a lower temperature since there is no heat dissipation from the other unit. Consequently the failure rate of the remaining unit decreases to $\lambda_s = 3 \times 10^{-6}$ failures per hour. Determine the reliability of the regulators at $t = 10{,}000$ hours. What is the MTTF?

SOLUTION

Let τ be the time to the first failure and $g_1(\tau)$ be the p.d.f. for the first failure $(0 < \tau < t)$. Thus,

$$g_1(\tau) = 2\lambda_b e^{-2\lambda_b \tau} = 12 \times 10^{-6}\, e^{-12\times 10^{-6}\tau} \qquad 0 < \tau < t$$

and

$$g_2(t/\tau) = \begin{cases} 3 \times 10^{-6}\, e^{-3\times 10^{-6}(t-\tau)} & 0 < \tau < t \\ 0 & \tau > t \end{cases}.$$

Since the two regulators are required for the battery protection, the j.d.f., $\phi(\tau, t)$, is

$$\phi(\tau, t) = g_1(\tau) g_2(t/\tau) \qquad 0 < \tau < t.$$

The reliability of the regulators is governed by the marginal density function, $f(t)$, which is obtained as

$$f(t) = \int_0^\infty \phi(\tau, t)\, d\tau$$

$$= \int_0^t \phi(\tau, \zeta)\, d\zeta$$

and

$$R(t) = 1 - \int_0^t f(\zeta)\, d\zeta$$

$$R(10{,}000) = \frac{12 \times 10^{-6}}{12 \times 10^{-6} - 3 \times 10^{-6}} e^{-3\times 10^{-6} \times 10^4}$$

$$- \frac{3 \times 10^{-6}}{12 \times 10^{-6} - 3 \times 10^{-6}} e^{-12\times 10^{-6} \times 10^4}$$

$$R(10{,}000) = 0.9983$$

and

$$MTTF = 3.888 \times 10^5 \text{ hours.}$$

Calculating the reliability for a system with a nonconstant failure rate becomes quite complex even when we deal with the simple linearly increasing failure rate model. Let us consider a simple system with two identical components connected in parallel. The failure rates of the components, when both are operating, are constant with parameter λ. When either of the components fails, the failure rate of the surviving component becomes $h(t) = kt$. The conditional densities are

$$g_1(\tau) = 2\lambda e^{-2\lambda\tau} \qquad 0 < \tau < t \tag{3.86}$$

$$g_2(t/\tau) = \begin{cases} [\lambda + k(t-\tau)]\, e^{-[\lambda + k(t-\tau)^2/2]} & 0 < \tau < t \\ 0 & \tau > 0 \end{cases}. \tag{3.87}$$

The joint density function is

$$\phi(\tau, t) = (2\lambda e^{-2\lambda\tau})\,[\lambda + k(t-\tau)]\, e^{-[\lambda + k(t-\tau)^2/2]}. \tag{3.88}$$

The marginal density function that governs the reliability of the system is obtained as

$$f(t) = \int_0^t \phi(\tau, t)\, d\tau$$

$$f(t) = 2\lambda \int_0^t e^{-\lambda(1+2\tau)} e^{-k(t-\tau)^2/2} [\lambda + k(t-\tau)]\, d\tau. \tag{3.89}$$

The reliability $R(t)$ is

$$R(t) = 1 - \int_0^t f(\zeta)\, d\zeta. \tag{3.90}$$

Approximate results of Eq. (3.89) can be obtained by expanding the exponentials in truncated series or by using numerical integration. Although the formulation of the joint density function is straightforward, the solution of the marginal density function over the time period of interest is computationally difficult.

Before closing this section, we briefly mention the *compound-events* approach. This approach is based on computing the state probabilities in terms of the system failure rates. It is similar to the Markov model approach with the exception that the failure rates are nonconstants. Although it shares the straightforwardness in model formulation with both the Markov model and the joint density function approach, it also shares the difficulty of obtaining the results with the joint density approach.

3.6. Redundancy and Standby

In Section 3.5 we presented reliability analysis approaches for systems with dependent failures. In such systems, when one of the components connected in parallel fails, the failure rates of the surviving components are affected. There is another type of failure dependency that arises when a component fails and a standby component replaces the failed one without affecting the failure rate of the standby component. In this section, we evaluate the reliability and availability of different standby and redundant systems.

Reliability of a system (or a component) can be improved by using redundant or standby systems (or components). The use of the factor of safety in engineering designs provides, in effect, some level of redundancy. The higher the factor of safety, the higher the level of redundancy. An example of this concept is the case of the supporting cables of suspension bridges. The cable contains thousands of wires arranged in a specific pattern. The number of wires required to carry the static load of the bridge, the wind effect, and the maximum applied dynamic load is significantly less than what the final design of the cable contains. If, for example, a factor of safety of two is used, the number of wires in the final design will be twice the amount as the number required to carry the maximum design load. This results in an implied redundancy in the system. Indeed, the reliability of a cable that contains n wires can be estimated by using the same procedure presented in Chapter 2 for k-out-of-n systems.

Other engineering designs include explicit component redundancies such as the number of components connected in parallel or the number of tires of a large transporter. In these cases the failure of one or more components or a tire will not necessarily result in the system failure. Redundancy can also be achieved by requiring *system redundancy*—that is, having one or more systems capable of performing the same function such as the brake system of a car where two redundant brake systems are always in operation.

We classify redundancy as *active* or *inactive*. If the redundant systems are continuously energized and are sharing a portion of the load, there is *active redundancy*. If the redundant systems do not perform any function unless the primary system fails, there is a *standby redundancy*.

We further classify the standby redundancy according to failure characteristics as follows:

- *Hot standby:* Standby components have the same failure rate as the primary component. Since the failure rate of one component is not affected by the other components, the hot standby redundancy consists of statistically independent components (Henley and Kumamoto, 1981).
- *Cold standby:* Standby components do not fail when they are in standby. The failure of the primary component results in the standby component being a primary component and in its failure rate becoming nonzero.
- *Warm standby:* A standby component has a smaller failure rate than the primary component but is greater than zero.

If the primary component has a failure rate λ, a hot standby component experiences a failure rate $\lambda_{hot} = \lambda$; a cold standby component has a failure rate $\lambda_{cold} = 0$; and a warm standby experiences a failure rate $\lambda_{warm} < \lambda$.

We can also classify redundant and standby systems as repairable or nonrepairable. Examples of nonrepairable systems include satellites and devices of an integrated circuit. Repairable standby systems include electric power generators, automotive brake systems, and airplane jet engines.

In repairable standby systems, when the primary unit fails, it undergoes repair and the standby unit assumes the functions of the primary unit. When the primary unit is repaired it assumes the position of the standby unit. The units alternate positions as failures and repairs occur.

Far-reaching decisions on the use of standby redundancy to assure product reliability are typically made in early phases of a project, well before design details required for the usual reliability predictions are available. Three major considerations are (1) number of standbys to be provided, (2) efficacy of the activation process, and (3) status of the standby when not used (Sears, 1990). In the following sections, we present methods for reliability and availability estimations of different redundant and standby systems.

3.6.1. NONREPAIRABLE SIMPLE STANDBY SYSTEMS

The simplest nonrepairable standby system is a two-unit system that functions successfully when the primary unit (Unit 1) does not fail, or if the primary unit fails during operating time t and the standby unit (Unit 2) assumes the function of the primary unit. The reliability of the system is the sum of the probability that Unit 1 does not fail until time t and the probability that Unit 1 fails at some time τ, $0 < \tau < t$, and the standby unit functions successfully from τ to time t. In other words,

$$R_{sb}(t) = R_1(t) + \int_{\tau=0}^{t} f_1(\tau)\, R_2(t-\tau)\, dt, \tag{3.91}$$

where

$R_{sb}(t)$ = the reliability of the standby system at t,

$R_1(t), R_2(t)$ = the reliabilities of the primary Unit 1 and the standby Unit 2 at time t, and

$f_1(t)$ = the p.d.f. of the failure time distribution of the first unit.

Assume that the failure rates of the primary and the standby units are constant with parameters λ_1 and λ_2, respectively. Then

$$R_{sb}(t) = e^{-\lambda_1 t} + \int_{\tau=0}^{t} \lambda_1 e^{-\lambda_1 \tau} e^{-\lambda_2 (t-\tau)}\, d\tau$$

$$= e^{-\lambda_1 t} + \lambda_1 e^{-\lambda_2 t} \int_{\tau=0}^{t} e^{-(\lambda_1 - \lambda_2)\tau}\, d\tau$$

or

$$R_{sb}(t) = e^{-\lambda_1 t} + \frac{\lambda_1 e^{-\lambda_2 t}}{\lambda_1 - \lambda_2}(1 - e^{-(\lambda_1 - \lambda_2)t}). \tag{3.92}$$

If $\lambda_1 = \lambda_2 = \lambda$, we use L'Hospital's rule and differentiate the second term of Eq. (3.92) with respect to λ_2 to obtain

$$R_{sb}(t) = (1 + \lambda t)e^{-\lambda t}. \tag{3.93}$$

The MTTF of the two-unit standby unit is

$$MTTF = \int_0^\infty R_{sb}(t)\,dt = \frac{1}{\lambda} + \frac{\lambda}{\lambda^2} = \frac{2}{\lambda}. \tag{3.94}$$

3.6.2. NONREPAIRABLE MULTIUNIT STANDBY SYSTEMS

We extend the standby system presented in Section 3.6.1 by allowing $(n - 1)$ units in standby for the case where the failure rates are constant with parameter λ. When the primary unit fails, one of the $(n - 1)$ standby units assumes the functions of the primary unit. When the second unit fails, one of the remaining $(n - 2)$ units assumes the functions of the system. The replacements are repeated until the failure of the last unit. The reliability of the multiunit standby system is given by

$$R_{sb}(t) = e^{-\lambda t}\left[1 + \lambda t + \frac{(\lambda t)^2}{2!} + \ldots + \frac{(\lambda t)^{n-1}}{(n-1)!}\right]. \tag{3.95}$$

The MTTF is

$$MTTF = \int_0^\infty R_{sb}(t)\,dt = \frac{n}{\lambda}. \tag{3.96}$$

From Eq. (3.95) it is evident that the reliability of the system increases as the number of standby units increases. However, the rate of improvement of the system reliability decreases exponentially as the number of standby units increases. Hence a decision regarding the number of standby units should consider the economics of adding standby units and the required reliability level of the system.

EXAMPLE 3.22

A thermocouple consists basically of two dissimilar metals, such as iron and constantan wires, joined to produce a thermal electromotive force when the junctions are at different temperatures. The measuring, or hot, junction is inserted into the medium where the temperature is to be measured. The reference, or cold, junction is the open end that is normally connected to the measuring instrument terminals. The electromagnetic force of a thermocouple increases as the difference in junction temperatures increases. Therefore, a sensitive instrument, capable of measuring electromagnetic force, can be calibrated and used to read temperature directly.

In order to measure the temperature around a retort, any number of thermocouples may be used in parallel connections as shown in Figure 3.7. All thermocouples must be of the same type and must be connected by the proper wires.

A producer of canned food uses thermocouples arranged in either a parallel or a standby configuration to ensure that the temperature of a retort is within an acceptable range (a lower temperature may result in a high microbial count while a higher temperature may result in a loss of food nutrition).

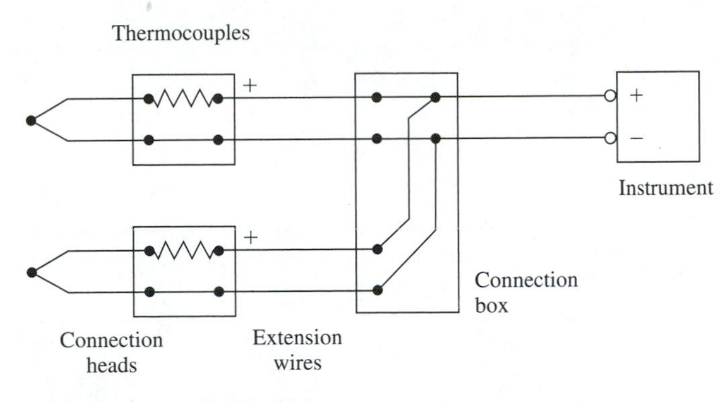

Figure 3.7. Thermocouples in Parallel

The thermocouples are identical and each has a constant failure rate $\lambda = 0.5 \times 10^{-6}$ failures per hour. Graph the reliability $R(t)$ for the parallel and the standby configuration. Determine the number of thermocouples needed in each configuration that ensures a system reliability of 0.999866 or higher at $t = 10^5$ hours.

SOLUTION

The reliability of the system for the parallel configuration is given by Eq. (3.97):

$$R_{\text{parallel}}(t) = 1 - (1 - e^{-\lambda t})^n. \tag{3.97}$$

The reliability of the multiunit standby system is given by Eq. (3.95) as

$$R_{sb}(t) = e^{-\lambda t}\left[1 + \lambda t + \frac{(\lambda t)^2}{2!} + \ldots + \frac{(\lambda t)^n}{n!}\right]. \tag{3.98}$$

Figures 3.8 and 3.9 show that the reliability of the parallel system is slightly lower than that of the standby system when $n \leq 3$ units. Indeed, the standby and the parallel systems require 2 and 3 units, respectively, to achieve a reliability of 0.999866 at 10^5 hours. The parallel system shows higher reliability values than the standby systems when $n > 3$.

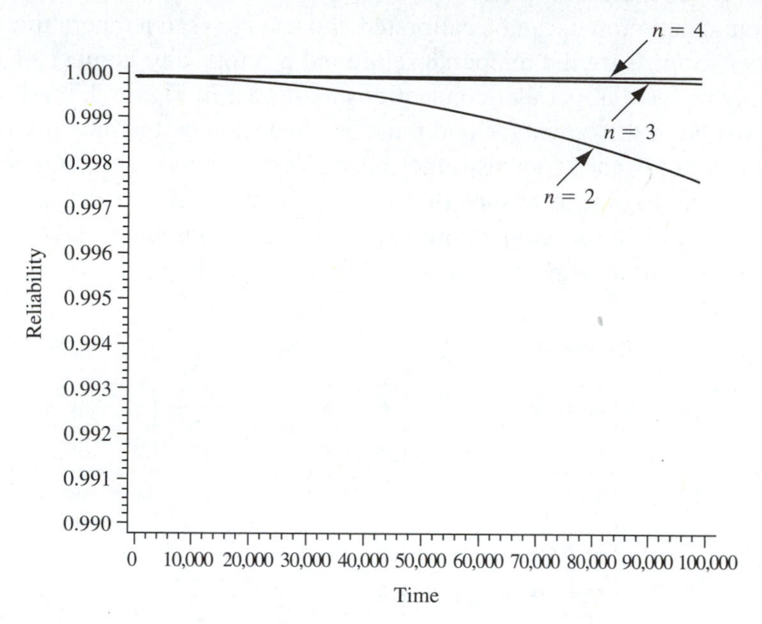

Figure 3.8. Reliability of the Parallel System

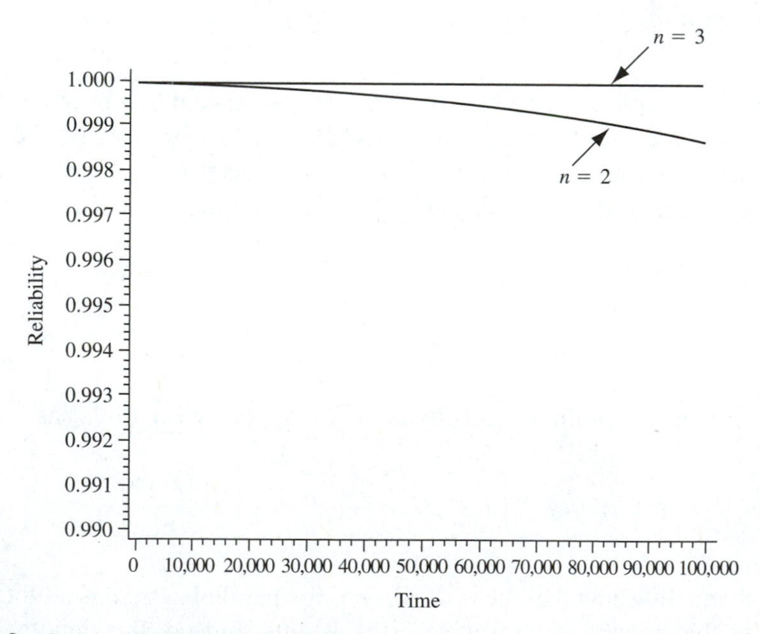

Figure 3.9. Reliability of the Standby System

3.6.3. REPAIRABLE STANDBY SYSTEMS

When the primary unit of a two-unit standby system fails, the standby unit assumes its functions while the primary unit undergoes repair. Upon completion of repairs, the primary unit returns to the system as a standby unit. The two units alternate positions as failures, and repairs occur. Examples of such repairable standby systems include two mainframe computers that share the same software and are connected in parallel, electric power generators, pumps in a chemical plant, and scrubbers in coal mining. In all these examples, the repair rate of the failed unit has a major effect on the instantaneous availability of the system as illustrated below.

Electrical-discharge machining (EDM) is a method of removing metal by a series of rapidly recurring electrical discharges between an electrode (the cutting tool) and the workpiece in the presence of a liquid (usually hydrocarbon dielectric). Minute particles of metal or *chips* are removed by melting and vaporization, and are flushed from the gap between the tool and the workpiece (Dallas, 1976).

EDM usually requires a liquid to provide a path for the discharge of electric current to remove metal particles produced from the gap and to cool the tool and workpiece. The liquid is circulated through the system by two hydraulic pumps connected in parallel. When either of the pumps fails, it is repaired while the surviving pump provides the necessary functions. Pumps A and B are shown in Figure 3.10 as well as the basic components of the electrical discharge machine.

The pumps are identical and their failure rates are constant. The repair rate is also constant with parameter μ. We are interested in estimating the instantaneous availability of the two-pump system under three conditions of standby: hot, warm, and cold.

There are five possible states for the two pumps as shown in Figure 3.11. They are

- *s1* Pump B is in operation, pump A is in standby, and neither is experiencing failure;
- *s2* Pump A is in operation, pump B is in standby, and neither is experiencing failure;
- *s3* Transition from state s_1 when B fails, A assumes B's functions, and B undergoes repair;
- *s4* Transition from state s_2 when A fails, B assumes A's functions, and A undergoes repair;
- *s5* Transition from either state s_3 or s_4 when the two pumps are under repair.

We follow the same steps presented in Section 3.3.2 to develop the state-transition equations. Let $P_{si}(t)$ be the probability that the two-pump system is in state *si*

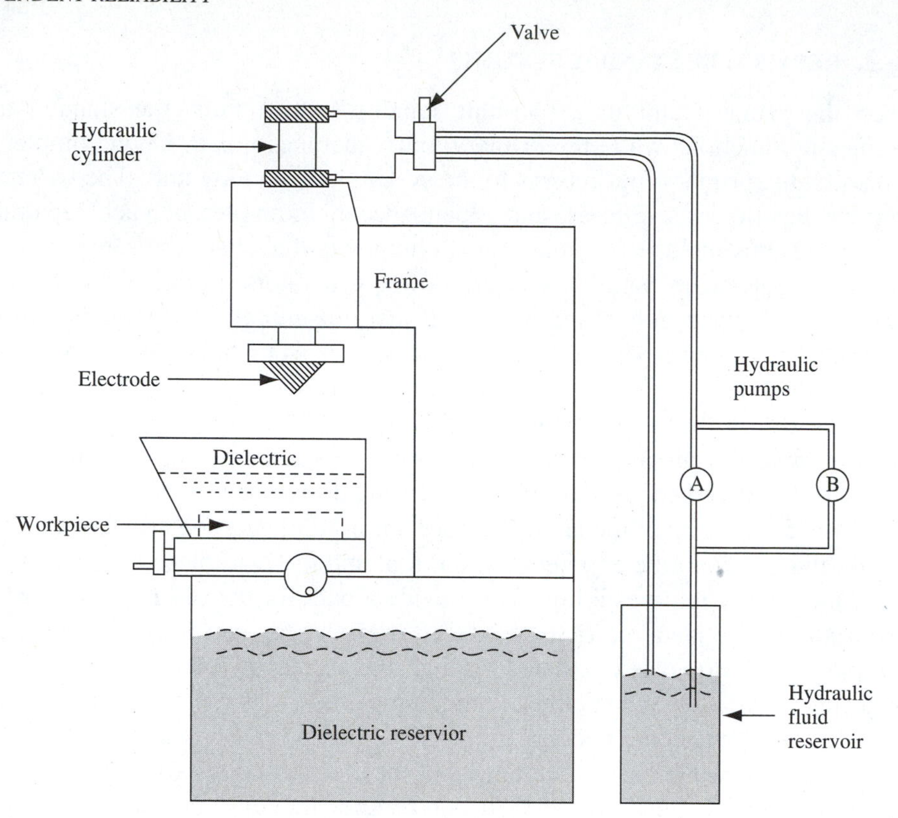

Figure 3.10. The Basic Components of an EDM

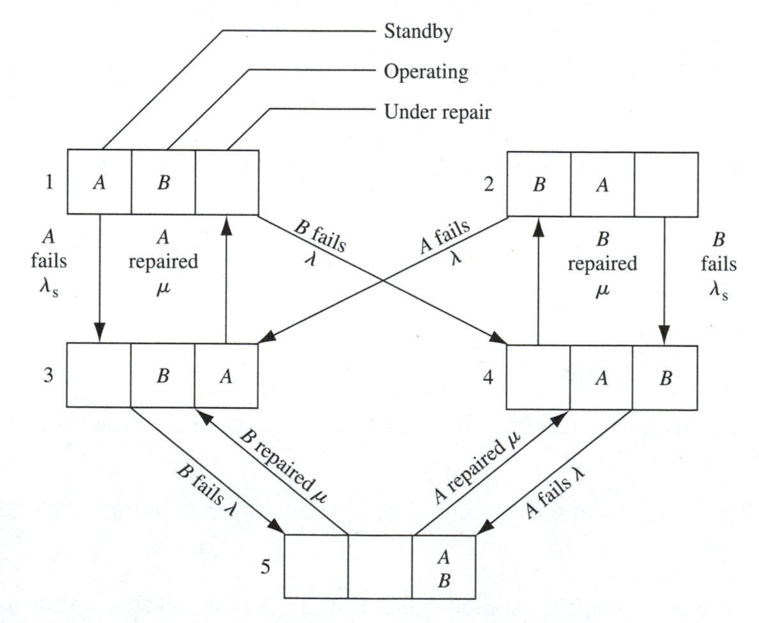

Figure 3.11. State-transition Diagram of the Two-Pump System

($i = 1, 2, \ldots$). Assume that the failure rate of the operating pump is λ and that of the standby pump is λ_s where

$$\lambda_s = \lambda_h = \lambda \text{ for hot standby,}$$

$$\lambda_s = \lambda_w \ (0 < \lambda_w < \lambda) \text{ for warm standby, and}$$

$$\lambda_s = \lambda_c = 0 \text{ for cold standby.}$$

The state-transition equations are

$$\dot{P}_{s1}(t) = -(\lambda + \lambda_s)P_{s1}(t) + \mu P_{s3}(t) \tag{3.99}$$

$$\dot{P}_{s2}(t) = -(\lambda + \lambda_s)P_{s2}(t) + \mu P_{s4}(t) \tag{3.100}$$

$$\dot{P}_{s3}(t) = -(\lambda + \mu)P_{s3}(t) + \lambda_s P_{s1}(t) + \lambda P_{s2}(t) + \mu P_{s5}(t) \tag{3.101}$$

$$\dot{P}_{s4}(t) = -(\lambda + \mu)P_{s4}(t) + \lambda P_{s1}(t) + \lambda_s P_{s2}(t) + \mu P_{s5}(t) \tag{3.102}$$

$$\dot{P}_{s5}(t) = -2\mu P_{s5}(t) + \lambda P_{s3}(t) + \lambda P_{s4}(t) \tag{3.103}$$

The initial conditions of the two-pump system are $P_{s1}(0) = 1$, $P_{si}(0) = 0$ ($i = 2, 3, 4, 5$). Equations (3.99) through (3.103) can be solved numerically or simplified as follows:

Since the states *s1* and *s2* are similar (exchange *A* and *B*) and the states *s3* and *s4* are also similar in that *A* and *B* can be exchanged, we add Eq. (3.99) to Eq. (3.100), and Eq. (3.101) to Eq. (3.102) to obtain

$$\frac{d[P_{s1}(t) + P_{s2}(t)]}{dt} = -(\lambda + \lambda_s)\,[P_{s1}(t) + P_{s2}(t)] + \mu[P_{s3}(t) + P_{s4}(t)] \tag{3.104}$$

$$\frac{d[P_{s3}(t) + P_{s4}(t)]}{dt} = -(\lambda + \mu)\,[P_{s3}(t) + P_{s4}(t)] + (\lambda + \lambda_s)\,[P_{s1}(t) + P_{s2}(t)] + 2\mu P_{s5}(t) \tag{3.105}$$

$$\frac{dP_{s5}(t)}{dt} = -\,2\mu P_{s5}(t) + \lambda[P_{s3}(t) + P_{s4}(t)]. \tag{3.106}$$

Define

$$P_1(t) = P_{s1}(t) + P_{s2}(t)$$

$$P_2(t) = P_{s3}(t) + P_{s4}(t)$$

$$P_3(t) = P_{s5}(t)$$

Substituting in Eqs. (3.104) through (3.106), we obtain

$$\dot{P}_1(t) = -(\lambda + \lambda_s)P_1(t) + \mu P_2(t) \tag{3.107}$$

$$\dot{P}_2(t) = -(\lambda + \mu)P_2(t) + (\lambda + \lambda_s)P_1(t) + 2\mu P_3(t) \tag{3.108}$$

$$\dot{P}_3(t) = -2\mu P_3(t) + \lambda P_2(t) \tag{3.109}$$

The new initial conditions are $P_1(0) = 1$, and $P_2(0) = P_3(0) = 0$.

Equations (3.107) through (3.109) can be solved by substituting $P_3(t) = 1 - P_1(t) - P_2(t)$ into Eq. (3.108), which results in

$$\dot{P}_2(t) = -(\lambda + 3\mu)P_2(t) + (\lambda + \lambda_s - 2\mu)P_1(t) + 2\mu. \tag{3.110}$$

Equations (3.107) and (3.110) can be written as

$$\dot{P}_1(t) = -(\lambda + \lambda_s)P_1(t) + \mu P_2(t) \tag{3.111}$$

$$\dot{P}_2(t) = -(\lambda + 3\mu)P_2(t) + (\lambda + \lambda_s - 2\mu)P_1(t) + 2\mu. \tag{3.112}$$

The instantaneous availability of the two-pump system, $A(t) = P_1(t) + P_2(t)$ is obtained by solving Eqs. (3.111) and (3.112) simultaneously.

EXAMPLE 3.23

Estimate the instantaneous availability for the two-pump system described above when the standby unit is considered hot, cold, or warm. Graph the availability for different λ and μ.

SOLUTION

Hot standby: In hot standby configurations, the failure rate of the standby unit equals that of the operating unit—that is, $\lambda_s = \lambda_h = \lambda$. Assume $\lambda = 5 \times 10^{-5}$ failures per hour and $\mu = 0.008$ repairs per hour. Substituting in Eqs. (3.111) and (3.112), we obtain

$$\dot{P}_1(t) = -(10 \times 10^{-5})P_1(t) + 0.008P_2(t) \tag{3.113}$$

$$\dot{P}_2(t) = -(0.02405)P_2(t) - 0.0159P_1(t) + 0.016. \tag{3.114}$$

Taking the Laplace transform of Eqs. (3.113) and (3.114) results in

$$(s + 10 \times 10^{-5})P_1(s) = 1 + 0.008P_2(s)$$

$$(s + 0.02405)P_2(s) = \frac{-0.0159}{(s + 10 \times 10^{-5})} - \frac{0.0001272}{(s + 10 \times 10^{-5})}P_2(s) + \frac{0.016}{s}$$

or

$$P_1(s) = \frac{(s + 0.02405)}{(s + 0.00805)(s + 0.0161)} + \frac{0.000128}{s(s + 0.00805)(s + 0.0161)} \tag{3.115}$$

and

$$P_2(s) = \frac{0.16 \times 10^{-5}}{s(s + 0.00805)(s + 0.0161)}. \tag{3.116}$$

We obtain the Laplace inverse of Eqs. (3.115) and (3.116) as

$$P_1(t) = 0.987616 + 0.0247675e^{-0.00805t} - 0.012384e^{-0.0161t} \tag{3.117}$$

and

$$P_2(t) = 0.0123452 - 0.0246904e^{-0.00805t} + 0.0123452e^{-0.0161t}. \tag{3.118}$$

The availability of the system is obtained by adding Eqs. (3.117) and (3.118) as

$$A(t) = 0.999961203 + 0.000077095e^{-0.00805t} - 0.000038797e^{-0.0161t}. \tag{3.119}$$

In order to compare the availability of the system for different values of λ, we consider the hot standby pump by setting $\lambda_s = \lambda$, the warm standby pump by setting $0 < \lambda_s < \lambda$ $(\lambda = 0.0002)$, and the cold standby system by setting $\lambda_s = 0$. As shown in Figure 3.12, the availability of the cold standby system is greater than the warm standby, which is greater than the hot standby system.

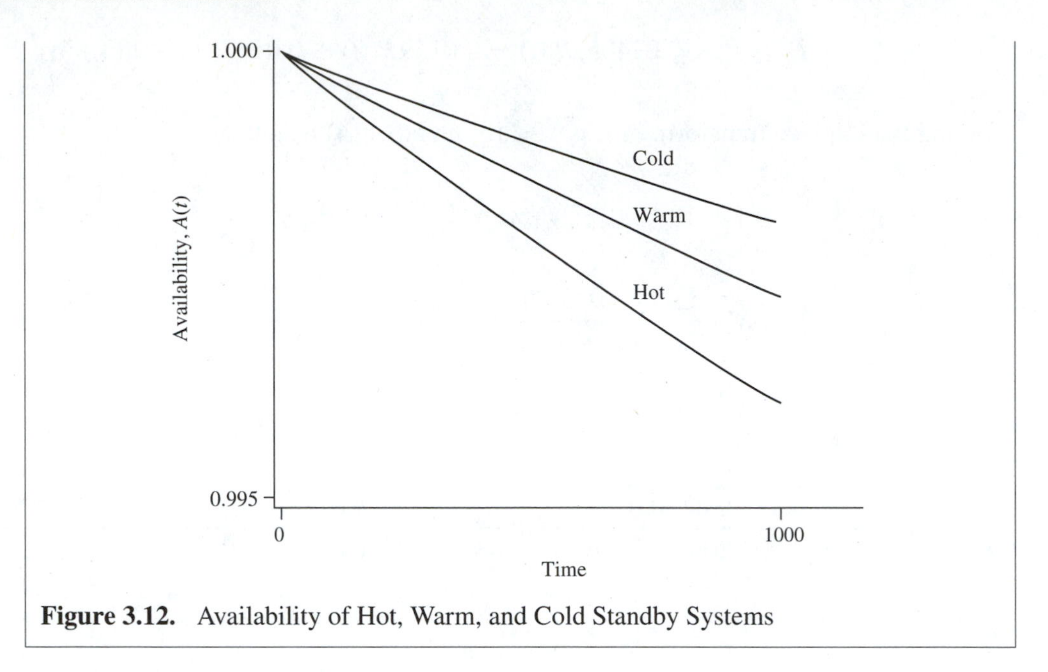

Figure 3.12. Availability of Hot, Warm, and Cold Standby Systems

PROBLEMS

3-1. Consider a consecutive-2-out-of-n: F system. What is the MTTF if the components are i.i.d. and each has a constant hazard rate λ?

3-2. Solve Problem 3-1 for both linearly increasing and Weibull hazard rates.

3-3. Solve Problem 3-1 for a consecutive-k-out-of-n: F system.

3-4. A system consists of three series components that are connected in series with four components in parallel. The components are identical, and the failure rate of each component follows a Weibull model with parameters $\gamma = 1.2$ and $\theta = 1.5 \times 10^5$. Derive a reliability expression for the system. What is the MTTF and the effective hazard rate of the system?

3-5. The main components of an undersea lightwave communication system are found in part under water, called the *wet plant*, and in part on land, the *dry plant*. The wet plant components consist of a cabled fiber transmission medium and repeaters containing optical amplifiers. The cable also contains a copper conductor to carry electrical power to the repeaters. Moreover, the system contains a branching unit, which provides for greater flexibility in undersea network architecture by allowing traffic to be split or switched. The dry plant components consist of terminal transmitter equipment (TTE), line monitoring equipment (LME), and power feed equipment (PFE). The TTE provides communication between the "dry" land communication network and the "wet" undersea transmission link. The LME monitors the transmission system and locates failures and faults, and the PFE energizes the link, providing power to the repeaters (Mortenson, Jackson, Shapiro, and Sirocky, 1995).

Consider a point-to-point undersea repeater system as shown in Figure 3.13. The system has 150 repeaters, and the failure of two consecutive repeaters interrupts the communication between the two points. The failure rate of each repeater is constant with $\lambda = 8.5 \times 10^{-7}$ failures per hour. The failure rates of the dry plant components are

$$h_{TTE}(t) = 5 \times 10^{-4} t^{1.25}$$

$$h_{LME}(t) = 12 \times 10^{-5} t$$

$$h_{PFE}(t) = 1.8 \times 10^{-6}.$$

Derive a reliability expression for the system. What is the MTTF? What do you recommend to improve the reliability of the system during a thirty-year life by 20 percent?

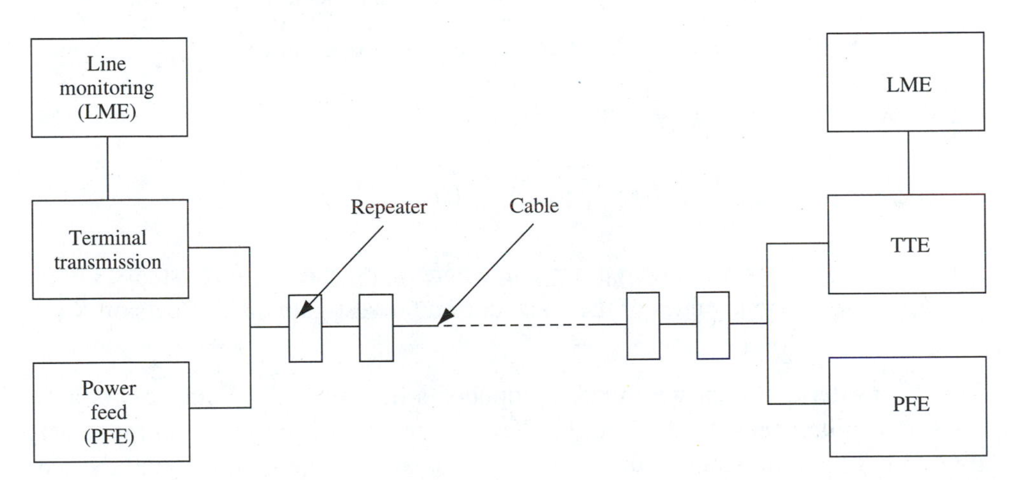

Figure 3.13. Point-to-Point Undersea Cable

3-6. One of the important components of the wet plant described in Problem 3-5 is the repeater. It is a collection of optical and electrical components in a beryllium-copper (BeCu) housing. The housing can support up to four amplifier pairs that support transmission on two communication lines simultaneously. Assume that the failure rate of an amplifier is linearly increasing with time with a parameter 6×10^{-12} failures per hour. A minimum of one pair of the amplifiers in each repeater is required to operate properly in order to ensure the transmission between end points. Derive a reliability expression for each repeater. What is the MTTF of the total system? Graph $R(t)$ for different values of the failure rate parameter.

3-7. A nonredundant subsystem has 100 units, each having a constant failure rate and a MTTF of 5×10^5 hours. What is the minimum number of units to be connected in parallel in order to achieve a system MTTF of two years? Provide alternative configurations that maintain a minimum reliability level of 0.999 after 3×10^4 hours.

3-8. One of the well-known approaches for improving system reliability is to add redundant components. This may be true for components with one type of failure mode. For components with multiple failure modes there exists an optimum number of redundant components that maximizes the total system reliability. Consider a system that consists of m components in parallel. Each component has three modes: normal (operational), fail open, and fail short. Let

$q_{si}(t)$ be the probability that the component fails short at time t and

$q_{oi}(t)$ be the probability that the component fails open at time t.

a. Show that the reliability of the system is

$$R_s(t) = \prod_{i=1}^{m} (1 - q_{si}(t)) - \prod_{i=1}^{m} q_{oi}(t).$$

b. Assume $q_{oi}(t) = a_i t^{bi}$ and $q_{si}(t) = \alpha_i t$, where a_i, b_i, and α_i are constants. Graph $R_s(t)$ for different values of the constants and investigate their effects on $R_s(t)$.

3-9. Consider a system with three components in series. The components have constant failure rates λ_1, λ_2, and λ_3. The failure rate of the third component is three times the failure rate of the second component, and the failure rate of the second component is twice that of the first component. It is desired to achieve a system reliability of 0.95 at time $t = 100$ hours. Determine

a. The failure rates of the components.

b. The MTTF of the system.

c. The probability of having 0, 1, and 2 failures in 100 hours of operation.

d. The failure rates of the components if a reliability of 0.95 is desired at the MTTF.

3-10. The Special Erlangian distribution is useful in modeling the failure rate of many electronic components. The p.d.f. of the distribution is

$$f(t) = \frac{t}{\lambda^2}\exp\left(-\frac{t}{\lambda}\right) \qquad t \geq 0.$$

a. Determine the MTTF of a component that exhibits such a failure time distribution. What is the variance of the TTF?

b. Assume that at $t = 1{,}000$ hours, the hazard rate of a component is 5×10^{-4} failures per hour. What is the reliability of a system composed of three similar components connected in parallel at time $t = 10^4$? What is the MTTF of such a system?

3-11. A system is configured using four components as shown in Figure 3.14:

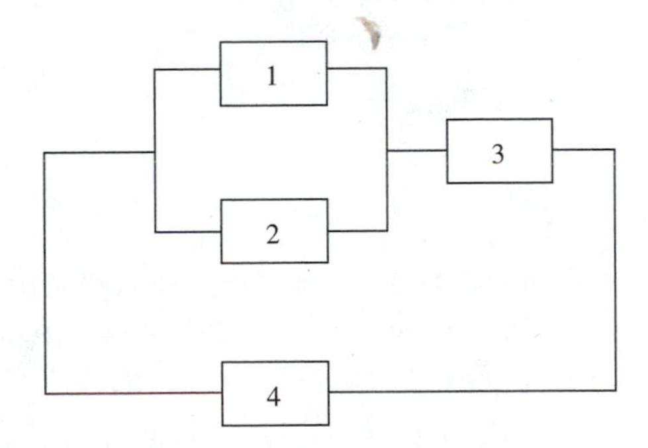

Figure 3.14. A Four Component System for Problem 3-11

Assume that the components have the following hazard rates:

$$h_1(t) = 0.5 \times 10^{-7} \text{ failures per hour,}$$

$$h_2(t) = 0.5 \times 10^{-7} \times t \text{ failures per hour,}$$

$$h_3(t) = 2.5 \times 10^{-6} \times t^{1.2} \text{ failures per hour, and}$$

$$h_4(t) = 0.5 \times 10^{-8} \times t^{1.3} \text{ failures per hour.}$$

Derive the reliability expression of the system. What is the reliability of the system at $t = 1{,}000$ hours? What is the MTTF?

3-12. Consider two nonrepairable systems, A and B, as shown in Figure 3.15. Each has the same number of components, n. The failure rate of component i is $h_i(t) = k_i t^{m_i}$.

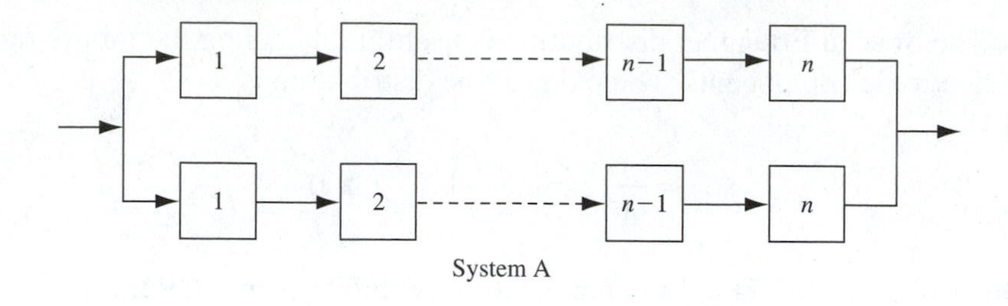

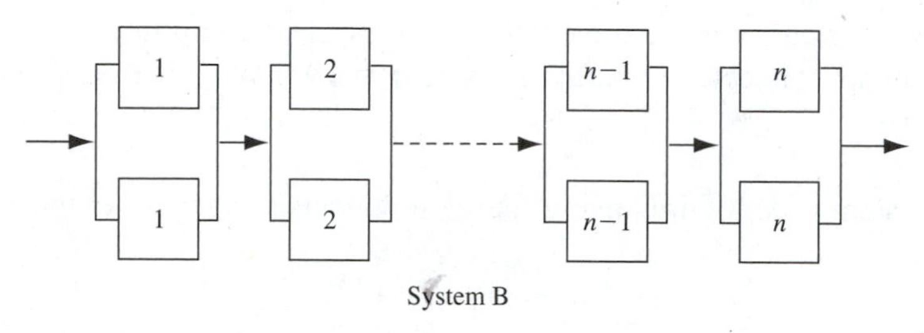

Figure 3.15. Two Nonrepairable Systems

a. Show that $R_A(t) \geq R_B(t)$.

b. Assume that the components of System *B* can be replaced with better components that exhibit the same failure rate. What are the parameters of the new components that result in $R_A(t) \leq R_B(t)$?

c. Derive expressions for the MTTF for both systems in (a) and (b).

3-13. Beam lead bonds of large-scale bonds integrated circuits (ICs) were observed. Two types of opens were observed on the failed ICs. The first type was a combination of silicon to beam interface separation and broken beam on the edge of the silicon chip. The second type was that of a broken beam at the heel or midspan. The failure rate of the first type is constant with parameter λ and the failure rate of the second type is a Weibull model with parameters γ and θ. Graph the reliability of the ICs for different values of λ, γ, and θ.

3-14. Assume that the failure rates of the ICs described in Problem 3-13 are constant with parameters λ_1 and λ_2 and the failed ICs are repaired with constant repair rates μ_l and μ_2 for the first and second type of failures, respectively.

a. Derive an expression for the ICs reliability at time *t*.

b. What is the steady-state availability of an IC?

c. If

$$\lambda_1 = 5 \times 10^{-6} \text{ failures per hour,}$$

$$\lambda_2 = 6 \times 10^{-7} \text{ failures per hour,}$$

$$\mu_1 = 0.5 \text{ repairs per hour, and}$$

$$\mu_2 = 1 \text{ repair per hour,}$$

what is the availability at time $t = 10^6$ hours?

d. What is the ratio between μ_1 and μ_2 that ensures a minimum availability of 0.9999 at time $t = 10^5$ hours?

e. What is the mean time between failures?

3-15. Further analysis of the integrated circuits given in Problem 3-13 shows that indeed the number of failure types is significantly more than two—for example, failure Type III consisted of darkened inclusions in the bond area, which acted as stress concentration centers, and resulted in degraded bond strength. In order to generalize the reliability and availability analysis of the ICs, the reliability engineers made the following assumptions:

- A typical IC may fail in any of the N failure modes. The failure rate of Type i is constant with parameter λ_i.
- The engineers also recommended that the failed IC be repaired and the repair rate of failure Type i is constant with parameter μ_i.

a. Derive expressions for $R(t)$ and $A(t)$.

b. For a wide range of λ_i and μ_i, determine the length of time which makes A_T equivalent to $A(t)$.

3-16. Pumping stations are considered major components of water supply systems. A typical pumping station consists of one or more pumping units supported by appropriate electrical, piping, control, and structural subsystems. The pumping unit is the primary subsystem, which includes four components: pump, driver (motor), power transmission, and controls. A common design practice is to install sufficient pumps to handle peak flows and include a spare pump of equal size to accommodate any down time of other pumps. Thus, the mechanical failure of the pumping station could be defined as the simultaneous failure of two or more pumping units while peak capacity is required (Mays, 1989).

The individual pumping units have two possible operating states: working and not working. The failure rates of the components of the individual units are

Component	*Failure Rate*
Pump	Constant with parameter λ_p
Drive	Increasing with $h_d(t) = \lambda_d t$
Power transmission	Weibull with $h_s(t) = \frac{\gamma}{\theta} t^{\gamma-1}$
Controls	Exponential with $h_c(t) = be^{\alpha t}$

The configuration of the components of the individual pump unit can be considered as a series system.

In order to meet the peak lead requirements of water supply, the planner of a city recommends the installation of four individual pump units in parallel in the major pumping station. Analysis of the data from an actual pumping station shows that

$$\lambda_p = 0.00133 \text{ failures per year,}$$

$$\lambda_d = 0.00288,$$

$$\gamma = 1.30,$$

$$\theta = 2.3 \times 10^3 \text{ per year,}$$

$$b = 1{,}000, \text{ and}$$

$$\alpha = 0.3.$$

The time t is expressed in years. Derive an expression for the reliability of the system. What are the two most critical components in an individual pump unit? Recommend two methods that improve the overall reliability of the pump station. Explain the advantages and disadvantages of each method.

3-17. A high-voltage system consisting of a power supply and two transmitters, A and B, uses mechanically tuned magnetrons. When the two transmitters are used in parallel, each transmitter tunes one-half of the desired frequency range; however, if one transmitter fails, the other tunes the entire range with a resultant change in the expected time to failure. Suppose that in order for this system to work, the power supply and at least one of the transmitters must operate properly (Pham, 1992). Let λ_A be the constant failure of transmitter A when transmitter B is operating in parallel with A. Let λ'_A be the failure rate when B fails. Similarly, suppose that transmitter B has a constant failure rate λ_B when A is operating in parallel with B and has a constant failure rate λ'_B when A fails.

a. Obtain an explicit reliability expression for the system.

b. Graph $R(t)$ for different ratios of λ_A / λ_B and λ'_A / λ'_B.

c. What is the MTTF of the system?

3-18. A repairable system consists of a primary unit and a standby unit. They alternate positions as failures and repairs occur. The units are identical, and each has two failure modes: open and short. The failure and repair rates are constant with the following parameters:

λ_o = the failure rate of the open mode failure,

λ_s = the failure rate of the short mode failure,

μ_o = the repair rate of the open failure, and

μ_s = the repair rate of the short failure.

When the primary unit fails in either mode, it is immediately replaced with the standby unit at a constant rate α (Elsayed and Dhillon, 1979).

a. Derive the state-transition equations.

b. Solve (a) for $P_i(t)$ (probability that the system is in state i at time t).

c. What is the instantaneous availability of the system?

d. Investigate the effect of μ_o / μ_s and α on $A(t)$.

3-19. A maintained system with two components in parallel each has a failure rate of $\lambda = 0.001$ failures per day independent of the number of components in operation. At $t = 0$, the two components are in an operative state (state zero). If it is desired to have the system in this state at least 50 percent of the time and to be in an operative state with one component working 95 percent of the time, what repair rates should be provided?

a. Assuming that you could only work on one component at a time and the repair rate was therefore independent of the number of failed components, what is the required repair rate now, assuming the above availabilities? What are the repair rates if the repair rate is a function of the number of failed components?

b. If it costs \$1 for each percent decrease in failure rate and \$2 for each percent of increase in repair rate determine the optimum policy that minimizes the total cost and maintains an availability of 0.95.

3-20. In a 3-out-of-n system with components having a linearly increasing hazard rate $h(t) = 0.5 \times 10^{-8}t$ failures per hour, determine the number of components for the system such that a reliability of 0.98 is achieved at $t = 10^3$ hours. What is the MTTF?

3-21. Determine the number of components that can be connected in parallel and results in the same reliability values as that of Problem 3-20.

3-22. A computer chip has 200,000 transistors connected in parallel and k transistors are required to operate properly for the chip to perform its function. Assuming that each transistor has a constant hazard model $h(t) = \lambda$, what is the value of k that ensures a chip reliability of .95 at $t = 10{,}000$ hours?

3-23. *Partial redundancy* is defined as the configuration where at least k out of m ($k < m$) possible paths must be successful. Consider a system of m diodes (a diode is an electronic device that allows current to pass in one direction and prevents it from passing in the other direction) connected in parallel and each diode is subject to either failures: (1) open, the current cannot pass in through the diode, and (2) short, the diode allows current to pass either way in the circuit. Assume that the probability of a diode failing in the open failure mode is q_o, the probability failing in the short mode is q_s, and the probability of working properly is p. Note that $q_o + q_s + p = 1$. Assuming that the system is successful if at least k of the m diodes are successful, develop an expression to describe the system's reliability. If $m = 6$, $k = 3$, $q_o = 0.02$, and $q_s = 0.05$, what is the reliability of the system?

3-24. In many pharmaceutical applications, the control system of the processes is crucial to ensure that the products meet quality requirements. Therefore, controllers of critical processes are designed such that some level of redundancy is provided. Figure 3.16 represents two simple configurations for a control system (System A and System B). The difference between the configurations is the addition of a redundant central processor for control Module I and a redundant communications network for System B. In these systems, the failure of the system occurs if there is a loss of communications network, loss of control Module I, and the loss of both operators' consoles (Renner, 1988). The failure rate of the operators' consoles is constant with parameter λ_c and their repair rate is also constant with parameter μ_c. The failure rate of the communication network is λ_n and its repair rate is μ_n. Similarly, the failure and repair rates for Module I are λ_I, μ_I, and those for Module II are λ_{II} and μ_{II}.

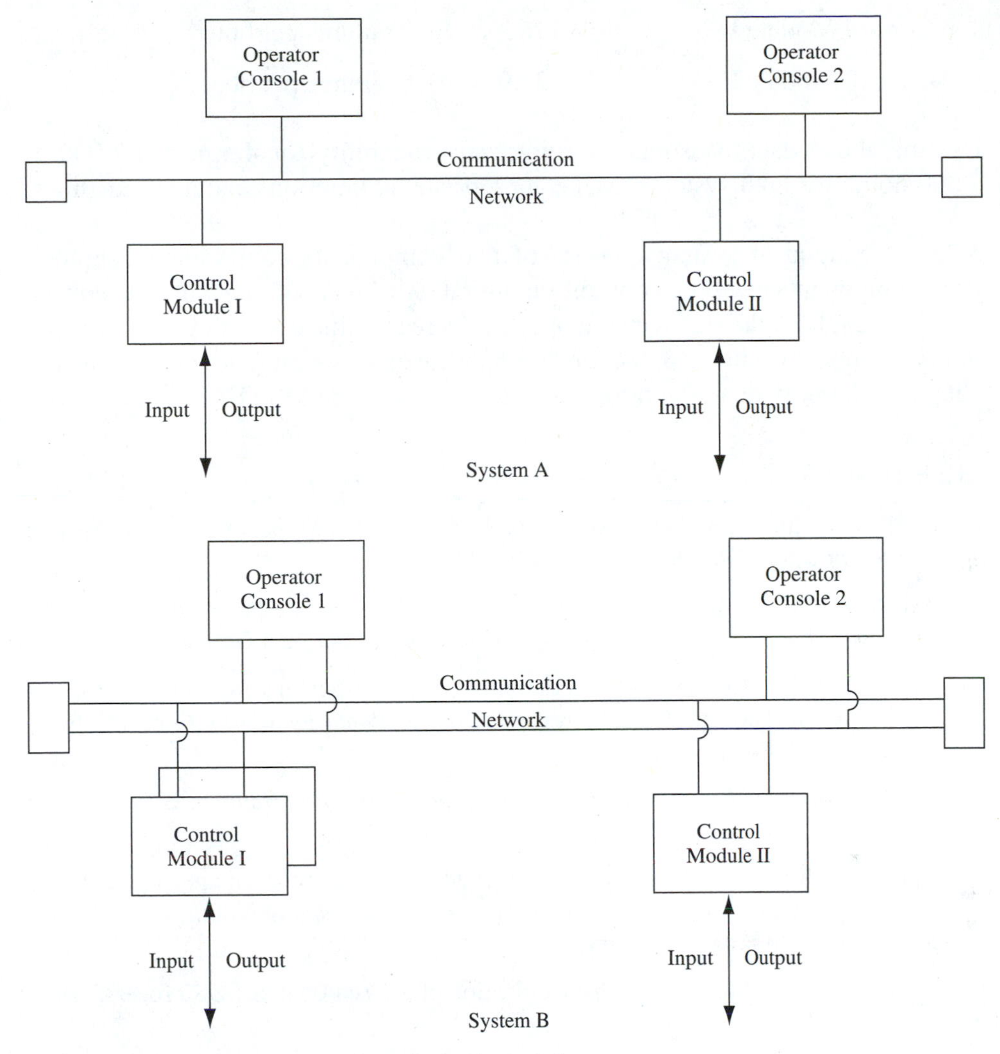

Figure 3.16. Two Configurations for a Control System

a. Derive the state transition equations of the system.

b. Derive expressions for the instantaneous availabilities of the two systems. What are the conditions that make the two systems equivalent?

c. Given the same number of components as that of System *B*, design a new control system that shows better performance than configuration *B*.

d. The failure rates of the components are

Operator Consoles 1 and 2	24×10^{-6} failures per hour
Communication links	0.25×10^{-6} failures per hour

Control Module I	116.2×10^{-6} failures per hour
Control Module II	130.0×10^{-6} failures per hour

Use the above data to estimate the interval availability for the period 2,000 to 7,000 hours for both systems. Make appropriate assumptions about repair times.

3-25. A redundant system consists of two components connected in parallel. Both components exhibit a constant failure rate λ when operating simultaneously. The failure rate of the surviving component increases linearly with time when one of the components fails. Use the joint probability distribution function approach to obtain a reliability expression for the system. What is the MTTF?

REFERENCES

Baruh, H., and Altiok, T. (1991). "Analytical Perturbations in Markov Chains." *European Journal of Operational Research* 51, 210–222.

Birolini, A., *On the Use of Stochastic Processes in Modeling Reliability Problems: Lecture Notes in Economic and Mathematical Systems*. Berlin: Springer-Verlag.

Crook, D. L. (1979). "Method of Determining Reliability Screens for Time Dependent Dielectric Breakdown." *Proceedings of the International Reliability Physics Symposium*, pp. 1–7. Piscataway: IEEE.

Dallas, D. (Ed.). (1976). *Tool and Manufacturing Engineers Handbook*. New York: McGraw-Hill.

Domangue, E., Rivera, R., and Shepard, C. (1984). "Reliability Prediction Using Large MOS Capacitors." *Proceedings of the Twenty-second Annual Reliability Physics Conference*, pp. 140–145. Las Vegas: IEEE.

Dugan, M. P. (1986). "Reliability Characterization of a 3-μm CMOS/SOS Process." *RCA Review* 47(2), 138–153.

Edwards, D. G. (1982). "Testing for MOS IC Failure Modes." *IEEE Transactions on Reliability* R-31(1), 9–17.

Elsayed, E. A., and Dhillon, B. (1979). "Repairable Systems with One Standby Unit." *Microelectronics Reliability* 19, 243–245.

Fostner, F. (1994). "A DSP-Based Test Solution for Regulators." *Evaluation Engineering* (November), 46–51.

Hawkins, C. F., and Soden, J. M. (1986). "Reliability and Electrical Properties of Gate Oxide Shorts in CMOS ICs." *Proceedings of the International Testing Conference*, pp. 443–451. Piscataway: IEEE.

Henley, E. J., and Kumamoto, H. (1981). *Reliability Engineering and Risk Assessment*. Englewood Cliffs: Prentice-Hall.

Holcomb, D. P. (1981). "Availability Estimates During Early Life." Holmdel: Technical Memorandum 81-59566-4, Bell Laboratories, Basking Ridge, NJ.

Johns, D. (1995). "Membrane Versus Mechanical Keyboards." *Electronic Products* (June), 43–45.

Lie, C. H., Hwang, C. L., and Tillman, F. A. (1977). "Availability of Maintained Systems: A State-of-the-Art Survey." *AIIE Transactions* 9(3), 247–259.

Mays, L. W. (1989). *Reliability Analysis of Water Distribution Systems.* New York: American Society of Civil Engineers.

Mortenson, R. L., Jackson, B. S., Shapiro, S., and Sirocky, W. F. (1995). "Undersea Optically Amplified Repeatered Technology, Products, and Challenges." *AT&T Technical Journal* (January/February), 33–46.

O'Connor, L. (1995). "Pumping Up Backstage." *Mechanical Engineering* (June), 130.

Pham, H. (1992). "Reliability Analysis of a High Voltage System with Dependent Failures and Imperfect Coverage." *Reliability Engineering and System Safety* 37, 25–28.

Renner, K. M. (1988). "Evaluating Process Control Systems: Availability, Reliability, and Redundancy." *Pharmaceutical Technology* (November), 24–34.

Schendel, U. (1989). *Sparse Matrices: Numerical Aspects with Applications for Scientists and Engineers*. Chichester: Ellis Horwood.

Sears, R. W. (1990). "Practical Models for Determining Standby Redundancy Levels." *Proceedings of the Annual Reliability and Maintainability Symposium*, pp. 120–126. Los Angeles: IEEE.

Shooman, M. (1968). *Probabilistic Reliability: An Engineering Approach.* New York: McGraw-Hill.

Swartz, G. A. (1986). "Gate Oxide Integrity of MOS/SOS Devices." *IEEE Transactions on Electron Devices* ED-33(11), 1826–1829.

PART II

Parameter Estimation and Reliability Testing

CHAPTER 4

Estimation Methods of the Parameters of Failure-Time Distributions

Reliability estimation requires the knowledge of the underlying failure-time distribution of the component or the system being modeled. Also, in order to predict the reliability or estimate the MTTF of components or systems subjected to an accelerated life test, we need to estimate the parameters of the probability distribution that describes the failure time of the population subjected to the test.

Clearly, the accuracy of the estimate of the parameters depends on the sample size and the method used for estimating the parameters. The statistics, calculated from the samples that are used to estimate population parameters, are called *estimators*. A good estimator should have the following properties:

- *Unbiased:* The estimator $\hat{\theta}$ is an unbiased estimator for a parameter θ if and only if $E[\hat{\theta}] = \theta$. In other words, an unbiased estimator should not consistently underestimate nor overestimate the true value of the parameter.

- *Consistent:* A consistent estimator is one that is unbiased and converges more closely to the true value of the population parameter as the sample size increases. In other words, the estimator $\hat{\theta}$ is said to be a consistent estimator of θ if the probability of making errors of any given size ε, tends to zero as n (sample size) tends to infinity—that is,

$$P[|\hat{\theta} - \theta| > \varepsilon] \rightarrow 0 \text{ as } n \rightarrow \infty, \text{ for any fixed positive } \varepsilon.$$

- *Efficient:* An efficient estimator is a consistent estimator whose standard deviation is smaller than the standard deviation of any other estimator for the same population parameter. We measure efficiency by

  Relative efficiency $= V(\hat{\theta}_2)/V(\hat{\theta}_1)$, where $V(\hat{\theta}_1)$ and $V(\hat{\theta}_2)$ are the variances of the two estimators $\hat{\theta}_1$ and $\hat{\theta}_2$ for the same population parameter: the better estimator has the smaller variance; and

  The asymptotic relative efficiency (ARE): the relative efficiency is a function of n and to avoid this we use the asymptotic relative efficiency as

$$ARE = \lim_{n \to \infty} \frac{V(\hat{\theta}_2)}{V(\hat{\theta}_1)}.$$

- *Sufficient:* A sufficient estimator is an estimator that utilizes all the information about the parameter that the sample possesses.

The statistic used to estimate the parameter of the population, θ, is called a *point estimator* for θ and is denoted by $\hat{\theta}$. The properties of the point estimator were discussed earlier.

Three of the most widely used methods for estimating the parameters of the population are *the method of moments*, *the maximum likelihood method*, and *the least-squares method*. It should be made clear that the estimate of the parameters regardless of the method being used depends on the quality of the data. This means that the engineer should check for outliers, such as abnormally short or abnormally long failure times. There are many statistical tests that identify outliers in a data set such as the Natrella-Dixon test (Dixon and Massey, 1957; Natrella, 1963) and the Grubbs test (1958). A comprehensive study of the outliers identification methods is presented in Hawkins (1980).

4.1. Method of Moments

The main idea of the method of moments is to equate certain sample characteristics such as mean and variance to the corresponding population expected values and then solve the resulting equations to obtain the estimates of the unknown parameter values.

If $x_1, x_2, \ldots, x_n$ represent a set of data, then the kth sample moment is

$$M_k = \frac{1}{n} \sum_{i=1}^{n} x_i^k.$$

If $\theta_1, \theta_2, \ldots, \theta_m$ are the unknown parameters of the population, then the *moment estimators* $\hat{\theta}_1, \hat{\theta}_2, \ldots, \hat{\theta}_m$ are obtained by equating the first m sample moments to the corresponding first m population moments and solving for $\theta_1, \theta_2, \ldots, \theta_m$.

EXAMPLE 4.1

Assume that $x_1, x_2, \ldots, x_n$ represent a random sample from an exponential distribution with parameter λ. What is the estimate of λ?

SOLUTION

The p.d.f. of the exponential distribution is

$$f(x) = \lambda e^{-\lambda x}$$

and

$$E[X] = \frac{1}{\lambda}.$$

Using the sample's first moment,

$$M_1 = \frac{\sum_{i=1}^{n} x_i}{n} = E[X] = \frac{1}{\lambda},$$

or the estimate of λ is

$$\hat{\lambda} = \frac{n}{\sum_{i=1}^{n} x_i}.$$

EXAMPLE 4.2

A manufacturer of a wireless data system uses infrared beams transmitted between devices mounted on the outside of buildings to provide a high-speed link data. The size of the infrared beam has a direct effect on the system reliability and its ability to reduce the effect of weather conditions such as snow and fog that obstruct the beam's path. Data are transmitted continuously using the infrared beams, and the times to failure in hours (not receiving the transmitted data) are recorded as follows:

47, 81, 127, 183, 188, 221, 253, 311, 323, 360, 489, 496, 511, 725, 772, 880, 1,509, 1,675, 1,806, 2,008, 2,026, 2,040, 2,869, 3,104, 3,205.

Assuming that the failure times follow an exponential distribution, determine the parameter of the distribution using the method of moments. Estimate the reliability of the system at time = 1,000 hours. (Note that the above data are generated from an exponential distribution with parameter $1/\lambda = 1{,}000$.)

SOLUTION The parameter of the exponential distribution is

$$\hat{\lambda} = \frac{n}{\sum_{i=1}^{n} x_i}$$

$$\hat{\lambda} = \frac{25}{26{,}209} = 0.00095387$$

$$\frac{1}{\hat{\lambda}} = 1{,}048.36.$$

This is very close to the same parameter value used in generating the data. Clearly, as the number of observations increases, the estimated parameter ($\hat{\lambda}$) quickly approaches the parameter of the actual distribution of the failure times.

The following example illustrates the method of moments in estimating the parameters of a two-parameter distribution such as a gamma distribution.

EXAMPLE 4.3 Let $x_1, x_2, \ldots, x_n$ be a random sample from a gamma distribution whose p.d.f. is

$$f(x) = \frac{1}{\Gamma(\alpha)\beta^{\alpha}} x^{\alpha-1} e^{-x/\beta} \qquad x > 0, \alpha \geq 0, \beta > 0.$$

Use the method of moments to obtain estimates of the parameters α and β.

SOLUTION As shown in Chapter 1, the mean and variance of the gamma distribution, respectively, are

$$E[X] = \alpha\beta$$

$$V(X) = \alpha\beta^2 = E[X^2] - (E[X])^2.$$

We replace $E[X]$ and $E[X^2]$ by their estimators M_1 and M_2, respectively, to obtain

$$M_1 = \hat{\alpha}\hat{\beta}$$

$$M_2 - M_1^2 = \hat{\alpha}\hat{\beta}^2.$$

Solving the above two equations simultaneously yields

$$\hat{\beta} = \frac{(M_2 - M_1^2)}{M_1} \tag{4.1}$$

$$\hat{\alpha} = \frac{M_1^2}{(M_2 - M_1^2)}. \tag{4.2}$$

EXAMPLE 4.4

A manufacturer of personal computers performs a burn-in test on twenty computer monitors and obtains the following failure times (in hours):

130, 150, 180, 40, 90, 125, 44, 128, 55, 102, 126, 77, 95, 43, 170, 130, 112, 106, 93, 71.

Assume that the main population of the failure times follows a gamma distribution with parameters α and β. What are the estimates of these parameters?

SOLUTION

We first determine M_1 and M_2 as

$$M_1 = \frac{\sum_{i=1}^{n} x_i}{n} = \frac{2{,}067}{20} = 103.35$$

$$M_2 = \frac{1}{20}\sum x_i^2 = \frac{1}{20} \times 244{,}823 = 12{,}241.15.$$

Using Eqs. (4.1) and (4.2), we obtain

$$\hat{\beta} = \frac{12{,}241.15 - (103.35)^2}{103.35} = 15.09$$

$$\hat{\alpha} = \frac{(103.35)^2}{12{,}241.15 - (103.35)^2} = 6.847.$$

The expected mean life of a monitor is $\hat{\alpha}\hat{\beta} = 103.3$ hours.

The following is another example that illustrates the use of the method of moments in estimating the parameters of a two-parameter distribution.

EXAMPLE 4.5 Use the method of moments to estimate the parameters μ and σ^2 of the normal distribution.

SOLUTION The p.d.f. of the normal distribution is

$$f(x) = \frac{1}{\sigma\sqrt{2\pi}} e^{-\frac{1}{2}\left(\frac{x-\mu}{\sigma}\right)^2}.$$

The first moment M_1 about the origin is

$$M_1 = \int_{-\infty}^{\infty} \frac{x}{\sigma\sqrt{2\pi}} e^{-\frac{1}{2}\left(\frac{x-\mu}{\sigma}\right)^2} dx.$$

Let

$$z = \frac{x-\mu}{\sigma}.$$

Then $x = \mu + \sigma z$ and $dx = \sigma\, dz$, and the first moment becomes

$$M_1 = \int_{-\infty}^{\infty} \frac{(\mu + \sigma z)}{\sqrt{2\pi}} e^{-\frac{z^2}{2}} dz$$

$$M_1 = \mu \int_{-\infty}^{\infty} \frac{1}{\sqrt{2\pi}} e^{-\frac{z^2}{2}} dz + \sigma \int_{-\infty}^{\infty} \frac{z}{\sqrt{2\pi}} e^{-\frac{z^2}{2}} dz.$$

Since

$$\int_{-\infty}^{\infty} \frac{1}{\sqrt{2\pi}} e^{-\frac{z^2}{2}} dz = 1$$

and

$$\int_{-\infty}^{\infty} \frac{z}{\sqrt{2\pi}} e^{-\frac{z^2}{2}} dz = \int_{-\infty}^{\infty} \frac{e^{-\frac{z^2}{2}}}{\sqrt{2\pi}} d\left(\frac{z^2}{2}\right) = \frac{-1}{\sqrt{2\pi}} e^{-\frac{z^2}{2}} \Big|_{-\infty}^{\infty} = 0,$$

then

$$M_1 = \mu = \frac{1}{n}\sum_{i=1}^{n} x_i. \tag{4.3}$$

The second moment, M_2, about the origin is obtained as

$$M_2 = \int_{-\infty}^{\infty} \frac{x^2}{\sigma\sqrt{2\pi}} e^{-\frac{1}{2}\left(\frac{x-\mu}{\sigma}\right)^2} dx.$$

Again, let $z = (x - \mu)/\sigma$, then

$$M_2 = \int_{-\infty}^{\infty} \frac{1}{\sqrt{2\pi}} (\mu + \sigma z)^2 e^{-\frac{z^2}{2}} dz$$

$$M_2 = \mu^2 \int_{-\infty}^{\infty} \frac{1}{\sqrt{2\pi}} e^{-\frac{z^2}{2}} dz + 2\sigma\mu \int_{-\infty}^{\infty} \frac{z}{\sqrt{2\pi}} e^{-\frac{z^2}{2}} dz$$

$$+ \sigma^2 \int_{-\infty}^{\infty} \frac{z^2}{\sqrt{2\pi}} e^{-\frac{z^2}{2}} dz.$$

The integral parts of the first two terms in the above equation have been obtained earlier. The integration of the third term is obtained by integration by parts:

$$\int_{-\infty}^{\infty} \frac{z^2}{\sqrt{2\pi}} e^{-\frac{z^2}{2}} dz = \frac{-1}{\sqrt{2\pi}} \int_{-\infty}^{\infty} z d\left(e^{-\frac{z^2}{2}}\right)$$

$$= \frac{-1}{\sqrt{2\pi}} z e^{-\frac{z^2}{2}} \Big|_{-\infty}^{\infty} + \frac{1}{\sqrt{2\pi}} \int_{-\infty}^{\infty} e^{-\frac{z^2}{2}} dz = 0 + 1$$

$$= 1.$$

Therefore,

$$M_2 = \mu^2 + \sigma^2 = \frac{1}{n}\sum_{i=1}^{n} x_i^2. \tag{4.4}$$

From Eqs. (4.3) and (4.4) the estimated parameters of the normal distribution are

$$\hat{\mu} = \frac{1}{n}\sum_{i=1}^{n} x_i$$

$$\hat{\sigma}^2 = \frac{1}{n}\sum_{i=1}^{n} x_i^2 - \left(\frac{1}{n}\sum_{i=1}^{n} x_i\right)^2$$

or

$$\hat{\sigma}^2 = \frac{1}{n}\sum_{i=1}^{n} (x_i - \bar{x})^2.$$

The method of moments is a simple method for estimating the parameters of the failure-time distribution provided that the underlying distribution is known. The errors in estimating the parameters are minimum when the underlying distribution is symmetric with no skewness and when the failure times are not censored or truncated.

4.1.1. CONFIDENCE INTERVALS

After the determination of the point estimate of the parameters of the distribution, we may be interested in determining a confidence interval for which the estimated parameters are close to the true values of the population. This can be accomplished by defining two limits—the lower confidence limit (LCL) and the upper confidence limit (UCL)—that form an interval that has a probability $1 - \alpha$ of capturing the true value of parameters, where $1 - \alpha$ is called the confidence coefficient. In other words, the confidence interval for the parameter θ is

$$P[LCL \leq \theta \leq UCL] = 1 - \alpha. \tag{4.5}$$

For brevity, we shall limit our presentation to the general distribution case. Other distributions can be easily treated in a similar fashion.

Suppose that a random sample $x_1, x_2, \ldots, x_n$ is taken from a population with mean μ and variance σ^2. Let $\bar{x}$ be the point estimator of μ. If n is large ($n \geq 30$), then $\bar{x}$ has approximately a normal distribution with mean μ and variance σ^2/n, or

$$Z = \frac{\bar{x} - \mu}{\sigma/\sqrt{n}}$$

has a standard normal distribution. For any value of α we can find (using standard normal tables) a value of $Z_{\alpha/2}$ such that

$$P[-Z_{\alpha/2} \leq Z \leq +Z_{\alpha/2}] = 1 - \alpha.$$

Rewriting the above expression, we obtain

$$1 - \alpha = P\left[-Z_{\alpha/2} \leq \frac{\bar{x} - \mu}{\sigma / \sqrt{n}} \leq + Z_{\alpha/2}\right]$$

$$= P\left[-Z_{\alpha/2}\frac{\sigma}{\sqrt{n}} \leq \bar{x} - \mu \leq + Z_{\alpha/2}\frac{\sigma}{\sqrt{n}}\right]$$

$$= P\left[\bar{x} - Z_{\alpha/2}\frac{\sigma}{\sqrt{n}} \leq \mu \leq \bar{x} + Z_{\alpha/2}\frac{\sigma}{\sqrt{n}}\right].$$

Thus the interval

$$\left(\bar{x} - Z_{\alpha/2}\frac{\sigma}{\sqrt{n}}, \bar{x} + Z_{\alpha/2}\frac{\sigma}{\sqrt{n}}\right)$$

forms the confidence interval of the estimated parameter $\bar{x}$ for μ with confidence coefficient of $(1 - \alpha)$.

EXAMPLE 4.6

Consider the failure times of Example 4.4. Find a confidence interval for the mean failure time with confidence coefficient of 0.95.

SOLUTION

From the data, we obtain

$$\bar{x} = 103.35$$

$$s = 40.52$$

s is the estimate of the standard deviation σ.

Since the sample size is small (< 30), it is more appropriate to use the t distribution than the standard normal in determining the confidence interval. Thus the confidence interval is

$$\bar{x} \pm t_{\alpha/2}\frac{\sigma}{\sqrt{n}} \quad \text{with } (1 - \alpha) = 0.95,$$

we obtain $t_{0.025} = 2.093$ and substitute s for σ to get

$$103.35 \pm 2.093 \times \frac{40.52}{\sqrt{20}}$$

$$103.35 \pm 18.96$$

$$(84.39, 122.31).$$

In other words, we have 95 percent confidence that the true mean failure time lies between 84.39 and 122.31 hours.

4.2. The Likelihood Function

The second most widely used method for estimating the parameters of a probability distribution is based on the likelihood function. This method plays a fundamental role in statistical inference and is applied in many practical problems. We first present the concept of the likelihood function, and we then follow it by a description of the maximum likelihood method. Other likelihood methods such as the marginal and partial likelihood methods are variants of the maximum likelihood and will not be discussed in this book.

Consider a manufacturer who checks the quality of the products by taking a sample of fifteen random products and inspects them for defective units. Assuming that θ is the proportion of defective units in the total production, then the probability of having x defective units in the sample is binomial:

$$P(x) = \binom{15}{x} \theta^x (1-\theta)^{15-x} \qquad x = 0, 1, \ldots, 15. \tag{4.6}$$

The probability of having two defective units in the sample is

$$P(2) = \binom{15}{2} \theta^2 (1-\theta)^{13}.$$

This probability is a function of θ and a plot of the $P(2)$ for different values of θ is shown in Figure 4.1. The numerical values of the probability, $P(2)$, and θ are shown in Table 4.1. The graph in Figure 4.1 is referred to as the likelihood function. One can deduce that the *likelihood function* is the joint probability of an observed sample as a function of the unknown parameters.

In situations where the sample size is very large, we may find that it is more convenient to calculate the logarithmic values of the likelihood function than to calculate the function itself. Therefore, plot of the likelihood function will be

Table 4.1. Values of θ and $P(2)$

θ	$P(2)$	θ	$P(2)$
0.02	0.0323	0.22	0.2010
0.04	0.0988	0.24	0.1707
0.06	0.1691	0.26	0.1416
0.08	0.2273	0.28	0.1150
0.10	0.2669	0.30	0.0916
0.12	0.2870	0.32	0.0715
0.14	0.2897	0.34	0.0547
0.16	0.2787	0.36	0.0411
0.18	0.2578	0.38	0.0303
0.20	0.2309	0.40	0.0219

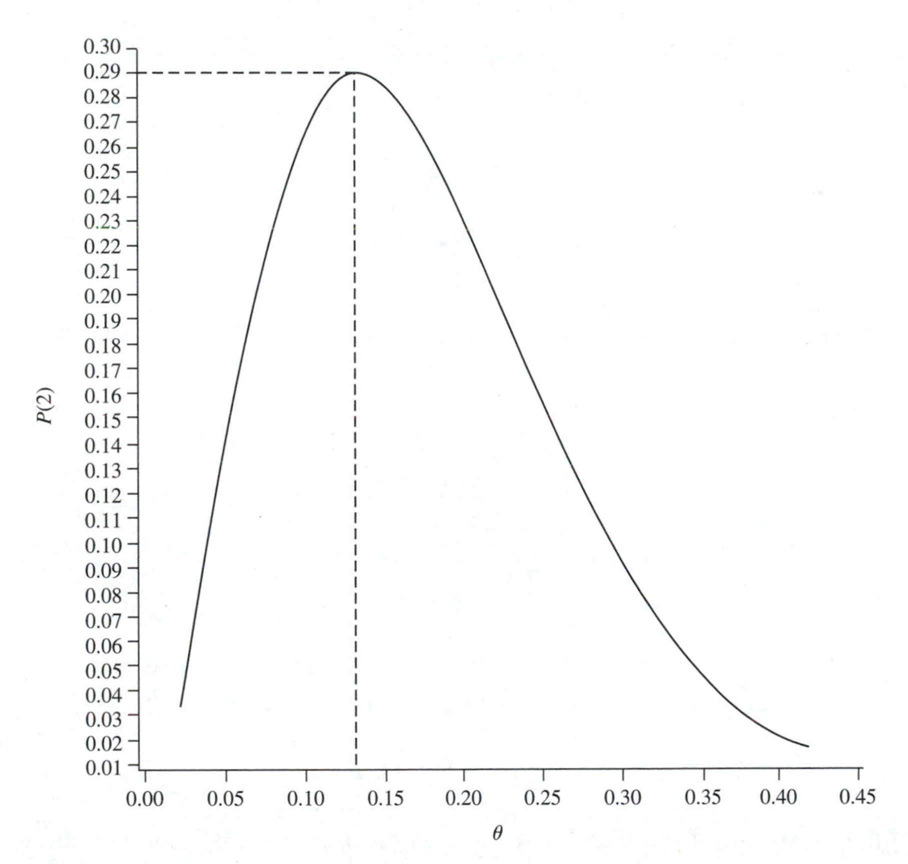

Figure 4.1. Plot of $P(2)$ Versus θ

greatly simplified since the likelihoods are usually obtained by multiplying the probabilities of independent events and by considering the logarithm of the function we can eliminate (or use as a scale) the constant term of the logarithm. This is illustrated in the following example.

EXAMPLE 4.7

The number of defectives in a production line is found to follow a Poisson distribution with an unknown mean μ. Two random samples are taken and the numbers of the defective units found are ten and twelve. What is the likelihood function of having ten and twelve defective units?

SOLUTION

The probability of having x units from a Poisson distribution is

$$P(x) = \frac{e^{-\mu}\mu^x}{x!} \qquad x = 0, 1, 2, \ldots$$

The probabilities of having ten and twelve defectives, respectively, are

$$P(10) = \frac{e^{-\mu}\mu^{10}}{10!}$$

and

$$P(12) = \frac{e^{-\mu}\mu^{12}}{12!}.$$

The likelihood function $[l(x, \mu)]$ is the product of $P(10)$ and $P(12)$—that is,

$$l(x, \mu) = \frac{e^{-\mu}\mu^{10}}{10!} \times \frac{e^{-\mu}\mu^{12}}{12!} \qquad (x = 10, 12)$$

$$= \frac{e^{-2\mu}\mu^{22}}{(10!\ 12!)}. \tag{4.7}$$

Evaluation of Eq. (4.7) for different values of μ can be simplified by taking the logarithm of $l(x, \mu)$. Let $L(x, \mu)$ be the logarithm of $l(x, \mu)$—that is,

$$L(x, \mu) = \log l(x, \mu)$$

and the logarithm of the likelihood function given in Eq. (4.7) can now be written as

$$L(\mu) = 22 \log \mu - 2\mu - \log(10!\ 12!). \tag{4.8}$$

Since the last term in Eq. (4.8) is constant, we may drop it and plot the relative values of the log likelihood function as shown in Figure 4.2. The values of $L(\mu)$ corresponding to the figure are shown in Table 4.2. It is obvious from Figure 4.2 that the probability of having ten and twelve defectives is maximum when the mean of the Poisson distribution is 11.

Table 4.2. Values of $L(\mu)$ for Figure 4.2

μ	$L(\mu)$
6	27.4187
7	28.8100
8	29.7477
9	30.3389
10	30.6569
11	30.7537
12	30.6679
13	30.4289
14	30.0593
15	29.5771
16	28.9970

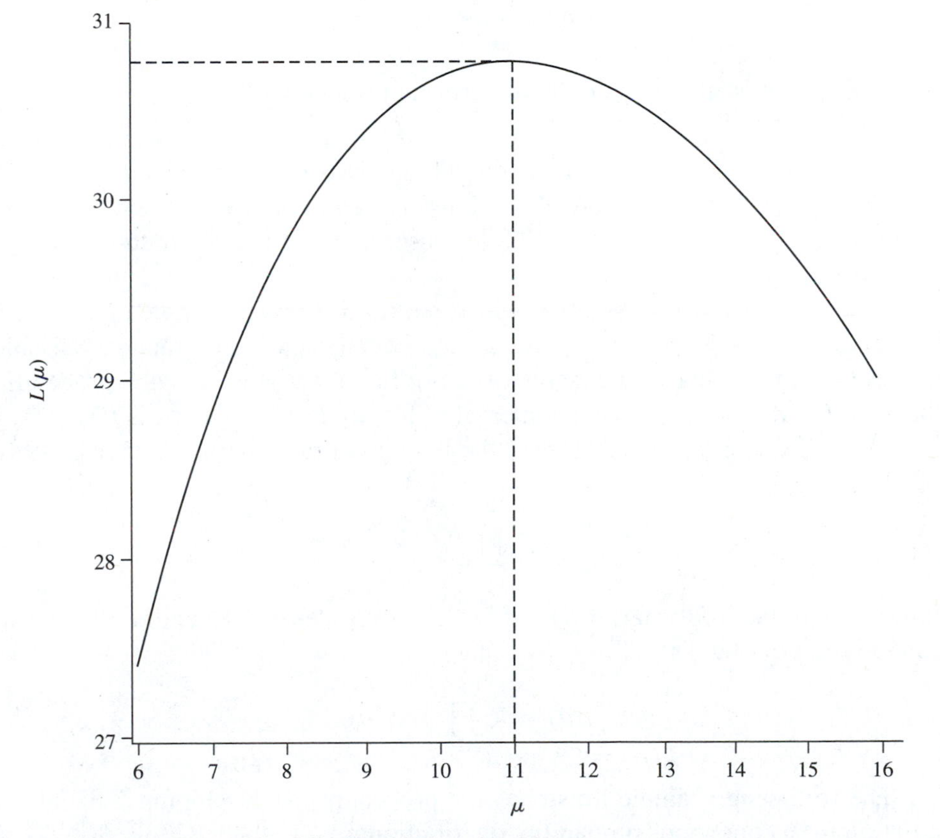

Figure 4.2. The Log of the Likelihood Function Versus μ

EXAMPLE 4.8

Suppose that the manufacturer of integrated circuits (ICs) takes three random samples from the same batch of sizes ten, fifteen, and twenty-five units. On inspection, it is found that these samples have two, three, and five defectives, respectively. What is the likelihood function for these probabilities?

SOLUTION

Since the three samples are taken from the same production batch, the underlying probability distribution has the same parameter θ. The probabilities of the three results are

$$\binom{10}{2}\theta^2(1-\theta)^8; \quad \binom{15}{3}\theta^3(1-\theta)^{12}; \quad \text{and} \quad \binom{25}{5}\theta^5(1-\theta)^{20}.$$

The likelihood function is simply the product of the three probabilities:

$$l(\theta) = \binom{10}{2}\theta^2(1-\theta)^8\binom{15}{3}\theta^3(1-\theta)^{12}\binom{25}{5}\theta^5(1-\theta)^{20}$$

$$l(\theta) = K\theta^{10}(1-\theta)^{40},$$

where K is a constant that includes all terms not involving θ.

So far, we have shown how a likelihood function can be developed for discrete probability distributions (binomial and Poisson). One can use the same procedure for developing the likelihood function for observations from continuous probability distributions.

Assume we have a distribution with a p.d.f. $f(x, \theta)$ with a single parameter θ, and let the observed results be $x_1, x_2, \ldots, x_n$. For any continuous random variable, the probability of obtaining exactly x is zero, for any x. However, the probability that an observation x occurs in an interval of length dx centered at x is $f(x, \theta)dx$. If $x_1, x_2, \ldots, x_n$ are independent, then the likelihood function is (Wetherill, 1981)

$$l(\theta) = \prod_{i=1}^{n} f(x_i, \theta)dx_i. \tag{4.9}$$

Since the product of the terms $dx_1, dx_2, \ldots, dx_n$ does not depend on θ, then we can rewrite Eq. (4.9) as

$$l(\theta) = K\prod_{i=1}^{n} f(x_i, \theta).$$

The following example illustrates the procedure for developing a likelihood function for a continuous probability distribution.

EXAMPLE 4.9

Assume that the manufacturer in Example 4.8 randomly selects three samples having the same size and observes that they have five, seven, and nine defectives. It is also observed that the defectives in a production batch follow a normal distribution with an unknown mean, μ, but the variance equals 1. What is the likelihood function?

SOLUTION

The p.d.f. of observation x_i is

$$\frac{1}{\sqrt{2\pi}} \exp[-1/2(x_i - \mu)^2] \qquad i = 1, 2, 3.$$

The likelihood function is

$$l(\mu) = \frac{1}{2\pi\sqrt{2\pi}} \exp[-1/2(5 - \mu)^2 - 1/2(7 - \mu)^2 - 1/2(9 - \mu)^2]. \qquad (4.10)$$

Expanding the squared terms will result in terms which include μ^2, μx_i, and x_i^2. The latter term can be dropped since it does not involve μ.

We can simplify Eq. (4.10) by rewriting the $x_i - \mu$ term as

$$x_i - \mu = x_i - \bar{x} + \bar{x} - \mu.$$

By squaring the above expression and adding, we have

$$\sum_{i=1}^{n} (x_i - \mu)^2 = \sum_{i=1}^{n} (x_i - \bar{x})^2 + n(\bar{x} - \mu)^2.$$

Since the mean of the observations is 7 and the term $\exp(-1/2 \sum (x_i - \bar{x})^2)$ may be dropped since it does not include θ, we can rewrite Eq. (4.10) as

$$L(\mu) = K - 2(7 - \mu)^2,$$

where K is a constant.

Clearly, if the probability distribution has more than one unknown parameter, we can develop a likelihood function in terms of these parameters as shown below.

EXAMPLE 4.10

Assume n observations $x_1, x_2, \ldots, x_n$ are randomly taken from a normal distribution with unknown mean μ and unknown variance σ^2. What is the likelihood function (Wetherill, 1981)?

SOLUTION Following the same procedure of Example 4.9, we obtain the p.d.f. of observation x_i as

$$\frac{1}{\sigma\sqrt{2\pi}} \exp\left[-1/2\left(\frac{x_i - \mu}{\sigma}\right)^2\right] \qquad i = 1, 2, \ldots, n.$$

The likelihood function is the product of these p.d.f.'s—that is,

$$l(\mu, \sigma^2) = \frac{(2\pi)^{\frac{-n}{2}}}{\sigma^n} \exp\left[-\frac{1}{2\sigma^2}\sum_{i=-1}^{n}(x_i - \mu)^2\right].$$

The log of the likelihood function is

$$L(\mu, \sigma^2) = -n \log\sqrt{2\pi} - n \log \sigma - \frac{1}{2\sigma^2}\sum_{i=1}^{n}(x_i - \bar{x})^2 - \frac{n}{2\sigma^2}(\bar{x} - \mu)^2. \tag{4.11}$$

4.2.1. THE METHOD OF MAXIMUM LIKELIHOOD

As discussed earlier, the likelihood function usually has a maximum at specific values of the distribution parameters. These values of the parameters are more likely to give rise to the observed data than other values. If we require a single value of the parameter to use as an estimate for the distribution, then it is clear that the value of the parameter that gives the maximum of the likelihood is the "best" estimate.

The objective is then to determine the best estimates of the parameters using the likelihood function. This can be accomplished by developing the likelihood function for the observations and obtaining its logarithmic expression. This expression is then differentiated with respect to the parameters, and the resulting equations are set to equal to zero. These equations are then solved simultaneously to obtain the best estimates of the parameters that maximize the likelihood function.

It should be noted that it is not necessary in all cases to obtain the logarithmic expression of the likelihood function. When it is possible, the likelihood function itself can be maximized without resorting to its logarithmic expression.

EXAMPLE 4.11

What is the maximum likelihood estimate of μ for the Poisson distribution given in Example 4.7?

SOLUTION

Using the logarithm of the maximum likelihood function given by Eq. (4.8),

$$L(\mu) = 22 \log \mu - 2\mu - \log(10!\ 12!).$$

The derivative of $L(\mu)$ with respect to μ is

$$\frac{dL(\mu)}{d\mu} = \frac{22}{\mu} - 2 = 0.$$

The "best" estimate of $\hat{\mu}$ is $22/2 = 11$.

EXAMPLE 4.12

What is the maximum likelihood estimate of θ in Example 4.8?

SOLUTION

$$l(\theta) = K\theta^{10}(1-\theta)^{40}$$

$$L(\theta) = \log K + 10 \log \theta + 40 \log(1-\theta)$$

$$\frac{dL(\theta)}{d\theta} = 0 + \frac{10}{\theta} - \frac{40}{1-\theta} = 0$$

and the "best" estimate of $\hat{\theta}$ is 1/5.

We now consider the maximum likelihood estimators (MLE) for the parameters of the exponential, Rayleigh, and normal distributions.

4.2.2. EXPONENTIAL DISTRIBUTION

The p.d.f. of the exponential distribution with parameter λ is

$$f(x, \lambda) = \lambda e^{-\lambda x}.$$

The p.d.f. of n observations $x_1, x_2, \ldots, x_n$ is

$$f(x_i, \lambda) = \lambda e^{-\lambda x_i} \qquad i = 1, 2, \ldots, n.$$

The likelihood function $l(x_1, x_2, \ldots, x_n; \lambda)$ is

$$l(x_1, x_2, \ldots, x_n; \lambda) = f(x_1, \lambda) f(x_2, \lambda) \ldots f(x_n, \lambda)$$

$$= \prod_{i=1}^{n} f(x_i, \lambda)$$

$$= \lambda^n \prod_{i=1}^{n} e^{-\lambda x_i}$$

$$= \lambda^n e^{-\lambda \sum_{i=1}^{n} x_i}.$$

The logarithm of the likelihood function is

$$L(x_1, x_2, \ldots, x_n; \lambda) = n \log \lambda - \lambda \sum_{i=1}^{n} x_i$$

and

$$\frac{\partial L(x_1, x_2, \ldots, x_n; \lambda)}{\partial \lambda} = \frac{n}{\lambda} - \sum_{i=1}^{n} x_i = 0.$$

The "best" estimate of λ is $n/\sum_{i=1}^{n} x_i$.

This is the same estimate as that obtained by the method of moments.

EXAMPLE 4.13 A sample of six electronic components is subjected to a reliability test to estimate the mean time to failure. The following are the times to failure of the components: 25, 75, 150, 230, 430, and 700 hours. What is the failure rate? Estimate the parameter(s) of the underlying failure-time distribution.

SOLUTION The mean of the time to failure is 260 hours and the standard deviation is 232 hours. Since the mean and the standard deviation are approximately equal, then the exponential distribution can be used to represent the failure-time distribution.

Therefore, the "best" estimate of $\hat{\lambda}$ (the parameter of the exponential distribution) as determined by the maximum likelihood method is

$$\hat{\lambda} = \frac{n}{\sum_{i=1}^{n} x_i},$$

where x_i is the time of the ith failure:

$$\hat{\lambda} = \frac{6}{1{,}610} = 3.726 \times 10^{-3} \text{ failures per hour.}$$

4.2.3. THE RAYLEIGH DISTRIBUTION

As explained in Chapter 1, the Rayleigh distribution is used to represent the failure-time distribution of components that exhibit linearly increasing failure rates. The p.d.f. of the Rayleigh distribution is

$$f(x) = \lambda x e^{-\frac{\lambda x^2}{2}},$$

where λ is the parameter of the Rayleigh distribution.

The likelihood function for n observations is

$$l(x_1, x_2, \ldots, x_n; \lambda) = f(x_1, \lambda) f(x_2, \lambda), \ldots, f(x_n, \lambda)$$

or

$$l(x_1, x_2, \ldots, x_n; \lambda) = \prod_{i=1}^{n} \lambda x_i e^{-\frac{\lambda x_i^2}{2}}.$$

Let

$$\prod_{i=1}^{n} x_i = X.$$

We can rewrite the likelihood function as

$$l(x_1, x_2, \ldots, x_n; \lambda) = \lambda^n X e^{-\frac{\lambda}{2}\sum_{i=1}^{n} x_i^2}.$$

The logarithm of the above function is

$$L(x_1, x_2, \ldots, x_n; \lambda) = n \log \lambda + \log X - \frac{\lambda}{2} \sum_{i=1}^{n} x_i^2. \quad (4.12)$$

Taking the derivative of Eq. (4.12) with respect to λ and equating the resultant to zero, we obtain

$$\frac{\partial L(x_1, x_2, \ldots, x_n; \lambda)}{\partial \lambda} = \frac{n}{\lambda} - \frac{1}{2}\sum_{i=1}^{n} x_i^2 = 0$$

or

$$\hat{\lambda} = \frac{2n}{\sum_{i=1}^{n} x_i^2}. \tag{4.13}$$

EXAMPLE 4.14

The following failure times are observed while conducting a reliability test: 15, 21, 30, 39, 52, and 68 hours. Assume that a Rayleigh distribution is considered an appropriate distribution to represent these failure times. Determine the parameter of the distribution. What are the mean and standard deviation of the failure time?

SOLUTION

Using Eq. (4.13), the parameter of the Rayleigh distribution is

$$\hat{\lambda} = \frac{2 \times 6}{10{,}415} = 0.00115 \text{ failures per hour.}$$

The mean and standard deviation of the failure times are

$$\hat{\mu} = \sqrt{\frac{\pi}{2\hat{\lambda}}} = 36.92 \text{ hours}$$

$$\hat{\sigma} = \sqrt{\frac{2}{\hat{\lambda}}\left(1 - \frac{\pi}{4}\right)} = 19.3 \text{ hours.}$$

4.2.4. THE NORMAL DISTRIBUTION

The p.d.f. of an observation x from a normal distribution with unknown mean μ and unknown variance σ^2 is

$$f(x) = \frac{1}{\sigma\sqrt{2\pi}} e^{-\frac{1}{2}\left(\frac{x-\mu}{\sigma}\right)^2}.$$

The likelihood function for n observations is

$$l(x_1, x_2, \ldots, x_n; \mu, \sigma) = \left(\frac{1}{\sigma\sqrt{2\pi}}\right)^n \prod_{i=1}^{n} e^{-\frac{1}{2}\left(\frac{x_i-\mu}{\sigma}\right)^2}$$

and the logarithm of the above function is

$$L(x_1, x_2, \ldots, x_n; \mu, \sigma) = n \log \frac{1}{\sigma\sqrt{2\pi}} - \frac{1}{2}\sum_{i=1}^{n}\left(\frac{x_i - \mu}{\sigma}\right)^2. \tag{4.14}$$

Taking the derivative of Eq. (4.14) with respect to μ results in

$$\frac{\partial L(x_1, x_2, \ldots, x_n; \mu, \sigma)}{\partial \mu} = \frac{1}{\sigma^2}\left(\sum_{i=1}^{n} x_i - n\mu\right) = 0$$

$$\hat{\mu} = \frac{1}{n}\sum_{i=1}^{n} x_i.$$

Similarly, taking the derivative of Eq. (4.14) with respect to σ, we obtain

$$\frac{\partial L(x_1, x_2, \ldots, x_n; \mu, \sigma)}{\partial \sigma} = \frac{\partial}{\partial \sigma}\left[n \log \frac{1}{\sqrt{2\pi}} - n \log \sigma - \frac{1}{2}\sum_{i=1}^{n}\left(\frac{x_i - \mu}{\sigma}\right)^2\right]$$

$$= -\frac{n}{\sigma} - \sum \frac{(x_i - \mu)^2}{2\sigma^3}(-2)$$

$$= \frac{1}{\sigma}\left[-n + \sum_{i=1}^{n} \frac{(x_i - \mu)^2}{\sigma^2}\right] = 0$$

$$\sigma^2 = \frac{1}{n}\sum_{i=1}^{n}(x_i - \mu)^2.$$

The estimate of σ^2 is

$$\hat{\sigma}^2 = \frac{1}{n}\sum_{i=1}^{n}(x_i - \hat{\mu})^2.$$

These are the same results as those obtained by the method of moments.

EXAMPLE 4.15

Assume that applied stresses on components and the corresponding failure times form paired observations $(x_1, y_1), \ldots, (x_n, y_n)$ that follow the model

$$E(Y) = \alpha + \beta x$$

$$\mathrm{Var}(Y) = \sigma^2,$$

where Y is an independently and normally distributed random variable. Use the maximum likelihood approach to estimate the parameters α and β.

SOLUTION

Since Y is independently and normally distributed, then the log likelihood is obtained using Eq. (4.14) as

$$L[(x_1, y_1), \ldots, (x_n, y_n)] = \frac{-n}{2} \log 2\pi - n \log \sigma - \frac{1}{2\sigma^2} \sum (y_i - \alpha - \beta x_i)^2. \tag{4.15}$$

The first two terms of the right side of Eq. (4.15) are independent of α and β. Hence, to minimize the log likelihood, it is sufficient to minimize the term

$$K = \sum_{i=1}^{n} (y_i - \alpha - \beta x_i)^2. \tag{4.16}$$

Taking the partial derivatives of K with respect to α and β and equating the derivatives to zero results in two linear equations in α and β. Their solutions yield

$$\hat{\beta} = \frac{\sum y_i (x_i - \bar{x})}{\sum (x_i - \bar{x})^2} \tag{4.17}$$

$$\hat{\alpha} = \bar{y} - \hat{\beta}\bar{x}, \tag{4.18}$$

where

$$\bar{x} = \frac{1}{n} \sum x_i$$

and

$$\bar{y} = \frac{1}{n} \sum y_i.$$

As we discussed earlier, in order to find the maximum likelihood estimator we set the derivatives of the log likelihood function with respect to the parameter being estimated to zero and solve the resulting equation for the value of the parameter. Unfortunately, sometimes there are no closed-form expressions for the estimated parameter(s), and to obtain an estimate of the parameter we may need to employ other methods, such as the following (Wetherill, 1981):

- *The gradient of the likelihood method:* This method is very effective when there is only one unknown parameter, θ. We simply calculate $dL/d\theta$ at various values of θ and plot $dL/d\theta$ versus θ to obtain a line which intersects with θ axis ($dL/d\theta = 0$) at the estimated value of θ. Drawing such a line rarely requires more than three or four calculations. The slope of the line is the second derivative of the likelihood function with respect to θ, ($d^2L/d\theta^2$), which is an approximate estimate of the variance of the estimator.

- *Newton's iterative method:* Newton's iterative method for finding the roots of $f(x) = 0$ is well known. We apply Newton's method to solve the derivative of the log likelihood function with respect to θ:

$$f(\theta) = \left\{\frac{dL}{d\theta}\right\}_{\theta=\theta} = 0. \tag{4.19}$$

If $\hat{\theta}_1$ is any rough estimate of θ, then using Taylor's expansion of Eq. (4.19) about $\hat{\theta}_1$ we obtain

$$\left\{\frac{dL}{d\theta}\right\}_{\hat{\theta}} = \left\{\frac{dL}{d\theta}\right\}_{\hat{\theta}_1} + (\hat{\theta} - \hat{\theta}_1)\left\{\frac{d^2L}{d\theta^2}\right\}_{\hat{\theta}_1} + \ldots = 0. \tag{4.20}$$

If $\hat{\theta}_1$ is sufficiently close to $\hat{\theta}$, we can simply ignore the higher terms of the expansion in Eq. (4.20). Thus,

$$\left\{\frac{dL}{d\theta}\right\}_{\hat{\theta}_1} + (\hat{\theta} - \hat{\theta}_1)\left\{\frac{d^2L}{d\theta^2}\right\}_{\hat{\theta}_1} \cong 0, \tag{4.21}$$

and a new estimate of $\hat{\theta}_2$ can be made as

$$\hat{\theta}_2 = \hat{\theta}_1 - \left\{\frac{dL}{d\theta}\right\}_{\hat{\theta}_1} \Bigg/ \left\{\frac{d^2L}{d\theta^2}\right\}_{\hat{\theta}_1}. \tag{4.22}$$

We can use the above expression recursively until the estimate of the parameter has converged. Clearly the rate of convergence depends on the selection of the initial value of $\hat{\theta}_1$. Unfortunately, there is no general method that enables us to provide a good initial estimate of $\hat{\theta}$ but a rough plot of the log likelihood function may provide an acceptable initial value for $\hat{\theta}_1$.

The maximum likelihood estimators are consistent, efficient, and unbiased. The bias of the estimators decreases as the number of observations increases. The method requires simple calculations for single parameter distributions but may require extensive computations for two or more parameter distributions. Moreover, the method is applicable for both censored (or truncated) and noncensored data.

4.2.5. INFORMATION MATRIX AND THE VARIANCE-COVARIANCE MATRIX

One of the major benefits of the use of the maximum likelihood estimator (MLE) to obtain the parameters of distribution(s) is that the logarithm of the likelihood function can be utilized in constructing the so-called *Fisher information matrix* (or *Hessian matrix*). The inverse of the matrix results in the well-known variance-covariance matrix.

Before we introduce the procedure for constructing the information matrix, we first need to define the variance-covariance matrix (or simply covariance matrix). If $X_1, X_2, \ldots, X_k$ are k mutually independent and identically distributed random variables within a p.d.f. $f(x, \theta_0)$, where θ_0 has k components and is the true value of θ, then the covariance matrix is defined as

$$\begin{bmatrix} \mathrm{Var}(\theta_1) & \mathrm{Cov}(\theta_1, \theta_2) & \mathrm{Cov}(\theta_1, \theta_3) & \ldots & \mathrm{Cov}(\theta_1, \theta_k) \\ \mathrm{Cov}(\theta_1, \theta_2) & \mathrm{Var}(\theta_2) & \mathrm{Cov}(\theta_2, \theta_3) & \ldots & \mathrm{Cov}(\theta_2, \theta_k) \\ \cdot & & & & \\ \cdot & & & & \\ \cdot & & & & \\ \mathrm{Cov}(\theta_1, \theta_k) & \mathrm{Cov}(\theta_2, \theta_k) & \ldots & \ldots & \mathrm{Var}(\theta_k) \end{bmatrix},$$

where $\mathrm{Cov}(\theta_i, \theta_j)$ is the covariance of θ_i and θ_j, and $\mathrm{Var}(\theta_i)$ is the variance of θ_i. This covariance matrix can be obtained from the information matrix as follows.

As was stated earlier in this chapter, when the sample size of data increases, the bias of the MLEs decreases, and they become asymptotically unbiased. In other words,

$$\lim_{n \to \infty} E[\hat{\theta}_i] = \theta_i, \qquad i = 1, 2, \ldots, k.$$

To find the asymptotic variances and covariances of the estimators, we first construct the information matrix $\boldsymbol{I}$, regarding the likelihood as a function of random variables observed in a given sample.

The (ij)th element of the information matrix $\boldsymbol{I}$ is

$$\boldsymbol{I}_{ij} = E\left[-\frac{\partial^2 L(\theta; X)}{\partial\theta_i \, \partial\theta_j}\right]. \tag{4.23}$$

The inverse matrix, $\boldsymbol{I}^{-1}$, with the (ij)th element denoted by I^{ij}, is the variance-covariance matrix of the $\hat{\theta}$'s, so that (Elandt-Johnson and Johnson, 1980)

$$\mathrm{Var}(\hat{\theta}_i) = I^{ii} \text{ and } \mathrm{Cov}(\hat{\theta}_i, \hat{\theta}_j) = I^{ij}. \tag{4.24}$$

EXAMPLE 4.16

A random sample $x_1, x_2, \ldots, x_n$ follows a normal distribution with parameters μ and σ^2. Use the information matrix to obtain the variances of $\hat{\mu}$ and $\hat{\sigma}$.

SOLUTION

The logarithm of the likelihood function of the normal distribution is given by Eq. (4.14) as

$$L(x_1, x_2, \ldots, x_n; \mu, \sigma) = n \log \frac{1}{\sigma\sqrt{2\pi}} - \frac{1}{2}\sum_{i=1}^{n}\left(\frac{x_i - \mu}{\sigma}\right)^2.$$

The partial derivatives of L with respect to μ and σ are

$$\frac{\partial L}{\partial \mu} = \frac{1}{\sigma^2}\left(\sum_{i=1}^{n} x_i - n\mu\right)$$

$$\frac{\partial L}{\partial \sigma} = \frac{1}{\sigma}\left[-n + \sum_{i=1}^{n}\frac{(x_i - \mu)^2}{\sigma^2}\right]$$

$$\frac{\partial^2 L}{\partial \mu^2} = \frac{-n}{\sigma^2} \tag{4.25}$$

$$\frac{\partial L^2}{\partial \mu\, \partial \sigma} = \frac{-2}{\sigma^3}\left(\sum_{i=1}^{n} x_i - n\mu\right) \tag{4.26}$$

$$\frac{\partial L^2}{\partial \sigma^2} = \frac{n}{\sigma^2} - \frac{3}{\sigma^4}\sum_{i=1}^{n}(x_i - \mu)^2. \tag{4.27}$$

In order to construct the information matrix, we obtain the expectations of Eqs. (4.25), (4.26), and (4.27). Then

$$E\left[\frac{\partial^2 L}{\partial \mu^2}\right] = \frac{-n}{\sigma^2} = -I^{11}$$

$$E\left[\frac{\partial L^2}{\partial \mu\, \partial \sigma}\right] = 0 = -I^{12} = -I^{21}$$

$$E\left[\frac{\partial L^2}{\partial \sigma^2}\right] = \frac{-2n}{\sigma^2} = -I^{22}.$$

Thus the information matrix $\boldsymbol{I}$ is constructed as

$$\boldsymbol{I} = \begin{pmatrix} n/\sigma^2 & 0 \\ 0 & 2n/\sigma^2 \end{pmatrix} \quad \text{and} \quad \boldsymbol{I}^{-1} = \begin{pmatrix} \sigma^2/n & 0 \\ 0 & \sigma^2/2n \end{pmatrix}.$$

The variance-covariance matrix, $\boldsymbol{I}^{-1}$, is

$$\begin{pmatrix} \mathrm{Var}(\hat{\mu}) & \mathrm{Cov}(\hat{\mu}, \hat{\sigma}) \\ \mathrm{Cov}(\hat{\mu}, \hat{\sigma}) & \mathrm{Var}(\hat{\sigma}) \end{pmatrix} = \begin{pmatrix} \sigma^2/n & 0 \\ 0 & \sigma^2/2n \end{pmatrix}.$$

EXAMPLE 4.17

A check-weigher is a piece of equipment that has three major components: a scale, a controller, and a diverter. In typical high-speed production systems such as those found in the canned food industry or the pharmaceutical manufacturing industry, one or more check-weighers are usually installed in the system to ensure that the weights of the products are within acceptable specification limits. If a product fails to meet the specifications, it is diverted away from the acceptable products. The diverter, being a mechanical system, is the most susceptible component to failure. The following times to failure (in weeks) of a diverter are observed:

14, 18, 18, 20, 21, 22, 22, 20, 17, 17, 15, 13.

Assume that the observations follow a normal distribution with mean μ and variance σ^2. Determine $\hat{\mu}$, $\hat{\sigma}$, and the variance-covariance matrix.

SOLUTION

From Section 4.2.4 we obtain $\hat{\mu}$ and $\hat{\sigma}$ as follows:

$$\hat{\mu} = \frac{1}{n}\sum_{i=1}^{n} x_i = \frac{1}{12} \times 217 = 18.08$$

and

$$\hat{\sigma}^2 = \frac{1}{n}\sum_{i=1}^{n} (x_i - \hat{\mu})^2 = 8.409$$

or

$$\hat{\sigma} = 2.9.$$

The variance-covariance matrix as shown in Example 4.17 is

$$\boldsymbol{I}^{-1} = \begin{pmatrix} \sigma^2/n & 0 \\ 0 & \sigma^2/2n \end{pmatrix} = \begin{pmatrix} 0.700 & 0 \\ 0 & 0.350 \end{pmatrix}.$$

Thus the variance of $\hat{\mu}$ is 0.700 and that of $\hat{\sigma}$ is 0.350.

4.3. Method of Least Squares

The method of least squares provides an efficient and unbiased estimator of the distribution parameters. The method defines the best fit as one that minimizes the sum of squared errors between the observed data and the fitted distribution (Elsayed and Boucher, 1994). Although the method is general and can be used for simple linear, multiple linear, and nonlinear regression models, we will limit our presentation to linear models.

Consider a set of data that may include extreme data points (or noise). We are interested in finding a function that reflects the pattern in the data and reduces the errors to a minimum. A simple plot of the data may reveal the underlying data-generating process as being linear or nonlinear. Assume that the data-generating process can be represented by the following linear model

$$f(x) = \alpha + \beta x + \varepsilon_x, \tag{4.28}$$

where

$f(x)$ = observed value of the function at x,

α, β = intercept and slope, respectively,

x = independent variable such as time, and

ε_x = random noise in the process at time x.

It is assumed that ε_x is normally and independently distributed, with mean $\bar{\varepsilon}_x = 0$ and $\text{Var}(\varepsilon_x) = \sigma^2$. Based on the assumption of a linear process, we propose to fit a model of the form

$$\hat{f}(x) = \hat{\alpha} + \hat{\beta}x, \tag{4.29}$$

where

$\hat{f}(x)$ = the estimated value of the function at x, and

$\hat{\alpha}, \hat{\beta}$ = estimates of α and β.

Let $e(x) = \hat{f}(x) - f(x)$ be the value of the error between the proposed polynomial $\hat{f}(x)$ and the actual data $f(x)$. Then we define the sum of squares of errors, SS_E, as

$$SS_E = \sum_{x=1}^{n} e^2(x), \tag{4.30}$$

where n is the total number of data points used for estimating $\hat{f}(x)$. Equation (4.30) can be rewritten as

$$SS_E = \sum_{x=1}^{n} [\hat{f}(x) - f(x)]^2. \tag{4.31}$$

The minimization of SS_E is accomplished by taking the partial derivatives of SS_E with respect to $\hat{\alpha}$ and $\hat{\beta}$ and setting the resulting equations to zero:

$$SS_E = \sum_{x=1}^{n} [f(x) - \hat{\alpha} - \hat{\beta}x]^2$$

$$\frac{\partial SS_E}{\partial \hat{\alpha}} = 2\sum_{x=1}^{n} [f(x) - \hat{\alpha} - \hat{\beta}x] = 0 \tag{4.32}$$

$$\frac{\partial SS_E}{\partial \hat{\beta}} = 2\sum_{x=1}^{n} [f(x) - \hat{\alpha} - \hat{\beta}x]\, x = 0. \tag{4.33}$$

Rewriting Eqs. (4.32) and (4.33) gives us

$$\sum_{x=1}^{n} f(x) = n\hat{\alpha} + \hat{\beta}\sum_{x=1}^{n} x \tag{4.34}$$

$$\sum_{x=1}^{n} x f(x) = \hat{\alpha}\sum_{x=1}^{n} x + \hat{\beta}\sum_{x=1}^{n} x^2, \tag{4.35}$$

which yields

$$\hat{\alpha} = \frac{\sum x^2 \sum f(x) - \sum x \sum x f(x)}{n \sum x^2 - \left(\sum x\right)^2} \quad (4.36)$$

$$\hat{\beta} = \frac{n \sum x f(x) - \sum x \sum f(x)}{n \sum x^2 - \left(\sum x\right)^2}. \quad (4.37)$$

After estimating the parameters of the model, one may wish to know how well the proposed model fits the data. The coefficient of determination and the coefficient of correlation are typical criteria that can be used for that purpose. They are, respectively, given by

$$r^2 = \frac{\sum (\hat{f}(x) - \bar{f}(x))^2}{\sum (f(x) - \bar{f}(x))^2} \quad (4.38)$$

and

$$\rho = \frac{\sigma_{x,f(x)}}{\sigma_x \, \sigma_{f(x)}}, \quad (4.39)$$

where

r^2 = coefficient of determination $0 \le r^2 \le 1$,

ρ = coefficient of correlation $0 \le \rho \le 1$,

$\sigma_{x,f(x)}$ = covariance of x and $f(x)$,

$= \Sigma_{x=1}^{n} (x - \bar{x}) (f(x) - \bar{f}(x))$,

$\sigma_x = \Sigma (x - \bar{x})^2$, and

$\sigma_{f(x)} = \Sigma (f(x) - \bar{f}(x))^2$.

Derivations of Eqs. (4.38) and (4.39) are given in Elsayed and Boucher (1994). A coefficient of determination of 0 indicates that the model does not fit the data, whereas when $r^2 = 1$, the model represents an ideal fit. Similarly when $r = 1$ or -1 the model indicates that there is a perfect positive correlation or a perfect negative correlation, respectively. When $r = 0$, then there is no correlation between the $f(x)$ and x. Therefore, when r^2 approaches 1 or r approaches ± 1, the model is a good fit of the data.

EXAMPLE 4.18

The surface mount technology (SMT) enables the manufacturers of electronic components to create printed circuits assemblies with high component density. One problem with SMT, though, is that a surface mount device contact is attached to a printed circuit board (PCB) with only solder. As a result, the SMT attachment is only as reliable as the solder. Hence, manufacturers are required to perform accelerated reliability testing to determine the long-term reliability at normal operating conditions. The following failure-time data is obtained from such a test: 10, 20, 30, 40, 50, 60, 70, 80, 93, and 111 hours. Assume that the failure rate is linearly increasing with t in the form $h(t) = \alpha + \beta t$. Determine the constants α and β and estimate the reliability at $t = 30$ hours.

SOLUTION

We construct Table 4.3 to estimate the values of $h(t)$. Using the least-squares method we obtain the following hazard-rate expression:

$$h(t) = -0.000498 + 0.000429t$$

and the reliability at $t = 30$ hours is

$$R(t) = e^{-\{0.0002145t^2 - 0.000498t\}}$$

$$R(30) = 0.83685.$$

The least-squares method provides an efficient, consistent, and unbiased estimate of the distribution parameters. The least-squares method is simple and computationally efficient and can be used for simple linear, multiple linear, and nonlinear

Table 4.3. Failure-Rate Calculations

t	$h(t) \times 10^{-3}$ *Failures per Hour*
10	10.00
20	11.11
30	12.50
40	14.28
50	16.66
60	20.00
70	25.00
80	33.33
93	38.40
111	55.45

modeling. Moreover, many nonlinear forms can be transformed to linear expressions by using simple transformations. For example,

$$f(x) = ax^b \tag{4.40}$$

can be linearized by taking the logarithm of both sides of the equation, which results in

$$\log f(x) = \log a + b \log x. \tag{4.41}$$

Let

$Y = \log f(x)$,

$X = \log x$, and

$A = \log a$.

Then Eq. (4.41) can be written in the linear form of

$$Y = A + bX. \tag{4.42}$$

In summary, the least squares estimators of a linear model (1) are unbiased, (2) have minimum variance among linear unbiased estimators, and (3) are obtained such that the residuals and estimators are uncorrelated (Wetherill, 1981).

PROBLEMS

4-1. Given are the following failure-time data—

40, 45, 55, 68, 78, 85, 94, 99, 120, 140, 160, and 175 hours.

a. Assuming that the data follow an exponential distribution, derive an expression for the failure-rate function.

b. Use the methods of moments to estimate the parameter of the exponential distribution.

c. What is the reliability of a component that belongs to the same population of tested units at time $t = 49$ hours?

d. Plot the reliability of the component against time.

4-2. The range of the parameters of the beta distribution enables it to model a wide variety of failure-time data. Hence, beta distribution is used widely in many reliability engineering applications. The probability density function of the beta distribution is

$$f(t) = \frac{\Gamma(\alpha + \beta + 2)}{\Gamma(\alpha + 1)\,\Gamma(\beta + 1)}\, t^{\alpha}(1 - t)^{\beta},$$

where $0 < t < 1$, $\alpha > -1$, and $\beta > -1$. α, β are shape parameters. The mean and the standard deviation of the distribution are

$$\mu = \frac{\alpha + 1}{\beta + \alpha + 2}$$

$$\sigma = \left[\frac{(\alpha + 1)(\beta + 1)}{(\alpha + \beta + 2)^2\,(\alpha + \beta + 3)}\right]^{\frac{1}{2}}.$$

The following failure-time data are obtained from a laboratory test

50, 100, 130, 140, 142, 150, 160, 172, 179, 200, 220 days.

a. Use the method of moments to estimate the parameters of the beta distribution.

b. Derive expressions for $f(t)$, $F(t)$, $R(t)$, and $h(t)$.

c. Plot $R(t)$ and $h(t)$ against time. *Hint:* Since $0 < t < 1$, use one year to represent one unit of time. Also, $\Gamma(11.807) = 24{,}952{,}916$.

4-3. The following failure times of the Weibull distribution are obtained from a reliability test:

320, 370, 410, 475, 562, 613, 662, 770, 865, 1,000 hours.

a. Use the method of moments to determine the parameters that fit the above data.

b. Use the maximum likelihood method to obtain the parameters of the Weibull distribution. Use Newton's approach to solve for the values of the parameters. (Use γ and θ obtained from (a) as starting values for Newton's method.)

c. Solve (b) using the least-squares method.

d. Compare the results obtained from a, b, and c. Draw your conclusions.

4-4. Use the maximum-likelihood method to obtain the parameters of a two-parameter exponential distribution having a p.d.f. of

$$f(t) = \lambda e^{-\lambda(t-\gamma)}, \quad f(t) \geq 0, \quad \lambda > 0, \quad t \geq \gamma,$$

where γ is the location parameter of the distribution.

4-5. Solve Problem 4-4 using the least-squares method.

4-6. Plot the contours of the likelihood function for the two-parameter exponential distribution for different values of λ and γ.

4-7. Plot the contours of the likelihood function for the normal distribution for different μ and σ.

4-8. Use the maximum-likelihood approach to estimate the parameters of the log-normal distribution whose p.d.f. is

$$f(t) = \frac{1}{\sigma t \sqrt{2\pi}} \exp\left[-\frac{1}{2}\left(\frac{\ln t - \mu}{\sigma}\right)^2\right].$$

a. Construct the information matrix and determine the covariance matrix.

b. Construct the confidence limits for μ.

4-9. Consider a system whose components fail if they enter either of two stages of failure mechanisms: the first mechanism is due to excessive voltage, and the second is due to excessive temperature. Suppose that the failure mechanism enters the first stage with probability θ, and the p.d.f. of the failure time is $\lambda_1 e^{-\lambda_1 t}$. It enters the second stage with probability $(1-\theta)$, and the p.d.f. of its failure time is $\lambda_2 e^{-\lambda_2 t}$. The failure of a component occurs at the end of either stage. Hence, the p.d.f. of the failure time is

$$f(t) = \theta\lambda_1 e^{-\lambda_1 t} + (1-\theta)\lambda_2 e^{-\lambda_2 t}.$$

a. Use the three methods: the method of moments, the maximum-likelihood approach, and the least-squares method to obtain the parameters of the above distribution.

b. Which method do you prefer? Why?

4-10. Construct the Fisher information matrix for the p.d.f. of the failure time given in Problem 4-9. Determine the variance of λ_1 and λ_2.

4-11. A producer of light emitting diodes (LED) subjects twenty-five units to reliability test conditions similar to those of the normal operating conditions. They are subjected to a temperature of 70 °F and an electric field of 5V. The failure-time data are recorded and observed to follow a special Erlangian distribution of the form

$$f(t) = \frac{t}{\lambda^2} \exp\left(\frac{-t}{\lambda}\right) \qquad t \geq 0.$$

a. Use the method of moments to estimate λ.

b. Use the method of the maximum likelihood to estimate λ.

c. Compare the results obtained in a and b.

4-12. A typical proportional, integral, and derivative (PID) controller consists of a stand-alone regulator (which adjusts the control variable of a process), a front end where the controller constants are manually entered, and a processor (or a computer) where the control algorithm is implemented. When a controller observes deviations in a process output from predefined reference values, the regulator automatically adjusts the process parameter to compensate for such deviations. The regulator is a mechanical, electrical, or an electromechanical system that implements the appropriate control action on the process parameter. Twenty controllers are placed in service and the times to failure of the regulators are recorded as follows:

551, 571, 571, 575, 583, 588, 590, 592, 594, 598, 606, 610, 611, 611, 613, 615, 615, 626, 629, 637.

a. Assuming that the failure data follow an exponential distribution, use the method of moments to obtain its parameter.

b. Assume that the failure data follow a Weibull distribution. Use the maximum-likelihood approach to estimate the parameters of the distribution.

c. Compare the results of a and b. What do you conclude about the failure-time distribution? Which method is preferred?

4-13. A manufacturer of hydraulic turbomachinery produces turbines, impellers, pumps, and similar equipment. The manufacturer is interested in estimating the expected life of the components of power generating turbines. The manufacturer subjects the turbines to high-speed flows through the components resulting in pressure differences that can cause the flow to vaporize and form bubbles. When the bubbles collapse because of a change in pressure, liquid particles bombard the surface of the machinery at high velocities. Such high-velocity, high-pressure liquid particles can chip metal out of the structure and create local fatigue regions in the equipment, which eventually results in the failure of the machinery.

The manufacturer subjects fifteen turbines to a high-speed flow test and obtains the following failure times:

46, 70, 76, 78, 81, 86, 87, 92, 93, 95, 101, 105, 148, 154, 158.

Assume that the failure-time data follow a log-logistic distribution of the form

$$f(t) = \lambda p(\lambda t)^{p-1}[1 + (\lambda t)^p]^{-2},$$

where $\lambda = e^{-\alpha}$ and $p = 1/\sigma$. Use the method of moments to estimate the parameters α and σ. Estimate the reliability at $t = 200$ hours.

4-14. The following are the probability density functions for four failure-time distributions:

- Cauchy

$$f(x) = \left\{\pi\beta\left[1 + \left(\frac{x-\alpha}{\beta}\right)^2\right]\right\}^{-1},$$

where $-\infty < \alpha < \infty, \quad \beta > 0, \quad -\infty < x < \infty;$

- Gumbel (or extreme value)

$$f(x) = \frac{1}{\beta}\exp\left[-e^{-(x-\alpha)/\beta} + (x-\alpha)/\beta\right],$$

where $-\infty < \alpha < \infty, \quad \beta > 0, \quad -\infty < x < \infty;$

- Logistic

$$f(x) = \frac{(1/\beta)e^{-(x-\alpha)/\beta}}{(1 + e^{-(x-\alpha)/\beta})^2},$$

where $-\infty < \alpha < \infty, \quad \beta > 0, \quad -\infty < x < \infty;$

- Pareto

$$f(x) = \frac{\alpha_2 c^{\alpha_2}}{x^{\alpha_2+1}},$$

where $c > 0, \quad \alpha_2 > 0, \quad x > c.$

a. Use an appropriate method to estimate the parameters of the above distributions. You may or may not use the same methods for the four functions.

b. Explain the situations and conditions under which each one of the above distributions can be used in reliability modeling.

4-15. The p.d.f. of the gamma distribution is given by

$$f(x) = \frac{1}{\Gamma(\alpha)\beta^{\alpha}} x^{\alpha-1} e^{-x/\beta} \qquad x > 0.$$

The mean and variance of the gamma distribution are $\alpha\beta$ and $\alpha\beta^2$, respectively.

a. Use the method of moments to obtain the parameters of the distribution.

b. Develop the likelihood function for the distribution and plot it against the parameters α and β.

4-16. The p.d.f. of the chi-squared distribution is

$$f(t) = \frac{t^{1/2\nu - 1}}{\Gamma(1/2\nu)\, 2^{1/2\nu}} e^{-1/2t} \qquad t > 0,$$

where ν is the number of degrees of freedom of the chi-squared distribution.

a. Use the method of moments to obtain the parameter of the chi-squared distribution if the mean and the variance are ν and 2ν, respectively.

b. Use the least-squares method to obtain the parameter of the chi-squared distribution.

REFERENCES

Dixon, W. J., and Massey, F. J., Jr. (1957). *Introduction to Statistical Analysis*. New York: McGraw-Hill.

Elandt-Johnson, R. C., and Johnson, N. L. (1980). *Survival Models and Data Analysis*. New York: John Wiley.

Elsayed, E. A., and Boucher, T. O. (1994). *Analysis and Control of Production Systems*. Englewood Cliffs, NJ: Prentice-Hall.

Grubbs, F. E. (1950). "Sample Criteria for Testing Outlying Observations." *Annals of Mathematical Statistics*, 21, 27–58.

Hawkins, D. M. (1980). *Identification of Outliers*. London: Chapman and Hall.

Natrella, M. G. (1963). *Experimental Statistics*. Washington, DC: National Bureau of Standards Handbook, Vol. 91.

Wetherill, G. B. (1981). *Intermediate Statistical Methods*. London: Chapman and Hall.

CHAPTER 5

Parametric Reliability Models

One of the most important factors that influence the design process of a product or a system is the reliability values of its components. For example, a modern piece of equipment, such as a personal computer, often uses a large number of integrated circuits in a single system. Clearly, a small percentage of the integrated circuits that exhibit early life failures can have a significant impact on the reliability of the system. It is commonly observed that the failure rate of semiconductor devices (components of the integrated circuits) is initially high and then decreases to a steady-state value.

There are many examples that illustrate that the reliability of the system is highly dependent on the reliability of the individual components that comprise the system. Submarine optical fiber transmission systems, weather satellites, telecommunication networks, and supercomputers are typical systems that require high-reliability components in order to achieve the reliability goals of the total system.

In order to estimate the reliability of the individual components or the entire system, we may follow one or more of the following approaches.

5.1. Approach 1: Historical Data

The failure data for components can be found in data banks such as GIDEP (Government-Industry Data Exchange Program), MIL-HDBK-217D (which includes failure data for components as well as procedures for reliability prediction), *AT&T Reliability Manual* (Klinger, Nakada, and Menendez, 1990), and *Bell Communications Research Reliability Manual* (Bell Communications Research, 1986). In such data banks and manuals, the failure data are collected from different manufacturers and presented with a set of multiplying factors that relate the failure rates to different manufacturer's quality levels and environmental conditions. For example, the general expression used to determine steady-state failure rate λ_{ss} of an electronic or electrical component that includes different devices is

$$\lambda_{ss} = \pi_E \left[\sum_{i=1}^{n} N_i (\lambda_G \pi_Q \pi_S \pi_T)_i \right], \tag{5.1}$$

where

π_E = environmental factor for a component,

N_i = quantity of the ith device,

n = number of different N_i devices in the system,

λ_G = generic failure rate for the ith device,

π_Q = quality factor for the ith device,

π_S = stress factor for the ith device, and

π_T = temperature factor for the ith device.

The factors π_E, π_Q, π_S, and π_T are estimated empirically and are found in Klinger, Nakada, and Menendez 1990 and Bell Communications Research, 1986.

It should be mentioned that the collection (and analysis) of field data poses challenging and important problems, and yet it has not been discussed much in the statistical literature (Lawless, 1983).

5.2. Approach 2: Operational Life Testing

An *operational life test* (OLT) is one in which prototypes of a product—whether it is a single component product such as a telephone pole or multi-component products such as cars and computers—are subjected to stresses and environmental conditions at typical operating conditions.

The duration of the test is determined by the number of products under test (sample size) and the number of failures. In all cases, the test should be terminated when its duration reaches the expected life of the product.

An example of the operational life testing is the testing of telephone poles by taking a sample and placing it under the same environmental and weather conditions and observing the failures times. Similar operational life testings are performed on electric switching systems and mechanical testing machines.

Usually, the OLT equipment is designed to be capable both of operating components and of testing them on a scanning basis. As mentioned earlier, the test conditions are not accelerated but rather designed to simulate the field operating conditions (such as temperature fluctuations, power on/off, and so on).

Analyses of test results are used to monitor and estimate the reliabilities and failure rates of products in order to achieve the desired specifications.

Although the results obtained from OLT are the most useful among other tests, the duration of the test is relatively long, and the costs associated with the tests may make them prohibitive to run. Indeed, this test may not be classified as an accelerated life testing since no real acceleration of time or stress is performed.

5.3. Approach 3: Burn-In Testing

It is often found that in a large population of components (or products) some individual components have quality defects that considerably affect the component life. In order to "weed out" these individual components, a *burn-in test* is performed at stressed conditions—that is, the time is accelerated. It is important to note that the test conditions must be determined such that the majority of failures are detected without significantly overstressing the remaining components. Additionally, an optimal burn-in period should be estimated such that the total cost to the producer and the user of the product is minimized. There are two cost elements that should be considered in estimating the optimal burn-in period. They are (1) cost per unit time of the test (long test periods are costly to the producer) and (2) cost of premature failure since short test periods may not completely weed out the defective components, which in turn results in significant costs for both producers and consumers. Mathematical models for estimating the optimal burn-in period are given in the literature of Jensen and Petersen (1982) and Bergman (1985).

5.4. Approach 4: Accelerated Life Testing

Accelerated life testing (ALT) is used to obtain information quickly on life distributions, failure rates, and reliabilities. ALT is achieved by subjecting the test units to conditions such that the failures occur sooner. Thus, prediction of long-term reliability can be made within a short period of time. Results from the ALT are used to extrapolate the unit characteristic at any future time t and at given normal operating conditions. There are two methods used for conducting an accelerated life test. In the first method, it is possible to accelerate the test by using the product more intensively than in normal use. For example, in evaluating the life distribution of a lightbulb of a telephone set which is used on the average one hour a day, a usage of the bulb during its expected life of forty years can be compressed into eighteen months by cycling the power on/off continuously during the test period. Another example, the endurance limit of a crankshaft of a car with an expected life of fifteen years (three hours of driving per day), can be obtained by compressing the test into two years. However, such time compression (accelerating time) may not be possible for a product that is in constant use, such as a mainframe computer.

When time cannot be compressed, the test is usually conducted at higher stress levels than those at normal use. For example, assuming the normal operating temperature of a computer is 25°C, we may accelerate the test by subjecting the critical components of the computer to a temperature of 100°C or higher. This causes the failure of the components to occur in a shorter time. Obviously, the higher the stress, the shorter the time needed for the failures to occur.

Another approach for accelerating life testing is to perform the test at very high stress levels in order to induce failures in very short times. We refer to this approach as *highly accelerated life testing* (HALT). This test attempts to greatly reduce the test time for both burn-in and life test. The ceramic capacitor is a good example for using HALT to evaluate both life test and production burn-in. Generally, the duration of the burn-in of the ceramic capacitor is about 100 hours. However, the use of the HALT approach can reduce the burn-in time significantly and in turn increase the throughput of the production facility. To apply HALT for ceramic capacitors, one or both of the two factors—temperature and voltage—may be used. Obviously, there are maximum stress levels beyond which the tested product will be damaged. Moreover, there are other dangers associated with accelerated tests. For example, voltage increases can create dangerous situations for personnel and equipment since the fuses, used to protect the bias supply, often explode.

The results at the accelerated stress conditions are then related to those at the design stress levels (that is, normal operating conditions) using a physically appropriate statistical model (Meeker and Hahn, 1985).

Two important statistical problems in accelerated life testing are *model identification* and *parameter estimation*. While model identification is the more difficult of the two, they are interrelated. Lack of fit of a model can be due in part to the use of an inefficient method of parameter estimation.

The model usually portrays a valid relationship between the results at accelerated conditions and the normal conditions when the failure mechanisms are the same at both conditions. Once an appropriate model has been identified, it is reasonable to ask which method of parameter estimation is better in terms of such criteria as root mean square (RMS) error and bias. The methods commonly used for parameter estimation are maximum likelihood estimator (MLE) and best linear unbiased estimator (BLUE). Methods for parameters estimation are discussed in Chapter 4.

A valid statistical analysis does not require that all test units fail. This is especially true in situations where the accelerated stress conditions are close to the normal operating conditions and failures may not occur during the predetermined test time. The information about nonfailed units at such stress levels is more important than the information about failed units, which are tested at much higher stress levels than the operating conditions. Therefore, the information about the nonfailed units (censored) must be incorporated into the analysis of the data. Recognizing this fact, it is important to present the different types of censoring used in the accelerated life testing.

5.5. Types of Censoring

One common aspect of reliability data that causes difficulties in the analysis is the censoring that takes place, since not all units under test will fail before the test-stopping criteria is met. There are many types of censoring, but we limit our presentation to the widely used types of censoring.

5.5.1. TYPE 1 CENSORING

Suppose we place n units under test for a period of time T. We record the failure times of r failed units as $t_1, t_2, t_3, \ldots, t_r \leq T$. The test is terminated at time T with $n - r$ surviving (nonfailed) units. The number of failures, r, is a random variable that depends on the duration of the test and the applied stress level.

Analysis cannot be performed about the reliability and failure rate of the units if no failures occur during T. Therefore, it is important to determine T such that at least some units fail during the test. The time T at which the test is terminated is referred to as the test censoring time, and this type of censoring is referred to as type 1 censoring.

5.5.2. TYPE 2 CENSORING

Suppose we place n units under test and the exact failure times of failed units are recorded. The test continues to run until exactly r failures occur.

The test is terminated at t_r. Since we specify r failures in advance, we know exactly how much data will be obtained from the test. It is obvious that this type of testing guarantees that r failure times will occur, and reliability analysis of the data is assured. Of course, the accuracy of reliability analysis is dependent on the number of failure times recorded. The test duration, T, is a random variable that depends on the same factors discussed in type 1 censoring.

In this type of test the censoring parameter is the number of failures, r, during the test. It is usually preferred to type 1 censoring.

5.5.3. RANDOM CENSORING

Random censoring arises when, for example, n units (devices) are divided among two or more independent pieces of test equipment. Suppose after time t_f has elapsed, we observe a failure of one of the pieces of test equipment. The units placed on this test equipment are removed from the test, while the remaining units on the other test equipment continue until the test is completed.

The time at which we observe the failure of the test equipment is called the censoring time of units. Since the failure time of the test equipment is a random variable, we refer to this type of censoring as *random censoring*.

In the following sections, we analyze the failure data obtained from the reliability testing. We start by using parametric fittings for the data when failure times of all units under test are known and when censoring exists.

5.5.4. HAZARD-RATE CALCULATIONS UNDER CENSORING

As discussed in Chapter 1, the hazard rate for a time interval is the ratio between the number of failures occurring during the time interval and the number surviving at the beginning of the interval divided by the length of the interval. Censored units during an interval should not be counted as part of the failed units during that interval. Otherwise, the hazard rate will be inflated. The following example illustrates the necessary calculations for both the hazard rate and the cumulative hazard.

EXAMPLE 5.1

Two hundred ceramic capacitors are subjected to a highly accelerated life test. The failure times of some of the capacitors are censored since the equipment used during testing these capacitors failed during the test. The number of survivors and censored units is shown in Table 5.1. Compute both the hazard rate and the cumulative hazard.

SOLUTION

The hazard rate at time t_i is computed as

$$h(t_i) = \frac{N_f(\Delta t_i)}{N_s(t_{i-1})\,\Delta t_i}, \tag{5.1}$$

where

$h(t_i)$ = the hazard rate at time t_i,

$N_f(\Delta t_i)$ = the number of failed units during the interval Δt_i,

$N_s(t_{i-1})$ = the number of survivors at the beginning of the interval Δt_i, and

Δt_i = the length of the time interval (t_{i-1}, t_i).

It should be noted that $N_f(\Delta t_i)$ does not include censored units. The calculations for the hazard rate and the cumulative hazard are given in Table 5.1. It is apparent that the hazard rate is constant with a mean value of 0.0319. Therefore, the mean time to failure is 31.35 hours.

Table 5.1. Hazard Rate and Cumulative Hazard

Time Interval	*Number of Failed Units*	*Number of Censored Units*	*Survivors at End of Interval*	*Hazard Rate* $\times 10^{-1}$	*Cumulative* $\times 10^{-1}$
0–10	0	3	197	0.0000	0.0000
10–20	6	8	183	0.0304	0.0304
20–30	7	9	167	0.0382	0.0686
30–40	6	8	153	0.0359	0.1045
40–50	6	15	132	0.0392	0.1437
50–60	5	20	107	0.0373	0.1810
60–70	4	18	85	0.0373	0.2183
70–80	3	20	62	0.0352	0.2535
80–90	2	30	30	0.0322	0.2857
90–100	1	29	0	0.0333	0.3190

In the following sections, we present parametric models to fit reliability failure data from reliability testings or field data to known failure-time distributions such as exponential, Weibull, lognormal, and gamma.

5.6. The Exponential Distribution

Since the exponential distribution has a constant failure rate and is commonly used in practice, we shall illustrate how to assess the validity of using the exponential distribution as a failure-time model. Let

t_1 = the time of the first failure,

t_i = the time between $i - 1$th and ith ($i = 2, 3, \ldots$) failures or the time to failure i (depending on the time being observed),

r = the total number of failures during the test (assuming no censoring),

T = the sum of the times between failures, $T = \sum_{i=1}^{r} t_i$, and

X = a random variable to represent time to failure.

In order to check whether or not the failure times follow an exponential distribution, we use the Bartlett's test whose statistic is

$$B_r = \frac{2r\left[\ln\left(\frac{T}{r}\right) - \frac{1}{r}\left(\sum_{i=1}^{r} \ln t_i\right)\right]}{1 + (r+1)/6r}, \tag{5.2}$$

where B_r is chi-square distributed statistics with $r - 1$ degrees of freedom.

The Bartlett's test does not contradict the hypothesis that the exponential distribution can be used to model a given time to failure data if the value of B_r lies between the two critical values of a two-tailed chi-square test with a $100(1-\alpha)$ percent significance level. The lower critical value is $\chi^2_{(1-\alpha/2),r-1}$ and the upper critical value is $\chi^2_{\alpha/2,r-1}$.

EXAMPLE 5.2

Twenty transistors are tested at five volts and 100°C. When a transistor fails, its failure time is recorded and the failed unit is replaced by a new one. The times between failures (in hours) are recorded in an increasing order as shown in Table 5.2. Test the validity of using a constant hazard rate for these transistors.

Table 5.2. Times Between Failures of the Transistors

Times Between Failures in Hours (t_i)	
200	32,000
400	34,000
2,000	36,000
6,000	39,000
9,000	42,000
13,000	43,000
20,000	48,000
24,000	50,000
26,000	54,000
29,000	60,000

SOLUTION

Since all the transistors failed during the test, the failure times can be assumed to come from a sample of twenty transistors, and each t_i is a value of the random variable, X, time to failure.

$$\sum_{i=1}^{20} \ln t_i = 193.28$$

$$T = \sum_{i=1}^{20} t_i = 567{,}600$$

$$B_{20} = \frac{2 \times 20 \left[\ln\left(\frac{567600}{20}\right) - \frac{1}{20} \times 193.28 \right]}{1 + (21)/(6 \times 20)}$$

$$B_{20} = 20.065.$$

The critical values for a two-tailed test with $\alpha = 0.10$ are

$$\chi^2_{0.95,\,19} = 10.117 \quad \text{and} \quad \chi^2_{0.05,\,19} = 30.144.$$

Therefore, B_{20} does not contradict the hypothesis that the failure times can be modeled by an exponential distribution.

The following example includes type 2 censoring.

EXAMPLE 5.3

In a test similar to the previous example, twenty transistors are subjected to an accelerated failure test (temperature 200°C and 2.0 volts). The test is discontinued when the tenth failure occurs. Determine whether the failure data in Table 5.3 follow an exponential distribution.

Table 5.3. Times Between Failures in Hours

600	2,800
700	3,000
1,000	3,100
2,000	3,300
2,500	3,600

SOLUTION

$$\sum_{i=1}^{10} \ln t_i = 75.554$$

$$T = \sum_{i=1}^{10} t_i = 22{,}600$$

$$B_{10} = \frac{2 \times 10 \left[\ln\left(\frac{22600}{10}\right) - \frac{1}{10}(75.554) \right]}{1 + (10+1)/(6 \times 10)}$$

$$B_{10} = 2.834.$$

The critical values for a two-sided test with $\alpha = 0.10$ are

$$\chi^2_{0.95,\,9} = 3.325 \quad \text{and} \quad \chi^2_{0.05,\,9} = 16.919.$$

Therefore, B_{10} contradicts the hypothesis that the failure data can be modeled by an exponential distribution. However, at a significance level of 98 percent, the critical values of the two-sided test become

$$\chi^2_{0.99,\,9} = 2.088 \quad \text{and} \quad \chi^2_{0.01,\,9} = 21.666,$$

and the test does not contradict the hypothesis that the failure times can be modeled by an exponential distribution.

5.6.1. TESTING FOR ABNORMALLY SHORT FAILURE TIMES

Short failure times may occur due to manufacturing defects such as the case of *freak* failures. These failure times do not actually represent the true failure times of the population. Therefore, it is important to determine whether the failure times are abnormally short before fitting the data to an exponential distribution. If the failure times are abnormally short, they should be discarded and not considered in determining the parameters of the failure time distribution.

Let $(t_1, t_2, \ldots, t_r)$ be a sequence of r independent and identically distributed exponential random variables that represent the time between failures for the first r failures. Then the quantity $2t_i/\theta$ is chi-square distributed with two degrees of freedom, where θ is the mean of the exponential distribution.

Therefore, if t is the time to first failure, which follows an exponential distribution with mean $= \theta$—that is,

$$f(t) = \frac{1}{\theta} e^{-t/\theta}$$

—then the random variable $y = 2t/\theta$ is a χ^2 with two degrees of freedom. This is explained as follows.

The density function of the random variable $\mathbf{t}$ is known and is given above. Our objective is to find the density function $g(y)$ for the random variable $\mathbf{y}$. We have

$$dy = \frac{2}{\theta}\, dt \quad \text{and} \quad t = \frac{\theta}{2} y.$$

Using the random variable transformation $g(y)\, dy = f(t)\, dt$, we write $g(y)$ as

$$g(y) = \frac{\theta}{2} f\left(\frac{\theta}{2} y\right),$$

which yields

$$g(y) = \frac{1}{2} e^{-\frac{y}{2}} \qquad y \geq 0.$$

This expression is indeed the probability density function of a χ^2 distribution with 2 degrees of freedom. Note that the general expression for the probability density function of χ^2 distribution with ν degrees of freedom is

$$g(y) = \frac{e^{\frac{-y}{2}} y^{\frac{\nu}{2}-1}}{2^{\frac{\nu}{2}} \Gamma\left(\frac{\nu}{2}\right)}.$$

The sum of two or more independent χ^2 distribution random variables is a new variable that follows χ^2 with a degree of freedom equal to the sum of the degrees of freedom of the individual random variables (Kapur and Lamberson, 1977).

So

$$\frac{2}{\theta} \sum_{i=2}^{r} t_i \text{ is } \chi^2 \text{ with } 2r - 2 \text{ degrees of freedom.}$$

Thus

$$F = \frac{\frac{2}{\theta} t_1 / 2}{\frac{2}{\theta} \sum_{i=1}^{r} t_i / (2r - 2)}.$$

This means that the F-distribution can be formed as

$$F_{2,\,2r-2} = \frac{(r-1) t_1}{\sum_{i=2}^{r} t_i},$$

where t_1, the short failure time, follows F distribution with degrees of freedom 2 and $2r - 2$.

If t_1 is small, then the F ratio becomes very small—that is,

$$F_{1-\alpha,\,2,\,2r-2} > \frac{(r-1)t_1}{\sum_{i=2}^{r} t_i}.$$

This inequality is equivalent to

$$F_{\alpha,\,2r-2,\,2} < \frac{\sum_{i=2}^{r} t_i}{(r-1)t_1}.$$

It is important to note that failure data should be ordered according to an increasing failure-time arrangement. In other words, the shortest failure time is listed first, followed by the second shortest time, and so forth.

EXAMPLE 5.4

Consider the failure data that represent cycles to failure for twenty turbine blades. The test is performed by subjecting a turbine to an accelerated load, replacing it by a new turbine upon failure, and recording the time to failure. Is the first failure time abnormally short (see Table 5.4).

Table 5.4. Failure Data of Turbine Blades

120	2,112	2,689	4,256
1,300	2,192	2,892	4,368
1,680	2,215	2,999	4,657
1,990	2,290	3,565	4,933
2,010	2,581	3,873	5,832

SOLUTION

The total failure time (except the first one) is

$$\sum_{i=2}^{20} t_i = 58{,}554$$

$$t_1 = 120$$

$$F_{\text{calculated}} = \frac{58{,}554}{(19)120} = 25.69.$$

The critical value of F at 95 percent confidence is

$$F_{0.05,\,38,\,2} = 19.47.$$

Thus the first failure is not a representative of the rest of the data. In other words, the hypothesis that the first failure time is not abnormally short should be rejected.

5.6.2. TESTING FOR ABNORMALLY LONG FAILURE TIMES

Following the above procedure, the failure time t_{ab} is considered to be an abnormally long failure time if

$$F_{\alpha,\,2,\,2r-2} < \frac{(r-1)t_{ab}}{\sum\limits_{i \neq ab}^{r} t_i}.$$

If t_r is the largest failure time, then the above equation can be rewritten as

$$F_{\alpha,\,2,\,2r-2} < \frac{(r-1)t_r}{\sum\limits_{i=1}^{r-1} t_i}.$$

EXAMPLE 5.5

Consider the failure time data in Table 5.5. Test whether the last failure time is abnormally long at 5 percent level of significance.

Table 5.5. Failure Times

30,000	46,585	63,200	77,990
34,500	49,970	66,600	80,330
37,450	54,430	70,000	84,450
39,950	57,600	73,120	88,960
43,760	59,990	75,690	99,550

SOLUTION

To check if the failure time of 99,550 time units is abnormally long, we obtain

$$\sum_{i=1}^{19} t_i = 1{,}134{,}575$$

$$F_{\text{calculated}} = \frac{19 \times 99{,}550}{1{,}134{,}575} = 1.667.$$

Since $F_{calculated} < F_{0.05,2,38} = 3.25$, then the last failure time (99,550) is not abnormally long.

Suppose that n units are placed under test and the exact failure times of the units are recorded. They are $t_1, t_2, \ldots, t_n$. Since all units have failed and no censoring exists, the maximum likelihood estimate (MLE) of λ is

$$\hat{\lambda} = \frac{n}{\sum_{i=1}^{n} t_i}, \tag{5.3}$$

where $\hat{\lambda}$ is the maximum likelihood estimator of the failure rate.

The mean μ of the exponential distribution is $1/\lambda$ and the MLE of μ is

$$\hat{\mu} = \frac{1}{\hat{\lambda}} = \frac{\sum_{i=1}^{n} t_i}{n} = \bar{t}. \tag{5.4}$$

$\hat{\mu}$ is referred to as the maximum likelihood estimator of the mean life.

EXAMPLE 5.6 Assume that the data given in Example 5.2 represent the time failure of the units under test. Determine the mean life of a transistor from this population.

SOLUTION Using Eq. (5.3), we obtain

$$\bar{t} = \frac{567{,}600}{20} = 28{,}380 \text{ hours.}$$

It can be shown that $2n\hat{\mu}/\mu$ has an exact chi-square distribution with $2n$ degrees of freedom. Since $\lambda = 1/\mu$ and $\hat{\lambda} = 1/\hat{\mu}$, then a $100(1 - \alpha)$ percent confidence interval for $\hat{\lambda}$ (assuming zero minimum life) is

$$\frac{\hat{\lambda}\chi^2_{1-\alpha/2,\,2n}}{2n} < \lambda < \frac{\hat{\lambda}\chi^2_{\alpha/2,\,2n}}{2n}, \tag{5.5}$$

where $\chi^2_{\alpha,\,2n}$ is the 100α percentage point of the chi-square distribution with $2n$ degrees of freedom—that is,

$$P[\chi^2_{2n} > \chi^2_{\alpha,\,2n}] = \alpha.$$

The confidence interval of the mean life corresponding to Eq. (5.5) is

$$\frac{2n\hat{\mu}}{\chi^2_{\alpha/2,\,2n}} < \mu < \frac{2n\hat{\mu}}{\chi^2_{1-\alpha/2,\,2n}}.$$

When n is large ($n \geq 25$), we can obtain an approximate interval for λ by approximating $\hat{\lambda}$ by a normal distribution with mean λ and variance λ^2/n. Thus,

$$\hat{\lambda} - \frac{\hat{\lambda} Z_{\alpha/2}}{\sqrt{n}} < \lambda < \hat{\lambda} + \frac{\hat{\lambda} Z_{\alpha/2}}{\sqrt{n}}, \tag{5.6}$$

where $Z_{\alpha/2}$ is $100(\alpha/2)$ percent point $[P(Z > Z_{\alpha/2}) = \alpha/2]$ of the standard normal distribution.

EXAMPLE 5.7

Determine a 95 percent two-sided confidence interval for the mean life of the transistors given in Example 5.6.

SOLUTION

Since $\hat{\mu} = 28{,}380$ hours, then at 95 percent confidence level, $\chi^2_{0.975,\,40} = 24.423$ and $\chi^2_{0.025,\,40} = 59.345$, the limits of μ are

$$\frac{2 \times 567{,}600}{59.345} < \mu < \frac{2 \times 567{,}600}{24.423}$$

or

$$19{,}218 \leq \mu \leq 46{,}480 \text{ hours.}$$

EXAMPLE 5.8

A mechanical engineer conducts a fatigue test to determine the expected life of rods made of a specific type of steel by subjecting twenty-five specimens to an axial load that causes a stress of 9,000 pounds per square inch (psi). The number of cycles are recorded at the time of failure of every specimen. Assuming that the test is run at ten cycles per minute, determine the reliability of a rod made of this steel at ten hours. Results of the test are shown in Table 5.6.

Table 5.6. Number of Cycles to Failures

Cycles to failure				
200	720	1,950	5,570	10,660
280	850	2,460	6,590	11,670
340	990	2,590	7,600	12,680
460	1,200	3,520	8,630	13,685
590	1,420	4,560	9,650	14,690

SOLUTION Using B_{25} we first check if the failure data can be represented by an exponential distribution. This requires the determination of the cycles between failures from the data given above as shown in Table 5.7.

Table 5.7. Number of Cycles to Failures and Between Failures

Rod No.	*Cycles to Failure (CTF)*	*Cycles Between Failures (CBF)*
1	200	200
2	280	80
3	340	60
4	460	120
5	590	130
6	720	130
7	850	130
8	990	140
9	1,200	210
10	1,420	220
11	1,950	530
12	2,460	510
13	2,590	130
14	3,520	930
15	4,560	1,040
16	5,570	1,010
17	6,590	1,020
18	7,600	1,010
19	8,630	1,030
20	9,650	1,020
21	10,660	1,010
22	11,670	1,010
23	12,680	1,010
24	13,685	1,005
25	14,690	1,005

$$T = \sum_{i=1}^{25} CBF_i = 14{,}690$$

$$\sum_{i=1}^{25} \ln CBF_i = 149.211$$

$$B_{25} = \frac{2 \times 25 \left[\ln \dfrac{14{,}690}{25} - \dfrac{1}{25} \times 149.211 \right]}{1 + \dfrac{26}{6 \times 25}}$$

$$B_{25} = 17.370.$$

The critical values for a two-tailed test with $\alpha = 0.10$ are

$$\chi^2_{0.95,\,24} = 13.848.$$

and

$$\chi^2_{0.05,\,24} = 36.418.$$

Hence, the B_{25} statistic does not contradict the hypothesis that the failure times can be modeled by an exponential distribution.

The reliability of a rod at ten hours is obtained as

$$R(t = 10 \text{ hours}) = e^{-\hat{\lambda} t},$$

where

$$\hat{\lambda} \quad \text{is} \quad \frac{25}{14{,}690} \quad \text{failures per cycle.}$$

$$R(t = 10 \text{ hours}) = e^{\frac{-25}{14{,}690} \times 60 \times 10 \times 10}$$

$$R(t = 10 \text{ hours}) = 0.3676 \times 10^{-4}.$$

We now consider the effect of censoring on the estimation of λ. First, we present type 1 censoring to be followed by type 2 censoring.

5.6.3. DATA WITH TYPE 1 CENSORING

Assume that n units are placed under test and the failure times t_i's of the failed units are recorded and reordered in an increasing order. Let T be the censoring time of the test. Thus, $t_1 \leq t_2 \leq t_3 \leq \ldots \leq t_r \leq t_1^+ = \ldots = t_{n-r}^+ = T$ where t_i^+ is the censoring time of censored unit i. The likelihood function is

$$L = \prod_{i=1}^{r} \lambda e^{-\lambda t_i} \prod_{i=1}^{n-r} e^{-\lambda t_i^+}. \tag{5.7}$$

Taking the logarithm of the above equation yields (we use l to refer to $\ln L$),

$$l = r \ln \lambda - \sum_{i=1}^{r} \lambda t_i - \sum_{i=1}^{n-r} \lambda t_i^+.$$

The derivative of the l with respect to λ is

$$\frac{dl}{d\lambda} = \frac{r}{\lambda} - \sum_{i=1}^{r} t_i - \sum_{i=1}^{n-r} t_i^+.$$

Equating the derivative to zero results in the MLE of λ as

$$\hat{\lambda} = \frac{r}{\sum_{i=1}^{r} t_i + \sum_{i=1}^{n-r} t_i^+}, \tag{5.8}$$

and the mean life of units can be estimated as

$$\hat{\mu} = \frac{1}{\hat{\lambda}} = \frac{1}{r}\left[\sum_{i=1}^{n} t_i + \sum_{i=1}^{n-r} t_i^+\right]. \tag{5.9}$$

Again, it can be shown that the statistic $2r\lambda/\hat{\lambda}$ has a chi-square distribution with $2r$ degrees of freedom. The mean and variance of $\hat{\lambda}$ are $r\hat{\lambda}/(r-1)$ and $\hat{\lambda}^2/(r-1)$, respectively (Lee, 1992). The $100(1-\alpha)$ percent confidence interval for λ is

$$\frac{\hat{\lambda}\chi^2_{1-\alpha/2,\,2r}}{2r} < \lambda < \frac{\hat{\lambda}\chi^2_{\alpha/2,\,2r}}{2r}. \tag{5.10}$$

The confidence interval for the mean life, μ, is

$$\frac{2r\hat{\mu}}{\chi^2_{\alpha/2,\,2r}} < \mu < \frac{2r\hat{\mu}}{\chi^2_{1-\alpha/2,\,2r}}.$$

When n is large ($n \geq 25$), the distribution of $\hat{\lambda}$ can be approximated by a normal distribution with mean λ and variance $\lambda^2/(r-1)$. The $100(1-\alpha)$ percent confidence interval is

$$\hat{\lambda} - \frac{\hat{\lambda}Z_{\alpha/2}}{\sqrt{r-1}} < \lambda < \hat{\lambda} + \frac{\hat{\lambda}Z_{\alpha/2}}{\sqrt{r-1}}. \tag{5.11}$$

EXAMPLE 5.9

A manufacturer of end mill cutters introduces a new ceramic cutter material. In order to estimate the expected life of a cutter, the manufacturer places ten units under continuous test and monitors the tool wear. A failure of the cutter occurs when the wear out exceeds a predetermined value. Because of budgeting constraints, the manufacturer decides to run the test for 50,000 minutes. The times to failure are recorded as shown in Table 5.8. Determine the mean life of a cutter

Table 5.8. Times to Failure of Cutters

Cutter's Life in Minutes
3,000
7,000
12,000
18,000
20,000
30,000

made from this material. What is the 90 percent confidence interval for the expected life? What is the reliability at 60,000 minutes?

SOLUTION

We check if the failure data follow an exponential distribution by estimating B_6 using the time between failures as shown in Table 5.9. We calculate

$$T = \sum_{i=1}^{6} TBF_i = 30{,}000$$

$$\sum_{i=1}^{6} \ln TBF_i = 50.33$$

$$B_6 = 1.2973.$$

Table 5.9. Failure Data of the Cutters

Cutter Number	*Time to Failure (TTF)*	*Time Between Failures (TBF)*
1	3,000	3,000
2	7,000	4,000
3	12,000	5,000
4	18,000	6,000
5	20,000	2,000
6	30,000	10,000

The chi-squared statistics are

$$\chi^2_{0.95,\,5} = 1.145 \quad \text{and} \quad \chi^2_{0.05,\,5} = 11.070.$$

Thus, the data follow an exponential distribution.

Using Eq. (5.9) we estimate $\hat{\mu}$ as

$$\hat{\mu} = \frac{1}{6}[30{,}000 + 4 \times 50{,}000] = 38{,}333 \text{ minutes},$$

and the 90 percent confidence interval for μ is

$$\frac{2 \times 6 \times 38{,}333}{18.307} < \hat{\mu} < \frac{2 \times 6 \times 38{,}333}{3.940}$$

or

$$25{,}127 < \hat{\mu} < 116{,}750.$$

The probability that a cutter will survive for 60,000 minutes is

$$\hat{R}(60{,}000) = e^{-\hat{\lambda}t} = e^{\frac{-1}{38{,}333} \times 60{,}000}$$

$$\hat{R}(60{,}000) = 0.209,$$

where $\hat{R}(t)$ is the estimated reliability at time t.

5.6.4. DATA WITH TYPE 2 CENSORING

Suppose that n units are placed under test at time zero and their failure times are recorded in an increasing order. Suppose that the test is terminated when r of the n units fail. The failure times of the n units are $t_1 \leq t_2 \leq t_3 \leq \ldots \leq t_r, t_1^+ = \ldots = t_{n-r}^+$, where t_i is the failure time of unit i and t_i^+ is the censoring time of censored unit i which is the censoring time of the test.

Following the same procedure of data analysis with type 1 censoring, we obtain the same equations for the MLE of λ and μ. Thus, there is no difference in results when either type 1 or type 2 censoring is applied.

We now examine the random censoring situation when n units undergo a reliability test at time zero. The test is terminated at time T. Let r be the number of units that fail before T and $n - r$ be the number of units that either survive the test time T or their test equipment fails during T while the units are operating. The data collected may be observed as follows: $t_1, t_2, \ldots, t_r, t_1^+, t_2^+, \ldots, t_{n-r}^+$. The + sign indicates censoring. We now order the failure times and the censoring times in ascending order $t_1 \leq t_2 \leq \ldots \leq t_r, t_1^+, t_2^+, \ldots, t_{n-r}^+$. Using the MLE method we obtain

$$\hat{\lambda} = \frac{r}{\sum_{i=1}^{r} t_i + \sum_{i=1}^{n-r} t_i^+}.$$

This is the same result as those for type 1 censoring.

What happens when all observations are censored? In this case, one estimates $\hat{\mu}$ as the sum of the censored time of all units:

$$\hat{\mu} = \sum_{i=1}^{n} t_i^+.$$

Clearly, this estimate of μ has little practical value, and this reliability test is considered to be poorly designed. Methods for handling all censored data will be discussed in the next chapter.

When the number of units under test, n, is large (≥ 25), the distribution of $\hat{\lambda}$ is approximately normal with mean λ and variance (Lee, 1980):

$$\mathrm{Var}(\hat{\lambda}) = \frac{\lambda^2}{\sum_{i=1}^{n} \left(1 - e^{-\lambda T_i}\right)},$$

where T_i is the time that the ith component is under observation (time until failure or end of test). If T_i is unknown due to an abnormal termination of the test, then the variance can be approximated as

$$\hat{\mathrm{Var}}(\hat{\lambda}) \cong \frac{\hat{\lambda}^2}{r},$$

where r is the number of failed components before the termination of the test.

The $100(1 - \alpha)$ percent confidence interval is

$$\hat{\lambda} - Z_{\alpha/2}\sqrt{\hat{\mathrm{Var}}(\hat{\lambda})} < \lambda < \hat{\lambda} + Z_{\alpha/2}\sqrt{\hat{\mathrm{Var}}(\hat{\lambda})},$$

and distribution of $\hat{\mu}$ is approximated by a normal distribution with mean μ and an estimated variance of

$$\hat{\mathrm{Var}}(\hat{\mu}) = \frac{\hat{\mu}^2}{\sum_{i=1}^{n} \left(1 - e^{-\lambda T_i}\right)}.$$

If T_i is unknown, then

$$\hat{\text{Var}}(\hat{\mu}) = \frac{\hat{\mu}^2}{r}.$$

Bounds on μ are

$$\hat{\mu} - Z_{\alpha/2}\sqrt{\hat{\text{Var}}(\hat{\mu})} < \mu < \hat{\mu} + Z_{\alpha/2}\sqrt{\hat{\text{Var}}(\hat{\mu})}.$$

5.7. The Rayleigh Distribution

The Rayleigh distribution exhibits a linearly increasing hazard function with time. This implies that when the time to failure follows the Rayleigh distribution, an intense aging of the equipment takes place, and the failures do not satisfy the conditions of a stationary random process. More important, during the early life of a product where the hazard rate is significant, the probability of failure-free operation of the product (or system) decreases with time more slowly than in the case of the exponential distribution. However, as the time increases, the probability of failure-free operation decreases with time at a faster rate than the exponential distribution. Rayleigh distribution is useful in modeling rapidly fading communication channels where the amplitude of the signal can be described by such a distribution. The probability density function of the Rayleigh distribution is

$$f(t) = \lambda t e^{\frac{-\lambda t^2}{2}}. \tag{5.12}$$

Following the exponential distribution, we estimate the parameter of the Rayleigh distribution for both noncensored and censored failure data as shown below.

5.7.1. ESTIMATION OF RAYLEIGH'S PARAMETER FOR DATA WITHOUT CENSORED OBSERVATIONS

Suppose that n devices are subjected to an accelerated life test and that the exact failure time of every device is recorded. The failure times are $t_1, t_2, \ldots, t_n$. Since all devices have failed, then the maximum likelihood estimate (MLE) of Rayleigh's parameter is obtained as follows:

$$L(\lambda, t) = \prod_{i=1}^{n} f(t_i)$$

$$= \prod_{i=1}^{n} \lambda t_i e^{\frac{-\lambda t_i^2}{2}}$$

or

$$L(\lambda, t) = \lambda^n \prod_{i=1}^{n} t_i e^{\frac{-\lambda t_i^2}{2}}. \tag{5.13}$$

The logarithm of Eq. (5.13) is

$$l(\lambda, t) = n \ln \lambda + \sum_{i=1}^{n} \ln t_i - \frac{\lambda}{2} \sum_{i=1}^{n} t_i^2. \tag{5.14}$$

In order to estimate λ, we take the derivative of Eq. (5.14) with respect to λ and equate the resultant equation to zero. Thus,

$$\frac{dl(\lambda, t)}{d\lambda} = \frac{n}{\lambda} - \frac{1}{2} \sum_{i=1}^{n} t_i^2$$

or

$$\hat{\lambda} = \frac{2n}{\sum_{i=1}^{n} t_i^2}. \tag{5.15}$$

The variance of Rayleigh distribution is

$$\text{Var}(t) = \frac{2}{\hat{\lambda}} \left(1 - \frac{\pi}{4}\right).$$

EXAMPLE 5.10

A manufacturer of an automotive speed sensor subjects ten sensors to a reliability test that simulates the environmental conditions (temperature and speed) at which the sensors will normally operate. A sensor is classified failed when its output falls outside 5 percent tolerance. The number of miles accumulated before the failures of the sensors are

110,000, 130,000, 150,000, 155,000, 159,000, 163,000, 166,000, 168,000, 169,000, 170,000.

Assume that the miles to failure follow a Rayleigh distribution. Determine the parameter of the distribution, the mean life of a sensor, and the variance of its life.

SOLUTION The parameter of the Rayleigh distribution is obtained using Eq. (5.15) as

$$\hat{\lambda} = \frac{2n}{\sum_{i=1}^{n} t_i^2}$$

$$\hat{\lambda} = \frac{2 \times 10}{2.40616 \times 10^{11}} = 8.31199 \times 10^{-11}.$$

The mean life is

$$\text{Mean life} = \sqrt{\frac{\pi}{2\hat{\lambda}}} = \sqrt{\frac{\pi}{2 \times 8.31199 \times 10^{-11}}} = 137{,}470 \text{ miles},$$

and the variance of the life is

$$\text{Variance} = \frac{2}{\hat{\lambda}}\left(1 - \frac{\pi}{4}\right) = 5.1636 \times 10^9.$$

The standard deviation is 71,859.

5.7.2. ESTIMATION OF RAYLEIGH'S PARAMETER FOR DATA WITH CENSORED OBSERVATIONS

Suppose that n devices are subjected to a test and that the failure times of the r failed units are recorded and listed in an ascending order as $t_1 \leq t_2 \leq \ldots \leq t_r$. The remaining $n - r$ units are censored—that is, these units have not failed before the test is terminated. We assume that the censoring is either of type 1 or type 2 only and the censored times are $t_1^+ = t_2^+ = \ldots t_{n-r}^+$.

The likelihood function is

$$L(\lambda, t) = \prod_{i=1}^{r} \lambda t_i e^{\frac{-\lambda t_i^2}{2}} \prod_{i=1}^{n-r} e^{\frac{-\lambda t_i^{+2}}{2}},$$

and the logarithm of the likelihood function is

$$l(\lambda, t) = r \ln \lambda + \sum_{i=1}^{r} \ln t_i - \frac{\lambda}{2}\left(\sum_{i=1}^{r} t_i^2 + \sum_{i=1}^{n-r} t_i^{+2}\right).$$

The estimated value of the parameter $\hat{\lambda}$ is obtained as

$$\hat{\lambda} = \frac{2r}{\sum_{i=1}^{r} t_i^2 + \sum_{i=1}^{n-r} t_i^{+2}}. \tag{5.16}$$

EXAMPLE 5.11

As an alternative to an automobile air-bag crash testing, a test engineer develops a sensor test system that uses a mechanical vibration shaker to replay measured actual crashes. The sensors are subjected to the same conditions measured during a crash test. Ten sensors are placed under test for fifty hours, and the following failure times are recorded:

10, 20, 30, 35, 39, 42, 44, 50^+, 50^+, 50^+.

Determine Rayleigh's parameter, the mean life of the sensors, and the standard deviation of the life.

SOLUTION

Using Eq.(5.16) we obtain

$$\hat{\lambda} = \frac{2 \times 7}{7{,}846 + 7{,}500} = 9.12289 \times 10^{-4}.$$

The mean life is

$$\text{Mean life} = \sqrt{\frac{\pi}{2 \times 9.12289 \times 10^{-4}}} = 41.49 \text{ hours.}$$

The standard deviation of the life is

$$\text{Standard deviation} = \sqrt{\frac{2}{\hat{\lambda}}\left(1 - \frac{\pi}{4}\right)} = 21.70 \text{ hours.}$$

5.7.3. BEST LINEAR UNBIASED ESTIMATE FOR RAYLEIGH'S PARAMETER FOR DATA WITH AND WITHOUT CENSORED OBSERVATIONS

The maximum likelihood estimator of the Rayleigh's parameter is biased when the number of observations is small. The bias increases as the number of observations decreases. If the p.d.f. of the failure time can be linearized, the bias in estimating the parameter(s) of the distribution can be decreased when the least square method is used in estimating the parameters. We refer to such an estimate as the best linear unbiased estimate (BLUE). In this section, we obtain the best linear unbiased estimate of Rayleigh's parameter for both censored and noncensored observations.

5.7.3.1. *BLUE for Rayleigh's Parameter*

Suppose that the failure times for n devices subjected to a reliability test are $t_{(1)} \leq t_{(2)} \leq \ldots \leq t_{(n)}$, where $t_{(i)}$ is the ith order statistics. Assume that the n ordered failure times follow the Rayleigh distribution with

$$f(t) = \frac{1}{\theta_2^2}(t - \theta_1)e^{-(t-\theta_1)^2/2\theta_2^2} \qquad (t > \theta_1 \geq 0, \theta_2 > 0) \quad \text{and} \tag{5.17}$$

$$f(t) = 0 \quad \text{elsewhere,}$$

where θ_1 is the location (threshold) parameter and θ_2 is the scale parameter. Note that Eq. (5.17) is identical to Eq. (5.12) when $\theta_1 = 0$ and $\lambda = 1/\theta_2^2$.

The BLUE θ_2^* of θ_2 when θ_1 is known can be estimated by

$$\theta_2^* = \sum_{i=1}^{n} b_i t_{(i)} - \theta_1 \frac{K_3}{K_2}. \tag{5.18}$$

If the location parameter θ_1 = zero, the density function, given by Eq. (5.17), becomes

$$f(t) = \frac{1}{\theta_2^2} t e^{-t^2/2\theta_2^2}, \tag{5.19}$$

and the estimate θ_2^* becomes

$$\theta_2^* = \sum_{i=1}^{n} b_i t_{(i)}. \tag{5.20}$$

The coefficients b_i's are given in Appendix B for $i = 1, \ldots, n$ for noncensored samples, and for censored samples with r largest observations censored, where $r = 0, 1, 2, \ldots, (n-2)$ and n is the sample size for $n = 5(1)25(5)45$. The variance of θ_2^* in terms of θ_2^2/n and K_3/K_2 is given in Appendix C.

EXAMPLE 5.12

A manufacturer of biosensors produces an electrochemical sensor array that is small enough to fit inside a blood vessel. The device is inserted into an artery within a catheter that has an inside diameter of 650 microns. It measures the levels of oxygen, carbon dioxide, and pH in the blood. The producer subjects twenty sensors to a functional test and observes the following failure times in hours:

0.9737	8.0327	13.1911	19.4369
1.0590	8.0833	13.4695	22.5168
3.3152	8.1957	14.0578	24.4470
3.3161	9.3706	14.8812	24.9225
5.2076	11.1886	17.9624	30.0000

(Note: These data are actually generated randomly from a Rayleigh distribution with θ_1 = zero and $\theta_2 = 10.5$.) Estimate the Rayleigh parameter and its variance.

SOLUTION

To estimate the scale parameter θ_2, we use the coefficients b_i's in Appendix B and K_3/K_2 in Appendix C for $n = 20$.

$$\begin{aligned}\theta_2^* &= (0.00767)(0.9737) + \ldots + (0.072142)(30.0000) - (0)(0.63995) \\ &= 10.7303 \\ &\cong 10.73.\end{aligned}$$

Using Appendix C, for $n = 20$ and $r = 0$, where the variances are given in terms of θ_2^2, we obtain

$$\begin{aligned}\text{Var}(\theta_2^*) &= 0.01260 \times \theta_2^{*2} = 0.01260 \times (10.73)^2 \\ &= 1.4507 \\ &\cong 1.45.\end{aligned}$$

Standard deviation of θ_2^* is $= \sqrt{1.45} = 1.20$.

EXAMPLE 5.13

A manufacturer of an automotive speed sensor subjects ten sensors to a reliability test that simulates the environmental conditions (temperature and speed) at which the sensors will normally operate. A sensor is classified failed when its output falls outside 5 percent tolerance. The number of miles accumulated before the failures of the sensors are

110,000, 130,000, 150,000, 155,000, 159,000, 163,000, 166,000, 168,000, 169,000, 170,000.

Assume that the miles to failure follow a Rayleigh distribution with a location parameter = 0. Find an estimate for the parameter θ_2, the mean life of the sensor, and the standard deviation of the estimate of θ_2.

SOLUTION Using Appendix B for $n = 10$ and $r = 0$, we get

$$\theta_2^* = (110{,}000)(0.02149) + (130{,}000)(0.03171) + \ldots$$

$$+ (170{,}000)(0.13149) - (0)(0.65170)$$

$$= 105{,}061.18$$

$$\cong 105{,}061 \text{ hours.}$$

The mean life is

$$\text{Mean life} = \theta_2^* \sqrt{\frac{\pi}{2}} = 131{,}674 \text{ miles.}$$

Using Appendix C, for $n = 10$ and $r = 0$, where the variances are given in terms of θ_2^2, we find

$$\text{Var}(\theta_2^*) = (0.02537)\theta_2^2.$$

Substituting the estimate θ_2^* for θ_2, we get

$$\text{Var}(\theta_2^*) = (0.02537)(105{,}061)^2$$

$$= 2.8 \times 10^8.$$

The standard deviation of the estimate θ_2^* is 16,733 hours.

EXAMPLE 5.14 Considering the data in Example 5.13, assume that the six largest values of failure times are censored. The miles accumulated before the failures of the sensors are 110,000, 130,000, 150,000, and 155,000. Estimate the parameter and its standard deviation.

SOLUTION Using Appendix B for $n = 10$ and $r = 6$ (where r in this Appendix refers to the number of censored observations from the right) we find

$$\theta_2^* = (110{,}000)(0.05301) + (130{,}000)(0.07777) + (150{,}000)(0.09821) + (155{,}000)(0.90085) - (0)(1.12984)$$

$$= 170{,}304.5.$$

Using Appendix C for $n = 10$ and $r = 6$, we find

$$\text{Var}(\theta_2^*) = 0.06434\theta^{*2}$$

$$= 0.06434 \times (170{,}304)^2$$

$$= 1.866 \times 10^9.$$

Standard deviation of the estimate θ_2^* is 43,198 miles.

5.7.3.2. *Confidence Interval Estimate for θ_2^2 for Noncensored Observations*

If $t_{(1)} \leq \ldots \leq t_{(n)}$ are order statistics from Rayleigh distribution

$$f(t) = \frac{1}{\theta_2^2} t e^{-t^2/2\theta_2^2},$$

then $y_{(1)} = t_{(1)}^2/2, y_{(2)} = t_{(2)}^2/2, \ldots, y_{(n)} = t_{(n)}^2/2$ are order statistics from exponential distribution of the form

$$f(y) = \frac{1}{\theta_2^2} e^{-y/\theta_2^2}.$$

Also, $2/\theta_2^2 \sum_{i=1}^n y_{(i)}^2 = 1/\theta_2^2 \sum_{i=1}^n t_{(i)}^2$ follows a χ^2 distribution with $(2n)$ degrees of freedom.
Thus, $100(1 - \alpha)$ percent confidence interval for θ_2^2

$$\frac{1}{\chi_{\alpha/2}^2} \sum_{i=1}^n t_{(i)}^2 \leq \theta_2^2 \leq \frac{1}{\chi_{1-\alpha/2}^2} \sum_{i=1}^n t_{(i)}^2. \tag{5.21}$$

Also, $100(1-\alpha)$ percent confidence interval for θ_2 is

$$\sqrt{\frac{1}{\chi^2_{\alpha/2}}\cdot\sum_{i=1}^{n}t^2_{(i)}}\le\theta_2\le\sqrt{\frac{1}{\chi^2_{1-\alpha/2}}\cdot\sum_{i=1}^{n}t^2_{(i)}},\tag{5.22}$$

where $\chi^2_{\alpha/2}$ and $\chi^2_{1-\alpha/2}$ are respectively the lower and upper $\alpha/2$ points of χ^2 with $2n$ degrees of freedom.

5.7.3.3. *Confidence Interval Estimate for θ_2^2 for Censored Observations*

Suppose that the sample size is n with r largest censored observations. The $100(1-\alpha)$ percent confidence interval for θ_2^2, using $(n-r)$ noncensored observations is

$$\frac{1}{\chi^2_{\alpha/2}}\sum_{i=1}^{(n-r)}t^2_{(i)}\le\theta_2^2\le\frac{1}{\chi^2_{1-\alpha/2}}\sum_{i=1}^{(n-r)}t^2_{(i)}.\tag{5.23}$$

Also $100(1-\alpha)$ percent confidence interval for θ_2 is

$$\sqrt{\frac{1}{\chi^2_{\alpha/2}}\cdot\sum_{i=1}^{(n-r)}t^2_{(i)}}\le\theta_2\le\sqrt{\frac{1}{\chi^2_{1-\alpha/2}}\cdot\sum_{i=1}^{(n-r)}t^2_{(i)}},\tag{5.24}$$

where $\chi^2_{\alpha/2}$ and $\chi^2_{1-\alpha/2}$ are respectively the lower and upper $\alpha/2$ points of χ^2 distribution with $2(n-r)$ degrees of freedom.

EXAMPLE 5.15 Determine the 95 percent confidence interval (C.I.) for the BLUE of θ_2^* for the data given in Example 5.13.

SOLUTION We are using noncensored data in this example, therefore we utilize Eq. (5.22)

$$\sum_{i=1}^{10}t_i^2=(2{,}406.16)10^8.$$

From the tables of percentiles of the χ^2 distribution for 20 degrees of freedom, we find

$$\chi^2_{0.025}=34.17,\qquad \chi^2_{0.975}=9.59.$$

Therefore a 95 percent C.I. for θ_2^* is

$$\sqrt{\frac{1}{34.17} \times 2{,}406.16 \times 10^8} \leq \theta_2^* \leq \sqrt{\frac{1}{9.59} \times 2{,}406.16 \times 10^8}$$

$$88{,}181.17 \leq \theta_2^* \leq 158{,}399.18.$$

EXAMPLE 5.16

For the data given in Example 5.14 where the six largest values of failure times are censored, find a 95 percent confidence interval for the estimate θ_2^*.

SOLUTION

We are using censored data in this example. We utilize Eq. (5.24)

$$\sum_{i=1}^{4} t_i^2 = (755.25)10^8.$$

From the tables of percentiles of the χ^2 distribution for eight degrees of freedom, we find

$$\chi^2_{0.025} = 17.53, \qquad \chi^2_{0.975} = 2.18.$$

Therefore a 95 percent C.I. for θ_2^* is

$$\sqrt{\frac{1}{17.53} \times 755.25 \times 10^8} \leq \theta_2^* \leq \sqrt{\frac{1}{2.18} \times 755.25 \times 10^8}$$

$$65{,}637.86 \leq \theta_2^* \leq 186{,}130.31.$$

5.8. The Weibull Distribution

In Chapter 1, we presented the probability density function of the Weibull distribution as

$$f(t) = \frac{\gamma}{\theta} t^{\gamma-1} \exp\left[-\left(\frac{t^\gamma}{\theta}\right)\right], \qquad t \geq 0, \gamma \geq 1, \theta > 0, \tag{5.25}$$

where γ and θ are the shape and scale parameters, respectively.

When the failure data are assumed to follow Weibull distribution, the estimated parameters of the distribution, $\hat{\theta}$ and $\hat{\gamma}$, can be obtained by using the MLE procedures proposed by Cohen (1965), Harter and Moore (1965), and discussed in

Lee (1980, 1992). This, as with the exponential distribution, will be presented for two cases.

5.8.1. FAILURE DATA WITHOUT CENSORING

The exact failure times of n units under test are recorded as $t_1, t_2, \ldots, t_n$. Assume that the failure data follow a Weibull distribution. The likelihood function is

$$L(\gamma, \theta, t) = \left(\frac{\gamma}{\theta}\right)^n \prod_{i=1}^{n} t_i^{\gamma-1} e^{\frac{-t_i^\gamma}{\theta}}. \tag{5.26}$$

Following the same procedure as the exponential and Rayleigh cases, we take the logarithm of Eq. (5.26). We then take the derivatives of the logarithmic function with respect to γ and θ. This results in the following two equations.

$$\frac{n}{\hat{\gamma}} + \sum_{i=1}^{n} \ln t_i - \frac{1}{\hat{\theta}} \sum_{i=1}^{n} t_i^{\hat{\gamma}} \ln t_i = 0 \tag{5.27}$$

$$-\frac{n}{\hat{\theta}} + \frac{1}{\hat{\theta}^2} \sum_{i=1}^{n} t_i^{\hat{\gamma}} = 0. \tag{5.28}$$

The MLE of γ and θ can be obtained by solving Eqs. (5.27) and (5.28) simultaneously. Substituting $\hat{\theta}$ obtained from Eq. (5.28) into (5.27), we obtain a difference $D(\hat{\gamma})$:

$$D(\hat{\gamma}) = \frac{\sum_{i=1}^{n} t_i^{\hat{\gamma}} \ln t_i}{\sum_{i=1}^{n} t_i^{\hat{\gamma}}} - \frac{1}{\hat{\gamma}} - \frac{1}{n} \sum_{i=1}^{n} \ln t_i = 0. \tag{5.29}$$

The above equation in $\hat{\gamma}$ can be solved numerically by using the Newton-Raphson method (described in Appendix I) or by trial and error. Once $\hat{\gamma}$ is estimated, we obtain $\hat{\theta}$ as

$$\hat{\theta} = \sum_{i=1}^{n} \frac{t_i^{\hat{\gamma}}}{n}.$$

Similar to the exponential distribution and assuming a large number of failure data, the $100(1-\alpha)$ percent confidence intervals for γ and θ are

$$\hat{\gamma} - Z_{\alpha/2}\sqrt{\mathrm{Var}(\hat{\gamma})} < \gamma < \hat{\gamma} + Z_{\alpha/2}\sqrt{\mathrm{Var}(\hat{\gamma})} \tag{5.30}$$

$$\hat{\theta} - Z_{\alpha/2}\sqrt{\mathrm{Var}(\hat{\theta})} < \theta < \hat{\theta} + Z_{\alpha/2}\sqrt{\mathrm{Var}(\hat{\theta})}, \tag{5.31}$$

where Var $(\hat{\gamma})$ and Var $(\hat{\theta})$ for large n are obtained as follows. Define

$$S_0 = \sum_{i=1}^{n} t_i^{\hat{\gamma}}$$

$$S_1 = \sum_{i=1}^{n} t_i^{\hat{\gamma}}(\ln t_i)$$

$$S_2 = \sum_{i=1}^{n} t_i^{\hat{\gamma}}(\ln t_i)^2.$$

Then

$$\mathrm{Var}(\hat{\gamma}) \cong \frac{\hat{\gamma}^2 S_0^2}{n(S_0^2 + \hat{\gamma}^2\, S_0 S_2 - \hat{\gamma}^2 S_1^2)}$$

$$\mathrm{Var}(\hat{\theta}) \cong \frac{S_0}{n^2}\left(\frac{S_0}{\hat{\gamma}^2} + S_2\right)\mathrm{Var}(\hat{\gamma})$$

$$\mathrm{Cov}(\hat{\gamma}, \hat{\theta}) \cong \frac{S_1}{n}\mathrm{Var}(\hat{\gamma}).$$

Unbiased estimates of the parameters $\hat{\theta}$, $\hat{\gamma}$, Var$(\hat{\theta})$, and Var$(\hat{\gamma})$ are discussed in Section 5.8.

EXAMPLE 5.17

Ten diodes are tested to failure at accelerated conditions. The failure times (in minutes) are recorded in Table 5.10. Suppose that the data follow Weibull distribution. Find the parameters $\hat{\gamma}$ and $\hat{\theta}$.

Table 5.10. Failure Data of the Diodes

31,000	51,000
36,000	51,500
40,000	54,000
44,000	57,000
50,000	63,000

SOLUTION The first step is to obtain a good initial value for $\hat{\gamma}$ to be substituted in Eq. (5.29). This can be achieved by using the relationship developed by Cohen (1965) between $\hat{\gamma}$ and CV (coefficient of variation), which is the ratio of sample standard deviation and mean. We may also use the following approximation to obtain $\hat{\gamma}$:

$$\hat{\gamma} = \frac{1.05}{\text{CV}}. \tag{5.32}$$

From the data:

$$\bar{t} = \frac{477{,}500}{10} = 47{,}750 \text{ hours}.$$

The sample standard deviation, s, is

$$s = 9{,}886$$

and

$$\text{CV} = 0.207.$$

Using Eq. (5.32) we obtain an initial value for $\hat{\gamma}$ as 5. Substituting in Eq. (5.29),

$$\sum_{i=1}^{10} t_i^{\hat{\gamma}} \ln t_i = 3.74 \times 10^{25}$$

$$\sum_{i=1}^{n} \ln t_i = 107.53$$

$$\sum_{i=1}^{10} t_i^{\hat{\gamma}} = 3.43 \times 10^{24}$$

$$D(5) = -0.04921.$$

We now try $\hat{\gamma} = 2.1$

$$\sum_{i=1}^{10} t_i^{\hat{\gamma}} \ln t_i = 75.758 \times 10^{10}$$

$$\sum_{i=1}^{10} \ln t_i = 107.53$$

$$\sum_{i=1}^{10} t_i^{\hat{\gamma}} = 6.99 \times 10^{10},$$

and

$$D(2.1) = -0.397345.$$

As $\hat{\gamma}$ decreases, the value of $D(\hat{\gamma})$ decreases and moves further away from zero. Therefore, we try a higher value of $\hat{\gamma}$. We now try $\hat{\gamma} = 3.64$:

$$\sum_{i=1}^{10} t_i^{\hat{\gamma}} \ln t_i = 13.676 \times 10^{18}$$

$$\sum_{i=1}^{10} \ln t_i = 107.53$$

$$\sum_{i=1}^{10} t_i^{\hat{\gamma}} = 1.2575 \times 10^{18}$$

$$D(3.64) = -0.15381.$$

We now try $\hat{\gamma} = 5.907$:

$$\sum_{i=1}^{10} t_i^{\hat{\gamma}} \ln t_i = 74.636 \times 10^{28}$$

$$\sum_{i=1}^{10} \ln t_i = 107.53$$

$$\sum_{i=1}^{10} t_i^{\hat{\gamma}} = 6.836 \times 10^{28}$$

$$D(5.907) = 0.00396145.$$

We now try $\hat{\gamma} = 6.0$:

$$\sum_{i=1}^{10} t_i^{\hat{\gamma}} \ln t_i = 20.519 \times 10^{29}$$

$$\sum_{i=1}^{10} \ln t_i = 107.53$$

$$\sum_{i=1}^{10} t_i^{\hat{\gamma}} = 1.8791 \times 10^{29}$$

$$D(6.0) = 0.0001322.$$

The exact value as obtained by the Newton-Raphson method (see the computer listing in Appendix J) is $\hat{\gamma} = 5.99697278$. Thus, using this value of $\hat{\gamma}$ we obtain

$$\sum_{i=1}^{10} t_i^{\hat{\gamma}} \ln t_i = 19.852 \times 10^{29}$$

$$\sum_{i=1}^{10} \ln t_i = 107.53$$

$$\sum_{i=1}^{10} t_i^{\hat{\gamma}} = 1.817998 \times 10^{29}$$

$$D(5.99697278) = 2.69800180 \times 10^{-16}.$$

Thus, an approximate value of $\hat{\gamma}$ is 6 and $\hat{\theta}$ is obtained as

$$\hat{\theta} = \frac{1}{n} \sum_{i=1}^{n} t_i^{\hat{\gamma}}$$

$$\hat{\theta} = \frac{1}{n} \sum_{i=1}^{n} t_i^{\hat{\gamma}} = \frac{1}{10} \times 1.8791 \times 10^{29} = 1.8791 \times 10^{28}.$$

Thus,

$$R(t) = e^{\frac{-t^{\gamma}}{\theta}}.$$

The reliability at $t = 40{,}000$ hours is

$$R(40{,}000) = 0.8041.$$

5.8.2. FAILURE DATA WITH CENSORING

Assume that the units under test are subjected to censoring of type 1 or type 2. The failure data can be represented by

$$t_1 \leq t_2 \leq t_3 \ldots \leq t_r = \overset{+}{t}_{r+1} = \ldots = \overset{+}{t}_n.$$

Suppose that the failure data follow a Weibull distribution. Following Eqs. (5.27) and (5.28), we obtain

$$\frac{r}{\hat{\gamma}} + \sum_{i=1}^{r} \ln t_i - \frac{1}{\hat{\theta}} \left[\sum_{i=1}^{r} t_i^{\hat{\gamma}} \ln t_i + (n-r) t_r^{\hat{\gamma}} \ln t_r \right] = 0 \tag{5.33}$$

$$-\frac{r}{\hat{\theta}} + \frac{1}{\hat{\theta}^2} \left[\sum_{i=1}^{r} t_i^{\hat{\gamma}} + (n-r) t_r^{\hat{\gamma}} \right] = 0. \tag{5.34}$$

Again, substituting $\hat{\theta}$ from Eq. (5.34) into Eq. (5.33), we obtain $D(\hat{\gamma})$ as

$$D(\hat{\gamma}) = \frac{\sum_{i=1}^{r} t_i^{\hat{\gamma}} \ln t_i + (n-r) t_r^{\hat{\gamma}} \ln t_r}{\sum_{i=1}^{r} t_i^{\hat{\gamma}} + (n-r) t_r^{\hat{\gamma}}} - \frac{1}{r} \sum_{i=1}^{r} \ln t_i - \frac{1}{\hat{\gamma}} = 0. \tag{5.35}$$

By trial and error or by using the Newton-Raphson method, we can estimate $\hat{\gamma}$ from Eq. (5.35). The estimate of $\hat{\theta}$ is

$$\hat{\theta} = \frac{1}{r} \left[\sum_{i=1}^{r} t_i^{\hat{\gamma}} + (n-r) t_r^{\hat{\gamma}} \right]. \tag{5.36}$$

We now present a procedure for obtaining unbiased estimates of $\hat{\theta}$ and $\hat{\gamma}$.

5.8.3. UNBIASED ESTIMATES OF THE WEIBULL PARAMETERS

Assume that the time to failure for a group of components under test follows a Weibull distribution. Moreover, we rewrite the p.d.f. of the Weibull distribution given in Eq. (5.25) as

$$f(t) = \frac{\gamma}{\theta_1}\left(\frac{t}{\theta_1}\right)^{\gamma-1} \exp\left[-\left(\frac{t}{\theta_1}\right)\right]^{\gamma}, \tag{5.37}$$

where

$$F(t) = 1 - \exp\left[-\left(\frac{t}{\theta_1}\right)^{\gamma}\right] \quad \text{and} \quad h(t) = \frac{\gamma}{\theta_1^{\gamma}} t^{\gamma-1}.$$

Note that θ_1 relates to θ of Eq. (5.25) by $\theta = \theta_1^{\gamma}$. Using the maximum likelihood procedure, it can be shown that $\hat{\gamma}$ is the solution of the equation

$$\frac{\sum_{i=1}^{r} t_i^{\hat{\gamma}} \ln t_i + (n-r)t_r^{\hat{\gamma}} \ln t_r}{\sum_{i=1}^{r} t_i^{\hat{\gamma}} + (n-r)t_r^{\hat{\gamma}}} - \frac{1}{\hat{\gamma}} = \frac{1}{r}\sum_{i=1}^{r} \ln t_i. \tag{5.38}$$

This equation can be solved using a numerical method such as the Newton-Raphson procedure as mentioned earlier in this chapter. Once $\hat{\gamma}$ is determined, $\hat{\theta}_1$ can be estimated as

$$\hat{\theta}_1 = \left[\frac{\sum_{i=1}^{r} t_i^{\hat{\gamma}} + (n-r)t_r^{\hat{\gamma}}}{r}\right]^{1/\hat{\gamma}}. \tag{5.39}$$

Equations (5.38) and (5.39) are general and can be used for both censored or complete data. In case of complete data, $r = n$ and all terms that include $(n - r)$ are eliminated.

5.8.4. VARIANCE OF THE MLE ESTIMATES

Since the MLE cannot be presented in a closed-form expression, determining properties of the estimators such as their bias, distribution, and so on is not straightforward. However, Bain and Engelhardt (1991) address these properties through Monte Carlo simulation. Following the procedure for the construction of the information matrix as presented in Section 4.2.5, the asymptotic variances and covariances of the MLE for complete or censored sampling are obtained as

$$\begin{bmatrix} \mathrm{Var}(\hat{\theta}_1) & \mathrm{Cov}(\hat{\theta}_1, \hat{\gamma}) \\ \mathrm{Cov}(\hat{\theta}_1, \hat{\gamma}) & \mathrm{Var}(\hat{\gamma}) \end{bmatrix} = \begin{bmatrix} c_{11}\hat{\theta}_1^2/n\hat{\gamma}^2 & c_{12}\hat{\theta}_1/n \\ c_{12}\hat{\theta}_1/n & c_{22}\hat{\gamma}^2/n \end{bmatrix}, \tag{5.40}$$

where c_{11}, c_{22}, and c_{12} depend on $p = r/n$ and are shown in Table 5.11.

Table 5.11. Asymptotic Value of the Coefficients to Be Used for the Calculations of Variances and Covariances of the MLE for Complete and Censored Sampling

p	c_{11}	c_{22}	c_{12}
1.0	1.108665	0.607927	0.257022
0.9	1.151684	0.767044	0.176413
0.8	1.252617	0.928191	0.049288
0.7	1.447258	1.122447	−0.144825
0.6	1.811959	1.372781	−0.446603
0.5	2.510236	1.716182	−0.935766
0.4	3.933022	2.224740	−1.785525
0.3	7.190427	3.065515	−3.438610
0.2	16.478771	4.738764	−7.375310
0.1	60.517110	9.744662	−22.187207

5.8.5. UNBIASED ESTIMATE OF $\hat{\gamma}$

The MLE may be used to provide a point estimate of $\hat{\gamma}$, but it is quite biased for a small n, particularly when heavy censoring occurs. Bain and Engelhardt (1991) suggest the use of an *unbiasing factor* G_n. Using this factor, the unbiased estimation of $\hat{\gamma}$ is

$$\hat{\gamma} = G_n \hat{\gamma}_{MLE}. \tag{5.41}$$

Tables for determining G_n are available in the literature. Alternatively, G_n can be computed using the following approximation:

$$G_n = 1.0 - 1.346/n - 0.8334/n^2. \tag{5.42}$$

For complete sampling, the asymptotic results for $\mathrm{Var}(\hat{\gamma})$ when $n \to \infty$ is $\mathrm{Var}(\hat{\gamma}) = c_{22}\hat{\gamma}^2 = 0.6079\hat{\gamma}^2$. However, if $n < 100$, instead of using $c_{22} = 0.6079$, a more accurate estimate of Var($\hat{\gamma}$) is obtained using C_n. Again, tables for C_n can be found in the literature or, alternatively, C_n can be computed using

$$C_n = 0.617 + \frac{1.8}{n} + \frac{78.25}{n^3} \tag{5.43}$$

and

$$\mathrm{Var}(\hat{\gamma}) = \frac{C_n \hat{\gamma}^2}{n}. \tag{5.44}$$

EXAMPLE 5.18 (Complete Sample)

Ten units are tested until failure. The data (time to failure) are

20, 22, 24, 25, 26, 27, 30, 35, 42, 52.

Fit a Weibull distribution to the data.

SOLUTION

Using Eqs. (5.38) and (5.39), we obtain the MLE estimates

$$\hat{\gamma} = 3.275$$

and

$$\hat{\theta}_1 = 33.75.$$

The corresponding variances can be calculated using Eq. (5.40) and Table 5.11 for $p = r/n = 1$ (no censoring):

$$\mathrm{Var}(\hat{\gamma}) = c_{22}\hat{\gamma}^2/n = 0.608(3.275)^2/10 = 0.6521$$

$$\mathrm{Var}(\hat{\theta}_1) = c_{11}\hat{\theta}_1^2/(\hat{\gamma}^2 n) = 1.109\,(33.75)^2/(3.275^2 \times 10) = 11.78$$

$$\mathrm{Cov}(\hat{\theta}_{1,}\,\gamma) = c_{12}\hat{\theta}_1/n = 0.257\,(33.75)/10 = 0.8674.$$

Now, using Eqs. (5.37) and (5.36) we obtain the unbiased estimate of $\hat{\gamma}$ as

$$G_n = 1.0 - (1.346/10) - (0.8334/10^2) = 0.857$$

$$\hat{\gamma} = G_n \hat{\gamma}_{MLE} = 0.857 \times 3.275 = 2.81$$

and the corresponding value of $\hat{\theta}_1$ is

$$\hat{\theta}_1 = \left(\sum_{i=1}^{r} \frac{t_i^{\hat{\gamma}}}{r} \right)^{1/\hat{\gamma}} = 33.01.$$

Also, we can calculate the variance of the unbiased estimate of $\hat{\gamma}$ using Eqs. (5.43) and (5.44):

$$C_n = 0.617 + (1.8/10) + (78.25/10^3) = 0.875$$

$$\text{Var}(\hat{\gamma}) = \frac{0.875 \times 2.81^2}{10} = 0.691.$$

EXAMPLE 5.19 (Censored Sample)

Thirty units are under test which is terminated after twenty-two failures occur. The times to failure are

18.5, 20, 20.5, 21.5, 22, 22.5, 23.5, 24, 24.3, 24.6, 25, 25.3, 25.6, 26, 26.3, 26.7, 27, 28, 29, 30, 32, 33.

Fit a Weibull distribution to the data.

SOLUTION

Using Eqs. (5.38) and (5.39), the MLE estimates are

$$\hat{\gamma} = 5.106$$

and

$$\hat{\theta}_1 = 30.58.$$

The corresponding variances can be calculated using Eq. (5.40) and interpolating from Table 5.11 for $p = 22/30 = 0.73$:

$$\text{Var}(\hat{\gamma}) = c_{22}\hat{\gamma}^2/n = 1.06(5.106)^2/30 = 0.921$$

$$\text{Var}(\hat{\theta}_1) = c_{11}\hat{\theta}_1^2/(\hat{\gamma}^2 n) = 1.38(30.58)^2/(5.106^2 \times 30) = 1.65,$$

and the unbiased estimate of $\hat{\gamma}$ is

$$G_n = 1.0 - (1.346/30) - (0.8334/30^2) = 0.954$$

$$\hat{\gamma} = G_n \hat{\gamma}_{MLE} = 4.87.$$

The corresponding value of $\hat{\theta}_1$ is

$$\hat{\theta}_1 = \left[\frac{\sum_{i=1}^{r} t_i^{\hat{\gamma}} + (n-r)t_r^{\hat{\gamma}}}{r}\right]^{\frac{1}{\hat{\gamma}}} = 30.59.$$

5.8.6. CONFIDENCE INTERVAL FOR $\hat{\gamma}$

Asymptotic results derived by Bain and Engelhardt (1991) indicate that for heavy censoring $\hat{\gamma}$ approximately follows a chi-squared distribution with $2(r-1)$ *df* (degrees of freedom), and it follows a chi-squared distribution with $(n-1)$ *df* when the sample is complete.

In order to take into account this transition, Bain and Engelhardt (1991) suggest the following approximation:

$$df = c(r-1), \tag{5.45}$$

where

$$c = 2/[(1+p^2)^2 p c_{22}].$$

Once the *df* has been calculated using these expressions, the $100(1-\alpha)$ percent confidence interval for $\hat{\gamma}$ can be computed using

$$\hat{\gamma}^L = \hat{\gamma}\left[\frac{\chi^2_{(1-\alpha),\, df}}{cr}\right]^{\frac{1}{1+p^2}} \tag{5.46}$$

$$\hat{\gamma}^U = \hat{\gamma}\left[\frac{\chi^2_{\alpha,\, df}}{cr}\right]^{\frac{1}{1+p^2}}. \tag{5.47}$$

The superscripts L and U denote lower and upper limits, respectively.

EXAMPLE 5.20 (Complete Sample)

Find the 90 percent confidence interval for $\hat{\gamma}$ estimated in Example 5.18.

SOLUTION

Since this is a complete sample, then $\hat{\gamma}$ approximately follows a chi-squared distribution with $(n-1)$ degrees of freedom:

$$\chi^2_{0.95,\,9} = 3.33 \quad \text{and} \quad \chi^2_{0.05,\,9} = 16.92$$

$$c = 2/[(1+1^2)^2(1) \times 0.608] = 0.822$$

$$\hat{\gamma}^L = \hat{\gamma}\left[\frac{\chi^2_{0.95,\,9}}{cr}\right]^{1/(1+p^2)}$$

$$\hat{\gamma}^L = 2.81\left[\frac{3.33}{0.822 \times 10}\right]^{1/2} = 1.788$$

$$\hat{\gamma}^U = \hat{\gamma}\left[\frac{\chi^2_{0.05,\,9}}{cr}\right]^{1/(1+p^2)}$$

$$\hat{\gamma}^U = 2.81\left[\frac{16.92}{0.822 \times 10}\right]^{1/2} = 4.032.$$

The following example illustrates the procedure for calculating the confidence interval for $\hat{\gamma}$ when some of the failure times are censored.

EXAMPLE 5.21 (Censored Sample)

Find the 90 percent confidence interval for $\hat{\gamma}$ estimated in Example 5.19.

SOLUTION

$\hat{\gamma}$ approximately follows a chi-squared distribution with *df* given by Eq. (5.45):

$$c = 2/[(1 + p^2)^2 p c_{22}] = 2/[(1 + 0.73^2)^2 \times (0.73) \times 1.06] = 1.10$$

$$df = c(r - 1) = 1.10(21) = 23$$

$$\chi^2_{0.95,\,23} = 13.09 \quad \text{and} \quad \chi^2_{0.05,\,23} = 35.17.$$

Using Eqs. (5.46) and (5.47) we obtain the upper and lower limits of the confidence interval for $\hat{\gamma}$ as

$$\hat{\gamma}^L = 4.87\left[\frac{13.09}{(1.10 \times 22)}\right]^{1/(1+p^2)} = 3.28$$

$$\hat{\gamma}^U = 4.87\left[\frac{35.17}{(1.10 \times 22)}\right]^{1/(1+p^2)} = 6.24.$$

5.8.7. INFERENCES ON $\hat{\theta}_1$

The bias of $\hat{\theta}_1$ is a function of both θ_1 and γ, and it is not easily assessed. Fortunately, in general, θ_1 is not very biased and the use of an unbiased $\hat{\gamma}$ in Eq. (5.39) provides a reasonable estimate of $\hat{\theta}_1$.

Confidence intervals for $\hat{\theta}_1$ can be constructed using the distribution of $U = \sqrt{n}\,\hat{\gamma}\ln(\hat{\theta}_1/\theta_1)$. It can be shown that the $100(1-\alpha)$ percent confidence intervals for $\hat{\theta}_1$ are

$$\theta_1^L = \hat{\theta}_1 \exp\left(-U_{1-\alpha/2}/(\sqrt{n}\,\hat{\gamma})\right) \tag{5.48}$$

$$\theta_1^U = \hat{\theta}_1 \exp\left(-U_{\alpha/2}/(\sqrt{n}\cdot\hat{\gamma})\right). \tag{5.49}$$

Bain and Engelhardt (1991) provide tables with the percentage points U_α such that $p(U \le U_\alpha) = \alpha$. Alternatively, $U_{0.05}$ and $U_{0.95}$ can be computed using the following approximation:

$$U_{0.05} = -1.715 - (3.868/n) - (44.23/\exp(n)) \tag{5.50}$$

$$U_{0.95} = 1.72 + (3.163/n) + (18.25/\exp(n)). \tag{5.51}$$

These expressions hold for complete samples only.

EXAMPLE 5.22 (Complete Sample)

Find the 90 percent confidence interval for $\hat{\theta}_1$ estimated in Example 5.18.

SOLUTION

Using Eqs. (5.50) and (5.51), we calculate the values of the U distribution:

$$U_{0.05} = -1.715 - (3.868/10) - (44.23/\exp(10)) = -2.09$$

and

$$U_{0.95} = 1.72 + (3.163/10) + (18.25/\exp(10)) = 2.04.$$

Now using Eqs. (5.48) and (5.49), the lower and upper limits for $\hat{\theta}_1$ are estimated as

$$\hat{\theta}_1^L = \hat{\theta}_1 \exp\left(-U_{0.95}/(\sqrt{n}\times\hat{\gamma})\right)$$

$$\hat{\theta}_1^L = 33.01 \exp\left(-2.04/(\sqrt{10}\times 2.81)\right) = 26.2$$

and

$$\hat{\theta}_1^U = \hat{\theta}_1 \exp\left(-U_{0.05}/(\sqrt{n}\times\hat{\gamma})\right)$$

$$\hat{\theta}_1^U = 33.01 \exp\left(2.09/(\sqrt{10}\times 2.81)\right) = 41.8.$$

For censored samples, U_α is a function of $p = r/n$ and n and some tabulated results for U_α are provided by Bain and Engelhardt (1991). Alternatively, $U_{0.05}$ and $U_{0.95}$ for censored samples can be computed using the following approximations ($p = r/n$):

$$U_{0.05} = -7.72 + 12.99p - 7.02p^2 + \frac{24.83}{n} + \frac{47.72}{n^2} - \frac{26.57}{np} - \frac{66.46}{(np)^2} \quad (5.52)$$

$$U_{0.95} = 4.08 - 4.76p + 2.43p^2 + \frac{11.41}{n} - \frac{9.85}{np} + \frac{10.46}{(np)^2}. \quad (5.53)$$

These expressions are valid for $5 \leq n < 120$ and $0.5 \leq p \leq 1.0$.

EXAMPLE 5.23 (Censored Sample)

Find the 90 percent confidence interval for $\hat{\theta}_1$ estimated in Example 5.19.

SOLUTION

In Example 5.19, we have $n = 30$ and $p = 22/30 = 0.733$. Using Eqs. (5.52) and (5.53), we calculate the values of the U distribution as

$$U_{0.05} = -2.44$$

$$U_{0.95} = 1.85.$$

Then, using Eq. (5.48) and (5.49), the lower and upper limits for $\hat{\theta}_1$ are

$$\hat{\theta}_1^L = \hat{\theta}_1 \exp\left(-U_{.95}/(\sqrt{n}\,\hat{\gamma})\right)$$

$$\hat{\theta}_1^L = 26.24 \exp\left(-1.85/(\sqrt{30} \times 4.87)\right) = 24.5$$

$$\hat{\theta}_1^U = \hat{\theta}_1 \exp\left(-U_{.05}/(\sqrt{n}\,\hat{\gamma})\right)$$

$$\hat{\theta}_1^U = 26.24 \exp\left(+2.44/(\sqrt{30} \times 4.87)\right) = 28.7.$$

5.9. Lognormal Distribution

When the failure times are assumed to follow a lognormal distribution, the p.d.f. $f(t)$ is given by

$$f(t) = \frac{1}{\sigma t \sqrt{2\pi}} \exp\left[-\frac{1}{2}\left(\frac{\ln t - \mu}{\sigma}\right)^2\right] \qquad t \geq 0.$$

Let $x = \ln t$, where x is normally distributed with a mean μ and standard deviation σ—that is,

$$E[x] = E[\ln t] = \mu$$

$$V[x] = \text{Var}[\ln t] = \sigma^2.$$

Since $t = e^x$,

$$E[t] = E[e^x] = \int_{-\infty}^{\infty} \frac{1}{\sigma\sqrt{2\pi}} \exp\left[x - \frac{1}{2}\left(\frac{x-\mu}{\sigma}\right)^2\right] dx$$

$$E[t] = \exp\left[\mu + \frac{\sigma^2}{2}\right] \int_{-\infty}^{\infty} \frac{1}{\sigma\sqrt{2\pi}} \exp\left[-\frac{1}{2\sigma^2}[x - (\mu + \sigma)]^2\right] dx$$

$$E[t] = \exp\left[\mu + \frac{\sigma^2}{2}\right]$$

$$E[t^2] = E[e^{2x}] = \exp[2(\mu + \sigma^2)]$$

$$\text{Var}[t] = [e^{2\mu+\sigma^2}][e^{\sigma^2} - 1].$$

But

$$F(t) = \int_0^t \frac{1}{\tau\sigma\sqrt{2\pi}} \exp\left[-\frac{1}{2}\left(\frac{\ln\tau - \mu}{\sigma}\right)^2\right] d\tau$$

$$F(t) = P(\mathbf{t} \le t) = P\left[z \le \frac{\ln t - \mu}{\sigma}\right]$$

$$R(t) = P[\mathbf{t} > t] = P\left[z > \frac{\ln t - \mu}{\sigma}\right]$$

$$h(t) = \frac{f(t)}{R(t)} = \frac{\dfrac{\phi\left(\dfrac{\ln t - \mu}{\sigma}\right)}{\sigma}}{t\sigma R(t)}.$$

Estimations of the parameters of the lognormal distribution when the failure data are not censored and when the failure data are censored are discussed next.

5.9.1. FAILURE DATA WITHOUT CENSORING

When the failure time T follows a lognormal distribution with p.d.f. $f(t)$,

$$f(t) = \frac{1}{t\sigma\sqrt{2\pi}} \exp\left[-\frac{1}{2\sigma^2}(\ln t - \mu)^2\right] \tag{5.54}$$

with a mean of

$$\exp\left(\mu + \frac{\sigma^2}{2}\right)$$

and a variance of

$$[e^{\sigma^2} - 1][e^{2\mu+\sigma^2}],$$

the estimation of the parameters $\hat{\mu}$ and $\hat{\sigma}$ can be obtained directly from Eq. (5.54). However, one of the simplest ways to obtain μ and σ^2 with optimum properties is by considering the distribution of $Y = \ln T$.

Assume that the failure times $t_1, t_2, \ldots, t_n$ are the exact failure times of n units that are subjected to a test. The MLE of μ and σ^2 of Y are

$$\hat{\mu} = \frac{1}{n}\sum_{i=1}^{n} \ln t_i \tag{5.55}$$

$$\hat{\sigma}^2 = \frac{1}{n}\left[\sum_{i=1}^{n}(\ln t_i)^2 - \frac{\left(\sum_{i=1}^{n}\ln t_i\right)^2}{n}\right]. \tag{5.56}$$

The estimate of $\hat{\mu}$ is unbiased. However, the estimate of $\hat{\sigma}^2$ is not unbiased. Therefore, to ensure that $\hat{\sigma}^2$ is unbiased, we use

$$s^2 = \hat{\sigma}^2\,[n/(n-1)],$$

where s^2 is the sample variance.

Obviously, $s^2 \approx \hat{\sigma}^2$ when n is large. The estimates of the mean and variance of T are

$$\exp(\hat{\mu} + \hat{\sigma}^2/2) \quad \text{and} \quad [e^{\hat{\sigma}^2} - 1][e^{2\hat{\mu} + \hat{\sigma}^2}],$$

respectively, and the $100(1 - \alpha)$ percent confidence interval for μ is obtained as

$$\hat{\mu} - Z_{\alpha/2}\frac{\sigma}{\sqrt{n}} < \mu < \hat{\mu} + Z_{\alpha/2}\frac{\sigma}{\sqrt{n}}. \tag{5.57}$$

If σ is unknown or when the sample size is relatively small ($n < 25$), then we replace it by s and use the student t-distribution instead. Thus,

$$\hat{\mu} - t_{\alpha/2,\,\mathrm{n}-1}\frac{s}{\sqrt{n}} < \mu < \hat{\mu} + t_{\alpha/2,\,\mathrm{n}-1}\frac{s}{\sqrt{n}}. \tag{5.58}$$

Similarly, the $100(1 - \alpha)$ percent confidence interval for $\hat{\sigma}^2$ [$n\hat{\sigma}^2/\sigma^2$ has a chi-square distribution with $(n - 1)$ degrees of freedom] is

$$\frac{n\hat{\sigma}^2}{\chi^2_{\alpha/2,\,(n-1)}} < \sigma^2 < \frac{n\hat{\sigma}^2}{\chi^2_{1-\alpha/2,\,(n-1)}}. \tag{5.59}$$

Engineers are normally interested in estimating the mean time to failure and confidence intervals for components whose failure time distribution is lognormal. Indeed, this is a main concern in many applications. To obtain a $100(1 - \alpha)$ percent confidence interval for the mean time to failure, $\hat{\tau}$, of the lognormal, we let

$$\tau = \mu + \frac{\sigma^2}{2} \tag{5.60}$$

and

$$\hat{\tau} = \hat{\mu} + \frac{n}{(n-1)}\frac{\hat{\sigma}^2}{2}, \tag{5.61}$$

where $\hat{\mu}$ and $\hat{\sigma}$ are obtained from Eqs. (5.55) and (5.56), respectively. For large samples, $\hat{\tau}$ is approximated by a normal distribution with variance $\sigma^2_{\hat{\tau}}$ as given in (Shapiro and Gross, 1981):

$$\sigma^2_{\hat{\tau}} = \mathrm{Var}(\hat{\mu}) + \mathrm{Var}[n\hat{\sigma}^2/(n-1)]/4. \tag{5.62}$$

However, $\text{Var}[n\hat{\sigma}^2/(n-1)] = n^2\hat{\sigma}^4/(n-1)^3$ and $\text{Var}(\hat{\mu}) = \hat{\sigma}^2/(n-1)$. Therefore we rewrite Eq. (5.62) as

$$\hat{\sigma}_{\hat{\tau}}^2 = \frac{\hat{\sigma}^2}{n-1} + \frac{n^2\hat{\sigma}^4}{4(n-1)^3}. \tag{5.63}$$

Once $\hat{\tau}$ and $\hat{\sigma}_{\hat{\tau}}^2$ are obtained using Eqs. (5.61) and (5.63), respectively, the $100(1-\alpha)$ percent confidence interval for T (mean time to failure of the population) can be determined as

$$\exp(\hat{\tau} - Z_{1-\alpha/2}\hat{\sigma}_{\hat{\tau}}) < T < \exp(\hat{\tau} + Z_{1-\alpha/2}\hat{\sigma}_{\hat{\tau}}). \tag{5.64}$$

EXAMPLE 5.24

A production engineer performs a burn-in test on eight video display terminals (VDTs). The following failure times (in hours) are recorded:

20, 28, 35, 39, 42, 44, 46, 47.

Suppose that the failure times follow a lognormal distribution. Determine the mean failure time and its standard deviation. What are the 95 percent confidence intervals for μ and σ^2?

SOLUTION

In order to calculate the parameters of the distribution we construct Table 5.12. Using Eqs. (5.55) and (5.56), we obtain

$$\hat{\mu} = \frac{\sum \ln t_i}{n} = \frac{28.747}{8} = 3.593$$

$$\hat{\sigma}^2 = \frac{1}{8}\left[103.912 - \frac{28.747^2}{8}\right] = 0.0766.$$

Table 5.12. Failure Times of the VDT

t_i	$\ln t_i$	$(\ln t_i)^2$
20	2.995	8.974
28	3.332	11.103
35	3.555	12.640
39	3.663	13.421
42	3.737	13.970
44	3.784	14.320
46	3.828	14.658
47	3.850	14.823
	28.747	103.912

The mean failure time is

$$\exp\left[\hat{\mu} + \frac{\hat{\sigma}^2}{2}\right] = \exp\,[3.6313] = 37.76 \text{ hours.}$$

The standard deviation of the failure time is

$$\sigma = \{[\exp(\hat{\sigma}^2) - 1]\exp[2\hat{\mu} + \hat{\sigma}^2]\}^{1/2}$$

$$\sigma = \{[\exp(0.0766) - 1]\exp[2 \times 3.593 + 0.0766]\}^{1/2}$$

$$\sigma = 10.654 \text{ hours.}$$

We now determine the 95 percent confidence interval for μ. Since n is less than 25, we estimate the variance σ^2 as

$$s^2 = \frac{8\hat{\sigma}^2}{8 - 1} = 0.0876$$

$$\hat{\mu} - t_{\alpha/2,\,n-1}\frac{s}{\sqrt{n}} < \mu < \hat{\mu} + t_{\alpha/2,\,n-1}\frac{s}{\sqrt{n}}$$

$$3.593 - 2.365\sqrt{8} < \mu < 3.593 + 2.365\sqrt{8}$$

$$2.893 < \mu < 4.293.$$

The 95 percent confidence interval for σ^2 is

$$\frac{n\hat{\sigma}^2}{\chi^2_{\alpha/2,\,n-1}} < \sigma^2 < \frac{n\hat{\sigma}^2}{\chi^2_{1-\alpha/2,\,n-1}}$$

$$\frac{8 \times 0.0766}{16.013} < \sigma^2 < \frac{8 \times 0.0766}{1.689}$$

or

$$0.0382 < \sigma^2 < 0.3628.$$

The 95 percent confidence interval for the mean life T can be estimated using Eq. (5.64). We first estimate $\hat{\tau}$ using Eq. (5.61), then we estimate $\sigma_{\hat{\tau}}^2$ using Eq. (5.63). Thus,

$$\hat{\tau} = 3.593 + \frac{7}{8} \times \frac{0.0766}{2} = 3.62651$$

$$\hat{\sigma}_{\hat{\tau}}^2 = \frac{0.0766}{7} + \frac{8^2 \times 0.0766^2}{4(8-1)^3} = 0.0112165.$$

The 95 percent confidence interval for the mean time to failure is

$$e^{(3.62651 - 1.96 \times 0.0112165)} < T < e^{(3.62651 + 1.96 \times 0.0112165)}$$

$$36.7642 < T < 38.416.$$

5.9.2. FAILURE DATA WITH CENSORING

Consider the placement of n units under test where the exact failure times of r units are

$$t_1 \leq t_2 \leq \ldots \leq t_r.$$

The test is censored after the occurrence of the rth failure or at time T_c. (We can assume r failures occurred within T_c.) Thus, we have either type 2 or type 1 censoring. As we discussed earlier, we use the fact that $Y = \log_e T$ has a normal distribution with mean μ and variance σ^2. We can estimate μ and σ^2 from the transformed data $y_i = \ln t_i$. We use the method of Sarhan and Greenberg (1956, 1957, 1958, 1962) to estimate μ and σ^2. They propose that the best estimates are linear combinations of the logarithms of the r exact failure times:

$$\hat{\mu} = \sum_{i=1}^{r} a_i \ln t_i \tag{5.65}$$

and

$$\hat{\sigma} = \sum_{i=1}^{r} b_i \ln t_i, \tag{5.66}$$

where a_i and b_i are given by Sarhan and Greenberg for $n \leq 20$ as well as the variance and covariance of $\hat{\mu}$ and $\hat{\sigma}$.

If the sample size is greater than twenty, the MLEs for normal distribution can be utilized in estimating the parameters of the lognormal (with censoring) as shown below:

$$\bar{y} = \frac{1}{r}\sum_{i=1}^{r} \ln t_i \tag{5.67}$$

$$s^2 = \frac{1}{r}\left[\sum_{i=1}^{r} (\ln t_i)^2 - \left(\sum_{i=1}^{r} \ln t_i\right)^2 \middle/ r\right], \tag{5.68}$$

and the MLEs of $\hat{\mu}$ and $\hat{\sigma}^2$ are

$$\hat{\mu} = \bar{y} - \lambda(\bar{y} - \ln t_r) \tag{5.69}$$

$$\hat{\sigma}^2 = s^2 + \lambda(\bar{y} - \ln t_r)^2. \tag{5.70}$$

The coefficient λ is a function of α and β (Cohen, 1961), where

$$\alpha = s^2/(\bar{y} - \ln t_r)^2 \tag{5.71}$$

$$\beta = (n - r)/n. \tag{5.72}$$

As shown in Eq. (5.72), 100β is the percentage of censored units. (Cohen, 1961) provides tabulated results for λ as a function of α and β. Alternatively, λ can be calculated using the following approximation:

$$\lambda = [1.136\alpha^3 - \ln(1 - \alpha)]\,[1 + 0.437\beta - 0.250\alpha\beta^{1.3}] + 0.08\alpha(1 - \alpha). \tag{5.73}$$

Equation (5.73) provides a good approximation of λ with a maximum error of 5 percent.

The asymptotic variances of $\hat{\mu}$ and $\hat{\sigma}$ can be estimated as

$$\mathrm{Var}(\hat{\mu}) = m_1\hat{\sigma}^2/n \tag{5.74}$$

$$\mathrm{Var}(\hat{\sigma}) = m_2\hat{\sigma}^2/n. \tag{5.75}$$

Cohen also provides tabulated values of m_1 and m_2 as a function of $\hat{c}$, where $\hat{c} = (\ln t_r - \hat{\mu})/\hat{\sigma}$. Alternatively, m_1 and m_2 can be calculated using the following approximation:

For $y < 0$,

$$m_1 = 1 + 0.51e^{2.5y} \tag{5.76}$$

$$m_2 = 0.5 + 0.74e^{1.6y}. \tag{5.77}$$

For $y > 0$,

$$m_1 = 0.52 + e^{(1.838y+0.354y^2)} - 0.391y - 0.676y^2 \tag{5.78}$$

$$m_2 = 0.24 + e^{(y+0.384y^2)} + 0.0507y + 0.2735y^2, \tag{5.79}$$

where $y = -\hat{c}$.

EXAMPLE 5.25

Ten units are subjected to a fatigue test. The test is terminated when seven units fail and their failure times (in weeks) are

30, 37, 42, 45, 47, 48, 50.

Suppose that the failure times follow a lognormal distribution. Determine $\hat{\mu}$ and $\hat{\sigma}$.

SOLUTION

In this case, $n = 10$, $r = 7$, and $n - r = 3$. Using Eqs. (5.65), (5.66), and Appendix D, we obtain

$$\hat{\mu} = 0.0244 \ln 30 + 0.0636 \ln 37 + 0.0818 \ln 42 + 0.0962 \ln 45 + 0.1089 \ln 47$$

$$+ 0.1207 \ln 48 + 0.5045 \ln 50 = 3.8447$$

$$\hat{\sigma} = -.3252 \ln 30 - 0.1758 \ln 37 - 0.1058 \ln 42 - 0.0502 \ln 45 - 0.0006 \ln 47$$

$$+ 0.0469 \ln 48 + 0.6107 \ln 50 = 0.2409.$$

The estimated mean failure time is

$$\mu = \exp\,[\hat{\mu} + \hat{\sigma}^2/2] = \exp\left[3.8447 + \left(\frac{0.2409^2}{2}\right)\right]$$

$$\mu = 48.12 \text{ weeks}.$$

And its estimated standard deviation is

$$\sigma = [(e^{\hat{\sigma}^2} - 1)\,(e^{2\hat{\mu}+\hat{\sigma}^2})]^{1/2}$$

$$\sigma = [(e^{0.2409^2} - 1)\,(e^{2\times 3.8447+0.2409^2})]^{1/2}$$

$$\sigma = [0.05975 \times 2315.6201\,]^{1/2} = 11.76.$$

5.10. The Gamma Distribution

The Gamma hazard model is useful in estimating the reliability of many practical situations. The two parameter Gamma density is given by

$$f(t;\theta,\gamma) = \frac{t^{\gamma-1}}{\theta^{\gamma}\Gamma(\gamma)} e^{-\frac{t}{\theta}}, \tag{5.80}$$

where $\Gamma(x)$ denotes the Gamma function, γ is the shape parameter, and $\theta = 1/\lambda$ (λ is sometimes referred to as the scale parameter). It is worth noting that the chi-square distribution is a special case of the Gamma distribution when $\theta = 2$ and $\gamma = \nu/2$ (where ν is an integer and is also the number of degrees of freedom of the chi-square distribution). Also the exponential distribution is a special case of Gamma distribution, when $\gamma = 1$. The value of $\Gamma(x)$ can be found in Appendix A.

5.10.1. FAILURE DATA WITHOUT CENSORING

We consider the case where n units are subjected to a reliability test and the failure times of all units are recorded. We assume that the failure times can be expressed by a two-parameter gamma distribution, and its p.d.f. is given by Eq. (5.80). To estimate γ and θ we use either the method of moments or the maximum likelihood estimation. However, the maximum likelihood estimation provides a more accurate result and will now be considered.

The likelihood function for a complete sample is

$$L(\theta,\gamma) = \frac{1}{\theta^{n\gamma}\Gamma^{n}(\gamma)}\left[\prod_{i=1}^{n} t_i\right]^{(\gamma-1)} e^{-\sum_{i=1}^{n} t_i/\theta}. \tag{5.81}$$

Taking partial derivatives of the logarithm of Eq. (5.81) with respect to θ and γ and equating the resultant equations to zero we obtain

$$\hat{\theta} = \bar{t}/\hat{\gamma} \tag{5.82}$$

$$\ln \hat{\gamma} - \psi(\hat{\gamma}) - \ln \bar{t} + \ln \tilde{t} = 0, \tag{5.83}$$

where

$$\bar{t} = \text{the arithmetic mean}, \quad \bar{t} = \frac{\sum_{i=1}^{n} t_i}{n}$$

$$\tilde{t} = \text{the geometric mean}, \quad \tilde{t} = \left(\prod_{i=1}^{n} t_i\right)^{1/n},$$

and the ψ function is defined as $\psi(x) = \Gamma'(x)/\Gamma(x)$, where $\Gamma'(\hat{\gamma})$ is the derivative of $\Gamma(\hat{\gamma})$ or

$$\Gamma'(\hat{\gamma}) = \int_0^{\infty} x^{\hat{\gamma}-1} \ln x \, e^{-x} \, dx.$$

Using ψ function tables, Eq. (5.83) can be solved iteratively. However, an easier approach is to use the approximation suggested by Greenwood and Durand (1960):

$$\hat{\gamma} = (0.50009 + 0.16488M - 0.054427M^2)/M; \qquad 0 < M \leq 0.5772 \tag{5.84a}$$

$$\hat{\gamma} = \frac{8.8989 + 9.0599M + 0.97754M^2}{M(17.797 + 11.968M + M^2)}; \qquad 0.5772 < M \leq 17 \tag{5.84b}$$

$$\hat{\gamma} = 1/M; \qquad M > 17, \tag{5.84c}$$

where $M = \ln(\bar{t}/\tilde{t})$. This approximation will lead to good estimates of the parameters obtained by the MLE method. Once we estimate $\hat{\gamma}$ we can easily estimate the value of $\hat{\theta}$ using Eq. (5.82).

As usual, for small n, the estimates of the parameters obtained using the MLE method are noticeably biased. To reduce the bias, the following expressions are suggested to provide unbiased estimates of γ and θ:

$$\hat{\gamma}_{\text{unbiased}} = \hat{\gamma}\frac{(n-3)}{n} + \frac{2}{3n} \tag{5.85}$$

$$\hat{\theta}_{\text{unbiased}} = \frac{\bar{t}}{\hat{\gamma}_{\text{unbiased}}\,[1 - 1/(\hat{\gamma}_{\text{unbiased}}\, n)]}. \tag{5.86}$$

Equation (5.85) is based on Bain and Engelhardt (1991), while Eq. (5.86) is based on Lee (1992).

EXAMPLE 5.26

A mechanical engineer conducts a fatigue test by subjecting ten identical steel rods to a stress level significantly higher than the endurance limit of the rod material. The number of cycles to failure are recorded as

20,000, 35,000, 47,000, 58,000, 68,000, 77,000, 85,000, 92,000, 97,000, 102,000.

Assume that the cycles to failure follow gamma distribution. What are the parameters of the distribution?

SOLUTION

We calculate $\bar{t}$, $\tilde{t}$, and M:

$$\bar{t} = \frac{\sum_{i=1}^{10} t_i}{10} = 68{,}100$$

$$\tilde{t} = \left[\prod_{i=1}^{10} t_i\right]^{\frac{1}{10}} = 61{,}492.22$$

$$M = \ln\left(\frac{\bar{t}}{\tilde{t}}\right) = 0.102.$$

Using Eq. (5.84a), we obtain $\hat{\gamma}$ as

$$\hat{\gamma} = \frac{1}{0.102}(0.50009 + 0.16488 \times 0.102 - 0.05442 \times 0.102^2)$$

$$\hat{\gamma} = 5.0588.$$

The unbiased estimates of $\hat{\gamma}$ and $\hat{\theta}$ are

$$\hat{\gamma}_{\text{unbiased}} = 5.0588 \times \frac{7}{10} + \frac{2}{30}$$

$$\hat{\gamma}_{\text{unbiased}} = 3.60$$

$$\hat{\theta}_{\text{unbiased}} = \frac{68{,}100}{3.60\left(1 - \frac{1}{3.6 \times 10}\right)} = 19{,}457.$$

The mean life $= \hat{\theta}_{\text{unbiased}}\,\hat{\gamma}_{\text{unbiased}} = 70{,}045$ cycles.

5.10.2. FAILURE DATA WITH CENSORING

When there are censored observations in the failure data, the estimation of the parameters becomes considerably more difficult. Wilk, Gnanadesikan, and Huyett (1962) and Bain and Engelhardt (1991) provide tables to aid in computing $\hat{\gamma}$ and $\hat{\theta}$. Alternatively, an approximation can be obtained using the following algorithm:

1. Compute the arithmetic and geometric mean of the available observations in the censored sample:

$$\bar{t}_c = \frac{\sum_{i=1}^{r} t_i}{r} \tag{5.87}$$

$$\tilde{t}_c = \left(\prod_{i=1}^{r} t_i\right)^{\frac{1}{r}}. \tag{5.88}$$

2. Compute NR, S, and Q as

$$NR = \frac{n}{r}$$

$$S = \frac{\bar{t}_c}{t_r}$$

$$Q = \frac{1}{\left(1 - \frac{\tilde{t}_c}{\bar{t}_c}\right)}.$$

3. Finally, compute $\hat{\gamma}_{\text{unbiased}}$ as a function of NR, S, and Q:

   - If $S < 0.42$, use

$$\hat{\gamma}_{\text{unbiased}} = 1.061(1 - \sqrt{Q}) + 0.2522Q(1 + (\sqrt{S}/NR^4)) + 1.953(\sqrt{S} - 1/Q) - 0.220/NR^4 + 0.1308Q/NR^4 + 0.4292/(Q\sqrt{S}). \tag{5.89}$$

   - When $0.42 < S < 0.80$, use

$$\hat{\gamma}_{\text{unbiased}} = 0.5311Q((1/NR^2) - 1) + 1.436 \log Q + 0.7536(QS - S) - 2.040/NR - 0.260QS/NR^2 + 2.489/(Q/NR)^{1/2}. \tag{5.90}$$

- When $S > 0.80$, use

$$\hat{\gamma}_{\text{unbiased}} = 1.151 + 1.448(Q(1 - S)/NR) - 1.024(Q + S) + 0.5311 \log Q + 1.541QS - 0.515(Q/NR)^{1/2}. \quad (5.91)$$

Once $\hat{\gamma}$ is estimated, then $\hat{\theta}$ can be estimated using the following expression:

$$\hat{\theta} = \frac{\left(\sum_{i=1}^{r} t_i + (n - r)t_r\right)\Big/ n}{\hat{\gamma}_{\text{unbiased}}\left[1 - 1/(\hat{\gamma}_{\text{unbiased}} r)\right]}.$$

The expressions for estimating $\hat{\gamma}$ in step 3 provide good approximations when $S \geq 0.12$, $1.2 \leq Q \leq 12$, and $NR \leq 3.0$. The accuracy of the estimation was not verified outside these limits. The standard error is approximately 0.04. For small values of $\hat{\gamma}$ ($\hat{\gamma} < 1$), the maximum percentile error can be large, about 10 percent. For large values of $\hat{\gamma}$ ($\hat{\gamma} > 2$), the maximum percentile error is less than 5 percent.

EXAMPLE 5.27

Ten components are subjected to a test. Seven of the components fail during the test and three components survive the predetermined test period of 4,900 hours. The failure times are

1,000, 3,000, 4,000, 4,400, 4,700, 4,800, 4,900, $4{,}900^{+}$, $4{,}900^{+}$, $4{,}900^{+}$.

Fit a gamma distribution to these data points and estimate its parameters. What is the mean life of a component from this population?

SOLUTION

We calculate $\bar{t}_c$, $\tilde{t}_c$, NR, S, and Q:

$$\bar{t}_c = \frac{\sum_{i=1}^{7} t_i}{7} = 3{,}829$$

$$t_c = \left(\prod_{i=1}^{7} t_i\right)^{\frac{1}{7}} = 3{,}452$$

$$NR = \frac{10}{7}$$

$$S = \frac{\bar{t}_c}{t_r} = \frac{3{,}829}{4{,}900} = 0.7814$$

$$Q = \frac{1}{1 - 0.9015} = 10.15.$$

Since $0.42 < S < 0.8$, we use Eq. (5.90) to obtain the unbiased estimate of $\hat{\gamma}$:

$$\hat{\gamma}_{\text{unbiased}} = 0.5311 \times 10.15\left(\frac{1}{1.428^2} - 1\right) + 1.436 \log 10.15$$

$$+ 0.7536(10.15 \times 0.7814 - 0.7814) - \frac{2.040}{1.428}$$

$$- \frac{0.260 \times 10.15 \times 0.7814}{1.428^2} + \frac{2.489}{\sqrt{10.15/1.428}} = 2.58$$

and

$$\hat{\theta}_{\text{unbiased}} = \frac{\left(\sum_{i=1}^{7} t_i + 3 \times 4900\right)\Big/10}{2.85 \times 0.91445} = 1{,}759.$$

The mean life of a component from this population is $\hat{\gamma}_{\text{unbiased}}\ \hat{\theta}_{\text{unbiased}} =$ 4,538 hours.

5.10.3. VARIANCE OF $\hat{\gamma}$ AND $\hat{\theta}$

For the case of a complete sample, the variances of $\hat{\gamma}$ and $\hat{\theta}$ are functions of γ itself and n (sample size). Bain and Engelhardt (1991) provide a table with coefficients (say, C_γ and C_θ) that permit estimates of the corresponding variances as

$$\text{Var}(\hat{\gamma}) = C_\gamma \hat{\gamma}^2/n \tag{5.92}$$

$$\text{Var}(\hat{\theta}) = C_\theta \hat{\theta}^2/n. \tag{5.93}$$

Alternatively, these coefficients can be calculated using the following approximate expressions:

$$C_\gamma = 2.076A - 0.3697A^2 + 0.01654A^3 + 5.463B$$
$$- 0.3917B^2 - 7.274\sqrt{B} + 0.0006823BA^4 \tag{5.94}$$

$$C_\theta = 1.976 + \frac{0.608}{\hat{\gamma}_{\text{unbiased}}{}^{1.2}} - \frac{1.8942}{n}, \tag{5.95}$$

where

$$A = 8 - (1/\hat{\gamma}_{\text{unbiased}})$$

$$B = n/(n-6).$$

For the case of censored samples, the variances of $\hat{\gamma}$ and $\hat{\theta}$ also depend on $p = r/n$. Asymptotic results ($n \to \infty$, $r/n \to p$) provided by Harter (1969) indicate that C_γ and C_θ, and thus $\text{Var}(\hat{\gamma})$ and $\text{Var}(\hat{\theta})$, will always be larger when there is censoring (as would be expected).

Thus, for censored sample, $\text{Var}(\hat{\gamma})$ and $\text{Var}(\hat{\theta})$ can be calculated using

$$\text{Var}(\hat{\gamma}) = C_1 C_\gamma \hat{\gamma}^2/n \tag{5.96}$$

$$\text{Var}(\hat{\theta}) = C_2 C_\theta \hat{\theta}^2/n, \tag{5.97}$$

where C_1 and C_2 are coefficients greater than one. Based on limited results provided by Harter (1969), Table 5.13 presents results that concern the asymptotic case—that is, the case where $n \to \infty$. The results can also be used to obtain approximations for $\text{Var}(\hat{\gamma})$ and $\text{Var}(\hat{\theta})$ when n is small.

Alternatively, C_1 and C_2 can be calculated using the following (approximate) expressions:

$$C_1 = 1 + 0.2942(1-p) + 0.5744\frac{(1-p)}{p} + 0.1021\frac{(1-p)}{p}\hat{\gamma}_{\text{unbiased}} \tag{5.98}$$

$$C_2 = 1 + 2.848(1-p) - 6.736(1-p)^2 + 14.49(1-p)^3 + 0.3832\frac{(1-p)}{\hat{\gamma}_{\text{unbiased}}\,p^2}. \tag{5.99}$$

Table 5.13. Coefficients C_1 and C_2

p	C_1 $\gamma = 1$	C_1 $\gamma = 2$	C_1 $\gamma = 3$	C_2 $\gamma = 1$	C_2 $\gamma = 2$	C_2 $\gamma = 3$
1.00	1	1	1	1	1	1
0.75	1.293	1.343	1.370	1.691	1.600	1.573
0.50	1.806	1.944	2.027	3.337	2.921	2.799
0.25	3.237	3.592	3.843	10.069	7.696	7.040

Equations (5.94) and (5.95) are valid for $\hat{\gamma} > 0.2$ and $n > 8$, while Eqs. (5.98) and (5.99) are valid for $0.5 < \hat{\gamma} < 5$ and $p > 0.25$. For all these expressions, the maximum error of estimate is approximately 2 percent.

5.10.4. CONFIDENCE INTERVALS FOR γ

For large γ and a complete sample, a $100(1 - \alpha)$ percent confidence interval is given by (Bain and Englehart, 1991):

$$\gamma^L = \frac{\chi^2_{\alpha/2}(n-1)}{2nS} \quad \text{and} \quad \gamma^U = \frac{\chi^2_{1-\alpha/2}(n-1)}{2nS}, \tag{5.100}$$

where $S = \ln(\bar{t}/\tilde{t})$. The number of degrees of freedom approaches $2(n - 1)$ as γ approaches 0, and it approaches $n - 1$ as γ approaches ∞. In other words, for small values of γ, one may use $df = 2(n - 1)$. Otherwise, $n - 1$ degrees of freedom should be used instead.

For small values of $\hat{\gamma}$ the df(degrees of freedom) and the denominator of the expressions given by Eq. (5.100) must be corrected. The correction is done by following an iterative procedure. For example, in order to find γ^L, begin with $\gamma^L = \hat{\gamma}_{\text{unbiased}}$ and then

1. Calculate

$$c = c(\gamma^L, n) = \frac{n\phi_1(\gamma^L) - \phi_1(n\gamma^L)}{n\phi_2(\gamma^L) - \phi_2(n\gamma^L)} \tag{5.101}$$

$$\nu = \nu(\gamma^L, n) = [n\phi_1(\gamma^L) - \phi_1(n\gamma^L)]\, c, \tag{5.102}$$

where

$$\phi_1(x) = 1 + 1/(1 + 6x) \tag{5.103}$$

$$\phi_2(x) = \begin{cases} 1 + 1/(1 + 2.5x) & 0 < x < 2 \\ 1 + 1/(3x) & 2 \le x < \infty. \end{cases} \tag{5.104}$$

2. Update γ^L using

$$\gamma^L = \frac{\chi^2_{1-\alpha/2}(\nu)}{2ncS}. \tag{5.105}$$

3. Return to step 1.

These calculations are continued until convergence is obtained. Now, replacing $\chi^2_{1-\alpha/2}$ by $\chi^2_{\alpha/2}$, and following the same procedure, γ^U can be estimated.

In the case of censored sample, the same procedure can be applied but $\bar{t}$ and $\tilde{t}$ must be replaced by A_r and G_r, respectively, where

$$A_r = \left[\sum_{i=1}^{r} t_i + (n-r)t_r\right] \Big/ n \tag{5.106}$$

$$G_r = \left[\left(\prod_{i=1}^{r} t_i\right)(t_r^{n-r})\right]^{1/n}. \tag{5.107}$$

The correction provided by Eqs. (5.101) through (5.105) is necessary because it can be shown that while for large γ,

$$2n\hat{\gamma}S \simeq \chi^2(n-1),$$

but for small γ (that is, where $\gamma \to 0$),

$$2n\hat{\gamma}S \simeq \chi^2(2(n-1)).$$

5.11. The Extreme Value Distribution

The extreme value distribution is useful in modeling the reliability of components that experience significant wearout—that is, highly increasing failure rate. The extreme value distribution is derived from the Weibull distribution as follows.

The p.d.f. of the Weibull distribution is

$$f(t) = \frac{\gamma}{\theta} t^{\gamma-1} \exp\left(\frac{-t^\gamma}{\theta}\right) \qquad t \geq 0, \gamma \geq 1, \theta > 0,$$

where γ and θ are the shape and scale parameters of the distribution, respectively. The reliability function of the Weibull distribution is

$$R(t) = e^{\frac{-t^\gamma}{\theta}} \qquad t \geq 0.$$

Assume n units are subjected to a reliability test that is terminated after $r (r \leq n)$ failures. This is type 2 censoring. The failure times of the r units follow a Weibull distribution and their values are

$$t_1 \leq t_2 \leq t_3 \ldots \leq t_r \leq t_{r+1}^{+} = t_{r+2}^{+} = \ldots = t_n^{+}.$$

Let $x_1 \leq x_2 \leq x_3 \ldots \leq x_r \leq x_{r+1}^+ = x_{r+2}^+ = \ldots = x_n^+$ be the corresponding extreme value lifetime, where $x_i = \ln t_i (i = 1, 2, \ldots, r)$ and $x_i^+ = \ln t_i^+ (i = r, r+1, \ldots, n)$.

The p.d.f. and the reliability function of the extreme value distribution are

$$f(x, a, b) = \frac{1}{b} e^{\left(\frac{x-a}{b}\right)} e^{-e^{\left(\frac{x-a}{b}\right)}} \qquad -\infty < x < \infty,$$

and

$$R(x, a, b) = e^{-e^{\left(\frac{x-a}{b}\right)}} \qquad -\infty < x < \infty,$$

where $a = 1/\gamma \log \theta$ and $b = 1/\gamma$ are the location and scale parameters of the extreme value distribution, respectively.

Following the derivations in Section 5.7.2, the likelihood function for complete or type 2 censored lifetimes is obtained as

$$L(a, b) = \prod_{i=1}^{r} f(x_i, a, b) \prod_{i=1}^{n-r} R(x_i^+, a, b)$$

or

$$L(a, b) = \frac{1}{b^r} \exp\left[\sum_{i=1}^{r} \left(\frac{x_i - a}{b}\right) - \sum_{i=1}^{r} e^{\left(\frac{x_i - a}{b}\right)} - (n - r) e^{\left(\frac{x_{r+1}^+ - a}{b}\right)}\right].$$

The logarithm of the likelihood function is

$$l(a, b) = -r \ln b + \sum_{i=1}^{r} \left(\frac{x_i - a}{b}\right) - \sum_{i=1}^{r} e^{\left(\frac{x_i - a}{b}\right)} - (n - r) e^{\left(\frac{x_{r+1}^+ - a}{b}\right)} \tag{5.108}$$

The maximum-likelihood estimates of a and b are obtained by taking the derivatives of Eq. (5.108) with respect to a and b and equating the resulting equations to zero, and then solving them simultaneously. These two equations are

$$\frac{\partial l(a, b)}{\partial a} = \frac{-r}{b} + \frac{1}{b} \sum_{i=1}^{r} e^{\left(\frac{x_i - a}{b}\right)} + \frac{n - r}{b} e^{\left(\frac{x_{r+1}^+ - a}{b}\right)} = 0 \tag{5.109a}$$

$$\frac{\partial l(a,b)}{\partial b} = \frac{-r}{b} - \frac{1}{b}\sum_{i=1}^{r}\left(\frac{x_i - a}{b}\right) + \frac{1}{b}\sum_{i=1}^{r}\left(\frac{x_i - a}{b}\right)e^{\left(\frac{x_i - a}{b}\right)}$$

$$+\left(\frac{n-r}{b}\right)\left(\frac{\overset{+}{x}_{r+1} - a}{b}\right)e^{\left(\frac{\overset{+}{x}_{r+1} - a}{b}\right)}. \tag{5.109b}$$

Solving Eq. (5.109a) for $\hat{a}$ in terms of $\hat{b}$ (Leemis, 1995), we obtain

$$\hat{a} = \hat{b}\ln\left[\frac{1}{r}\sum_{i=1}^{r} e^{x_i/\hat{b}} + \left(\frac{n-r}{r}\right)e^{\overset{+}{x}_{r+1}/\hat{b}}\right]. \tag{5.110}$$

Substituting Eq. (5.110) into Eq. (5.109b), we obtain

$$-\hat{b} - \frac{1}{r}\sum_{i=1}^{r} x_i + \frac{\sum_{i=1}^{r} x_i\, e^{x_i/\hat{b}} + (n-r)\overset{+}{x}_{r+1}\, e^{\overset{+}{x}_{r+1}/\hat{b}}}{\sum_{i=1}^{r} e^{x_i/\hat{b}} + (n-r)\, e^{\overset{+}{x}_{r+1}/\hat{b}}} = 0. \tag{5.111}$$

Solving Eq. (5.111) results in the maximum-likelihood estimate of $\hat{b}$. Approximate estimates of variances of $\hat{a}$ and $\hat{b}$ and their covariance ($\mathrm{Cov}(\hat{a}, \hat{b})$) are given in Balakrishnan and Varadan (1991).

EXAMPLE 5.28

Manufacturers of flight data recorders conduct reliability testings by subjecting the recorders to extremely high impacts, pressures, and temperatures and to corrosive fluids. The last test is performed by completely immersing the recorder for forty-eight hours in each of the several different fluids found on an airplane, including hydraulic oil, jet fuel, and de-icing fluid. The recorder is also dipped in fire fighting fluid, such as Halon foam, for eight hours (O'Connor, 1995).

Thirteen recorders are subjected to the corrosive fluid test, which is terminated at the time of the tenth failure. The times to failure are

2.25, 5.6, 8.9, 10.6, 13.8, 13.9, 15.7, 17.4, 25.3, 30.5 hours.

Assuming that the data follow a Weibull distribution, obtain the parameters of the corresponding extreme value distribution. What is the reliability of a recorder immersed in such fluids for twenty hours?

SOLUTION

In order to obtain estimates of the parameters of the extreme value distribution, we transform the failure time data by taking the logarithms of the observations:

0.812, 1.723, 2.186, 2.361, 2.625, 2.632, 2.754, 2.856, 3.231, 3.418.

We also have

$$n = 13, r = 10, n - r = 3.$$

Substituting in Eq. (5.111), we obtain

$$-\hat{b} - \frac{1}{10} \times 24.598 + \frac{\sum_{i=1}^{10} x_i e^{x_i/\hat{b}} + 3 \times 3.418\, e^{3.418/\hat{b}}}{\sum_{i=1}^{10} e^{x_i/\hat{b}} + 3e^{3.418/\hat{b}}} = 0.$$

Using the Newton-Raphson method, we solve the above equation and obtain $\hat{b} = 0.692$. Substituting $\hat{b} = 0.692$ into Eq. (5.110), we obtain $\hat{a} = 3.143$. The reliability of a recorder immersed in the fluids for twenty hours is

$$R(20) = e^{-e^{\left(\frac{2.996-3.143}{0.692}\right)}} = 0.446.$$

5.12. The Half-Logistic Distribution

The half-logistic distribution is commonly used in modeling the failure times of components that exhibit increasing failure rates. A unique characteristic of the standard half-logistic distribution is that its hazard rate is a monotonically increasing function of x ($x = (t - \mu)/\sigma$) and tends to 1 as $x \to \infty$, where t is the failure time and μ and σ are the parameters of the distribution. The probability density function and the cumulative distribution function of the half-logistic distribution, respectively, are given by

$$g(y; \mu, \sigma) = \frac{2\exp\left(-\frac{y-\mu}{\sigma}\right)}{\sigma\left[1 + \exp\left(-\frac{y-\mu}{\sigma}\right)\right]^2} \tag{5.112}$$

$$G(y; \mu, \sigma) = \frac{1 - \exp\left(-\frac{y-\mu}{\sigma}\right)}{1 + \exp\left(-\frac{y-\mu}{\sigma}\right)} \qquad y \geq \mu, \sigma > 0. \tag{5.113}$$

The above half-logistic distribution can be transformed into a standard half-logistic distribution by letting the random variable $X = (Y - \mu)/\sigma$. Thus, the p.d.f. and c.d.f. of the standard half-logistic distribution are

$$f(x) = \frac{2e^{-x}}{(1 + e^{-x})^2} \tag{5.114}$$

$$F(x) = \frac{1 - e^{-x}}{1 + e^{-x}} \qquad 0 \le x < \infty. \tag{5.115}$$

The reliability and the hazard functions are

$$R(x) = \frac{2e^{-x}}{1 + e^{-x}} \tag{5.116}$$

and

$$h(x) = \frac{f(x)}{R(x)} = \frac{1}{1 + e^{-x}} \tag{5.117}$$

Figure 5.1 shows $f(x)$ and $h(x)$ for different values of μ and σ.

The rth moment of X can be found by direct integration

$$E[X^r] = 2\int_0^{\infty} \frac{x^r e^{-x}}{(1 + e^{-x})^2} dx$$

$$E[X^r] = r!2\sum_{i=1}^{\infty} (-1)^{i-1} i^{-r}.$$

The mean of the distribution is

$$E[X] = \ln 4.$$

The variance is obtained as

$$\mathrm{Var}[X] = E[X^2] - (\ln 4)^2 = \frac{\pi^2}{3} - (\ln 4)^2.$$

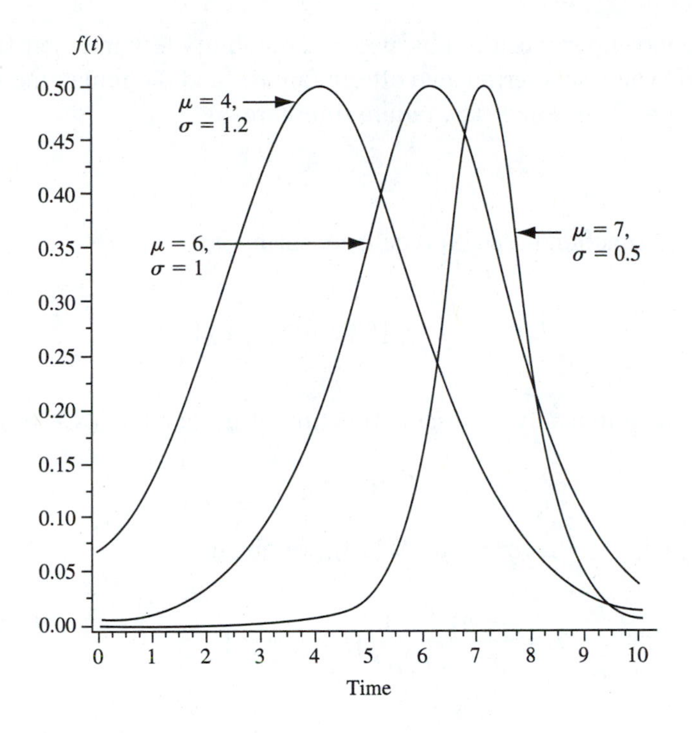

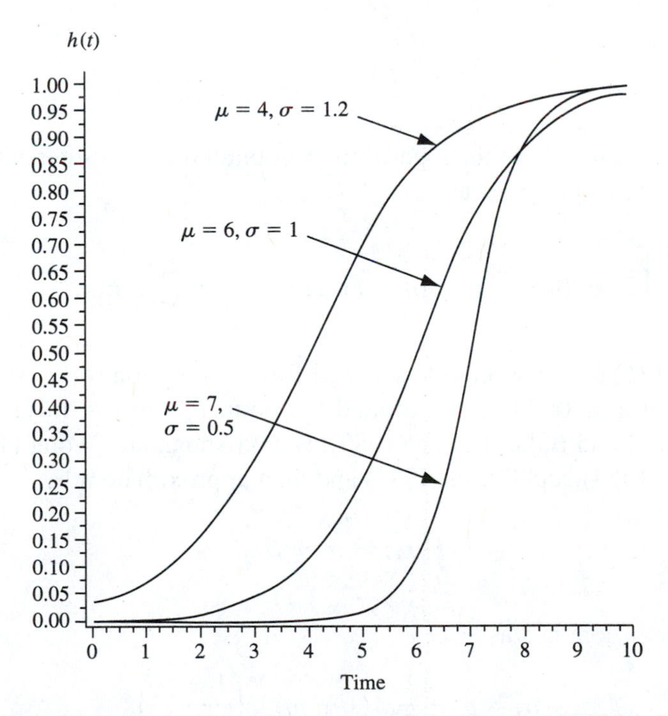

Figure 5.1. The p.d.f. and the Hazard Functions of the Standard Half-Logistic Distribution

Assume n components are subjected to a reliability test and their failure times are recorded. The test is terminated after r failures and the remaining $n - r$ components are type 2 censored. The failure times are

$$y_1 \leq y_2 \leq \ldots \leq y_r.$$

The likelihood function based on type 2 censoring is

$$L = \frac{n!}{(n-r)!}[\bar{G}(y_r)]^{n-r} \prod_{i=1}^{r} g(y_i). \tag{5.118}$$

Since L is a monotonically increasing function of μ, then the MLE of μ is

$$\hat{\mu} = y_1. \tag{5.119}$$

Substituting $x = (y - \mu)/\sigma$ in Eq. (5.118), we obtain

$$L = \frac{n!}{(n-r)!}\frac{1}{\sigma^r}[\bar{F}(x_r)]^{n-r} \prod_{i=1}^{r} f(x_i). \tag{5.120}$$

Substituting

$$f(x) = \frac{1}{2}\{\bar{F}(x)[1 + F(x)]\}$$

into Eq. (5.120) and taking the logarithm, we obtain the derivative of the logarithm l with respect to σ as

$$\frac{\partial l}{\partial \sigma} = \frac{-1}{2\sigma}\left[2r - (n-r)x_r - (n-r)x_r F(x_r) - 2\sum_{i=1}^{r} x_i F(x_i)\right] = 0. \tag{5.121}$$

Equation (5.121) does not provide a closed-form expression for σ. We expand the function $F(x_i)$ in a Taylor series around the point $F^{-1}(p_i) = \ln((1 + p_i)/q_i)$ as given in Arnold and Balakrishnan (1989), Balakrishnan and Wong (1991), David and Johnson (1954), and David (1981) and then approximate it by

$$F(x_i) \cong \alpha_i + \beta_i x_i,$$

where

$$\alpha_i = p_i - \frac{1}{2} q_i (1 + p_i) \ln\left(\frac{1 + p_i}{q_i}\right)$$

$$\beta_i = \frac{q_i(1 + p_i)}{2}$$

$$p_i = \frac{i}{n + 1}$$

$$q_i = 1 - p_i.$$

Following Balakrishnan and Wong (1991), we approximate Eq. (5.121) by

$$\frac{\partial l}{\partial \sigma} = \frac{-1}{2\sigma}\left[2r - (n - r)(1 + \alpha_r)x_r - 2\sum_{i=1}^{r} \alpha_i x_i - (n - r)\beta_r x_r^2 - 2\sum_{i=1}^{r} \beta_i x_i^2\right] = 0$$

or

$$\hat{\sigma} = \left[B + (B^2 + 8rC)^{\frac{1}{2}}\right] \Big/ 4r, \tag{5.122}$$

where

$$B = (n - r)(1 + \alpha_r)(y_r - y_1) + 2\sum_{i=2}^{r} \alpha_i (y_i - y_1)$$

$$C = (n - r)\beta_r (y_r - y_1)^2 + 2\sum_{i=2}^{r} \beta_i (y_i - y_1)^2.$$

The estimator of $\hat{\sigma}$ given in Eq. (5.122) remains the same when y_i is replaced by $y_i + \alpha$ $(1 \leq i \leq r)$, and it becomes $\beta\hat{\sigma}$ when y_i is replaced by $\beta y_i (1 \leq i \leq r)$. Therefore, realizing that the estimator of $\hat{\sigma}$ is statistically biased for small sample sizes, Balakrishnan and Wong (1991) simulated censored samples from the half logistic population and estimated values of the unbiased factor (b) of $\hat{\sigma}$ as shown in Table 5.14. Note that $s = n - r$.

The unbiased estimate of $\hat{\sigma}$ is referred to as σ and is expressed as

$$\sigma = b\hat{\sigma}. \tag{5.123}$$

We now need to determine the unbiased estimator of μ. From Eq. (5.119)

$$E[\hat{\mu}] = E[y_1] = \mu + \sigma E[x_1].$$

Table 5.14. Unbiasing Factor (b) for $\hat{\sigma}$

N	$s=0$	$s=1$	$s=2$	$s=3$	$s=4$	$s=5$	$s=6$	$s=7$	$s=8$	$s=9$	$s=10$	$s=11$	$s=12$	$s=13$	$s=14$	$s=15$	$s=16$	$s=17$	$s=18$
2	1.882																		
3	1.458	2.054																	
4	1.296	1.523	2.085																
5	1.209	1.333	1.536	2.082															
6	1.179	1.258	1.369	1.566	2.119														
7	1.147	1.203	1.267	1.376	1.563	2.117													
8	1.129	1.171	1.217	1.279	1.389	1.579	2.141												
9	1.115	1.144	1.182	1.225	1.290	1.387	1.583	2.127											
10	1.101	1.125	1.153	1.186	1.231	1.291	1.386	1.567	2.135										
11	1.090	1.112	1.134	1.157	1.185	1.226	1.284	1.376	1.565	2.117									
12	1.085	1.101	1.117	1.135	1.158	1.189	1.230	1.291	1.393	1.580	2.133								
13	1.076	1.090	1.103	1.121	1.139	1.160	1.189	1.228	1.288	1.378	1.556	2.121							
14	1.075	1.087	1.100	1.113	1.127	1.145	1.167	1.198	1.237	1.295	1.383	1.559	2.079						
15	1.067	1.079	1.089	1.101	1.114	1.128	1.145	1.167	1.193	1.232	1.290	1.382	1.560	2.081					
16	1.059	1.069	1.076	1.087	1.098	1.111	1.123	1.140	1.162	1.189	1.224	1.279	1.370	1.536	2.075				
17	1.061	1.069	1.076	1.085	1.095	1.104	1.116	1.130	1.150	1.172	1.198	1.233	1.288	1.373	1.540	2.067			
18	1.057	1.064	1.071	1.080	1.087	1.097	1.106	1.118	1.131	1.146	1.165	1.192	1.230	1.282	1.371	1.543	2.065		
19	1.056	1.064	1.069	1.075	1.081	1.089	1.097	1.107	1.118	1.135	1.149	1.173	1.201	1.238	1.290	1.384	1.559	2.070	
20	1.048	1.053	1.059	1.063	1.070	1.076	1.083	1.090	1.101	1.114	1.129	1.146	1.165	1.189	1.228	1.284	1.374	1.556	2.084

Source: Reprinted from *IEEE Transactions on Reliability*, Vol. 40, No. 2, June 1991 by N. Balakrishnan and K. H. T. Wong. © 1991 IEEE.

The unbiased estimator of μ is

$$\mu = \hat{\mu} - \sigma E[x_1]. \tag{5.124}$$

The value of $E[x_1]$ required for Eq. (5.124) can be obtained from Table 5.15, which is prepared by Balakrishnan (1985) for sample sizes up to 15, or can be found for larger sample sizes as discussed in Balakrishnan (1985).

Table 5.15. Means of Order Statistics

n	$E[x_1]$	n	$E[x_1]$
1	1.38629	9	0.20326
2	0.77259	10	0.18430
3	0.54518	11	0.16860
4	0.42369	12	0.15538
5	0.34738	13	0.14410
6	0.29475	14	0.13435
7	0.25617	15	0.12584
8	0.22663		

Source: Abridged from Balakrishnan (1985).

EXAMPLE 5.29

Flight recorders are insulated by a ceramic fiber impregnated with a phase-change material designed to control the memory module's temperature. By changing from a liquid to a gas, the material absorbs energy and delays a rise in temperature. Twelve flight recorders are subjected to 250°C to determine the effectiveness of its insulation in protecting the contents of the memory module. The failure times of the recorders' insulators, in minutes, are

18, 33, 37, 42, 64, 70, 105, 112, 144, 147, 208, 208^+.

Determine the parameters of the half-logistic distribution that fits the failure data. What is the reliability of a recorder after being subjected to this temperature for 2.5 hours?

SOLUTION

The censoring time is 208 since the test is terminated at the eleventh failure. We estimate the biased $\hat{\sigma}$ by using the calculations shown in Table 5.16:

$$B = 1 \times [1 + 0.4932] \times 190 + 2 \times 202.2549 = 688.2178$$

$$C = 1 \times [(0.1420)(190)^2 + 2 \times 20596.230] = 46319.66$$

$$\hat{\sigma} = [688.2178 + (688.2178^2 + 8 \times 11 \times 46319.66)^{\frac{1}{2}}]/44 = 64.119.$$

Table 5.16. Calculations for B and C

i	p_i	q_i	$y_i - y_1$	α_i	$\alpha_i(y_i - y_1)$	β_i	$\beta_i(y_i - y_1)^2$
1	0.0769	0.9230	0.0	0.0003	0.0000	0.4970	0.000
2	0.1538	0.8461	15.0	0.0024	0.0365	0.4881	109.837
3	0.2307	0.7692	19.0	0.0082	0.1573	0.4733	170.887
4	0.3076	0.6923	24.0	0.0198	0.4752	0.4526	260.733
5	0.3846	0.6153	46.0	0.0391	1.7999	0.4260	901.491
6	0.4615	0.5384	52.0	0.0686	3.5685	0.3934	1064.000
7	0.5384	0.4615	87.0	0.1110	9.6583	0.3550	2687.219
8	0.6153	0.3846	94.0	0.1695	15.9399	0.3106	2744.911
9	0.6923	0.3076	126.0	0.2484	31.3069	0.2603	4133.396
10	0.7692	0.2307	129.0	0.3534	45.5908	0.2041	3397.127
11	0.8461	0.1538	190.0	0.4932	93.7208	0.1420	5126.627
Total					202.2549		20596.230

Using Table 5.14, we obtain the unbiasing factor of 1.101, thus

$$\sigma = 1.101 \times 64.119 = 70.59.$$

The approximate MLE of μ is

$$\hat{\mu} = E[y_1] = 18.$$

Thus, the unbiased approximate MLEs of μ and σ are obtained using Eq. (5.124) as

$$\mu = 18 - 70.59 \times E[x_1],$$

where $E[x_1]$ is obtained from Table 5.15. Therefore

$$\mu = 18 - 70.59 \times 0.1553 = 7.037.$$

Using $\mu = 7.037$ and $\sigma = 70.59$, we obtain an unbiased estimate of the mean failure time as

$$\text{Mean failure time} = \mu + \sigma \ln 4 = 7.037 + 70 \ln 4 = 104 \text{ minutes}.$$

The reliability at $t = 2.5 \times 60 = 150$ minutes is

$$x = \frac{150 - 104}{70.59} = 0.6516$$

$$R(0.6516) = \frac{2e^{-0.6516}}{1 + e^{-0.6516}} = 0.685$$

Further details about the half-logistic and other continuous univariate distributions are given in Johnson, Kotz, and Balakrishnan (1994, 1995).

5.13. Linear Models

When several hazard-rate functions can fit the same data, it becomes necessary to discriminate among the functions to determine the "best" function. This can be achieved by substituting the reliability estimators obtained by the different hazard functions into a likelihood function. The "best" function is the one that maximizes the likelihood function. This can be easily accomplished when the hazard functions are linear. As we have seen in Chapter 1, most of the hazard functions are nonlinear. However, simple transformations can simply change some of the nonlinear functions to linear ones. In this section, we illustrate the use of linear hazard functions in conjunction with a likelihood function in determining the "best" hazard function that fits a given set of failure data.

Consider the following hazard-rate functions:

Constant hazard	$h(t) = \lambda$
Weibull hazard	$h(t) = \gamma/\theta\, t^{\gamma-1}$
Rayleigh hazard	$h(t) = \lambda t$
Gompertz hazard	$h(t) = \exp(a + bt)$
Linear exponential hazard	$h(t) = a + bt$

All the above hazard-rate functions are either already linear functions of time or can be linearized by taking the logarithm of both sides of the hazard-rate function. In all cases, we can write the linear hazard-rate function as

$$y_i = a + bx_i,$$

where y_i is the estimated hazard function or its logarithm at the ith interval, x_i is the midpoint of the time interval t_i or its logarithm, and a and b are constants. Using the weighted least-squares method, we express the weighted sum of squares (WSS) of the differences between the actual y_i and the estimated $\hat{y}_i = \hat{a} + \hat{b}x_i$ for N intervals as

$$WSS = \sum_{i=1}^{N} w_i(y_i - \hat{a} - \hat{b}x_i)^2, \qquad (5.125)$$

where w_i is the weight. Researchers considered the case where $w_i = 1$ or $w_i = n_i b_i$ where b_i and n_i are the width and number of components under test in the ith interval. Other weights can also be assigned. In order to minimize WSS, we take the derivatives of Eq. (5.125) with respect to $\hat{a}$ and $\hat{b}$ to obtain two equations. These resultant equations are set equal to zero and solved simultaneously to obtain $\hat{a}$ and $\hat{b}$ as follows:

$$\hat{b} = \frac{\sum_{i=1}^{N} w_i (x_i - \bar{x}_w)(y_i - \bar{y}_w)}{\sum_{i=1}^{N} w_i (x_i - \bar{x}_w)},$$

and

$$\hat{a} = \bar{y}_w - \hat{b}\bar{x}_w,$$

where $\bar{x}_w$ and $\bar{y}_w$ are the weighted averages of x_i's and y_i's, respectively. They are expressed as

$$\bar{x}_w = \frac{\sum_{i=1}^{N} w_i x_i}{\sum_{i=1}^{N} w_i}$$

$$\bar{y}_w = \frac{\sum_{i=1}^{N} w_i y_i}{\sum_{i=1}^{N} w_i}.$$

Having estimated $\hat{a}$ and $\hat{b}$ for each of the above hazard-rate functions, we estimate the corresponding reliabilities using

Constant	$R(t) = e^{-\lambda t}$
Weibull	$R(t) = e^{-t^{\gamma}/\theta}$
Rayleigh	$R(t) = e^{-\lambda t^2/2}$
Gompertz	$R(t) = \exp\left[\frac{-e^{a}}{b}(\exp(bt) - 1)\right]$

Linear exponential $\quad R(t) = \exp\left[-\left(at + \frac{1}{2}bt^2\right)\right].$

We then substitute in the logarithm of the likelihood function L:

$$L = \prod_{i=1}^{N-1} \left[1 - \frac{\hat{R}(t_{i+1})}{\hat{R}(t_i)}\right]^{r_i} \left[\frac{\hat{R}(t_{i+1})}{\hat{R}(t_i)}\right]^{n_i - r_i}.$$

The logarithm of the likelihood function is

$$l = \sum_{i=1}^{N-1} r_i \ln\left[1 - \frac{\hat{R}(t_{i+1})}{\hat{R}(t_i)}\right] + \sum_{i=1}^{N-1} (n_i - r_i) \ln\left[\frac{\hat{R}(t_{i+1})}{\hat{R}(t_i)}\right], \tag{5.126}$$

where n_i and r_i are the number of units under test and the number of failed units in the interval i, respectively. We finally compare the log-likelihood values of the observed data under the various hazard-rate functions. We chose the hazard-rate function that results in the maximum value of l as the "best" function.

5.14. Multicensored Data

So far we have dealt with parametric fitting of failure time data for type 1, type 2, and random censoring. There are situations when a combination of censoring may occur during the same test. For example, consider a reliability test where some of the units under test are removed due to the malfunction of the test equipment and the remaining units continue their testing until the test is terminated (when a specified number of failures occurs or when a specified test time is reached). The test results contain multicensored data. There is no known parametric method that can accommodate such multicensored data. However, there are two well-known simple approaches that can easily deal with such data. They are referred to as the *product-limit estimator*, which is developed by Kaplan and Meier (1958), and the *cumulative-hazard estimator* which is developed by Nelson (1979, 1982). In the following two sections, we present these estimators and compare their reliability estimates.

5.14.1. PRODUCT-LIMIT ESTIMATOR (PLE)

As stated earlier, the main advantage of this estimator lies in its ability of handling multicensored data and in the simplicity of calculations. The estimator relies on the fact that the probability of a component surviving during an interval of time

(t_i, t_{i+1}) is estimated as the ratio between the number of units that did not fail during the interval and the units that were under the reliability test at the beginning of the interval (time t_i). The reliability estimate at that interval is then obtained as the product of all ratios from time zero until time t_{i+1}. In other words, the reliability function using the product-limit estimator at the distinct lifetime t_j is

$$\hat{R}_{pl}(t_j) = \prod_{l=1}^{j} (n_l - d_l)/n_l = \prod_{l=1}^{j} (1 - x_l), \tag{5.127}$$

where t_j is ordered distinct lifetime at time j, $j = 1, 2, \ldots, k$, and t_0 is the start time of the reliability test, n_j and d_j are the number of units under test and the number of failed units at time j, respectively. Some of the units may be right-censored (type 2 censored) during the test interval $[t_j, t_{j+1})$; we refer to the censored observations during this interval as e_j. Thus,

$$n_j = \sum_{l=j}^{k} (d_l + e_l), \qquad n_0 = n,$$

where

n = the total number of units under test,

k = the distinct failure times, and

$x_j = d_j/n_j$.

Equation (5.127) is a nonparametric maximum likelihood estimator of the reliability function.

5.14.2. CUMULATIVE-HAZARD ESTIMATOR (CHE)

This is an alternative procedure for dealing with multicensored data. The estimates of the hazard-rate and cumulative-hazard functions are

$$\hat{h}(t_j) = \frac{d_j}{n_j} = x_j \tag{5.128}$$

$$\hat{H}(t_j) = \sum_{l=1}^{j} \hat{h}(t_l) = \sum_{l=1}^{j} x_l. \tag{5.129}$$

From the relationships in Chapter 1, we calculate the reliability using the CHE as

$$\hat{R}_{CH}(t_j) = \exp\left(-\sum_{l=1}^{j} x_l\right)$$

or

$$\hat{R}_{CH}(t_j) = \prod_{l=1}^{j} \exp(-x_l) \qquad \text{for all } j \in \{1, 2, \ldots, k\}. \tag{5.130}$$

Bohoris (1994) shows that the reliability estimates obtained using CHE are larger than those obtained by the product limit estimator. The following example confirms the observation.

EXAMPLE 5.30

Nonmetallic bearings (dry bearings) are made from polymers and polymer composites. They are preferred in operating environments where there is no adequate lubrication present or where a combination of high-load, low-speed, or intermittent motion makes lubrication difficult. The main disadvantages of such bearings are their poor creep strength, low softening temperature, high thermal expansion coefficients, and the ability to absorb liquids. Therefore, producers of the nonmetallic bearings perform extensive reliability experiments to determine the failure rates at different operating conditions.

A manufacturer of dry bearings subjects twenty-five units to a creep test and observes the following failure times (in hours):

70, 180, 190^+, 200, 210, 230, 275, 295, 310, 370^+, 395, 420, 480, 495, 560, 600^+, 620^+, 680, 750, 780, 800, 900, 980^+, $1{,}010^+$, $1{,}020^+$.

The "+" sign indicates censoring. Calculate reliability functions using the product limit and the cumulative-hazard estimators.

SOLUTION

We use Eqs. (5.127), (5.128), (5.129), and (5.130) to obtain the estimates shown in Table 5.17. As shown in the table, the reliability estimates obtained by the product-limit method is always less than those obtained by the cumulative-hazard method. These two methods, though simple, are quite useful in many applications that have multicensored data. The reliability can be estimated at any time t by fitting a parametric exponential function to the reliability values given in the last two columns of Table 5.17.

Table 5.17. Estimates of the Reliability Functions

i	t_i	j	t_j	n_j	d_j	e_j	$\hat{h}(t_j)$	$\hat{H}(t_j)$	$\hat{R}_{CH}(t_j)$	$\hat{R}_{pl}(t_j)$
1	70	1	70	25	1	0	0.040	0.040	0.961	0.960
2	180	2	180	24	1	1	0.042	0.082	0.921	0.920
3	190^+	—	—	—	—	—	—	—	—	—
4	200	3	200	22	1	0	0.045	0.127	0.881	0.879
5	210	4	210	21	1	0	0.048	0.175	0.840	0.840
6	230	5	230	20	1	0	0.050	0.225	0.799	0.798
7	275	6	275	19	1	0	0.053	0.278	0.757	0.756
8	295	7	295	18	1	0	0.056	0.334	0.716	0.714
9	310	8	310	17	1	1	0.059	0.393	0.675	0.672
10	370^+	—	—	—	—	—	—	—	—	—
11	395	9	395	15	1	0	0.067	0.460	0.631	0.627
12	420	10	420	14	1	0	0.071	0.531	0.588	0.582
13	480	11	480	13	1	0	0.077	0.608	0.544	0.537
14	495	12	495	12	1	0	0.083	0.691	0.501	0.492
15	560	13	560	11	1	2	0.091	0.782	0.457	0.447
16	600^+	—	—	—	—	—	—	—	—	—
17	620^+	—	—	—	—	—	—	—	—	—
18	680	14	680	8	1	0	0.125	0.907	0.404	0.391
19	750	15	750	7	1	0	0.143	1.050	0.350	0.335
20	780	16	780	6	1	0	0.167	1.217	0.296	0.279
21	800	17	800	5	1	0	0.200	1.417	0.242	0.223
22	900	18	900	4	1	3	0.250	1.667	0.189	0.167
23	980^+	—	—	—	—	—	—	—	—	—
24	1010^+	—	—	—	—	—	—	—	—	—
25	1020^+	—	—	—	—	—	—	—	—	—

PROBLEMS

5-1. The following failure data are obtained from a fatigue test of helical gears (the times between failures are in hours):

Times Between Failures in Hours	
220	25,000
400	28,000
590	31,000
790	35,000
1,200	40,500
1,900	45,000
3,000	49,000
4,900	53,000
6,500	58,000
9,000	64,000
14,000	68,000
19,000	75,000
22,000	100,000

a. Does the exponential distribution fit this data?

b. Is the first failure abnormally short?

c. Is the last failure abnormally long?

d. If the data fit an exponential distribution, what is the estimated MTBF?

5-2. Consider the following failure times of a wear test of composite tires:

4,000
4,560
5,800
7,900
10,000
12,000
15,000
17,000
23,000
26,000
30,000
36,000
40,000
48,000

52,000
70,000

a. Check if an exponential distribution can be used to fit the failure data (failure data are measured in miles).

b. What are the parameters of the exponential distribution?

c. Set a 95 percent confidence limit on the parameters of the distribution.

d. Set a 90 percent confidence limit on the reliability at 30,000 miles.

e. Would you buy a tire from this population? If yes, state why. If not, state why not.

5-3. Power supplies are major units for most electronic products. The manufacturers usually use a reliability demonstration test to establish a measure of reliability. For example, to demonstrate a 20,000 hr MTTF, thirteen power supplies must be operated at full load for sixty days without observing any failure. To extend the demonstrated MTTF to 200,000 hours, 127 units must be operated over the same duration (Eimar, 1990). A manufacturer subjects ten power supplies to a reliability test and observes the following times to failure:

10,000, 18,000, 21,000, 22,000, 22,500, 23,000, 25,000, 30,000, 40,000, 70,000 hours.

a. Fit an exponential distribution to the above failure-time data (check for abnormal failure times).

b. Estimate the parameters of the distribution using three different methods.

c. If you were to own a power supply produced by this manufacturer, what would the reliability of your unit be at t = 20,000 hours?

d. Do the units meet the conditions for the reliability demonstration test?

5-4. Assume that the manufacturer in the above problem conducts a reliability test using fifteen power supplies instead. The times to failure of ten units are identical to those obtained in the above problem. However, the manufacturer terminates the test for the remaining units at t = 80,000 hours. Answer questions (a) through (d) in the above problem using the results of the new test.

5-5. Twenty-one units are subjected to a fatigue test. The times to failure in hours are

8, 8, 8, 9, 13, 15, 18, 25, 26, 8^+, 10^+, 13^+, 19^+, 22^+, 29^+, 33^+, 36^+, 40^+, 45^+, 47^+, 49^+.

a. Plot the hazard function of this data.

b. Assuming an exponential failure time distribution, estimate the failure rate and the mean time to failure.

c. What is the reliability of a unit at time $t = 52$ hours?

d. Assuming that the observations fit a Rayleigh distribution, estimate its parameter using both the maximum-likelihood method and the best linear unbiased estimate (BLUE). Compare the results and explain the causes of differences, if any.

5-6. Fit a Weibull distribution for the data given in problem 5-5 and estimate the reliability of a unit at time $t = 52$ hours. Compare the results of the reliability at $t = 52$ hours with that obtained in the above problem. Explain the difference in the results.

5-7. Suppose that a manufacturer of tires makes a new prototype and provides 20 customers with a pair of these tires. The failure times measured in miles of driving are

3,000, 4,000, 6,000, 9,000, 9,000, 11,000, 12,000, 14,000, 16,000, 18,000, 30,000, 35,000, 38,000, $8{,}000^+$, $13{,}000^+$, $22{,}000^+$, $28{,}000^+$, $36{,}000^+$, $45{,}000^+$, and $46{,}000^+$.

The "+" sign indicates that the customers left the study at the indicated miles.

a. Fit a Weibull distribution to this data. Determine the parameters of the distribution.

b. What is the reliability of a tire at $t = 50{,}000$ miles?

5-8. A manufacturer of long life-cycle toggle switches observes fifteen switches under test and records the number of switch activations to failure as follows:

Failure Number	*Number of Activations*
1	50,000
2	51,000
3	60,000
4	72,000
5	80,000
6	85,000
7	89,000
8	94,000
9	97,000
10	99,000
11	110,000
12	115,000
13	$116{,}000^{+}$
14	$117{,}000^{+}$
15	$118{,}000^{+}$

The "+" sign indicates censoring.

a. Assuming that the activations to failure follow a lognormal distribution, determine the parameters of the distribution.

b. Determine the 95 percent confidence interval for the parameters of the distribution.

c. What are the variances of the parameters?

d. Suppose that the manufacturer ignores the censored observations and limits the analysis to the noncensored data only. Compare the mean lives when the noncensored and censored data are included in the analysis.

5-9. Being unsure of the failure distribution, the manufacturer in the above problem wishes to use a gamma distribution instead.

a. Solve items (a) through (d) of the above problem using gamma distribution.

b. Compare the mean lives obtained from the gamma and the lognormal distribution.

c. Obtain the variances of the parameters of the gamma distribution.

5-10. A manufacturer of utility power tubings analyzes the failure of two superheater tubes that are operating under conditions of 540°C. The analyses reveal creep voids near the rupture of the tubes. Therefore, the manufacturer designs an accelerated stress test where fifteen tubes are tested at 750°C. The following failure times are observed:

Failure Number	*Failure Time*
1	173.902965
2	188.913180
3	124.104797
4	177.705833
5	105.309583
6	45.437542
7	101.241912
8	243.574609
9	34.538182
10	269.868394
11	85.672150
12	134.730817
13	42.700680
14	258.385793
15	29.749256

The manufacturer also accumulates the following failure times from units operating under normal conditions:

Failure Number	*Failure Time*
1	867.202829
2	1681.223701
3	1785.556757
4	1088.084422
5	347.900778
6	819.299461
7	1035.156087
8	816.992475
9	1214.372096
10	1094.081077
11	1453.072243
12	715.786344
13	294.697870
14	867.420406
15	434.520703

Assume that the failure mechanism at the accelerated stresses (failure due to creep voids) is the same as that occurring at normal conditions. Determine the parameters of the failure time distributions at the accelerated and normal operating conditions (assume Weibull distribution). What is the ratio between the mean lives at the accelerated conditions and the normal conditions?

5-11. In order to reduce the test time, twelve ceramic capacitors are subjected to a highly accelerated life testing (HALT). The test is terminated after nine capacitors fail. The survival times in hours are

$6, 9, 10, 11, 13, 16, 22, 23, 27, 27^{+}, 27^{+}, 27^{+}.$

The "+" sign indicates censoring.

a. Assume that the failure times follow the lognormal distribution. Determine the parameters of the distribution and their 95 percent confidence intervals.
b. A Weibull distribution can also fit the failure times of the capacitors. What are the parameters of the Weibull distribution and their 95 percent confidence intervals?

5-12. Suppose that twenty products are placed under a vibration test and the time to failure (in months) is recorded as follows:

1, 2, 3, 3, 4, 4, 4, 5, 5, 6, 7, 7, 8, 9, 9, 10, 15, 16, 20, 25.

The time to failure can be described by a Weibull distribution. Determine the parameters of the distribution and the mean time to failure. What are the variances of the parameters?

5-13. An automated testing laboratory conducts an experiment using a sample of ten devices. The failure rate of the units is observed to follow a linear model:

$$h(t) = a + bt.$$

The failure-time data are

$20, 50, 80, 110, 130, 150^{+}, 150^{+}, 150^{+}, 150^{+}, 150^{+}.$

The "+" sign indicates censoring.

a. Use the maximum likelihood estimation procedure to estimate the parameters a and b.

b. Use the following failure time data to estimate the reliability of a device at time = 100 hours.

5-14. The most frequently employed environmental test is the 85/85 temperature and humidity stress test. The purpose of the test is to determine the ability of the device to withstand long-term exposure to warm, humid environments. The test involves subjecting a sample of devices to 85°C and unsaturated moisture of 85 percent RH under static electrical bias for 1,000 to 2,000 hours. The devices are then analyzed to determine whether the metallic wire bonds have corroded. The test usually lasts for about twelve weeks. A manufacturer of high-capacity hard disk drives uses the test to demonstrate that the mean time between failures of the drives is greater than 100,000 hours at the 85/85 test.

A sample of twenty-five drives is subjected to this test and the following failure-time data are obtained:

1,000, 1,100, 1,300, 1,450, 1,520, 1,600, 1,720, 1,750, 1,800, 1,910, 2,000, $2{,}000^{+}$ hours.

The "+" sign indicates censoring time of the remaining devices. The manufacturer is not sure which failure-time distribution should be used to model the failure times. Since the Weibull and the Gamma distributions are widely used, the manufacturer decides to use both distributions.

a. The p.d.f. of the Weibull model is

$$f(t) = \frac{\gamma}{\theta} t^{\gamma-1} \exp\left[-\left(\frac{t^{\gamma}}{\theta}\right)\right] \qquad t \geq 0, \gamma \geq 1, \theta > 0.$$

Estimate the parameters of the Weibull distribution for the failure time data. What is the MTBF?

b. The p.d.f. of the Gamma model is

$$f(t) = \frac{\lambda}{\Gamma(\gamma)} (\lambda t)^{\gamma-1} \exp(-\lambda t) \qquad t \geq 0, \gamma, \lambda > 0.$$

Estimate the parameters of the Gamma model. What is the MTBF?

c. What is the probability that the MTBF is greater than 100,000 hours for the following cases?

- Weibull is used.
- Gamma is used.

5-15. A mining company owns a 1,400-car fleet of 80-ton high-side, rotary-dump gondolas. A car will accumulate about 100,000 miles per year. In their travels from the mines to a power plant, the cars are subjected to vibrations due to track input in addition to the dynamic effects of the longitudinal shocks coming through the couplers. As a consequence the couplers encounter high dynamic impacts and experience fatigue failures and wear. Twenty-eight cars are observed, and the miles driven until the coupler is broken are recorded as follows:

Car	*Number of Miles*	*Car*	*Number of Miles*
1	131,375	12	199,284
2	153,802	13	202,996
3	167,934	14	203,754
4	171,842	15	204,356
5	178,770	16	209,866
6	184,104	17	213,354
7	189,838	18	218,898
8	193,242	19	226,196
9	196,150	20	234,634
10	198,949	21	233,567
11	199,986	22	235,987

The remaining six cars left the service after 151,345, 154,456, 161,245, 167,876, 175,547, and 177,689 miles. None of them experienced a broken coupler.

a. Fit a Weibull distribution to the failure miles and determine the parameters of the distribution.

b. Obtain unbiased estimates of the parameters and their variances.

c. Construct 90 percent confidence intervals for the parameters.

d. What is the probability that a car's coupler will break after 150,000 miles have been accumulated?

5-16. A manufacturer of resistors conducts an accelerated test on ten resistors and records the following failure times (in days):

2, 3.8, 6, 9, 12, 15, 20, 33, 45, 60.

a. Assume that the failure times follow the gamma distribution. Determine the parameters of the distribution and their 95 percent confidence intervals.

b. A Weibull distribution can also fit the failure times of the resistors. What are the parameters of the Weibull distribution and their 95 percent confidence intervals? Determine the variances of the parameters.

5-17. Assume n units are subjected to a test and r different failure times are recorded as $t_1 \leq t_2 \leq \ldots \leq t_r$. The remaining $n - r$ units are censored and their censoring times are $t_r = t_1^+ = t_2^+ = \ldots = t_{n-r}^+$. Assuming that the failure times follow the special Erlang distribution, whose probability density function is

$$f(t) = \frac{t}{\lambda^2} e^{-\frac{t}{\lambda}} \qquad t \geq 0,$$

estimate the distribution's parameter.

5-18. The following failure times are recorded:

200, 300, 390, 485, 570, 640, 720, 720^+, 720^+, 720^+.

The "+" indicates censoring. What is the Erlang's parameter? What is the mean time to failure? What is the reliability of a component from this population at $t =$ 500 hours?

5-19. One of the techniques for performing stress screening is referred to as highly accelerated stress screening (HASS), which uses the highest possible stresses beyond "qualification" level to attain time compression on the tested units. Assume that fifteen units are subjected to a HASS and the failure times of the first eleven units are recorded in minutes. The remaining four units are still operating properly when the test is terminated. The failure times are

1.5, 4.0, 7.0, 11.0, 14.0, 16.5, 19.0, 22.0, 24.0, 26.4, 28.5.

The test is terminated after thirty minutes.

a. Assume that the engineer in-charge suspects that the data follow an exponential distribution. In order not to limit the analysis, the engineer suggests that the Weibull distribution would be a better fit for the data. Fit both the exponential and the Weibull distributions to the data and estimate their parameters.

b. Determine the 90 percent confidence intervals for all the parameters obtained above.

c. Obtain the reliability of a unit from the above population using both distributions at time = 16 minutes. What do you conclude?

5-20. Prove that the reliability estimates obtained by the CHE are larger than those obtained by the PLE.

5-21. Repeaters are used to connect two or more ethernet segments of any media type. As segments exceed their maximum number of nodes or maximum length, signal quality begins to deteriorate. Repeaters provide the signal amplification and retiming required to connect segments. It is, therefore, necessary that repeaters used in high traffic networks have low failure rates. A manufacturer subjects twenty repeaters to a reliability vibration test and obtains the following failure times (in hours):

25, 50, 89, 102, 135, 136, 159, 179, 254, 300, 360, 395, 460, 510, 590, 670, 699, 780^+, 780^+, 780^+.

The "+" indicates censoring. The manufacturer believes that an extreme value distribution of the form

$$f(t; \mu, \sigma) = \frac{1}{\sigma} e^{(y-\mu)/\sigma} \exp\left(-e^{(y-\mu)/\sigma}\right)$$

is appropriate to fit the failure times where μ and σ are the parameters of the distribution. The cumulative distribution function is

$$F(t; \mu, \sigma) = 1 - \exp\left(-e^{(y-\mu)/\sigma}\right) \qquad -\infty < y < \infty, -\infty < \mu < \infty, \sigma > 0.$$

Determine the parameters of the distribution. Estimate the reliability of a repeater obtained from the same population as that of the test units at time $t = 500$ hours. Assume that your estimate of μ and σ has ± 20 percent error from the actual values. What is the range of the reliability estimate at $t = 500$ hours?

5-22. The most common choice of metallic spring material is carbon steel, either wire or flat form. The majority of spring applications require significant deflections, but however good the spring material, there are limits over which it cannot be expected to work consistently and will exhibit a reasonable fatigue life. Most spring failures result from too high forces creating too high material stresses for too many deflections. Manufacturers use different methods such as shot peening or vapor blasting to increase the working stresses under fatigue conditions.

A reliability test is conducted on twenty springs to determine the effect of shot peening on their expected lives. The following failure times (in hours) are obtained:

610, 1,090, 1,220, 1,430, 2,160, 2,345, 3,535, 3,765, 4,775, 4,905, 6,500, 7,250, 7,900, 8,348, 9,000, 9,650, 9,980, $11{,}000^+$, $11{,}000^+$, $11{,}000^+$.

The "+" indicates censoring. Assume that the failure data follow a half-logistic distribution. Determine the parameters of the distribution, the reliability of a spring at time $t = 9{,}700$ hours and the unbiased estimate of the mean failure time.

5-23. Assume that the failure data of Problem 5-22 can also fit a logistic distribution of the form

$$f(t) = \frac{\pi e^{-\pi(t-\mu)/\sqrt{3}\,\sigma}}{\sqrt{3}\,\sigma(1 + e^{-\pi(t-\mu)/\sqrt{3}\,\sigma})^2} \qquad t < \infty, 0 < \mu < \infty, 0 < \sigma < \infty.$$

The cumulative distribution function and the hazard function are

$$F(t) = \frac{1}{1 + e^{-\pi(t-\mu)/\sqrt{3}\,\sigma}}$$

$$h(t) = \frac{f(t)}{1 - F(t)} = \frac{\pi}{\sqrt{3}\,\sigma} F(t).$$

The density function is similar to the gamma density in that the hazard function approaches a constant, and thus it may be a useful alternative to the Weibull model. Determine the parameters (μ, σ) of the logistic distribution that fit the data of Problem 5-22. What is the mean life of a spring from the same population as that of the test units?

5-24. A manufacturer produces micromotors that rotate at hundreds of thousands of revolutions per minute. The medical devices that utilize such micromotors may eventually be used to perform neurosurgery, unclog arteries, and study abnormal cells. The reliability of the motors is of special concern for the users of such devices. Therefore, the manufacturer conducts a reliability test by subjecting the motors to 1.16 micronewton and observes the number of cycles to failure. The average rotation of a motor is 150,000 rpm (revolutions per minute). Twenty-five motors are subjected to the test and the following number of cycles multipled by 10^7 are observed:

150, 170, 180, 190^+, 195^+, 199, 210, 230, 260, 270^+, 295, 330, 380, 390^+, 420, 460, 500, 560^+, 590, 675, 725, 794, 830, 850, 950^+.

The sign "+" indicates censoring. Use both the CHE and PLE methods to develop reliability functions of the motors. Determine the reliability of a motor after 8.5×10^9 cycles of operations using both methods. What is the estimated mean time to failure?

5-25. Magnetic abrasive machining is used to achieve 2μm surface roughness of round steel bars. This process reduces the number of surface notches or scratches that contribute to the initiation of cracks when the bars are subjected to a fatigue test. Twenty bars are tested and the time to failure can be expressed by a special Erlangian distribution having the following p.d.f.:

$$f(t) = \frac{t}{\lambda^2} e^{-t/\lambda},$$

where λ is the parameter of the distribution. Derive expressions to estimate λ for both complete samples and right censored samples. The failure times of the units are

1,000, 1,500, 1,700, 1,900, 2,200, 2,350, 2,880, 3,309, 3,490, 3,600, 3,695, 3,825, 4,050, 5,000, 6,000, 6,750, $8{,}000^+$, $8{,}000^+$, $8{,}000^+$, $8{,}000^+$.

The sign "+" indicates censoring. What is the estimated value of λ? What is the variance of the estimator of λ? Determine the 95 percent confidence interval for λ. What is the mean time to failure?

5-26. Use the product limit estimator to develop a reliability expression for the failure data in Problem 5-25. Compare the mean time to failure obtained by using the developed reliability expression with that obtained from the special Erlangian distribution. Explain the source of differences.

5-27. Assume that a reliability engineer fits a half logistic distribution to the failure data given in Problem 5-25. An additional reliability test is conducted and the following data are obtained:

1,500, 2,000, 2,200, 2,800, 3,500, 3,900, 4,500, 4,900, 5,200, 5,750, 6,125, 6,680, 7,125, 7,795, 8,235, 8,699, $9{,}000^+$, $9{,}000^+$, $9{,}000^+$.

The sign "+" indicates censoring. The engineer fits an exponential distribution to the data. Realizing that some of the data fitted by the half logistic distribution exhibits an increasing hazard rate and that the data fitted by the exponential distribution exhibits a constant hazard rate, the engineer decides to combine the two data sets and fit them using one distribution. In doing so, the hazard rate of the mixture may exhibit a decreasing failure rate. Determine the conditions that result in a decreasing hazard rate (if it exists in this case). What is the estimated mean time to failure based on the mixed distribution?

REFERENCES

Arnold, B. C., and Balakrishnan, N. (1989). *Relations Bounds and Assumptions for Order Statistics*. Lecture Notes in Statistics No. 53. New York: Springer-Verlag.

Bain, L. J., and Engelhardt, M. (1991). *Statistical Analysis of Reliability and Life-Testing Models: Theory*. New York: Marcel Dekker.

Balakrishnan, N. (1985). "Order Statistics from the Half Logistic Distribution." *Journal of Statistical Computation and Simulation* 20, 287–309.

Balakrishnan, N., and Varadan, J. (1991). "Approximate MLEs for the Location and Scale Parameters of the Extreme Value Distribution with Censoring." *IEEE Transactions on Reliability* 24(2), 146–151.

Balakrishnan, N., and Wong, K. H. T. (1991). "Approximate MLEs for the Location and Scale Parameters of the Half-Logistic Distribution with Type II Right-Censoring." *IEEE Transactions on Reliability* 40(2), 140–145.

Bell Communications Research. (1986). *Bell Communications Research Reliability Manual*. SR-TSY-000385. Morristown, NJ: Bell Communications Research.

Bergman, B. (1985). "On Reliability Theory and Its Applications." *Scandinavian Journal of Statistics* 12, 1–41.

Bohoris, G. A. (1994). "Comparison of the Cumulative-Hazard and Kaplan-Meier Estimators of the Survivor Function." *IEEE Transactions on Reliability* 43 (2), 230–232.

Cohen, A. C., Jr. (1961). "Table for Maximum Likelihood Estimates: Singly Truncated and Singly Censored Samples." *Technometrics* 3, 535–541.

Cohen, A. C., Jr. (1965). "Maximum Likelihood Estimation in the Weibull Distribution Based on Complete and on Censored Samples." *Technometrics* 7, 579–588.

David, F. N., and Johnson, N. L. (1954). "Statistical Treatment of Censored Data: I Fundamental Formulae." *Biometrika* 41, 228–240.

David, H. A. (1981). *Order Statistics* (2nd ed.). New York: John Wiley.

Eimar, B. (1990). "Reliability Dictates True Cost of Ownership." *Engineering Evaluation* (January), 31.

Greenwood, J. A., and Durand, D. (1960). "Aids for Fitting the Gamma Distribution by Maximum Likelihood." *Technometrics* 2, 55–65.

Harter, H. L. (1969). *Order Statistics and Their Use in Testing and Estimation*. Aerospace Research Laboratory, Wright-Patterson AFB, Ohio, Vol. 2, Table C1, pp. 426–456.

Harter, H. L., and Moore, A. H. (1965). "Maximum Likelihood Estimation of the Parameters of Gamma and Weibull Populations from Complete and from Censored Samples." *Technometrics* 7, 1603–1617.

Jensen, F., and Petersen, N. E. (1982). *Burn-in: An Engineering Approach to the Design and Analysis of Burn-in Procedures*. West Sussex, England: Chichester.

Johnson, N. L., Kotz, S., and Balakrishnan, N. (1994). *Continuous Univariate Distributions* (vol. 1). New York: John Wiley.

Johnson, N. L., Kotz, S., and Balakrishnan, N. (1995). *Continuous Univariate Distributions* (vol. 2). New York: John Wiley.

Kaplan, E. L., and Meier, P. (1958). "Nonparametric Estimation from Censored Incomplete Observations." *Journal of American Statistical Association* 53, 457–481.

Kapur, K., and Lamberson, L. R. (1977). *Reliability in Engineering Design*. New York: John Wiley.

Klinger, D. J., Nakada, Y., and Menendez, M. A. (eds.) (1990). *AT&T Reliability Manual*. New York: Van Nostrand Reinhold.

Lawless, J. F. (1983). "Statistical Methods in Reliability." *Technometrics* 25, 305–316.

Lee, Elisa, T. (1980). *Statistical Methods for Survival Data Analysis*. Belmont, CA: Wadsworth.

Lee, Elisa, T. (1992). *Statistical Methods for Survival Data Analysis*. New York: John Wiley.

Leemis, L. M. (1995). *Reliability: Probabilistic Models and Statistical Methods*. Englewood Cliffs, NJ: Prentice-Hall.

Meeker, W. Q., and Hahn, G. J. (1985). *How to Plan an Accelerated Life Test: Some Practical Guidelines* (vol. 10). Milwaukee, WI: American Society for Quality Control.

Nelson, W. (1979). *How to Analyze Data with Simple Plots*. Milwaukee, WI: American Society for Quality Control.

Nelson, W. (1982). *Applied Life Data Analysis*. New York: John Wiley.

O'Connor, L. (1995). "Inside the Black Box." *Mechanical Engineering* (January), pp. 72–74.

Sarhan, A. E., and Greenberg, B. G. (1956). "Estimation of Location and Scale Parameters by Order Statistics from Singly and Doubly Censored Samples. Part I, The Normal Distribution up to Samples of Size 10." *Annals of Mathematical Statistics* 27, 427–451.

Sarhan, A. E., and Greenberg, B. G. (1957). "Estimation of Location and Scale Parameters by Order Statistics from Singly and Doubly Censored Samples, Part III." Technical Report No. 4-OOR Project 1597. U.S. Army Research Office.

Sarhan, A. E., and Greenberg, B. G. (1958). "Estimation of Location and Scale Parameters by Order Statistics from Singly and Doubly Censored Samples, Part II." *Annals of Mathematical Statistics* 29, 79–105.

Sarhan, A. E., and Greenberg, B. G. (1962). *Contributions to Order Statistics*. New York: John Wiley.

Shapiro, S. S., and Gross, A. J. (1981). *Statistical Modeling Techniques*. New York: Marcel Dekker.

Wilk, M. B., Gnanadesikan, R., and Huyett, M. J. (1962a). "Estimation of Parameters of the Gamma Distribution Using Order Statistics." *Biometrika* 49, 525–545.

Wilk, M. B., Gnanadesikan, R., and Huyett, M. J. (1962b). "Probability Plots for the Gamma Distribution." *Technometrics* 4, 1–20.

CHAPTER 6

Models for Accelerated Life Testing

In the previous chapter, different methods for estimating the parameters of the underlying failure-time distributions were presented. Confidence intervals for the values of the parameters were also discussed. Once the failure-time distribution is identified and its parameters are estimated, we can then estimate reliability measures of interest such as the expected number of failures during a specified time interval $[T_1, T_2]$ and the mean time to failure. Obviously, such measures are useful when they are estimated from failure data obtained at the normal operating conditions of the components, products, or systems of interest. However, in many situations—such as the case of the development of a new component or a product—failure data at normal operating conditions are lacking and the reliability measures become difficult, if not impossible, to estimate. Indeed, there are cases where the reliability of a component is "high" and failure data of the component when operating at normal conditions (design conditions) may not be attainable during its expected life. In such cases, accelerated life testing induces failures, and the failure data at the accelerated conditions are used to estimate the reliability at normal operating conditions. In this chapter, methods for accelerated life testing and the use of failure data obtained at accelerated testing in estimating reliability at normal operating conditions are presented.

6.1. Accelerated Testing

Accelerated life testing is usually conducted by subjecting the product (or component) to severer conditions than those that the product will be experiencing at normal conditions or by using the product more intensively than in normal use without changing the normal operating conditions. We refer to these approaches as *accelerated stress* and *accelerated failure time*, respectively. It is clear that the accelerated-failure-time approach is suitable for products or components that are used on a continuous time basis, such as tires, toasters, heaters, and light bulbs.

For example, in evaluating the failure-time distribution of light bulbs that are used on the average about six hours per day, one year of operating experience can be compressed into three months by using the light bulb twenty-four hours every day. Similarly the failure-time distribution of automobile tires that are used on the average about two hours per day (equivalent to 60 miles per day) can be obtained by observing the failure times during seventy days of continuous use (50,000 miles) at normal operating conditions. The accelerated failure time (AFT) testing is preferred to the accelerated stress testing since no assumptions need to be made about the relationship of the failure-time distributions at both accelerated and normal conditions. Of course, this is possible only if the time can be compressed as shown earlier. When it is not possible to compress the product life due to the constant use of the product—such as the case of components of a power-generating unit, communication satellites, and monitors of the air traffic controllers—then reliability estimates of such products or components can be obtained by conducting an accelerated test at stress (temperature, humidity, volt, vibration) levels higher than those of the normal operating conditions. The results at the accelerated-stress testing are then related to the normal conditions by using appropriate models as illustrated later in this chapter.

6.2. Accelerated-Failure-Data Models

There are three types of models for relating the failure data at accelerated conditions to reliability measures at normal (or design) stress conditions. The underlying assumption in relating the failure data when using any of the models is that the components (or products) operating under the normal conditions experience the same failure mechanism as those occurring at the accelerated stress conditions. For example, if the macroscopic examination of the fracture surface of the failed components indicates that fatigue cracking initiated at a corrosion pit is the cause of the failure at normal operating conditions, then the accelerated test should be designed so that the failure mechanism is identical to that of the normal conditions.

Models can be classified as statistics-based models (parametric and nonparametric), physics-statistics-based models, and physics-experimental-based models. In all of these models, we assume that the stress levels applied at the accelerated conditions are within a range of true acceleration—that is, if the failure-time distribution at a high stress level is known and time-scale transformation to the normal conditions is also known, we can mathematically derive the failure-time distributions at normal operating conditions (or any other stress level). For practical purposes, we assume that the time-scale transformation (also referred to as acceleration factor, $A_F > 1$) is constant, which implies that we have a true linear acceleration. Thus the relationships between the accelerated and normal conditions are summarized as follows (Tobias and Trindade, 1986). Let the subscripts o and s refer to the operating conditions and stress conditions, respectively. Thus,

- The relationship between the time to failure at operating conditions and stress conditions is

$$t_o = A_F \times t_s. \tag{6.1}$$

- The cumulative distribution functions are related as

$$F_o(t) = F_s\left(\frac{t}{A_F}\right). \tag{6.2}$$

- The probability density functions are related as

$$f_o(t) = \left(\frac{1}{A_F}\right) f_s\left(\frac{t}{A_F}\right). \tag{6.3}$$

- The failure rates are given by

$$h_o(t) = \frac{f_o(t)}{1 - F_o(t)}.$$

$$= \frac{\left(\frac{1}{A_F}\right) f_s\left(\frac{t}{A_F}\right)}{1 - F_s\left(\frac{t}{A_F}\right)}$$

$$h_o(t) = \left(\frac{1}{A_F}\right) h_s\left(\frac{t}{A_F}\right). \tag{6.4}$$

We now explain the accelerated models.

6.3. Statistics-Based Models: Parametric

Statistics-based models are generally used when the exact relationship between the applied stresses (temperature, humidity, voltage) and the failure time of the component (or product) is difficult to determine based on physics or chemistry principles. In this case, components are tested at different accelerated stress levels $s_1, s_2, \ldots$. The failure times at each stress level are then used to determine the most appropriate failure-time probability distribution along with its parameters.

As stated earlier, the failure times at different stress levels are linearly related to each other. Moreover, the failure-time distribution at stress level s_1 is expected to be the same at different stress levels $s_2, s_3, \ldots$ as well as at the normal operating conditions. The shape parameters of the distributions are the same for all stress levels (including normal conditions), but the scale parameters may be different.

When the failure-time probability distribution is unknown, we use the non-parametric models discussed later in this chapter. We now present the parametric models.

6.3.1. EXPONENTIAL DISTRIBUTION ACCELERATION MODEL

This is the case where the time to failure at an accelerated stress s is exponentially distributed with parameter λ_s. The hazard rate at the stress is constant. The CDF at stress s is

$$F_s(t) = 1 - e^{-\lambda_s t}. \tag{6.5}$$

Following Eq. (6.2), the CDF at the normal operating condition is

$$F_o(t) = F_s\left(\frac{t}{A_F}\right) = 1 - e^{\frac{-\lambda_s t}{A_F}}. \tag{6.6}$$

Similarly,

$$\lambda_o = \frac{\lambda_s}{A_F}. \tag{6.7}$$

The failure rate at stress level s can be estimated for both noncensored and censored failure data as follows:

$$\lambda_s = \frac{n}{\sum_{i=1}^{n} t_i} \quad \text{for noncensored data}$$

and

$$\lambda_s = \frac{r}{\sum_{i=1}^{r} t_i + \sum_{i=1}^{n-r} t_i^+} \quad \text{for censored data,}$$

where t_i is the time of the ith failure, t_i^+ is the ith censoring time, n is the total number of units under test at stress s, and r is the number of failed units at the accelerated stress s.

EXAMPLE 6.1

An accelerated life test is conducted on twenty integrated circuits (ICs) by subjecting them to 150°C and recording the failure times. Assume that the failure-time data exhibit an exponential distribution with a mean time to failure at stress condition, $MTTF_s = 6{,}000$ hours. The normal operating temperature of the ICs is 30°C and the acceleration factor is 40. What are the failure rate, the mean time to failure, and the reliability of an IC operating at the normal conditions at time = 10,000 hours (one year)?

SOLUTION

The failure rate at the accelerated temperature is

$$\lambda_s = \frac{1}{MTTF_s} = \frac{1}{6{,}000} = 1.666 \times 10^{-4} \text{ failures per hour.}$$

Using Eq. (6.7), we obtain the failure rate at the normal operating condition as

$$\lambda_o = \frac{\lambda_s}{A_F} = \frac{1.666 \times 10^{-4}}{40} = 4.166 \times 10^{-6} \text{ failures per hour.}$$

The mean time to failure at normal operation condition, $MTTF_o$, is

$$MTTF_o = \frac{1}{\lambda_o} = 240{,}000 \text{ hours.}$$

The reliability at 10,000 hours is

$$R(10{,}000) = e^{-\lambda_o t} = e^{-4.166 \times 10^{-6} \times 10^4} = 0.9591.$$

Typical accelerated-testing plans allocate equal units to the test stresses. However, units tested at stress levels close to the design or operating conditions may not experience enough failures that can be effectively used in the acceleration models. Therefore, it is preferred to allocate more test units to the low-stress conditions than to the high-stress conditions (Meeker and Hahn, 1985) so as to obtain an equal expected number of failures at each condition. When censoring occurs, we can use the methods discussed in Chapter 5 to estimate the parameters of the failure-time distribution. In the following example, we illustrate the use of failure data at accelerated conditions to predict reliability at normal conditions when the accelerated test is censored.

EXAMPLE 6.2

Highly accelerated stress testing (HAST) is an accelerated test where components are subjected to a high-temperature, high-humidity atmosphere under pressure to further accelerate the testing process. In order to observe the latch-up failure mode associated with complementary metal-oxide-silicon (CMOS) devices where the device becomes nonfunctional and draws excessive power supply current causing overheating and permanent device damage, a manufacturer subjects twenty devices to highly accelerated stress-testing conditions and observes the following failure times in minutes:

91, 145, 257, 318, 366, 385, 449, 576, 1,021, 1,141, 1,384, 1,517, 1,530, 1,984, 3,656, $4{,}000^+$, $4{,}000^+$, $4{,}000^+$, $4{,}000^+$, and $4{,}000^+$.

The sign "+" indicates censoring time. Assuming an acceleration factor of 100 is used, what is the MTTF at normal operating conditions? What is the reliability of a device from this population at $t = 10{,}000$ minutes?

SOLUTION

The Bartlett test does not reject the hypothesis that the above failure-time data follow an exponential distribution. Therefore, we estimate the parameter of the exponential distribution at the accelerated conditions as follows:

$$n = 20$$

$$r = 15$$

$$\hat{\lambda}_s = \frac{r}{\sum_{i=1}^{r} t_i + \sum_{i=1}^{n-r} t_i^+}$$

$$\hat{\lambda}_s = \frac{15}{14{,}820 + 20{,}000} = 4.3078 \times 10^{-4} \text{ failures per minute.}$$

The failure rate at normal operating conditions is

$$\lambda_o = \frac{\hat{\lambda}_s}{A_F} = 4.3078 \times 10^{-6} \text{ failures per minute,}$$

and the MTTF is 2.321×10^5 minutes or 3,868 hours (about five months).

The reliability at 10,000 minutes is

$$R(10{,}000) = e^{-4.3078 \times 10^{-6} \times 10^4} = 0.9578.$$

6.3.2. WEIBULL DISTRIBUTION ACCELERATION MODEL

Again, we consider the true linear acceleration case. Therefore, the relationships between the failure-time distributions at the accelerated and normal conditions can be derived using Eqs. (6.2) and (6.3). Thus,

$$F_s(t) = 1 - e^{-\left(\frac{t}{\theta_s}\right)^{\gamma_s}} \qquad t \geq 0, \gamma_s \geq 1, \theta_s > 0$$

and

$$F_o(t) = F_s\left(\frac{t}{A_F}\right) = 1 - e^{-\left(\frac{t}{A_F\theta_s}\right)^{\gamma_s}} = 1 - e^{-\left(\frac{t}{\theta_o}\right)^{\gamma_o}}. \tag{6.8}$$

The underlying failure-time distributions at both the accelerated stress and operating conditions have the same shape parameters—that is, $\gamma_s = \gamma_o$, and $\theta_o = A_F\theta_s$. If the shape parameters at different stress levels are significantly different, then either the assumption of true linear acceleration is invalid or the Weibull distribution is inappropriate to use for analysis of such data.

Let $\gamma_s = \gamma_o = \gamma \geq 1$. Then the probability density function at normal operating conditions is

$$f_o(t) = \frac{\gamma}{A_F\theta_s}\left(\frac{t}{A_F\theta_s}\right)^{\gamma-1} \exp -\left(\frac{t}{A_F\theta_s}\right)^{\gamma} \qquad t \geq 0, \theta_s \geq 0. \tag{6.9}$$

The MTTF at normal operating conditions is

$$MTTF_o = \theta_o \Gamma\left(1 + \frac{1}{\gamma}\right). \tag{6.10}$$

The hazard rate at the normal conditions is

$$h_o(t) = \frac{\gamma}{A_F\theta_s}\left(\frac{t}{A_F\theta_s}\right)^{\gamma-1} = \frac{h_s(t)}{A_F{}^{\gamma}}. \tag{6.11}$$

EXAMPLE 6.3

Gold bonding failure mechanisms are usually related to gold-aluminum bonds on the integrated circuit (IC) chip. When gold-aluminum beams are present in an IC, carbon impurities may lead to cracked beams. This is common with power transistors and with analog circuits due to the elevated temperature environment

(Christou, 1994). An accelerated life test is designed to cause gold bonding failures in a newly developed transistor. Three stress levels s_1, s_2, and s_3 (mainly temperature) are determined and s_1 is higher than s_2, which is higher than s_3. The sample sizes for s_1, s_2, and s_3 are 22, 18, and 22, respectively. The failure times at these stresses follow:

Stress Level	*Failure Times in Minutes*
s_1	438, 641, 705, 964, 1,136, 1,233, 1,380, 1,409, 1,424, 1,517, 1,614 1,751, 1,918, 2,044, 2,102, 2,440, 2,600, 3,352, 3,563, 3,598, 3,604, 4,473
s_2	427, 728, 1,380, 2,316, 3,241, 3,244, 3,356, 3,365, 3,429, 3,844, 3,955 4,081, 4,462, 4,991, 5,322, 6,244, 6,884, 8,053
s_3	1,287, 2,528, 2,563, 3,395, 3,827, 4,111, 4,188, 4,331, 5,175, 5,800 5,868, 6,221, 7,014, 7,356, 7,596, 7,691, 8,245, 8,832, 9,759, 10,259 10,416, 15,560

Assume an acceleration factor of 30 between the lowest stress level and the normal operating conditions. What is the MTTF at normal conditions? What is the reliability of a transistor at $t = 1{,}000$ minutes?

SOLUTION Using the maximum likelihood estimation procedure, the parameters of the Weibull distributions corresponding to the stress levels s_1, s_2, and s_3 are

For s_1: $\gamma_1 = 1.953$, $\theta_1 = 2260$

For s_2: $\gamma_2 = 2.030$, $\theta_2 = 4325$

For s_3: $\gamma_3 = 2.120$, $\theta_3 = 7302$.

Since $\gamma_1 = \gamma_2 = \gamma_3 \cong 2$, then the Weibull distribution model is appropriate to describe the relationship between failure times at accelerated stress conditions and normal operating conditions. Moreover, we have a true linear acceleration. Thus,

The A_F from s_3 to $s_2 = 1.68$

The A_F from s_2 to $s_1 = 1.91$

The A_F from s_3 to $s_1 = 3.24$.

The relationship between the scale parameter θ_s at s_3 and the normal operating conditions is

$$\theta_o = A_F \theta_3 = 30 \times 7{,}302 = 219{,}060.$$

The MTTF at normal conditions is

$$MTTF_o = 219{,}060\ \Gamma\left(\frac{3}{2}\right) = 194{,}130 \text{ minutes.}$$

The corresponding reliability of a component at 1,000 minutes is

$$R(1000) = e^{-\left(\frac{1{,}000}{219{,}060}\right)^2} = 0.999979161.$$

6.3.3. RAYLEIGH DISTRIBUTION ACCELERATION MODEL

The Rayleigh distribution appropriately describes linearly increasing failure-rate models. When the failure rates at two different stress levels are linearly increasing with time and have the same slopes, we may express the hazard rate at the accelerated stress s as

$$h_s(t) = \lambda_s t. \tag{6.12}$$

The p.d.f. for the normal operating conditions is

$$f_o(t) = \frac{\lambda_s t}{(A_F)^2} e^{\frac{-\lambda_s t^2}{2(A_F)^2}} = \lambda_o t e^{\frac{-\lambda_o t^2}{2}}, \tag{6.13}$$

where

$$\lambda_o = \frac{\lambda_s}{A_F^2}.$$

The reliability function at time t is

$$R(t) = e^{\frac{-\lambda_s t^2}{2(A_F)^2}}. \tag{6.14}$$

The mean time to failure at normal conditions is

$$MTTF = \sqrt{\frac{\pi}{2\lambda_o}}. \tag{6.15}$$

EXAMPLE 6.4

The failure of silicon and gallium arsenide substrate is the main cause for the reduction of yield and for the introduction of microcracks and dislocations during processing of integrated circuits. Thermal fatigue crack propagation in the substrate reduces the reliability levels of many electronic products that contain such integrated circuits as components. A manufacturer of integrated circuits (ICs) subjects a sample of fifteen units to a temperature of 200°C and records their failure times in minutes as

2,000, 3,000, 4,100, 5,000, 5,200, 7,100, 8,400, 9,200, 10,000, 11,500, 12,600, 13,400, $14{,}000^+$, $14{,}000^+$, and $14{,}000^+$.

The sign "+" indicates censored time. Assume that the acceleration factor between the accelerated stress and the operating condition is 20 and that the failure times follow a Rayleigh distribution. What is mean time to failure of components at normal operating conditions? What is the reliability at $t = 20{,}000$ minutes?

SOLUTION

The parameter of the Rayleigh distribution at the accelerated stress is obtained as

$$\lambda_s = \frac{2r}{\sum_{i=1}^{r} t_i^2 + \sum_{i=1}^{n-r} t_i^{+^2}}, \tag{6.16}$$

where r is the number of failed observations, n is the sample size, t_i is the failure time of the ith component, and t_i^+ is the censoring time of the ith component. Thus

$$\lambda_s = \frac{2 \times 12}{8.5803 \times 10^8 + 5.88 \times 10^8} = \frac{24}{14.4603 \times 10^8}$$

$$= 1.6597 \times 10^{-8} \text{ failures per minute.}$$

From Eq. (6.13), we obtain

$$\lambda_o = \frac{1}{(A_F)^2}\lambda_s = \frac{1.6597 \times 10^{-8}}{400} = 4.149 \times 10^{-11}.$$

The mean time to failure is

$$MTTF = \sqrt{\frac{\pi}{2\lambda_o}} = 194{,}575 \text{ minutes (about 4.5 months).}$$

The reliability at $t = 20{,}000$ minutes is

$$R(20{,}000) = e^{\frac{-\lambda_o t^2}{2}} = 0.8471.$$

6.3.4. LOGNORMAL DISTRIBUTION ACCELERATION MODEL

Lognormal distribution is widely used in modeling the failure times in the accelerated testing of electronic components when they are subjected to high temperatures, high electric fields, or a combination of both temperature and electric field. Indeed, the lognormal distribution is used for calculating the failure rates due to electromigration in discrete and integrated devices.

Another failure mechanism that results in failures of integrated circuits that can be modeled by a lognormal distribution is the fracture of the substrate. For example, in field usage of IC packages, power on and off of the device makes the junction temperature fluctuate (due to the differences in the coefficients of thermal expansion of material in the package). This temperature cycle develops stresses in the substrate, which in turn may develop microcracks and cause failure.

The probability density function of the lognormal distribution is

$$f(t) = \frac{1}{\sigma t\sqrt{2\pi}} e^{-\frac{1}{2}\left(\frac{\ln t-\mu}{\sigma}\right)^2} \qquad t \geq 0. \tag{6.17}$$

The mean and variance of the lognormal distribution are

$$\begin{aligned} \text{Mean} &= e^{\left(\mu+\frac{\sigma^2}{2}\right)} \\ \text{Variance} &= (e^{\sigma^2} - 1)(e^{2\mu+\sigma^2}). \end{aligned} \tag{6.18}$$

The p.d.f. at accelerated stress s is

$$f_s(t) = \frac{1}{\sigma_s t\sqrt{2\pi}} e^{-\frac{1}{2}\left(\frac{\ln t-\mu_s}{\sigma_s}\right)^2},$$

and the p.d.f. at normal conditions is

$$f_o(t) = \frac{1}{\sigma_o t\sqrt{2\pi}} e^{-\frac{1}{2}\left(\frac{\ln t-\mu_o}{\sigma_o}\right)^2} = \frac{1}{\sigma_s t\sqrt{2\pi}} e^{-\frac{1}{2}\left[\frac{\ln\left(\frac{t}{A_F}\right)-\mu_s}{\sigma_s}\right]^2}, \tag{6.19}$$

which implies that $\sigma_o = \sigma_s = \sigma$. This is similar to the Weibull model where the shape parameter is the same at all stress levels. The parameter σ for the lognormal distribution is equivalent to the shape parameter of the Weibull distribution. Therefore, when σ is the same at all stress levels, then we have a true linear acceleration. Equation (6.19) also implies that the parameters μ_s and μ_o are related as $\mu_o = \mu_s + \ln A_F$. The relationship between failure rates at different stress levels is time dependent and should be calculated at specified times.

EXAMPLE 6.5

Electric radiant element heaters are used in furnaces that hold molten potline aluminum before casting into ingots. Stainless steel sheet metal tubes are used to protect the elements from the furnace atmosphere and from splashes of molten aluminum and they are expected to survive for 2.5 years (Esaklul, 1992). The furnace temperatures range from 900 to 700°C with metal being cast at 710°C. In order to predict the reliability of the tubes, an accelerated test is performed at 1,000°C using sixteen tubes, and their failure times (in hours) are

2,617, 2,701, 2,757, 2,761, 2,846, 2,870, 2,916, 2,962, 2,973, 3,069, 3,073, 3,080, 3,144, 3,162, 3,180, 3,325.

Assume that the acceleration factor between the normal operating conditions and the acceleration conditions is 10. What is the mean life at normal conditions assuming a lognormal distribution?

SOLUTION

Using the MLE procedure, we obtain the parameters of the lognormal distribution at the accelerated conditions as follows:

$$\mu_s = \frac{1}{n}\sum_{i=1}^{16} \ln t_i = \frac{1}{16} \times 127.879 = 7.9925$$

$$\sigma_s^2 = \frac{1}{n}\left[\sum_{i=1}^{16} (\ln t_i)^2 - \frac{1}{n}\left(\sum_{i=1}^{16} \ln t_i\right)^2\right] = 0.0042.$$

The mean life and the standard deviation at the accelerated conditions are

$$\text{Mean life} = e^{\mu_s + \frac{\sigma_s^2}{2}} = 2{,}964.75 \text{ hours}$$

$$\text{Standard deviation} = \sqrt{(e^{\sigma_s^2} - 1)(e^{2\mu_s + \sigma_s^2})} = 192.39.$$

The parameters of the lognormal distribution at normal operating conditions are

$$\mu_o = \mu_s + \ln A_F = 7.9925 + \ln 10 = 10.295,$$

and $\sigma_s = \sigma_o = \sigma$. The mean life is then obtained as

$$e^{\mu_o + \frac{\sigma^2}{2}} = e^{10.2971} = 29{,}649 \text{ hours}$$

or

$$\text{Mean life} = 2.96 \text{ years.}$$

It is recognized that the lognormal distribution can be effectively used to model the failure times of metal-oxide-semiconductor (MOS) integrated circuits when they are subjected simultaneously to two types of stress accelerations—thermal acceleration and electric field acceleration. In this case the p.d.f. at the normal operating conditions can be expressed as

$$f_o(t) = \frac{1}{\sigma_o t\sqrt{2\pi}} \exp\left\{-\frac{1}{2}\left[\ln\left(\frac{t}{e^{\mu_o}/(A_T A_{EF})}\right)^{\frac{1}{\sigma_o}}\right]^2\right\}, \tag{6.20}$$

where $\sigma_o = \sigma_s = \sigma$—that is, the shape parameter of the lognormal is the same at all stress levels. Moreover, σ and $\mu_o - \ln A_T - \ln A_{EF}$ are the standard deviation and mean of the logarithmic failure time, respectively. A_T and A_{EF} are the thermal and electric field acceleration factors respectively. The quantity $e^{\mu}/(A_T A_{EF})$ can be viewed as a scale parameter. In spite of the popularity of the lognormal distribution as given in Eq. (6.20) its hazard-rate function is not a representative model of the device behavior. McPherson and Baglee (1985) advocate the use of the lognormal distribution as given in Eq. (6.20) to model the time-dependent dielectric breakdown (TDDB) of MOS integrated circuits. The thermal acceleration factor A_T is the ratio of the reaction rate at the stress temperature T_s to that at the normal operating temperature T_o. It is given by

$$A_T = \exp\left[\frac{E_a}{k}\left(\frac{1}{T_o} - \frac{1}{T_s}\right)\right], \tag{6.21}$$

where E_a is the activation energy and k is Boltzmann's constant ($k = 8.623 \times 10^{-5}$ eV/Kelvin). A more detailed explanation of the activation energy is based on the transition-state theory. The theory predicts approximately that the rate of reaction (or breakdown rate) is given by $E_a T/k$ where T is the temperature or the applied stress, E_a, and k are defined above. In effect, the activation energy represents the difference between the energy of reacting molecules at the final stress level and the energy of the molecules at the initial stress level.

The electric field acceleration factor, A_{EF}, between the accelerated electric field and the normal electric field is expressed as

$$A_{EF} = \exp\left(\frac{E_s - E_o}{E_{EF}}\right), \tag{6.22}$$

where E_s is the stress field in MV/cm, E_o is the normal operating field in MV/cm, and E_{EF} is the electric field acceleration parameter. Even though Eq. (6.22) is commonly used, other functional forms for the electric field acceleration factor may be used instead. For instance, Chen and Hu (1987) propose that the logarithm of the electric field acceleration is inversely proportional to the stress field E_s—that is,

$$A_{EF} = \exp\left[C_{EF}\left(\frac{1}{E_o} - \frac{1}{E_s}\right)\right], \tag{6.23}$$

where C_{EF} is the proportionality constant. Interaction terms between temperature and electric field may be included in Eq. (6.20). For example, the product $A_T A_{EF}$ of the acceleration factor given in Eq. (6.20) can be replaced by (McPherson and Baglee, 1985)

$$A_{\text{combined}} = A \exp\left[-\frac{Q}{kT_s}\right] \exp[-\gamma(T_s)E], \tag{6.24}$$

where A is a constant that normalizes the acceleration factor to 1 at the operating conditions, Q is an energy term associated with material breakdown, and $\gamma(T_s)$ is a temperature dependent parameter, given by

$$\gamma(T) = B + \frac{C}{T}, \tag{6.25}$$

where B and C are constants. Substitution of Eq. (6.25) into Eq. (6.24) shows that the combined acceleration factor is proportional to the exponential of

$$\left[\frac{C_1}{T_s} + C_2 E_s + C_3 \frac{E_s}{T_s}\right],$$

where C_1, C_2, and C_3 are constants. Thus an interaction term $C_3(E_s/T_s)$ is introduced.

EXAMPLE 6.6

Twenty-five long-life bipolar transistors for submarine cable repeaters are subjected to both temperature and electric field accelerated stresses in order to predict the expected mean life at normal operating conditions of 10°C and 5 eV. The accelerated stress conditions are 50°C and 15 eV. The following failure times are obtained from the accelerated test:

830, 843, 870, 882, 900, 932, 946, 953, 967, 992, 1,005, 1,010, 1,019, 1,023, 1,028, 1,035, 1,036, 1,044, 1,054, 1,064, 1,078, 1,099, 1,106, 1,115, 1,135.

Assume that the electric field acceleration parameter is 3.333 and the activation energy of the bipolar transistors is 0.07 eV. What are the mean life and standard deviation at the normal operating conditions?

SOLUTION

Similar to Example 6.5, we obtain the parameters of the lognormal distribution at the accelerated conditions as

$$\mu_s = \frac{1}{n}\sum_{i=1}^{25} \ln t_i = \frac{1}{25} \times 172.568 = 6.903$$

$$\sigma_s^2 = \frac{1}{n}\left[\sum_{i=1}^{25} (\ln t_i)^2 - \frac{1}{n}\left(\sum_{i=1}^{25} (\ln t_i)\right)^2\right] = 0.00765.$$

Therefore, the mean life and standard deviation at the accelerated conditions are

$$\text{Mean life} = e^{\mu_s + \frac{\sigma_s^2}{2}} = 999 \text{ hours}$$

$$\text{Standard deviation} = \sqrt{(e^{\sigma_s^2} - 1)(e^{2\mu_s + \sigma_s^2})} = 87.55 \text{ hours}.$$

The mean life at normal conditions is obtained as

$$\mu_o = \mu_s + \ln A_T + \ln A_{EF}.$$

With

$$A_T = \exp\left[\frac{E_a}{k}\left(\frac{1}{T_o} - \frac{1}{T_s}\right)\right]$$

$$A_T = \exp\left[\frac{0.07}{8.623 \times 10^{-5}}\left(\frac{1}{283} - \frac{1}{323}\right)\right] = 1.426$$

and

$$A_{EF} = \exp\left[\frac{15 - 5}{3.333}\right] = 20.0915.$$

Therefore,

$$\mu_o = 6.903 + \ln 1.426 + \ln 20.0915$$

$$\mu_o = 10.258.$$

The mean life at normal operating conditions is

$$\text{Mean life} = e^{\mu_o + \frac{\sigma^2}{2}} = e^{10.258+0.0038} = 28{,}633 \text{ hours.}$$

The mean life is approximately 2.9 years.

6.4. Statistics-Based Models: Nonparametric

When the failure-time data involve complex distributional shapes that are largely unknown or when the number of observations is small making it difficult to accurately fit a failure-time distribution and to avoid making assumptions that would be difficult to test, semiparametric or nonparametric statistics-based models appear to be a very attractive alternative to the parametric ones. There are several nonparametric models that can be used in modeling failure-time data.

This section presents two nonparametric models—a widely used multiple regression model referred to as the linear model and a second that is gaining acceptance in reliability modeling and is referred to as the proportional-hazards model.

6.4.1. THE LINEAR MODEL

The standard linear model is

$$T_i = \alpha + \beta x_i + e_i \tag{6.26}$$

or

$$T_i = \alpha + \boldsymbol{\beta}^T \mathbf{x}_i + e_i \qquad i = 1, 2, \ldots, n, \tag{6.27}$$

where

T_i = the time to failure of the ith unit,

$\mathbf{x}_i$ = the vector of the covariates (stresses) associated with time to failure T_i,

$\boldsymbol{\beta}^T$ = the vector of regression coefficients, and

$e_1, e_2, \ldots, e_n$ = identical and independent error coefficients with a common distribution.

Linear models are connected to hazard models through an accelerated time model (Miller, 1981). Suppose t_o is a survival time with hazard rate

$$\lambda_o(t) = \frac{f_o(t)}{1 - F_o(t)}. \tag{6.28}$$

Also, assume that the survival time of a component with stress vector $\mathbf{x}$ has the same distribution as

$$t_x = e^{\beta^T \mathbf{x}} t_o \tag{6.29}$$

If $\boldsymbol{\beta}^T\mathbf{x} < 0$, then $t_{\mathbf{x}}$ is shorter than t_o and the stress accelerates the time to failure and the acceleration factor is

$$A_F = e^{-\beta^T \mathbf{x}}. \tag{6.30}$$

The hazard rate of $t_{\mathbf{x}}$ is

$$\lambda_{\mathbf{x}}(t) = \frac{f_{\mathbf{x}}(t)}{1 - F_{\mathbf{x}}(t)} \tag{6.31}$$

or

$$\lambda_{\mathbf{x}}(t) = \lambda_o(e^{-\beta^T \mathbf{x}} t) e^{-\beta^T \mathbf{x}}. \tag{6.32}$$

EXAMPLE 6.7

The reliability modeling of computer memory devices such as dynamic random access memory device (DRAM) is of particular interest to manufacturers and consumers. To predict the reliability of a newly developed DRAM, the manufacturer subjects twenty-two devices to combined accelerated stresses testing of temperature and electric field. The test conditions and the time to failure are recorded in Table 6.1.

Table 6.1. Failure Times and Test Conditions

Failure Time (Hours)	*Temperature °C*	*Electric Field eV*
19.00	200	15
19.00	200	15
19.10	200	15
19.20	200	15
19.30	200	15
19.32	200	15
19.38	200	15
19.40	200	15
19.44	200	15
19.49	200	15
110.00	150	10
110.50	150	10
110.70	150	10
111.00	150	10
111.40	150	10
111.80	150	10
1,000.00	100	10
1,002.00	100	10
1,003.00	100	10
1,004.00	100	10
1,005.00	100	10
1,006.00	100	10

Determine the time to failure at normal operating conditions of 25°C and 5 eV. What is the acceleration factor between the normal conditions and the most severe stress conditions?

SOLUTION It is important to first convert the temperature from °C to K (Kelvin). Then we develop a multiple regression model in the form

$$t \text{ (time to failure)} = \alpha + \beta_1 T + \beta_2 E, \tag{6.33}$$

where

α = a constant,

β_1, β_2 = the coefficients of the applied stresses,

T = the temperature in Kelvin, and

E = the electric field in eV.

Using the standard multiple linear regression method, we obtain

$$t = 6{,}059.29 - 17.848T + 160.16E.$$

The time to failure at normal conditions is obtained as

$$t_o = 6{,}059.29 - 17.848 \times 298 + 160.16 \times 5$$

$$t_o = 1541.38 \text{ hours.}$$

The acceleration factor =

$$\frac{t_o}{t_s \text{ (at 200°C, 15 eV)}} = \frac{1541.38}{19.26} = 80.02.$$

6.4.2. PROPORTIONAL-HAZARDS MODEL

The second set of the nonparametric models is the proportional-hazards model, introduced by Cox (1972). The model is essentially "distribution-free" since no assumptions need to be made about the failure-time distribution. The only assumption that needs to be made about the failure times at the accelerated test is that the hazard-rate functions for different devices when tested at different stress levels must be proportional to one another. However, the need for proportionality can be relaxed by using time-dependent explanatory variables (time-dependent stress levels) or stratified baseline hazards. One more advantage of the proportional-hazards model is that it can easily accommodate the coupling effects between applied stresses.

Unlike standard regression models, the proportional-hazards model assumes that the applied stresses have a multiplicative (rather than additive) effect on the hazard rate—a much more realistic assumption in many cases (Dale, 1985). Moreover, the model can handle censored failure times, tied values, and failure times equal to zero. Each of these commonly occurring phenomena causes difficulty when using standard analyses. The proportional-hazards model has been widely used in the medical field to model the survival times of patients. Only recently has the model been used in the reliability field (Dale, 1985; Elsayed and Chan, 1990).

The basic proportional hazards model is given by

$$\lambda(t; z_1, z_2, \ldots, z_k) = \lambda_0(t) \exp(\beta_1 z_1 + \beta_2 z_2 \ldots + \beta_k z_k), \tag{6.34}$$

where

$\lambda(t; z_1, z_2, \ldots, z_k) =$ the hazard rate at time t for a device under test with regressor variables (covariates) $z_1, z_2, \ldots, z_k$,

$z_1, z_2, \ldots, z_k =$ regressor variables (these are also called explanatory variables or the applied stresses),

$\beta_1, \beta_2, \ldots, \beta_k =$ regression coefficients, and

$\lambda_o(t) =$ unspecified baseline hazard-rate function.

The explanatory variables account for the effects of environmental stresses (such as temperature, voltage, and humidity) on the hazard rate. We should note that the number of regressor variables may not correspond to the number of environmental stresses used in the accelerated life testing. For example, when a device is subjected to an accelerated temperature T and an electric field E, the hazard function may be explained by three regressors—$1/T$, E, and E/T. The first two terms refer to temperature and electric field effects, whereas the last term refers to the interaction between the two terms. To simplify our presentation, we assume that the number of regressors corresponds to the number of stresses used in the accelerated test.

It is important to reemphasize the fact that in the proportional-hazards model, it is assumed that the ratio of the hazard rates for two devices tested at two different stresses (such as two temperatures, T_o and T_s); $\lambda(t; T_o)/\lambda(t; T_s)$ does not vary with time. In other words, $\lambda(t; T_o)$ is directly proportional to $\lambda(t; T_s)$—hence the term *proportional-hazards model* (PHM).

The unknowns of the PHM are $\lambda_o(t)$ and β_i's. In order to determine these unknowns, we utilize Cox's partial likelihood function to estimate β_i's as follows. Suppose that a random sample of n devices under test gives d distinct observed failure times and $n - d$ censoring times. The censoring times are the times at which the functional devices are removed from test. The observed failure times are $t_{(1)} < t_{(2)} < \ldots < t_{(d)}$. To estimate β, we use the partial likelihood function $L(\beta)$ without specifying the failure-time distribution:

$$L(\beta) = \prod_{i=1}^{d} \frac{e^{(\beta z_{(i)})}}{\sum_{r} e^{(\beta z_{(r)})}}, \tag{6.35}$$

where $z_{(i)}$ is the regressor variable associated with the device that failed at $t_{(i)}$. The index, r, refers to the units under test at time $t_{(i)}$. We illustrate the construction of the likelihood function $L(\beta)$ by the following example.

EXAMPLE 6.8

In an accelerated life experiment, 100 devices are subjected to a temperature acceleration test at $T_1 = 130°C = 403$ K. One device fails at $t = 900$ hours, and the test is discontinued at 1,000 hours. In other words, the remaining ninety-nine devices survive the test. Another test is performed on five devices but at an elevated temperature of $T_2 = 250°C = 523$ K. Three devices fail at times 500, 700, and 950 hours. Two devices are removed from the test at 800 hours. The results of the experiments are summarized in Table 6.2.

Table 6.2. Failure Time Data of the Experiments

Time (hr)	*Temperature (°C)*	*Observation*
500	250	1 failed
700	250	1 failed
800	250	2 removed
900	130	1 failed
950	250	1 failed
1,000	130	99 removed

SOLUTION

In order to simplify the analysis, we determine a scale factor s for the regressor variables (Temperature T_1 and T_2) z_1 and z_2 such that $z_1 = 0$ and $z_2 = 1$. To do so, we use

$$z = s\left(\frac{1}{T_1} - \frac{1}{T}\right)$$

$$z_1 = 0$$

$$z_2 = s\left(\frac{1}{T_1} - \frac{1}{T_2}\right) = 1$$

$$z_2 = s\left(\frac{1}{403} - \frac{1}{523}\right) = 1$$

$$s = 1756K.$$

Now consider the total population of devices under test. There are 105 devices. The probability of failure occurring in a particular device of those tested at 130°C is

$$\frac{e^0}{100 + 5e^{\beta}}.$$

Similarly, the probability of the failure occurring in a particular device of those tested at 250°C is

$$\frac{e^{\beta}}{100 + 5e^{\beta}}.$$

When $\beta = 0$—that is, there is no activation energy, the two probabilities are equal to 1/105. This means that temperature has no effect on the hazard rate. When $\beta > 0$ we have $e^{\beta} > e^{0}$ and the probability of the failure occurring in a particular device of those tested at 250°C is higher than its 130°C counterpart.

We should note that the denominator $(100 + 5e^{\beta})$ used in computing the probability is a weighted sum in which the number of devices is weighted by a hazard coefficient—that is, 100 is weighted by 1, and 5 is weighted by e^{β}. If β is low (the activation energy is low), then it is more likely to observe the first failure from the 130°C group because there are more devices under test at 130°C. On the other hand, if β is high, then it is more likely that the first failure will be from the 250°C group because the hazard coefficient $e^{\beta} > 1$ is dominant. Given the first failure did occur at 250°C, the probability of the failure occurring in the 250°C group is $e^{\beta}/(100 + 5e^{\beta})$, this is the first term in the partial likelihood function:

$$L(\beta) = \left(\frac{e^{\beta}}{100 + 5e^{\beta}}\right)\left(\frac{e^{\beta}}{100 + 4e^{\beta}}\right)\left(\frac{1}{100 + e^{\beta}}\right)\left(\frac{e^{\beta}}{99 + e^{\beta}}\right).$$

As shown above, $L(\beta)$ is simply the product of the probabilities. To obtain an estimate of β, we take the natural logarithm of the partial likelihood $L(\beta)$ and equate it to zero.

Thus,

$$\ln L(\beta) = \sum_{i=1}^{d} z_{(i)}\beta - \sum_{i=1}^{d} \ln\left(\sum_{r} e^{\beta z_{(r)}}\right),$$

and

$$\frac{\partial \ln L(\beta)}{\partial \beta} = 0.$$

Since $\lambda(t; \mathbf{Z}) = \lambda_0(t)\, e^{\beta \mathbf{Z}}$, the reliability function $R(t; \mathbf{Z})$ is obtained as

$$R(t; \mathbf{Z}) = e^{-\int_0^t \lambda_0(t) e^{\beta \mathbf{Z}}\, dt}$$

$$= R_0(t) \exp(\mathbf{Z}\beta), \qquad (6.36)$$

where $\mathbf{Z}$ is the vector of the applied stresses, $\boldsymbol{\beta}$ is the vector of the regression coefficients, and $R_o(t)$ is the underlying reliability function when $\mathbf{Z} = 0$. To obtain $R_o(t)$ we utilize the life table method proposed by Kalbfleisch and Prentice (1980). We first group data into intervals $I_1, I_2, \ldots, I_k$ such that $I_j = (b_0 + \ldots + b_{j-1}, b_0 + \ldots + b_j), j = 1, \ldots, k$ is of width b_j with $b_0 = 0$ and $b_k = \infty$. The method then considers the hazard function to be a step function in the form

$$\lambda_0(t) = \lambda_j, \quad t \in I_j, \qquad j = 1, \ldots, k.$$

Take $\boldsymbol{\beta} = \hat{\boldsymbol{\beta}}$ as estimated from the partial likelihood to obtain the maximum likelihood estimate of λ_j as

$$\hat{\lambda}_j = \frac{d_j}{S_j}, \tag{6.37}$$

where d_j is the number of failures in I_j and

$$S_j = b_j \sum_{l \in R_j} e^{Z_l \beta} + \sum_{l \in D_j} (t_l - b_1 - \ldots - b_{j-1}) e^{Z_l \beta},$$

where R_j is the number of units under test at $b_0 + \ldots + b_j - 0$ and D_j is the set of units failing in I_j. The corresponding estimator of the baseline reliability function for $t \in I_j$ is

$$\hat{R}_0(t) = \exp\left[-\hat{\lambda}_j(t - \sum_0^{j-1} b_l) - \sum_1^{j-1} \hat{\lambda}_j b_i\right]. \tag{6.38}$$

The above estimator is a continuous function of time. The following example illustrates the use of the proportional-hazards model in using failure data from accelerated condition to estimate the hazard rate at normal operating conditions. This example is based on the data presented in Nelson and Hahn (1978) and the results obtained by Dale (1985).

EXAMPLE 6.9

An accelerated life test is conducted by subjecting motorettes to accelerated temperatures. Four temperature stress levels are chosen, and ten motorettes are tested at each level. The number of hours to failure at each stress level is shown in Table 6.3.* Estimate the hazard-rate function at normal operating conditions of 130°C.

* Reprinted from *Reliability Engineering* 10, C.J. Dale, "Application of Proportional Hazard Model in the Reliability Field," pp. 1–14, 1985 with kind permission from Elsevier Science Ltd, The Boulevard, Langford Lane, Kidlington OX5 1GB, UK.

Table 6.3. Hours to Failure of Motorettes

Temperature	*Hours to Failure*
150°C	10 motorettes without failure at 8,064 hr
170°C	1,764, 2,772, 3,444, 3,542, 3,780, 4,860, 5,196 3 motorettes without failure at 5,448 hr
190°C	408, 408, 1,344, 1,344, 1,440 5 motorettes without failure at 1,680 hr
220°C	408, 408, 504, 504, 504 5 motorettes without failure at 528 hr

SOLUTION The failure data exhibit severe censoring since only seventeen of the forty motorettes failed before the end of the test, and none of those tested at 150° experienced failure before the end of the test. There is also a number of tied values: seventeen failures occurred at only eleven distinct time values. Moreover, the number of failure-time observations at any stress level is too small to use parametric models to fit the data. Therefore, we use a proportional-hazards model of the form

$$\lambda(t; Z) = \lambda_0(t) \exp(\beta Z),$$

where Z is the reciprocal of the absolute temperature. Fitting the data in Table 6.3 as described earlier (using the SAS software or equivalent), the estimated value of β is $-19{,}725$. Thus, with $\lambda_0(t)$ representing the hazard-rate function applying to operating conditions at 130°C, the fitted model is

$$\lambda(t; 150°\text{C}) = 10\lambda_0(t)$$

$$\lambda(t; 170°\text{C}) = 83\lambda_0(t)$$

$$\lambda(t; 190°\text{C}) = 568\lambda_0(t)$$

$$\lambda(t; 220°\text{C}) = 7{,}594\lambda_0(t).$$

The base-line hazard-rate function $\lambda_0(t)$ can be estimated using a parametric model such as Weibull or a nonparametric method as discussed above. The nonparametric maximum likelihood estimates of the hazard-rate function at 130°C and 150°C are shown in Table 6.4.* Fitting a nonlinear model for the hazard values at 130°C results in

$$\lambda_{130°\mathrm{C}}(t) = 3.68 \times 10^{-9} t^{1.5866}$$

$$R_{130°\mathrm{C}}(t) = e^{\frac{-t^{2.5866}}{7.02\times10^{8}}}.$$

The reliability at 100 hours of operation is

$$R_{130°\mathrm{C}}(100\ \mathrm{hr}) = 0.9997.$$

Verification of the proportional-hazards assumption can be achieved by plotting $\ln[-\ln(\hat{R}(t))]$ versus $\ln t$ for different stress levels. Parallel lines indicate that the proportional-hazards assumption is satisfied.

Table 6.4. Hazard Rates at Two Temperatures

Failure Time (Hours)	*Temperature*	
	130°C	*150°C*
408	0.000054	0.001
504	0.000053	0.001
1,344	0.000405	0.004
1,440	0.000246	0.002
1,764	0.001124	0.011
2,772	0.001239	0.013
3,444	0.001382	0.014
3,542	0.001561	0.016
3,780	0.001794	0.018
4,680	0.002110	0.021
5,196	0.002558	0.026

* Reprinted from *Reliability Engineering* 10, C.J. Dale, "Application of Proportional Hazard Model in the Reliability Field," pp. 1–14, 1985 with kind permission from Elsevier Science Ltd, The Boulevard, Langford Lane, Kidlington OX5 1GB, UK.

The proportional-hazards model is also capable of modeling the hazard rates of accelerated life testing when the covariates (or applied stresses) are time dependent. Examples of time-dependent covariates include step stressing, linear increase of the applied electric field with time, and temperature cycling.

Another variant of the proportional-hazards models is the additive-hazards models (AHM). Under the AHM the effects of the explanatory variables (applied stresses) are assumed to be additive on the base-line hazard rather than multiplicative as is the case in the PHM approach (Wightman, Bendell, and McCollin, 1994). The most common and easiest to implement form of the covariate effects in an AHM formulation is to assume a linear function (other forms are discussed in Hastie and Tibshirani, 1990).

$$\lambda(t; z_1, z_2, \ldots, z_n) = \lambda_0(t) + \alpha_1 z_1 + \alpha_2 z_2 + \ldots + \alpha_n z_n, \tag{6.39}$$

where $\lambda_0(t)$ is the base-line hazard; $z_1, z_2, \ldots, z_n$ are the covariate values and $\alpha_1, \alpha_2, \ldots, \alpha_n$ are the parameters of the model. Like the PHM, the base-line hazard, $\lambda_0(t)$ can be estimated using either parametric or nonparametric approaches. The parameters $\alpha_1, \alpha_2, \ldots, \alpha_n$ can be estimated using the linear regression approach.

Clearly the choice between PHM and AHM depends on the effects of the covariates on the failure data. For example, Wightman, Bendell, and McCollin (1994) state that the PHM is inappropriate for modeling repairable software systems. However, the AHM is more applicable, robust, and easier to implement.

6.5. Physics-Statistics-Based Models

The physics-statistics-based models utilize the effect of the applied stresses on the failure rate of the units under test. For example, the failure rate of many integrated circuits is accelerated by temperature and the model that relates the failure rate with temperature should reflect the physical and chemical properties of the units. Moreover, since several units are usually tested at the same stress level and all times to failure are random events, the failure-rate expression should also reflect the underlying failure-time distribution. Thus, physics-statistics-based models are needed to describe the failure-rate relationships. The following sections present such models for both single and multiple stresses.

6.5.1. THE ARRHENIUS MODEL

Elevated temperature is the most commonly used environmental stress for accelerated life testing of microelectronic devices. The effect of temperature on the device is generally modeled using the Arrhenius reaction rate equation given by

$$r = Ae^{-(E_a/kT)}, \tag{6.40}$$

where

r = the speed of reaction,

A = an unknown nonthermal constant,

E_a = the activation energy (eV); energy that a molecule must have before it can take part in the reaction,

k = the Boltzmann Constant (8.623×10^{-5} eV/K), and

T = the temperature in Kelvin.

Activation energy (E_a) is a factor that determines the slope of the reaction rate curve with temperature—that is, it describes the acceleration effect that temperature has on the rate of a reaction and is expressed in electron volts (eV). For most applications, E_a is treated as a slope of a curve rather than a specific energy level. A low value of E_a indicates a small slope or a reaction that has a small dependence on temperature. On the other hand, a large value of E_a indicates a high degree of temperature dependence.

Assuming that device life is proportional to the inverse reaction rate of the process, then Eq. (6.40) can be rewritten as

$$L = Ae^{+(E_a/kT)}.$$

The lives of the units at normal operating temperature L_o and accelerated temperature L_s are related by

$$\frac{L_o}{L_s} = \frac{e^{(E_a/kT_o)}}{e^{(E_a/kT_s)}}$$

or

$$L_o = L_s \exp \frac{E_a}{k}\left(\frac{1}{T_o} - \frac{1}{T_s}\right). \tag{6.41}$$

When the mean life L_o at normal operating conditions is calculated and the underlying life distribution is exponential, then the failure rate at normal operating temperature is

$$\lambda_o = \frac{1}{L_o},$$

and the thermal acceleration factor is

$$A_T = \frac{L_o}{L_s}$$

or

$$A_T = \exp\left[\frac{E_a}{k}\left(\frac{1}{T_o} - \frac{1}{T_s}\right)\right]. \tag{6.42}$$

Equation (6.42) is the same as Eq. (6.21) and is similar to the proportional hazard when E_a/k is replaced by β.

EXAMPLE 6.10

An accelerated test is conducted at 200°C. Assume that the mean failure time of the microelectronic devices under test is found to be 4,000 hours. What is the expected life at a normal operating temperature of 50°C?

SOLUTION

The mean life at the accelerated conditions is $L_s = 4{,}000$ hours, the accelerated temperature is $T_s = 200 + 273 = 473\text{K}$, and the operating temperature is $T_o = 50 + 273 = 323\text{K}$. Assuming an activation energy of 0.191 eV (Blanks, 1980), then

$$L_o = 4{,}000 \exp\left[\frac{0.191}{8.623 \times 10^{-5}}\left(\frac{1}{323} - \frac{1}{473}\right)\right]$$

$$= 35{,}198 \text{ hours.}$$

The acceleration factor is

$$A_T = \exp\left[\frac{0.191}{8.63 \times 10^{-5}}\left(\frac{1}{323} - \frac{1}{473}\right)\right] = 8.78.$$

Simply, it is the ratio between L_o and L_s or $35{,}198/4{,}000 = 8.78$.

6.5.2. THE EYRING MODEL

The Eyring model is similar to the Arrhenius model. Therefore, it is commonly used for modeling failure data when the accelerated stress is temperature. It is more general than the Arrhenius model since it can model data from temperature acceleration testing as well as data from other single stress testing such as electric field. The Eyring model for temperature acceleration is

$$L = \frac{1}{T} \exp\left[\frac{\beta}{T} - \alpha\right], \tag{6.43}$$

where α and β are constants determined from the accelerated test data, L is the mean life, and T is the temperature in Kelvin. As shown in Eq. (6.43), the underlying failure time distribution is exponential. Thus the hazard rate λ is $1/L$. The relationship between lives at the accelerated conditions and the normal operating conditions is obtained as follows. The mean life at accelerated stress conditions is

$$L_s = \frac{1}{T_s} \exp\left[\frac{\beta}{T_s} - \alpha\right]. \tag{6.44}$$

The mean life at normal operating conditions is

$$L_o = \frac{1}{T_o} \exp\left[\frac{\beta}{T_o} - \alpha\right]. \tag{6.45}$$

Dividing Eq. (6.45) by Eq. (6.44), we obtain

$$L_o = L_s\left(\frac{T_s}{T_o}\right) \exp\left[\beta\left(\frac{1}{T_o} - \frac{1}{T_s}\right)\right]. \tag{6.46}$$

The acceleration factor is

$$A_F = \frac{L_o}{L_s}.$$

Equation (6.46) is identical to the result of the Arrhenius model given in Eq. (6.41) with the exception that the ratio (T_s/T_o) of the nonexponential curve in Eq. (6.46) is set to equal 1. In this case, β reduces to be the ratio between E_a and k (Boltzmann's constant).

The constants α and β can be obtained through the maximum likelihood method, by solving the following two equations for l samples tested at different stress levels and r_i failures ($i = 1, 2, \ldots, l$) are observed at stress level V_i. The equations are the resultants of taking the derivatives of the likelihood function with respect to α and β, respectively and equating them to zero (Kececioglu and Jacks, 1984).

$$\sum_{i=1}^{l} R_i - \sum_{i=1}^{l} [R_i/(\hat{\lambda}_i V_i)] \exp\,[\alpha - \beta(V_i^{-1} - \bar{V})] = 0 \tag{6.47}$$

$$\sum_{i=1}^{l} (R_i/(\hat{\lambda}_i V_i))(V_i^{-1} - \bar{V}) \exp\,[\alpha - \beta(V_i^{-1} - \bar{V})] = 0, \tag{6.48}$$

where

$\hat{\lambda}_i$ = the estimated hazard rate at stress V_i,

$$R_i = \begin{cases} r_i & \text{if the location of the parameter is known,} \\ r_i - 1 & \text{if the location of the parameter is unknown,} \end{cases}$$

$$\bar{V} = \frac{\sum_{i=1}^{l} \frac{R_i}{V_i}}{\sum_{i=1}^{l} R_i}.$$

V = stress variable. If temperature, then V is in Kelvin.

EXAMPLE 6.11 A sample of twenty devices is subjected to an accelerated test at 200°C. The failure times (in hours) shown in Table 6.5 are observed. Use the Eyring model to estimate the mean life at 50°C. What is the acceleration factor?

Table 6.5. Failure Data of 20 Devices

170.948	6,124.780
1,228.880	6,561.350
1,238.560	6,665.030
1,297.360	7,662.570
1,694.950	7,688.870
2,216.110	9,306.410
2,323.340	9,745.020
3,250.870	9,946.490
3,883.490	10,187.600
4,194.720	10,619.100

SOLUTION Using the data in Table 6.5, we estimate the mean life at the accelerated stress (200°C) as

$$L_s = \frac{1}{20} \sum_{i=1}^{20} t_i = 5{,}300.32 \text{ hours.}$$

The constant α is obtained by substituting in Eq. (6.47) as follows

$$20 - \frac{20 \times 5{,}300.32}{473} \exp\left[\alpha - \beta\left(\frac{1}{473} - \frac{1}{473}\right)\right] = 0$$

or

$$20 - 224.115e^{\alpha} = 0$$

$$\alpha = -2.416.$$

The constant β is obtained by substituting in Eq. (6.43) as follows:

$$5{,}300.32 = \frac{1}{473} \exp\left[\frac{\beta}{473} - \alpha\right]$$

or

$$\frac{\beta}{473} - \alpha = 14.735$$

$$\beta = 5{,}826.706.$$

The mean life at normal operating conditions of 50°C is

$$L_o = 5{,}300.32\left(\frac{473}{323}\right)\exp\left[5{,}826.706\left(\frac{1}{323} - \frac{1}{473}\right)\right]$$

$$L_o = 2.368 \times 10^6 \text{ hours.}$$

The Eyring model can be used effectively when multiple stresses are applied simultaneously at the accelerated test. For example, McPherson (1986) developed a generalized Eyring model to analyze thermally activated failure mechanisms. The general form of the Eyring model is

$$L_s = \frac{\alpha}{T_s} \exp\left(\frac{E_a}{kT_s}\right) \exp\left[\left(\beta + \frac{\gamma}{T_s}\right)s\right], \tag{6.49}$$

where E_a is the activation energy of the device under test, k is the Boltzmann's constant, T_s is the applied temperature stress in Kelvin, s is the applied physical stress (load/area) and α, β, and γ are constants. The model relates the time to failure (or life) to two different stresses: thermal and mechanical. It predicts a stress-activated energy, provided that two conditions are met: (1) the applied stress must be of the same order of magnitude as the strength of the material, and (2) a stress acceleration parameter must be a function of temperature (Christou, 1994).

6.5.3. THE INVERSE POWER RULE MODEL

The inverse power rule model is derived based on the Kinetic theory and activation energy. The underlying life distribution of this model is Weibull. The mean time to failure (life) decreases as the nth power of the applied stress (usually voltage). The inverse power law is expressed as

$$L_s = \frac{C}{V_s^n} \qquad C > 0, \tag{6.50}$$

where L_s is the mean life at the accelerated stress V_s and C and n are constants. The mean life at normal operating conditions is

$$L_o = \frac{C}{V_o^n}. \tag{6.51}$$

Thus,

$$L_o = L_s \left(\frac{V_s}{V_{\hat{o}}} \right)^n . \tag{6.52}$$

To obtain estimates of C and n, Mann, Schafer, and Singpurwalla (1974) amended Eq. (6.50) without changing its basic character to

$$L_i = \frac{C}{(V_i/\dot{V})^n}, \tag{6.53}$$

where L_i is the mean life at stress level V_i and $\dot{V}$ is the weighted geometric mean of the V_i's and is expressed as

$$\dot{V} = \prod_{i=1}^{k} (V_i)^{R_i / \sum_{i=1}^{k} R_i}, \tag{6.54}$$

where $R_i = \gamma_i$ (number of failures at stress V_i) or $R_i = \gamma_i - 1$ depending on whether or not the shape parameter of the failure time distribution is known. The likelihood function of C and n is

$$\prod_{i=1}^{k} \Gamma^{-1}(R_i) \left[\frac{R_i}{C} \left(\frac{V_i}{\dot{V}} \right)^n \right]^{R_i} (\hat{L}_i)^{R_i - 1} \exp\left[-\frac{R_i \hat{L}_i}{C} \left(\frac{V_i}{\dot{V}} \right)^n \right],$$

where $\hat{L}_i$ is the estimated mean life at stress V_i. The maximum likelihood estimators of $\hat{C}$ and $\hat{n}$ are obtained by solving the following two equations:

$$\hat{C} = \frac{\sum_{i=1}^{k} R_i \hat{L}_i (V_i/\dot{V})^{\hat{n}}}{\sum_{i=1}^{k} R_i}. \tag{6.55}$$

$$\sum_{i=1}^{k} R_i \hat{L}_i \left(\frac{V_i}{\dot{V}}\right)^{\hat{n}} \log \frac{V_i}{\dot{V}} = 0. \tag{6.56}$$

The asymptotic variances of $\hat{n}$ and $\hat{C}$ are

$$\sigma_{\hat{n}}^2 = \left[\sum_{i=1}^{k} R_i \left(\log \frac{V_i}{\dot{V}}\right)^2\right]^{-1} \tag{6.57}$$

$$\sigma_{\hat{C}}^2 = C^2 \left(\sum_{i=1}^{k} R_i\right)^{-1}. \tag{6.58}$$

EXAMPLE 6.12

CMOS integrated circuits suffer from a dielectric induced instability at negative bias, which eventually causes defects and breakdown of the device. A manufacturer subjects two samples of twenty devices to two electric field stresses of 25 eV and 10 eV, respectively. The failure times (in hours) are listed in Table 6.6.

Table 6.6. Failure Data of 40 Devices

25 eV Test		*10 eV Test*	
809.10	3,802.88	1,037.39	9,003.08
1,135.93	3,944.15	3,218.11	9,124.50
1,151.03	4,095.62	3,407.17	9,365.93
1,156.17	4,144.03	3,520.36	9,642.53
1,796.53	4,305.32	3,879.49	10,429.50
1,961.23	4,630.58	3,946.45	10,470.60
2,366.54	4,720.63	6,635.54	11,162.90
2,916.91	6,265.99	6,941.07	12,204.50
3,013.68	6,916.16	7,849.78	12,476.90
3,038.61	7,113.82	8,452.49	23,198.30

1. Assuming that the shape parameter of the failure time distribution is known, use the inverse power rule model to estimate the mean life at 5 eV. What are the variances of the model parameters?
2. Assuming that the failure times follow Weibull distribution, estimate the mean life at the normal operating condition of 5 eV.

SOLUTION

1. Define the 25 eV and 10 eV stress levels as s_1 and s_2, respectively. Thus

$$R_{s_1} = R_{s_2} = 20$$

$$L_{s_1} = \frac{\sum \text{failure times}}{20} = 3{,}464.25 \text{ hours}$$

$$L_{s_2} = 8{,}298.33$$

$$\dot{V} = (25)^{1/2}(10)^{1/2} = 15.81.$$

Using Eqs. (6.55) and (6.56) we obtain

$$\hat{C} = \frac{1}{2}\left[3{,}464.25\left(\frac{25}{15.81}\right)^{\hat{n}} + 8{,}298.33\left(\frac{10}{15.81}\right)^{\hat{n}}\right]$$

$$69{,}285\left(\frac{25}{15.81}\right)^{\hat{n}} \log 1.5812 + 165{,}966.6\left(\frac{10}{15.81}\right)^{\hat{n}} \log 0.6325 = 0,$$

which results in $\hat{n} = 0.95318$ and $\hat{C} = 5362.25$.

The mean life at 5 eV is

$$L_{5\text{eV}} = L_{25}\left(\frac{25}{5}\right)^{0.95318} = 16{,}065 \text{ hours}$$

or

$$L_{5\text{eV}} = \frac{\hat{C}}{(5/15.81)^{\hat{n}}} = 16{,}065 \text{ hours}.$$

The standard deviations of $\hat{n}$ and $\hat{C}$ are

$$\hat{\sigma}_n = 0.7946.$$

$$\hat{\sigma}_C = 847.84.$$

2. The shape parameters of the Weibull distributions at stresses s_1 and s_2 are obtained from fitting a Weibull distribution to each stress. This results in

$$\gamma_{s_1} = 1.98184, \qquad \theta_{s_1} = 3{,}916.97$$

$$\gamma_{s_2} = 1.83603, \qquad \theta_{s_2} = 9{,}343.58.$$

Let $\gamma_{s_1} = \gamma_{s_2} = \gamma_o \cong 2$, where γ_o is the shape parameter at the normal operating conditions. Assume an acceleration factor of 1.5. The mean time to failure is 12,140 hours.

6.5.4. COMBINATION MODEL

This model is similar to the Eyring multiple stress model when temperature and another stress such as voltage are used in the accelerated life test. The essence of the model is that the Arrhenius reaction model and the inverse power rule model are combined to form this combination model. It is valid when the shape parameter of the Weibull distribution is equal to one in the inverse power rule model (Kececioglu and Jacks, 1984). The model is given by

$$\frac{L_o}{L_s} = \left(\frac{V_o}{V_s}\right)^{-n} \exp\left[E_a/k\left(\frac{1}{T_o} - \frac{1}{T_s}\right)\right], \tag{6.59}$$

where

L_o = the life at normal operating conditions,

L_s = the life at accelerated stress conditions,

V_o = the normal operating volt,

V_s = the accelerating stress volt,

T_s = the accelerated stress temperature, and

T_o = the normal operating temperature.

EXAMPLE 6.13

Samples of long-life bipolar transistors for submarine cable repeaters are tested at accelerated conditions of both temperature and volt. The mean lives at the combinations of temperature and volt are given in Table 6.7. Assume an activation energy of 0.2 eV. Estimate the mean life at normal operating conditions of 30°C and 25 volt.

Table 6.7. Mean Lives in Hours at Stress Conditions

Temperature	*Applied Volt (V)*			
°C	*50*	*100*	*150*	*200*
60	1,800	1,500	1,200	1,000
70	1,500	1,200	1,000	800

SOLUTION Substitution in Eq. (6.59) using two stress levels results in

$$\frac{1{,}800}{1{,}200} = \left(\frac{50}{100}\right)^{-n} \exp\left[\frac{0.2}{8.623 \times 10^{-5}}\left(\frac{1}{333} - \frac{1}{343}\right)\right].$$

Solving the above equation, we obtain $n = 0.292$. Therefore,

$$L_o = L_s\left(\frac{V_o}{V_s}\right)^{-n} \exp\left[\frac{E_a}{k}\left(\frac{1}{T_o} - \frac{1}{T_s}\right)\right]$$

$$L_o = 1500\left(\frac{25}{50}\right)^{-0.292} \exp\left[\frac{0.2}{8.623 \times 10^{-5}}\left(\frac{1}{303} - \frac{1}{343}\right)\right]$$

$$L_o = 4484.11 \text{ hours.}$$

The acceleration factor =

$$\frac{4484.11}{1{,}500} \cong 3.0$$

6.6. Physics-Experimental-Based Models

The time to failure of many devices and components can be estimated based on the physics of the failure mechanism by either the development of theoretical basis for the failure mechanisms or the conduct of experiments using different levels of the parameters that affect the time to failure. There are many failure mechanisms resulting from the application of different stresses at different levels. For example, the time to failure (TTF) of packaged silicon integrated circuits due to the electromigration phenomenon is affected by the current density through the circuit and by the temperature of the circuit. Similarly, the time to failure of some components may be affected by relative humidity only.

The following sections present the most widely used models for predicting the time to failure as a function of the parameters that result in device or component failures.

6.6.1. ELECTROMIGRATION MODEL

Electromigration is the transport of microcircuit current conductor metal atoms due to electron wind effects. If, in an aluminum conductor, the electron current density is sufficiently high, an electron wind effect is created. Since the size and mass of an electron are small compared to the atom, the momentum imparted to an aluminum atom by an electron collision is small (Christou, 1994). If enough electrons collide with an aluminum atom, then the aluminum atom will move gradually causing a depletion at the negative end of the conductor. This will result in voids or hillocks along the conductor, depending on the local microstructure, causing a catastrophic failure. The median time to failure (MTF) in the presence of electromigration is given by Black's (1969) equation:

$$MTF = AJ^{-n}e^{E_a/kT}, \tag{6.60}$$

where A, n are constants, J is the current density, k is Boltzmann's constant, T is the absolute temperature, and E_a is the activation energy ($\simeq$0.6 eV for aluminum and $\simeq$0.9 eV for gold). The electromigration exponent n ranges from 1 to 6.

In order to determine the lives of components at normal operating conditions, we perform accelerated life testing on samples of these components by subjecting them to different stresses. In the case of electromigration, the stresses are the electric current and the temperature. Buehler, Zmani, and Dhiman (1991) use linear regression and propagation-of-errors analyses of the linearized equation to select the proper levels of the currents and temperatures. They show that from three or more stress conditions, the electromigration parameters such as E_a and n can be obtained.

For a fixed current, we can estimate the median life at the operating temperature as

$$\frac{t_{50}(T_o)}{t_{50}(T_s)} = \exp\left[\frac{E_a}{k}\left(\frac{1}{T_o} - \frac{1}{T_s}\right)\right], \tag{6.61}$$

where $t_{50}(T_i)$ is the median life at T_i ($i = o$ or s).

Similarly, we can fix the temperature and vary the current density. Thus,

$$\frac{t_{50}(J_o)}{t_{50}(J_s)} = \left(\frac{J_o}{J_s}\right)^{-n}.$$

6.6.2. HUMIDITY DEPENDENCE FAILURES

Corrosion in a plastic integrated circuit may deteriorate the leads outside the encapsulated circuit or the metallization interconnect inside the circuit. The basic ingredients needed for corrosion are moisture (humidity) and ions for the formation of an electrolyte, and metal for electrodes and an electric field. If any of these is missing, corrosion will not take place.

The general humidity model is

$$t_{50} = A(RH)^{-\beta} \quad \text{or} \quad t_{50} = Ae^{-\beta(RH)},$$

where t_{50} is the median life of the device, A and β are constants, and RH is the relative humidity. However, conducting an accelerated test for only humidity requires years before meaningful results are obtained. Therefore, temperature and humidity are usually combined for life testing, which is referred to as highly accelerated stress testing (HAST). The most common form of HAST is the 85/85 test where devices are tested at a relative humidity of 85 percent and a temperature of 85°C. Voltage stress is usually added to this stress in order to reduce the duration of the test further. The time to failure of a device operating under temperature, relative humidity, and voltage conditions is expressed as (Gunn, Camenga, and Malik, 1983)

$$t = ve^{\frac{E_a}{kT}} e^{\frac{\beta}{RH}}, \tag{6.62}$$

where

t = the time to failure,

v = the applied voltage,

E_a = the activation energy,

k = Boltzmann's constant,

T = the absolute temperature,

β = a constant, and

RH = the relative humidity.

Let the subscripts s and o represent the accelerated stress conditions and the normal operating conditions, respectively. The acceleration factor is obtained as

$$A_F = \frac{t_o}{t_s} = \frac{v_o}{v_s} e^{\frac{E_a}{k}\left[\frac{1}{T_o} - \frac{1}{T_s}\right]} e^{-\beta\left[\frac{1}{RH_o} - \frac{1}{RH_s}\right]}. \tag{6.63}$$

Changes in the microelectronics require that the manufacturers consider faster methodologies to detect failures caused by corrosion. Some manufacturers use pressure cookers to induce corrosion failures in a few days of test time. Studies showed that pressurized humidity test environments forced moisture into the plastic encapsulant much more rapidly than other types of humidity test methods.

6.6.3. FATIGUE FAILURES

When repetitive cycles of stresses are applied to material, fatigue failures usually occur at a much lower stress than the ultimate strength of the material due to the accumulation of damage. Fatigue loading causes the material to experience cycles of tension and compressions, which result in crack initiations at the points of discontinuity, defects in material, or notches or scratches where stress concentration is high. The crack length grows as the repetitive cycles of stresses continue until the stress on the remaining cross-section area exceeds the ultimate strength of the material. At this moment, sudden fracture occurs, causing instantaneous failure of the component or member carrying the applied stresses. It is important to recognize that the applied stresses are not only caused by applying physical load or force but also by temperature or voltage cycling. For example, *creep fatigue*, or the thermal expansion strains caused by thermal cycling, is the dominant failure mechanism causing breaks in surface mount technology (SMT)—solder attachments of printed circuits. Each thermal cycle produces a specific net strain energy density in the solder that corresponds to a certain amount of fatigue damage. The long-term reliability depends on the cyclically accumulated fatigue damage in the solder, which eventually results in fracture (Flaherty, 1994). The reliability of components or devices subject to fatigue failure is often expressed in number of stress cycles corresponding to a given cumulative failure probability. A typical model for fatigue failure of a solder attachment is given by (Engelmaier, 1993)

$$N_f(x\%) = \frac{1}{2}\left[\frac{2\varepsilon}{F}\frac{h}{L_D\,\Delta\alpha\,\Delta T_e}\right]^{\frac{-1}{c}}\left[\frac{\ln(1-0.01x)}{\ln(0.5)}\right]^{\frac{1}{\beta}}, \tag{6.64}$$

where

$N_f(x\%)$ = number of cycles (fatigue life) that correspond to x percent failures,

ε = the solder ductility,

F = an experimental factor (Engelmaier, 1993),

h and L_D = dimensions of the solder attachment,

$\Delta\alpha$ = a factor of the differences in the thermal expansion coefficient of component and substrate (that produces the stress),

ΔT_e = the effective thermal cycling range,

c = a constant that relates the average temperature of the solder joint and the time for stress relaxation and creep per cycle, and

$\beta = 4$ for the leadless surface mounted attachment.

6.7. Degradation Models

Most reliability data obtained from accelerated life testing are time-to-failure measurements obtained from testing samples of units at different stresses. However, there are many situations where the actual failure of the units, especially at stress levels close to the normal operating condition, may not fail catastrophically but degrade within the allotted test time. For example, a component may start a test with an acceptable resistance value reading, but during the test the resistance reading "drifts" (Tobias and Trinidade, 1986). As the test time progresses, the resistance eventually reaches an unacceptable level that causes the unit to fail. In such cases, measurements of the degradation of the characteristics of interest (those whose failure may cause catastrophic failure of the part) are frequently taken during the test. The degradation data are then analyzed and used to predict the time to failure at normal conditions. It is obvious that there is no general degradation model that can be used for all devices or parameters for a specific device. For example, the degradation in the resistance of a device requires a model different from the one that measures degradation in the output current of the same device. Therefore, this section is limited to a presentation of specific degradation models.

6.7.1. RESISTOR DEGRADATION MODEL

The thin film integrated circuit resistor degradation mechanism can be described by (Chan, Boulanger, and Tortorella, 1994):

$$\frac{\Delta R(t)}{R_0} = \left(\frac{t}{\tau}\right)^m, \tag{6.65}$$

where,

$\Delta R(t)$ = the change in resistance at time t,

R_0 = the initial resistance,

t = time,

τ = the time required to cause 100 percent change in resistance, and

m = a constant.

The temperature dependence is embedded in τ as

$$\tau = \tau_0 e^{\frac{E_a}{kT}}, \tag{6.66}$$

where τ_0 is constant.

Substituting Eq. (6.66) into Eq. (6.65) and taking the logarithm, we obtain

$$\ln\left(\frac{\Delta R(t)}{R_0}\right) = m\left[\ln(t) - \ln(\tau_0) - \frac{E_a}{kT}\right]$$

or

$$\ln(t) = \ln(\tau_0) + \frac{1}{m}\ln\left(\frac{\Delta R(t)}{R_0}\right) + \frac{E_a}{kT}. \tag{6.67}$$

Once the constants m and τ_0 are determined we can use Eq. (6.67) to calculate the change in resistance at any time. The above equation can also be used to predict the life of a device subject to electromigration failures. Recall that the time to failure due to electromigration is given by Eq. (6.60). Taking the natural logarithm of Eq. (6.60) results in

$$\ln(MTF) = \ln(A) - n\ln(J) + \frac{E_a}{kT}. \tag{6.68}$$

Note Eqs. (6.68) and (6.67) are identical.

The constants m and τ_0 can be obtained using the standard multiple regression procedure as shown in the following example.

EXAMPLE 6.14

The data shown in Table 6.8 represent sixteen measurements of $\Delta R(t)/R_0$ at different time intervals from one sample at two temperatures (100°C and 150°C). Both the change in resistance and the exact time of the measurements are multiplied by arbitrary scale factors. Determine the time at which $\ln(\Delta R(t)/R_0) = 8.5$ when the resistor is operating at 28°C. This change in resistor corresponds to a catastrophic failure of the device.

SOLUTION

Using the data given in Table 6.8, we develop a multiple linear regression model of the form

$$\ln(t) = \ln(\tau_0) + \frac{1}{m}\ln\left(\frac{\Delta R(t)}{R_0}\right) + \frac{E_a}{kT}.$$

Table 6.8. Degradation Data of the Resistor*

Time t in seconds	$\ln t$	$\ln\left(\frac{\Delta R(t)}{R_0}\right)$	$(kT)^{-1}$
5,000	8.517193	6.3969297	31.174013
15,000	9.615805	6.6846117	31.174013
50,000	10.819778	6.9077553	31.174013
150,000	11.918390	7.3132204	31.174013
500,000	13.122363	7.5286778	31.174013
1,500,000	14.220975	8.1992672	31.174013
5,000,000	15.424948	8.5003481	31.174013
10,000,000	16.118095	8.7809009	31.174013
5,000	8.517193	7.2412823	27.489142
15,000	9.615805	7.8913435	27.489142
50,000	10.819778	8.0966259	27.489142
150,000	11.918390	8.2540092	27.489142
500,000	13.122363	8.7809009	27.489142
1,500,000	14.220975	9.2256796	27.489142
5,000,000	15.424948	9.9034876	27.489142
10,000,000	16.118095	10.308953	27.489142

*Reprinted from *1994 Proceedings of the Annual Reliability and Maintainability Symposium*, "Analysis of Parameter Degradation Data Using Life-Data Analysis Program," C.K. Chan, M. Boulanger, and M. Tortorella, pp. 288–291. © 1994 IEEE.

The coefficients $\ln(\tau_0)$, $1/m$, and E_a are obtained from the regression model as −15.982, 2.785, and 0.24, respectively. The time required for a device operating at 28°C to reach $\ln[\Delta R(t)/R_0] = 8.5$ is

$$\ln(t) = -15.982 + 2.785 \times 8.5 + (0.24/8.623 \times 10^{-5} \times 301)$$

$$\ln(t) = 16.93719$$

or time $= 22.6845 \times 10^6$ seconds or 7.6 months.

6.7.2. LASER DEGRADATION

A laser diode is a source of radiation that utilizes simulated emission. Through high current density, a large excess of charge carriers is generated in the conduction band of the laser so that a strong simulated emission can take place. The performance of the laser diode is greatly affected by the driving current. Therefore, the degradation parameter that should be observed is the change in current with time. We utilize the degradation model developed by Takeda and Suzuki (1983) and modified by Chan, Boulanger, and Tortorella (1994):

$$\frac{D(t)}{D_0} = \exp\left[-\left(\frac{t}{\tau_d}\right)^p\right], \tag{6.69}$$

where

$D(t)$ = the change in degradation parameter at time t,

D_0 = the original value of the degradation parameter, and

τ_d, p = constants.

Again, we linearize Eq. (6.69) by taking its logarithm twice to obtain

$$\ln\left[\ln\left(\frac{D(t)}{D_0}\right)\right] = -p[\ln(t) - \ln(\tau_d)]. \tag{6.70}$$

The parameters p and τ_d can be obtained in a similar fashion as discussed above.

6.7.3. HOT-CARRIER DEGRADATION

Technological advances in very large-scale integrated (VLSI) circuits fabrication resulted in significant reductions in the device dimensions such as the channel length, the gate oxide thickness, and the junction depth without proportional reduction in the power supply voltage. This has resulted in a significant increase of both the horizontal and vertical electric fields in the channel region. Electrons and holes gaining high kinetic energies in the electric field (hot-carriers) may be injected into the gate oxide causing permanent changes in the oxide-interface charge distribution and degrading the current-voltage characteristics of the device. For example, the damage caused by hot-carrier injection affects the characteristics of the nMOS transistors by causing a degradation in transconductance, a shift in the threshold voltage, and a general decrease in the drain current capability (Leblebici and Kang, 1993).

This performance degradation in the devices leads to the degradation of circuit performance over time. Therefore, in order to estimate the reliability of a device that may fail due to hot-carrier effects, a degradation test may be conducted and changes in device characteristics with time should be recorded. A degradation model is then developed to relate the device life with the changes in a critical characteristic. The model can be used to estimate the time or life of the device when a specified value of change in the device characteristics occurs. For example, the device life τ can be defined as the time required for a 10 mV threshold voltage shift under stress bias conditions. An empirical formula that relates the amount of the normalized substrate current ($I_{\text{substrate}}/W$) is given by (Leblebici and Kang, 1993):

$$\tau = A\left[\frac{I_{\text{substrate}}}{W}\right]^{-n}, \tag{6.71}$$

where

$I_{\text{substrate}}$ = the substrate current,

W = the channel width of the transistor,

A = a process-dependent constant, and

n = an empirical constant.

Taking the logarithm of Eq. (6.71) results in

$$\ln \tau = \ln A - n \ln\left[\frac{I_{\text{substrate}}}{W}\right]. \tag{6.72}$$

The parameters A and n can be obtained using linear regression.

The degradation model given in Eq. (6.71) is simple. However other degradation models for hot-carrier can be quite complex, as discussed in Quader, Fang, Yue, Ko, and Hu (1994).

6.8. Accelerated Life Testing Plans

The type of the accelerated stress to be applied on the device, component, or part to be tested depends on the failure modes and the environment at which the device will normally operate. For example, if the component will be subjected to cycles of tension and compression, then an accelerated fatigue test deems appropriate to provide prediction of the life of the component at normal operating conditions. Similarly, if the part will operate in a hot and humid environment, then a test in a temperature and humidity chamber will simulate the environment at much higher stress levels.

The main questions that need to be addressed in order to effectively conduct an accelerated life test are as follows:

- What type of stresses should be applied on the device or component?
- At what stress levels should the device or component be tested?
- What is the number of units to be tested at each stress level?

As mentioned above, the type of stress to be applied depends on the actual functions that the device or component will be performing at the normal operating conditions. For example, if the device has many physical connections and will be used in an airplane cockpit, then a vibration test appears appropriate to conduct. Moreover, if the relative humidity level in the cockpit is greater than 30 percent, then humidity acceleration should be included in the test. Furthermore, the failure mechanism may also dictate the type of stress to be applied. Table 6.9 provides a summary of some failure mechanisms for electronic devices and the corresponding stresses that induce such mechanisms (Brombacher, 1992).

In choosing the stress levels, it is necessary to establish the highest stress to be used as the one that represents the most extreme conditions where the assumed failure model can still reasonably be expected to hold (Meeker and Hahn, 1985). The next step is to conduct tests at two or more stress conditions (including the high stress) in each case, at least $200P$ percent (P is the percentile of the time to failure distribution) of the units will fail within the duration of the test at the higher stress level. Moreover, at least $100P$ percent of the units at the lower stress level must fail within the test duration (P represents percentile). For example, if it is desired to estimate the tenth percentile of the time to failure distribution, tests should be conducted at least at two stresses (that is, the high and middle stress conditions) at which 20 percent (and, as a minimum at the middle stress, 10 percent) of the units tested should be expected to fail within the duration of the test. Finally, conduct a test at a third stress level that is as close as possible to the normal operating conditions but that will result in at least five failures (Meeker and Hahn, 1985). Clearly, other test plans that minimize cost or maximize the information obtained from the test can be used; see, for example, Barton (1980), Kielpinski and Nelson (1975), Meeker and Nelson (1975).

Table 6.9. Failure Mechanism and Corresponding Stress

Failure Mechanism	*Applied Stress*
Electromigration	Current density Temperature
Thermal cracks	Dissipated power Temperature
Corrosion	Humidity Temperature
Mechanical fatigue	Repeated cycles of lead Vibration
Thermal fatigue	Repeated cycle of temperature change

The number of units to be allocated to each stress level is inversely proportional to the applied stress. In other words, more test units should be allocated to low stress levels than to the high stress levels because of the higher proportion of failures at the high stress levels. When conducting an accelerated life test, arrangements should be made to ensure that the failures of the units are independent of each other and that the conditions of the test are the same for all units under test. For example, when conducting a temperature acceleration test, arrangements should be made to ensure that the temperature distribution is uniform within the test chamber.

PROBLEMS

6-1. The failure times of diode X (Schottky diode) at both the accelerated stress conditions and the normal operating conditions are found to follow gamma distributions having the same shape parameter γ. The following is the probability density function of the gamma distribution—

$$f(t) = \frac{t^{\gamma-1}}{\theta^{\gamma}\Gamma(\gamma)} e^{\frac{-t}{\theta}}.$$

Where θ is the scale parameter of the distribution,

a. Develop the relationship between the hazard rates at both the accelerated stress condition and the normal operating conditions.

b. Determine the reliability expression at the normal operating conditions as a function of the acceleration factor and the parameters of the gamma distribution at the accelerated conditions.

c. Develop an expression for the acceleration factor.

6-2. Electrolytic corrosion of metallization involves the transport of metallic ions across an insulating surface between two metals. The conductivity of the surface affects the rate of material transport and hence the device life. An accelerated life test is conducted on twenty integrated circuits by subjecting them to moisture with known relative humidity. The following failure times in years are obtained from the test:

0.0031667
0.0056359
0.0061977
0.0067325
0.0069382

0.0076820
0.0106340
0.0107340
0.0116650
0.0119780
0.0122230
0.0128110
0.0132280
0.0154420
0.0156650
0.0164890
0.0217990
0.0296007
0.0311001
0.0373216

Plotting of the hazard rate reveals that the failure times can be described by a gamma distribution. Assuming that an acceleration factor of 20 is used in the experimentation, estimate the parameters of the distribution at the normal operating conditions. What is the reliability of a device at a time of 0.6 years?

6-3. In performing the analysis of the data given in Problem 6-2, the analyst realizes that there are five more observations that are censored at time 0.0395460 years (termination time of the test). Rework the above analysis, and compare the results with those obtained in Problem 6-2.

6-4. Creep failure results whenever the plastic deformation in a machine members accrues over a period of time under the influence of stress and temperature until the accumulated dimensional changes cause the part not to perform its function or to rupture (part failure). It is clear that the failure of the part is due to stress-time-temperature effect. Fifteen parts made of 18-18 plus stainless steel are subjected to a mechanical acceleration method of creep testing, in which the applied stress levels are significantly higher than the contemplated design stress levels, so that the limiting design strains are reached in a much shorter time than in actual service. The times (in hours) to failure at an accelerated stress of 710 MPa are

30.80, 36.09, 65.68, 97.98, 130.97, 500.75, 530.22, 653.96, 889.91, 1173.76, 1317.08, 1490.44, 1669.33, 2057.95, 2711.36.

Assume that the failure times can be modeled by an exponential distribution and that the acceleration factor between the mean life at normal operating conditions and the accelerated stress condition is 20. Determine the parameter of the distribution at the normal conditions and the reliability that a part will survive to 10,000 hours.

6-5. Verification of the reliability of a new stamped suspension arm requires demonstration of R90 (reliability of 0.90) at 50,000 cycles on a vertical jounce to rebound test fixture (for testing in up and down directions). Eight suspension arm units are subjected to an accelerated test, and the following results are obtained (Allmen and Lu, 1994):

Cycles to Failure	*Status*
75,000	Failure
95,000	Failure
110,000	Failure
125,000	Failure
125,000	Censored
125,000	Censored
125,000	Censored
125,000	Censored

Assume that the acceleration factor is 1.4 and that a Weibull distribution represents the failure time probability distribution. Determine the parameters of the distribution. Does the test verify the reliability requirements at normal operating conditions?

6-6. Derive expressions that relate the hazard-rate functions, probability density functions, and the reliability functions at the accelerated stress and at the normal operating conditions when the failure-time distribution at the two conditions is special Erlang in the form

$$f(t) = \frac{t}{\lambda^2} e^{\frac{-t}{\lambda}},$$

where λ is the parameter of the distribution.

6-7. Optical cables are subject to a wide range of temperatures. Buried and duct cables experience small temperature variations, whereas aerial cables experience a much wider range in temperature variations. For each cable type the transmission properties of the optical fiber may change only within a limited range.

In order to determine temperature performance, the change in attenuation is measured as a function of the temperature. This is usually performed by placing the cable along with the drum on which it is wound in a computer-controlled temperature chamber with both ends attached to an attenuation measuring test set. The average attenuation change at a wavelength of 130 nm at −40°C is less than 0.1 dB/km (Mahlke and Gossing, 1987).

Two samples each having twenty-five cables are subjected to temperature acceleration stresses of −20°C and 70°C and the times until an attenuation change of 0.3 dB/km are recorded (an attenuation greater than 0.28 dB/km is deemed unacceptable) as shown below:

Failure Times at −20°C	*Failure Times at 70°C*
4,923.01	9,082.28
4,937.33	9,090.59
4,938.33	9,228.30
4,957.34	9,248.30
4,957.42	9,271.51
4,960.18	9,394.54
4,969.82	9,438.95
4,971.28	9,693.95
4,971.64	9,694.27
4,977.37	9,778.96
4,979.76	9,966.46
4,983.77	10,015.00
4,992.01	10,086.80
4,994.98	10,115.00
4,997.83	10,131.90
5,003.59	10,141.90
5,004.30	10,149.50
5,005.54	10,205.60
5,009.98	10,249.60
5,017.21	10,291.20
5,022.17	10,310.50
5,027.04	10,313.10
5,027.89	10,341.00
5,032.00	10,469.00
5,045.87	10,533.50

a. Assume that the failure times follow lognormal distributions. Estimate the parameters of the distributions at both stress levels.

b. Assume that the mean life of the cable is linearly related to the temperature of the environment. What is the mean life at 25°C?

6-8. The gate oxide in MOS devices is often the source of device failure, especially for high-density arrays that require thin gate oxides. Voltage and temperature are two main factors that cause breakdown of the gate oxide. Therefore, accelerated life testings for MOS devices usually include these two factors. An accelerated test is performed on two samples of fifteen n-channel transistors each by subjecting the first sample to a voltage of 20V and temperature of 200°C and the second sample to a voltage of 27V and 120°C. The normal operating conditions are 9V and 30°C. The following failure times (in seconds) are obtained:

Failure Times for the First Sample	*Failure Times for the Second Sample*
48.7716	10.4341
48.8160	12.4544
49.1403	12.9646
49.3617	13.0883
50.0852	13.1680
50.6413	13.1984
50.7534	13.6002
50.8506	14.1088
51.1490	14.8734
51.2638	15.1088
51.3007	15.4149
51.3085	16.0556
51.4376	16.2214
51.9868	18.2557
53.5653	18.2615

a. Calculate both the electric field and thermal acceleration factors for both stress levels (the activation energy is 0.12 eV and $E_{EF} = 3$).

b. Assuming that the data at each stress level can be modeled using a lognormal distribution as given in Eq. (6.20), determine the mean life at the normal operating conditions.

6-9. Use the data in Problem 6-8 to obtain a combined acceleration factor as given by Eq. (6.24).

6-10. The following is a subset of actual failure times at different accelerated stress conditions of experiments conducted on samples of an electronic device. The failure times are in seconds.

Failure Time	*Temp. °C*	*Volt*	*Failure Time*	*Temp. °C*	*Volt*	*Failure Time*	*Temp. °C*	*Volt*
1	25	27	1	225	26	1365	125	25.7
1	25	27	14	225	26	1401	125	25.7
1	25	27	20	225	26	1469	125	25.7
73	25	27	26	225	26	1776	125	25.7
101	25	27	32	225	26	1789	125	25.7
103	25	27	42	225	26	1886	125	25.7
148	25	27	42	225	26	1930	125	25.7
149	25	27	43	225	26	2035	125	25.7
153	25	27	44	225	26	2068	125	25.7
159	25	27	45	225	26	2190	125	25.7
167	25	27	46	225	26	2307	125	25.7
182	25	27	47	225	26	2309	125	25.7
185	25	27	53	225	26	2334	125	25.7
186	25	27	53	225	26	2556	125	25.7
214	25	27	55	225	26	2925	125	25.7
214	25	27	56	225	26	2997	125	25.7
233	25	27	59	225	26	3076	125	25.7
252	25	27	60	225	26	3140	125	25.7
279	25	27	60	225	26	3148	125	25.7
307	25	27	61	225	26	3736	125	25.7

Use a multiple linear model to develop a relationship between the failure time, temperature, and volt. What is the time to failure at normal operating conditions of 30°C and 5 volts? $E_a = 0.08$ eV.

6-11. Use the proportional-hazard model to estimate the failure rate at normal operating conditions of 30°C and 5 volts. Compare the estimate with that obtained in Problem 6-10. If there is a difference between the two estimates, explain why.

6-12. A temperature acceleration test is performed on twelve units at 300°C and the following failure times (in hours) are obtained:

200, 240, 300, 360, 390, 450, 490, 550, 590, 640, 680, 730.

a. Assume that an Eyring model describes the relationship between the mean life and temperature. What is the expected life at a temperature of 40°C?

b. Assume that the activation energy of the unit's material is $E_a = 0.04$ eV. Use the Arrhenius model to estimate the mean life at 40°C.

6-13. Use the Eyring model to obtain a relationship between the mean life, temperature, and volt. Compare the estimate of mean life at normal operating conditions (30°C and 5 volts) with the estimates obtained in Problems 6-10 and 6-11. Are the physics-statistics-based models more appropriate than the physics-experimental-based models when estimating the failure rate for the data given in Problem 6-10?

6-14. High-voltage power transistors are used in many applications where both high voltage and high current are present. Under these conditions catastrophic device failure can occur due to reverse-biased second breakdown (RBSB). This breakdown can occur when a power transistor switches off an inductive load. The voltage across the device can rise by several hundred volts within a few hundred nanoseconds causing the failure of the device (White, 1994). Therefore, the manufacturers of such transistors often run accelerated life and operational life testing to improve the design of the transistor and to ensure its reliability. A manufacturer conducts an accelerated test by subjecting transistor units to two voltages of 50 volts and 80 volts. The failure times (hours) are given below:

Failure Times at 50V	*Failure Times at 80V*
10.55	3.01
11.56	3.05
12.78	3.06
13.00	3.12
13.50	4.20
15.00	4.30
15.01	4.45
16.02	5.62
19.01	5.67
25.06	8.60
25.50	8.64
29.60	9.10
30.10	9.21
35.00	9.26
45.00	9.29
49.00	10.01
58.62	10.20

a. Use the inverse power rule model to estimate the mean life at the normal operating conditions of five volts. What is the reliability of a device operating at the normal conditions at a time of 10,000 hours? $E_a = 0.60$ eV.

b. What are the acceleration factors used in the test? Are they proper?

c. Solve part a above using the Eyring model. Explain the causes for the difference in results.

6-15. The manufacturer of transistors in Problem 6-14 provides the following additional information about the failure data:

- The accelerated life test using fifty volts is conducted at 90°C.
- The accelerated life test using eighty volts is conducted at 150°C.

a. Solve Problem 6-14 using the additional information assuming T_o = 30°C.

b. Compare the results obtained in a with those obtained from the combination model.

6-16. A specific type of device is susceptible to failure due to electromigration. An accelerated life test is conducted under the following conditions:

Failure Time (Hours)	*Current Intensity*	*Temperature °C*	*Failure Time (Hours)*	*Current Intensity*	*Temperature °C*
300	10	200	264	10	250
340	10	200	270	10	250
345	10	200	271	10	250
349	10	200	272	10	250
361	10	200	280	10	250
362	10	200	285	10	250
363	10	200	200	15	200
369	10	200	205	15	200
374	10	200	207	15	200
379	10	200	209	15	200
380	10	200	210	15	200
390	10	200	211	15	200
250	10	250	215	15	200
251	10	250	220	15	200
252	10	250	222	15	200
260	10	250	225	15	200
262	10	250	228	15	200
263	10	250	230	15	200

a. Use Black's equation to obtain a relationship among the median time to failure, the current intensity, and the temperature. E_a = 0.60 eV.

b. What is median life at J = 5 and T = 30°C?

c. Assume you were not aware of the electromigration model. Apply the Eyring model and compare the results.

6-17. Solve Problem 6-10 if the relative humidities (RH) at test conditions are

Temperature	*Voltage*	*RH*
25°C	27	70%
225°C	26	50%
125°C	25.7	40%

Assume that the device's normal operating conditions are temperature = 30°C, voltage = 5 volts, and RH = 30 percent.

6-18. Hale, Mactaggart, and Shaw (1986) report on the change in resistance of a cathode ray tube (CRT) bleed resistor. The resistor has the function of regulating the power supply to the CRT by providing a constant impedance across it. If the bleed resistance increases, the effect will be severe front-of-screen distortion, where the outside edges of a video image are curved. If the resistance decreases significantly, the circuit will be overloaded, and the display will power down. The main criterion for the performance of the resistor is the change in its resistance with time. The following degradation model describes the relationship between the change in resistance with time and the applied temperature:

$$\frac{dR}{dt} = Ae^{\frac{-E_a}{kT}}.$$

A manufacturer observes the resistance of two resistors tested at two different temperatures of 80°C and 120°C. The results of the test that is conducted for 1,000 hours are as follows:

Temperature 80°C		*Temperature 120°C*	
Time (hours)	*Resistance in MΩ*	*Time (hours)*	*Resistance in MΩ*
0	250	0	250
100	270	100	280
200	291	200	309
300	310	300	341
400	328	400	369
500	349	500	402
600	370	600	432
700	387	700	460
800	412	800	490
900	430	900	516
1000	448	1000	547

The normal value of the resistor is 250 MΩ. The edges of a video image become unacceptably curved when the resistor's value changes to 340 MΩ, and failure of the display occurs when the resistance's value decreases to 180 MΩ. Determine the time to system failure (distorted video or failed display) if the resistor is expected to operate normally at 30°C.

6-19. The breakdown strength of electrical insulation depends on age and temperature. The dielectric strength is measured in kV. Nelson (1981) reports the results of an accelerated test of 128 specimens and the strength of their electrical insulations. The test requires four specimens for each combination of four test temperatures (180, 225, 250, 275°C) and eight aging times (1, 2, 4, 8, 16, 32, 48, 64 weeks). The dielectric strengths for the 128 specimens are as follows:

Week	*Temp.*	*Strength (kV)*	*Week*	*Temp.*	*Strength (kV)*	*Week*	*Temp.*	*Strength (kV)*
1	180	15.0	4	250	13.5	32	225	11.0
1	180	17.0	4	275	10.0	32	225	11.0
1	180	15.5	4	275	11.5	32	250	11.0
1	180	16.5	4	275	11.0	32	250	10.0
1	225	15.5	4	275	9.5	32	250	10.5
1	225	15.0	8	180	15.0	32	250	10.5
1	225	16.0	8	180	15.0	32	275	2.7
1	225	14.5	8	180	15.5	32	275	2.7
1	250	15.0	8	180	16.0	32	275	2.5
1	250	14.5	8	225	13.0	32	275	2.4
1	250	12.5	8	225	10.5	48	180	13.0
1	250	11.0	8	225	13.5	48	180	13.5
1	275	14.0	8	225	14.0	48	180	16.5
1	275	13.0	8	250	12.5	48	180	13.6
1	275	14.0	8	250	12.0	48	225	11.5
1	275	11.5	8	250	11.5	48	225	10.5
2	180	14.0	8	250	11.5	48	225	13.5
2	180	16.0	8	275	6.5	48	225	12.0
2	180	13.0	8	275	5.5	48	250	7.0
2	180	13.5	8	275	6.0	48	250	6.9
2	225	13.0	8	275	6.0	48	250	8.8
2	225	13.5	16	180	18.5	48	250	7.9
2	225	12.5	16	180	17.0	48	275	1.2
2	225	12.5	16	180	15.3	48	275	1.5
2	250	12.5	16	180	16.0	48	275	1.0
2	250	12.0	16	225	13.0	48	275	1.5
2	250	11.5	16	225	14.0	64	180	13.0
2	250	12.0	16	225	12.5	64	180	12.5
2	275	13.0	16	225	11.0	64	180	16.5
2	275	11.5	16	250	12.0	64	180	16.0
2	275	13.0	16	250	12.0	64	225	11.0

(*continued*)

Week	Temp.	Strength (kV)	Week	Temp.	Strength (kV)	Week	Temp.	Strength (kV)
2	275	12.5	16	250	11.5	64	225	11.5
4	180	13.5	16	250	12.0	64	225	10.5
4	180	17.5	16	275	6.0	64	225	10.0
4	180	17.5	16	275	6.0	64	250	7.2
4	180	13.5	16	275	5.0	64	250	7.5
4	225	12.5	16	275	5.5	64	250	6.7
4	225	12.5	32	180	12.5	64	250	7.6
4	225	15.0	32	180	13.0	64	275	1.5
4	225	13.0	32	180	16.0	64	275	1.0
4	250	12.0	32	180	12.0	64	275	1.2
4	250	13.0	32	225	11.0	64	275	1.2
4	250	12.0	32	225	9.5			

Reprinted from *IEEE Transactions on Reliability*, Vol. R-30, No. 2, "Analysis of Performance Degradation Data from Accelerated Tests," W. Nelson p. 149. ©1981 IEEE.

Assume that the relationship between median (50 percent point) log breakdown voltage $V_{50\%}$, absolute temperature T, and exposure time t is

$$\log V_{50\%} = \alpha - \beta t \exp(-\gamma / T),$$

where α, β, and γ are constants. Estimate the time for the insulation to degrade below 2 kV breakdown strength at the normal operating temperature of 150°C.

6-20. *In situ* accelerated aging technique is based on the same idea of the accelerated life testing: most of the physical and chemical processes are thermally activated. In accelerated life testing, the purpose of the application of thermal stress is to induce a number of failures, so that a failure rate can be calculated, whereas in an *in situ* test, the thermal stress is applied to increase the rate at which physicochemical processes occur in the system, in order to measure their effect during the aging treatment on a parameter characterizing the performance of the system (DeSchepper et al., 1994). In other words, the main characteristic of the *in situ* technique is that the effect of the accelerated physicochemical processes on the relevant parameter is measured *during* thermal stress. For example, a simple model that represents the aging of a thin film resistor can be written in the form

$$\frac{dR(t)}{R_o} = kt^n,$$

where

R_o = the initial value of the resistor,

k, n = constants, and

t = the time corresponding to the change in the resistance.

Use the data of Example 6.14 to estimate the time to reach $d\,R(t)/R_o = 8.5$. Compare the solution with that obtained in Example 6.14.

REFERENCES

Allmen, C. R., and Lu, M-W. (1994). "A Reduced Sampling Approach for Reliability Verification," *Quality and Reliability Engineering International* 10, 71–77.

Barton, R. R. (1980). "Optimal Accelerated Life-Time Plans That Minimize the Maximum Test-Stress." *IEEE Transactions on Reliability* 40, 166–172.

Black, J. R. (1969). "Electromigration: A Brief Survey and Some Recent Results." *IEEE Transactions on Electron Devices* 16, 338.

Blanks, H. S. (1980). "The Temperature Dependence of Component Failure Rate." *Microelectronics Reliability*, 20, 297–307.

Brombacher, A. C. (1992). *Reliability by Design*. New York: John Wiley.

Buehler, M., Zmani, N., and Dhiman, J. (1991). "Electromigration Error Analysis for Optimal Experimental Design." NASA's Jet Propulsion Laboratory, Case No. NPO-18012, August.

Chan, C. K., Boulanger, M. and Tortorella, M. (1994). "Analysis of Parameter-Degradation Data Using Life-Data Analysis Programs." *1994 Proceedings of the Annual Reliability and Maintainability Symposium* (pp. 288–291) Piscataway: IEEE.

Chen, I. C., and Hu, C. (1987). "Accelerated Testing of Time-Dependent Breakdown of SiO_2." *IEEE Electron Device Letters* 8(4), 140–142.

Christou, A. (1994). *Integrating Reliability into Microelectronics Manufacturing*. New York: John Wiley.

Cox, D. R. (1972). "Regression Models and Life Tables (With Discussion)." *Journal of Royal Statistical Society B*, 34, 187.

Dale, C. J. (1985). "Application of the Proportional Hazards Model in the Reliability Field." *Reliability Engineering* 10, 1–14.

DeSchepper, L., DeCeuninck, W., Lekens, G., Stals, L., Vanhecke, B., Roggen, J., Beyne, E., and Tielemans, L. (1994). "Accelerated Aging with *In Situ* Electrical Testing: A

Powerful Tool for the Building-in Approach to Quality and Reliability in Electronics." *Quality and Reliability Engineering International* 10, 15–26.

Elsayed, E. A., and Chan, C. K. (1990). "Estimation of Thin-Oxide Reliability Using Proportional Hazards Models." *IEEE Transactions on Reliability* 39(3), 329–335.

Engelmaier, W. (1993). "Reliability of Surface Mount Solder Joints: Physics and Statistics of Failure." *National Electronic Packaging and Production West Proceedings* (p. 1782), Ann Arbor: University of Michigan.

Esaklul, K. (Ed.). (1992). *Handbook of Case Histories in Failure Analysis*. Material Park, Ohio: ASM International.

Flaherty, J. M. (1994). "How to Accelerate SMT Attachment Reliability Testing." *Test and Measurement World* (January), 47–54.

Gunn, J. E., Camenga, R. E., and Malik, S. K. (1983). "Rapid Assessment of the Humidity Dependence of IC Failure Modes by Use of HAST." *Proceedings IEEE/IRPS* (pp. 66–72). Piscataway: IEEE.

Hale, P., Mactaggart, I., and Shaw, M. (1986). "Cathode Ray Tube Bleed Resistor Reliability: A Case Study." *Quality and Reliability Engineering International* 2, 165–170.

Hastie, T. J., and Tibshirani, R. J. (1990). *Generalized Additive Models*. New York: Chapman and Hall.

Kalbfleisch, J. D., and Prentice, R. L. (1980). *The Statistical Analysis of Failure Time Data*. New York: John Wiley.

Kececioglu, D., and Jacks, J. (1984). "The Arrhenius, Eyring, Inverse Power Law and Combination Models in Accelerated Life Testing." *Reliability Engineering* 8, 1–9.

Kielpinski, T. J., and Nelson, W. B. (1975). "Optimum Accelerated Life Tests for Normal and Lognormal Life Distributions." *IEEE Transactions on Reliability* (Vol. R-24) (pp. 310–320). Piscataway: IEEE.

Leblebici, Y., and Kang, S-M. (1993). *Hot-Carrier Reliability of MOS VLSI Circuits*. Boston: Kluwer.

Mahlke, G., and Gossing, P. (1987). *Fiber Optic Cables*. New York: John Wiley.

Mann, N. R., Schafer, R. E., and Singpurwalla, N. D. (1974). *Methods for Statistical Analysis of Reliability and Life Data*. New York: John Wiley.

McPherson, J. (1986). "Stress Dependent Activation Energy." *Proceedings IEEE/IRPS* (pp. 118–125). Piscataway: IEEE.

McPherson, J. W., and Baglee, D. A. (1985). "Accelerated Factors for Thin Gate Oxide Stressing." *Proceedings of the Twenty-third International Reliability Physics Symposium* (pp. 1–5). Piscataway: IEEE.

Meeker, W. Q., and Hahn, G. J. (1985). *How to Plan an Accelerated Life Test: Some Practical Guidelines* (Vol. 10). Milwaukee: ASQC.

Meeker, W. Q., and Nelson, W. B. (1975). "Optimum Accelerated Life-Tests for the Weibull and Extreme Value Distribution." *IEEE Transactions on Reliability* R-14, 321–332.

Miller, R. G., Jr. (1981). *Survival Analysis*. New York: John Wiley.

Nelson, W. (1981). "Analysis of Performance Degradation Data from Accelerated Tests." *IEEE Transactions on Reliability* R-30, 149–155.

Nelson, W., and Hahn, G. J. (1978). "Linear Estimation of a Regression Relationship from Censored Data, Part I—Simple Methods and Their Application." *Technometrics* 14, 214.

Quader, K. N., Fang, P., Yue, J. T., Ko, P. K., and Hu, C. (1994). "Hot-Carrier: Reliability Design Rules for Translating Device Degradation to CMOS Digital Circuit Degradation." *IEEE Transactions on Electron Devices* 41(5), 681–690.

Takeda, E., and Suzuki, N. (1983). "An Empirical Model for Device Degradation Due to Hot-Carrier Injection." *IEEE Transaction on Electron Device Letters*. 4, 111.

Tobias, P. A., and Trindade, D. (1986). *Applied Reliability*. New York: Van Nostrand Reinhold.

White, G. L. (1994). "Operational Life Testing of Power Transistors Switching Unclamped Inductive Loads." *Quality and Reliability Engineering International*. 10, 63–69.

Wightman, D., Bendell, T., and McCollin, C. (1994). "Comparison of Proportional Hazards Modelling, Proportional Intensity Modelling and Additive Hazards Modelling in a Software Reliability Context." Paper presented at the Second International Applied Statistics in Industry Conference, Witchita, Kansas, June 6–8.

PART III

Reliability Improvement: Warranty and Preventive Maintenance

CHAPTER 7

Renewal Processes and Expected Number of Failures

One of the most frequently sought after quantities is the expected number of failures of a system during a time interval $(0, t]$. This quantity is used to determine the optimal preventive maintenance schedule and as a criterion for reliability acceptance tests. The latter exemplifies a typical reliability test when n units (components or systems) are drawn at random from a production lot. They are subjected to specified test conditions, and the entire production lot is accepted if x $(x < n)$ or more units survive the test by time t.

More important, this quantity is extremely useful for manufacturers in estimating the cost of a warranty. As an example, consider the case when a manufacturer agrees to replace, free of charge, the product when it fails before a time period T (warranty period). Suppose that $M(T)$ is the expected number of replacements (renewals) during the warranty period. Then the expected warranty cost $C(T)$ is

$$C(T) = c \cdot M(T), \tag{7.1}$$

where c is the fixed cost per replacement. Clearly, the cost of warranty is greatly affected by the number of replacements, and if the manufacturer produces a very large number of units, it becomes crucial for the manufacturer to determine $M(T)$ with a much greater accuracy.

The role of $M(T)$ in estimating the warranty cost for a given warranty policy and in determining the optimal preventive replacement periods for repairable systems is emphasized in Chapters 8 and 9. The following sections present two different approaches for determining $M(T)$. The first approach, a *parametric approach*, is used when the failure-time distribution of the units is known. The second approach, a *nonparametric approach*, is used when the failure-time distribution is unknown or when the mean and the standard deviation of the failure times are the only known parameters. Because the estimation of $M(T)$ may be quite difficult to obtain, we present approximate methods for estimating $M(T)$.

7.1. Parametric Renewal Function Estimation

When the failure-time distribution is known, we can determine the expected number of failures (or renewals) during any time interval $(0, t]$ by using either the continuous-time or the discrete-time approaches given below.

7.1.1. CONTINUOUS TIME

This approach is also referred to as the *renewal theory approach*. Consider the case when a unit is operating until it fails. Upon failure, the unit is either replaced by a new identical unit or repaired to its original condition. This is considered a renewal process and can be formally defined as a nonterminating sequence of independent, identically distributed (i.i.d.) nonnegative random variables. To determine the expected number of failures in interval $(0, t]$, we follow Jardine's (1973) work and define the following notations as shown in Figure 7.1. Let

$N(t)$ = the number of failures in interval $(0, t]$,

$M(t)$ = the expected number of failures in interval $(0, t] = E[N(t)]$, where $E[\,]$ denotes expectations,

t_i = length of the time interval between failures $i - 1$ and i,

S_r = total time up to the rth failure $S_r = t_1 + t_2 + \ldots + t_r = \Sigma_{i=1}^{r} t_i$.

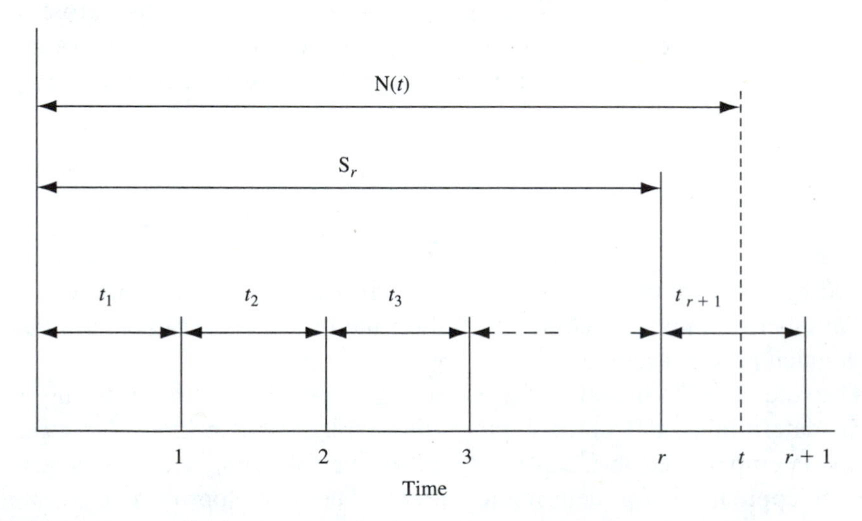

Figure 7.1. Failures in $(0,t]$

The probability that the number of failures $N(t) = r$ is the same as the probability that t lies between the rth and $(r + 1)$th failure. Thus,

$$P[N(t) < r] = 1 - F_r(t),$$

where $F_r(t)$ is the cumulative distribution function of S_r—that is, $F_r(t) = P[S_r \leq t]$, and

$$P[N(t) > r] = F_{r+1}(t).$$

But

$$P[N(t) < r] + P[N(t) = r] + P[N(t) > r] = 1.$$

Thus,

$$P[N(t) = r] = F_r(t) - F_{r+1}(t).$$

The expected value of $N(t)$ is then

$$M(t) = \sum_{r=0}^{\infty} r P[N(t) = r]$$

$$= \sum_{r=0}^{\infty} r\,[F_r(t) - F_{r+1}(t)]$$

or

$$M(t) = \sum_{r=1}^{\infty} F_r(t). \tag{7.2}$$

$M(t)$ is referred to as the *renewal function*. Eq. (7.2) can be written as

$$M(t) = F(t) + \sum_{r=1}^{\infty} F_{r+1}(t),$$

where $F_{r+1}(t)$ is the convolution of $F_r(t)$ and F. Let f be the p.d.f. of F, then

$$F_{r+1}(t) = \int_0^t F_r(t - x) f(x)\, dx$$

and

$$M(t) = F(t) + \sum_{r=1}^{\infty} \int_0^t F_r(t-x) f(x)\, dx$$

$$= F(t) + \int_0^t \left[\sum_{r=1}^{\infty} F_r(t-x) \right] f(x)\, dx$$

—that is,

$$M(t) = F(t) + \int_0^t M(t-x) f(x)\, dx. \tag{7.3}$$

We refer to Eq. (7.3) as the *fundamental renewal equation.*

By taking Laplace transforms of both sides of Eq. (7.3), we obtain

$$M^*(s) = \frac{f^*(s)}{s[1 - f^*(s)]}, \tag{7.4}$$

where

$$f^*(s) = E[e^{-sS_{N(t)}}] = \int_0^{\infty} e^{-st} f(t)\, dt$$

and M(t) $= \mathcal{L}^{-1} M^*$(s) is the Laplace inverse of M^*(s). The renewal density $m(t)$ is the derivative of $M(t)$ or

$$m(t) = \frac{dM(t)}{dt}.$$

$m(t)$ is interpreted as the probability that a renewal occurs in the interval $[t, t + \Delta t]$. Thus, in the case of a Poisson process, renewal density $m(t)$ is the Poisson rate λ.

We can also write

$$m(t) = \sum_{r=1}^{\infty} f_r(t)$$

or

$$m(t) = f(t) + \int_0^t m(t-x) f(x)\, dx. \tag{7.5}$$

Equation (7.5) is known as the *renewal density equation*. To solve the renewal equation, we use Laplace transforms:

$$\mathcal{L}f(t) = \int_0^\infty e^{-st} f(t)\, dt$$

and

$$\mathcal{L}m(t) = \int_0^\infty e^{-st} m(t)\, dt.$$

Using the convolution property of transforms,

$$\mathcal{L}m(t) = \mathcal{L}f(t) + \mathcal{L}m(t)\mathcal{L}f(t)$$

$$m^*(s) = f^*(s) + m^*(s) f^*(s)$$

$$m^*(s) = \frac{f^*(s)}{1 - f^*(s)}$$

$$M^*(s) = \frac{f^*(s)}{s[1 - f^*(s)]} \tag{7.6}$$

and

$$f^*(s) = \frac{m^*(s)}{1 + m^*(s)}.$$

EXAMPLE 7.1

A component that exhibits constant failure rate is replaced upon failure by an identical component. The p.d.f. of the failure-time distribution is

$$f(t) = \lambda e^{-\lambda t}.$$

What is the expected number of failures during the interval $(0, t]$?

SOLUTION

Taking the Laplace transform of the p.d.f. results in

$$f^*(s) = \frac{\lambda}{s + \lambda}.$$

The Laplace transform of the renewal density becomes

$$m^*(s) = \frac{\lambda}{s + \lambda - \lambda} = \frac{\lambda}{s}.$$

The inverse of the above expression is

$$m(t) = \lambda \qquad t \geq 0.$$

Thus,

$$M(t) = \lambda t \qquad t \geq 0$$

—that is, the number of failures in $(0, t]$ is λt.

We now consider a numerical example for the constant-failure-rate case.

EXAMPLE 7.2 A system is found to exhibit a constant failure rate of 6×10^{-6} failures per hour. What is the expected number of failures after one year of operation? Note that the system is repaired upon failure and is returned to its original condition.

SOLUTION The Laplace transform of the p.d.f. of the constant-hazard rate λ is

$$f^*(s) = \int_0^\infty \lambda e^{-\lambda t} e^{-st}\, dt = \frac{\lambda}{s + \lambda}.$$

Substituting in Eq. (7.4), we obtain

$$M^*(s) = \frac{\lambda/(\lambda + s)}{s[1 - \lambda/(\lambda + s)]} = \frac{\lambda}{s^2}.$$

The inverse of $M^*(s)$ to $M(t)$ is

$$M(t) = \mathscr{L}^{-1} \frac{\lambda}{s^2} = \lambda t.$$

The expected number of failures after one year of service (10^4 hours) is $6 \times 10^{-6} \times 10^4 = 0.06$ failures.

Let us consider another example. If $X_1, X_2, \ldots, X_n$ are n exponentially independent and identically distributed random variables having a mean of $1/\lambda$, then

$X_1 + X_2 + \ldots + X_n$ is a gamma distribution with parameters n and λ. Its p.d.f. is given by

$$f(t) = \lambda e^{-\lambda t} \frac{(\lambda t)^{n-1}}{(n-1)!}$$

or

$$f(t) = \frac{\lambda(\lambda t)^{n-1} e^{-\lambda t}}{\Gamma(n)}.$$

The Laplace transform of $f(t)$ is

$$f^*(s) = \frac{\lambda^n}{(\lambda + s)^n}.$$

The Laplace transform of the expected number of failures is

$$M^*(s) = \frac{f^*(s)}{s[1 - f^*(s)]} = \frac{\lambda^n}{s[(\lambda + s)^n - \lambda^n]}.$$

The inverse of the above transform is difficult to obtain. Therefore, we may obtain $M(t)$ by numerically computing it for discrete time intervals, by using nonparametric approaches, or by using approximate methods as discussed later in this chapter.

7.1.1.1. *Availability Analysis Under Renewals*

Consider the case when a failure occurs, it is repaired, and the component becomes "as good as new." Let T_i be the duration of the ith functioning period and D_i be the system downtime for the ith repair or replacement. We have a sequence of random variables $\{X_i = T_i + D_i\}$ $i = 1, 2, \ldots$ as shown in Figure 7.2.

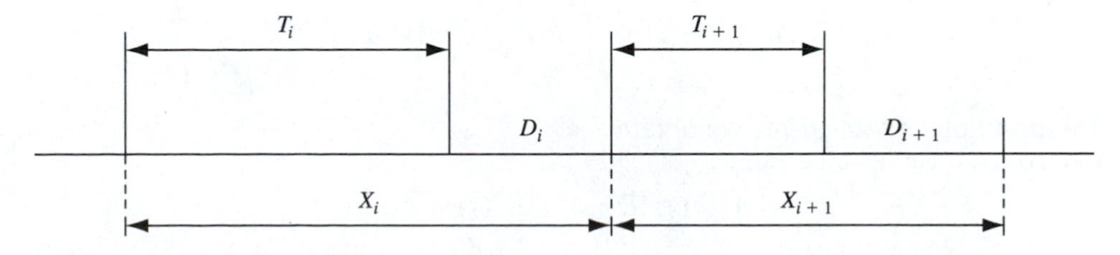

Figure 7.2. Renewal Processes Using Repairs

Assume T_i's are i.i.d. with cumulative distribution function $W(t)$ and probability density function $w(t)$. D_i's are i.i.d. with c.d.f. of $G(t)$ and p.d.f. $g(t)$. Then X_i's are i.i.d. The underlying density $f(t)$ of the renewal process is the convolution of w and g. Thus,

$$\mathscr{L}f(t) = \mathscr{L}w(t)\mathscr{L}g(t)$$

or

$$f^*(s) = w^*(s)g^*(s).$$

Therefore,

$$m^*(s) = \frac{w^*(s)g^*(s)}{1 - w^*(s)g^*(s)}. \tag{7.7}$$

As shown in section 3.3.1 $M^*(s)$ is obtained as

$$M^*(s) = \frac{w^*(s)g^*(s)}{s[1 - w^*(s)g^*(s)]}. \tag{7.8}$$

We define the availability $A(t)$ as the probability that the component is properly functioning at time t. If no repair is performed, then $R(t) = A(t) = 1 - W(t)$.

The component may be functioning at time t by reason of two mutually exclusive cases: either the component has not failed from the beginning (no renewals in $(0, t]$) with probability $R(t)$, or the last renewal (repair) occurred at time x, $0 < x < t$, and the component has continued to function since that time (Trivedi, 1982). The probability associated with the second case is

$$\int_0^t R(t - x)m(x)\,dx.$$

Thus,

$$A(t) = R(t) + \int_0^t R(t - x)m(x)\,dx.$$

Taking Laplace transforms we obtain

$$A^*(s) = R^*(s) + R^*(s)m^*(s)$$

$$A^*(s) = R^*(s)[1 + m^*(s)]. \tag{7.9}$$

Substituting Eq. (7.7) into Eq. (7.9) results in

$$A^*(s) = R^*(s)\left[1 + \frac{w^*(s)\,g^*(s)}{1 - w^*(s)\,g^*(s)}\right]$$

or

$$A^*(s) = \frac{R^*(s)}{1 - w^*(s)\,g^*(s)}.$$

But $R(t) = 1 - W(t)$ and its Laplace transform is

$$R^*(s) = \frac{1}{s} - W^*(s)$$

or

$$R^*(s) = \frac{1}{s} - \frac{w^*(s)}{s}$$

$$= \frac{1 - w^*(s)}{s}.$$

Thus,

$$A^*(s) = \frac{1 - w^*(s)}{s[1 - w^*(s)\,g^*(s)]}.$$

The steady-state availability A is

$$A = \lim_{t\to\infty} A(t) = \lim_{s\to 0} sA^*(s).$$

When s is small, we approximate $\mathrm{e}^{-\mathrm{st}} \simeq 1 - st$

or
$$w^*(s) = \int_0^\infty e^{-st}\,w(t)\,dt$$

$$\simeq \int_0^\infty w(t)\,dt - s\int_0^\infty t\,w(t)\,dt$$

$$\simeq 1 - \frac{s}{\alpha},$$

where $1/\alpha$ is the MTTF (mean time to failure). Also

$$g^*(s) \simeq 1 - \frac{s}{\beta},$$

where $1/\beta$ is the MTTR (mean time to repair):

$$A = \lim_{s \to 0} \frac{1 - \left[1 - \frac{s}{\alpha}\right]}{1 - \left[1 - \frac{s}{\alpha}\right]\left[1 - \frac{s}{\beta}\right]} = \frac{\frac{1}{\alpha}}{\frac{1}{\alpha} + \frac{1}{\beta}}$$

$$A = \frac{MTTF}{MTTF + MTTR}.$$

EXAMPLE 7.3 Consider the case of exponential failure and repair-time distributions. Derive an expression for the renewal density $m(t)$. What are the availability $A(t)$ and the steady-state availability $A(\infty)$?

SOLUTION Let $w(t)$ and $g(t)$ represent the failure-time and repair-time distributions. Then,

$$w(t) = \lambda e^{-\lambda t}$$

$$g(t) = \mu e^{-\mu t}.$$

The Laplace transforms of these two functions are:

$$w^*(s) = \frac{\lambda}{s + \lambda}$$

$$g^*(s) = \frac{\mu}{s + \mu}.$$

Using Eq. (7.7), the renewal density is obtained as

$$m^*(s) = \frac{w^*(s) g^*(s)}{1 - w^*(s) g^*(s)}$$

$$= \frac{\lambda\mu}{s[s + (\lambda + \mu)]}$$

or

$$m^*(s) = \frac{\lambda\mu}{(\lambda + \mu)s} - \frac{\lambda\mu}{(\lambda + \mu)^2} \frac{\lambda + \mu}{s + \lambda + \mu}.$$

The renewal density function is

$$m(t) = \frac{\lambda\mu}{\lambda + \mu} - \frac{\lambda\mu}{\lambda + \mu} e^{-(\lambda+\mu)t}$$

$$\lim_{t\to\infty} m(t) = \frac{\lambda\mu}{\lambda + \mu}$$

or

$$\lim_{t\to\infty} m(t) = \frac{1}{MTTF + MTTR}.$$

The availability of the system is obtained as

$$A^*(s) = \frac{1 - \dfrac{\lambda}{s + \lambda}}{s\left[1 - \dfrac{\lambda\mu}{(s + \lambda)(s + \mu)}\right]}$$

or

$$A^*(s) = \frac{s + \mu}{s[s + (\lambda + \mu)]}.$$

No transform for this function exists in Laplace transform tables. However, a partial-fraction algebra reduces the above expression to known results:

$$A^*(s) = \frac{\dfrac{\mu}{\lambda + \mu}}{s} + \frac{\dfrac{\lambda}{\lambda + \mu}}{s + (\lambda + \mu)}.$$

The Laplace inverse is

$$A(t) = \frac{\mu}{\lambda + \mu} + \frac{\lambda}{\lambda + \mu} e^{-(\lambda+\mu)t}$$

and

$$A = \lim_{t \to \infty} A(t) = \frac{\mu}{\lambda + \mu}.$$

EXAMPLE 7.4

Permanent magnet synchronous motor (PMSM) brushless DC (BLDC) servos are becoming attractive replacements for DC motors in industrial servo motors. The PMSM BLDC servo has higher torque and velocity bandwidth and does not require the regular brush and maintenance requirements of conventional motors.

A producer of the PMSMs designs a reliability test by subjecting a motor to a continuous load. Upon failure, the motor is immediately repaired and restored to its initial condition. The test is then continued and the above procedure is repeated. The failure and the repair time distributions are exponential with rates λ and μ with estimates of 6×10^{-5} failures per hour and 4×10^{-2} repairs per hour, respectively.

Determine the expected number of motor's failures during $(0, 2 \times 10^4$ hours) and the availability of the motor at the end of two years of testing. Plot $M(t)$ and $A(t)$ for different values of λ and μ.

SOLUTION

Since failure and repair times are exponential, following Example 7.3 we have

$$m^*(s) = \frac{\lambda\mu}{(\lambda + \mu)\, s} - \frac{\lambda\mu}{(\lambda + \mu)^2} \cdot \frac{1}{(s + \lambda + \mu)}$$

and

$$m(t) = \frac{\lambda\mu}{\lambda + \mu} - \frac{\lambda\mu}{(\lambda + \mu)} e^{-(\lambda+\mu)t}.$$

The expected number of renewals in $(0, t]$ is

$$M(t) = \frac{\lambda\mu}{(\lambda + \mu)} t - \frac{\lambda\mu}{(\lambda + \mu)^2} + \frac{\lambda\mu}{(\lambda + \mu)^2} e^{-(\lambda+\mu)t}.$$

Substitution of the values of λ and μ in the above expression results in

$$M(t) = 5.991 \times 10^{-5} t - 0.001495 + 1.495 \times 10^{-3} e^{-4.006 \times 10^{-2} t}.$$

The expected number of failures in a 2×10^4 hours interval is

$$M(2 \times 10^4) = 1.197 \text{ failures.}$$

The availability of the motor is

$$A(t) = \frac{\mu}{\lambda + \mu} + \frac{\lambda}{\lambda + \mu} e^{-(\lambda+\mu)t},$$

and the availability at the end of two years of testing is

$$A(2 \times 10^4) = \frac{4 \times 10^{-2}}{4.006 \times 10^{-2}} + \frac{6 \times 10^{-5}}{4.006 \times 10^{-2}} e^{-(4.006)\times 2 \times 10^2}$$

or

$$A(2 \times 10^4) = 0.9985.$$

The plots of $M(t)$ and $A(t)$ for different values of λ and μ are shown in Figures 7.3 and 7.4, respectively.

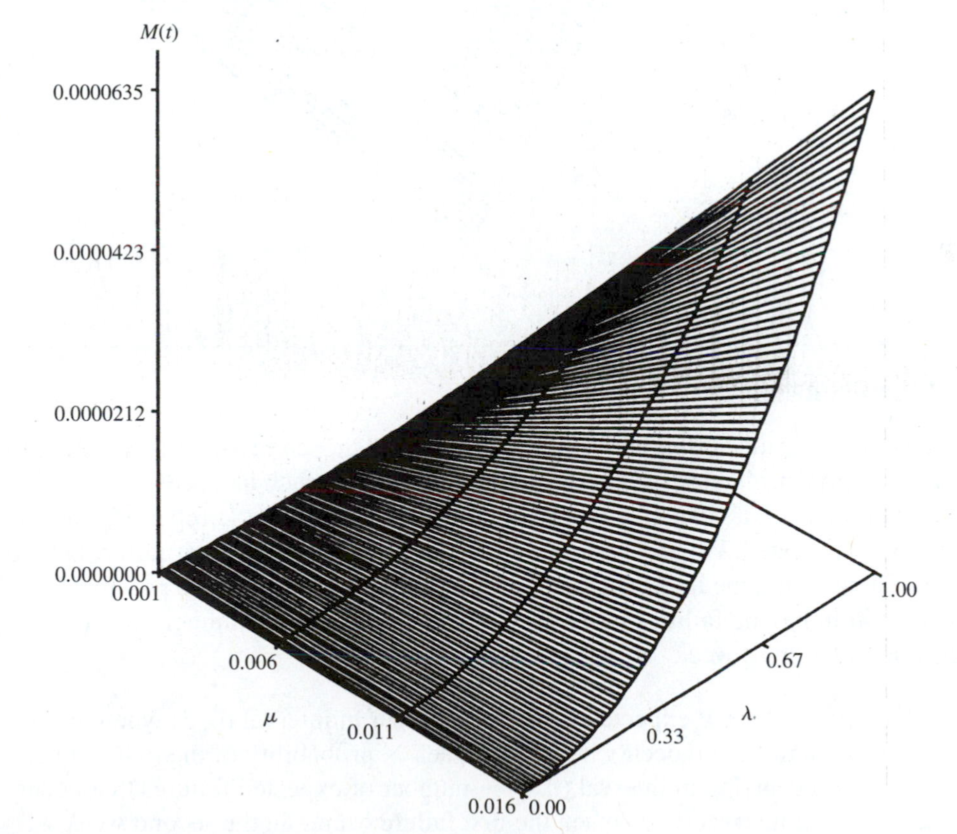

Figure 7.3. $M(t)$ for Different λ and μ

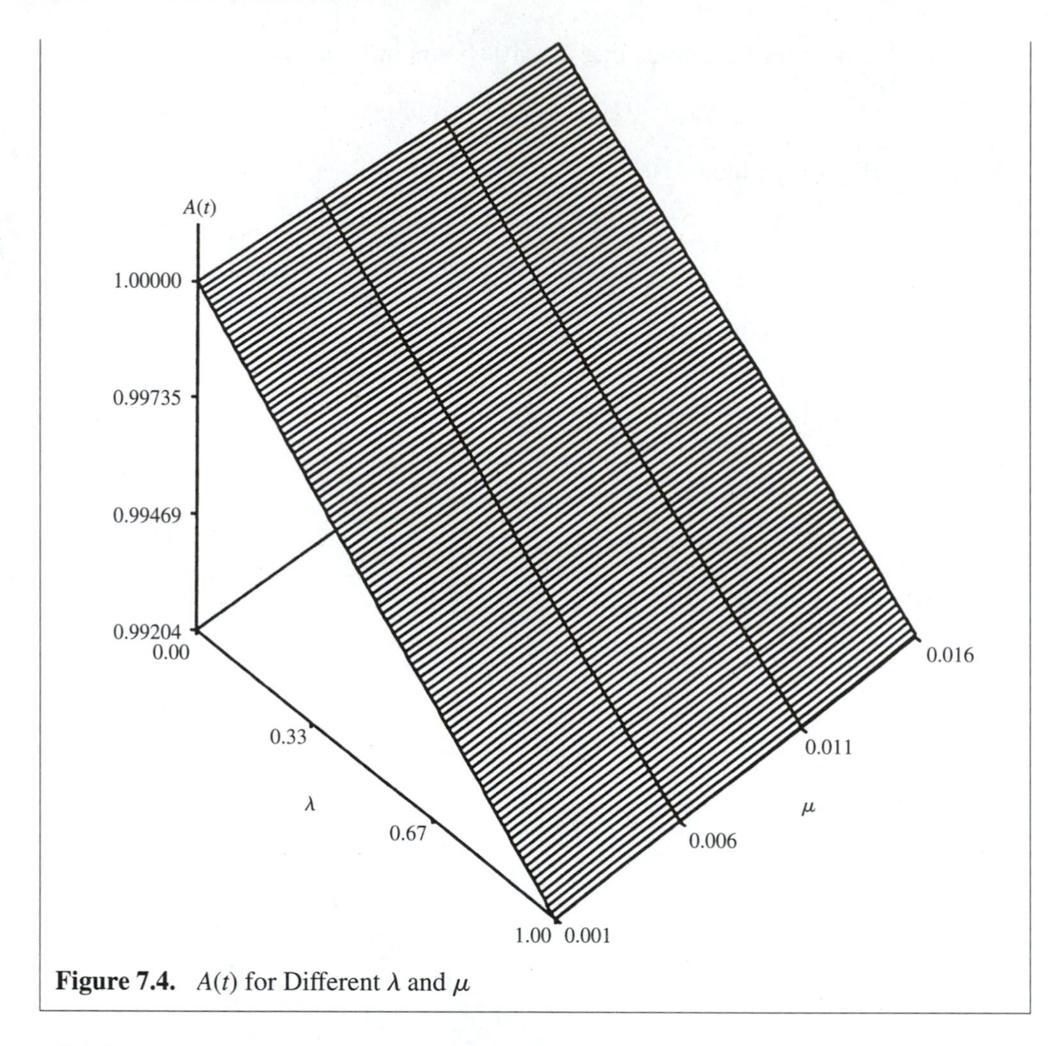

Figure 7.4. $A(t)$ for Different λ and μ

7.1.2. DISCRETE TIME

Let us consider the situation where the time scale is discrete—that is, the system (or a component) is observed at discrete time intervals such as once a week, once a month, or once a year. If a failure is observed, the system is repaired, and the process is repeated. We are interested in determining the number of failures at the end of a discrete time interval—say, the third week. There are three possible ways that result in having failures in the third week. The expected number of failures at the end of the third week, $M(3)$, is obtained as

$M(3) =$ number of expected failures that occur in interval (0, 3) when the first failure occurs in the first week $\times$ probability of the first failure occurring in interval (0, 1) + number of expected failures that occur in interval (0, 3) when the first failure occurs in the second week $\times$

probability of the first failure occurring in interval (1, 2) + number of expected failures that occur in interval (0, 3) when the first failure occurs in the third week $\times$ probability of the first failure occurring in interval (2, 3).

The expected number of failures which occur in the interval (0, 3) when the first failure occurs in the first week can be written as

$M_1(3) =$ the number of failures that occurred in the first week + the expected number of failures in the remaining two weeks

or

$M_1(3) = 1 + M(2)$, where $M_1(3)$ is the expected number of failures at the end of the three weeks, provided that the first failure occurred at the first week.

By definition, the expected number of failures in the remaining two weeks is $M(2)$, starting with a new component on replacing the failed one after the failure that occurred in the first week. In order to calculate the expected number of failures in any interval we need to calculate the probability that the first failure occurs in an interval (t_1, t_2) as follows.

$$\text{Probability that the first failure occurs in the interval } (t_1, t_2) = \int_{t_1}^{t_2} f(t)\,dt.$$

Thus, the expected number of failures by the third week is

$$M(3) = [1 + M(2)] \int_0^1 f(t)\,dt + [1 + M(1)] \int_1^2 f(t)\,dt + [1 + M(0)] \int_2^3 f(t)\,dt.$$

Since $M(0) = 0$, then the above equation can be rewritten as

$$M(3) = \sum_{i=0}^{2} [1 + M(2 - i)] \int_i^{i+1} f(t)\,dt.$$

In general, the number of failures at time period T is obtained as

$$M(T) = \sum_{i=0}^{T-1} [1 + M(T - i - 1)] \int_i^{i+1} f(t)\,dt \qquad T \geq 1, \tag{7.10}$$

with $M(0) = 0$.

EXAMPLE 7.5 The manufacturer of five-volt electric bulbs estimates the expected number of failures during a twenty week period by subjecting a bulb to ten volts. Upon failure, the bulb is replaced by a new one, and the process is repeated. The failure time for the bulbs is found to follow a uniform distribution between $0 \le t \le 20$ weeks with a $f(t) = 1/20$. Determine the expected number of failures in the fourth week.

SOLUTION Using Eq. (7.10) we obtain $M(4)$ as follows:

$$M(4) = \sum_{i=0}^{3} [1 + M(3 - i)] \int_{i}^{i+1} f(t)\, dt$$

$$M(4) = [1 + M(3)] \int_{0}^{1} \frac{1}{20} dt + [1 + M(2)] \int_{1}^{2} \frac{1}{20} dt + [1 + M(1)] \int_{2}^{3} \frac{1}{20} dt + [1 + M(0)] \int_{3}^{4} \frac{1}{20} dt$$

$$M(0) = 0$$

$$M(1) = [1 + M(0)] \int_{0}^{1} \frac{1}{20} dt = \frac{1}{20}$$

$$M(2) = [1 + M(1)] \int_{0}^{1} \frac{1}{20} dt + [1 + M(0)] \int_{1}^{2} \frac{1}{20} dt = \frac{41}{400}$$

$$M(3) = [1 + M(2)] \int_{0}^{1} \frac{1}{20} dt + [1 + M(1)] \int_{1}^{2} \frac{1}{20} dt + [1 + M(0)] \int_{2}^{3} \frac{1}{20} dt = \frac{1{,}261}{8{,}000}$$

$M(4) = 0.2155$ failures.

The following example illustrates the estimation of the expected number of failures when the failure time is normally distributed.

EXAMPLE 7.6

Consider the case when the system is observed every two weeks and the failure-time distribution is normal with a mean = 4 and a standard deviation = 1 week. Determine the expected number of failures at the end of two weeks.

SOLUTION

$$M(2) = [1 + M(1)] \frac{1}{\sqrt{2\pi}} \int_0^1 \exp\left[\frac{-(t-4)^2}{2}\right] dt$$

$$+ [1 + M(0)] \frac{1}{\sqrt{2\pi}} \int_1^2 \exp\left[\frac{-(t-4)^2}{2}\right] dt,$$

but

$$\frac{1}{\sqrt{2\pi}} \int_0^1 \exp\left[-\frac{(t-4)^2}{2}\right] dt = \Phi(1-4) - \Phi(0-4)$$

$$= \Phi(-3) - \Phi(-4),$$

where

$$\Phi(t) = \frac{1}{\sqrt{2\pi}} \int_\infty^t \exp\left[-\frac{t^2}{2}\right] dt$$

is the cumulative distribution function of the standard normal distribution. From the standard tables, we obtain

$$\Phi(-3) - \Phi(-4) \simeq 0.0014.$$

Meanwhile,

$$\frac{1}{\sqrt{2\pi}} \int_1^2 \exp\left[\frac{-(t-4)^2}{2}\right] dt = \Phi(-2) - \Phi(-3)$$

$$= 0.0228 - 0.0014 = 0.0214$$

$M(0) = 0$

$$M(1) = [1 + M(0)] \frac{1}{\sqrt{2\pi}} \int_0^1 \exp\left[\frac{-(t-4)^2}{2}\right] dt$$

$$= [1 + 0]0.0014 = 0.0014$$

$$M(2) = (1 + 0.0014)0.0014 + (1 + 0)(0.0214) = 0.0228 \text{ failures.}$$

7.2. Nonparametric Renewal Function Estimation

When it is difficult to determine the Laplace transform or its inverse for complex probability density functions or when the failure-time distribution is unknown but the mean and standard deviation are known, one may estimate the expected number of failures (or renewals) in interval $(0, t]$ when the time horizon is continuous or when the time interval is discrete as described below.

7.2.1. CONTINUOUS TIME

When the time horizon is continuous, one may use a general expression to determine the expected number of failures at time t. This expression is developed by Cox (1962), and its derivation is given below.

Consider the form of $M(t)$ as $t \to \infty$. Let us examine the behavior of $M^*(s)$ for small s. The Laplace transform of the p.d.f. of the failure time $f(t)$ is $f^*(s)$. From the properties of the Laplace transform, the mean (μ) and standard deviation (σ) of the failure time can be determined by using the following equations:

$$\left.\frac{df^*(s)}{ds}\right|_{s=0} = -\mu$$

$$\left.\frac{d^2f^*(s)}{ds^2}\right|_{s=0} = \sigma^2 + \mu^2, \qquad f^*(0) = 1.$$

From the above equations, we can express $f^*(s)$ as a Taylor series expansion around the point $s = 0$:

$$f^*(s) = 1 - s\mu + \frac{1}{2}s^2(\mu^2 + \sigma^2) + O(s^2), \tag{7.11}$$

where $O(s^2)$ denotes a function of s tending to zero as $s \to 0$ faster than s^2.

Substituting Eq. (7.11) into Eq. (7.8) we obtain

$$M^*(s) = \frac{1 - s\mu + \frac{1}{2}s^2(\mu^2 + \sigma^2) + O(s^2)}{s^2\mu - \frac{1}{2}s^3(\mu^2 + \sigma^2) + O(s^3)}.$$

The above equation can be simplified by using the partial-fraction-expansion formula given below:

$$g(s) = \frac{N(s)}{D(s)} = \frac{N(s)}{\prod_{i=1}^{n}(s + r_i)} = \frac{A_1}{s + r_1} + \frac{A_2}{s + r_2} + \ldots + \frac{A_n}{s + r_n},$$

where

$$A_i = \left[\frac{N(s)}{D(s)} \cdot (s + r_i)\right]_{s=-r_i}.$$

The above equation is valid when the roots of the $D(s)$ expression are all real and different. Clearly, the solution should be modified when we have repeated roots or some of the roots are imaginary.

The denominator of the $M^*(s)$ equation has two repeated roots $r_1 = r_2 = 0$ and a third real root r_3. Thus, we can rewrite $M^*(s)$ as

$$M^*(s) = \frac{A_1}{s^2 + 0} + \frac{A_2}{s + 0} + \frac{A_3}{s + r_3}.$$

The coefficients A_1, A_2, and A_3 can be obtained as follows:

$$A_1 = \left[\frac{(1 - s\mu + \frac{1}{2}s^2(\mu^2 + \sigma^2)) + O(s^2)}{s^2\mu - \frac{1}{2}s^3(\mu^2 + \sigma^2) + O(s^3)} \cdot s^2\right]_{s=0} = \frac{1}{\mu}.$$

Since the first two roots s_1 and s_2 are repeated, A_2 is obtained as

$$A_2 = \frac{d}{ds}\left[\frac{(1 - s\mu + \frac{1}{2}s^2(\mu^2 + \sigma^2)) + O(s^2)}{s^2\mu - \frac{1}{2}s^3(\mu^2 + \sigma^2) + O(s^3)} \cdot s^2\right]_{s=0} = \frac{-\mu^2 + \frac{1}{2}(\mu^2 + \sigma^2)}{\mu^2}$$

or

$$A_2 = \frac{(\sigma^2 - \mu^2)}{2\,\mu^2},$$

and the last term is a function of the order $O(1/s)$. This results in

$$M^*(s) = \frac{1}{s^2\mu} + \frac{1}{s}\frac{\sigma^2 - \mu^2}{2\mu^2} + O\left(\frac{1}{s}\right). \tag{7.12}$$

The inverse of Eq. (7.12) as $t \to \infty$ is

$$M(t) = \frac{t}{\mu} + \frac{\sigma^2 - \mu^2}{2\mu^2} + O(1). \tag{7.13}$$

The above equation holds true as σ^2 is finite. One should note that

- If $\sigma = \mu$, then $M(t) = t/\mu + O(1)$. For the exponential failure time $M(t) = t/\mu$ or λt.
- If $\sigma < \mu$, then $(\sigma^2 - \mu^2)/2\mu^2$ in Eq. (7.13) becomes negative, and if $\sigma << \mu$, then

$$M(t) \approx \frac{t - \frac{1}{2}\mu}{\mu} + O(1). \tag{7.14}$$

  This implies that to start with a new component rather than an *average* component is equivalent to saving one-half a failure (Cox, 1962).

- If $\sigma > \mu$, the second term in Eq. (7.13) becomes positive. This implies that when the coefficient of variation σ^2/μ^2 is greater than one, it is likely to have appreciable probability near zero failure-time and that to start with a new component is therefore worse than to start with an *average* component (Bartholomew, 1963).

The following two examples illustrate the use of the general equation to determine the expected number of failures (renewals) when the parameters of the failure time distribution are known and when the inverse of the Laplace transform is difficult to obtain.

EXAMPLE 7.7

Consider a component that fails according to a distribution with $\mu = 5$, $\sigma^2 = 1$. When the component fails it is immediately repaired and placed in service. Moreover, a preventive replacement is performed every 1,000 weeks. How many failures would have occurred before a preventive replacement is made?

SOLUTION

Using Eq. (7.13), we obtain the expected number of failures as

$$M(t) = \frac{t}{\mu} + \frac{\sigma^2 - \mu^2}{2\mu^2}$$

$$M(1{,}000) = \frac{1{,}000}{5} + \frac{1 - 25}{2 \times 25} = 199.5 \text{ failures.}$$

We now determine the expected number of failures for components whose failure time follows a Gamma distribution.

EXAMPLE 7.8

It is found that the failure-time distribution of a complex system with a large number of units each has an exponential failure time distribution that can be described by a Gamma distribution. The parameters of the distribution are $n = 50$, and $\lambda = 0.001$. Determine the expected number of failures after 10^5 hours of operation.

SOLUTION

The Gamma distribution has the following $f(t)$:

$$f(t) = \frac{\lambda(\lambda t)^{n-1} e^{-\lambda t}}{\Gamma(n)}$$

The Laplace transform of the expected number of failures is

$$M^*(s) = \frac{1}{s\left[\left(1 + \frac{s}{\lambda}\right)^n - 1\right]}.$$

The inverse of the Laplace transform of $M^*(s)$ is complex and difficult to obtain. Therefore, we utilize the general expression for the expected number of failures as shown below.

The mean is n/λ and the variance is n/λ^2. Thus, using Eq. (7.13), we obtain

$$M(10^5) = \frac{10^5}{n/\lambda} + \frac{(n/\lambda^2) - (n/\lambda)^2}{2(n/\lambda)^2}$$

$$= \frac{10}{5} - \frac{2{,}450}{5{,}000}$$

$$= 1.51 \text{ failures.}$$

Systems with Two Stages of Failure: Consider a system whose components fail if they enter either of two stages of failure mechanisms. The first mechanism is due to excessive voltage, and the second is due to excessive temperature. Suppose that the failure mechanism enters the first stage with probability θ, and the p.d.f. of the failure time is $\lambda_1 e^{-\lambda_1 t}$. It enters the second stage with probability $(1 - \theta)$, and the p.d.f. of its failure time is $\lambda_2 e^{-\lambda_2 t}$. The failure of a component occurs at the end of either stage. Hence, the p.d.f. of the failure time is

$$f(t) = \theta\lambda_1 e^{-\lambda_1 t} + (1 - \theta)\lambda_2 e^{-\lambda_2 t}.$$

The Laplace transform of $f(t)$ is

$$f^*(s) = \frac{\theta\lambda_1}{\lambda_1 + s} + \frac{(1 - \theta)\lambda_2}{\lambda_2 + s}$$

$$f^*(s) = \frac{\lambda_1\lambda_2 + \theta\lambda_1 s + (1 - \theta)\lambda_2 s}{(\lambda_1 + s)(\lambda_2 + s)},$$

but

$$M^*(s) = \frac{f^*(s)}{s[1 - f^*(s)]}.$$

Substitution of $f^*(s)$ into the above equation yields

$$M^*(s) = \frac{s(\theta\lambda_1 + (1 - \theta)\lambda_2) + \lambda_1\lambda_2}{s^2(s + (1 - \theta)\lambda_1 + \theta\lambda_2)}. \tag{7.15}$$

Equation (7.15) has double poles $r_1 = 0$ and $r_2 = 0$ and the root $r_3 = -\{(1 - \theta)\lambda_1 + \theta\lambda_2\}$. Thus, one can rewrite Eq. (7.15) as

$$M^*(s) = \frac{A_1}{s^2} + \frac{A_2}{s} + \frac{A_3}{s - r_3},$$

where A_1, A_2, and A_3 are obtained as follows:

$$A_1 = \left[\frac{s(\theta\lambda_1 + (1-\theta)\lambda_2) + \lambda_1\lambda_2}{s^2(s + (1-\theta)\lambda_1 + \theta\lambda_2)} \cdot s^2 \right]_{s=0} = \frac{\lambda_1\lambda_2}{(1-\theta)\lambda_1 + \theta\lambda_2}.$$

Since $f'^*(s = 0) = -\mu$, then

$$\frac{\lambda_1\lambda_2}{(1-\theta)\lambda_1 + \theta\lambda_2} = A_1 = \frac{1}{\mu}.$$

Similarly, A_2 can be obtained as

$$A_2 = \frac{\sigma^2 - \mu^2}{\mu^2}.$$

Finally, A_3 is obtained as

$$A_3 = \left[\frac{s(\theta\lambda_1 + (1-\theta)\lambda_2) + \lambda_1\lambda_2}{s^2(s + (1-\theta)\lambda_1 + \theta\lambda_2)} \cdot (s + (1-\theta)\lambda_1 + \theta\lambda_2) \right]_{s\,=\,-[(1-\theta)\lambda_1 + \theta\lambda_2]}$$

$$A_3 = \frac{-\theta(\lambda_1 - \lambda_2)^2 + \theta^2(\lambda_1 - \lambda_2)^2}{[(1-\theta)\lambda_1 + \theta\lambda_2]^2}.$$

Using the above $A_i\ i = 1, 2,$ and 3 in Eq. (7.15), $M^*(s)$ becomes

$$M^*(s) = \frac{1}{s^2\mu} + \frac{1}{s}\frac{\sigma^2 - \mu^2}{2\mu^2} - \frac{\theta(1-\theta)(\lambda_1 - \lambda_2^2)}{\{(1-\theta)\lambda_1 + \theta\,\lambda_2\}^2\ (s\ -\ r_3)}.$$

The inverse of the above equation is

$$M(t) = \frac{t}{\mu} + \frac{\sigma^2 - \mu^2}{2\mu^2} - \frac{\theta(1-\theta)(\lambda_1 - \lambda_2)^2}{\{(1-\theta)\lambda_1 + \theta\lambda_2\}^2} \exp\left[-\{(1-\theta)\lambda_1 + \theta\lambda_2\}t\right]. \tag{7.16}$$

EXAMPLE 7.9

The laser diodes (LD) used in a submarine optical fiber transmission system are composed of many kinds of circuit elements such as lenses, dielectric multilayer thin-film filters, and optical fibers as well as their holders, which are comprised of various materials such as metals, ceramics, and organic matter. The failure of the system is caused by the failure of the individual components as well as the thermal stresses in their joints. Suppose that the dielectric filters fail due to thermal stresses with a probability of 0.2 and the failure time follows a p.d.f. $\lambda_1 e^{-\lambda_1 t}$ with $\lambda_1 = 0.00001$. It may also fail due to voltage stresses with a probability of 0.8, and the failure-time distribution follows a p.d.f. $\lambda_2 e^{-\lambda_2 t}$ with $\lambda_2 = 0.00006$. The filter fails when either of the two events occur. Determine the expected number of failures during the first year of operation.

SOLUTION

The parameters of the system are

$$\theta = 0.20$$

$$1 - \theta = 0.80$$

$$\lambda_1 = 0.00001$$

$$\lambda_2 = 0.00006$$

$$\mu = \frac{(1-\theta)\lambda_1 + \theta\lambda_2}{\lambda_1\lambda_2} = 3.3333 \times 10^4$$

$$\sigma^2 = \frac{(2-\theta)\theta\lambda_2^2 + (1-\theta^2)\lambda_1^2 - 2\theta(1-\theta)\lambda_1\lambda_2}{\lambda_1^2\lambda_2^2} = 3.3333 \times 10^9.$$

At time $t = 10^4$, substituting the above parameters into Eq. (7.16), we obtain

$$M(10^4) = \frac{10^4}{3.3333 \times 10^4} + \frac{33.3333 - 11.1111}{22.2222} - e^{-0.20}$$

$$M(10^4) = 0.3000 + 1.0000 - 0.8187 = .4812 \text{ failures.}$$

7.2.2. DISCRETE TIME

The asymptotic result of Eq. (7.13) is very useful when large values of t are used. The meaning of *large* depends on the distribution of the failure times and the accuracy required, but the approximation is not usually acceptable unless at least, say, $t \geqq 2\mu$. The smaller values of t necessitate initially fitting some p.d.f., $f(t)$, to the data. However, the choice of the most appropriate p.d.f. is not always easy, especially when some of the data are censored or when the sample size is small. Therefore, it is more appropriate to consider a nonparametric approach for determining $M(t)$. The development of a nonparametric approach for $M(t)$ for discrete time intervals is now discussed.

Let $X_1, X_2, \ldots, X_n$ be independent and identically distributed (i.i.d.) nonnegative random variates, having a p.d.f. $f(x)$ with mean μ and variance σ^2. If $S_k = \Sigma_{i=1}^{k} X_i$, the renewal function may be written as

$$M(t) = \sum_{k=1}^{\infty} P(S_k \leqq t). \tag{7.17}$$

Frees's (1986b) estimator of $M(t)$ is

$$\hat{M}_p(t) = \sum_{k=1}^{p} \hat{F}_n^{(k)}(t), \tag{7.18}$$

where p is the cut-off, and $\hat{F}_n^{(k)}(t)$ is the distribution function of the permutation distribution of the sum of any k of the random variates X_i. $\hat{F}_n^{(k)}(t)$ can be written in terms of an indicator function I as

$$\hat{F}_n^{(k)}(t) = (C_k^n)^{-1} \sum_i I(X_{i1} + \ldots + X_{ik} \leqq t), \tag{7.19}$$

where the sum is over all C_k^n choices of k out of the n values of X_i, and $I(A)$ is 1 if event A occurs, and 0 otherwise. The choice of p is somewhat arbitrary, typically 5 or 10 (Baker, 1993). The following example illustrates the use of Eqs. (7.18) and (7.19) to obtain the expected number of failures in an interval $(0, T]$.

EXAMPLE 7.10

This example illustrates the use of the nonparametric discrete time approach to determine the renewal function. We use some of the data in Juran and Gryna (1993), which can also be found in Kolb and Ross (1980). Use the nonparametric expression of Eq. (7.13) to estimate $M(t)$ at $t = 20$ hours and 100 hours. The data are given in Table 7.1.

SOLUTION

The sample mean is $\bar{X} = 55.603$ hours and the sample standard deviation is $s = 43.926$ hours. Using Eq. (7.13), we obtain

$$M(20) = 0.17173 \text{ and } M(100) = 1.6105.$$

Table 7.1. Failure Data of Electronic Ground Support

1.0	1.2	1.3	2.0	2.4	2.9	3.0	3.1	3.3	3.5
3.8	4.3	4.6	4.7	4.8	5.2	5.4	5.9	6.4	6.8
6.9	7.2	7.9	8.3	8.7	9.2	9.8	10.2	10.4	11.9
13.8	14.4	15.6	16.2	17.0	17.5	19.2	28.1	28.2	29.0
29.9	30.6	32.4	33.0	35.3	36.1	40.1	42.8	43.7	44.5
50.4	51.2	52.0	53.3	54.2	55.6	56.4	58.3	60.2	63.7
64.6	65.3	66.2	70.1	71.0	75.1	75.6	78.4	79.2	84.1
86.0	87.9	88.4	89.9	90.8	91.1	91.5	92.1	97.9	100.8
102.6	103.2	104.0	104.3	105.0	105.8	106.5	110.7	112.6	113.5
114.8	115.1	117.4	118.3	119.7	120.6	121.0	122.9	123.3	124.5
125.8	126.6	127.7	128.4	129.2					

Source: The data are adapted from Juran and Gryna (1993).

Using Frees's estimator for $M(t)$,

$$M_p(t) = \sum_{k=1}^{p} F_n^{(k)}(t),$$

where $F_n^{(k)}(t)$ is defined by Eq. (7.19).

The values of $\Sigma_{k=1}^{p} F_{105}^{(k)}(t)$ for $t = 20$ and 100, and $p = 1, 2, \ldots, 8$ appear in Table 7.2. These values are based on Frees (1993). We now show how the values in column 2 for $p = 1$ and $p = 2$ are obtained.

Table 7.2. Calculations of $\sum_{k=1}^{p} F_{105}^{(k)}(t)$

p	$\sum_{k=1}^{p} F_{105}^{(k)}(20)$	$\sum_{k=1}^{p} F_{105}^{(k)}(100)$
1	0.35238	0.75238
2	0.44286	1.1775
3	0.45970	1.3729
4	0.46178	1.4514
5	0.46194	1.4798
6	0.46194	1.4890
7	0.46194	1.4917
8	0.46194	1.4924

For $p = 1$

For this case the $\Sigma_i I(X_{i1} \leqq 20)$ is the total number of observations whose individual values are $\leqq 20$. Thus,

$$F_{105}^{(1)}(20) = \frac{1}{C_1^{105}} \sum_i I(X_{i1} \leqq 20) = \frac{37}{105} = 0.35238.$$

Therefore, $M_1(20) = 0.35238$.

For $p = 2$

In this case the $\Sigma_i I(X_{i1} + X_{i2} \leqq 20)$ is the total number of cases where the sum of any two observations is $\leqq 20$. Thus,

$$F_{105}^{(2)}(20) = \frac{1}{C_2^{105}} \sum_i I(X_{i1} + X_{i2} \leqq 20) = \frac{494 \times 2}{104 \times 105} = 0.09048.$$

Therefore, $M_2(20) = 0.35238 + 0.09048 = 0.44286$.

From Table 7.2, using the recommended value of $p = 5$, the expected number of failures for times 20 and 100 hours are 0.46194 and 1.4798, respectively.

Baker (1993) develops a discretization approach to calculate $\hat{M}_n(t)$ by scaling up the X_i's and approximating them by integers. The p.d.f. of the permutation distribution of the sum of any k of the X_i—whose distribution function is $\hat{F}_n^{(k)}(t)$—is represented as a histogram, and the kth histogram has C_k^n partial sums of k of the X_i contributing to it. The set of n histograms is built up by adding the X_i successively. The error in estimating $\hat{M}(t)$ decreases as the number N of histogram intervals increases. The choice of N can be accomplished by running simulation experiments and choosing N that reduces the discretization error to an acceptably low value without increasing the computational difficulty. The following algorithm is developed by Baker (1993):

1. Sort the $X_i s$ into ascending order—that is, work with the order statistics $X_{(i)}$.
2. Find the sums of the first p order-statistics and hence the largest value of p such that $\Sigma_{i=1}^{p} X_{(i)} < t_0$, where t_0 is the largest value of t for which M must be estimated.
3. Zero the p histograms, and for i from 1 to n, add each $X_{(i)}$ in turn. All translations that add to array elements $> N$ are discarded.
4. Normalize each of the p histograms to unity by dividing the kth histogram by C_k^n.
5. Add the p calculated histograms (distribution functions) together to give $\hat{M}(t)$ for each of the N times, jh, where h is the step size, and $j = 1, 2, \ldots, N$.
6. Make a continuity correction to each value of $\hat{M}(t)$ by averaging it with the corresponding value for the previous time-point.

EXAMPLE 7.11 Use Baker's algorithm to determine $M(20)$ and $M(100)$ for the data given in Example 7.10.

SOLUTION The algorithm was coded by Baker in a computer program listed in Appendix K. The results obtained from the program are listed in Table 7.3. Note that we only listed those results in the neighborhood of $t = 20$ and $t = 100$. The results obtained from the program are $M(20) = 0.462972$ and $M(100) = 1.493999$, which approximately equal those obtained from Free's estimator.

Table 7.3. Expected Number of Failures Using Baker's Approach

Time	*Expected Number of Failures*
********	********
********	********
16.799999	0.415758
17.849999	0.432715
18.899999	0.449924
19.949999	0.462617
20.999999	0.470082
********	********
********	********
96.599996	1.443323
97.649996	1.460376
98.699996	1.477292
99.749995	1.489561
100.799995	1.506470

EXAMPLE 7.12 Fifty n-channel MOS transistor arrays are subjected to a voltage stress of 27V and 25°C to investigate the time-dependent dielectric breakdown (TDDB) behavior of such transistors (Swartz, 1986). The times to failure in minutes are given in Table 7.4. Use Baker's algorithm to determine the expected number of failures starting from $t = 100$ to $t = 150$ min. Compare these estimates with those obtained using Eq. (7.13).

Table 7.4. Failure-Time Data of MOS Transistor Arrays

1.0	60.0	73.0	74.0	75.0	90.0	101.0	103.0	113.0	117.0
131.0	132.0	135.0	148.0	149.0	150.0	152.0	153.0	155.0	159.0
160.0	161.0	163.0	167.0	171.0	176.0	182.0	185.0	186.0	194.0
197.0	211.0	214.0	215.0	220.0	233.0	235.0	236.0	237.0	241.0
252.0	268.0	278.0	279.0	292.0	307.0	344.0	379.0	445.0	465.0

SOLUTION

We use Baker's algorithm and Eq. (7.13) to obtain the results shown in Table 7.5. The first column in the table is the time, the second is the estimate of $M(t)$ using Baker's algorithm, and the third column is the estimate of $M(t)$ obtained using Eq. (7.13). As shown from Table 7.5, the estimates obtained using Eq. (7.13) are higher than those obtained by Baker's algorithm and that the difference between the two approaches increases with time.

Table 7.5. Estimates of $M(t)$

Time	*Baker's Estimate*	*Eq. (7.13) Estimate*
100	0.124082	0.138927
102	0.134082	0.149383
104	0.154490	0.159839
106	0.165306	0.170295
108	0.165714	0.180751
110	0.165714	0.191207
112	0.165714	0.201663
114	0.175714	0.212119
116	0.186122	0.222574
118	0.196531	0.233030
120	0.206939	0.243486
122	0.207347	0.253942
124	0.207347	0.264398
126	0.207347	0.274854
128	0.207347	0.285310
130	0.207347	0.295766
132	0.227347	0.306222
134	0.248980	0.316677
136	0.261071	0.327133
138	0.271964	0.337589
140	0.272398	0.348045
142	0.272398	0.358501
144	0.272398	0.368957
146	0.272398	0.379413
148	0.282806	0.389869
150	0.314872	0.400324

7.3. Alternating Renewal Process

Suppose a machine breaks down and is repaired as exhibited in Figure 7.2. The breakdown (or failure) of the machine and the repair of the failure are two processes that do not occur simultaneously but alternate with time. This process is called an *alternating renewal process*. Similarly, suppose a component of a system can be replaced upon failure by either a type A component or a type B component. If the replacement is done in such a way that when a type A component fails it is replaced by a type B component and vice versa, then we have an alternating renewal process. It should be clear that the alternating renewal process is not limited to two types of replacements. In fact, the above example can be generalized to k types of components following one another in a strict cyclic order, or it can be generalized by having a probability transition matrix with element p_{ij} specifying the probability that a type i component is replaced upon failure by a type j component. Such a system is called a semi-Markov process (Cox, 1962).

7.3.1. EXPECTED NUMBER OF FAILURES IN AN ALTERNATING RENEWAL PROCESS

Consider an alternating renewal process as depicted in Figure 7.2. Instead of T_i and D_i, which represent uptime and downtime of the machine, we consider the case when a type A component is replaced by a type B component upon failure and vice versa. Let $X_i = X_{Ai} + X_{Bi}$, $i = 1, 2, \ldots$ be the random variables that represent the renewal process, with X_{Ai} and X_{Bi} as the sequence of times during which the machine is up when type A component and type B component are used, respectively. Thus, if type A component was used at time $t = 0$, the first breakdown occurs at time X_{A1}, the second breakdown occurs at time $X_{A1} + X_{B1}$, and so on. Also, the first breakdown when type B component is in use occurs at time $X_{A1} + X_{B1}$, and the second failure when type B component is in use occurs at time $X_{A1} + X_{B1} + X_{A2} + X_{B2}$, and so on. Therefore, we can apply the results in Section 7.1.1 taking the distribution of failure time as the convolution of $f_A(x)$ and $f_B(x)$, with Laplace transform $f_A^*(s)f_B^*(s)$. The expected number of type B component failures in the interval $(0, t]$, $M_B(t)$, is obtained as a result of the Laplace inverse of

$$M_B^*(s) = \frac{f_A^*(s)f_B^*(s)}{s[1 - f_A^*(s)f_B^*(s)]}. \tag{7.20}$$

The expected number of type A component failures is obtained by modifying the renewal process such that the p.d.f. of the first failure time is $f_A(x)$ and the p.d.f. of the subsequent failure times is the convolution of $f_A(x)$ and $f_B(x)$ (Cox, 1962). Hence,

$$M_A^*(s) = \frac{f_A^*(s)}{s[1 - f_A^*(s)f_B^*(s)]}. \tag{7.21}$$

The renewal densities corresponding to $M_A^*(s)$ and $M_B^*(s)$ are

$$m_j^*(s) = sM_j^*(s), \qquad j = A, B. \tag{7.22}$$

7.3.2. PROBABILITY THAT TYPE j COMPONENT IS IN USE AT TIME t

One of the important criterion of component performance is the probability that it is in use at a specified time. For example, one may be interested in determining the probability $P_A(t)$ that type A component is in use at time t (when the machine is observed). This probability is obtained as the sum of the probabilities of two mutually exclusive events: in the first event the initial type A component has a failure time greater than t, and in the second event type B component fails in the time interval $(u, u + \delta u)$, for some time $u < t$, and is replaced by a type A component that does not fail during the interval $t - u$. Thus,

$$P_A(t) = R_A(t) + \int_0^t m_B(u) R_A(t-u)\,du, \tag{7.23}$$

where $R_A(t)$ is the reliability of component A at time t. By taking the Laplace transform of Eq. (7.23), we obtain

$$P_A^*(s) = [1 - f_A^*(s)][1 + m_B^*(s)]/s. \tag{7.24}$$

Substituting Eq. (7.20) into the above equation, we have

$$P_A^*(s) = \frac{1 - f_A^*(s)}{s[1 - f_A^*(s) f_B^*(s)]}. \tag{7.25}$$

Equation (7.25) implies that $P_A^*(s) = M_B^*(s) - M_A^*(s) + 1/s$. Thus,

$$P_A(t) = M_B(t) - M_A(t) + 1. \tag{7.26}$$

EXAMPLE 7.13

Microcasting is a droplet-based deposition process. The droplets of the molten material to be cast are relatively large (1 to 3 mm in diameter). They contain sufficient heat to remain significantly superheated until impacting the substrate and rapidly solidify due to significantly low substrate temperatures. By controlling the superheat of the droplets and the substrate temperature, conditions can be attained such that the impacting droplets superficially remelt the underlying material, leading to metallurgical interlayer bonding (Merz, Prinz, and Weiss, 1994).

The apparatus used for microcasting usually fails due to the clogging of the nozzle that controls the size of the droplets. Therefore, the manufacturer of such an apparatus includes two nozzles, A and B, which are alternatively changed. Nozzle A is made of material with a higher melting temperature than that of nozzle B. The failure times of nozzles A and B follow exponential distributions with parameters 1×10^{-5} and 0.5×10^{-5} failures per hour, respectively. What is the probability that nozzle A is in use at $t = 10^4$ hours?

SOLUTION The probability density functions of nozzles A and B are

$$f_A(t) = \lambda_A e^{-\lambda_A t} \quad \text{and} \quad f_B(t) = \lambda_B e^{-\lambda_B t}.$$

Using Eqs. (7.20) and (7.21) we obtain the expected number of failures as

$$M_B(t) = \frac{\lambda_A \lambda_B}{\lambda_A + \lambda_B} t + \left(-\frac{\lambda_A \lambda_B}{(\lambda_A + \lambda_B)^2}\right) + \frac{\lambda_A \lambda_B}{(\lambda_A + \lambda_B)^2} e^{-(\lambda_A + \lambda_B)t}.$$

$$M_A(t) = \frac{\lambda_A \lambda_B}{\lambda_A + \lambda_B} t + \frac{\lambda_A^2}{(\lambda_A + \lambda_B)^2} - \frac{\lambda_A^2}{(\lambda_A + \lambda_B)^2} e^{-(\lambda_A + \lambda_B)t}.$$

Substituting the above expressions into Eq. (7.26), we obtain

$$P_A(t) = M_B(t) - M_A(t) + 1.$$

$$\text{or } P_A(t) = \frac{\lambda_B}{\lambda_A + \lambda_B} + \frac{\lambda_A}{\lambda_A + \lambda_B} e^{-(\lambda_A + \lambda_B)t}.$$

Thus, the probability that nozzle A is in use at $t = 10^4$ hours is

$$P_A(10^4) = \frac{0.5}{1.5} + \frac{1}{1.5} e^{-0.15} = 0.907.$$

7.4. Approximations of $M(t)$

Estimating the expected number of renewals $M(t)$ using Eq. (7.3) is difficult since $M(t)$ appears on both sides of the equation. Therefore, researchers investigated approximate methods for the integral of Eq. (7.3) in order to obtain $M(t)$ by direct substitutions. In this section we summarize three approximations.

The first approximation is proposed by Bartholomew (1963) and is given by

$$M_b(t) = F(t) + \lambda \int_0^t [1 - F_e(t - x)]\, dx, \tag{7.27}$$

where

$$F_e(t) = \lambda \int_0^t [1 - F(x)]\, dx,$$

and $\lambda = 1/\mu$, μ is the expected value of the time between renewals.

The second approximation is proposed by Ozbaykal (1971), $M_o(t)$, which is given by

$$M_o(t) = \lambda t - F_e(t) + \int_0^t [1 - F_e(t - x)]\, dx. \tag{7.28}$$

The third approximation is proposed by Deligönül (1985), $M_d(t)$, and is derived as follows. The renewal function $M(t)$ is

$$M(t) = F(t) + \int_0^t M(t-x) f(x)\, dx.$$

The renewal density $m(t)$ is obtained as

$$m(t) = \frac{dM(t)}{dt}. \tag{7.29}$$

An equivalent equation to the renewal density can be written as

$$m(t) = f(t) + \int_0^t m(t-x) f(x)\, dx. \tag{7.30}$$

Karlin and Taylor (1975) provide an alternative expression of the renewal function $M(t)$ as

$$M(t) = \lambda t - F_e(t) + \int_0^t [1 - F_e(t-x)]\, dM(x). \tag{7.31}$$

As defined earlier, $F_e(t) = \lambda \int_0^t [1 - F(x)]\, dx$ and $\lambda = 1/\mu$ where μ is the expected value of the time between renewals.

Since $M(t)$ satisfies the renewal equation, it also satisfies the equation

$$F(x) = \int_0^x m(x-t)[1 - F(x)]\, dx. \tag{7.32}$$

Combining Eqs. (7.29) through (7.32) yields

$$M(t) = \lambda t - F_e(t) + \int_0^t [1 - F_e(t-x)] \left[\frac{F(x) \int_0^x m(x-t) f(t)\, dt}{\int_0^x m(x-t)[1 - F(t)]\, dt} + f(x) \right] dx \tag{7.33}$$

Deligönül (1985) approximates Eq. (7.33) by dropping out $m(x-t)$'s to obtain the following approximate estimate of $M_d(t)$:

$$M_d(t) = \lambda t - F_e(t) + \int_0^t [1 - F_e(t-x)] \left[f(x) + \frac{\lambda F^2(x)}{F_e(x)} \right] dx. \tag{7.34}$$

A comparison between $M_b(t)$ and $M_d(t)$ estimates for an increasing hazard-rate gamma distribution of the form $F(t) = 1 - (1 + 2t)e^{-2t}$ with mean = 1 and var = 1/2 is shown in Table 7.6.

Table 7.6. A Comparison Between $M_b(t)$ and $M_d(t)$ for Gamma Distribution (Mean = 1, Var = 1/2)

t	$M_b(t)$	$M_d(t)$	$M(t)$, exact
0.1	0.0176	0.0176	0.0176
0.2	0.0626	0.0626	0.0623
0.3	0.1264	0.1263	0.1253
0.4	0.2031	0.2029	0.2005
0.5	0.2888	0.2882	0.2838
0.6	0.3807	0.3794	0.3727
0.7	0.4768	0.4746	0.4652
0.8	0.5758	0.5723	0.5602
0.9	0.6766	0.6717	0.6568
1.0	0.7787	0.7718	0.7546
1.5	1.2946	1.2745	1.2506
2.0	1.8074	1.7720	1.7501
2.5	2.3148	2.6624	2.2500
3.0	2.8185	2.7605	2.7500
3.5	3.3202	3.2562	3.2500
4.0	3.8210	3.7534	3.7500
5.0	4.8215	4.7509	4.7500
6.0	5.8215	5.7502	5.7500
7.0	6.8215	6.7500	6.7500
8.0	7.8215	7.7500	7.7500
9.0	8.8215	8.7500	8.7500
10.0	9.8215	9.7500	9.7500
11.0	10.8215	10.7500	10.7500
12.0	11.8215	11.7500	11.7500
13.0	12.8215	12.7500	12.7500
14.0	13.8215	13.7500	13.7500
15.0	14.8215	14.7500	14.7500

Note: $F(t) = 1 - (1 + 2t)e^{-2t}$ (Deligönül, 1985).

Before we present other important characteristics of the expected number of renewals such as the variance, the confidence interval for $M(t)$, and the residual life, we briefly discuss other types of renewal processes.

7.5. Other Types of Renewal Processes

So far, we have only considered the case where the times to renewal (failure or failure and repair) are nonnegative identical and independent. We refer to this type of renewal processes as the *ordinary renewal process*. There are two slightly different renewal processes: the *modified renewal process* (or the *delayed renewal process*) and the *equilibrium renewal process*. In the modified renewal process the time to the first failure T_1, has a p.d.f. $f_1(t)$ and failure times between two successive failures beyond the first all have the same p.d.f. $f(x)$. In other words, the conditions for the modified renewal process are the same as those of the ordinary renewal process, except that the time from the origin to the first failure has a different distribution from the other failure times (Cox, 1962).

The equilibrium renewal process is a special case of the modified renewal process where the time to the first failure has the p.d.f. $R(t)/\mu$, where $R(t)$ is the reliability function at time t and μ is the mean failure time.

We now give typical examples of these three types of renewal processes. The ordinary renewal process is exemplified by replacements of an electric light bulb upon failure, the air filter and the brake pads of an automobile, and the spark plugs of an engine. The modified renewal process arises when the life of the original part or component is significantly different from that of its replacements. For example, when a customer acquires a new vehicle, the oil and air filters are usually replaced after 1,000 miles, whereas subsequent replacements of the oil filters occur at approximately equal intervals of 3,000 miles. The equilibrium renewal process can be regarded as an ordinary renewal process in which the system or component has been operating for a long time before it is first observed.

It should be noted that the renewal density for the modified renewal process is similar to that of the ordinary renewal process given by Eq. (7.5). It is expressed as

$$m_m(t) = f_1(t) + \int_0^t m_m(t-x) f(x)\, dx, \tag{7.35}$$

where $m_m(t)$ is the renewal density of the modified renewal process and $f_1(t)$ is the p.d.f. of the time to the first failure (or renewal).

Finally, the expected number of renewals of the equilibrium renewal process is

$$M_e(t) = \frac{t}{\mu}, \tag{7.36}$$

where μ is the mean failure (renewal) time.

7.6. The Variance of the Number of Renewals

As shown later in Chapters 8 and 9, the warranty cost, the length of a warranty policy, and the optimal maintenance schedule for a component (replacement or repair) are dependent on the expected number of failures (or renewals) during the warranty period and length of the maintenance schedule. Moreover, the variance of the number of renewals has a more significant impact on the choice of the appropriate warranty policy. Indeed, when two warranty policies have the same expected warranty cost for the same warranty length, the variance of the warranty cost would be the deciding factor in preferring one policy to another. Hence, it is important to determine the variance of the number of renewals Var$[N(t)]$ in the interval $(0, t]$.

From the definition of the variance, we obtain

$$\mathrm{Var}[N(t)] = E[N^2(t)] - E[N(t)]^2. \tag{7.37}$$

But

$$E[N(t)] = M(t) = \sum_{r=0}^{\infty} rP[N(t) = r],$$

which is expressed in Eq. (7.2) as

$$E[N(t)] = M(t) = \sum_{r=1}^{\infty} F_r(t). \tag{7.38}$$

Similarly,

$$E[N^2(t)] = \sum_{r=0}^{\infty} r^2 P[N(t) = r]$$

$$= \sum_{r=0}^{\infty} r^2 [F_r(t) - F_{r+1}(t)]$$

or

$$E[N^2(t)] = \sum_{r=1}^{\infty} (2r - 1)F_r(t). \tag{7.39}$$

Substituting Eqs. (7.38) and (7.39) into Eq. (7.37) results in

$$\mathrm{Var}[N(t)] = \sum_{r=1}^{\infty} (2r - 1)F_r(t) - [M(t)]^2. \tag{7.40}$$

Equation (7.40) is computationally difficult to evaluate. Therefore we follow Cox's (1962) work and obtain a simpler algebraic form of Var$[N(t)]$ by using $\psi(t)$, which is defined as

$$\psi(t) = E[N(t)(N(t) + 1)]. \tag{7.41}$$

Equation (7.41) represents the sum of $E[N^2(t)] + E[N(t)]$. Thus, the variance of $N(t)$ can be expressed in terms of $\psi(t)$ as

$$\mathrm{Var}[N(t)] = \psi(t) - M(t) - M^2(t). \tag{7.42}$$

Equation (7.41) can be written as

$$\psi(t) = \sum_{r=0}^{\infty} r(r+1)P[N(t) = r]. \tag{7.43}$$

But

$$P[N(t) = r] = F_r(t) - F_{r+1}(t). \tag{7.44}$$

Substituting Eq. (7.44) into Eq. (7.43) results in

$$\psi(t) = \sum_{r=0}^{\infty} r(r+1)[F_r(t) - F_{r+1}(t)]. \tag{7.45}$$

Taking the Laplace transform of Eq. (7.45) yields

$$\psi^*(s) = \frac{1}{s}\sum_{r=0}^{\infty} r(r+1)[f_r^*(s) - f_{r+1}^*(s)]$$

or

$$\psi^*(s) = \frac{1}{s}[0 + 2F_1^*(s) - 2f_2^*(s) + 6f_2^*(s) - 6f_3^*(s) + 12f_3^*(s) - 12f_4^*(s) + \ldots]$$

$$\psi^*(s) = \frac{2}{s}\sum_{r=1}^{\infty} rf_r^*(s). \tag{7.46}$$

For an ordinary renewal process $f_r^*(s) = [f^*(s)]^r$. Thus,

$$\psi_o^*(s) = \frac{2f^*(s)}{s[1 - f^*(s)]^2}. \tag{7.47}$$

For an equilibrium renewal process $F_r^*(s) = [f^*(s)]^{r-1}[1 - f^*(s)]/s\mu$, or

$$\psi_e^*(s) = \frac{2}{s^2\mu[1 - f^*(s)]}. \tag{7.48}$$

Cox (1962) shows that there is a relationship between Eq. (7.48) and the renewal function of the ordinary renewal process as

$$\psi_e^*(s) = \frac{2}{s\mu}\left[M_o^*(s) + \frac{1}{s}\right] \tag{7.49}$$

or

$$\psi_e(t) = \frac{2}{\mu}\int_0^t M_o(x)\,dx + \frac{2t}{\mu}. \tag{7.50}$$

Substituting Eqs. (7.36) and (7.50) into Eq. (7.42), we obtain the variance of the number of renewals of the equilibrium process as

$$\text{Var}\,[N_e(t)] = \frac{2}{\mu}\int_0^t \left[M_o(x) - \frac{x}{\mu} + \frac{1}{2}\right] dx. \tag{7.51}$$

EXAMPLE 7.14

Most machine parts are subjected to fluctuating or cyclic loads that induce fluctuating or cyclic stresses that often result in failure by fatigue. Fatigue may be characterized by a progressive failure phenomenon that proceeds by the *initiation* and propagation of cracks to an unstable size. Thus, the time to failure can be represented by a two-stage process that can be modeled as a special two-stage Erlangian distribution with a parameter λ failures per hour.

1. Assuming an ordinary renewal process, graph $M_o(t)$ and $\text{Var}[N_o(t)]$ for different values of λ and t.
2. Repeat 1 under the equilibrium renewal process assumption.

SOLUTION

The probability density function of the special Erlangian distribution is

$$f(t) = \frac{t}{\lambda^2} e^{-t/\lambda}.$$

The Laplace transform of $f(t)$ is

$$f^*(s) = \frac{1}{(1 + s\lambda)^2}. \tag{7.52}$$

1. The expected number of renewals of the ordinary renewal process is obtained by substituting Eq. (7.52) into Eq. (7.6). Thus,

$$M_o^*(s) = \frac{f^*(s)}{s[1 - f^*(s)]}$$

$$M_o^*(s) = \frac{\dfrac{1}{(1 + s\lambda)^2}}{s\left[1 - \dfrac{1}{(1 + s\lambda)^2}\right]} = \frac{1/\lambda^2}{s^2(s + 2/\lambda)}. \tag{7.53}$$

To obtain the inverse of $M_o^*(s)$, we rewrite Eq. (7.53) as

$$M_o^*(s) = \frac{1}{2\lambda s^2} - \frac{1}{4s} + \frac{1}{4\left(s + \dfrac{2}{\lambda}\right)}. \tag{7.54}$$

Thus, the inverse is

$$M_o(t) = \frac{t}{2\lambda} - \frac{1}{4} + \frac{1}{4}e^{-2t/\lambda}. \tag{7.55}$$

To obtain the variance of the number of renewals of the ordinary renewal process, we utilize Eq. (7.42):

$$\mathrm{Var}[N_o(t)] = \psi_o(t) - M_o(t) - M_o^2(t). \tag{7.56}$$

We first estimate $\psi_o(t)$ by obtaining the inverse of $\psi_o^*(s)$:

$$\psi_o^*(s) = \frac{2f^*(s)}{s[1 - f^*(s)]^2}$$

or

$$\psi_o^*(s) = \frac{2\left[\dfrac{1}{(1 + s\lambda)^2}\right]}{s\left[1 - \dfrac{1}{(1 + s\lambda)^2}\right]} = \frac{2(1 + s\lambda)^2}{\lambda s^3(2 + s\lambda)^2}$$

and

$$\psi_o(t) = \frac{t^2}{4\lambda^2} + \frac{t}{2\lambda} - \frac{1}{8} + \frac{1}{8} e^{\frac{-2t}{\lambda}} - \frac{t}{4\lambda} e^{\frac{-2t}{\lambda}}. \tag{7.57}$$

Substituting Eqs. (7.55) and (7.57) into Eq. (7.56), we obtain

$$\text{Var}[N_o(t)] = \frac{t}{4\lambda} + \frac{1}{16} - \frac{t}{2\lambda} e^{\frac{-2t}{\lambda}} - \frac{1}{16} e^{\frac{-4t}{\lambda}}. \tag{7.58}$$

Figures 7.5 and 7.6 show the effect of λ and t on $M_o(t)$ and on $\text{Var}[N_o(t)]$, respectively.

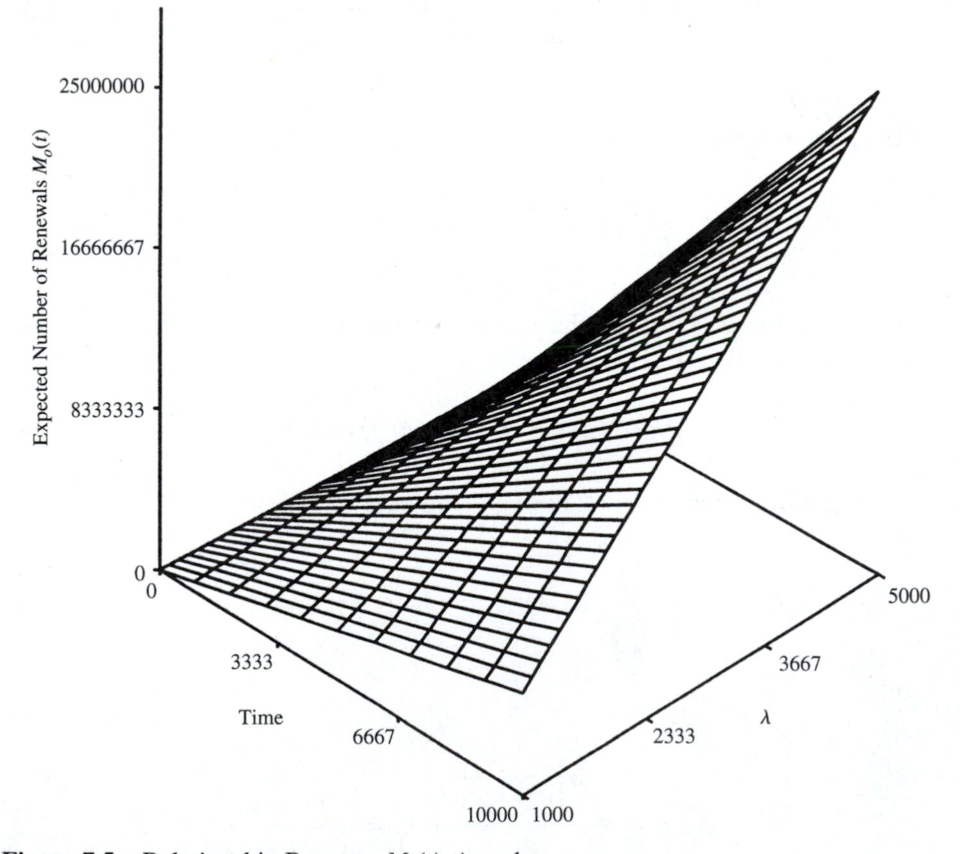

Figure 7.5. Relationship Between $M_o(t)$, λ, and t

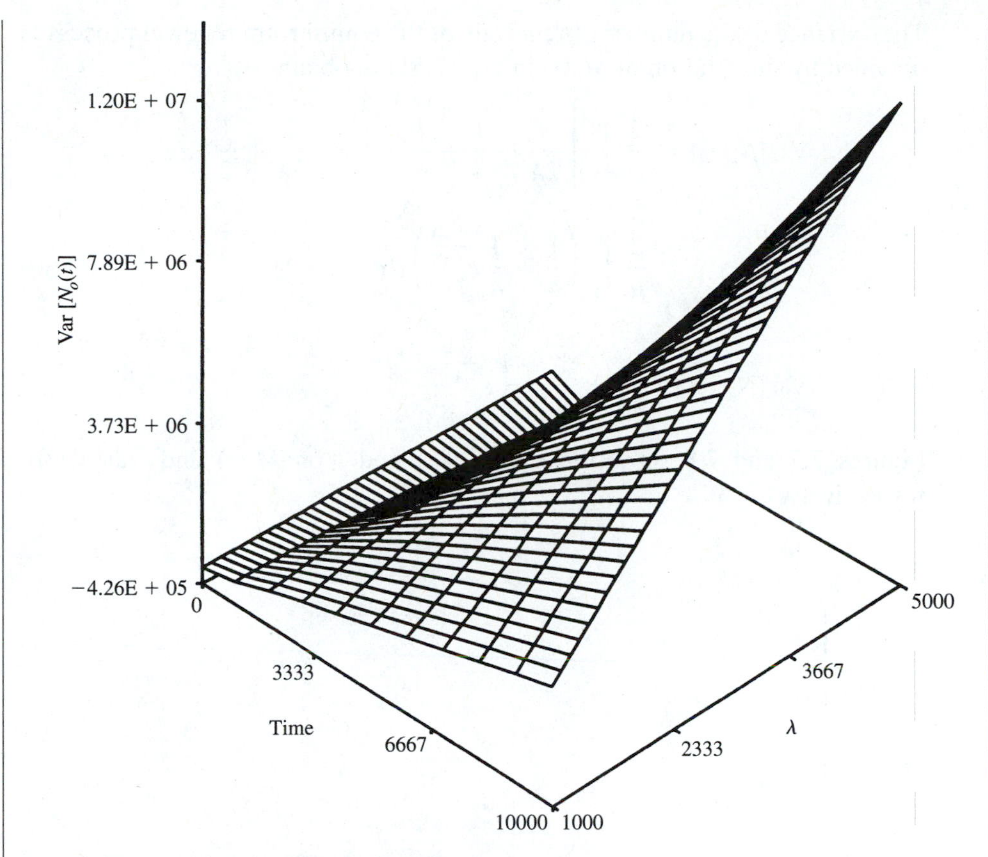

Figure 7.6. Effect of λ and t on Var$[N_o(t)]$

2. The expected number of renewals of the equilibrium renewal process is given by Eq. (7.36) as

$$M_e(t) = \frac{t}{\mu},$$

where μ of the special Erlang distribution is 2λ, thus

$$M_e(t) = \frac{t}{2\lambda}. \tag{7.59}$$

The variance of the number of renewals of the equilibrium renewal process is obtained by substitution of $M_o(t)$ in Eq. (7.51) to obtain

$$\begin{aligned}\operatorname{Var}[N_e(t)] &= \frac{2}{\mu}\int_0^t \left[\frac{x}{2\lambda} - \frac{1}{4} + \frac{1}{4}e^{\frac{-2x}{\lambda}} - \frac{x}{2\lambda} + \frac{1}{2}\right] dx \\ &= \frac{2}{\mu}\int_0^t \left(\frac{1}{4} + \frac{1}{4}e^{\frac{-2x}{\lambda}}\right) dx \qquad (7.60) \\ \operatorname{Var}[N_e(t)] &= \frac{t}{4\lambda} + \frac{1}{8} - \frac{1}{8}e^{\frac{-2t}{\lambda}}.\end{aligned}$$

Figures 7.7 and 7.8 show the effect of λ and t on $M_e(t)$ and $\operatorname{Var}[N_e(t)]$, respectively.

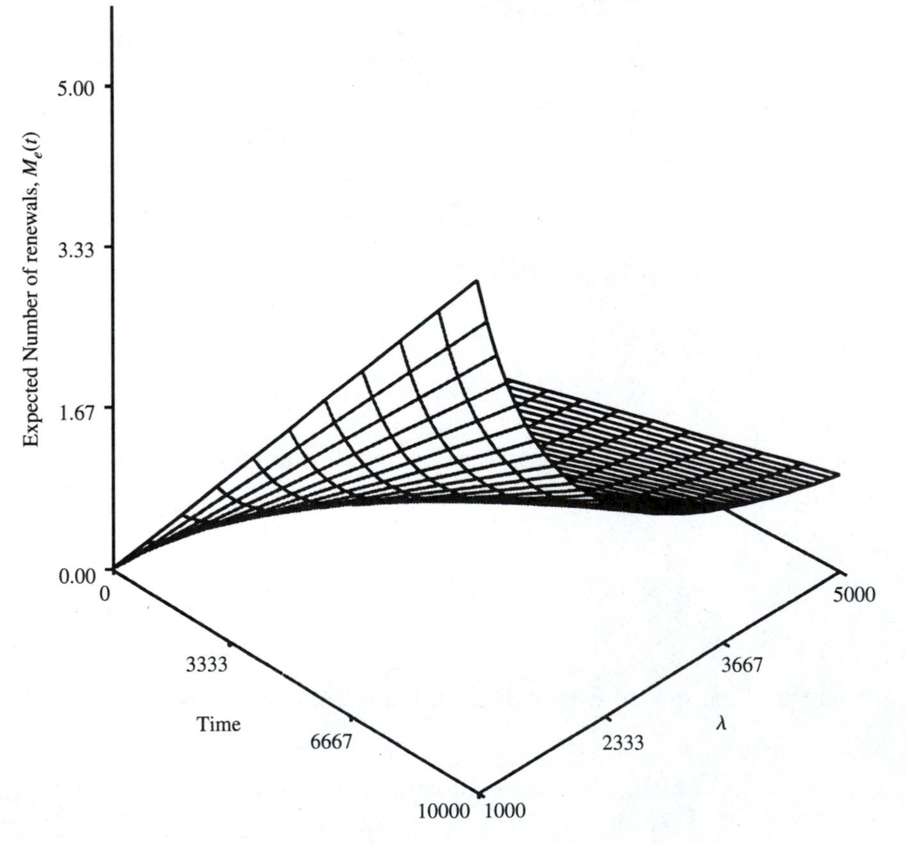

Figure 7.7. Effect of λ and t on $M_e(t)$

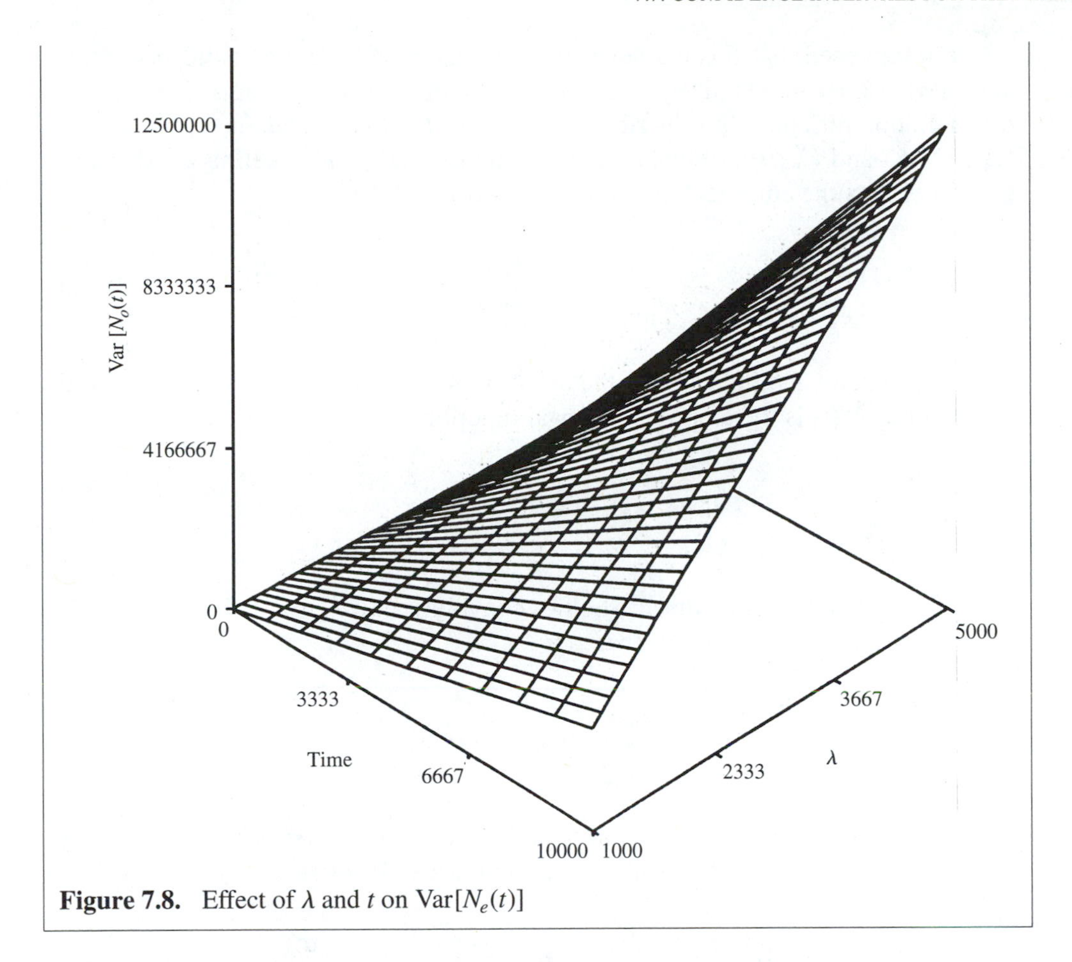

Figure 7.8. Effect of λ and t on $\mathrm{Var}[N_e(t)]$

7.7. Confidence Intervals for the Renewal Function

The point estimate of $M(t)$ was derived earlier in this chapter. Approximate confidence intervals may be calculated when the parameter estimates are asymptotically normally distributed. If the functional forms of the underlying distribution functions are unknown, a nonparametric approach is required. Frees (1986a, 1986b, 1988) presents nonparametric estimators of the renewal function and constructs a nonparametric confidence interval for $M(t)$.

In this section, we present an alternative nonparametric confidence interval, based on Baxter and Li (1994), for the renewal function, which is easier to compute than that of Frees (1986a) and which is appreciably narrower. The approach is based on the assumption that the empirical renewal function converges weakly to a Gaussian process as the sample size increases.

When $F(t)$ is known, the renewal function is given by Eq. (7.2). When $F(t)$ is unknown, we follow the same derivations given in Section 7.2.2 to obtain an alternative estimate of $M(t)$.

It is supposed that F is unknown and we wish to calculate a confidence interval for $M(t)$ for a fixed t given $x1, x2, \ldots, xn$, a random sample of n observations of a random variable with distribution function F. Baxter and Li (1994) utilize Eqs. (7.18) and (7.19) to obtain a nonparametric maximum likelihood estimator (MLE) of F as the empirical distribution function (EDF):

$$\hat{F}_n(t) = \frac{1}{n} \sum_{i=1}^{n} I_{\{xi \leq t\}}, \tag{7.61}$$

where I_A denotes the indicator of the event A (see Section 7.2.2). Thus, a natural estimator of $M(t)$ is the empirical renewal function

$$\hat{M}_n(t) = \sum_{k=1}^{\infty} \hat{F}_n^{(k)}(t), \tag{7.62}$$

where $\hat{F}_n^{(k)}$ is the k-fold recursive Stieltjes convolution of $\hat{F}_n$.

Baxter and Li (1994) prove that as $n \rightarrow \infty$,

$$\frac{\sqrt{n}}{\hat{\sigma}_n(t)} [\hat{M}_n(t) - M(t)]$$

converges in distribution to a standard normal variate. Hence, for $\alpha \in (0, 1)$, an approximate $100(1 - \alpha)$ percent confidence interval for $M(t)$ is

$$\hat{M}_n(t) - z_{\alpha/2} \frac{\hat{\sigma}_n(t)}{\sqrt{n}} \leq M(t) \leq \hat{M}_n(t) + z_{\alpha/2} \frac{\hat{\sigma}_n(t)}{\sqrt{n}}, \tag{7.63}$$

where $z_\alpha/2$ denotes the upper $\alpha/2$ quantile of the standard normal distribution. An alternative procedure for calculating $\hat{M}_n(t)$ rather than using Eq. (7.62), requires the partitioning of the interval $(0, t]$ into k subintervals of equal width—say, $0 = t_0 < t_1 < \ldots < t_k = t$—where the value of k depends on t and on the actual observations. $\hat{M}_n(t_i)$ $(i = 1, 2, \ldots, k)$ can then be recursively calculated as

$$\hat{M}_n(t_i) = \hat{F}_n(t_i) + \sum_{j=1}^{i} \hat{M}_n(t_i - t_j) [\hat{F}_n(t_j) - \hat{F}_n(t_{j-1})].$$

Clearly, if F is known, we utilize Eq. (7.4) or its approximations (Eqs. (7.27), (7.28), and (7.34)) for the ordinary renewal process or Eqs. (7.35) and (7.36) for the modified renewal and equilibrium renewal process respectively to estimate

$M(t)$. We then use Eq. (7.42) to obtain the corresponding estimate of the variance. Finally, assuming $n = 25$, we substitute these estimates in Eq. (7.63) to obtain the confidence interval for $M(t)$.

EXAMPLE 7.15

Determine the 95 percent confidence intervals for $M(t)$ of the ordinary renewal process and the equilibrium renewal process for $t = 100$ to 1,000 (increments of 100) and for $t = 2{,}000$ to 10,000 (increments of 1,000) for the machine parts given in Example 7.14. Assume $\lambda = 5 \times 10^3$ hours between failures.

SOLUTION

Substitute $\lambda = 5 \times 10^3$ in Eqs. (7.55), (7.58), (7.59), and (7.60) to obtain $M_o(t)$, $\text{Var}[N_o(t)]$, $M_e(t)$, and $\text{Var}[N_e(t)]$, respectively. The confidence intervals for $M_o(t)$ and $M_e(t)$ are obtained by using Eq. (7.63) and substituting $n = 30$. The results are shown in Table 7.7.

Table 7.7. $M_o(t)$, $M_e(t)$, $\text{Var}[N_o(t)]$, $\text{Var}[N_e(t)]$, and Bounds for $M(t)$

Time	$\text{Var}[N_o(t)]$	*Lower* $M_o(t)$	$M_o(t)$	*Upper* $M_o(t)$	$\text{Var}[N_e(t)]$	*Lower* $M_e(t)$	$M_e(t)$	*Upper* $M_e(t)$
100	0.0002	0.0001	0.0002	0.0003	0.0099	0.0065	0.0100	0.0135
200	0.0008	0.0005	0.0008	0.0011	0.0196	0.0130	0.0200	0.0270
300	0.0017	0.0011	0.0017	0.0023	0.0291	0.0196	0.0300	0.0404
400	0.0030	0.0020	0.0030	0.0041	0.0385	0.0262	0.0400	0.0538
500	0.0047	0.0030	0.0047	0.0064	0.0477	0.0329	0.0500	0.0671
600	0.0066	0.0043	0.0067	0.0090	0.0567	0.0397	0.0600	0.0803
700	0.0089	0.0058	0.0089	0.0121	0.0655	0.0466	0.0700	0.0934
800	0.0115	0.0074	0.0115	0.0156	0.0742	0.0534	0.0800	0.1066
900	0.0143	0.0093	0.0144	0.0195	0.0828	0.0604	0.0900	0.1196
1,000	0.0174	0.0114	0.0176	0.0238	0.0912	0.0674	0.1000	0.1326
2,000	0.0600	0.0409	0.0623	0.0838	0.1688	0.1396	0.2000	0.2604
3,000	0.1165	0.0836	0.1253	0.1670	0.2374	0.2151	0.3000	0.3849
4,000	0.1792	0.1364	0.2005	0.2646	0.2998	0.2927	0.4000	0.5073
5,000	0.2437	0.1966	0.2838	0.3710	0.3581	0.3719	0.5000	0.6281
6,000	0.3076	0.2626	0.3727	0.4827	0.4137	0.4520	0.6000	0.7480
7,000	0.3697	0.3329	0.4652	0.5975	0.4674	0.5327	0.7000	0.8673
8,000	0.4298	0.4064	0.5602	0.7140	0.5199	0.6140	0.8000	0.9860
9,000	0.4879	0.4823	0.6568	0.8314	0.5716	0.6955	0.9000	1.1045
10,000	0.5442	0.5599	0.7546	0.9493	0.6227	0.7772	1.0000	1.2228

7.8. Remaining Life at Time t

In Section 7.1.1, we defined the total time up to the rth failure as $S_r = t_1 + t_2 + \ldots t_r = \Sigma_{i=1}^{r} t_i$, where t_i is the interval between failures $i-1$ and i. In this section we are interested in estimating the time from t until the next renewal—that is, the excess life or remaining life at time t. Let $L(t)$ represent the remaining life at t—that is,

$$L(t) = S_{N(t)+1} - t. \tag{7.64}$$

The distribution function of $L(t)$ is given for $x \geq 0$ by

$$P(L(t) \leq x) = F(t + x) - \int_0^t [1 - F(t + x - y)] \, dM(y). \tag{7.65}$$

In general, it is difficult to solve Eq. (7.65) analytically. When $t \to \infty$, then

$$\lim_{t \to \infty} P(L(t) \leq x) = \frac{\int_0^x [1 - F(y)] \, dy}{\mu}, \tag{7.66}$$

where $\mu = E[X]$.

Consider a renewal process that has been "running" for a very long time but that is observed beginning at time $t = 0$. Let T_1 denote the time to the first renewal after time $t = 0$. Then T_1 is the remaining life of the unit that was in operation at time $t = 0$ (Hoyland and Rausand, 1994). From Eq. (7.66), the distribution of T_1 is

$$F_{T_1}(t) = \frac{1}{\mu} \int_0^t [1 - F_T(y)] \, dy. \tag{7.67}$$

When the age of the unit (or component) that is in operation at time $t = 0$ is greater than 0, we have a modified renewal process. The distribution of the remaining lifetime $L(t)$ becomes

$$P(L(t) \leq x) = F_{T_1}(t + x) - \int_0^t [1 - F_T(t + x - y)] \, dM_m(y), \tag{7.68}$$

where $M_m(y)$ is the renewal function of the modified renewal process.

The mean remaining lifetime in a stationary renewal process is

$$E[L(t)] = \int_0^\infty [1 - P(L(t)) \leq x] \, dx, \tag{7.69}$$

and

$$\lim_{t\to\infty} E[L(t)] = \frac{E[X^2]}{2\mu},$$

where X is the time between renewals.

EXAMPLE 7.16

Communications cables are drawn in cable duct plants (conduits), either by using a cable grip when the diameter of the cable beneath the sheath is less than 50 mm or by a drawing ring applied to the cable when the diameter is greater than 50 mm. A cable production facility has a drawing machine with two identical rings. The machine stops production only when the two rings fail. The failure time follows an exponential distribution with parameter $\lambda = 0.002$ failures per hour. Determine the mean remaining lifetime of the drawing machine.

SOLUTION

Since the drawing machine stops production only when the two rings fail, the machine reliability is therefore estimated as

$$R(t) = 2e^{-\lambda t} - e^{-2\lambda t}.$$

The mean time to failure, μ, is

$$\mu = \int_0^\infty R(t)\,dt = \frac{3}{2\lambda}.$$

Using Eq. (7.66), we obtain

$$P(L(t) \le x) = \frac{1}{\mu}\int_0^x R(t)\,dt$$

$$= \frac{2\lambda}{3}\left[\frac{-2}{\lambda}e^{-\lambda t} + \frac{2}{2\lambda}e^{-2\lambda t}\right]_0^x$$

or

$$P(L(t) \le x) = \frac{2\lambda}{3}\left[\frac{-2}{\lambda}e^{-\lambda x} + \frac{1}{2\lambda}e^{-2\lambda x} + \frac{3}{2\lambda}\right].$$

The mean remaining lifetime is

$$E[L(t)] = \int_0^{\infty} [1 - P(L(t) \le x)]\, dx$$

$$= \int_0^{\infty} \left(\frac{4}{3} e^{-\lambda x} - \frac{1}{3} e^{-2\lambda x}\right) dx$$

or

$$E[L(t)] = \frac{7}{6\lambda} = 583 \text{ hours.}$$

7.9. Poisson Processes

Two important point processes are commonly used in modeling repairable systems. A *repairable system* is a system that can be repaired when failures occur, such as cars, airplanes, and computers. A *nonrepairable system* is a system that is discarded or replaced upon failure, such as electronic chips and inexpensive calculators. The point processes to be discussed are the homogeneous Poisson process (HPP) and the nonhomogeneous Poisson process (NHPP).

7.9.1. HOMOGENEOUS POISSON PROCESS (HPP)

Before defining the homogeneous Poisson process (HPP), we introduce the counting process $N(t)$, $t \ge 0$. It represents the total number of events (such as failures and repairs) that have occurred up to time t. The counting process $N(t)$ must satisfy the following:

1. $N(t) \ge 0$,
2. $N(t)$ is integer valued,
3. If $t_1 < t_2$ then $N(t_1) \le N(t_2)$, and
4. The number of events that occur in the interval $[t_1, t_2]$ where $t_1 < t_2$ is $N(t_2) - N(t_1)$.

For an HPP, condition 4 is modified such that the number of events (failures) in the interval $[t_1, t_2]$ has a Poisson distribution with mean $\lambda(t_2 - t_1)$ where λ is the failure rate and as additional conditions $N(0)=0$ and the numbers of events in nonoverlapping intervals are independent—that is, the process has independent increments. Thus, for $t_2 > t_1 \ge 0$, the probability of having n failures in the interval $[t_1, t_2]$ is

$$P\{N(t_2) - N(t_1) = n\} = \frac{e^{-\lambda(t_2-t_1)}\{\lambda(t_2 - t_1)\}^n}{n!}$$

for $n \geq 0$.

It follows from condition 4 that a Poisson process has an expected number of failures (events) as

$$E[N(t_2 - t_1)] = \lambda(t_2 - t_1).$$

The Poisson process is referred to as homogeneous when λ is not time dependent—that is, the number of events in an interval depends only on the length of the interval (process has stationary increments). Hence, the reliability function $R(t_1, t_2)$, for the interval $[t_1, t_2]$ is

$$R(t_1, t_2) = e^{-\lambda(t_2-t_1)}.$$

7.9.2. NONHOMOGENEOUS POISSON PROCESS (NHPP)

The nonhomogeneous Poisson process (NHPP) is similar to the HPP with the exception that the failure rate (occurrence rate of the event) is time dependent. Thus, the process is nonstationary. In other words, we modify condition 4 as follows:

4. The number of events that occur in the interval $[t_1, t_2]$ where $t_1 < t_2$ has a Poisson distribution with mean $\int_{t_1}^{t_2} \lambda(t)\, dt$.

Therefore, the probability of having n failures in the interval $[t_1, t_2]$ is

$$P[N(t_2) - N(t_1) = n] = \frac{e^{-\int_{t_1}^{t_2} \lambda(t)\, dt}\left[\int_{t_1}^{t_2} \lambda(t)\, dt\right]^n}{n!},$$

and the expected number of failures in $[t_1, t_2]$ is

$$E[N(t_2) - N(t_1)] = \int_{t_1}^{t_2} \lambda(t)\, dt.$$

The reliability function of the NHPP for the interval $[t_1, t_2]$, is

$$R(t_1, t_2) = e^{-\int_{t_1}^{t_2} \lambda(t)\, dt}.$$

EXAMPLE 7.17

Determine $M(t)$ when

1. $F(t) = 1 - e^{-t/\mu_1}$
2. $f(t)$ is a gamma density of order k:

$$f(t) = \frac{\lambda(\lambda t)^{k-1}}{(k-1)!} e^{-\lambda t}.$$

SOLUTION

1. The distribution function

$$F(t) = 1 - e^{-t/\mu_1}$$

has a probability density function

$$f(t) = \frac{1}{\mu_1} e^{-t/\mu_1}$$

$$\text{Set } \lambda = \frac{1}{\mu_1}.$$

But

$$m^*(s) = \frac{f^*(s)}{1 - f^*(s)}$$

or

$$m^*(s) = \frac{\dfrac{\lambda}{s+\lambda}}{1 - \dfrac{\lambda}{s+\lambda}} = \frac{\lambda}{s}.$$

The inverse of $m^*(s)$ is

$$m(t) = \lambda$$

and

$$M(t) = \int_0^t \lambda \, dt = \lambda t$$

or

$$M(t) = \frac{t}{\mu_1}.$$

Hence, for the HPP, the expected number of renewals in an interval of length t is simply t divided by the mean life.

2. It is known that the p.d.f. of a gamma distribution of order k is the convolution of k exponentials with parameter λ. Therefore, the probability of n renewals in $(0, t]$ for a renewal process defined by $f(t)$ is equal to the probability of either $nk, nk + 1, \ldots$ or $nk + k - 1$ events occurring in $(0, t]$ for a Poisson process with parameter λ. Therefore,

$$P[N(t) = n] = \frac{(\lambda t)^{nk}}{(nk)!} e^{-\lambda t} + \frac{(\lambda t)^{nk+1}}{(nk+1)!} e^{-\lambda t} + \ldots$$

$$+ \frac{(\lambda t)^{nk+k-1}}{(nk+k-1)\,!} e^{-\lambda t}.$$

Let $m(t)$ be the renewal density for a gamma density of order k. For $k = 1$ (exponential density), $m(t) = \lambda$ as shown in part (1) of this example. Since $m(t)\,dt$ is the probability of a renewal in $[t, t + dt]$, we can interpret this probability for the gamma density of order k as (Barlow, Proschan, and Hunter, 1965)

$$m(t)\,dt = \sum_{j=1}^{\infty} \left[\frac{(\lambda t)^{kj-1}}{(kj-1)!} e^{-\lambda t} \right] \lambda\,dt. \tag{7.70}$$

The right side is the probability of $kj-1$ events occurring in $(0, t]$ from a Poisson process with parameter λ times the probability of an additional event occurring in $[t, t + dt]$ and summed over all permissible j.

When $k = 2$,

$$m(t) = \frac{\lambda}{2} - \frac{\lambda}{2} e^{-2\lambda t}$$

and

$$M(t) = \frac{\lambda t}{2} - \frac{1}{4} + \frac{1}{4} e^{-2\lambda t}.$$

The expected number of failures during an interval $(0, t]$ for a component whose failure time exhibits an Erlang distribution with k stages is obtained by integrating Eq. (7.67) to obtain (Parzen, 1962):

$$M(t) = \frac{\lambda t}{k} + \frac{1}{k} \sum_{j=1}^{k-1} \frac{\theta^j}{1 - \theta^j} [1 - e^{-\lambda t(1-\theta^j)}],$$

where

$$\theta = e^{(2\pi i/k)}$$

and

$$i = \sqrt{-1}.$$

Details of the above derivation are given in Barlow, Proschan, and Hunter (1965).

PROBLEMS

7-1. Assume $f(t) = 1/4, 0 \leq t \leq 4$ (uniform distribution). Determine the expected number of failurės if a replacement occurs every two weeks.

7-2. Capacitors are used in electrical circuits whenever radio interference needs to be suppressed. Most radio frequency interference (RFI) capacitors are made from either a metalized plastic film or metalized paper. Metalized paper capacitors often fail after short circuits. But the major advantage of metalized paper over metalized plastic capacitors is their superior self-healing capability under dry conditions. A manufacturer develops capacitors with different structures that result in better performance than both the metalized paper and the metalized plastic capacitors. The manufacturer subjects a capacitor to a transient voltage of 1.2kV and 10 pulses per day. The duration of each pulse is 10 μs. When a capacitor fails, it is repaired and immediately placed under the same test conditions. Assume that the failure time of the capacitor follows an Erlang distribution with a p.d.f. of

$$f(t) = \frac{t}{\lambda^2} e^{-t/\lambda},$$

where λ is 2×10^3 hours. The repair time is exponentially distributed with a repair rate of 4×10^2 repairs per hour. Determine the expected number of failures during one year of testing $(0, 10^4$ hours] and the availability of the capacitor at the end of the testing period.

7-3. Consider a component whose failure time exhibits a shifted exponential distribution as shown below:

$$F(t) = \begin{cases} 0, & t < \beta \\ 1 - e^{-\lambda(t-\beta)}, & t \geq \beta \end{cases}.$$

Determine the expected number of failures in the interval (0, t].

7-4. Given the following failure times, what are the expected number of failures during the periods (0, 20] and (0, 40]? Compare your results with those obtained using the asymptotic equation of the expected number of failures. Failure times in hours are

1.20, 3.5, 4.5, 6.0, 7.9, 12.8, 15.9, 17.9, 22.7, 26.9, 29.8, 30.5, 37.8, 39.0, 48.0, 58.0, 67.0, 75.0.

7-5. Use the following Laplace expression for the expected number of failures in the interval $(0, t]$—

$$M^*(s) = \frac{f^*(s)}{s[1 - f^*(s)]}$$

—to estimate the expected number of failures at time t for n components in operation beginning at time 0 when

a. The failure time distribution follows the special Erlang given by

$$f(t) = \frac{t}{\lambda^2} e^{\frac{-t}{\lambda}} \qquad \lambda > 0.$$

b. The failure time distribution follows the normal distribution given by

$$f(t) = \frac{1}{\sigma\sqrt{(2\pi)}} \exp\left[-(t-\mu)^2/2\sigma^2\right].$$

7-6. The following failure data are given:

1.0, 1.5, 2.0, 2.3, 2.5, 3.1, 3.7, 4.2, 4.8, 5.6, 5.9, 6.2, 6.7, 8.9, 10, 12, 15.2, 17.0, 18.9, 20.3, 21.5, 24.5, 26.8, 29.1, 34.6, 44.5, 47.8, 50, 52.7, 55.5, 59.3.

a. Use Frees' method to estimate the expected number of failures at time $t = 25$.

b. Use Baker's approach to obtain the expected number of failures at times $t =$ 25, 40, 50.

7-7. Electromagnetic (EM) sensors and actuators are replacing many of the mechanical components in automobiles. An example of such replacements is the antilock braking systems that replace traditional hydraulic components with EM sensors and actuators. As a result, an accurate estimate and prediction of the reliability of the EM components is of a high importance for the automobile's manufacturer. A producer of EM sensors subjects fifty units to an electric field and obtains the following failure times:

Failure Times × 100				
0.076196	0.480874	0.745838	1.08577	1.57504
0.145768	0.512149	0.774938	1.12684	1.62709
0.24849	0.547918	0.832483	1.12863	1.67457
0.268816	0.556499	0.863123	1.20560	1.68656
0.278996	0.599449	0.926084	1.20575	1.73760
0.292879	0.614937	0.926734	1.31209	1.80716
0.322036	0.633408	0.973047	1.40185	1.94672
0.371150	0.636191	0.988017	1.43525	2.08160
0.393230	0.680449	1.022900	1.44437	2.23592
0.462698	0.719642	1.057490	1.49011	2.40073

a. Determine the expected number of failures at $t = 200$ hours.

b. Solve a using Baker's approximation.

c. Fit the above data to a Weibull distribution and determine the expected number of failures at $t = 200$ hours.

d. Compare the results obtained from a, b, and c. What do you conclude?

7-8. Recent advances in semiconductor integration, motor performance, and reliability have resulted in the development of inexpensive electronics that have brushless DC (BLDC) motors. Unlike a brush-type motor, the brushless DC motor has a wound stator, a permanent-magnet rotor, and internal or external devices to sense rotor position. The sensing devices can be optical encoders or resolvers providing signals for electronically switching the stator windings in the proper sequence to maintain the rotation of the magnet. The elimination of brushes reduces maintenance due to arcing and dust, reduces noise, and increases life and reliability. A manufacturer of motors wishes to replace its product from brush-type to brushless motors if it is shown to be economically feasible. Assuming that the current facility and equipment can be used to produce either type of motor and that the cost of producing a brushless motor is $\$x$ higher than the cost of a brush-type motor but the warranty cost will decrease by $\$y$ per motor per year. The manufacturer's experience with brush-type motors reveals that their failure time distribution is given by a mixture of two exponential distributions

$$f(t) = \theta\lambda_1 e^{-\lambda_1 t} + (1 - \theta)\lambda_2 e^{-\lambda_2 t},$$

where $\theta = 0.2$, $\lambda_1 = 0.6 \times 10^{-4}$, and $\lambda_2 = 1.8 \times 10^{-4}$ failures per hour. On the other hand, the failure time of the BLDC motors can be expressed by a special Erlang distribution with the following p.d.f.:

$$f(t) = \frac{t}{\lambda^2} e^{-t/\lambda},$$

where

$$\lambda = 0.45 \times 10^4.$$

Determine the relationship between x and y that will make the production of the brushless motors feasible.

7-9. Solve Problem 7-8 using Frees's estimator and Baker's estimator. Compare the relationships obtained using these estimators with the relationship obtained from the exact solution of the renewal density function. What are your conclusions?

7-10. A telephone switching system uses two types of exchangeable modules A and B. When module A fails, it is instantaneously replaced by module B and module A undergoes repair. When B fails, it is instantaneously replaced by A and B undergoes repair. Assume that the repair time is significantly less than the time to failure and that the p.d.f. of the failure-time distributions for modules A and B are as follows. The p.d.f. for A is

$$f_A(t) = \frac{\beta t^{\beta-1}}{\lambda^\beta} \exp\left[-\left(\frac{t}{\lambda}\right)^\beta\right], \quad \text{where } t \geq 0,\ \beta,\ \lambda > 0.$$

The p.d.f. for B is

$$f_B(t) = \frac{t^{\beta-1}}{\lambda^\beta \, \Gamma(\beta)} \exp\left[-\frac{t}{\lambda}\right], \quad \text{where } t \geq 0,\ \beta,\ \lambda > 0.$$

It is found that $\lambda_A = 1{,}000$, $\beta_A = 3$, $\lambda_B = 2{,}000$, and $\beta_B = 2$. Note that λ_A and β_A are the parameters of module A, whereas λ_B and β_B are the parameters of module B. Determine the following:

a. Expected number of failures in the interval (0, 200 hours) for both modules A and B.

b. What is the probability that module A is functioning at $t = 200$ hours?

Also note that

$$f_A^*(s) = \sum_{j=0}^{\infty} (-1)^j \frac{(\lambda s)^j}{j!} \Gamma\left(\frac{j+\beta}{\beta}\right)$$

and

$$f_B^*(s) = \frac{1}{(1 + \lambda s)^\beta}.$$

7-11. Recent developments in the area of microelectromechanical systems (MEMS) have resulted in the construction of microgrippers, which are capable of handling microsized objects and have wide applications in biomedicals and microtelerobotics. A typical microgripper consists of a fixed closure driver and two movable jaws that are closed by an electrostatic voltage applied across them and the closure driver. A typical gripper can exert 40 nN of force on the object between its jaws, with an applied voltage of 40V. A microtelerobot (MT) is used in experimental medical applications where microgrippers are attached to the MT. The microgrippers exert repeated forces on objects clogging a pathway. The time to failure of the grippers follows a Weibull distribution with a p.d.f. of the form

$$f(t) = \frac{\beta t^{\beta-1}}{\lambda^\beta} \exp\left[-\left(\frac{t}{\lambda}\right)^\beta\right] \qquad t \geq 0, \beta > 0, \lambda > 0,$$

and its

$$f^*(s) = \sum_{j=0}^{\infty} (-1)^j \frac{(\lambda s)^j}{j!} \Gamma\left(\frac{j+\beta}{\beta}\right).$$

The parameters β and λ are 2 and 2,000, respectively. When the grippers fail, they are replaced by a new set of grippers, and the replacement time follows an exponential distribution of the form

$$f(t) = \theta \exp(-\theta t) \qquad t \geq 0,$$

with a parameter $\theta = 500$.

Assuming that the sequence of the grippers' failure and replacement follows an alternating renewal process, determine the following:

a. The expected number of the grippers' failures in 10,000 hours.

b. The availability of the grippers at 10,000 hours.

c. The steady-state availability of the grippers.

d. The probability that the grippers will fail during a medical operation of an expected length of eight hours.

e. A way to improve the availability of such grippers?

7-12. Paper stock consists of cellulose fibers suspended in water. Once the stock has been washed and screened to remove unwanted chemicals and impurities, it is refined to improve the quality of the paper sheets. Additives such as starch, alum, and clay fillers are then introduced to develop required characteristics of the paper product. The paper stock is then pumped to different tanks and processed. There are two preferred types of pumps for that purpose: the reciprocating suction pumps and the centrifugal pumps. The latter are frequently clogged with high-density paper stock. A paper producer uses a centrifugal pump in order to pump the paper stock from the main tank to the next process. When the pump fails (mainly due to clogging), it is replaced by a reciprocating suction pump and vice versa. The following failure times (in hours) are observed for the centrifugal pump:

38.93	443.61	1352.84	2728.80	4064.14
79.38	447.44	1375.61	2755.42	4074.49
89.63	558.08	1492.33	2890.38	4335.69
117.39	682.27	1525.85	2891.07	4337.81
274.81	898.10	1559.99	2999.77	5078.74
299.70	946.85	1662.41	3108.44	5418.34
326.80	1013.81	1763.87	3458.03	6659.95
417.36	1157.73	2060.99	3529.30	7038.81
421.82	1285.96	2122.60	3754.15	7762.72
432.78	1326.85	2297.35	3780.32	7859.20

Similarly, the following failure times (in hours) are observed for the reciprocating pump:

9.87	259.64	592.30	934.95	1630.92
67.20	330.22	592.44	1018.67	1661.09
77.58	337.38	643.88	1140.36	1821.21
80.43	366.20	649.49	1153.35	1885.12
85.48	381.19	657.77	1260.16	2470.53
127.74	412.98	672.60	1361.80	2697.63
142.53	457.82	674.74	1421.64	2862.29
146.49	538.24	679.83	1425.61	3356.10
157.99	553.35	710.15	1488.32	3372.39
206.87	565.57	783.42	1493.67	3878.58

a. What is the expected number of failures for each type of pump in the interval (0, 10^4 hours]?

b. What is the probability that the centrifugal pump is in use at time $t = 10^4$ hours?

c. Graph the above probability over the interval (0, 10^4 hours].

7-13. A producer of motor control boards uses a surface-mount chip resistor subassembly as a part of the board's assembly. The chip substrate is high-purity alumina, and the resistive element is a sintered thick film that is coated with a protective glass film after laser trimming and is finished with an epoxy coating. Continuity through the resistive element is established by solder attachments to the subassembly lead frame through edge terminations. Field results show that the resistor exhibits a failure mode characterized by an increase in resistance beyond the system's tolerance. Therefore, the producer develops a thermal shock test (from −85 to 200 °F) and obtains the following failure times (in hours):

0.90	25.95	59.25	93.50	163.00
6.75	33.05	59.45	102.00	166.20
7.80	33.80	64.40	115.20	182.20
8.00	36.60	64.90	115.80	188.50
8.65	38.15	65.75	126.50	247.00
12.80	41.30	67.25	136.20	269.75
14.26	45.80	67.45	142.60	286.30
14.61	53.85	68.00	142.90	335.40
15.80	55.00	71.05	148.90	336.90
20.65	56.60	78.30	150.20	390.00

a. Compare the estimates of the expected number of failures in the interval (0, 5000 hours] using M_b, M_d, and M_o.

b. Fit an exponential distribution to the failure data and obtain its parameter.

c. Compare the results obtained from a and b.

d. Assume that when the resistor subassembly fails, it is replaced by a new one. Thus, the failure replacement sequence can be represented by an ordinary renewal process. Calculate its variance and the 95 percent confidence interval for the number of renewals in the interval (0, 5,000 hours].

7-14. The producer of the resistor subassembly in Problem 7-13 modifies the resistor but observes that the time to the first failure has a distinct distribution of the form

$$f_1(t) = \lambda_1 e^{-\lambda_1 t},$$

where $\lambda_1 = 0.0133$ failures per hour.

The failure times between any two successive failures beyond the first have the same p.d.f. with a p.d.f. obtained from the failure data given in Problem 7-13. When a resistor subassembly fails, it is replaced by a new subassembly and subsequent failures are replaced accordingly.

a. Estimate the expected number of replacements during the interval (0, 10^4 hours]. What is its variance?

b. Construct a 90 percent confidence interval for the expected number of replacements obtained in a.

7-15. Consider a NHPP with the following hazard function:

$$h(t) = \frac{t^{\beta-1}}{\lambda^{\beta}\,\Gamma(\beta)\displaystyle\sum_{j=0}^{\beta-1}\left(\frac{t}{\lambda}\right)^{j}\frac{1}{\Gamma(j+1)}}.$$

Assume $\beta = 3$ and $\lambda = 500$.

a. Determine the probability that five failures occur in the interval (0, 6,000 hours].

b. What is the expected number of failures during the same interval? Plot the expected number of failures in the interval $(0, t]$ versus t.

c. What are your conclusions regarding the hazard-rate function?

7-16. A system composed of 100 identical components and its operation is independent of the number of failed units at any time t. Assume the failures constitute a nonhomogeneous Poisson process with failure rate function $\lambda(t)$ given by

$$\lambda(t) = \begin{cases} 0.001 + 0.0002t & 0 \le t \le 200 \text{ hours} \\ 0.041 & 200 \le t \le 300 \text{ hours} \\ 0.044 - 0.00001t & 300 \le t \le 400 \text{ hours} \end{cases}$$

and

$$\lambda(t) = 0.007t \qquad t > 400.$$

a. What is the expected number of failures in the interval (200, 600]?

b. What is the reliability of the system at time $t = 600$ hours?

REFERENCES

Baker, Rose, D. (1993). "A Nonparametric Estimation of the Renewal Function." *Computers and Operations Research* 20(2), 167–178.

Barlow, R. E., Proschan, F., and Hunter, L. C. (1965). *Mathematical Theory of Reliability*. New York: John Wiley.

Bartholomew, D. J. (1963). "An Approximate Solution of the Integral Equation of Renewal Theory." *Journal of Royal Statistical Society* 25B, 432–441.

Baxter, L. A., and Li, L. (1994). "Non-Parametric Confidence Intervals for the Renewal Function and the Point Availability." *Scandinavian Journal of Statistics* 21, 277–287.

Cox, D. R. (1962). *Renewal Theory*. London: Methuen.

Deligönül, Z. (1985). "An Approximate Solution of the Integral Equation of Renewal Theory." *Journal of Applied Probability* 22, 926–931.

Frees, E. W. (1986a). "Nonparametric Renewal Function Estimation." *Annals of Statistics* 14, 1366–1378.

Frees, E. W. (1986b). "Warranty Analysis and Renewal Function Estimation." *Naval Research Logistics Quarterly* 33, 361–372.

Frees, E. W. (1988). "Correction: Nonparametric Renewal Function Estimation." *Annals of Statistics* 16, 1741.

Hoyland, A., and Rausand, M. (1994). *System Reliability Theory*. New York: John Wiley.

Jardine, A. K. S. (1973). *Maintenance, Replacement, and Reliability*. New York: John Wiley.

Juran, J. M., and Gryna, F. M. (1993). *Quality Planning and Analysis* (3rd ed.). New York: McGraw-Hill.

Karlin, S., and Taylor, H. M. (1975). *A First Course in Stochastic Processes*. New York: Academic Press.

Kolb, J., and Ross, S. S. (1980). *Product Safety and Liability* (3rd ed.). New York: McGraw-Hill.

Merz, R., Prinz, F. B., and Weiss, L. E. (1994). "Method and Apparatus for Depositing Molten Metal." U.S. Patent 5,286,573, 1994.

Ozbaykal, T. (1971). "Bounds and Approximations of the Renewal Function." Unpublished M.S. thesis, Naval Postgraduate School, Department of Operations Research and Administrative Science, Monterey, California.

Parzen, E. (1962). *Stochastic Processes*. San Francisco: Holden-Day.

Swartz, G. A. (1986). "Gate Oxide Integrity of NMOS Transistor Arrays." *IEEE Transactions on Electron Devices* ED-33(11), 1826–1829.

Trivedi, K. S. (1982). *Probability and Statistics with Reliability, Queueing, and Computer Science Applications*. Englewood Cliffs, NJ: Prentice-Hall.

CHAPTER 8

Warranty Models

Behold the warranty: the bold print giveth, and the fine print taketh away.

—Anonymous

8.1. Introduction

The increasing worldwide competition is prompting manufacturers to introduce innovative approaches in order to increase their market shares. In addition to improving quality and reducing prices, they also provide attractive warranties for their products. In other words, warranties are becoming an important factor in the consumer's decision-making process. For example, when several products that perform the same function are available in the market and their prices are essentially equal, the customer's deciding factor of preference for one product over the other includes the manufacturer's reputation and the type and length of the warranty provided with the product. Because of the impact of the warranty on future sales, manufacturers who traditionally did not provide warranties for some products and services are now providing or required to provide some type of warranty. For example, there were no warranties on the weapon systems until the Defense Procurement Reform Act was established in 1985, requiring the prime contractor for the production of weapon systems to provide written guarantees for such systems. The Defense Procurement Reform Act also lists the types of coverage required, the possible remedies and specific reasons for securing a waiver, and actions to be taken in the event a waiver is sought. Thus, the warranty is becoming increasingly important for both consumer and military products.

A warranty is a contract or an agreement under which the manufacturer of a product or service must agree to repair, replace, or provide service when the product fails or the service does not meet the customer's requirements before a specified time (length of warranty). This specified "time" may be measured in calendar time units such as hours, months, and years, or in usage units such as miles, hours of operation, number of times the product has been used (for example, number of copies made by a copier or number of pages printed by a printer) or both. Other warranties have no specified "times" and are referred to as lifetime warranties.

Three types of warranties are commonly used for consumer goods: the *ordinary free replacement* warranty, the *unlimited free replacement* warranty, and the *pro-rata* warranty. Under an ordinary free replacement warranty, if an item fails before the end of the warranty length, it is replaced or repaired at no cost to the consumer. The repaired or replaced item is then covered by an ordinary free replacement warranty with a length equal to the remaining length of the original warranty. Such warranty ensures that the consumer will receive as many free repairs or replacements as needed during the original length of the warranty. The ordinary free replacement warranty is the most common type of warranty; it is most often used to cover consumer durables such as cars and kitchen appliances (Mamer, 1987).

The second type of warranty, the unlimited free replacement, is identical to the ordinary replacement warranty except that each replacement item carries an identical warranty to the original purchase warranty. Such warranties are used only for small electronic appliances that have high early failure rates and are usually limited to very short periods of time.

Thus, under the free replacement warranty, whether it is ordinary or unlimited, a long warranty period will result in a very large warranty cost. Furthermore, an increase in the warranty period will reduce the number of replacement purchases over the life cycle of the product, which consequently reduces the total profit of the manufacturer. Clearly, the free replacement policy is more beneficial to the consumer than to the manufacturer. Therefore, it is extremely important for the manufacturer to determine the optimal price of the product and the optimal warranty length such that the total cost over the product life cycle is minimized.

The third type of warranty is the pro-rata warranty. Under this warranty, if the product fails before the end of the length of the warranty, it is replaced at a cost that depends on the age of the item at the time of failure and the replacement item is covered by an identical warranty. Typically, a discount proportional to the remaining length of the warranty is given on the purchase price of the replacement item. For example, if the length of the warranty is w and the item fails at a time $t < w$, the consumer pays the proportion t/w of the cost of the replacement items (automobile tires represent an ideal product for which the pro-rata warranty policy is appropriate), and the remaining cost of the product replacement is covered by the manufacturer. Unlike the free replacement warranty, the pro-rata warranty is more beneficial to the manufacturer than the consumer.

As presented above, a pure free replacement warranty favors the consumer, whereas a pure pro-rata warranty favors the manufacturer. Therefore, a mix of these policies may present an alternative warranty that is fair to both the consumer and manufacturer. For example, a policy that involves an initial free replacement period followed by a pro-rata period is a sensible one, being fair to both the manufacturer and the consumer (Nguyen and Murthy, 1984b). It has a promotional appeal to attract consumers and at the same time keeps the warranty cost for the manufacturer within a reasonable amount. There are other types of warranty policies such as reliability improvement warranty where the manufacturer provides guaranteed mean time between failure (MTBF) or provides support for engineering changes (Blischke and Murthy, 1994).

In deciding which warranty policy should be used, the manufacturer should consider the type of repair or replacements to be made. We classify the products (items) into two types: repairable and nonrepairable. Repairable products are those for which repair cost is significantly less than the cost of replacing the products with new ones, such as copying machines, printers, computers, large appliances, and automobiles. Nonrepairable products are those for which the repair cost is close to the replacement cost or those that cannot be repaired due to the difficulty of accessing the product. Typical examples of nonrepairable products are small appliances, radios, inexpensive watches, and satellites. The rebate warranty policy is commonly used for nonrepairable products. Under the rebate policy, the consumer is refunded some proportion of the sale price if the product fails before the warranty period expires (Nguyen and Murthy, 1984).

The manufacturers face two main problems in planning a warranty program. These problems require the determination of

- The type of warranty policy, length of warranty period, and its cost; and
- the amount of capital that must be allocated to cover future expenses for failures during a specified warranty period—that is, determination of the allocation for future warranty expenses. Too large a warranty reserve might make the sale price noncompetitive, thus reducing sales volume and profit. On the other hand, too little warranty reserve results in hidden losses that impact future profits.

This chapter discusses different warranty policies and addresses the above two problems for both repairable and nonrepairable products.

8.2. Warranty Models for Nonrepairable Products

In this section, we determine the optimal warranty reserve fund for nonrepairable products when a pro-rata warranty policy is used. Under this policy, a product is replaced by a new product when the warranty is invoked, and a rebate, which decreases linearly with time or product use, is subtracted from the replacement price of the product. As mentioned in Section 8.1, this type of rebate is applicable to consumer products like batteries and tires.

8.2.1. WARRANTY COST FOR NONREPAIRABLE PRODUCTS

Consider a product whose hazard-rate function, $h(t)$, is constant—that is,

$$h(t) = \lambda, \tag{8.1}$$

where λ is a constant failure rate. The probability of failure at any time less than or equal to t, $F(t)$, is

$$F(t) = 1 - e^{-\lambda t}. \tag{8.2}$$

The mean time to failure, m, obtained from Eq. (8.1) is $1/\lambda$. We rewrite Eq. (8.2) as

$$F(t) = 1 - e^{-\frac{t}{m}}. \tag{8.3}$$

The procedure for determining the warranty reserve fund is as follows: knowing the MTTF, determine the expected number of products that will fail in any small time interval dt. Then multiply the expected number of failures by the cost of replacement at time t to estimate the increment of warranty reserve that must be set aside for failures during the interval. Finally, add the incremental warranty cost for all increments dt from $t = 0$ to $t = w$ (end of warranty period). This results in the total warranty reserve fund. It is assumed that all failures during the warranty period are claimed. To transform this procedure into mathematical expressions, we define the following notations (Menke, 1969).

c = constant unit product price, including warranty cost,

t = time,

m = MTTF of the product,

w = duration of warranty period,

L = product lot size for warranty reserve determination,

R = total warranty reserve fund for L units,

$C(t)$ = pro-rata customer rebate at time t, $C(t) = c[1 - (t/w)]$, $0 < t < w$,

r = warranty reserve cost per unit product, $r = R/L$, and

E = expected number of failures at time t.

Using Eq. (8.3), we obtain the expected number of failures occurring at any time t:

$$E[N(t)] = LP[\text{product failure before or at time } t] = L\left[1 - e^{\frac{-t}{m}}\right].$$

The total number of failures in the interval t and $t + dt$ is

$$dE[N(t)] = \frac{\partial E\,[N(t)]}{\partial t}\,dt = (L/m)e^{-t/m}\,dt.$$

The cost for the failures in t and $t + dt$ is

$$d(R) = C(t)\,dE\,[N(t)] = c\left(1 - \frac{t}{w}\right)(L/m)e^{-t/m}\,dt. \tag{8.4}$$

The total cost for all failures occurring in $t = 0$ to $t = w$ is

$$R = \int_0^w \frac{Lc}{m}\left(1 - \frac{t}{w}\right)e^{-t/m}\,dt$$

or

$$R = Lc\left[1 - \left(\frac{m}{w}\right)(1 - e^{-w/m})\right]. \tag{8.5}$$

The warranty reserve fund per unit is

$$r = \frac{R}{L} = c\left[1 - \left(\frac{m}{w}\right)(1 - e^{-w/m})\right].$$

Thus,

$$\frac{r}{c} = 1 - \left(\frac{m}{w}\right)(1 - e^{-w/m}). \tag{8.6}$$

Let c' be the unit price before warranty cost is added. Then,

$$c = c' + r$$

and

$$c' = c\left(1 - \frac{r}{c}\right)$$

or

$$c = \frac{c'}{1 - \frac{r}{c}}. \tag{8.7}$$

The total warranty reserve fund to be allocated for L units of production is obtained from Eq. (8.5).

The ratio between the warranty reserve cost and the product cost increases as the ratio between the warranty length and the MTTF increases as shown in Figure 8.1.

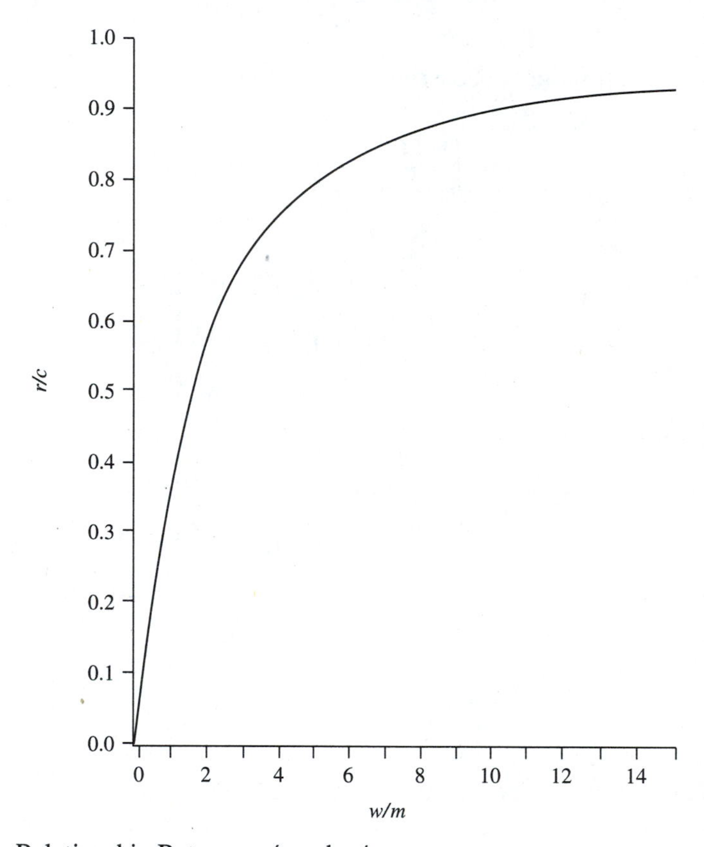

Figure 8.1. Relationship Between r/c and w/m

EXAMPLE 8.1

Assume the manufacturer of shortwave radios wishes to extend a twelve-month warranty on new types of radios. An accelerated life test was performed that indicated that the failure time of the radio follows an exponential distribution with parameter $\lambda = 0.01$ failures per month. The manufacturer's cost of a radio (not including warranty cost) is \$45. Assuming a total production run of 4,000 radios, determine the warranty reserve fund and the adjusted price of the radio.

SOLUTION

Following are the data for the radios:

$w = 12$ months,

$c' = \$45$,

$m = 1/\lambda = 100$ months,

$L = 4{,}000$ units.

Using Eq. (8.6) we obtain r/c as

$$\frac{r}{c} = 1 - \frac{100}{12}\left(1 - e^{-\frac{12}{100}}\right) = 0.0576.$$

Using Eq. (8.7) we obtain the adjusted price of the radio

$$c = \frac{c'}{1 - \frac{r}{c}} = \frac{45}{1 - 0.0576} = \$47.75.$$

The warranty reserve fund for the 4,000 radios is

$$R = 0.05767 \times 4{,}000 \times 47.75 = \$11{,}014.$$

8.2.2. WARRANTY RESERVE FUND: LUMP-SUM REBATE

If the administrative cost and the errors in estimating pro-rata claims are too expensive, the manufacturer may wish to consider an alternative warranty plan by paying a fixed or lump-sum rebate to the customer for any failure occurring before the warranty expires. Again, we are interested in determining the adjusted price of the product and the warranty reserve fund that meets customer claims.

Let k be the proportion of the unit cost to be refunded as a lump-sum rebate and S be the unit lump-sum rebate ($S = kc$).

Substituting $C(t) = kc$ in the pro-rata warranty model, we obtain

$$r_s = kc\left(1 - e^{-\frac{w}{m}}\right),$$

where r_s is the warranty reserve cost per unit under the lump-sum warranty plan.

If it is desirable to make the warranty cost per unit of production equal for both the pro-rata and the lump-sum plan, then

$$1 - \frac{m}{w}\left(1 - e^{-\frac{w}{m}}\right) = k\left(1 - e^{-\frac{w}{m}}\right)$$

or

$$k = \frac{1}{1 - e^{-w/m}} - \frac{m}{w}. \tag{8.8}$$

The proportion of the unit cost to be refunded as a lump-sum rebate as a function of ratio w/m is shown in Figure 8.2. The lump-sum rebate per unit becomes

$$S = kc.$$

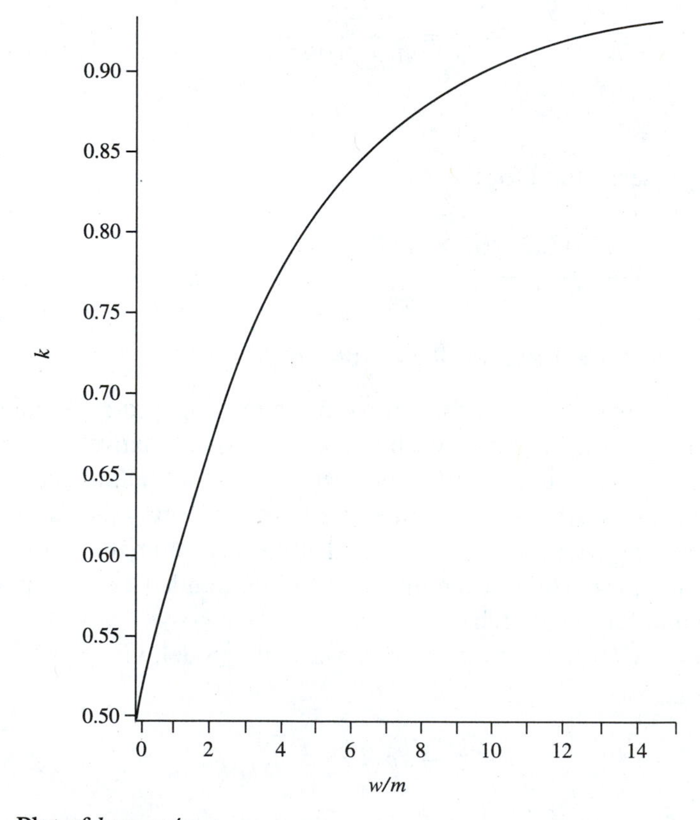

Figure 8.2. Plot of k vs. w/m

The total warranty reserve fund R_s is (Menke, 1969)

$$R_s = LS\left(1 - e^{-\frac{w}{m}}\right). \tag{8.9}$$

EXAMPLE 8.2

Assume that the radio manufacturer in Example 8.1 wishes to adopt a lump-sum warranty plan equivalent to the pro-rata plan by providing a lump sum of the initial price to customers whose radios fail during the warranty period. Determine the portion of the price to be refunded upon failure before the warranty expires.

SOLUTION

Using the same data for the radio:

$w = 12$ months,

$c' = \$45$,

$m = 100$ months, and

$L = 4{,}000$ units.

We first determine the proportion of the unit cost to be returned as a lump sum by using Eq. (8.8):

$$k = \frac{1}{1 - e^{-w/m}} - \frac{m}{w}$$

$$= \frac{1}{1 - e^{-\frac{12}{100}}} - \frac{100}{12} = 0.5099.$$

In other words, the manufacturer should pay 51 percent of the initial price to customers whose radios fail before the expiration of the warranty period. The warranty cost per unit is

$$c = c' + r = \$47.75$$

and

$$S = ck = \$24.35.$$

The total warranty reserve fund is

$$R_s = 4{,}000 \times 24.35\left(1 - e^{-\frac{12}{100}}\right)$$

$$R_s = \$11{,}014.$$

It is important to consider the situations for which a pro-rata or a lump-sum warranty policy can be used. For example, a manufacturer should consider the use of a lump-sum plan when it is possible to determine that the product failed before its warranty period but the exact time of failure is not possible to determine (Menke, 1969).

The model discussed in Sections 8.2.1 and 8.2.2 overstates the required warranty reserve fund since the discounting of future warranty claim costs for the time value of money, and changes in the general price level due to inflation (or deflation) have been ignored.

Let θ be the rate of return earned through the investment of the warranty reserve fund and ϕ be the expected change per period in the general price level. Then, the real present value of the warranty claims (Amato and Anderson, 1976) in the t and $t + dt$ period is obtained by rewriting Eq. (8.4) as

$$d(R^*) = c^*\left(1 - \frac{t}{w}\right)(1 + \theta + \phi)^{-t}\left(\frac{L}{m}\right)e^{-\frac{t}{m}}\,dt, \tag{8.10}$$

where R^* and c^* are the present values of the warranty reserve fund and the price of the product, respectively, and $(1 + \theta)^n (1 + \phi)^n \simeq (1 + \theta + \phi)^n$ for small θ and ϕ. Equation (8.10) can be rewritten as

$$\begin{aligned}
R^* &= \frac{Lc^*}{m}\int_0^w \left(1 - \frac{t}{w}\right)\left[(1 + \theta + \phi)e^{\frac{1}{m}}\right]^{-t} dt \\
&= \left\{\frac{Lc^*}{1 + m\ln(1 + \theta + \phi)}\right\} \times \\
&\quad \left\{1 - \left(\frac{m}{w}\right)\frac{1}{1 + m\ln(1 + \theta + \phi)} \times \left[1 - (1 + \theta + \phi)^{-w} e^{-\frac{w}{m}}\right]\right\}.
\end{aligned} \tag{8.11}$$

Again, the manufacturer can assign the following per unit price to its product in order to incorporate the warranty cost

$$c^* = c' + r^*,$$

where

$$\frac{r^*}{c^*} = [1 + m \ln(1 + \theta + \phi)]^{-1} \times$$

$$\left\{1 - \frac{m}{w}[1 + m \ln(1 + \theta + \phi)]^{-1}\left[1 - (1 + \theta + \phi)^{-w} e^{-\frac{w}{m}}\right]\right\} \tag{8.12}$$

and

$$r^* = \frac{R^*}{L}. \tag{8.13}$$

EXAMPLE 8.3

The manufacturer of the radios in Example 8.1 intends to invest the warranty reserve fund to earn an interest rate of 5 percent and to increase the price of the radio the following year by 6 percent. Determine the warranty reserve fund and the price of the radio after adjustments.

SOLUTION

The following data were provided in Example 8.1:

$m = 100$ months,

$w = 12$ months,

$c' = \$45$, and

$L = 4{,}000$ units.

We now include the effect of θ and ϕ

$$(1 + \theta + \phi) = 1 + 0.05 + 0.06 = 1.11.$$

We obtain r^*/c^* from Eq. (8.12):

$$\frac{r^*}{c^*} = (1 + 100 \ln 1.11)^{-1}\left\{1 - \frac{100}{12}(1 + 100 \ln 1.11)^{-1}\left[1 - (1.11)^{-12} e^{-\frac{12}{100}}\right]\right\}$$

$$= 0.09495\,\{1 - 0.791\,[1 - 0.2858 \times 0.8869]\} = 0.03888.$$

But

$$c^* = \frac{c'}{1 - \frac{r^*}{c^*}} = \frac{45}{1 - 0.03888} = \$46.82$$

and

$$R^* = L \times \frac{r^*}{c^*} \times c^* = 4{,}000 \times 0.0388 \times 46.82 = \$7{,}281.$$

These two estimates are smaller than those of Example 8.1.

8.2.3. MIXED WARRANTY POLICIES

In the previous two sections we presented the pro-rata and the lump-sum rebate warranty policies. In this section, we present and compare two warranty policies. The first policy, which we refer to as *full rebate policy*, occurs when a manufactured product is sold under full warranty. If a failure occurs within w_0 units of time, the product is replaced at no cost to the consumer and a new warranty is issued. The second policy is a *mixed warranty policy* where a full compensation is provided to the consumer if the product fails before time w_1, followed by a linear pro-rated compensation up to the end of the warranty coverage period w_2. In order to simplify the analysis, we define the following notations (Ritchken, 1985).

w_0 = warranty length of the full rebate policy,

w_1 = length of the full compensation period for the mixed policy,

$w_2 - w_1$ = length of the pro-rated period for the mixed policy,

c_0 = unit cost of replacement,

ϕ = notation for the full rebate policy,

ψ = notation for the mixed policy,

X_i = time between failures i and $i - 1$; $X_i > 0$,

$I(X_i)$ = cost of a failure to the manufacturer under a given policy (ϕ or ψ),

V = random variable representing the total warranty cost accumulated per product,

$F(X_i), R(X_i)$ = cumulative distribution function and the reliability function of X; $R(X_i) = 1 - F(X_i)$,

$F^{(2)}(x) = \int_0^x F(w)\,dw$,

$F^{(3)}(x) = \int_0^x F^{(2)}(w)\,dw$, and

N' = number of failures that occur until a failure time exceeds the warranty period.

The two warranty policies are shown in Figure 8.3. The cost associated with the full rebate policy is

$$I_\phi(X_i) = \begin{cases} c_0 & 0 \le X_i \le w_0 \\ 0 & \text{otherwise.} \end{cases} \tag{8.14}$$

The cost associated with the mixed policy is

$$I_\psi(X_i) = \begin{cases} c_0 & 0 \le X_i \le w_1 \\ c_0(w_2 - X_i)/(w_2 - w_1) & w_1 \le X_i \le w_2 \\ 0 & \text{otherwise.} \end{cases} \tag{8.15}$$

The total warranty expenses accumulated per product sold is

$$V = \sum_{i=1}^{N'} I(X_i). \tag{8.16}$$

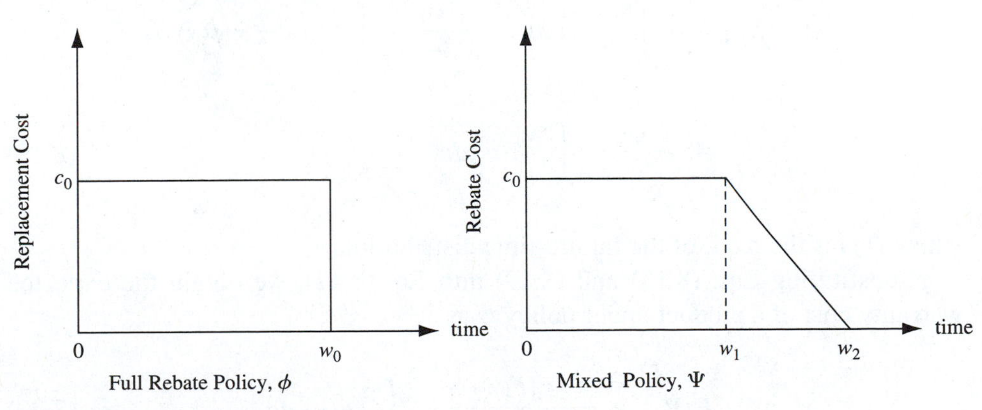

Figure 8.3. Two Warranty Policies

The decision variable for the full rebate policy ϕ is w_0, whereas the decision variables for the mixed policy ψ are the time of full compensation and the total time of the warranty length.

We now derive the expected total warranty cost for a product under policy ψ:

$$E[V_\psi] = E\left\{E\left[\sum_{i=1}^{N'} I_\psi(X_i)\right]\right\}.$$

Since N' is a stopping time,

$$E[V_\psi] = E[N']E[I_\psi(X_i)]. \tag{8.17}$$

But N' is geometric—that is,

$$P[N' = k] = [1 - F(w_2)]\,[F(w_2)]^k \qquad k = 0, 1, 2, \ldots$$

or

$$P[N' = k] = R(w_2)F(w_2)^k.$$

Thus,

$$E[N'] = \sum_{k=0}^{\infty} kR(w_2)F(w_2)^k = F(w_2)/R(w_2). \tag{8.18}$$

Moreover, from Ritchken (1985) and Thomas (1983):

$$E[I_\psi(X_i)] = c_0\int_0^{w_1} f(x)\,dx + \frac{c_0}{w_2 - w_1}\int_{w_1}^{w_2} (w_2 - x)f(x)\,dx$$

$$= \frac{c_0}{w_2 - w_1}\int_{w_1}^{w_2} F(u)\,du, \tag{8.19}$$

where $f(x)$ is the p.d.f. of the failure-time distribution.

Substituting Eqs. (8.18) and (8.19) into Eq. (8.17), we obtain the expected warranty cost of a product under policy ψ as

$$E[V_\psi] = \frac{c_0 F(w_2)}{(w_2 - w_1)R(w_2)}\int_{w_1}^{w_2} F(u)\,du. \tag{8.20}$$

EXAMPLE 8.4

Consider a product that exhibits a constant failure rate with mean time to failure of sixty months. It is intended to use a mixed policy with $w_1 = 3$ months, and $w_2 = 12$ months. The cost of a replacement is \$120. What is the expected warranty cost?

SOLUTION

Since the product exhibits constant failure rate, then

$$F(x_i) = 1 - e^{\frac{-x_i}{60}} \qquad x_i \geq 0$$

$$R(x_i) = e^{\frac{-x_i}{60}}.$$

Using Eq. (8.20), we obtain

$$E[V_\psi] = \frac{120F(w_2)}{(12-3)R(w_2)} \int_3^{12} 1 - e^{-\frac{x}{60}}\, dx$$

$$= \$3.0997.$$

For each mixed policy ψ, there is a full rebate policy ϕ that yields the same cost. The expected cost of a failure under a full rebate policy is

$$E[I_\phi(X_i)] = c_0 F(w_0). \tag{8.21}$$

Using Eqs. (8.17), (8.18), and (8.21) we obtain

$$E[V_\phi] = \frac{c_0 F(w_0)^2}{R(w_0)}. \tag{8.22}$$

We equate Eqs. (8.20) and (8.22) so that the two policies will have the same cost. Thus,

$$\frac{F(w_0)^2}{R(w_0)} = \frac{F(w_2)}{R(w_2)} \frac{1}{(w_2 - w_1)} \int_{w_1}^{w_2} F(u)\, du,$$

which is reduced to

$$\frac{F(w_0)^2}{R(w_0)} = \frac{F(w_2)}{R(w_2)} \frac{F^{(2)}(w_2) - F^{(2)}(w_1)}{(w_2 - w_1)}. \tag{8.23}$$

If we consider a linear pro-rata warranty policy only ($w_1 = 0$), then the above equation can be rewritten as

$$\frac{F(w_0)^2}{R(w_0)} = \frac{F(w_2)}{R(w_2)} \frac{F^{(2)}(w_2)}{w_2}. \tag{8.24}$$

In other words, the two policies are equivalent if w_0 is chosen such that Eq. (8.24) is satisfied.

EXAMPLE 8.5

Using the data of Example 8.4, determine the warranty length (w_0) for the full rebate policy, which makes it equivalent to the mixed policy.

SOLUTION

Substituting in Eq. (8.23), we obtain

$$\frac{F(w_0)^2}{R(w_0)} = \frac{0.22140(1.12384 - 0.07376)}{12 - 3} = 0.0258$$

$$\frac{\left(1 - e^{-\frac{w_0}{60}}\right)^2}{e^{-\frac{w_0}{60}}} = 0.0258$$

$$w_0 = 9.6 \text{ months.}$$

EXAMPLE 8.6

Determine the warranty length (w_0) for the full rebate policy that makes it equivalent to a pro-rata policy with $w_2 = 12$.

SOLUTION

Using $w_2 = 12$ and substituting into Eq. (8.24), we obtain

$$\frac{F(w_0)^2}{R(w_0)} = \frac{F(w_2)}{R(w_2)} \frac{F^{(2)}(w_2)}{w_2}$$

$$w_0 = 8.63 \text{ months.}$$

It is not sufficient to compare two policies based only on the expected cost since the variance of the cost (or the distribution of the cost) may influence the choice of the warranty policy. For example, the manufacturer may prefer a warranty policy with smaller cost variance. The manufacturer may also compare different warranty policies using the mean-variance orderings. Therefore, the variances of the warranty cost need to be determined. Ritchken (1985) derives the following expressions for the variances of total warranty cost—

- For the linear pro-rated policy ψ, with $w_1 = 0$,

$$\text{Var}(V_\psi) = c_0^2 F(w_2)[2R(w_2)F^{(3)}(w_2) + F^{(2)}(w_2)^2 F(w_2)]/w_2^2 R(w_2)^2. \tag{8.25}$$

- For the full rebate policy ϕ, the variance is

$$\text{Var}(V_\phi) = c_0^2 F(w_0)^2\, [R(w_0)^2 + F(w_0)]/R(w_0)^2 \tag{8.26}$$

—when

$$F(w) = 1 - e^{-\lambda w} \qquad w \geq 0,$$

and λ is the failure rate, then

$$F^{(2)}(w) = [\lambda w - F(w)]/\lambda \tag{8.27}$$

$$F^{(3)}(w) = (\lambda w - 1)^2 + (2F(w) - 1)/2\lambda^2. \tag{8.28}$$

Substituting Eqs. (8.27) and (8.28) into Eq. (8.25) results in

$$\text{Var}(V_\psi) = \frac{c_0^2 F(w_2)}{[\lambda w_2 R(w_2)]^2}[(\lambda w_2 - 1)^2 + 2F(w_2) - 1]R(w_2). \tag{8.29}$$

The expected time to the first failure that is not covered by the warranty cost is

$$E[T] = E[X_i]/R(w_2), \tag{8.30}$$

where $E[X_i]$ is the mean time to failure.

EXAMPLE 8.7

Using the warranty lengths of $w_2 = 12$ and $w_0 = 8.63$ that make the pro-rata policy equivalent to the full rebate policy, determine the variances of the total warranty cost for each policy. Which policy do you prefer?

SOLUTION

From Eq. (8.29), we obtain the variance for the pro-rata policy as

$$\text{Var}(V_\psi) = \frac{120^2\left(1 - e^{-\frac{12}{60}}\right)}{\left(\frac{12}{60} \times e^{-\frac{12}{60}}\right)^2}\left[\left(\frac{12}{60} - 1\right)^2 + 2\left(1 - e^{-\frac{12}{60}}\right) - 1\right] e^{-\frac{12}{60}}$$

$$= 167.44.$$

Using Eq. (8.26), the variance for the full rebate policy is

$$\mathrm{Var}(V_\phi) = \frac{120^2\left[1 - e^{-\frac{8.63}{60}}\right]^2}{\left(e^{-8.63/60}\right)^2}\left[\left(e^{-8.63/60}\right)^2 + \left(1 - e^{-8.63/60}\right)\right] = 304.61.$$

Since the two policies are equivalent, the manufacturer should adopt the pro-rata policy in order to reduce the variability in the total warranty cost.

8.2.4. OPTIMAL REPLACEMENTS FOR ITEMS UNDER WARRANTY

A typical age replacement policy of items calls for an item replacement upon failure or at a fixed time, whichever comes first. Clearly, such a policy is applicable only for items whose failure rates increase with age. In this section, we develop a model for the determination of the optimal age replacement policies for warrantied items, such that the average cost is minimized. We summarize an age replacement policy as follows (Ritchken and Fuh, 1986):

Assume a warrantied nonrepairable item is installed at time zero. If the item fails during its warranty period, it is replaced at a cost shared by both the manufacturer and the customer (such as a linear pro-rata policy) in accordance with a rebate policy. After the warranty expires, an age replacement policy is followed, with the item being replaced after an additional fixed time or upon failure—whichever comes first. We define the following notation:

ϕ = warranty policy,

w = length of the warranty period,

X_i = time to the ith failure,

$F(t)$ = c.d.f. of time to failure,

$R(t)$ = reliability function at time t,

$h(t)$ = hazard rate at time t,

$F_r()$ = c.d.f. of the residual lifetime beyond w,

$f_r()$ = p.d.f. of the residual lifetime beyond w,

$[N(t), t > 0]$ = number of times an item fails in the time interval $(0, t)$,

$m(t) = E[N(t)]$ = renewal function,

Y = time for which first failure occurs outside warranty period—that is, $Y \equiv \inf\,[t \mid N(t) = N(w) + 1]$,

$r(w)$ = residual life of the functioning item at time w,
$r(w) = Y - w$,

T = age replacement parameter measured from the end of the warranty period,

T_r = time between replacements outside the warranty interval,
$T_r \equiv \min\{T + w, Y\}$,

$G(T) = \int_0^T x f_r(x)\,dx$ partial mean of time to replacement beyond w,

$\hat{C}(T)$ = mean cost incurred between replacements outside the warranty interval,

c_1 = cost of replacing a failed item,

c_2 = cost of replacing a functioning item,

$I_j(\phi)$ = cost to the consumer for replacement j under the warranty policy ϕ,

W_ϕ = total mean cost of replacements over the warranty period.

The age replacement policy under warranty is illustrated in Figures 8.4 and 8.5.

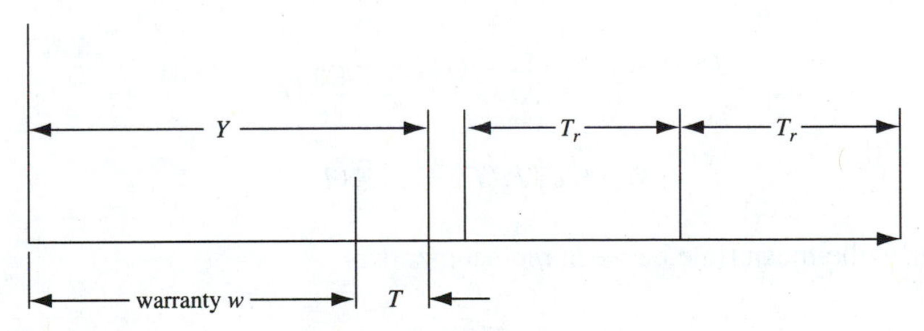

Figure 8.4. Replacement of an Item that Fails at Y

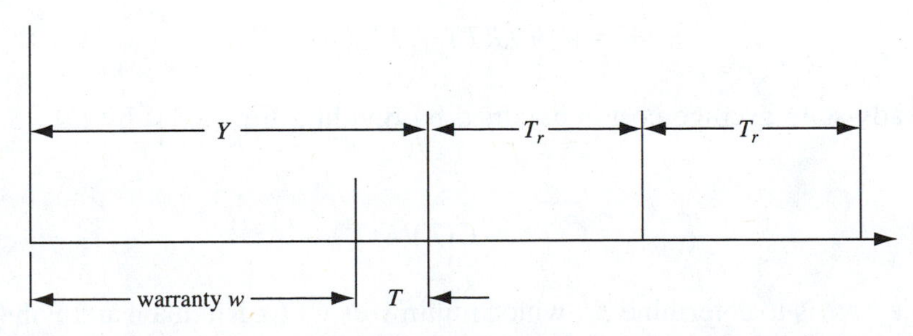

Figure 8.5. Replacement of an Item that Survives Until the Replacement Interval

Consider a linear pro-rata policy. Then the cost of replacing an item j that fails before the warranty period w is

$$I_j(\phi) = X_j c_1/w \quad \text{for } X_j < w. \tag{8.31}$$

The mean cost of replacements to the customer over the warranty period is

$$W_\phi = E\left[\sum_{j=1}^{N(w)} I_j(\phi)\right]. \tag{8.32}$$

If the item survives beyond the warranty period, the residual life of the item, $r(w)$, is a random variable with c.d.f. given by (Ross, 1970):

$$F_r(t) = F(w+t) - \int_0^w R(w+t-x)\,dm(x). \tag{8.33}$$

The cost incurred over the full replacement cycle T_r is the sum of the warranty expenses over the warranty period w, together with the replacement cost of either a failed item at time Y (as shown in Figure 8.4) or a functioning item at time $T + w$. The expected cost over the cycle is

$$\begin{aligned}\hat{C}(T) &= W_\phi + c_1\int_0^T f_r(x)\,dx + c_2\int_T^\infty f_r(x)\,dx \\ &= W_\phi + c_1 F_r(T) + c_2\bar{F}_r(T).\end{aligned} \tag{8.34}$$

Similarly, the mean time between replacements is

$$\begin{aligned}E[T_r] &= w + \int_0^T t f_r(t)\,dt + T\bar{F}_r(T) \\ &= w + G(T) + T\bar{F}_r(T).\end{aligned} \tag{8.35}$$

The steady-state average cost is obtained by dividing Eq. (8.34) by Eq. (8.35) as follows:

$$\bar{C}(T) = \hat{C}(T)/E(T_r). \tag{8.36}$$

The objective is to determine T^*, which minimizes $\bar{C}(T)$. Ritchken and Fuh (1986) prove that if $h(t)$ is continuous and monotonically nondecreasing, then a unique solution exists that minimizes Eq. (8.36).

EXAMPLE 8.8

Consider an item that exhibits a constant failure rate λ. What is the optimal replacement interval?

SOLUTION

Since the failure rate is constant, then

$$h(t) = \lambda$$

and

$$F_r(t) = 1 - e^{-\lambda t} \qquad 0 \leq t < \infty.$$

Substituting in Eq. (8.36),

$$\bar{C}(T) = \frac{W_\phi + c_1(1 - e^{-\lambda T}) + c_2 e^{-\lambda T}}{w + (1 - e^{-\lambda T})/\lambda}.$$

$\bar{C}(T)$ is monotonically decreasing in T. Hence, the optimal policy is that items should not be replaced before failure.

EXAMPLE 8.9

A hot standby system consists of two components, in parallel (1-out-of-2 system). The cost of replacing a failed unit is \$11, the failure rates of the components are identical, $\lambda = 0.2$ failures per month, $W_\phi = 1$, and the warranty length is five months. What is the optimal replacement interval?

SOLUTION

The failure distribution of two components in parallel is

$$F(x) = \prod_{i=1}^{2} F_i(x).$$

Let

$$F_i(x) = 1 - e^{-\lambda x} \qquad i = 1, 2.$$

Then

$$F(x) = 1 - 2e^{-\lambda x} + e^{-2\lambda x}.$$

The Laplace transform of $F()$ is

$$F^*(s) = \int_0^\infty e^{-sx}\, dF(x) \qquad x > 0$$

and

$$m(x) = \sum_{n=1}^{\infty} F_n(x).$$

The Laplace transform of $m()$ is

$$m^*(s) = \sum_{n=1}^{\infty} F_n^*(s) = \sum_{n=1}^{\infty} [F^*(s)]^n$$

$$m^*(s) = \frac{F^*(s)}{1 - F^*(s)}.$$

From the above equations, we obtain

$$F^*(s) = \frac{2\lambda^2}{(s+\lambda)(s+2\lambda)}$$

$$m^*(s) = \frac{2\lambda}{3s} - \frac{2\lambda}{3s + 9\lambda}.$$

Hence,

$$m'(x) = \frac{2\lambda}{3} - \frac{2\lambda}{3} e^{-3\lambda x}.$$

From Eq. (8.33), we obtain the residual life distribution after time t as

$$F_r(t) = 1 - 2e^{-\lambda(w+t)} + e^{-2\lambda(w+t)} + \frac{4}{3} e^{-\lambda(w+t)} [e^{\lambda w} - 1]$$

$$- \frac{1}{3} e^{-2\lambda(w+t)} - \frac{2}{3} e^{-\lambda(w+t)} [1 - e^{-2\lambda w}]$$

$$+ \frac{2}{3} e^{-2\lambda(w+t)} [1 - e^{-\lambda w}]$$

or

$$F_r(t) = 1 - 1.25e^{-0.2\text{t}} + 0.15e^{-0.4\text{t}}$$

and

$$h(t) = \frac{0.25e^{-0.2t} - 0.06e^{-0.4t}}{1.25e^{-0.2t} - 0.15e^{-0.4t}}.$$

$h(t)$ is monotonically nondecreasing. Hence, there exists a finite T^* that minimizes $\bar{C}(t)$. Assuming $c_2 = \$5$ and substituting in Eq. (8.36), we obtain

$$\bar{C}(T) = \frac{e^{-0.4T} - 8\frac{1}{3}e^{0.2T} + 13\frac{1}{3}}{e^{0.4T} - 16\frac{2}{3}e^{-0.2T} + 2\frac{1}{3}}.$$

$$T^* = 3.849 \text{ months.}$$

8.3. Warranty Models for Repairable Products

Most products are repairable upon failure. A warranty for such products may have a fixed duration in terms of calendar time or other measures of usage. Such products may also have a lifetime warranty, which means that the manufacturer must repair or replace the failed product during the consumer ownership of the product. The lifetime of the product may terminate due to technological obsolescence, changes in design, change in the ownership of the product, or failure of a critical component that is not under warranty. In this section, we present warranty models for repairable products.

8.3.1. WARRANTY COST FOR REPAIRABLE PRODUCTS

Consider a product that is subject to failure. Minimum repair is performed to return the product to an average condition for a working product of its age. In other words, repair is performed to restore the unit to its operational conditions (Park, 1979). The product is warranted for a warranty length w. If the product fails at any time before w, it is minimally repaired to bring it to an operational condition comparable to other products having the same age. No warranty extension beyond w is provided after repair. We now develop a model to determine the present worth of the repairs during w.

We define the following notation after Park and Yee (1984):

$R(t)$ = reliability of the product at time t,

λ = Weibull scale parameter,

β = Weibull shape parameter,

$h(t)$ = hazard rate of the product at time t,

$H(t) = \int_0^t h(t)\,dt$, cumulative hazard function,

$f_n(t)$ = p.d.f. of failure n,

r = average cost per repair,

i = nominal interest rate for discounting the future cost,

C_w = present worth of repair during w,

C_∞ = present worth of repair for a product with lifetime warranty, and

poim $(k; \mu)$ = Poisson p.m.f.; $\mu^k e^{-\mu}/k!$

Since minimal repair is performed upon failure, and the hazard rate resumes at $h(t)$ instead of returning to $h(0)$, the system failure times are not renewal points but can be described by a nonhomogeneous Poisson process (NHPP). The probability density of the time to the nth failure is (Park, 1979)

$$f_n(t) = h(t)\ \text{poim}\ (n - 1; H(t))$$

or

$$f_n(t) = \lambda\beta(\lambda t)^{\beta-1}\ \{\exp\left[-(\lambda t)^\beta\right](\lambda t)^{(n-1)\beta}/\Gamma(n)\} \tag{8.37}$$

for a Weibull distribution, where $H(t) = (\lambda t)^\beta$.

The present worth of the repairs during the warranty period is

$$C_w = \sum_{n=1}^{\infty} \int_0^w re^{-it} f_n(t)\ dt$$

$$= r\beta \int_0^{\lambda w} \exp\left[-\ iu/\lambda\right] u^{\beta-1}\ du \tag{8.38}$$

$$= r\beta(\lambda w)^\beta \exp\left(-iw\right) \sum_{k=0}^{\infty} \frac{(iw)^k}{\beta(\beta + 1)\ldots(\beta + k)}$$

or

$$C_w = r\beta\ (\lambda/i)^\beta \exp\left(-\ iw\right) \sum_{k=0}^{\infty} \frac{(iw)^{\beta+k}}{\beta(\beta + 1)\ldots(\beta + k)}. \tag{8.39}$$

For a lifetime warranty, the cost is obtained as

$$C_\infty = r\int_0^\infty h(t)e^{-it}\,dt$$

$$C_\infty = r\left(\frac{\lambda}{i}\right)^\beta \Gamma(\beta + 1). \tag{8.40}$$

EXAMPLE 8.10

The major component of a product experiences constant failure rate of 0.4 failures per year. The average repair cost is $r = \$12$, and the nominal interest rate is 5 percent per year. What is the expected warranty cost for one year?

SOLUTION

Since the component exhibits a constant failure rate, then

$$R(t) = e^{-\lambda t}.$$

Substituting $\beta = 1$ in Eq. (8.39), we obtain

$$C_w = \frac{r\lambda}{i}[e^{-iw}(e^{iw} - 1)].$$

Set $w = 1$. Then

$$C_1 = \frac{12 \times 0.4}{0.05}[1 - e^{-0.05\times 1}] = \$4.68.$$

The lifetime warranty cost is obtained by

$$C_\infty = \frac{r\lambda}{i} = \frac{12 \times 0.4}{0.05} = \$96.$$

EXAMPLE 8.11

A producer of nondestructive testing equipment is manufacturing a new ultrasonic testing unit that assesses the quality of concrete. Testing is confined to measurement of the time-of-flight of an ultrasonic pulse through the concrete from a transmitting to a receiving transducer. The pulse velocity value represents the quality of the concrete between the two transducers. Analysis of the measurements can detect the number and size of the voids in the concrete.

The producer warrants the product for a period of two years. If the product fails at any time before two years, it is minimally repaired to bring it to an age comparable to other products produced at the same time. The producer does not provide any warranty beyond the two years. The average cost per repair is \$80, and the interest rate is 5 percent per year. The cumulative hazard function of the products is expressed as

$$H(t) = (\lambda t)^{\beta},$$

where

$\lambda = 3$ years, and

$\beta = 2.5$.

Determine the expected value of the repair cost during the warranty period. Also, determine the expected value of the repair cost if the producer extends the lifetime warranty for the product.

SOLUTION The parameters of the product and the warranty policy are

$\beta = 2.5,$

$\lambda = 3,$

$w = 2,$

$r = \$80$, and

$i = 0.05.$

Using Eq. (8.39) we obtain the present value of the repair cost during the warranty period as

$$C_2 = 5.046 \times 10^6 \sum_{k=0}^{\infty} \frac{(0.1)^{3.5+k}}{(2.5)(3.5)\ldots(2.5+k)}$$

or

$$C_2 = 5.046 \times 10^6 \times 15.56 \times 10^{-6} = \$78.52.$$

The repair cost for the lifetime warranty is

$$C_\infty = r\left(\frac{\lambda}{i}\right)^{\beta}\Gamma(\beta+1)$$

or

$$C_\infty = 80\left(\frac{3}{0.05}\right)^{2.5} \Gamma(3.5) = \$7{,}413{,}847.$$

Clearly, this warranty cost is excessive, due to the increasing failure rate of the product. In such a situation, the producer may wish to redesign the product in order to reduce its failure rate.

The failure-free warranty policy is commonly used for repairable products. Under this policy the manufacturer agrees to pay the repair cost for all failures occurring during the warranty period. We first develop a general warranty model that estimates the expected warranty cost per product for a warranty length w when the failure-time distribution is arbitrary and the repair cost depends on the number of repairs carried out. We then develop models for different repair policies.

8.3.2. WARRANTY MODELS FOR A FIXED-LOT SIZE: ARBITRARY FAILURE-TIME DISTRIBUTION

Again, when a product fails, it is restored to its operating condition by repair. This model considers the case when the failure-time distribution is arbitrary. To simplify the analysis, we assume that the repair time is negligible. In addition to other notations presented earlier in this chapter, we define the following notation:

S_n = total time to the nth failure (random),

$f^{(n)}(t)$ = p.d.f. of S_n,

$F^{(n)}(t)$ = cumulative distribution function of S_n,

$N(t)$ = number of failures in $[0, t]$,

$M(t)$ = expected number of failures in $[0, t]$ (renewal function),

c_i = expected cost of the ith *repair*,

C_w = expected warranty cost per product for a warranty period w,

C_n = warranty cost given when there are exactly n failures in $[0, w]$, $C_n = \Sigma_{i=1}^{n} c_i$,

$\sigma^2(w)$ = variance of the warranty cost per product for a warranty period w, and

L = number of products sold.

The expected warranty cost per product is

$$C_w = \sum_{n=1}^{\infty} C_n P[N(w) = n], \tag{8.41}$$

where $P[N(w) = n]$ is the probability of having n failures during the warranty period $[0, w]$. Also,

$$P[N(t) = n] = F^{(n)}(t) - F^{(n+1)}(t), \tag{8.42}$$

with $F^{(0)}(t) = 1$. Substituting Eq. (8.42) into Eq. (8.41), we obtain

$$C_w = \sum_{n=0}^{\infty} C_n F^{(n)}(w). \tag{8.43}$$

The total warranty cost is equal to LC_w. Similarly, $\sigma^2(w)$ is given as (Nguyen and Murthy, 1984a)

$$\sigma^2(w) = \sum_{n=1}^{\infty} C_n^2 P[N(w) = n] - [C_w]^2$$

or

$$\sigma^2(w) = \sum_{n=1}^{\infty} [C_n^2 - C_{n-1}^2] F^{(n)}(w) - [C_w]^2. \tag{8.44}$$

The expected number of failures during the warranty period is

$$M(w) = \sum_{n=0}^{\infty} nP[N(w) = n].$$

Using Eq. (8.42), we rewrite $M(w)$ as

$$M(w) = \sum_{n=1}^{\infty} F^{(n)}(w), \tag{8.45}$$

which is the definition of the renewal function.

If the repair cost is independent of the number of failed units $C_n = C$, then the repair cost during the warranty period is

$$C_w = CM(w)$$

and the variance of the warranty cost per product becomes

$$\sigma^2(\mathrm{w}) = C^2 \,\mathrm{Var}[N(w)],$$

where

$$\mathrm{Var}[N(w)] = \sum_{n=1}^{\infty} n^2[F^{(n)}(w) - F^{(n+1)}(w)]$$

or

$$\mathrm{Var}[N(w)] = \sum_{n=1}^{\infty} (2n-1)F^{(n)}(w) - [M(w)]^2.$$

The variance for the total warranty cost is $L^2\sigma^2(w)$.

We next consider three different repair policies: the minimal repair policy, the "good-as-new" repair policy, and a mixture of these two policies. We use the subscripts 1 and 2 to refer to the first and second repair policies, respectively.

8.3.3. WARRANTY MODELS FOR A FIXED-LOT SIZE: MINIMAL REPAIR POLICY

Under this repair policy, when an item fails, it is repaired and restored to the same failure rate at the time of failure. This is the case of repairing components of large and complex systems. Clearly, repairing one or more components will not affect the total failure rate of the system since the aging of the other components will ensure that the system failure rate will remain unchanged.

The model can be characterized by a counting process $\{N(t), t \geq 0\}$ and the probability of having exactly one failure in $[t, t + dt]$ is $h(t)\,dt$. Ross (1970) shows that this process is a nonhomogeneous Poisson process (NHPP) since the failure rate does not change with time, and

$$M_1(w) = \int_0^w h(t)\,dt = -\ln F(w) \tag{8.46}$$

and

$$P[N_1(w) = n] = \frac{[M_1(w)]^n e^{-M_1(w)}}{n!}. \tag{8.47}$$

Using Eqs. (8.45) and (8.46) we show that $F_1^{(1)}(w) = F(w)$ and

$$F_1^{(n)}(w) = 1 - \sum_{i=0}^{n-1} \frac{[M_1(w)]^i e^{-M_1(w)}}{i!} \qquad n > 1. \tag{8.48}$$

8.3.4. WARRANTY MODELS FOR A FIXED-LOT SIZE: GOOD-AS-NEW REPAIR POLICY

This type of repair is usually performed for simple products where the product is completely overhauled after a failure. It is assumed that the repair will return the product to its "new" condition—that is, the failure rate after repair is significantly lower than the failure rate at the time of failure. Unlike the minimal repair policy, the good-as-new repair policy is a renewal process $\{N_2(t), t \geq 0\}$. Therefore,

$$F_2^{(1)}(w) = F(w)$$

$$F_2^{(n)}(w) = \int_0^w F_2^{(n-1)}(w - t) f(t)\, dt, \qquad n > 1, \tag{8.49}$$

and $M_2(w)$ is given by the standard renewal function

$$M_2(w) = F(w) + \int_0^w M_2(w - t) f(t)\, dt. \tag{8.50}$$

The values of $M_2(w)$ and $F_2^{(n)}(w)$ can be analytically obtained for the mixed exponential and the Erlang distributions. However, their values for a general failure-time distribution can be obtained by numerical methods.

EXAMPLE 8.12

The failure time of a product follows an Erlang distribution with k stages and its cumulative distribution function is given by

$$F(t) = 1 - e^{-\lambda t} \sum_{i=0}^{k-1} \frac{(\lambda t)^i}{i!}.$$

Assume $\lambda = 2$ failures per year and the repair cost $C_n = n$. What is the total warranty cost for a fixed production lot of 1,000 products assuming either repair policies (minimal repair or good-as-new repair)? Assume $w = 0.5$ and 2 years.

SOLUTION

For $k = 2$, the cumulative distribution function of the Erlang distribution becomes

$$F(t) = 1 - e^{-\lambda t}(1 + \lambda t).$$

We obtain the expected number of failures during the warranty period for the minimum repair policy by using Eq. (8.46)

$$M_1(t) = \lambda t - \ln(1 + \lambda t).$$

For $w = 0.5$ years,

$$M_1(0.5) = 1 - \ln 2 = 0.307.$$

Using $F_1^{(1)}(w) = F(w)$ and Eq. (8.48), we obtain the following $F_1^{(n)}(0.5)$ for different values of n:

$$F_1^{(1)}(0.5) = 1 - 2e^{-1} = 0.2642$$

$$F_1^{(2)}(0.5) = 1 - [e^{-0.307} + (0.307)e^{-0.307}] = 0.0385$$

$$F_1^{(3)}(0.5) = 1 - \left[e^{-0.307} + 0.307e^{-0.307} + \frac{(0.307)^2}{2!}e^{-0.307}\right] = 0.0038$$

$$F_1^{(4)}(0.5) = 0.000252.$$

Higher orders of $F_1^{(n)}(0.5)$ will rapidly approach zero. Therefore, without significant loss in accuracy, we stop at $F_1^{(4)}(w)$:

$$C_{0.5} = \sum_{n=0}^{\infty} nF_1^{(n)}(0.5)$$

$$C_{0.5} = 0.353.$$

The total warranty cost at $w = 0.5$ years is $0.353 \times 1{,}000 = \$353$. Similarly, for $w = 2$, we obtain

$$M_1(2) = 4 - \ln 5 = 2.39$$

and

$$C_2 = \$5{,}264.$$

The total warranty cost for $w = 2$ is \$5,264.

For the good-as-new repair policy the cumulative distribution function of the Erlang distribution is

$$F(t) = 1 - e^{-\lambda t}(1 + \lambda t)$$

and $F_2^{(n)}(t)$ and $M_2(t)$ are given in Barlow and Proschan (1965) as follows:

$$F_2^{(1)}(t) = F(t)$$

$$F_2^{(n)}(t) = 1 - e^{-\lambda t} \sum_{i=0}^{nk-1} \frac{(\lambda t)^i}{i!}$$

$$M_2(t) = \frac{\lambda t}{k} + \frac{1}{k} \sum_{j=1}^{k-1} \frac{\theta^j}{1 - \theta^j} [1 - \exp[-\lambda t(1 - \theta^j)]],$$

where $\theta = \exp(2\pi i/k)$ is a kth root of unit.

For $k = 2$, $w = 0.5$, and $\lambda = 2$ failures per year,

$$F(t) = 1 - e^{-\lambda t}(1 + \lambda t)$$

$$M_2(t) = [2\lambda t - 1 + e^{-2\lambda t}]/4.$$

Therefore,

$$F_2^{(1)}(0.5) = 0.2642$$

$$F_2^{(2)}(0.5) = 1 - e^{-1.0}\left[\frac{1}{0!} + \frac{1}{1!} + \frac{1}{2!} + \frac{1}{3!}\right]$$

$$= 0.018988$$

$$F_2^{(3)}(0.5) = 1 - e^{-1.0}\left[\frac{1}{0!} + \frac{1}{1!} + \frac{1}{2!} + \frac{1}{3!} + \frac{1}{4!} + \frac{1}{5!}\right]$$

$$= 0.0006216$$

Higher values of $F_2^{(n)}(w)$ will rapidly approach zero, and the warranty cost per product at $w = 0.5$ for good-as-new repair policy is

$$C_{0.5} = \sum_{n=1}^{\infty} nF_2^{(n)}(0.5)$$

$$= 0.30404.$$

For a lot of 1,000 units the total warranty cost is $304.04.

The expected number of failures during the warranty period is

$$M_2(0.5) = [2 \times 1 - 1 + e^{-2}]/4 = 0.284.$$

Similarly, the warranty cost per product for $w = 2$ years is

$$C_2 = \sum_{n=1}^{\infty} nF_2^{(n)}(2)$$

$$= 2.93686.$$

The total warranty cost for a lot size of 1,000 products is $2,937.

It is clear that the expected warranty cost for the minimal repair policy is always higher than the cost for the good-as-new repair policy. This is to be expected, since the product has an increasing failure-rate distribution when the minimal repair policy is used. Moreover, the rate of increase of the warranty cost as the warranty length increases is significantly much higher for the minimal repair policy when compared with the rate of increase for the good-as-new policy.

8.3.5. WARRANTY MODELS FOR A FIXED-LOT SIZE: MIXED REPAIR POLICY

We now consider the case where a repair can be either minimal or good-as-new depending on the type of failure of the product. For example, there are many components in large systems—such as modular electronic components—that must be replaced by new components upon failure. However, there are other components that require minimum repair upon failure. Indeed, these types of components are commonplace. In other situations, the same component, depending on its age, may require minimal repair or may require a replacement with a new component. In this section, we discuss the latter situation and assume that a component may experience two types of failures: type 1 requires good-as-new repair, and type 2 requires minimal repair. A product of age t experiences type 1 failure with probability $p(t)$ and type 2 with probability $1 - p(t)$. We now derive expressions for the expected number of failures of each type.

Since the failure rate after a minimal repair remains unchanged, and since for an age t the probability of good-as-new repair at failure is $p(t)$, we can define a good-as-new repair rate of the product as $p(t)\ h(t)$. After repair, the product continues to function until the next failure. The process is repeated and the intervals between good-as-new repairs are independent and identically distributed with distribution function $\mathfrak{R}(t)$ given by Nguyen and Murthy (1984a) as

$$\mathfrak{R}(t) = 1 - e^{-\int_0^t p(x)h(x)\,dx} \tag{8.51}$$

and

$$\mathfrak{R}'(t) = p(t)h(t)\bar{\mathfrak{R}}(t),$$

where

$$\bar{\mathfrak{R}}(t) = 1 - \mathfrak{R}(t).$$

The sequence of good-as-new repairs is a renewal process whose expected number of repairs during the warranty period $[0, w]$ is $M_1(w)$:

$$M_1(w) = \mathfrak{R}(w) + \int_0^w M_1(w - t)\,d\mathfrak{R}(t). \tag{8.52}$$

The expected number of minimal repairs at time t given that the age of the product is x can be expressed as

$$m_2(x) = \bar{p}(x)h(x). \tag{8.53}$$

Using the distribution function of x, we rewrite Eq. (8.53) as

$$m_2(t) = \bar{p}(t)h(t)\bar{\mathfrak{R}}(t) + \int_0^t \bar{p}(x)h(x)\bar{\mathfrak{R}}(x)\,dM_1(t - x). \tag{8.54}$$

The expected number of minimal repairs during the warranty period $[0, w]$ is obtained by integrating Eq. (8.54) with respect to t over the warranty period—that is,

$$M_2(w) = \int_0^w m_2(t)\,dt$$

or

$$M_2(w) = \int_0^w [1 + M_1(w - t)]h(t)\bar{\mathfrak{R}}(t)\,dt - M_1(w). \tag{8.55}$$

Now we can determine the total expected number of repairs during the warranty period by adding the expected number of each type of repair:

$$M(w) = M_1(w) + M_2(\mathrm{w}).$$

Add Eq. (8.52) and Eq. (8.55) to obtain

$$M(w) = \int_0^w [1 + M_1(w - t)]h(t)\bar{\mathfrak{R}}(t)\,dt. \tag{8.56}$$

Assuming that c_1 and c_2 are the expected repair costs for the good-as-new and the minimal repair policies, respectively, then

$$\begin{aligned} C_w &= c_1 M_1(w) + c_2 M_2(w) \\ &= (c_1 - c_2)M_1(w) + c_2 \int_0^w [1 + M_1(w - t)]h(t)\bar{\mathfrak{R}}(t)\,dt. \end{aligned} \tag{8.57}$$

If $p(t) = 1$, then the repair policy is good-as-new only and when $p(t) = 0$ it becomes a minimal repair policy only. Also, if $p(t) =$ constant, p, then $\bar{\mathfrak{R}}(t) = [\bar{F}(t)]^p$ and Eq. (8.56) and Eq. (8.57) reduce to (Nguyen and Murthy, 1984a)

$$M(w) = M_1(w)/p \tag{8.58}$$

$$C_w = (pc_1 + \bar{p}c_2)M_1(w)/p. \tag{8.59}$$

In some situations, a repair may result in an increase, a decrease, or a constant failure rate of the product. Under such situations, we consider the repair to be imperfect—that is, the failure-time distribution changes after each repair and the failure-time distribution of a product depends on the number of repairs performed. It is possible that the mixed repair policy discussed earlier in this chapter may experience an imperfect repair that impacts the warranty cost of the product. Therefore, the failure-time distribution of the nth failure needs to be modified, to reflect the effect of imperfect repairs, as follows.

As presented earlier, $F^{(n)}(t)$ and $f^{(n)}(t)$ are the failure-time distribution function and failure-time density function for the nth failure, respectively, and

$$F^{(1)}(w) = F(w)$$

$$F^{(n)}(w) = \int_0^w F^{(n-1)}(w - t) f_n(t)\,dt \qquad \text{for } n > 1. \tag{8.60}$$

Consider the situation where the failure-time distributions are exponential with different means—that is, $F_i(t) = 1 - e^{-\lambda_i t}$ with $\lambda_1 < \lambda_2 < \lambda_3 \ldots$, indicating a decrease in the mean time to failure. Nguyen and Murthy (1984a) illustrate that by taking Laplace transform of Eq. (8.60) and solving for $F^{(n)}(w)$, we obtain

$$F^{(1)}(w) = 1 - e^{-\lambda_1 w} \tag{8.61}$$

$$F^{(n)}(w) = \sum_{i=1}^{n} \left[\prod_{\substack{j=1 \\ j \neq i}}^{n} \left(\frac{\lambda_j}{\lambda_j - \lambda_i} \right) \right] [1 - e^{-\lambda_i w}]. \tag{8.62}$$

Once $F^{(n)}(w)$ is obtained, we can easily obtain the total warranty cost using Eq. (8.43).

EXAMPLE 8.13

A manufacturer wishes to estimate the warranty cost for 2,000 products. Assume that every time a repair is performed, it decreases the mean time to the next failure. The field data show that the failure-time distribution function is exponential with different means—that is, $F_i(t) = 1 - e^{-\lambda_i t}$ with $\lambda_1 < \lambda_2 < \lambda_3 < \ldots < \lambda_w$, where λ_i is the failure rate of the ith failure and λ_w is the failure rate of the last failure before the expiration of the warranty length w.

The manufacturer wishes to extend the warranty for two years. The failure rates of the first five failures are 0.5, 0.8, 1, 1.2, and 3 failures per year. The corresponding repair costs are 20, 19, 18, 18, and 18. Determine the total warranty cost.

SOLUTION

Since $F_i(t) = 1 - e^{-\lambda_i t}$ with $\lambda_1 < \lambda_2 < \lambda_3 \ldots$, then by using Laplace transform of the following expression

$$F^{(n)}(w) = \int_0^w F^{(n-1)}(w - t) f_n(t)\, dt, \quad \text{for } n > 1,$$

we obtain

$$F^{(1)}(w) = 1 - e^{-\lambda_1 w}$$

and

$$F^{(n)}(w) = \sum_{i=1}^{n} \left[\sum_{\substack{j=1 \\ j \neq i}}^{n} \left(\frac{\lambda_j}{\lambda_j - \lambda_i} \right) \right] [1 - e^{-\lambda_i w}]$$

$\lambda_1 = 0.5$ failures per year,

$\lambda_2 = 0.8$ failures per year,

$$\lambda_3 = 1 \text{ failure per year,}$$

$$\lambda_4 = 1.2 \text{ failures per year,}$$

$$\lambda_5 = 3 \text{ failures per year,}$$

$$w = 2 \text{ years,}$$

$$F^{(1)}(2) = 0.6321,$$

$$F^{(2)}(2) = 0.3554,$$

$$F^{(3)}(2) = 1.2869,$$

$$F^{(4)}(2) = 2.9383, \text{ and}$$

$$F^{(5)}(2) = 5.8716.$$

$$C_w = \sum_{n=0}^{\infty} C_n F^{(n)}(w)$$

$$C_w = \$201.$$

Manufacturers usually perform burn-in on new products to ensure that the product, when acquired by the customer, has already survived beyond the "infant mortality" or the decreasing failure-rate region. Thus, the number of repairs and the warranty cost during the early period of the customer's ownership of the product are minimized. However, ensuring that all products marketed have survived beyond the failure-rate region is a difficult, if not impossible, task to achieve. Moreover, most warranty periods for nonrepairable products are short. They are indeed shorter than the "infant mortality" region. Therefore, manufacturers place more emphasis on the warranty cost during the decreasing failure-rate region.

The Weibull distribution is often used to model the failure-times during this region. However, it is not an analytically tractable model. Researchers hypothesize that a failure distribution of a mixture of two or more exponential densities would exhibit the desired failure-rate characteristics. The following example shows how the warranty cost is estimated during the decreasing failure-rate region for a mixture of exponential densities.

EXAMPLE 8.14

This problem is based on the warranty model developed by Karmarkar (1978). The failure distribution of a product consists of a mixture of two exponential densities that exhibits the desired failure-rate characteristics. This can be interpreted as having a mixture of two kinds of units—a proportion p of defectives with a high failure rate λ_1 and a proportion $1 - p$ of "normal" units with a lower failure rate λ_2. The p.d.f. of the model is

$$f(t) = p\lambda_1 e^{-\lambda_1 t} + (1 - p)\lambda_2 e^{-\lambda_2 t}.$$

1. Show that the failure rate is monotone decreasing in t.
2. Assume $\lambda_1 = 4$ failures per year, $\lambda_2 = 2$ failures per year, the cost per repair is \$100, and $p = 0.4$. Determine the warranty cost for a warranty length of five years.

SOLUTION

1. The product has a mixed failure rate with

$$f(t) = p\lambda_1 e^{-\lambda_1 t} + (1 - p)\lambda_2 e^{-\lambda_2 t}$$

$$F(t) = 1 - pe^{-\lambda_1 t} - (1 - p)e^{-\lambda_2 t},$$

and

$$R(t) = pe^{-\lambda_1 t} + (1 - p)e^{-\lambda_2 t}.$$

The failure rate is

$$h(t) = \frac{f(t)}{1 - F(t)} = \frac{p\lambda_1 e^{-\lambda_1 t} + (1 - p)\lambda_2 e^{-\lambda_2 t}}{pe^{-\lambda_1 t} + (1 - p)e^{-\lambda_2 t}}.$$

At $t = 0$, $h(0) = p\lambda_1 + (1 - p)\lambda_2$ and

$$\frac{dh(t)}{dt} = \frac{-(\lambda_1 - \lambda_2)^2 p(1 - p)e^{-(\lambda_1 + \lambda_2)t}}{[pe^{-\lambda_1 t} + (1 - p)e^{-\lambda_2 t}]^2} < 0.$$

Therefore, $h(t)$ is monotone decreasing in t.

2. In order to determine the warranty cost for a five-year warranty period, we first determine the expected number of failures during the warranty period as follows. The Laplace transform of the density function is

$$f^*(s) = [p\lambda_1/(\lambda_1 + s)] + [(1 - p)\lambda_2/(\lambda_2 + s)].$$

The Laplace transform of the renewal function is

$$M(s) = \frac{f^*(s)}{s[1 - f^*(s)]}$$

$$M(s) = \frac{\lambda_1\lambda_2 + s[p\lambda_1 + (1 - p)\lambda_2]}{s^2\{s + [(1 - p)\lambda_1 + p\lambda_2]\}}.$$

Following Karmarkar (1978), we define

$$\Lambda_1 = p\lambda_1 + (1-p)\lambda_2, \Lambda_2 = (1-p)\lambda_1 + p\lambda_2 = (\lambda_1 + \lambda_2) - \Lambda_1.$$

Taking the inverse transformation of $M(s)$, we obtain

$$M(t) = ([\Lambda_1\Lambda_2 - \lambda_1\lambda_2]/\Lambda_2^2)[1 - e^{-\Lambda_2 t}] + (\lambda_1\lambda_2/\Lambda_2)t.$$

Substituting $\lambda_1 = 4$, $\lambda_2 = 2$, $p = 0.4$, and $t = 5$ in the above expression, we obtain

$$\Lambda_1 = 2.8, \Lambda_2 = 3.2, \text{ and}$$

$$M(5) = ([8.96 - 8]/3.2^2)[0.9999998875] + 12.5$$

$$M(5) = 12.5937$$

The expected warranty cost $= 100 \times 12.5937 = \$1{,}259.37$.

8.4. Warranty Claims

In the preceding sections different warranty policies and methods were presented for determining warranty length, warranty cost per unit, and warranty reserve fund that the manufacturer should allocate to cover the warranty claims during the service life of the product. It is more beneficial to the manufacturer to allocate warranty cost as a function of the age of the products, the number of claims during any time, and the number of products in service at that time. Therefore, continuous analysis of the claims data enables the manufacturer to more accurately predict the future warranty claims and to compare claim rates and cost for different product lines, different components of a product, and units from the same product that are manufactured at different times. Continuous analysis of claims data may also enable the manufacturer to assess product performance that may possibly lead to product improvement.

This section discusses methods for analyzing warranty claims in order to estimate the expected number of warranty claims per unit in service as a function of the time in service. Moreover, forecasts of the number and cost of claims on the population of all units in service along with standard error of the forecasts are also presented (Kalbfleisch, Lawless, and Robinson, 1991).

We first determine the number of claims at time t. Assume that units are sold to the customers on days x ($0 \leq x \leq \tau$). The number of claims for a unit at t days later is assumed to be Poisson with mean λ_t ($t = 0, 1, \ldots$). Since the expected number of claims λ_t is small for most situations, λ_t can be interpreted as the probability of a claim at age t. The prediction of cost of claims requires that any repair claim be immediately entered into the claims database and momentarily used in the analysis. However, repair claims are usually entered using one or more of the following procedures: (1) claims are entered as soon as they occur, (2) claims are individually entered after the lapse of time l, and (3) claims are accumulated and entered as a group at a later date.

Suppose N_x identical products are sold on day x. Repair claims enter the database after a time lag l. We define N_{xtl} to be the number of claims for products sold on day x having an age t and repair claims time lag l. The distribution of N_{xtl} is Poisson with mean $\mu_{xtl} = N_x \lambda_t f_l$, where f_l is the probability that a repair claim enters the database after a time lag l. The expected number of claims for a product up to and including time t is $\Lambda_t = \Sigma_{u=0}^{t} \lambda_u$.

Thus, the average number of claims at time t for products sold (or put in service) over the period $(0, \tau)$ is

$$m(t) = \frac{\sum_{x=0}^{\tau} \sum_{l=0}^{\infty} N_{xtl}}{\sum_{x=0}^{\tau} N_x}, \qquad t = 0, 1, \ldots \tag{8.63}$$

and

$$M(t) = \sum_{u=0}^{t} m(u). \tag{8.64}$$

We follow Kalbfleisch, Lawless, and Robinson (1991) and assume that the data are available over the calendar time 0 to T. All the claims that entered into the database by time T are included in the analysis and the counts N_{xtl} for x, t, l, such that $0 < x + t + l \leq T$ are observed. This makes the estimation of $m(t)$ and $M(t)$ a prediction problem that requires the prediction of N_{xtl}'s. Once $m(t)$ and $M(t)$ are estimated, an estimate of the cost of warranty claims can be easily made. In the following sections, we present two models for estimating the number and cost of warranty claims. The first model operates under the assumption that the probabilities of the lag time l for entering (or reporting) claims into the database are known. The second model considers the case when claims are entered as groups into the database.

8.4.1. WARRANTY CLAIMS WITH LAG TIMES

We assume that the probability of entering a warranty claim into the database after a time lag l since the claim took place, f_l, is known. Let $F_l = f_0 + f_1 + \ldots + f_l$. Moreover, the number of products (identical units of the same product) that are sold on day x, N_x, is known for $x = 0, 1, \ldots, T$, where T is the current date. Thus, the likelihood function for the claim frequency N_{xtl} is

$$L = \prod_{x+t+l \leq T} \prod \prod \frac{(N_x \lambda_t f_l)^{N_{xtl}} e^{-N_x \lambda_t f_l}}{N_{xtl}!}. \tag{8.65}$$

The maximum likelihood estimators obtained from Eq. (8.65) are

$$\hat{\lambda}_t = \frac{N_e(t)}{R_{T-t}} \qquad t = 0, 1, \ldots, T, \tag{8.66}$$

where

$$N_e(t) = \sum_{x+l \leq T-t} \sum N_{xtl} \tag{8.67}$$

is the total number of claims that have occurred at time (or age) t, and

$$R_{T-t} = \sum_{x=0}^{T-t} N_x F_{T-t-x} \tag{8.68}$$

is the adjusted count of the number of products at risk at time t. The number of products (units) sold on day x is adjusted by the probability that for a product in this group, a claim at age t would be reported by time T. In other words, to account for those claims that occurred before time T and would not be included in the analysis at T, we multiply N_x by a corresponding probability of reporting the claim before T. The average number of claims at time t for products put in service is

$$\hat{m}(t) = \hat{\lambda}_t \tag{8.69}$$

and

$$\hat{M}(t) = \sum_{u=0}^{t} \hat{\lambda}_u = \hat{\Lambda}_t. \tag{8.70}$$

It is important to note that if the time lag l is ignored or if the entering of the claims into the database is instantaneous, then the estimates of $\hat{m}(t)$ and $\hat{M}(t)$ are obtained with all of the F_l's ($l = 0, 1, \ldots$) equal to 1. Moreover, R_{T-t} is, in effect, the total number of products sold that have an age of at least t at time T. Clearly, if there is a time lag l and if it is purposely ignored in the analysis, then the estimates of λ_t are biased downward, resulting in serious errors in claim predictions.

It is also important to note that true age of a product at time t is greater than t, since products, in most cases, are temporarily stored in a warehouse as soon as they are produced until they are sold. Although the products are not in use while in the warehouse, their failure rates are affected. The longer the storage period, the higher the number of warranty claims during the warranty period, since the warranty period starts from the time the product is sold regardless of the age of the product at that time. In this case, manufacturers may reduce such claims by either adjusting the production rate, such that the total inventory and claims cost are minimized, or by redesigning the product to significantly reduce its early failure rate.

The total cost of warranty claims can be estimated by multiplying the average number of claims at time t, $M(t)$, by the average cost of a claim. It can also be estimated by grouping the claims according to the cost as follows: suppose that claim costs are indexed by $c = 1, 2, \ldots, m$ and $k(c)$ is the cost of a claim in the cth group. Also, suppose that $\lambda_t^{(c)}$ is the expected number of claims of cost $k(c)$ for a product at age t, and that $N_{xtl}^{(c)}$ is Poisson $(N_x \lambda_t^{(c)} f_l)$ independently for x, t, and l. Following the derivation of Eq. (8.66) we obtain

$$\lambda_t^{(c)} = N_e^{(c)}(t)/\mathrm{R}_{T-t}. \tag{8.71}$$

Similarly $m^{(c)}(t)$ and $M^{(c)}(t)$ are natural extensions of Eqs. (8.69) and (8.70) representing the average number of claims of cost $k(c)$ at age t and up to age t for products sold over the period $0, 1, \ldots, \tau$. The average cost of all claims up to age t for all products sold in $t = 0, 1, \ldots, \tau$ is

$$K(t) = \sum_{c=1}^{m} k(c) M^{(c)}(t). \tag{8.72}$$

EXAMPLE 8.15

A manufacturer produces temperature and humidity chambers that are used for performing accelerated life testings. The chambers are introduced over a sixty-day period with equal numbers of chambers being introduced every day. The warranty length of the chamber is one year. The true claim rate is 0.004 per chamber per day. Suppose that reporting lags of the claims are distributed over zero to fifty-nine days with probabilities $f_l = 1/80$ for $l = 0, 1, \ldots, 19$, and $40, 41, \ldots, 59$ days, and $f_l = 1/40$ for $l = 20, 21, \ldots, 39$ days. The average cost per claim is \$45. Determine the total warranty claims over a two-month period.

SOLUTION

The estimate of the claim rate at time t is

$$\hat{\lambda}_t = \frac{N_e(t)}{\sum_{x=0}^{T-t} N_x}.$$

The expected value of $N_e(t)$ is

$$E[N_e(t)] = \lambda_t R_{T-t},$$

where R_{T-t} is given by Eq. (8.68).
Thus,

$$\hat{\lambda}_t = \frac{\lambda_t R_{T-t}}{\sum_{x=0}^{T-t} N_x} = \frac{\lambda_t \sum_{x=0}^{T-t} N_x F_{T-t-x}}{\sum_{x=0}^{T-t} N_x}.$$

Since $N_x = N$ for $x = 0, 1, \ldots$, we rewrite the above expression as

$$\hat{\lambda}_t = \frac{\lambda_t \sum_{x=0}^{T-t} F_{T-t-x}}{(T-t)}.$$

Substituting the values of λ_t and F_{T-t-x}, we obtain $\hat{\lambda}_t$ for $t = 0, 1, 2, \ldots, 59$ as shown in Table 8.1.

Table 8.1. $\hat{\lambda}_t$ and F_{T-t-x} for Example 8.15

t	F_{T-t-x}	$\sum_{x=0}^{T-t} F_{T-t-x}$	$\hat{\lambda}_t$
0	0.01250	30.48749	0.00207
1	0.01250	29.48749	0.00203
2	0.01250	28.49999	0.00200
3	0.01250	27.52499	0.00197
4	0.01250	26.56249	0.00193
5	0.01250	25.61250	0.00190
6	0.01250	24.67500	0.00186
7	0.01250	23.75000	0.00183
8	0.01250	22.83750	0.00179
9	0.01250	21.93750	0.00175
10	0.01250	21.05000	0.00172
11	0.01250	20.17500	0.00168
12	0.01250	19.31250	0.00164
13	0.01250	18.46250	0.00161
14	0.01250	17.62500	0.00157
15	0.01250	16.80000	0.00153

Table 8.1. (*continued*)

t	F_{T-t-x}	$\sum_{x=0}^{T-t} F_{T-t-x}$	$\hat{\lambda}_t$
16	0.01250	15.98750	0.00149
17	0.01250	15.18750	0.00145
18	0.01250	14.40000	0.00140
19	0.02500	13.62500	0.00136
20	0.02500	12.86250	0.00132
21	0.02500	12.11250	0.00128
22	0.02500	11.38750	0.00123
23	0.02500	10.68750	0.00119
24	0.02500	10.01250	0.00114
25	0.02500	9.36250	0.00110
26	0.02500	8.73750	0.00106
27	0.02500	8.13750	0.00102
28	0.02500	7.56250	0.00098
29	0.02500	7.01250	0.00094
30	0.02500	6.48750	0.00089
31	0.02500	5.98750	0.00086
32	0.02500	5.51250	0.00082
33	0.02500	5.06250	0.00078
34	0.02500	4.63750	0.00074
35	0.02500	4.23750	0.00071
36	0.02500	3.86250	0.00067
37	0.02500	3.51250	0.00064
38	0.02500	3.18750	0.00061
39	0.01250	2.88750	0.00058
40	0.01250	2.61250	0.00055
41	0.01250	2.36250	0.00053
42	0.01250	2.12500	0.00050
43	0.01250	1.90000	0.00048
44	0.01250	1.68750	0.00045
45	0.01250	1.48750	0.00043
46	0.01250	1.30000	0.00040
47	0.01250	1.12500	0.00038
48	0.01250	0.96250	0.00035
49	0.01250	0.81250	0.00033
50	0.01250	0.67500	0.00030
51	0.01250	0.55000	0.00028
52	0.01250	0.43750	0.00025
53	0.01250	0.33750	0.00023
54	0.01250	0.25000	0.00020
55	0.01250	0.17500	0.00018
56	0.01250	0.11250	0.00015
57	0.01250	0.06250	0.00013
58	0.01250	0.02500	0.00010

The expected value for estimate $\hat{\Lambda}_{58} = \Sigma_{i=0}^{58} \lambda_i$, or $\hat{\Lambda}_{58} = 0.05928$ claims. Assume that ten chambers are introduced every day. The expected warranty cost for the claims after two months is

$$\text{Claim cost} = 0.05928 \times 10 \times 58 \times 45$$

or

$$\text{Claim cost} = \$1{,}547.$$

8.4.2. WARRANTY CLAIMS FOR GROUPED DATA

In this section we estimate the total warranty claims when the claims are grouped based on the age of the product. For example, we may know the total number of claims for all products whose ages fall between t_1 and t_2 days, $t_2 + 1$ and t_3, and so forth. All the claims for the products whose ages are within a time interval are reported as a group at a future time t—that is, all the products in the group have the same reporting lag. Again, we assume that the reporting lag distribution f_l is known.

Consider some age interval $t = [a, b]$, inclusive. The average number of claims per product for this age interval is

$$M(a, b) = \sum_{t=a}^{b} \lambda_t. \tag{8.73}$$

Using Eqs. (8.66) and (8.73), we estimate the average number of claims per product for the age interval $[a, b]$ as

$$\sum_{t=a}^{b} \hat{\lambda}_t = \sum_{t=a}^{b} \frac{N_e(t)}{R_{T-t}}. \tag{8.74}$$

If we only observe the total number of claims that have occurred during the interval $[a, b]$—that is, if we observe only $\Sigma_{t=a}^{b} N_e(t)$, then we approximate Eq. (8.74) by

$$M(a, b) = \frac{\sum_{t=a}^{b} N_e(t)}{R(a, b)}, \tag{8.75}$$

where $M(a, b)$ is an approximation of $M(a, b)$, and $R(a, b)$ is an estimate of the product-days in service. An approximation of $R(a, b)$ is

$$R(a, b) = \frac{1}{2}(R_{T-a} + R_{T-b}) \tag{8.76}$$

or

$$R(a, b) = \frac{1}{b - a + 1} \sum_{t=a}^{b} R_{T-t}. \tag{8.77}$$

The expected warranty claim cost C_{total} is

$$C_{\text{total}} = \bar{k} M(a, b) N_{a,b}, \tag{8.78}$$

where

$\bar{k}$ = the average cost per claim, and

$N_{a,b}$ = the number of products whose ages are between a and b, inclusive.

The approximation (8.78) becomes more accurate as the interval $[a, b]$ decreases.

PROBLEMS

8-1. A manufacturer of medical devices uses shape memory alloys (notably nickel-titanium) to manufacture novel devices. The alloys can be heated at one temperature and then heated to recover their original shape. They can be elastically deformed 10 to 20 percent more than the conventional materials. The manufacturer produces a "micro vessel correction" device that, when inserted into blood vessels, is warmed by the blood and expands outward to maintain the desired vessel shape.

Experimental results show that the device experiences a constant failure rate of 0.008333 failures per month. The price of the device is $1,250, and the yearly production of the device is 3,000 devices. Assume that the manufacturer wishes to extend a five-year warranty for this device. Determine the warranty reserve fund and the adjusted price of the device.

8-2. The producer of high-precision instruments needs to extend a warranty for a new sensor that is capable of measuring temperature accurately in the range of 1,500°F to 2,000°F. Historical data show that the sensor's hazard rate can be expressed as

$$h(t) = kt,$$

where $k = 0.000085$. The cost of the sensor, not including warranty, is \$80. Assume that the manufacturer is limiting the selling price to \$90 after inclusion of the warranty cost. Determine the warranty length and the total warranty reserve fund for 3,000 sensors.

8-3. Determine the lump-sum value to be paid to the customer when the product fails during the warranty period for Problem 8-2.

8-4. A manufacturer wishes to estimate the warranty cost for 3,000 products. Assume that every time a repair is performed, it decreases the mean time to the next failure. The field data show that the failure-time distribution function is exponential with different means—that is, $F_i(t) = 1 - e^{-\lambda_i t}$ with $\lambda_1 < \lambda_2 < \lambda_3 < \ldots < \lambda_w$, where λ_i is the failure rate of the ith failure and λ_w is the failure rate of the last failure before the expiration of the warranty length, w. The manufacturer wishes to extend the warranty for three months. The failure rate of the first five failures are 0.5, 0.8, 1, 1.2, and 3 failures per year. The corresponding repair costs are 20, 19, 18, 18, 18. Determine the total warranty cost.

8-5. The failure distribution of a product consists of a mixture of two exponential densities that exhibit the desired failure-rate characteristics. This can be interpreted as having a mixture of two kinds of units—a proportion p of defectives with a high failure rate λ_1, and a proportion $1 - p$ of "normal" units with a lower failure rate λ_2. The p.d.f. of the model is

$$f(t) = p\lambda_1 e^{-\lambda_1 t} + (1 - p)\lambda_2 e^{-\lambda_2 t}.$$

Assume $\lambda_1 = 4$ failures per year, $\lambda_2 = 3$ failures per year, and the cost per repair is \$80. Determine the warranty cost for a warranty length of three years. Also, assume $p = 0.3$.

8-6. Consider the following notation.

c = product price including warranty cost,

w = length of warranty,

m = MTTF of the product,

$C(t)$ = pro-rata customer rebate at time t; $C(t) = C(1-t/w)$, $0 < t < w$, and

r = warranty reserve cost per unit.

Derive an expression for the total warranty reserve fund for L units of production assuming that the product exhibits the following hazard-rate function:

$$h(t) = \frac{\beta}{\lambda^{\beta}} t^{\beta-1}.$$

8-7. Determine the proportion of the unit cost to be refunded as a lump-sum rebate that makes both the pro-rata and lump-sum plan equivalent for Problem 8-6.

8-8. Develop the confidence interval for the expected cost of the pro-rata warranty when the failure-time distribution is given by

$$f(t) = \lambda e^{-\lambda t} \qquad \lambda > 0.$$

8-9. Develop the confidence interval for the expected cost of the full replacement warranty policy (FRW) when the failure time is given by

$$f(t) = \frac{\gamma}{\theta} t^{\gamma-1} \exp\left[\frac{-t^{\gamma}}{\theta}\right] \qquad t \geq 0, \gamma \geq 1, \theta > 0.$$

8-10. A car insurance company wishes to estimate the average warranty claims per year for a newly introduced car model. In collaboration with the manufacturer, the insurance company obtained the failure data shown in Table 8.2 from the laboratory testing of different components of the car. The claims are approximately equal in value regardless of the type of failure. This implies that the failure data of all the components can be analyzed as if they came from one type of failure. The table shows data obtained from subjecting eighty-four cars to continuous testing. Assume that the failure time follows a Weibull distribution; the cost of a claim is \$85; the average miles per car is 15,000 per year; and 20,000 cars were introduced into the market. What is the total cost of claims per year for the next five years?

Table 8.2. Failure Times in Hours

Failure Time	*Number of Failed Units*
1,000	10
2,100	12
3,400	9
4,400	11
5,800	10
7,000	14
8,200	10
10,000	8

8-11. A manufacturer of portable telephones intends to extend a thirty-six month warranty on the new product. An accelerated test showed that the failure time of such products at normal operating conditions exhibits a Weibull distribution with a shape parameter of 2.2 and a scale parameter of 10,000. The price of a telephone unit is $120 (not including warranty cost). Assuming a total production of 15,000, 20,000, and 25,000 in years 1, 2, and 3, respectively, determine the warranty reserve fund and the adjusted price of the telephone unit.

8-12. The manufacturer of the telephone units wishes to offer the customer a choice of one of the following warranties:

a. A full rebate policy for the duration of the warranty length.

b. A mixed policy that offers a full compensation if the product fails before time w_1, followed by a linearly pro-rated compensation up to the end of the warranty service of thirty-six months.

Design a mixed warranty policy whose reserve fund is equivalent to that of the full rebate policy.

8-13. Consider a good-as-new repair policy where the product is completely overhauled after a failure. The repair returns the product to its "new" condition. Assume that the failure-time distribution is

$$f(t) = \begin{cases} \dfrac{t^{\alpha_1 - 1}(1 - t)^{\alpha_2 - 1}}{B(\alpha_1, \alpha_2)} & \text{if } 0 < x < 1 \\ 0 & \text{otherwise,} \end{cases}$$

where $B(\alpha_1, \alpha_2)$ is the beta function defined by

$$B(\alpha_1, \alpha_2) = \int_0^1 t^{\alpha_1 - 1}(1 - t)^{\alpha_2 - 1}\, dt$$

for any real numbers $\alpha_1 > 0$ and $\alpha_2 > 0$. Also,

$$B(\alpha_1, \alpha_2) = \frac{\Gamma(\alpha_1)\Gamma(\alpha_2)}{\Gamma(\alpha_1 + \alpha_2)}.$$

The parameters of the failure distribution are $\alpha_1 = 1.5$ and $\alpha_2 = 3.0$. Assume $w =$ 1 (one year) and the repair cost $c_n = n$. What is the total warranty cost for a fixed production lot of 2,000 units?

8-14. A warm standby system consists of two components in parallel. The cost of replacing a failed unit is \$20. The failure rates of the components are $\lambda_1 = 0.3$ and $\lambda_2 = 0.1$ failures per month. Assuming $W_\phi = 1$ and a warranty length of six months, what is the optimal replacement interval?

8-15. Burrs are considered a major problem in machining operations, punching, or casting processes. Many applications require that all burrs and sharp edges be removed to the extent that material fragments are not visible and sharpness cannot be felt. A manufacturer produces a cost-effective deburring tool that removes burrs and sharpness. The manufacturer intends to sell the tool (excluding warranty cost) for \$22 and provides a lump-sum warranty for a six-month duration. The annual production is 6,000 tools and the manufacturer intends to invest the warranty fund at an interest rate of 4 percent per year and to increase the price by 3 percent after six months. Assume that the tools experience a constant failure rate of 0.006 failures per month. Determine the price of the tool including warranty cost, the proportion of the lump-sum rebate to be paid to the customer when the tool fails before six months, and the total warranty reserve fund.

8-16. A manufacturer wishes to change the current warranty policy on one of the products from being a pure pro-rata rebate policy with a duration of twelve months to a full rebate policy. The full rebate consists of a lump-sum equivalent to the initial cost of the product if it fails before W_0 months from the date of purchase. The failure time of the product follows an Erlangian distribution with three stages ($k = 3$). The parameter λ of the distribution is 0.005 failures per hour. The cost of the product is \$120, and its failure-time distribution function is

$$F(t) = 1 - e^{-\lambda t}\left\{\sum_{j=0}^{k-1} \frac{(\lambda t)^j}{j!}\right\}.$$

a. What is the expected number of failures during a twelve-month period?

b. Determine the length of the full rebate that makes it equivalent, in cost, to the current warranty policy.

8-17. A manufacturer introduces a new thermal printer over a six-month period and the number of printers sold at day x equals 5 when x is even and equals 10 when x is odd. Assume that the claims are entered into the database with time lags distributed over zero to twenty-nine days. The probabilities associated with the time lags are $f_l = 1/20$ for $l = 0, 1, \ldots, 9$, $f_l = 1/30$ for $l = 10, 11, \ldots, 19$, and $f_l = 1/60$ for $l = 20, 21, \ldots, 29$. Assume that the average claim rate is 0.001 per printer per day and that the average claim cost is \$35. Determine the total warranty claims after two months of introducing the printers (the length of warranty is six months).

8-18. Assume that the manufacturer in Problem 8-17 decides to group the claims of the printers based on their age. In doing so, the manufacturer groups all the claims for printers whose ages fall within the same fifteen-day interval. After two months, the following claims were accumulated:

Age group	*0–14 days*	*15–30 days*	*31–45 days*	*46–60 days*
Number of Claims	12	10	9	5

The probability distribution of the time lag for reporting the claims of a group is $f_l = 1/20$ for $l = 0, 1, \ldots, 9$, $f_l = 1/30$ for $l = 10, 11, \ldots, 19$, and $f_l = 1/60$ for $l = 20, 21, \ldots, 29$.

Determine the total cost of warranty claims over a two-month period.

REFERENCES

Amato, H. N., and Anderson, E. E. (1976). "Determination of Warranty Reserves: An Extension." *Management Science* 22(2), 1391–1394.

Barlow, R. E., and Proschan, F. (1965). *Mathematical Theory of Reliability*. New York: John Wiley.

Blischke, W. R., and Murthy, D. N. P. (1994). *Warranty Cost Analysis*. New York: Marcel Dekker.

Kalbfleisch, J. D., Lawless, J. F., and Robinson, J. A. (1991). "Methods for the Analysis and Prediction of Warranty Claims." *Technometrics* 33(3), 273–285.

Karmarkar, U. S. (1978). "Future Costs of Service Contracts for Consumer Durable Goods," *AIIE Transactions* 10(4), 380–387.

Mamer, J. W. (1987). "Discounted and Per Unit Costs of Product Warranty." *Management Science* 33(7), 916–930.

Menke, W. W. (1969). "Determination of Warranty Reserves." *Management Science* 15(10), B542–B549.

Nguyen, D. G., and Murthy, D. N. P. (1984a). "A General Model for Estimating Warranty Costs for Repairable Products." *IIE Transactions* 16(4), 379–386.

Nguyen, D. G., and Murthy, D. N. P. (1984b). "Cost Analysis of Warranty Policies." *Naval Research Logistics Quarterly* 31, 525–541.

Park, K. S. (1979). "Optimal Number of Minimal Repairs Before Replacement." *IEEE Transactions on Reliability* R-28, 137–140.

Park, K. S., and Yee, S. R. (1984). "Present Worth of Service Cost for Consumer Product Warranty," *IEEE Transactions on Reliability* R-33(5), 424–426.

Ritchken, P. H. (1985). "Warranty Policies for Non-Repairable Items Under Risk Aversion." *IEEE Transactions on Reliability* R-34(2), 147–150.

Ritchken, P. H., and Fuh, D. (1986). "Optimal Replacement Policies for Irrepairable Warrantied Items." *IEEE Transactions on Reliability* R-35(5), 621–623.

Ross, S. M. (1970). *Applied Probability Models with Optimization*. San Francisco: Holden-Day.

Thomas, M. U. (1983). "Optimum Warranty Policies for Non-Repairable Items." *IEEE Transactions on Reliability* R-32(3), 282–287.

CHAPTER 9

Preventive Maintenance and Inspection

A component's degree of reliability is directly proportional to its ease of accessibility; i.e., the harder it is to get to, the more often it breaks down.

—Jonathan Waddell, crew member of the oil tanker Exxon New Orleans

Reliability of a system is greatly affected by its structural design, the quality and reliability of its components, and the implementation of an effective preventive maintenance and inspection program, when applicable. Earlier chapters of this book presented methods for estimating reliability of different structural designs such as series, parallel, parallel-series, series-parallel, k-out-of-n, and complex networks. We also presented methods for estimating reliability of components using accelerated and operational life testing. In this chapter we introduce models for optimum preventive maintenance, replacements, and inspection (PMRI) schedules. The term *optimum* arises from the fact that high frequency of PMRI increases the total cost of maintenance and reduces the cost due to the downtime of the system, whereas low frequency of PMRI decreases the cost of maintenance but increases the cost due to the downtime of the system. Hence, depending on the type of failure-time distribution, an optimum PMRI may exist. Preventive maintenance may imply minimal repairs, replacements, or inspection of the components. In some systems, such as a microprocessor of a programmable logical controller, minimal repairs or inspections are not applicable. Similarly, there are systems whose status can be determined only by inspection or testing—as in "one-shot" devices, such as military explosives. In this case, replacement is the only possible alternative.

The primary function of the preventive maintenance and inspections is to control the condition of the equipment and ensure its availability. Doing so requires the determination of the following:

- Frequency of the preventive maintenance, replacements, and inspections,
- Replacement rules for components,
- Effect of technological changes on the replacement decisions,
- The size of the maintenance crew,
- Optimum inventory levels of spare parts,
- Sequencing and scheduling rules for maintenance jobs, and
- Number and type of machines available in the maintenance workshop.

The above topics are a partial list of what constitutes a comprehensive preventive maintenance, replacement, and inspection system. This chapter presents analytical models that address some of these topics. More specifically, we present different approaches for determining the optimum frequency to perform preventive maintenance, replacements, and inspections for systems operating under different conditions. Methods for determining the optimum inventory levels of spare parts also are discussed.

9.1. Preventive Maintenance and Replacement Models: Cost Minimization

Preventive maintenance and replacements are maintenance actions that are performed on the system by making minimal repairs or full replacements of some of the system's components or the entire system. Before presenting the analytical models for preventive maintenance and replacements, it is important to note that most, if not all, models available in the literature reasonably assume the following:

- The total cost associated with failure replacement is greater than that associated with a preventive maintenance action whether it is a repair or replacement. In other words, the cost to repair the system after its failure is greater than the cost of maintaining the system before its failure. For example, replacement of a cutting tool in a milling operation before the breakage of the tool may result in a reduced total cost of the milling operation since a sudden tool breakage may cause damage to the workpiece.
- The system's failure-rate function is monotonically increasing with time. Clearly, if the system's failure rate is decreasing with time, then the system is likely to improve with time and any preventive maintenance action or replacement is considered a waste of resources. Likewise, if the equipment or system has a constant failure rate, then any preventive maintenance action is also a waste of resources. This can be attributed to the fact that when the failure rate is constant, replacing equipment before failure does not affect the probability that the equipment will fail in the next instant, given that it is now good (Jardine and Buzacott, 1985).

- Minimal repairs do not change the failure rate of the system. Even though a component in a system may be replaced with a new component, the complexity of the system and the large number of components in the system make the effect of such replacement negligible or nonexistent.

In the following sections, we examine common policies for preventive maintenance and replacements.

9.1.1. THE CONSTANT INTERVAL REPLACEMENT POLICY (CIRP)

The constant interval replacement policy (CIRP) is the simplest preventive maintenance and replacement policy. Under this policy, two types of actions are performed. The first type is the preventive replacement that occurs at fixed intervals of time. Components or parts are replaced at predetermined times regardless of the age of the component or the part being replaced. The second type of action is the failure replacement where components or parts are replaced upon failure. This policy is illustrated in Figure 9.1 and is also referred to as *block replacement policy*.

As mentioned earlier, the objective of the PMRI models is to determine the parameters of the preventive maintenance policy that optimize some criterion. The most widely used criterion is the total expected replacement cost per unit time. This can be accomplished by developing a total expected cost function per unit time as follows.

Let $c(t_p)$ be the total replacement cost per unit time as a function of t_p. Then

$$c(t_p) = \frac{\text{Total expected cost in interval } (0, t_p]}{\text{Expected length of the interval}}. \tag{9.1}$$

The total expected cost in the interval $(0, t_p]$ is the sum of the expected cost of failure replacements and the cost of the preventive replacement. During the interval $(0, t_p]$, one preventive replacement is performed at a cost of c_p and $M(t_p)$ failure

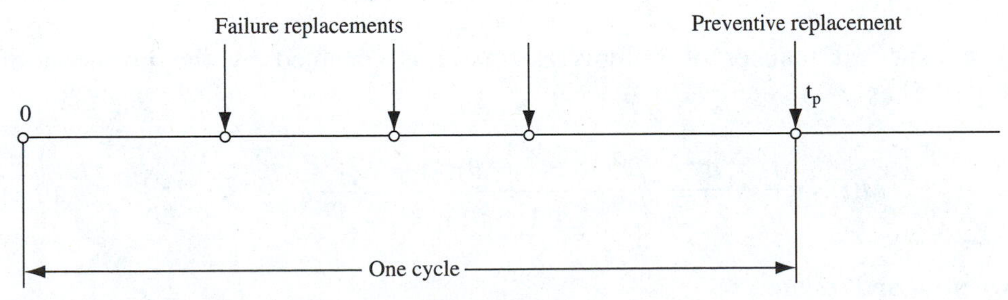

Figure 9.1. Constant Interval Replacement Policy

replacements at a cost of c_f each, where $M(t_p)$ is the expected number of replacements (or renewals) during the interval $(0, t_p]$. The expected length of the interval is t_p. Equation (9.1) can be rewritten as

$$c(t_p) = \frac{c_p + c_f M(t_p)}{t_p}. \tag{9.2}$$

The expected number of failures, $M(t_p)$, during $(0, t_p]$ may be obtained using any of the methods discussed earlier in this book (see Chapter 7).

EXAMPLE 9.1

A critical component of a complex system fails when its failure mechanism enters one of two stages. Suppose that the failure mechanism enters the first stage with probability θ and that it enters the second stage with probability $1 - \theta$. The probability density functions of failure time for the first and second stage are $\lambda_1 e^{-\lambda_1 t}$ and $\lambda_2 e^{-\lambda_2 t}$, respectively. Determine the optimal preventive replacement interval of the component for different values of λ_1, λ_2, and θ if c_p = \$100 and c_f = \$300.

SOLUTION

The p.d.f. of the failure time of the component is

$$f(t) = \theta\lambda_1 e^{-\lambda_1 t} + (1 - \theta)\lambda_2 e^{-\lambda_2 t}, \tag{9.3}$$

and the Laplace transform is

$$f^*(s) = \frac{\lambda_1\lambda_2 + \theta\lambda_1 s + (1 - \theta)\lambda_2 s}{(\lambda_1 + s)(\lambda_2 + s)}. \tag{9.4}$$

The above equation has roots 0 and $s_1 = -[(1 - \theta)\lambda_1 + \theta\lambda_2]$. The expansion of the Laplace transform equation of the expected number of failures is

$$M^*(s) = \frac{1}{s^2\mu} + \frac{1}{s}\frac{\sigma^2 - \mu^2}{2\mu^2} - \frac{\theta(1 - \theta)(\lambda_1 - \lambda_2)^2}{[(1 - \theta)\lambda_1 + \theta\lambda_2]^2 (s - s_1)}. \tag{9.5}$$

The expected number of failures at time t is obtained by the inversion of Eq. (9.5) as

$$M(t) = \frac{t}{\mu} + \frac{\sigma^2 - \mu^2}{2\mu^2} - \frac{\theta(1 - \theta)(\lambda_1 - \lambda_2)^2}{[(1 - \theta)\lambda_1 + \theta\lambda_2]^2} e^{-[(1-\theta)\lambda_1 + \theta\lambda_2]t}, \tag{9.6}$$

where μ and σ^2 are

$$\mu = E(t) = \int_0^\infty t f(t)\, dt$$

$$\sigma^2 = E(t^2) - [E(t)]^2.$$

Without solving this example numerically, it is clear that when $\lambda_1 = \lambda_2$ the component has a constant failure rate. Therefore, the optimal preventive replacement is to replace the component upon failure.

When the preventive replacement is performed at discrete time intervals, such as every four weeks, it is then more appropriate to estimate the expected number of failures during the time interval by using the discrete time approach discussed earlier in this book.

EXAMPLE 9.2

A sliding bearing of a high-speed rotating shaft wears out according to a normal distribution with mean of 1,000,000 cycles and standard deviation of 100,000 cycles. The cost of preventive replacement is $50 and that of the failure replacement is $100. Assuming that the preventive replacements can be performed at discrete time intervals equivalent to 100,000 cycles per interval, determine the optimum preventive replacement interval.

SOLUTION

Substituting the cost elements in Eq. (9.2) results in

$$c(t_p) = \frac{50 + 100M(t_p)}{t_p}.$$

Using the discrete-time approach for calculating the expected number of failures and equating 100,000 cycles to one time interval, then

$$M(0) = 0$$

$$M(1) = [1 + M(0)] \frac{1}{\sqrt{2\pi}} \int_0^1 \exp\left[\frac{-(t-10)^2}{2}\right] dt$$

$$M(1) = [1 + M(0)]\phi(1 - 10) = [1 + 0]0 = 0,$$

where

$$\phi(t) = \frac{1}{\sqrt{2\pi}} \int_{-\infty}^{t} \exp\left[\frac{-t^2}{2}\right] dt$$

is the cumulative distribution function of the standardized normal distribution with mean = 0 and standard deviation = 1:

$$M(2) = [1 + M(1)]\frac{1}{\sqrt{2\pi}}\int_0^1 \exp\left[\frac{-(t-10)^2}{2}\right]dt$$

$$+ [1 + M(0)]\frac{1}{\sqrt{2\pi}}\int_1^2 \exp\left[\frac{-(t-10)^2}{2}\right]dt$$

$$M(2) = [1 + 0]0 + [1 + 0][\phi(-8) - \phi(-9)] = 0.$$

Similarly,

$$M(3) = 0, \quad M(4) = 0, \quad M(5) = 0, \quad M(6) = 0$$

$$M(7) = [1 + M(6)]\frac{1}{\sqrt{2\pi}}\int_0^1 \exp\left[\frac{-(t-10)^2}{2}\right]dt$$

$$+ [1 + M(5)]\frac{1}{\sqrt{2\pi}}\int_1^2 \exp\left[\frac{-(t-10)^2}{2}\right]dt$$

$$+ [1 + M(4)]\frac{1}{\sqrt{2\pi}}\int_2^3 \exp\left[\frac{-(t-10)^2}{2}\right]dt$$

$$+ [1 + M(3)]\frac{1}{\sqrt{2\pi}}\int_3^4 \exp\left[\frac{-(t-10)^2}{2}\right]dt$$

$$+ [1 + M(2)]\frac{1}{\sqrt{2\pi}}\int_4^5 \exp\left[\frac{-(t-10)^2}{2}\right]dt$$

$$+ [1 + M(1)]\frac{1}{\sqrt{2\pi}}\int_5^6 \exp\left[\frac{-(t-10)^2}{2}\right]dt$$

$$+ [1 + M(0)]\frac{1}{\sqrt{2\pi}}\int_6^7 \exp\left[\frac{-(t-10)^2}{2}\right]dt$$

$$M(7) = [1 + 0]\,[\phi(-3) - \phi(-4)] = 0.0014$$

$$M(8) = 0 + [1 + M(1)]\frac{1}{\sqrt{2\pi}}\int_6^7 \exp\left[\frac{-(t-10)^2}{2}\right]dt$$

$$+ [1 + M(0)]\frac{1}{\sqrt{2\pi}}\int_7^8 \exp\left[\frac{-(t-10)^2}{2}\right]dt = 0.00275$$

$$M(9) = 0.15875$$

$$M(10) = 0.50005$$

$$M(11) = 0.84135.$$

The summary of the calculations is shown in Table 9.1. From the table, the minimum cost per cycle corresponds to 800,000 cycles. Therefore, the optimum preventive replacement length is equivalent to 800,000 cycles of the sliding bearing.

Table 9.1. Calculations for the Optimal Preventive Interval

Interval, t_p	$M(t_p)$	$c(t_p)$
100,000	0	0.000500
200,000	0	0.000250
300,000	0	0.000166
400,000	0	0.000125
500,000	0	0.000100
600,000	0	0.000083
700,000	0.0014	0.000072
800,000	0.00275	0.000063*
900,000	0.15875	0.000073
1,000,000	0.50005	0.000100
1,100,000	0.84135	0.0001219

*Indicates minimum cost

9.1.2. REPLACEMENT AT PREDETERMINED AGE

The disadvantage of the constant interval replacement policy is that the units or components are replaced at failures and at a constant interval of time since the last preventive replacement. This may result in performing preventive replacements on units shortly after failure replacements. Under the replacement at predetermined age policy, the units are replaced upon failure or at age t_p, whichever occurs first. The models of Barlow and Hunter (1960), Senju (1957), and Jardine (1973), *inter alia*, apply. Following Blanks and Tordan (1986) and Jardine (1973), if the part's operating cost is independent of time, the cost per unit time is

$$c(t_p) = \frac{\text{Total expected replacement cost per cycle}}{\text{Expected cycle length}}. \tag{9.7}$$

To calculate both numerator and denominator of Eq. (9.7) we need first to discuss a typical cycle. There are two possible cycles of operation. The first is when the equipment reaches its planned preventive replacement age t_p, and the second is when the equipment fails before the planned replacement age. Hence, the numerator of the above equation can be calculated as (Jardine and Buzacott, 1985)

Numerator Cost of preventive replacement × Probability the equipment survives to the planned replacement age + Cost of failure replacement × Probability of equipment failure before t_p:

$$= c_p R(t_p) + c_f[1 - R(t_p)]. \tag{9.8}$$

Similarly, the denominator is obtained as

Denominator Length of a preventive cycle × Probability of a preventive cycle + Expected length of a failure cycle × Probability of a failure cycle:

$$= t_p R(t_p) + \int_{-\infty}^{t_p} t f(t)\, dt. \tag{9.9}$$

Dividing Eq. (9.8) by (9.9), we obtain

$$c(t_p) = \frac{c_p R(t_p) + c_f[1 - R(t_p)]}{t_p R(t_p) + \int_{-\infty}^{t_p} t f(t)\, dt}. \tag{9.10}$$

The optimum value of the length of the preventive replacement cycle is obtained by determining t_p that minimizes Eq. (9.10). This can be achieved by taking the partial derivative of Eq. (9.10) with respect to t_p and equating the resultant equation to zero as shown below:

$$\frac{\partial c(t_p)}{\partial t_p} = \frac{[-c_p f(t_p) + c_f f(t_p)] \cdot \int_0^{t_p} R(t)\, dt - [c_p R(t_p) + c_f F(t_p)] R(t_p)}{\left[\int_0^{t_p} R(t)\, dt\right]^2} = 0.$$

The optimal preventive replacement cycle t_p^* is obtained by simple algebraic manipulations of the above expression as follows:

$$f(t_p^*)[c_f - c_p] \int_0^{t_p^*} R(t)\,dt = [c_p R(t_p^*) + c_f F(t_p^*)]R(t_p^*)$$

$$\frac{f(t_p^*)}{R(t_p^*)} \int_0^{t_p^*} R(t)\,dt = \frac{1}{c_f - c_p}[c_p R(t_p^*) + c_p F(t_p^*) + c_f F(t_p^*) - c_p F(t_p^*)]$$

$$h(t_p^*) \int_0^{t_p^*} R(t)\,dt = \frac{1}{c_f - c_p}[c_p + (c_f - c_p)F(t_p^*)]$$

$$\text{or} \quad h(t_p^*) \int_0^{t_p^*} R(t)\,dt = \frac{c_p}{c_f - c_p} + F(t_p^*).$$

EXAMPLE 9.3

Assume that constant interval replacement policy in Example 9.2 is to be compared with an age replacement policy using the same cost values. Determine the optimum preventive replacement interval for the age replacement policy. Which policy is preferred?

SOLUTION

We evaluate Eq. (9.10) for different values of t_p:

$$c(t_p) = \frac{100 - 50\,R(t_p)}{t_p R(t_p) + \int_{-\infty}^{t_p} t f(t)\,dt}, \tag{9.11}$$

where

$$R(t_p) = 1 - \int_{-\infty}^{t_p} f(t)\,dt = \int_{t_p}^{\infty} f(t)\,dt$$

or

$$R(t) = \frac{1}{\sqrt{2\pi}} \int_{t_p}^{\infty} \exp\left[\frac{-(t-10)^2}{2}\right] dt.$$

Equivalently,

$$R(t_p) = \frac{1}{\sqrt{2\pi}} \int_{t_p-10}^{\infty} \exp\left[\frac{-t^2}{2}\right] dt. \tag{9.12}$$

The second term in the denominator of Eq. (9.11) is obtained as follows:

$$\int_{-\infty}^{t_p} tf(t)\,dt = \int_{-\infty}^{t_p} \frac{t}{\sigma\sqrt{2\pi}} \exp\left[\frac{-(t-\mu)^2}{2\sigma^2}\right] dt$$

$$= \frac{1}{\sigma\sqrt{2\pi}} \int_{-\infty}^{t_p} [(t-\mu)+\mu] \exp\left[\frac{-(t-\mu)^2}{2\sigma^2}\right] dt$$

$$= \frac{1}{\sigma\sqrt{2\pi}} \int_{-\infty}^{t_p} (t-\mu) \exp\left[\frac{-(t-\mu)^2}{2\sigma^2}\right] dt$$

$$+ \frac{1}{\sigma\sqrt{2\pi}} \int_{-\infty}^{t_p} \mu \exp\left[\frac{-(t-\mu)^2}{2\sigma^2}\right] dt$$

$$= \frac{\sigma}{\sqrt{2\pi}} \int_{-\infty}^{t_p} -d\left(\exp\left[\frac{(t-\mu)^2}{2\sigma^2}\right]\right) + \mu\Phi\left(\frac{t_p-\mu}{\sigma}\right)$$

$$= \frac{-\sigma}{\sqrt{2\pi}} \exp\left[\frac{-(t_p-\mu)^2}{2\sigma^2}\right] + \mu\Phi\left(\frac{t_p-\mu}{\sigma}\right)$$

or

$$\int_{-\infty}^{t_p} tf(t)\,dt = -\sigma\phi\left(\frac{t_p-\mu}{\sigma}\right) + \mu\Phi\left(\frac{t_p-\mu}{\sigma}\right), \tag{9.13}$$

where

$$\phi(t) = \frac{1}{\sqrt{2\pi}} \exp\left[\frac{-t^2}{2}\right] \quad \text{and} \quad \Phi(t) = \frac{1}{\sqrt{2\pi}} \int_{-\infty}^{t} \exp\left[\frac{-t^2}{2}\right] dt$$

are referred to as the ordinate and cumulative distribution functions of the standard normal distribution $N(0, 1)$, respectively, and their values are shown in Appendix E. The summary of the calculations for different values of t_p is shown in Table 9.2. The optimal preventive replacement interval for the age replacement is 800,000 cycles. The results of this policy are identical to the constant interval replacement policy. This is due to the fact that the time is incremented by 100,000 cycles. Smaller increments of time may result in a significant difference between these two policies.

Table 9.2. Optimal Age Replacement Policy

t_p	$R(t_p)$	$\phi\left(\frac{t_p - \mu}{\sigma}\right)$	$\Phi\left(\frac{t_p - \mu}{\sigma}\right)$	$c(t_p)$ *per cycle*
100,000	1.00	0	0	0.000500
200,000	1.00	0	0	0.000250
300,000	1.00	0	0	0.000166
400,000	1.00	0	0	0.000125
500,000	1.00	0	0	0.000100
600,000	1.00	0	0	0.000083
700,000	0.9987	0.004	0.0013	0.000072
800,000	0.9773	0.054	0.0227	0.000064*
900,000	0.8413	0.242	0.1587	0.000065
1,000,000	0.5000	0.398	0.5000	0.00016
1,100,000	0.1587	0.242	0.8413	0.00009

*Indicates minimum cost

9.2. Preventive Maintenance and Replacement Models: Downtime Minimization

The models discussed in Section 9.1 determine the optimum preventive maintenance interval that minimizes the total cost per unit time. There are many situations where the availability of the equipment is more important than the cost of repair or maintenance. Indeed, the consequences of the downtime of equipment may exceed any measurable cost. In such cases, it is more appropriate to minimize the downtime per unit time than to minimize the total cost per unit time. In the following sections, we present two preventive replacement policies with the objective of minimizing the total downtime per unit time.

9.2.1. THE CONSTANT INTERVAL REPLACEMENT POLICY (CIRP)

This is the simplest preventive maintenance and replacement policy. It is identical to the policy discussed in Section 9.1.1 with the exception that the objective is to minimize the total downtime per unit time—that is, minimize the unavailability of the equipment. Under this policy replacements are performed at predetermined times regardless of the age of the equipment being replaced. In addition, replacements are performed upon failure of the equipment. Following Jardine (1973) and Blanks and Tordan (1986), we rewrite Eq. (9.1) as follows:

$$D(t_p) = \frac{\text{Total downtime}}{\text{Cycle length}}, \tag{9.14}$$

where

Total downtime = Downtime due to failure + Downtime due to preventive replacement

= Expected number of failures in $(0, t_p] \times$ Time to perform a failure replacement + T_p

or

$$\text{Total downtime} = M(t_p)T_f + T_p,$$

where

T_f = time to perform a failure replacement,

T_p = time to perform a preventive replacement, and

$M(t_p)$ = expected number of failures in the interval $(0, t_p]$.

The cycle length is the sum of the time to perform preventive maintenance and the length of the preventive replacement cycle $= T_p + t_p$. Thus, Eq. (9.14) becomes

$$D(t_p) = \frac{M(t_p)T_f + T_p}{T_p + t_p}. \tag{9.15}$$

9.2.2. PREVENTIVE REPLACEMENT AT PREDETERMINED AGE

Again, this policy is similar to that discussed in Section 9.1.2. Under this policy, preventive replacements are performed upon equipment failure or when the equipment reaches age t_p. The objective is to determine the optimal preventive replacement age t_p that minimizes the downtime per unit time:

$$D(t_p) = \frac{\text{Total expected downtime/cycle}}{\text{Expected cycle length}}. \tag{9.16}$$

Total expected downtime per cycle is the sum of the downtime due to a preventive replacement $\times$ the probability of a preventive replacement and the downtime due to a failure cycle $\times$ the probability of a failure cycle. The numerator of Eq. (9.16) is

$$T_pR(t_p) + T_f[1 - R(t_p)].$$

Similarly, the expected cycle length (Jardine, 1973) is

$$(t_p + T_p)R(t_p) + \left[\int_{-\infty}^{t_p} tf(t)\,dt + T_f\right][1 - R(t_p)].$$

Therefore,

$$D(t_p) = \frac{T_pR(t_p) + T_f[1 - R(t_p)]}{(t_p + T_p)R(t_p) + \left[\int_{-\infty}^{t_p} tf(t)\,dt + T_f\right][1 - R(t_p)]}. \qquad (9.17)$$

It is important to note that the conditions for the cost minimization models are also applicable to the downtime minimization models. Moreover, we replace the cost constraint by a replacement time constraint—that is, the time to perform failure replacements is greater than the time to perform preventive replacements or $T_f > T_p$.

EXAMPLE 9.4

Assume that $T_f = 50{,}000$ cycles and $T_p = 25{,}000$ cycles. Determine the parameters of the constant preventive replacement interval policy and the age replacement policy for the equipment given in Example 9.2.

SOLUTION

We calculate $M(t_p)$, $\int_{-\infty}^{t_p} f(t)\,dt$ and $R(t_p)$ as shown in Table 9.3. For the constant interval replacement policy (CIRP) we substitute the known parameter of the policy into Eq. (9.15) to obtain

$$D_{\text{CIRP}}(t_p) = \frac{25{,}000\,[1 + 2M(t_p)]}{25{,}000 + t_p}. \qquad (9.18)$$

Table 9.3. $M(t_p)$, $R(t_p)$, and $\int_{-\infty}^{t_p} t\,f(t)\,dt$

t_p	$M(t_p)$	$R(t_p)$	$1 - R(t_p)$	$\phi(t_p)$	$\Phi(t_p)$	$\int_{-\infty}^{t_p} t\,f(t)\,dt$
100,000	0	1.00	0	0	0	0
200,000	0	1.00	0	0	0	0
300,000	0	1.00	0	0	0	0
400,000	0	1.00	0	0	0	0
500,000	0	1.00	0	0	0	0
600,000	0	1.00	0	0	0	0
700,000	0.00140	0.9987	0.0013	0.004	0.0013	900
800,000	0.00275	0.9773	0.0227	0.054	0.0227	17300
900,000	0.15875	0.8413	0.1587	0.242	0.1587	134500
1,000,000	0.50050	0.5000	0.5000	0.398	0.5000	460110
1,100,000	0.84135	0.1587	0.8413	0.242	0.8413	817100

The downtime policy per cycle for the age replacement policy (ARP) is obtained using

$$D_{\mathrm{ARP}}(t) = \frac{25{,}000[2 - R(t_p)]}{(25{,}000 + t_p)R(t_p) + \left[\int_{-\infty}^{t_p} tf(t)\,dt + 50{,}000\right][1 - R(t_p)]}. \quad (9.19)$$

The summary of the calculations is shown in Table 9.4. The two policies result in the same optimal preventive replacement interval of 800,000 cycles.

Table 9.4. Summary of $D(t_p)$ Calculations

t_p	$D_{CIRP}(t_p)$	$D_{ARP}(t_p)$
100,000	0.2000	0.2000
200,000	0.1111	0.1111
300,000	0.0769	0.0769
400,000	0.0588	0.0588
500,000	0.0476	0.0476
600,000	0.0400	0.0400
700,000	0.0346	0.0346
800,000	0.0305*	0.0316*
900,000	0.0356	0.0362
1,000,000	0.0488	0.0505
1,100,000	0.0596	0.0532

*Indicates minimum downtime

9.3. Minimal Repair Models

Maintaining a complex system that is composed of many components may be achieved by replacing, repairing, or adjusting the components of the system. The replacements, repairs, or adjustments of the components usually restore function to the entire system, but the failure rate of the system remains unchanged, as it was just before failure. This type of repair is called *minimal repair*. Since the failure rate of complex systems increases with age, it would become increasingly expensive to maintain operation by minimal repairs (Valdez-Flores and Feldman, 1989). The main decision variable is the optimal time to replace the entire system instead of performing minimal repairs.

Minimal repair models generally assume that the system's failure-rate function is increasing and that the minimal repairs do not affect the failure rate. Like the preventive replacement models, the cost of minimal repair c_f is less than the cost of replacing the entire system c_r. The expected cost per unit time at age t is

$$c(t) = \frac{c_f M(t) + c_r}{t}, \quad (9.20)$$

where $M(t)$ is the expected number of minimal repairs during the interval $(0, t]$. This model is similar to the preventive replacement model given by Eq. (9.1).

Tilquin and Cléroux (1975, 1985) add cost of adjustments to the numerator of Eq. (9.20). The adjustment cost $c_a(ik)$ at age ik, $i = 1, 2, 3$, and $k > 0$ is added to the cost of minimal repair and the cost of system replacement. This model is closer to reality since the adjustment cost $c_a(ik)$ can be used to reflect the actual operating cost of the system such as periodic adjustment costs, depreciation costs, or interest charges. Rewriting Eq. (9.20) to include the adjustment costs, we obtain

$$c(t) = \frac{c_f M(t) + c_r + \overset{*}{c}_a(\nu(t))}{t}, \tag{9.21}$$

where $\overset{*}{c}_a(\nu(t)) = \Sigma_{i=0}^{\nu(t)} c_a(ik)$ and $\nu(t)$ is the number of adjustments in the interval $(0, t]$. This model can be extended to modify the minimal repair cost c_f to include two parts: the first part represents a fixed charge or setup a, and the second part represents a variable cost that depends on the number of minimal repairs that occurred since the last replacement—that is, $c_r = a + bk$, where $a > 0$ and $b \geq 0$ are constants and k represents the kth minimal repair.

9.3.1. OPTIMAL REPLACEMENT UNDER MINIMAL REPAIR

As mentioned earlier, most repair models assume that repairs result in making the system functions "as good as new." In other words, the system is renewed after each failure. Although this is true for some situations, as in the case of replacing the entire brake system of a vehicle with a new one, there are situations where the failed system will function again after repair but will have the same failure rate and the same effective age at the time of failure. Clearly, when a machine has an increasing failure rate, the duration of its function after repairs will become shorter and shorter, resulting in a finite functioning time. Similarly, as the system ages, its repair time will become longer and longer and will tend to infinity—that is, the system becomes nonrepairable. Thus, in an appropriate model for such systems, successive survival times are stochastically decreasing, each survival time is followed by a repair time, and the repair times are stochastically increasing.

This problem can be modeled using the nonhomogeneous Poisson process as described in Ascher and Feingold (1984), Barlow, Proschan, and Hunter (1965), Downton (1971), and Thompson (1981). Lam (1988, 1990) modeled the problem using geometric processes. We consider a more general replacement model based on Stadje and Zuckerman (1990) and Lam (1990). Assume that the successive survival (operational) times of the system (X_n, $n = 1, 2, \ldots$) form a stochastically decreasing process, that each survival time has an increasing failure rate (IFR) and the consecutive repair times (Y_n, $n = 1, 2, \ldots$) constitute a stochastically increasing process, and that each repair time has the property that new is better than used in expectation (NBUE). A replacement policy T is considered. Under this policy the system is replaced (repaired) after the elapse of time T from the last replacement. Assume that the repair cost rate is c and the replacement cost during an operating interval is c_o. We also assume that the replacement cost is c_f if the system is replaced upon failure or during repair and $c_f \geq c_o$. The reward or profit per unit time of system is R.

Theorem 1 (Stadje and Zuckerman, 1990): If

1. X_n and Y_n are both nonnegative random variables, $\forall n \geq 1$, $\lambda_n = E(X_n)$ is nonincreasing and $\mu_n = E(Y_n)$ is nondecreasing,
2. $\lim_{n\to\infty} \lambda_n = 0$, or $\lim_{n\to\infty} \mu_n = \infty$,
3. $(X_n, n = 1, 2, \ldots)$ and $(Y_n, n = 1, 2, \ldots)$ are two independent sequences of independent random variables, also X_n has IFR and Y_n is NBUE, $\forall n \geq 1$, and
4. $c_o = c_f$,

then the optimum replacement policy is

$$T^* = \sum_{i=1}^{n_0} X_i + \sum_{i=1}^{n_0-1} Y_i, \tag{9.22}$$

where

$$n_0 = \min \{n \geq 1/(c + \phi^*)\mu_n \geq (R - \phi^*)\lambda_{n+1}\} \tag{9.23}$$

and ϕ^* is the optimal value of the long-run average reward (profit).

An equivalent replacement policy is that to replace the system after the Nth failure where the time at which the Nth failure occurs is near time T^*. We refer to this policy as policy N. Once N^* is determined, we can immediately evaluate the corresponding T^* and ϕ^* as shown below.

We now consider a policy N that operates under the following assumptions:

1. When the system fails, it is either repaired or replaced by a new and identical system.
2. Similar to policy T, the survival (or operating) time X_k after the $(k - 1)$th repair forms a sequence of nonnegative random variables with nonincreasing means $E[X_k] = \lambda_k$ and the repair time Y_k after the kth failure forms a sequence of nonnegative random variables with nondecreasing means $E[Y_k] = \mu_k$.

3. The repair cost is c, the replacement cost under this policy is c_f and the reward (or profit) per unit time of system operation is R.

From renewal theory, the average reward per unit time until the Nth failure $R(N)$ is (Lam, 1990):

$$R(N) = \frac{R \sum_{k=1}^{N} \lambda_k - c \sum_{k=1}^{N-1} \mu_k - c_f}{\sum_{k=1}^{N} \lambda_k + \sum_{k=1}^{N-1} \mu_k} \tag{9.24}$$

or

$$R(N) = R - c(N), \tag{9.25}$$

where R is the reward per unit time of system operation and $c(N)$ is the cost per unit time and is given by

$$c(N) = \frac{(c + R) \sum_{k=1}^{N-1} \mu_k + c_f}{\sum_{k=1}^{N} \lambda_k + \sum_{k=1}^{N-1} \mu_k}, \qquad N = 1, 2, \ldots \tag{9.26}$$

The optimum replacement policy N^* is obtained by finding N that maximizes $R(N)$ or minimizes $c(N)$. This can be accomplished by using Eq. (9.26) and subtracting $c(N^*)$ from $c(N^* + 1)$ as shown:

$$c(N^* + 1) - c(N^*) = \{(c + R) f_N - c_f(\lambda_{N+1} + \mu_N)\}/\Delta_N,$$

where

$$\Delta_N = \left(\sum_{k=1}^{N+1} \lambda_k + \sum_{k=1}^{N} \mu_k\right)\left(\sum_{k=1}^{N} \lambda_k + \sum_{k=1}^{N-1} \mu_k\right)$$

and

$$f_N = \mu_N \sum_{k=1}^{N} \lambda_k - \lambda_{N+1} \sum_{k=1}^{N-1} \mu_k. \tag{9.27}$$

Define

$$g_N = (\lambda_{N+1} + \mu_N)/f_N. \tag{9.28}$$

Hence,

$$g_{N+1} - g_N = (\lambda_{N+2}\mu_N - \lambda_{N+1}\mu_{N+1})\left(\sum_{k=1}^{N+1}\lambda_k + \sum_{k=1}^{N}\mu_k\right)\Big/(f_N f_{N+1}) \leq 0.$$

But g_N is a nonincreasing sequence from $g_1 = (\lambda_2 + \mu_1)/(\lambda_1\mu_1)$ to $g_\infty = \lim_{N\to\infty} g_N$ (Lam, 1990) and $c(N+1) \leq$ or $\geq c(N)$ if $g_N \geq$ or $\leq (c+R)/c_f$. Therefore, the optimal policy N^* is determined by

$$N^* = \min\{N \geq 1/g_N \leq (c+R)/c_f\}. \tag{9.29}$$

The optimal replacement policy N^* given by (9.29) is exactly the same as policy T^* given by Eq. (9.22), and the optimal ϕ^* is

$$\phi^* = R(N^*) = R - c(N^*). \tag{9.30}$$

EXAMPLE 9.5

A turbine is used to derive power at a natural gas letdown station. The impeller is the most critical component of the turbine. To reduce the effect of impeller fracture and crack propagation a replacement policy N is implemented. When the impeller fails, it is replaced or repaired at a cost c = \$100. After N replacements (repairs) of the impeller, the turbine is replaced by a new one at a cost c_f = \$12,000. Assume that the reward rate is \$20 per hour. The operating time λ_k and the repair time μ_k after the kth failure are, respectively,

$$\lambda_k = \frac{5000}{2^{k-1}}$$

$$\mu_k = 100 \times 2^{k-1}.$$

Determine the parameters of the optimum replacement policy N^*.

SOLUTION We calculate

$$\frac{c+R}{c_f} = \frac{100+20}{12{,}000} = 0.01.$$

We also calculate g_N and f_N for different N as shown in Table 9.5. From Table 9.2 the optimal replacement policy is $N^* = 2$. In other words, the turbine should be replaced after two failures or 7,500 hours. The optimal ϕ^* is obtained by substituting corresponding values in Eq. (9.30):

$$\phi^* = R - c(N^*)$$

or

$$\phi^* = 20 - 3.077 = \$16.923.$$

Table 9.5. Calculations for N^*

N	λ_N	μ_N	f_N	g_N
1	5000	100	5×10^5	0.0102
2	2500	200	13.75×10^5	0.00196

9.4. Optimum Replacement Intervals for Systems Subject to Shocks

The preventive maintenance and replacement models discussed so far assume that the components or the systems exhibit wear or gradual deterioration—that is, increasing failure rate.

There are many situations where the system is subject to shocks that cause it to deteriorate. For example, the hydraulic and electrical systems of airplanes are subject to shocks that occur at takeoff and landing. Likewise, a breakdown of an insert of a multiinsert cutting tool may subject the tool to a sudden shock. Clearly, systems that are subject to such shocks will eventually deteriorate and fail. In this section, we discuss an optimum replacement policy for such systems.

Assume that the normal cost of running the system is a per unit time and that each shock increases the running cost by c per unit time. The system is entirely replaced at times T, $2T$, . . . at a cost of c_0 per complete system replacement. This is referred to as a periodic replacement policy of length T. The only parameter of such policy is T, and reliability engineers usually seek the optimal value of T that optimizes some criterion such as the minimization of the long-run average cost per unit of time or the maximization of the system availability during T. Clearly, if the length of the period T is rather long, then the cost of system operation and replacement will vary from one period to another due to the changes in labor and material cost with time. Therefore, we present two periodic replacement policies where the first policy considers all cost components of the system to be time-independent and the second policy considers some of the cost components to be time-dependent. These policies are based on Abdel-Hameed's (1986) work.

9.4.1. PERIODIC REPLACEMENT POLICY: TIME-INDEPENDENT COST

As mentioned in the previous section, the system is subject to repeated shocks and is entirely replaced after a fixed period of time T has elapsed. Let $N(t)$ be the number of shocks that the system is subject to during the interval $(0, t]$, and let $n = (N(t), t \geq 0)$. To simplify the analysis, we assume that the jumps of N are of one unit magnitude and that τ_n is the sequence of the jump times of the process N.

The total cost of running the system per period T for a given realization of the sequence τ_n is

$$aT + c(\tau_2 - \tau_1) + \ldots + c(N(T) - 1)(\tau_{N(T)} - \tau_{N(T)-1}) + cN(T)(T - \tau_{N(T)}) + c_0. \tag{9.31}$$

The above expression can be rewritten as

$$aT + c\int_0^T N(t)\,dt + c_0. \tag{9.32}$$

Utilizing Fubini's Theorem (Heyman and Sobel, 1982), the expected total cost of running the system per period is given by

$$aT + c\int_0^T M(t)\,dt + c_0, \tag{9.33}$$

where $M(t)$ is the expected number of shocks in $(0, t]$. The long-run average cost per unit time, $C(T)$, is obtained by dividing Eq. (9.33) by the length of the replacement period T. In other words,

$$C(T) = [aT + c\int_0^T M(t)\,dt + c_0]/T. \tag{9.34}$$

The objective is to determine T that minimizes Eq. (9.34). Since $C(T)$ is a differential function of T and the first-order derivative of $C(T)$ is given by

$$C'(T) = [c\int_0^T [M(T) - M(t)]\,dt - c_0]/T^2,$$

and since $\int_0^T [M(T) - M(t)]\,dt$ is positive and increasing, then the optimal value of the periodic replacement time always exists and is equal to the unique solution of

$$\int_0^T [M(T) - M(t)]\,dt = \frac{c_0}{c}. \tag{9.35}$$

Moreover, the value of T^* is finite if and only if (Abdel-Hameed, 1986)

$$\lim_{T\to\infty} \int_0^T [M(T) - M(t)]\, dt > c_0/c. \tag{9.36}$$

Abdel-Hameed (1986) develops an expression to estimate $M(t)$ when the shocks occur according to a nonstationary pure birth process. If the shock occurrence rate is λ, the probability of a shock occurring in $[t, t + \Delta)$ given that k shocks occurred in $(0, t]$ is $\lambda_k(t)$.

Assume that $\{N(t); t > 0\}$ counts the number of shocks. When $\lambda_k = k\lambda(t)$ and $N(0) > 0$, the counting process is called *Yule process* (Heyman and Sobel, 1982); when $\lambda_k \equiv \lambda$, it is called a Poisson process. Since $N(t)$ is a nonstationary Yule process—that is, for $k = 1, 2, \ldots$

$$\lambda_k = k\lambda(t)$$

—then we have that

$$M(t) - 1 = \int_0^t \lambda(x) M(x)\, dx. \tag{9.37}$$

Equation (9.37) has the solution

$$M(t) = e^{\int_0^t \lambda(x)\, dx}. \tag{9.38}$$

As discussed earlier in this chapter, the optimal value of the periodic replacement policy (T^*) exists only if $\lambda(t)$ is an increasing function of t.

EXAMPLE 9.6

Consider a component whose $\lambda(t) \equiv \lambda$. Assume that the normal cost of running the system is \$0.50 per hour, the increase in running cost due to each shock is \$0.055 per hour, and the cost of replacing the entire component is \$15,000. Determine the optimal value of T and the corresponding long-run average costs of replacement per hour for different values of shock rates.

SOLUTION

From the above description we list

$a = \$0.5$,

$c = \$0.055$, and

$c_0 = \$15{,}000$.

Using Eq. (9.38), we obtain

$$M(t) = e^{\lambda t}.$$

Substituting in Eq. (9.35), we obtain

$$e^{\lambda T}[\lambda T - 1] = \frac{\lambda c_o}{c} - 1.$$

Solving the above equation for different values of λ, we obtain the optimum replacement interval and the long-range average costs of replacement per unit time $C(T)$ as shown in Table 9.6.

Table 9.6. Optimum Replacement Interval

λ	T^*	$C(T^*)$
1×10^{-4}	27,239.267	1.0507
2×10^{-4}	15,972.763	1.4391
4×10^{-4}	9,231.254	2.1251
6×10^{-4}	6,657.679	2.7534
8×10^{-4}	5,266.755	3.3487
1×10^{-3}	4,385.343	3.9214
6×10^{-3}	970.945	15.9680
8×10^{-3}	758.135	20.3165
1×10^{-2}	625.206	24.5376
6×10^{-2}	129.794	117.0886
8×10^{-2}	100.487	151.4683
1×10^{-1}	82.347	185.1727

9.4.2. PERIODIC REPLACEMENT POLICY: TIME-DEPENDENT COST

This policy is similar to that presented in Section 9.4.1 with the exception that the replacement cost per unit time is dependent on the number of shocks and the time at which shocks occur. Following the policy in Section 9.4.1 we assume that the shock process $N = (N(t),\ t \geq 0)$ has jumps of size 1. Let τ_n be the sequence describing the jump times of the shock process N. The additional cost of operating the system per unit time due to every additional shock in the interval $[\tau_i, \tau_i + 1)$ is $c_i(u)$, $i = 0, 1, \ldots$, and u is the state space at which the periodic replacement can be performed. We assume that $\tau_0 = 0$. The normal cost of running the system is a per unit time, and the cost of completely replacing the system is c_0 (Abdel-Hameed, 1986). Similar to the periodic replacement policy with constant cost structure, the system is completely replaced at $T, 2T, \ldots$ and the process is reset to time zero at every replacement. The expected total cost of running the system per period is the sum of the operating cost (aT), expected additional cost due to shock $\int_0^T Ec_{N(t)}(t)\,dt$, and the cost of completely replacing the system. This is expressed as

$$aT + \int_0^T Ec_{N(t)}(t)\,dt + c_0, \tag{9.39}$$

where $Ec_{N(t)}$ is the expectation of the additional cost function due to shocks at time t. The long-run average cost per unit time is obtained by dividing Eq. (9.39) by the length of the period T. Abdel-Hameed (1986) defines $h(t) = E(c_{N(t)})$ and shows that if h is continuous and increasing, then the optimal value of the periodic replacement time exists and is the unique solution of Eq. (9.40):

$$\int_0^T [h(T) - h(t)]\,dt = \frac{c_0}{c}. \tag{9.40}$$

The period T is finite if and only if

$$\lim_{T\to\infty} \int_0^T [h(T) - h(t)]\,dt > \frac{c_0}{c}. \tag{9.41}$$

9.5. Preventive Maintenance and Number of Spares

In Section 9.1.1, the total expected cost per unit time is expressed as

$$c(t_p) = \frac{c_p + c_f M(t_p)}{t_p}, \tag{9.42}$$

where c_p and c_f are the cost per preventive replacement (or planned replacement) and the cost per failure replacement, respectively. $M(t_p)$ is the renewal function and is related to the failure-time distribution $F(t)$ by the integral renewal equation.

One of the most important decisions regarding preventive maintenance function is the determination of the number of spare units to carry in the spares inventory. Clearly, if the number of spares on hand is less than what is needed during the preventive maintenance cycle, the system to be repaired will experience unnecessary downtime until the spares become available. On the other hand, if more spares are carried in the spares inventory than what is needed during the preventive maintenance cycle, an unnecessary inventory carrying cost will incur. Ideally, the number of spares carried in the inventory should equal the number of repairs during the preventive cycle. However, the number of repairs (failures) is a random variable that makes the determination of the exact number of spares a difficult task. In this section, we determine the optimal number of spares to carry at the beginning of the preventive maintenance cycle such that the total cost during the cycle is minimized.

The number of spares needed per preventive maintenance cycle is $[1 + N_1(t_p)]$, where $N_1(t_p)$ is a random variable that represents the number of replacements due to failure in the cycle t_p. Assume that the initial inventory of the spares is L units. At the end of the preventive cycle the inventory cost equals zero if $L = N_1(t_p) + 1$; otherwise, a carrying cost or shortage cost (cost due to unavailability of spares when needed) will incur. Define $g(L, N_1(t_p))$ as the penalty function which increases as $[L - (N_p(t_p) + 1)]$ deviates from zero. Following Taguchi, Elsayed, and Hsiang's (1989) loss function and Murthy's (1982) function and assuming that the inventory carrying cost per excess unit equals the shortage cost per unit, we express the penalty function g as

$$g[L, N_1(t_p)] = [L - (N_1(t_p) + 1)]^2. \tag{9.43}$$

Since $N_1(t_p)$ is a random variable, then the expected value of the penalty function $D(L, t_p)$ is

$$D(L, t_p) = E\{g[L, N_1(t_p)]\}$$

or

$$D(L, t_p) = \sum_{n=0}^{\infty} g(L, n)p_n, \tag{9.44}$$

where p_n is the probability that $N_1(t_p) = n$. The overall cost function is the sum of two terms: the first is the average cost of the system per unit time given by Eq. (9.42), and the second is the penalty cost associated with the number of spares. Thus, we express the overall cost as

$$TC = c(t_p) + \alpha D(L, t_p), \tag{9.45}$$

where α is a scaling factor. If $\alpha = 0$, then there is no penalty cost, and a large value of α implies a very high penalty cost for both shortage and excess inventory. It is obvious that the optimum preventive maintenance cycle t_p is a function of α; hence, it is more appropriate to represent it as $t_{p\alpha}$. The optimal values of $t_{p\alpha}^*$ and L^* are obtained by minimizing TC with respect to $t_{p\alpha}$ and L.

We rewrite $D(L, t_p)$ as

$$\begin{aligned} D(L, t_p) &= E[L - (M(t_p) + 1) + M(t_p) - N_1(t_p)]^2 \\ &= [L - (M(t_p + 1))]^2 + E[N_1(t_p) - M(t_p)]^2 \end{aligned}$$

or

$$D(L, t_p) = [L - (M(t_p) + 1))]^2 + \text{Var}\,[N_1(t_p)], \tag{9.46}$$

where $M(t_p) = E[N_1(t_p)]$ and Var $[N_1(t_p)]$ is the variance of $N_1(t_p)$. Substituting Eqs. (9.46) and (9.42) into Eq. (9.45) we obtain

$$TC = [c_f M(t_p) + c_p]/t_p + \alpha\, \text{Var}\,[N_1(t_p)] + \alpha[L - (M(t_p) + 1)]^2. \tag{9.47}$$

The optimal number of spares at the beginning of the preventive cycle is obtained by setting $\partial TC/\partial L = 0$ as follows:

$$\frac{\partial TC}{\partial L} = 2\alpha[L - (M(t_p) + 1)] = 0$$

or

$$L^* = 1 + M(t_p). \tag{9.48}$$

This is an intuitive and expected result, since it states that the optimal number of spares must equal the expected number of repairs (failures) during the preventive maintenance cycle.

Similarly, the optimal length of the preventive maintenance cycle for a given α is obtained by setting $\partial TC/\partial t_p = 0$, which results in

$$\frac{\partial c(t_p)}{\partial t_p} + \alpha \frac{\partial V(t_p)}{\partial t_p} = 0, \tag{9.49}$$

where

$$V(t_p) = \text{Var}(N_1(t_p)).$$

The optimal value of $t^*_{p\alpha}$ is obtained by solving Eq. (9.49).

EXAMPLE 9.7

The blades of a high-pressure compressor used in the first stage of an aeroengine are subject to fatigue cracking. The fatigue life of a blade is evaluated as the product of the amplitude and frequency (AF value) during vibratory fatigue testing. If a blade exhibits unusually low AF values, it is replaced in order to avoid the fatigue cracking of the blade. Since the cost of conducting the fatigue test on a regular basis is high, the users of such compressors usually use a preventive maintenance schedule to replace the blades based on the number of operating hours. The cost of replacing a blade at the end of the preventive maintenance cycle is \$250 whereas the cost of replacing a blade during the cycle is \$1,000. The time between successive failures is expressed by a two stage Erlang distribution with a parameter of $\lambda = 0.005$ failures per hour. Assuming $\alpha = 0.8$, what are the optimal number of spares and preventive maintenance intervals that minimize the total cost?

SOLUTION The p.d.f. of the two stage Erlang distribution is

$$f(t; \lambda) = \lambda^2 t e^{-\lambda t}.$$

The expected number of failures during the preventive maintenance cycle t_p is

$$M(t_p) = \frac{1}{2}\lambda t_p - \frac{1}{4} + \frac{1}{4}e^{-2\lambda t_p}. \tag{9.50}$$

Consequently, $c(t_p)$ is

$$c(t_p) = \frac{c_p}{t_p} + \frac{1}{2}\lambda c_f - \frac{1}{4}\frac{c_f}{t_p} + \frac{c_f}{4t_p}e^{-2\lambda t_p}. \tag{9.51}$$

The variance of the number of failures is obtained as discussed in Cox (1962):

$$\text{Var}\,(N(t_p)) = \frac{1}{4}\lambda t_p + \frac{1}{16} - \frac{1}{2}\lambda t_p e^{-2\lambda t_p} - \frac{1}{16}e^{-4\lambda t_p}. \tag{9.52}$$

Substituting Eqs. (9.51) and (9.52) into Eq. (9.49) we obtain

$$\frac{-c_p}{t_p^2} + \frac{c_f}{4t_p^2} - \frac{c_f}{4}e^{-2\lambda t_p}\left(\frac{2\lambda t_p + 1}{t_p^2}\right)$$

$$+ \alpha\left[\frac{\lambda}{4} - \frac{1}{2}\lambda e^{-2\lambda t_p}(1 - 2\lambda t_p) + \frac{1}{4}\lambda e^{-4\lambda t_p}\right] = 0$$

or

$$e^{-0.001t_p}[t_p^3 - 10t_p^2 - 12{,}500t_p - 1{,}250{,}000] + 5t_p^2(1 + e^{-0.02t_p}) = 0.$$

The solution of the above equation is

$$t_{p\alpha}^* = 140.568 \text{ hours}.$$

The corresponding expected number of failures during the preventive maintenance cycles is 0.163. Therefore, the optimal number of spares is 1.163.

The difficulty in using Eq. (9.49) is due to the estimation of the variance of the expected number of failures during the preventive cycle t_p. Cox (1962) derives asymptotic results for the variance as a function of the mean and standard deviation of the failure time distribution. The asymptotic variance is

$$\text{Var}\ [N(t_p)] = \frac{\sigma^2}{\mu^3} t_p. \tag{9.53}$$

We now illustrate the use of Eq. (9.53) when closed-form expressions for $M(t_p)$ and Var $(N(t_p))$ are difficult to attain.

EXAMPLE 9.8

High-pressure ball valves made from martensitic stainless steel are usually used in chemical plants to control the flow of dry synthetic gas (a three to one mixture of nitrogen and hydrogen with 4 percent ammonia). Assume that the failure times follow a Weibull distribution of the form

$$f(t) = \frac{\gamma}{\theta} t^{\gamma-1} e^{\frac{-t^\gamma}{\theta}} \qquad t > 0, \theta > 0.$$

The estimated values of γ and θ are 2 and 50, respectively (measurements in hundreds). The cost of a failure replacement is $500 and that of a preventive replacement is $300. The scale of the penalty function is 3.5. Determine the optimal preventive maintenance interval and the optimal number of spares during the maintenance interval.

SOLUTION

Since it is difficult to estimate the expected number of failures during t_p, we utilize the asymptotic form of the renewal function

$$M(t_p) = \frac{t_p}{\mu} + \frac{\sigma^2 - \mu^2}{2\mu^2},$$

where μ and σ are the mean and the standard deviation of the failure-time distribution. The mean and variance of the Weibull distribution are

$$\mu = \theta^{\frac{1}{\gamma}} \Gamma\left(1 + \frac{1}{\gamma}\right) = (50)^{\frac{1}{2}} \Gamma\left(\frac{3}{2}\right) = 6.27$$

$$\sigma^2 = \theta^{\frac{2}{\gamma}} \left[\Gamma\left(1 + \frac{2}{\gamma}\right) - \left(\Gamma\left(1 + \frac{1}{\gamma}\right)\right)^2\right]$$

or

$$\sigma^2 = 50[1 - 0.7854] = 10.73$$

and

$$\sigma = 3.275 \text{ hours.}$$

Thus,

$$M(t_p) = \frac{t_p}{6.27} - 0.363. \tag{9.54}$$

The variance of the expected number of failures is

$$\text{Var}\,[N(t_p)] = 0.0435 t_p. \tag{9.55}$$

Substituting Eq. (9.54) into Eq. (9.42) we obtain

$$c(t_p) = \frac{c_p}{t_p} + \frac{c_f}{t_p}\left[\frac{t_p}{6.27} - 0.363\right]$$

$$c(t_p) = \frac{118.5}{t_p} + 79.74. \tag{9.56}$$

Substituting Eqs. (9.55) and (9.56) into Eq. (9.49) results in

$$\frac{-118.5}{t_p^2} + (3.5) \times 0.0435 = 0$$

or

$$t^*_{p\alpha} = 27.898 \quad \text{or} \quad 2{,}789.8 \text{ hours.}$$

The optimal number of spares is

$$L^* = 1 + M(t^*_{p\alpha}) = 1 + 4.086 = 5.086 \text{ valves.}$$

9.5.1. NUMBER OF SPARES AND AVAILABILITY

When systems provide critical services such as the computer systems of the Federal Reserve Bank, it is important to stock spare parts on hand to ensure a specified availability level of the system. In this case, we utilize Erlang's loss formula to estimate the number of spares during the constant failure-rate region. The formula is given as follows (Cooper, 1972):

$$\bar{A}(s, a) = \frac{a^s/s!}{\sum_{k=0}^{s} (a^k/k!)},$$

where

$\bar{A}(s, a)$ = steady-state unavailability of the system when the number of spares on hand is s and the number of units under repair is a,

a = number of units under repair,

s = number of spares on hand, and

$A(s, a) = 1 - \bar{A}(s, a)$ = availability of the system.

The number of units under repair depends on the total number of units in service, the repair rate, and the average lead time for obtaining spares. Let N be the total number of units for which we need to provide spares in order to maintain a specified availability level. Assume that the repair rate is R units per unit time and l is the lead time. Then a, number of units under repair, is in effect the product NRl. Thus, the number of spares s required to obtain a specified availability level can be obtained by substituting the values of a and $A(s, a)$ into the Erlang's loss formula and solving for s.

EXAMPLE 9.9

A telephone company maintains a large communications network that contains 2,200 repeaters (devices capable of receiving one or two communication signals and delivering corresponding signals). Each repeater experiences a constant failure rate of 3,000 FITs (one FIT is 10^{-9} failure per hour). Assume that the company's standard repair rate = 1.70 FITs and the average lead time is forty-eight hours. Determine the required number of spares that maintains an availability of 0.998.

SOLUTION

The repair rate is $1.7 \times 3{,}000 \times 10^{-9} = 5{,}100 \times 10^{-9}$ replacements per hour. The product $NRl = 2{,}200 \times 5{,}100 \times 10^{-9} \times 48 = 0.539$.

Using Erlang's loss function,

$$\bar{A}(s, a) = \frac{a^s/s!}{\sum_{k=0}^{s} (a^k/k!)}$$

$$\bar{A}(s, a) = 0.002.$$

For $s = 2$,

$$\bar{A}(2, a) = \frac{0.1450}{1 + 0.539 + 0.145} = 0.08 > 0.002.$$

For $s = 3$,

$$\bar{A}(3, a) = \frac{0.02609}{1 + 0.539 + 0.145 + 0.02609} = 0.015 > 0.002.$$

For $s = 4$,

$$\bar{A}(4, a) = \frac{0.003516}{1 + 0.539 + 0.145 + 0.02609 + 0.00351} = 0.00208 \cong 0.002.$$

Therefore, the number of spares required to achieve an availability level of 0.998 is 4.

The relationship among the availability, the product NRl, and the number of spares is shown in Figure 9.2. The number of spares can be obtained from this

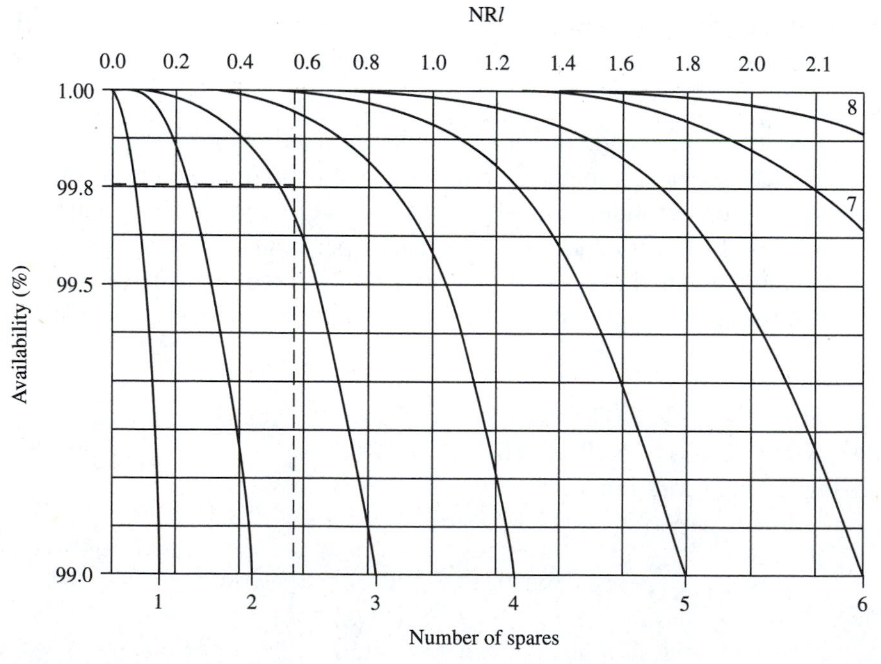

Figure 9.2. Relationship Between the Availability and Number of Spares

figure by observing where the intersection of NRl and $A(s, a)$ occurs. Similar graphs can be developed for different ranges of NRls and availabilities (AT&T, 1983).

9.6. Group Maintenance

The preventive maintenance schedules presented so far in this chapter are limited to the replacement of the components of the system one at a time at predetermined time intervals that either minimize the cost of maintenance or maintain an acceptable level of system availability. By studying the tradeoff between preventive maintenance and corrective maintenance (failure replacements) costs, it is shown (Barlow, Proschan, and Hunter, 1965) that the optimum scheduled time for preventive maintenance is nonrandom and there exists a unique optimum policy if the distribution of time to failure has IFR (increasing failure rate).

In situations where many similar products or machines perform the same function—such as the copiers in a copy center or a fleet of identical vehicles that transport passengers from one location to another regularly—it is perhaps more economical to perform preventive maintenance on a group of the products, machines, or vehicles at the same time. Similar to the single component or product preventive maintenance models, we are interested in determining the optimum preventive maintenance interval for the group.

Consider a group of N independent operating machines that are subject to failure. The repair cost is composed of a fixed cost for each repair and a variable cost per machine. The repair cost per machine decreases as the number of machines requiring repairs increases while the production loss due to machine breakdowns increases. Let $N(t)$ represent the number of machines operating at time $t (0 \leq N(t) \leq N)$, and let the machines have identical failure distributions $F(t)$. The distribution of $N(t)$ is

$$P[N(t) = n] = \binom{N}{n} [1 - F(t)]^n \, [F(t)]^{N-n}, \tag{9.57}$$

where $P\,[N(t) = n]$ is the probability that the number of machines operating at time t equals n. The distribution of $N(t)$ is binomial with a mean of

$$E[N(t)] = N[1 - F(t)]. \tag{9.58}$$

Suppose that the failed machines are repairable at a fixed c_o and a variable cost c_1 per machine. If a failed machine is not repaired upon failure, then a production loss of c_2 per unit time per machine is incurred. Since the production will increase as the scheduled time for repair increases and the repair cost per machine decreases, then there exists an optimum scheduled time (maintenance time) that minimizes the expected total cost per unit time.

Following Okumoto and Elsayed (1983), we consider a random maintenance scheduling policy that states that repairs are undertaken whenever the number of operating machines reaches a certain level n. The time to reach this level is a random variable T with c.d.f. of $G(t)$. T represents the nth-order statistic of N random variables. The expected repair cost per cycle R_c is

$$R_c = c_0 + c_1 \int_0^{\infty} [N - E[N(t)]] \, dG(t)$$

or

$$R_c = c_0 + c_1 N \int_0^{\infty} F(t) \, dG(t). \tag{9.59}$$

The expected production loss per cycle P_c is

$$P_c = c_2 \int_0^{\infty} [N - E[N(t)]] \bar{G}(t) \, dt$$

or

$$P_c = c_2 N \int_0^{\infty} F(t) \bar{G}(t) \, dt, \tag{9.60}$$

where $\bar{G}(t) = 1 - G(t)$. The total expected cost per unit time is

$$c[G(t)] = \frac{R_c + P_c}{\int_0^{\infty} t \, dG(t)}. \tag{9.61}$$

Barlow, Proschan, and Hunter (1965) show that the optimum scheduling policy that minimizes Eq. (9.61) is deterministic. In other words,

$$G(t) = \begin{cases} 0 & \text{if } t \leq t_0 \\ 1 & \text{if } t > t_0, \end{cases}$$

where t_0 is the scheduled time for group maintenance. Thus, Eq. (9.61) can be rewritten as

$$c(t_0) = \frac{1}{t_0} [c_0 + c_1 N F(t_0) + c_2 N \int_0^{t_0} F(t) \, dt]. \tag{9.62}$$

From Eq. (9.62), $c(0) = \infty$ and $c(\infty) = c_2N$. This implies that the cost per unit time for the optimum schedule is less than c_2N. The optimum schedule policy is summarized as follows:

Assume that $F(t)$ is continuous, the derivative of $f(t)$ exists, and the failure rate per machine is λ. Suppose $-f'(t)/f(t) < c_2/c_1$ for $t \geq 0$. Then,

1. if $c_2/\lambda > c_0/N + c_1$, then there exists a unique and finite optimum scheduling time t_0^* that satisfies the following equation:

$$c_1 t_0 f(\overset{*}{t_0}) + c_2 t_0 F(\overset{*}{t_0}) - c_1 F(\overset{*}{t_0}) - c_2 \int_0^{t_0} F(t)\, dt = c_0/N. \qquad (9.63)$$

   The minimum cost per unit time is obtained by

$$c(\overset{*}{t_0}) = c_1 N f(\overset{*}{t_0}) + c_2 N F(\overset{*}{t_0}). \qquad (9.64)$$

2. Otherwise, $t_0^* = \infty$.

The condition in (1) is realistic. For instance, if the failure-time distribution is assumed to be exponential with a rate of λ, then the condition $-f'(t)/f(t) < c_2/c_1$ becomes $(c_2/\lambda) > c_1$, which translates to the average production loss per machine. Furthermore, the condition $c_2/\lambda > c_0/N + c_1$ implies that the average production loss per machine is more than the total repair cost per machine when group repair of N machines is performed. Therefore, it is reasonable to schedule the repair before the failure of all machines (Okumoto and Elsayed, 1983).

We have shown earlier in this chapter that when a machine exhibits constant failure rate, the optimal preventive maintenance policy is to replace the component upon failure. However, when N identical machines each exhibit a constant failure rate λ, the condition for the existence of an optimum policy is given by $c_2/\lambda > c_0/N + c_1$. By substituting the p.d.f. of the exponential distribution into Eq. (9.63), we obtain

$$(X^* + 1)e^{-X^*} = A, \qquad (9.65)$$

where

$$X^* = \lambda \overset{*}{t_0} \qquad (9.66)$$

and

$$A = \frac{c_2/\lambda - (c_0/N + c_1)}{c_2/\lambda - c_1}. \qquad (9.67)$$

Equation (9.65) represents the expected number of machines to be repaired under the optimum policy. The expected cost for the optimum policy is

$$c(\overset{*}{t_0}) = c_2 N + N\lambda(c_1 - c_2/\lambda)e^{-\lambda \overset{*}{t_0}}. \tag{9.68}$$

EXAMPLE 9.10

The milling department in a large manufacturing facility has ten identical computer numerically controlled (CNC) milling machines. Each machine exhibits a constant failure rate of 0.0005 failures per hour. The cost of lost production is \$200 per machine per hour. The repair cost consists of two components: a fixed cost of \$150 and a variable cost of \$100 per machine. Determine the optimum preventive maintenance schedule and the corresponding cost.

SOLUTION

The machines exhibit constant failure rates that follow exponential distributions. The p.d.f. is

$$f(t) = \lambda e^{-\lambda t}.$$

The repair cost has two elements $c_0 = \$150$ and $c_1 = \$100$. The production loss $c_2 = \$200$ per hour. We check the condition $(c_2/\lambda) > c_0/N + c_1$ for the existence of an optimum policy.

Since

$$\frac{200}{0.0005} > \frac{150}{10} + 100,$$

then an optimum policy exists. The optimum repair time of the machines is obtained by using Eqs. (9.65) through (9.67). From Eq. (9.67), we obtain

$$A = \frac{c_2/\lambda - (c_0/N + c_1)}{c_2/\lambda - c_1}$$

or

$$A = \frac{400{,}000 - 115}{400{,}000 - 100} = \frac{399{,}885}{399{,}900} = 0.9999624.$$

From Eq. (9.66), $X^* = 0.0005\, t_0^*$. Thus, Eq. (9.65) becomes

$$(0.0005\overset{*}{t_0} + 1)e^{-0.0005\overset{*}{t_0}} = 0.9999624,$$

and the optimal time to repair the failed machines is 17.5 hours. The corresponding cost is

$$c(17.5) = c_2 N + N\lambda(c_1 - c_2/\lambda)e^{-\lambda t_0^*}$$

$$= 2{,}000 + 10 \times 0.0005\left(100 - \frac{200}{0.0005}\right)e^{-0.0005\times 17.5}$$

or

$$c(17.5) = \$17.92.$$

9.7. Periodic Inspection

Performing preventive maintenance at specified schedules will certainly improve the reliability of the system. To further improve reliability, especially for critical systems, a preventive maintenance schedule is usually coupled with a periodic inspection schedule. In such situations, the status of the systems is determined by inspection; such is the case of bridges and structures. Continuous monitoring of the system is an alternative to periodic inspection. Methods for condition monitoring are presented in Section 9.8. In this section we discuss two inspection policies.

9.7.1. AN OPTIMUM INSPECTION POLICY

Under this policy, the state of the equipment is determined by inspection. For example, the quality of the products produced by a machine may fall outside the acceptable control limits indicating machine degradation (assuming other factors affecting product quality have not been changed). When a failure or degradation is detected, the equipment is repaired or adjusted and is returned to its original condition before the failure or the degradation. If the inspection fails to detect the failure or the degradation of the equipment, then unnecessary cost associated with equipment failure will incur. We refer to this cost as *nondetection cost*. The objective is to determine an optimum inspection schedule that minimizes the total cost per unit time associated with inspection, repair, and the nondetection cost.

The inspection policy is to perform inspections at times $x_1, x_2, x_3, \ldots$ until a failed or degraded equipment is detected. Repairs are immediately performed upon the detection of a failure or degradation. The inspection intervals are not necessarily equal but may be reduced as the probability of failure increases (Jardine, 1973). We use the following definitions:

c_i = the inspection cost per inspection,

c_u = the cost per unit time of undetected failure or degradation,

c_r = the cost of a repair,

T_r = the time required to repair a failure (or degradation), and

$f(t)$ = the p.d.f. of the equipment's time to failure.

The expected total cost per unit time is

$$c(x_1, x_2, x_3, \ldots) = \frac{E_c}{E_l}, \tag{9.69}$$

where E_c and E_l are the total expected cost per cycle and expected cycle length, respectively.

We now illustrate the estimation of E_c and E_l as follows. Assume that failure of the equipment occurs between any pair of inspection times. If the failure occurs at time t_1 between time 0 and x_1, then the cost of the cycle would be

$$c_i(1) + c_u(x_1 - t_1) + c_r.$$

The expected value of this cost is

$$\int_0^{x_1} [c_i(1) + c_u(x_1 - t) + c_r] f(t)\, dt. \tag{9.70}$$

Similarly, if the failure occurs between the inspection times x_1 and x_2, the expected value of the cost would be

$$\int_{x_1}^{x_2} [c_i(2) + c_u(x_2 - t) + c_r] f(t)\, dt. \tag{9.71}$$

Thus, the total expected cost per cycle is

$$\begin{aligned} E_c = &\int_0^{x_1} [c_i(0 + 1) + c_u(x_1 - t) + c_r] f(t)\, dt \\ &+ \int_{x_1}^{x_2} [c_i(1 + 1) + c_u(x_2 - t) + c_r] f(t)\, dt \\ &+ \int_{x_2}^{x_3} [c_i(2 + 1) + c_u(x_3 - t) + c_r] f(t)\, dt \\ &+ \ldots + \int_{x_j}^{x_{j+1}} [c_i(j + 1) + c_u(x_{j+1} - t) + c_r] f(t)\, dt + \ldots \end{aligned} \tag{9.72}$$

Equation (9.72) can be written as

$$E_c = \sum_{k=0}^{\infty} \int_{x_k}^{x_{k+1}} [c_i(k+1) + c_u(x_{k+1} - t) + c_r] f(t)\, dt$$

or

$$E_c = c_r + \sum_{k=0}^{\infty} \int_{x_k}^{x_{k+1}} [c_i(k+1) + c_u(x_{k+1} - t)] f(t)\, dt. \qquad (9.73)$$

We estimate the expected cycle length E_l by following the same steps of estimating the expected total cost per cycle. Hence, the expected cycle length is

$$\int_0^{x_1} [t + (x_1 - t) + T_r] f(t)\, dt + \int_{x_1}^{x_2} [t + (x_2 - t) + T_r] f(t)\, dt + \ldots$$

$$+ \int_{x_j}^{x_{j+1}} [t + (x_{j+1} - 1) + T_r] f(t)\, dt + \ldots$$

or

$$E_l = \mu + T_r + \sum_{k=0}^{\infty} \int_{x_k}^{x_{k+1}} (x_{k+1} - t) f(t)\, dt, \qquad (9.74)$$

where μ is the mean time to failure of the equipment.

Substituting Eqs. (9.73) and (9.74) into Eq. (9.69) we obtain

$$c(x_1, x_2, x_3, \ldots) = \frac{c_r + \sum_{k=0}^{\infty} \int_{x_k}^{x_{k+1}} [c_i(k+1) + c_u(x_{k+1} - t)] f(t)\, dt}{\mu + T_r + \sum_{k=0}^{\infty} \int_{x_k}^{x_{k+1}} (x_{k+1} - t) f(t)\, dt}. \qquad (9.75)$$

The optimal inspection schedule is obtained by taking the derivatives of Eq. (9.75) with respect to $x_1, x_2, x_3, \ldots$ and equating the resulting equations to zero, and then solving the resulting equations simultaneously.

Following Brender (1962), Barlow and Proschan (1965), and Jardine (1973), we present the following procedure to determine the optimum inspection schedule.

Define a residual function as

$$R(L; x_1, x_2, x_3, \ldots) = LE_l - E_c, \qquad (9.76)$$

where L represents either an initial estimation of the minimum cost $c(x_1, x_2, x_3, \ldots)$ or a value of $c(x_1, x_2, x_3, \ldots)$ obtained from a previous cycle of an iteration process. The schedule that minimizes $R(L; x_1, x_2, x_3, \ldots)$ is the same schedule that minimizes $c(x_1, x_2, x_3, \ldots)$. The following procedure determines $x_1, x_2, x_3, \ldots$:

Step 1. Choose a value of L.

Step 2. Choose a value of x_1.

Step 3. Generate a schedule $x_1, x_2, x_3, \ldots$ using the following relationship:

$$x_{i+1} = x_i + \frac{F(x_i) - F(x_{i-1})}{f(x_i)} - \frac{c_i}{c_u - L}. \tag{9.77}$$

Step 4. Compute R using Eq. (9.76).

Step 5. Repeat steps 2 through 4 with different values of x_1 until $R_{\max}$ is obtained.

Step 6. Repeat steps 1 through 5 with different values of L until $R_{\max} = 0$.

A procedure for adjusting L until it is identical with the minimum cost can be obtained from

$$c(L; x_1, x_2, x_3, \ldots) = L - \frac{R_{\max}}{E_l}.$$

EXAMPLE 9.11

The strain-gauge technique is used to measure the stress-strain fields in the pipes of chemical plants. The stress data measured by the strain-gauge technique are compared with the results of the thermomechanical and sectional flexibility analyses. If inconsistencies exist between the measured data and the results of the analyses, then further examinations of the pipes using ultrasonic or acoustic emission techniques are warranted. This, of course, will eliminate possible leaks in the pipes or ruptures of their walls.

The reliability engineer of this plant recommends an inspection policy where inspections of the strain-gauge's measurements are analyzed at times $x_1, x_2, x_3, \ldots$ until an inconsistency exists between the results of thermomechanical analysis and the data from the strain gauges. At that time, preventive maintenance or repair is performed on the pipes. The inspection cost per inspection is \$80, the cost per unit time of undetected failure or degradation of the pipe is \$9 per hour, and the cost of repair is \$2,000. The time required to repair a failure is ninety hours. The time to failure of the pipes follows a gamma distribution with the following p.d.f.:

$$f(t) = \frac{\alpha(\alpha k)^{k-1} \exp[-\alpha t]}{(k-1)!},$$

where $\alpha = 1/b$, b is a scale parameter. The mean of the distribution, $\mu = kb$. Assuming $k = 3$ and $\mu = 1{,}000$ hours, determine the first four points of the optimal schedule.

SOLUTION

By substituting the distribution parameters into the p.d.f. of the gamma distribution, we obtain

$$f(t) = \frac{12.65}{10^9} e^{\frac{-3}{1{,}000}t}.$$

Using Brender's algorithm, we obtain the results shown in Table 9.7. No schedule exists for $x_1 < 20$. The optimum inspection schedule for the pipes is $x_1 = 20$, $x_2 = 41$, $x_3 = 64$, and $x_4 = 89$, and the corresponding cost is \$95.5.

Table 9.7. Optimum Inspection Schedule

x_1	x_2	x_3	x_4	*Total Cost*
100	216	356	530	1,581.5
80	170	274	396	1,083.5
60	125	198	280	648.5
40	82	128	178	312.5
20	41	64	89	95.5*

9.7.2. PERIODIC INSPECTION AND MAINTENANCE

Periodic inspection coupled with preventive maintenance is usually performed on critical components and systems such as airplane engines, standby power generators for hospitals, missiles and weaponry systems, and backup computers for banking and airline passenger reservation systems. In all these cases, inspection is periodically performed. When failures of the components are detected, the components are repaired or replaced with new ones. If the failed components are not detected during inspection and the system fails after the inspection is performed, a nondetection cost (or loss) will incur. Obviously, more frequent inspections will reduce the nondetection cost but will increase the cost of inspections. Therefore, an inspection schedule that minimizes the expected cost until detection of failure and also minimizes the expected cost assuming renewal at detection of failure is desirable.

We consider a system that is periodically inspected to determine whether or not it requires repair (or replacement) and that at the same time provides preventive maintenance if needed. Let us assume that after inspection, the unit (or component) has the same age as before with probability p and is as good as new with probability q (Nakagawa, 1984). We are interested in estimating the mean time to failure and the expected number of inspections before failure. We are also interested in estimating the total expected cost and the expected cost per unit of time until detection of failure. Furthermore, we seek the optimum number of checks that minimize the expected cost.

Let us assume that the system to be inspected begins operating at time $T = 0$ and that inspection is performed at times kT ($k = 1, 2, \ldots$), where $T > 0$ is constant and previously determined. The cumulative distribution function of the time to failure of the system is $F(t)$ with a finite mean μ. The failure of the system is detected only by inspection and the time to perform inspection is negligible when compared with the length of time between two successive inspections. Following Nakagawa (1984), we estimate the mean time to failure of the system $\gamma(T, p)$ as

$$\gamma(T, p) = \sum_{j=1}^{\infty} \left\{ p^{j-1} \int_{(j-1)T}^{jT} t\, dF(t) + p^{j-1} q \bar{F}(jT)[jT + \gamma(T, p)] \right\}, \tag{9.78}$$

where $\bar{F}(t) = 1 - F(t)$. The first term in Eq. (9.78) represents the mean time until the system fails between the $(j-1)$th and the jth inspections. The second term represents the mean time until the system becomes new by the jth check; after that it fails. By solving and rearranging terms of Eq. (9.78), we obtain

$$\gamma(T, p) = \frac{\sum_{j=0}^{\infty} p^j \int_{jT}^{(j+1)T} \bar{F}(t)\, dt}{\sum_{j=0}^{p} p^j \{\bar{F}(jT) - \bar{F}[(j+1)\,T]\}}. \tag{9.79}$$

If $p = 0$, then the system is as good as new after each inspection and

$$\gamma(T, 0) = \frac{\int_0^T \bar{F}(t)\, dt}{F(t)}. \tag{9.80}$$

On the other hand, if $p = 1$, then the system has the same age as before inspection and

$$\gamma(T, 1) = \mu. \tag{9.81}$$

Following Eq. (9.79), the expected number of inspections before failure $M(T, p)$ is obtained as follows:

$$M(T, p) = \frac{\sum_{j=0}^{\infty} p^j \bar{F}[(j+1)T]}{\sum_{j=0}^{\infty} p^j \{\bar{F}(jT) - \bar{F}[(j+1)T]\}}. \tag{9.82}$$

When $p = 0$, then

$$M(T, 0) = \frac{\bar{F}(T)}{F(T)}. \tag{9.83}$$

When $p = 1$, then

$$M(T, 1) = \sum_{j=0}^{\infty} \bar{F}[(j+1)T]. \tag{9.84}$$

Assume that c_1 is the cost of each inspection and c_2 is the cost of undetection of failure—that is, the cost associated with the elapsed time between failure and its detection per unit of time. The total expected cost until a failure is detected $c(T; p)$ can be expressed as

$$c(T; p) = (c_1 + c_2 T)[M(T, p) + 1] - c_2 \gamma(T, p). \tag{9.85}$$

Substituting Eqs. (9.79) and (9.82) into Eq. (9.85) results in

$$c(T; p) = \frac{(c_1 + c_2 T)\left\{\sum_{j=0}^{\infty} p^j \bar{F}[(j+1)T] + \sum_{j=0}^{\infty} p^j \bar{F}(jT) - \sum_{j=0}^{\infty} p^j \bar{F}(j+1)T\right\}}{\sum_{j=0}^{\infty} p^j \{\bar{F}(jT) - \bar{F}[(j+1)T]\}}$$

$$- \frac{c_2 \sum_{j=0}^{\infty} p^j \int_{jT}^{(j+1)T} \bar{F}(t)\, dt}{\sum_{j=0}^{\infty} p^j \{\bar{F}(jT) - \bar{F}[(j+1)T]\}}$$

or

$$c(T;p) = \frac{(c_1 + c_2T)\sum_{j=0}^{\infty} p^j \bar{F}(jT) - c_2 \sum_{j=0}^{\infty} p^j \int_{jT}^{(j+1)T} \bar{F}(t)\,dt}{\sum_{j=0}^{\infty} p^j\{\bar{F}(jT) - \bar{F}[(j+1)T]\}}. \tag{9.86}$$

From Eq. (9.86), $\lim_{T\to 0} c(T;p) = \lim_{T\to\infty} c(T;p) = \infty$, which implies that there exists a finite optimal value T^* that minimizes the total expected cost $c(T;p)$. Nakagawa (1984) shows that

$$M(T,p) \le \frac{\gamma(T,p)}{T} \le [1 + M(T,p)].$$

Consequently,

$$c_1 \frac{\gamma(T,p)}{T} \le c(T,p) \le c_1[1 + M(T,p)] + c_2T. \tag{9.87}$$

It may be of interest to seek the optimum inspection interval that minimizes the expected cost per unit time until detection of failure; $c_d(T,p)$. In this case, we follow the same derivation as that of Eq. (9.85) to obtain

$$c_d(T;p) = \frac{(c_1 + c_2T)[M(T,p) + 1] - c_2\gamma(T,p)}{T\,[M(T,p) + 1]}$$

or

$$c_d(T;p) = \frac{c_1}{T} + c_2 \left\{ 1 - \frac{\sum_{j=0}^{\infty} p^j \int_{jT}^{(j+1)T} \bar{F}(t)\,dt}{T \sum_{j=0}^{\infty} p^j \bar{F}(jT)} \right\}. \tag{9.88}$$

The bounds of $c_d(T;p)$ are $\lim_{T\to 0} c_d(T;p) = \infty$ and $\lim_{T\to\infty} c_d(T;p) = c_2$. From Eq. (9.88) and for a given value of T, the expected cost $c_d(T;p)$ is an increasing function of p for an increasing failure rate. Therefore,

$$\frac{c_1 + c_2 \int_0^T F(t)\,dt}{T} \le c_d(T;p) \le \frac{(c_1 + c_2T)\sum_{j=0}^{\infty} \bar{F}(jT) - c_2\mu}{T \sum_{j=0}^{\infty} \bar{F}(jT)}. \tag{9.89}$$

EXAMPLE 9.12

In a copper mining operation, the copper ore bodies are mined about a half mile below the surface. The mining operation consists of drilling, blasting, and transportation processes. The diesel loaders are considered an essential part of the mining operation since they carry both drilling equipment and the exploded ore. Because of the high cost of lost production, the loaders are subject to an extensive inspection program: more specifically, the wheel axles of the loaders are inspected for possible cracks. The failure time of the axles is exponential with a failure rate of 0.0005 failures per hour. The cost per inspection is $120, and the cost of an undetected failure is $80. Determine the following:

1. The optimum inspection interval that minimizes the total expected cost per unit time.
2. The optimum inspection interval that minimizes the expected cost per unit time until detection of failure.

SOLUTION

Since the failure time is exponentially distributed, $[F(t) = 1 - e^{-\lambda t}]$, then for any value of p, Eq. (9.86) reduces to

$$c(T; p) = \frac{c_1 + c_2 T}{1 - e^{-\lambda T}} - \frac{c_2}{\lambda} \tag{9.90}$$

and Eq. (9.88) reduces to

$$c_d(T; p) = \frac{c_1 + c_2 T - (c_2/\lambda)(1 - e^{-\lambda T})}{T}. \tag{9.91}$$

1. The optimum inspection interval that minimizes the total expected cost per unit time is the value of T^* that minimizes Eq. (9.90). In other words,

$$c(T^*; p) = \frac{120 + 80T^*}{1 - e^{-0.0005T^*}} - \frac{80}{0.0005}$$

and

$$T^* = 78 \text{ hours.}$$

2. The optimum inspection interval that minimizes the expected cost per unit time until detection of failure is obtained using Eq. (9.91) as given below:

$$c_d(T^*; p) = \frac{120 + 80T^* - \left(\dfrac{80}{0.0005}\right)(1 - e^{-0.0005T^*})}{T^*}$$

or

$$T^* = 77 \text{ hours.}$$

In other words, the axles should be inspected periodically at fixed intervals of 77 hours.

9.8. On-Line Surveillance and Monitoring

Thus far, we have presented preventive maintenance and inspection models that determine the optimum preventive maintenance and inspection intervals that minimize the total expected cost and downtime, and maximize the system availability. In many practical situations, it is critical to continuously monitor the condition of the system in order to prevent its failure for any period of time. The recent developments in sensors, chemical and physical nondestructive testing (NDT), and sophisticated measurement techniques have facilitated the continuous monitoring of the system performance. When linked with microcomputer-based control equipment, performance monitoring and predictive control of complete systems have become more practical and affordable. Moreover, sensors for monitoring system components eliminate the time for diagnostics, thus reducing the time to perform the actual repair and testing in order to verify the repair's quality.

Data needed to diagnose the condition of equipment or a system include noise level, speed, flow rate, temperature, differential expansion, vibration, position, accuracy, repeatability, and others. The majority of sensors and monitoring devices are based on vibration, acoustic, electrical, hydraulic, pneumatic, corrosion, wear, vision, and motion patterns. In this section, we briefly discuss some of the most commonly used diagnostic systems for component or system monitoring.

9.8.1. VIBRATION ANALYSIS

Machines or equipment produce vibration when in operation. Each machine has a characteristic vibration or "signature" composed of a large number of harmonic vibrations of different amplitudes. The effects of component wear and failure on these harmonic vibrations differ widely depending on the contribution made by a particular component to the overall "signature" of the machine. For example, in reciprocating engines and compressors, the major forcing is produced by harmonic gas forcing torques, which are functions of the thermodynamic cycle on which the machine operates. In a multicylinder engine, the major harmonic vibrations are calculated from the number of working strokes per revolution. Thus, misfiring of one or more cylinders would produce significantly different vibrations than the original "signature" of the engine, which can be easily detected by an accelerometer. The accelerometer is an electromechanical transducer that produces an electric output proportional to the vibratory acceleration to which it is exposed (Niebel, 1994).

The vibration "signature" of equipment is dependent on the frequency, amplitude, velocity, and acceleration or wave slope of the vibration wave. In general, there is no single transducer that is capable of the extreme wide range of signatures. Therefore, transducers are developed for different frequency, amplitude, velocity, and acceleration ranges. For example, the bearing probe, being a displacement measuring device, is sensitive only to low-frequency large-amplitude vibrations. This makes it useful only as a vibration indicator for gear box vibrations and turbine blades.

On the other hand, seismic accelerometers can be used for monitoring vibrations of electrical machines that are characterized by the usual bearing and balance vibrations in addition to high-frequency vibrations that are functions of the electrical geometry via number of starter and rotor poles (Downham, 1975).

9.8.2. ACOUSTIC EMISSION AND SOUND RECOGNITION

Acoustic emission (AE) can be defined as the transient elastic energy spontaneously released from materials undergoing deformation or fracture or both. The released energy produces high-frequency acoustic signals. The strength of the signals depend on parameters such as the rate of deformation, the volume of the participating material, and the magnitude of the applied stress. The signals can be detected by sensors often placed several feet away from the source of signal generation.

Most of the AE sensors are broad-band or resonant piezoelectric devices. Optical transducers for AE are in the very early stage of development. They have the advantages of being used as contacting and noncontacting measurement probes and of the flat frequency response over a large bandwidth. Acoustic emission is used in many applications such as tool wear monitoring, material fatigue, and weld defects.

Sound recognition is used to detect a wide range of abnormal occurrences in manufacturing processes. The sound recognition system recognizes various operational sounds, including stationary and shock sounds, using a speech recognition technique and then compares them with the expected normal operational sounds (Takata and Ahn, 1987).

The operational sound is collected by a unidirectional condenser microphone that is set near the component or machine to be monitored. When the sound of an abnormal operation is generated, such as the sound of tool breakage or the sound of worn-out motor bearings, the features of the sound signal are extracted and a sound pattern is formed. The sound pattern is then compared with the standard patterns through pattern matching techniques, and the most similar standard pattern is selected. The failure or fault corresponding to this category is then recognized and diagnosed.

9.8.3. TEMPERATURE MONITORING

Elevation in component or equipment temperature is frequently an indication of potential problems. For example, most of the failures of electric motors are attributed to excessive heat that is generated by antifriction bearings. The bearing lives are dependent on the preventive maintenance schedule and their operating conditions. Similarly, hot spots in electric boards indicate that failure is imminent. The hot spots are usually caused by excessive currents.

Therefore, a measure of temperature variation can be effectively used in monitoring components and equipment for preventive maintenance purposes. There is a wide range of instruments for measuring variations in temperature—such as mercury thermometers, which are capable of measuring temperatures in the range of −35 to 900°F, and thermocouples, which can provide accurate measurements up to about 1,400°F. Optical pyrometers, where the intensity of the radiation is compared optically with a heated filament, are useful for the measurements of very high temperatures (1,000 to 5,000°F).

Recent advances in computers made the use of the infrared temperature measure possible for many applications that are difficult and impractical to contact with other instruments (Niebel, 1994).

The infrared emissions are the shortest wavelengths of all radiant energy and are visible with special instrumentation. Clearly, the intensity of the infrared emission from an object is a function of its surface temperature. Therefore, when a sensing head is aimed at the object whose surface temperature is being measured, the computer calculates the surface temperature and provides a color graphic display of temperature distribution. This instrument is practical and useful in monitoring temperature of controllers and detecting heat loss in pipes.

9.8.4. FLUID MONITORING

Analysis of equipment fluids such as oil can reveal important information about the equipment wear and performance. It can also be used to predict the reliability and expected remaining life of parts of the equipment. As the equipment operates, minute particles of metal are produced from the oil-covered parts. The particles remain in suspension in the oil and are not removed by the oil filters due to their small size. The particle count will increase as equipment parts wear out. There are several methods that can identify the particle count and the types of particles in the oil. The two most commonly used methods are *atomic absorption* and *spectrographic emission.*

With the atomic absorption method, a small sample of oil is burnt, and the flame is analyzed through a light source that is particular for each element. The method is very accurate and can obtain particle count as low as 0.1 parts per million (ppm). However, the analysis is tedious and time consuming except when the type of particle is known (Cumming, 1990).

Spectrographic emission is similar to the atomic absorption method in burning a small sample of oil. It has the advantage that all quantities of all the materials can be read at one burn. However, it is capable only of detecting particle counts of 1 ppm or higher. Moreover, spectrometry is unable to give adequate warning in situations when the failure mode is characterized by the generation of large particles from rapidly deteriorating surfaces (Eisentraut et al., 1978).

9.8.5. CORROSION MONITORING

Corrosion is a degradation mechanism of many metallic components. Clearly, monitoring the rate of degradation—the amount of corrosion—has a major impact on the preventive maintenance schedule and the availability of the system. There are many techniques for monitoring corrosion, such as visual, ultrasonic thickness monitoring, electrochemical noise, impedance measurements, and thin-layer activation. We briefly describe one of the most effective on-line corrosion monitoring techniques—*thin layer activation* (TLA). The principle of TLA is that trace quantities (1 in 10^{10}) of a radioisotope are generated in a thin surface layer of the component under study by an incident high-energy ion beam. Loss of the material (due to corrosion) from the surface of the component can be readily detected by a simple γ-ray monitor (Asher et al., 1983). The reduction in activity is converted to give a depth of corrosion directly, and, provided that the corrosion is not highly localized, this gives a reliable measurement of the average loss of material over the surface.

9.8.6. OTHER DIAGNOSTIC METHODS

Components and systems can be monitored in order to perform maintenance and replacements by observing some of the critical characteristics using a variety of sensors or microsensors. For example, pneumatic and hydraulic systems can be monitored by observing pressure, density of the flow, rate of flow, and temperature change. Similarly, electrical components or systems can be monitored by observing the change in resistance, capacitance, volt, current, temperature, and magnetic field intensity. Mechanical components and systems can be monitored by measuring velocities, stress, angular movements, shock impulse, temperature, and force.

Recent technological advances in measurements and sensors resulted in observing characteristics that were difficult or impossible to observe, such as odor sensing. At this point in time, silicon microsensors have been developed that are capable of mimicking the human sense of sight (such as a CCD), touch (such as a tactile sensor array), and hearing (such as a silicon microphone). Sensors to mimic the human sense of smell to discriminate between different odor types or notes are at the early stage of development. Nevertheless, some commercial odor-discriminating sensors are now available such as the Fox 2000 or Intelligent Nose (Alpha

MOS, France). The instrument is based on an array of six sintered metal oxide gas sensors that respond to a wide range of odorants. The array signals are processed using an artificial neural network (ANN) technique. The electronic nose is first trained on known odors; then the ANN can predict the nature of the unknown odors with a high success rate (Gardner, 1994).

The improvements in sensors' accuracy and the significant reduction in their cost have resulted in their use in a wide variety of applications. For example, most of the automobiles are now equipped with electronic diagnostic systems that provide signals indicating the times to service the engine, replace the oil filter, and check engine fluids.

Most important, the advances in microcomputers, microprocessors, and sensors can now offer significant benefits to the area of preventive maintenance and replacements. Many components, systems, and entire plants can now be continuously monitored for sources of disturbances and potential failures. Moreover, online measurements, analysis, and control of properties and characteristics, which have been traditionally performed off-line, result in monitoring of a wider range of components and systems than ever before.

PROBLEMS

9-1. In a block replacement policy, the cost of a failure replacement is \$150 while the cost of a preventive replacement is \$80. Assume that the failure times of a component that is replaced based on the block replacement policy follow a beta distribution. The parameters of the distribution are $\alpha = 4$ and $\beta = 3$ hours. The p.d.f. of the beta distribution is

$$f(t) = \frac{\Gamma(\alpha + \beta)}{\Gamma(\alpha)\Gamma(\beta)} t^{\alpha-1}(1 - t)^{\beta-1}, \qquad 0 < t < 1.$$

a. What is the optimal replacement interval?

b. Assume that the penalty factor for surplus or shortage inventory (amount of stock that meets the expected failures during the replacement interval) is \$20. Assuming that there are 200 units in operation, what are the optimal interval T and stock level L?

9-2. Consider a block replacement policy, where the failing unit is replaced by a new one at time instants $t = KT$, $K = 1, 2, \ldots$ and at failure. The cost of replacing a failed unit is \$80, and the cost of replacing nonfailed units is \$50. The penalty for both shortage or excess spares in the inventory is \$25 per unit. There are 100 units in operation at the beginning of a replacement cycle, and the failure density function is given by

$$f(t) = \begin{cases} \dfrac{1}{\Gamma(\alpha)\beta^{\alpha}} t^{\alpha-1} e^{-t/\beta} & t > 0 \\ 0 & \text{otherwise,} \end{cases}$$

where α and β are parameters that determine the specific shape of the curve. For $\alpha = 2$ and $\beta = 400$, determine the optimal replacement period and the optimal inventory level of the spares.

9-3. Consider a replacement policy where the time at which preventive replacement occurs depends on the age of the equipment; failure replacements are made when failures occur. Let t_p be the age of the equipment at which a preventive replacement is made, c_p the cost of a preventive replacement, c_f the cost of a failure replacement. It is given that $c_p = \$50$, $c_f = \$100$, $\lambda = 1$, and the p.d.f. of the failure-time distribution is given by the special Erlangian function

$$f(t) = \frac{t}{\lambda^2} \exp(-t/\lambda) \qquad t \geq 0.$$

a. What is the optimal replacement interval?

b. Repeat part a for $f(t) = \lambda e^{-\lambda t}$.

9-4. A typical twin-turboprop transport aircraft has an average gross landing weight of 40,000 pounds and a tricycle landing gear. The main landing gear is equipped with two wheels on each side. It is the principal support for the aircraft and has many components, including air/oil shock struts to absorb landing impact and taxiing loads, alignment and support units, retraction mechanisms and safety devices, auxiliary gear protective devices, wheels, tires, tubes, and braking systems.

The landing gear system (main nosewheel, left landing gear, right landing gear) is periodically inspected and preventive maintenance or replacements are carried out. The cost of preventive maintenance or replacement is \$5,000, whereas the cost of a failure replacement depends on where the failure occurs. The failure of either the left or right landing gear may result in right- or left-wing tipping, which causes the propeller blades of the engines and the lower portion of the rear fuselage to scrape along the runway, resulting in a substantial damage of \$25,000. The failure of nosewheel landing gear results in a damage worth \$45,000. The landing gears have equal probabilities of failure. The time to failure follows a normal distribution with mean of 500 aircraft landings and standard deviation of twenty landings. Determine the optimum preventive constant replacement intervals.

9-5. Assume that the constant interval replacement policy in Problem 9-4 is to be compared with an age replacement policy using the same cost values. Determine the optimum preventive replacement interval for the age replacement policy.

9-6. Transport aircraft with a steel piston engine crankshaft may fail catastrophically during flight if the crankshaft fails. The massive and complex-shaped crankshaft is usually produced by forging a triple alloy steel. During flight, the crankshaft experiences significant and complex stresses including bending and torsion. It is periodically checked for crack indications and defects using the magnetic particle nondestructive test method. If cracks or defects are found, the crankshaft is repaired or replaced. The time to perform preventive replacement, T_p, is twenty hours and the time to perform failure replacement is fifty hours. Assume that failure times of the crankshaft follow a Weibull distribution having a p.d.f. as

$$f(t) = \frac{\gamma}{\theta} t^{\gamma-1} e^{-t^{\gamma}/\theta} \qquad t > 0.$$

The shape parameter γ is found to be 2.7, and scale parameter θ is 250. Determine the optimum preventive replacement age t_p that minimizes the downtime per unit time when the following preventive replacement policy is implemented: perform preventive replacement when the equipment reaches age t_p.

9-7. In the optimal replacement policy under minimal repair (Lam, 1990), a critical component is observed. When the component fails, it is replaced or minimally repaired at a cost of \$1,200. After N minimal repairs (or replacements) of the component, the entire system is replaced by a new one at a cost of \$42,000. Assume that the reward rate is \$100 per hour and the operating time λ_k and the repair time μ_k after the kth failure are

$$\lambda_k = \frac{9{,}000}{3^{k-1}}$$

$$\mu_k = 50 \times 3^{k-1}.$$

Determine the parameters of the optimum replacement policy N^*.

9-8. Consider a periodic replacement policy where the component is subject to shock. Assume that the normal cost of running the system is a per unit of time, that each shock to the system increases its running cost by c per unit of time, and the cost of completely replacing the system is c_0. Prove that, when the shock rate is

$$\lambda(t) = \begin{cases} 0, & t < 1 \\ \dfrac{1}{t} & t \geq 1, \end{cases}$$

the optimal value of the period replacement time is

$$T^* = \sqrt{\frac{2c_0}{c} + 1}.$$

9-9. Determine the optimal preventive maintenance interval and the optimal number of spares when the penalty function g is expressed as

$$g(L, N_1(t_p)) = |L - (N_1(t_p) + 1)|.$$

9-10. Consider a group maintenance policy for N machines, each having a failure rate λ. Assume that $F(t)$ is continuous, $f(t)$ exists, and $-f'(t)/f(t) < c_2/c_1$ for $t \geq 0$. Prove that if $c_2/\lambda > c_0/N + c_1$, then there exists a unique and finite optimum scheduling time t_0^* that satisfies Eq. (9.63).

9-11. A group of production machines consists of N machines. Each machine exhibits the same failure-time distribution with the following p.d.f.:

$$f(t) = kte^{\frac{-kt^2}{2}},$$

where $k = 0.01$. Assume that $c_0 = \$300$, $c_1 = \$150$, and $c_2 = \$250$. Determine the optimum group replacement interval.

9-12. Consider an inspection policy where inspections of a power generator unit are performed at times x_1, x_2, x_3, and x_4. The power generator unit exhibits a failure-time distribution with a p.d.f. given by

$$f(t) = \frac{\alpha(\alpha k)^{k-1} \exp(-\alpha t)}{(k-1)!},$$

where $\alpha = 1/b$, b is a scale parameter. The mean of the distribution, $\mu = kb$. Assume $k = 3$, $\mu = 1{,}000$ hours, cost of inspection $= \$150$, cost of undetected failure $c_u = \$300$, and the cost of repair is $c_r = \$2{,}000$. Determine the first four points of the optimal inspection schedule.

9-13. An aircraft maintenance group uses a thermographic detection method to detect corrosion over a large surface area of the aircraft. The detection method uses a noncontact device. The inspection involves heating the surface with flash or quartz lamps and then measuring the temperature with an infrared camera over a set time span. Heating the surfaces creates temperature differences that indicate disbonds or corrosion.

The aircraft is periodically inspected to determine whether or not corrosion is formed. The corroded components or surfaces are repaired or replaced, and at the same time, preventive maintenance is provided if needed. We assume that after inspection the component (or surface) has the same age as before with probability p and that the component is as good as new with probability q. The times to the formation of corrosion follow a log logistic distribution with the following p.d.f.:

$$f(t) = \frac{\lambda k(\lambda t)^{k-1}}{[1 + (\lambda t)^k]^2} \qquad 0 \leq t < \infty,$$

where $k = 2$ and $\lambda = 0.008$ failures per hour. The cost per inspection is \$2,500 and the cost of undetected corrosion is \$500. Determine the following:

a. The optimum inspection interval that minimizes the total expected cost per unit time.

b. The optimum inspection interval that minimizes the expected cost per unit time until detection of failure.

c. Solve a and b if the true cost of an undetected failure is \$3,000.

9-14. Leaf springs are attached to the undercarriage assemblies of trains in order to provide a smooth ride to the passengers. The repeated loads on a leaf spring result in subjecting both surfaces of the spring to cycles of tension and compression stress. Such repeated loads coupled with crack initiation and shocks may result in the spring failure.

Consider a leaf spring whose $\lambda(t) = \lambda t$ with $\lambda = 10 \times 10^{-5}$. Assume that the normal cost of running the system is \$50 per hour, the increase in running cost due to each shock is \$6 per hour, and the cost of replacing the entire spring system is \$20,000. Determine the optimal value of T (the length of the periodic replacement policy) that minimizes the long-run average cost per unit time.

9-15. Assume that the replacement cost of the leaf springs in Problem 9-14 is a function of time (or number of shocks). In other words, the operating cost per unit time due to every additional shock in the interval $[\tau_i, \tau_{i+1})$ is $c_i(u)$, $i = 0, 1, \ldots,$ and u is the state space at which the periodic replacement can be performed. Let

$c_{N(t)}$ represent the additional operating cost per unit as a function of the number of shocks at time t; $N(t)$:

$$c_{N(t)} = 5.0 + 2N(t) + 1.5(N(t))^2.$$

Determine the optimum replacement interval T^*.

9-16. Consider a maintenance policy in which a component (or a system) is minimally repaired at equal intervals of time. The minimal repairs involve adjustments, cleaning, and replacement of nonessential parts. After a minimum repair, the component (or system) has the same age as before the repair. Moreover, the component is replaced by a new one upon failure and when the ratio between the failure rates $r + 1$ minimal repairs and r is greater than $\phi(\phi > 1)$. Assume that the hazard rate of the component is given by

$$h(t) = \delta k^t,$$

where δ and k are positive constants. The cost per minimal repair is c_m, the cost of the scheduled replacement is c_r and the cost due to the failure replacement is c_f. It should be noted that $c_f > c_r > c_m$. Determine the optimum minimal repair interval and the optimum replacement time that minimize the total expected cost per unit time.

9-17. Define, e_π, the effectiveness of a preventive maintenance or replacement policy π, as the ratio between the expected number of failures avoidable by the implementation of the policy and the total expected number of failures under the failure replacement policy (FRP). Under the FRP, components are only replaced upon failure (Al-Najjar, 1991). Derive e_π for the policy stated in Problem 9-16.

9-18. The efficiency of a preventive maintenance or replacement policy, η, can be measured as the ratio between the total expected cycle cost when the policy is in effect and the total expected cycle cost when replacements are made only when failures occur. Derive an expression for η for the policy stated in Problem 9-16.

9-19. The Kurtosis method is a statistical means of studying the time domain signal generated due to vibrations caused by running a machine. The principle of the Kurtosis method is to take observations in a suitable frequency range. The Kurtosis of the signal is calculated as

$$K = \frac{1}{\sigma^4} \sum_{i=1}^{N} \frac{(x_i - \bar{x})^4}{N},$$

where σ^2 is the variance, N is the number of observations, $\bar{x}$ is the mean value of the observations, and x_i is the observed value i. The use of K to monitor the condition of rotating machinery is based on the fact that a rolling bearing in normal operating conditions exhibits a Kurtosis value of about 3. However, the value of K increases rapidly as the wear of the parts increases.

Reliability engineers use the Kurtosis method to determine whether or not to perform preventive maintenance or replacement of the sliding bearings of high-pressure presses. In other words, the Kurtosis method is used as an inspection tool. Under this policy, inspections are performed at times $x_1, x_2, x_3, \ldots$ until a failed or degraded bearing is detected. Repairs are immediately performed upon the detection of a failure or degradation. The inspection intervals are not necessarily equal as they may be reduced as the probability of failure increases.

In order to accurately predict the optimal inspection intervals, observations from the last eleven inspection intervals are shown in Table 9.8. Using the information in Tables 9.8 and 9.9, obtain the p.d.f. of the failure times. The cost per

Table 9.8. Observations from Eleven Inspection Intervals

Obs. A	*Obs. B*	*Obs. C*	*Obs. D*	*Obs. E*	*Obs. F*	*Obs. G*	*Obs. H*	*Obs. I*	*Obs. J*	*Obs. K*
830	850	830	835	855	860	855	860	864	860	834
853	863	853	853	863	873	873	873	874	874	844
880	870	885	886	886	880	886	883	887	884	857
892	882	892	893	893	892	893	895	898	896	878
980	900	984	985	985	990	995	996	997	997	987
999	932	995	996	996	992	997	997	998	997	998
999	936	999	999	999	1,000	999	1,002	999	1,004	999
1,000	953	10	1,001	1,001	1,001	1,011	1,003	1,001	1,002	1,001
1,002	969	1,002	1,002	1,002	1,002	1,002	1,004	1,012	1,005	1,012
1,004	972	1,005	1,004	1,003	1,003	1,003	1,005	1,013	1,005	1,023
1,005	996	1,007	1,007	1,006	1,004	1,016	1,006	1,015	1,006	1,025
1,010	1,005	1,010	1,011	1,015	1,005	1,015	1,007	1,016	1,008	1,016
1,019	1,009	1,019	1,018	1,018	1,009	1,018	1,009	1,017	1,009	1,017
1,023	1,013	1,025	1,023	1,022	1,013	1,032	1,012	1,021	1,011	1,021
1,028	1,018	1,028	1,026	1,023	1,016	1,023	1,015	1,024	1,013	1,034
1,035	1,025	1,035	1,038	1,034	1,027	1,025	1,025	1,025	1,023	1,035
1,036	1,026	1,036	1,034	1,035	1,029	1,036	1,029	1,026	1,025	1,026
1,044	1,034	1,043	1,042	1,042	1,034	1,044	1,035	1,031	1,032	1,041
1,054	1,034	1,054	1,050	1,051	1,039	1,055	1,038	1,054	1,036	1,054
1,064	1,044	1,064	1,064	1,064	1,044	1,044	1,046	1,044	1,048	1,064
1,078	1,058	1,078	1,078	1,075	1,053	1,065	1,056	1,064	1,055	1,074
1,099	1,069	1,099	1,099	1,094	1,060	1,074	1,061	1,075	1,063	1,095
1,106	1,076	1,106	1,106	1,104	1,066	1,100	1,064	1,098	1,067	1,098
1,115	1,089	1,115	1,115	1,116	1,069	1,110	1,065	1,100	1,066	1,110
1,135	1,099	1,135	1,135	1,125	1,079	1,115	1,076	1,105	1,073	1,115

Table 9.9. Times at Which the Observations Are Taken

Obs. A	*Obs.* B	*Obs.* C	*Obs.* D	*Obs.* E	*Obs.* F	*Obs.* G	*Obs.* H	*Obs.* I	*Obs.* J	*Obs.* K
58	60	61	81	82	94	107	123	127	134	146

inspection is \$200, cost of undetected failure is \$20, and the cost of repair is \$1,800. The repair time is 30 hours. Determine the first three optimum inspection intervals that minimize the expected total cost per unit time.

9-20. Exhaust gases from paper plants usually contain fine particles that are removed by electrostatic precipitators. A typical precipitator contains thin wires that are charged to several thousands of volts. When the gases pass through the wires, the fine particles are attracted to the wires and to the dust-collector plates. The wires are vibrated periodically to remove the particles, which are then concentrated in receptacles, collected, and disposed of. Since the wires are subject to thermal stresses, they may experience breaks that reduce its effectiveness in removing the particles. Therefore, a periodic inspection policy is implemented. Under this policy, the wires are inspected at times $x_1, x_2, x_3, x_4, \ldots$ until a broken wire is detected. Repairs are immediately performed when breaks are found.

The cost per inspection is \$150; the cost of undetected failure is \$12; and the cost of repair is \$225. The repair time is 12 hours. The p.d.f. of the wire breaks follow

$$f(t) = kte^{\frac{-kt^2}{2}},$$

where $k = 0.00004$. Determine the first five inspection intervals that minimize the total cost.

REFERENCES

Abdel-Hameed, M. (1986). "Optimum Replacement of a System Subject to Shocks." *Journal of Applied Probability* 23, 107–114.

Al-Najjar, B. (1991). "On the Selection of Condition-Based Maintenance for Mechanical Systems." In Holmberg, K., and Folkeson, A. (Eds.), *Operational Reliability and Systematic Maintenance* (pp. 153–174). London: Elsevier Applied Science.

Ascher, H., and Feingold, H. (1984). *Repairable Systems Reliability*. New York: Marcel Dekker.

Asher, J., Conlon, T. W., Tofield, B. C., and Wilkins, N. J. M. (1983). "Thin-Layer Activation: A New Plant Corrosion-Monitoring Technique." In Butcher, D.W. (Ed.),

On-Line Monitoring of Continuous Process Plants (pp. 95–105). West Sussex, England: Ellis Horwood.

AT&T. (1983). *Reliability Manual*. Basking Ridge, NJ: AT&T.

Barlow, R. E., and Hunter, L. (1960). "Optimum Preventive Maintenance Policies." *Operations Research* 8, 90–100.

Barlow, R. E., Proschan, F., and Hunter, L. C. (1965). *Mathematical Theory of Reliability*. New York: John Wiley.

Blanks, H. S., and Tordan, M. J. (1986). "Optimum Replacement of Deteriorating and Inadequate Equipment." *Quality and Reliability Engineering International* 2, 183–197.

Brender, D. M. (1962). "A Surveillance Model for Recurrent Events." Research Report RC-837, IBM Corporation, Watson Research Center, Yorktown Heights, New York.

Cooper, R. B. (1972). *Introduction to Queueing Theory*. New York: MacMillan.

Cox, D. R. (1962). *Renewal Theory*. New York: Methuen.

Cumming, A. C. D. (1990). "Condition Monitoring Today and Tomorrow: An Airline Perspective." In Rao, R., Au, J., and Griffiths, B. (Eds.), *Condition Monitoring and Diagnostic Engineering Management*. London: Chapman and Hall.

Downham, E. (1975). "Vibration Monitoring and Wear Prediction." In *Terotechnology: Does it Work in the Process Industries?* (pp. 29–34). New York: Mechanical Engineering Publications Limited for the Institute of Mechanical Engineers.

Downton, F. (1971). "Stochastic Models for Successive Failures." *Proceedings of the Thirty-eighth Sessional International Statistical Institute* 44(1), 667–694. The Hague: International Statistical Institute.

Eisentraut, K. J., Thornton, T. J., Rhine, W. E., Constandy, S. B., Brown, J. R., and Fair, P. S. (1978). "Comparison of the Analysis Capability of Plasma Source Spectrometers vs. Rotating Disc Atomic Emission and Atomic Absorption Spectrometry of Wear Particles in Oil: Effect of Wear Metal Particle Size." Paper presented at the First International Symposium on Oil Analysis, Erding, Germany, July 4–6.

Gardner, J. W. (1994). *Microsensors Principles and Applications*. New York: John Wiley.

Heyman, D. P., and Sobel, M. J. (1982). *Stochastic Models in Operations Research* (Vol. 1). New York: McGraw-Hill.

Jardine, A. K. S. (1973). *Maintenance, Replacement, and Reliability*. New York: Pitman/Wiley. Reprinted edition available from International Academic Services, PO Box 2, Kingston, Ontario, Canada, K7L 4V6.

Jardine, A. K. S., and Buzacott, J. A. (1985). "Equipment Reliability and Maintenance." *European Journal of Operational Research* 19, 285–296.

Lam, Y. (1988). "A Note on the Optimal Replacement Policy." *Advanced Applied Probability* 20, 479–482.

Lam, Y. (1990). "A Repair Replacement Model." *Advanced Applied Probability* 22, 494–497.

Murthy, D. N. P. (1982). "A Note on Block Replacement Policy." *Journal of Operational Research Society* 33, 481–483.

Nakagawa, T. (1984). "Periodic Inspection Policy with Preventive Maintenance." *Naval Research Logistics Quarterly* 31, 33–40.

Niebel, B. W. (1994). *Engineering Maintenance Management*. New York: Marcel Dekker.

Okumoto, K., and Elsayed, E. A. (1983). "An Optimum Group Maintenance Policy." *Naval Research Logistics Quarterly* 30, 667–674.

Senju, S. (1957). "A Probabilistic Approach to Preventive Maintenance." *Journal of Operations Research Society of Japan* 1, 49–58.

Stadje, W., and Zuckerman, D. (1990). "Optimal Strategies for Some Repair Replacement Models." *Advances in Applied Probability* 22(3), 641–656.

Taguchi, G., Elsayed, E. A., and Hsiang, T. (1989). *Quality Engineering in Production Systems*. New York: McGraw-Hill.

Takata, S., and Ahn, J. H. (1987). "Overall Monitoring System by Means of Sound Recognition." In Milacic, V. R., and McWaters, J. F. (Eds.), *Diagnostic and Preventive Maintenance Strategies in Manufacturing Systems* (pp. 99–111). Amsterdam: North-Holland.

Thompson, W. A. (1981). "On the Foundations of Reliability." *Technometrics* 23, 1–13.

Tilquin, C., and Cléroux, R. (1975). "Periodic Replacement with Minimal Repair at Failure and Adjustment Costs." *Naval Research Logistics Quarterly* 22, 243–254.

Tilquin, C., and Cléroux, R. (1985). "Periodic Replacement with Minimal Repair at Failure and General Cost Function." *Journal of Statistical Computer Simulation* 4, 63–77.

Valdez-Flores, C., and Feldman, R. M. (1989). "A Survey of Preventive Maintenance Models for Stochastically Deteriorating Single-Unit Systems." *Naval Research Logistics* 36, 419–446.

CHAPTER 10

Case Studies

Case 1: A Crane Spreader Subsystem*

Introduction

Cranes are considered the primary method of transferring loads in ports. The operating conditions of a crane depend on the type and weight of the loads to be transferred, the coordinates of the points of pickup and discharge, the intensity of the flow of the loads, the location of the crane, and the effects of the environment (such as temperature, wind, snow, humidity, and dust content).

The coordinates of the points of pickup and discharge as well as the overall dimensions of the loads determine the principal dimensions of the crane (Kogan, 1976). The coordinates are given with known tolerances that determine the accuracy with which the loads are transferred. The tolerances affect the drive mechanisms and their operation.

Ports are usually equipped with several container handling gantry type cranes. Figure 10.1 is a diagrammatic sketch of a typical gantry type crane. During the unloading of a cargo ship, the crane picks up one container at a time from the ship and places it on a chassis of a transporter, which is then moved away to a designated location in the port. Loading a ship with containers is performed when a transporter carrying the container arrives at a specific location within reach of the crane. The container is then lifted by the crane and placed in a proper position on the ship.

Container handling gantry cranes are usually equipped with automated spreaders in order to permit rapid, safe, and efficient loading and unloading of containers. A remotely controlled, telescoping spreader is used to lift loads safely and transmit them vertically through the cornerposts of the container. The spreader depends on hydraulic pumps to provide the power required for most of its operations. These pumps activate the hydraulic cylinders of the telescoping system, flippers, and twist locks.

The telescoping system of the spreader is used to adjust the length of the spreader to accommodate containers of different lengths (20-foot, 35-foot, or 40-foot containers). The flippers (or "gather guides") are retractable corner guides

* Based on actual operation of a major shipping company.

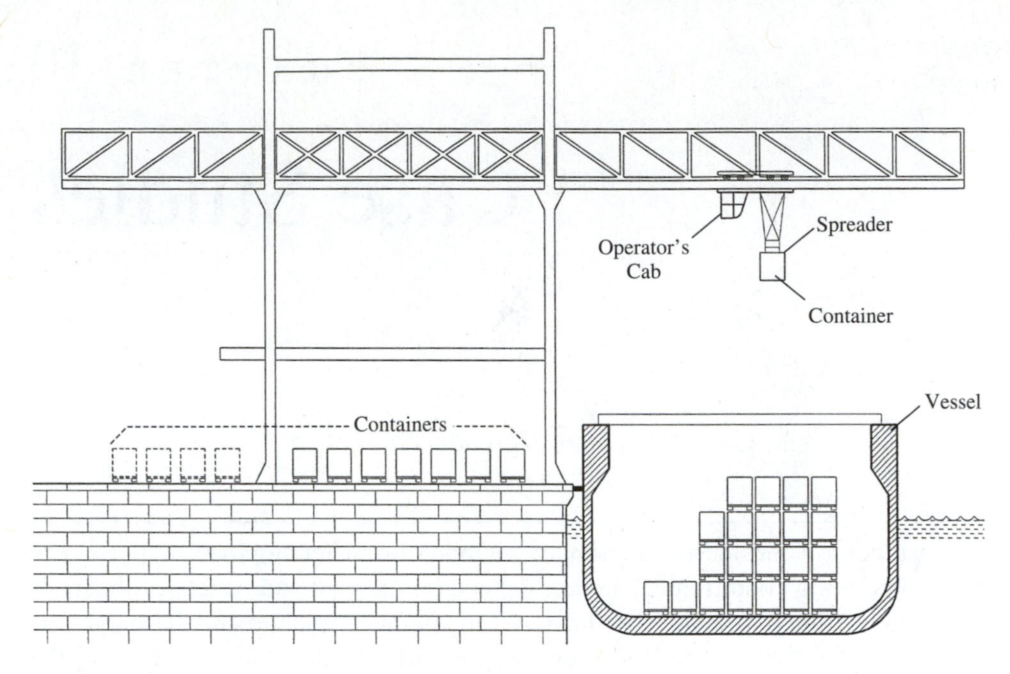

Figure 10.1. Diagrammatic Sketch of a Gantry Type Crane

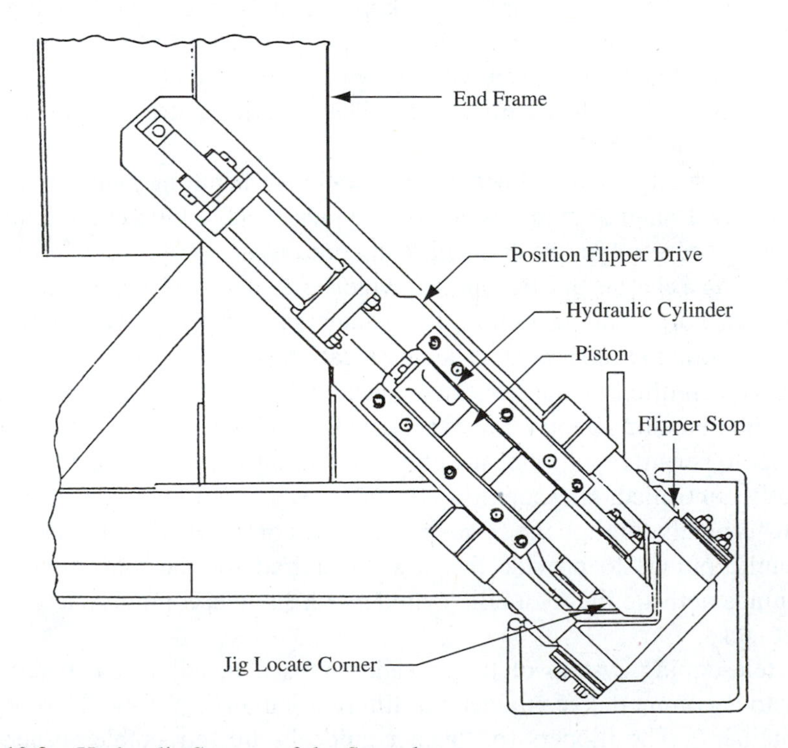

Figure 10.2. Hydraulic System of the Spreader

that help the operator in lowering the spreader onto the container. When a container is lifted, four twist locks must be engaged (one twist lock at each corner casting of the container). If one of the twist locks does not operate or function properly, none of the other three twist locks can be engaged as a safety precaution. The twist locks are turned into locked or unlocked positions through a hydraulic power unit.

Another critical component of the spreader is the limit switch. The function of the limit switch is to alter the electrical circuit of a machine or piece of equipment so as to limit its motion. The twist lock limit switches are located on the hydraulic cylinder. There are two limit switches per twist lock to terminate the action of the hydraulic cylinder whenever the locked or unlocked positions are reached. Limit switches are also used by the telescoping system to terminate the expansion or contraction of the spreader whenever the desired length is reached. Figure 10.2 shows the plan view of a corner of the hydraulic system of a spreader subsystem.

The maintenance records for six cranes operated by a worldwide company for shipping and receiving containers show that failures involving the spreader and its components account for approximately 65 percent of all crane failures. The basic components of the spreader are shown in Table 10.1. The table also shows the

Table 10.1. Failure-Time Distributions of the Components

Type of Failure	*Components*	*Number of Failures*	*Failure-Time Distribution*	*Parameter(s) of the Distribution*
A. Electrical	1. Loose connections	8[a]	—	—
	2. Short or open circuit	62	Exponential	7.14×10^{-4} failures/hr
	3. Wires parted	10[a]	—	—
B. Flippers	1. Damaged (replaced)	129	Exponential	1.03×10^{-3} failures/hr
	2. Hydraulic cylinder	29	Exponential	3.571×10^{-4} failures/hr
	3. Flipper mechanism	44	Exponential	8.333×10^{-4} failures/hr
C. Twist locks	1. Locks	85	Exponential[b]	6.666×10^{-4} failures/hr
	2. Cylinder	55	Exponential[b]	4.640×10^{-4} failures/hr
	3. Limit switches	178	Exponential[b]	5.319×10^{-4} failure/hr
D. Telescoping system	1. Cylinders	12[a]	—	—
	2. Limit switches	112	Exponential	11.764×10^{-4} failures/hr
E. Hydraulic system	1. Power unit	146	Exponential	12.5×10^{-4} failures/hr
	2. Hydraulic piping	45	Exponential	8.33×10^{-4} failures/hr
	3. Fittings (other than those on the power unit)	32	Exponential	9.756×10^{-4} failures/hr

(continued)

[a] There is insufficient data to determine the failure-time distribution (rare events during the nine-year period of the study).
[b] Both the Weibull and the exponential distributions appropriately fit the failure data. Comparisons of the sum of squares of errors show that the exponential distribution yields slightly lower values than the Weibull distribution.

Table 10.1. *(continued)*

Type of Failure	*Components*	*Number of Failures*	*Failure-Time Distribution*	*Parameter(s) of the Distribution*
F. Frame (structural)	1. Main structure	5[a]	—	—
	2. Corners	7[a]	—	—
	3. Expanding trays	6[a]	—	—
	4. Corner trays	11[a]	—	—
G. Head block	1. Frame	6[a]	—	—
	2. Twist locks	10[a]	—	—
	3. Limit switches	18[a]	—	—

[a] There is insufficient data to determine the failure-time distribution (rare events during the nine-year period of the study).

number of failures, the failure-time distribution, and the parameter(s) of the distribution as estimated from the failure data.

Statement of the Problem

Cranes operate continuously for a period of ten hours per day including weekends. When a ship arrives at the port for loading or unloading, depending on the work load, one or more cranes immediately proceed with the task. Although spreaders are interchangeable between cranes, it is customary that only one spreader is assigned to each crane. The spreader requires an average repair time of two hours when it fails, independent of the type of failure—that is, repair rate is constant. A failure of the spreader results in delaying the ship at a cost of $10,000 per hour. The cost of repairs is $200 per hour.

The management of the company is interested in choosing one of the following alternatives in order to minimize the total cost of the system:

1. Acquire additional spreaders at a cost of $100,000 per unit and assign two spreaders to each crane. The expected life of a spreader is five years.
2. Increase the crew size in order to reduce the repair time to one hour. This will increase the repair cost to $400 per hour.

We will investigate these two alternatives and make a recommendation.

SOLUTION The failure data of each critical component of the spreader are analyzed, and the failure-time distributions and their parameters are shown in Table 10.1. We construct the reliability block diagram of the spreader as shown in Figure 10.3. We use the notation *Xn* to refer to component *n* of failure type *X*. For example, *C3* refers

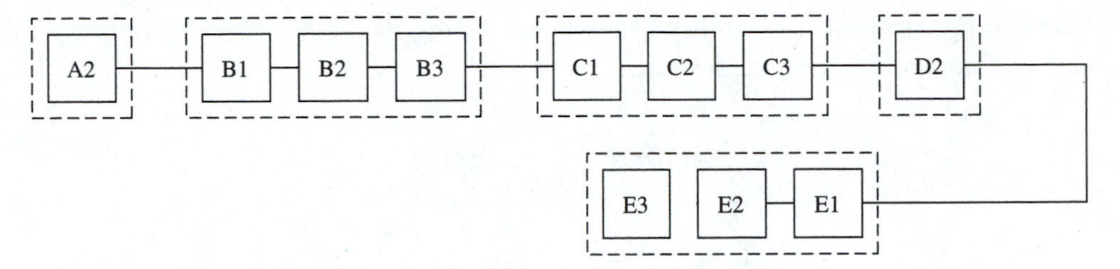

Figure 10.3. Block Diagram of the Spreader Components

to the limit switches of the twist locks (see Table 10.1). In constructing the block diagram, we drop those components that exhibit rare failures.

Using Eq. (3.7) we develop an expression for the reliability of the spreader as

$$R_s(t) = \exp[-(\lambda_{A2} + \sum_{i=1}^{3} \lambda_{Bi} + \sum_{i=1}^{3} \lambda_{Ci} + \lambda_{D2} + \sum_{i=1}^{3} \lambda_{Ei})t] \tag{10.1}$$

or

$$R_s(t) = \exp[-\,8.831 \times 10^{-3} t]. \tag{10.2}$$

The hazard rate of the spreader subsystem is

$$h_s(t) = 8.831 \times 10^{-3} \text{ failures per hour.}$$

Assuming that the preventive maintenance is performed at scheduled times that do not interrupt the normal operation of the crane, then the availability of the spreader is obtained by substituting $w^*(s) = \lambda/(s + \lambda)$ and $g^*(s) = \mu/(s + \mu)$ into Eq. (3.41) or by using Eq. (3.63) directly. Thus,

$$A(t) = \frac{\mu}{\lambda + \mu} + \frac{\lambda}{\lambda + \mu} e^{-(\lambda+\mu)t}$$

or

$$A(t) = \frac{0.5}{8.831 \times 10^{-3} + 0.5} + \frac{8.831 \times 10^{-3}}{8.831 \times 10^{-3} + 0.5} e^{-(8.831\times 10^{-3}+0.5)t}$$

$$A(t) = 0.9826 + 0.017355 e^{-0.50883t}. \tag{10.3}$$

The average up-time availability of the crane during its ten hours of operation per day is obtained from Eq. (3.69) as

$$A(10) = \frac{1}{10}\int_0^{10} A(t)\,dt$$

$$A(10) = \frac{1}{10}[0.9826t - 0.034107659e^{-0.50883t}]\Big|_0^{10}$$

$$A(10) = 0.9859897.$$

If the average repair time is two hours, the down-time cost of the current system in which one spreader is assigned to each crane can be calculated as

$$\text{Current downtime cost} = [1 - A(10)] \times 10{,}200 = \$142.905 \text{ per hour.} \qquad (10.4)$$

We now investigate alternatives 1 and 2. The repair rate of alternative 1 is 0.5 repairs per hour. The two spreaders will function as a cold standby system—that is, one spreader is used while the second spreader is not in operation, and its failure rate is zero. When the first spreader fails, it undergoes repairs, and the second spreader becomes the primary unit. Therefore, we use Eqs. (3.111) and (3.112) to obtain

$$\dot{P}_1(t) = -8.831 \times 10^{-3}P_1(t) + 0.5P_2(t) \qquad (10.5)$$

$$\dot{P}_2(t) = -1.508831P_2(t) - 0.991169P_1(t) + 1. \qquad (10.6)$$

Solving Eqs. (10.5) and (10.6) results in

$$P_1(t) = 0.982636 + 0.00868e^{-0.50015t}$$

$$P_2(t) = 0.016329 - 0.016037e^{-0.50015t}.$$

The availability of the spreader is

$$A(t) = P_1(t) + P_2(t) = 0.998965 - 0.007357e^{-0.50015t} \qquad (10.7)$$

The average up-time availability of the crane is

$$A(10) = \frac{1}{10}\int_0^{10} A(t)\,dt = \frac{1}{10}[0.998965t + 0.014709e^{-0.50015t}]\Big|_0^{10}$$

$$A(10) = 0.997509.$$

Therefore, the loss due to downtime of the crane during the ten hours of operation is

$$\text{Downtime cost} = [1 - A(10)]10{,}200 = \$25.45 \text{ per hour.}$$

The cost due to the acquisition of the crane is

$$\text{Cost of acquisition per hour} = \frac{100{,}000}{365 \times 10 \times 5} = \$5.47.$$

Therefore, the total cost of alternative 1 is

$$\text{Total cost} = 25.45 + 5.47 \times 10 = \$80.15 \text{ per hour.} \tag{10.8}$$

Alternative 2 increases the repair rate to one repair per hour. Substituting into Eq. (3.41), we obtain

$$A(t) = \frac{1}{8.831 \times 10^{-3} + 1} + \frac{8.831 \times 10^{-3}}{8.831 \times 10^{-3} + 1} e^{-(8.831\times10^{-3}+1.0)t}$$

$$A(t) = 0.9912463 + 0.0087536e^{-1.008831t}$$

The average up-time availability for alternative 2 is

$$A(10) = \frac{1}{10}\int_0^{10} [0.9912463 + 0.0087536e^{-1.008831t}]\, dt$$

$$A(10) = 0.9921140.$$

The downtime cost for alternative 2 is

$$\text{Downtime cost} = [1 - A(10)] \times 10{,}400 = \$82 \text{ per hour.} \tag{10.9}$$

Comparison of Eqs. (10.4), (10.8), and (10.9) shows that alternative 1 is the preferred alternative since it results in the least down-time cost per hour.

Case 2: Design of a Production Line*

Introduction

A food processing line that is used to fill ingredients into packages and to seal and label those packages is shown in Figure 10.4. The food product in this case study is beef stew. The operation of the system is summarized below.

Raw material is manually moved into the production area, and a paper record is made of the lot number and the time and date when the material is placed into the product feeders. This is shown at the upper left of Figure 10.4. Federal government regulations for the food industry requires that the material lots be traceable to the production lots in which they are produced. This is required in the event of product recall. The materials of the beef stew product are beef and mixed vegetables. The beef product feeder, which is a hopper with a conveyor, feeds the volumetric filler, shown at the top of Figure 10.4, which fills cups volumetrically. The transport system from the beef filler includes an in-line checkweigher that weights the contents of the cup and recycles cups back to the filler if they are outside the weight specification (Boucher et al., 1996).

Acceptable cups move onto the vegetable filler where they receive a vegetable fill. When they arrive at the filling station of the packaging machine, they are moved into a cup dumper that overturns the cups into packages. Gravy is added to the package separately by the gravy filler, as shown at the bottom left of Figure 10.4. Through this series of events an automatic fill is achieved.

The package is a polymer pouch, which is formed from roll stock on the packaging machine at the forming station, which is just prior to the cup dumper (filling) station. The packaging machine, which runs horizontally along the bottom of Figure 10.4, is a horizontal form-fill-seal (F/F/S) machine. The following steps are performed on that machine: after the materials are dumped into the package and gravy is automatically dispensed, the package is indexed forward through a sealing station, where a top layer of polymer film is heat sealed to the package. At the next forward index, a Videojet printer labels the package for product name and time of production, and a slitter cuts the roll stock into individual packages. Finally, pouches are inspected and loaded into racks to be taken to the retort station, where they are subjected to a temperature of 250 °F for 30 minutes.

The coordination of the cup filling and cup transport system with the packaging line is achieved using the form-fill-seal (F/F/S) packaging machine controller. Each index of the F/F/S presents six pouches for filling. Upon completion of the index, the F/F/S controller signals the product transfer system controller, which in

* Based on an actual production line initiated by the Defense Logistics Agency and the Combat Rations Advanced Manufacturing Technology Demonstration, Piscataway, New Jersey. The author acknowledges Thomas Boucher of Rutgers University for providing a summary of this case.

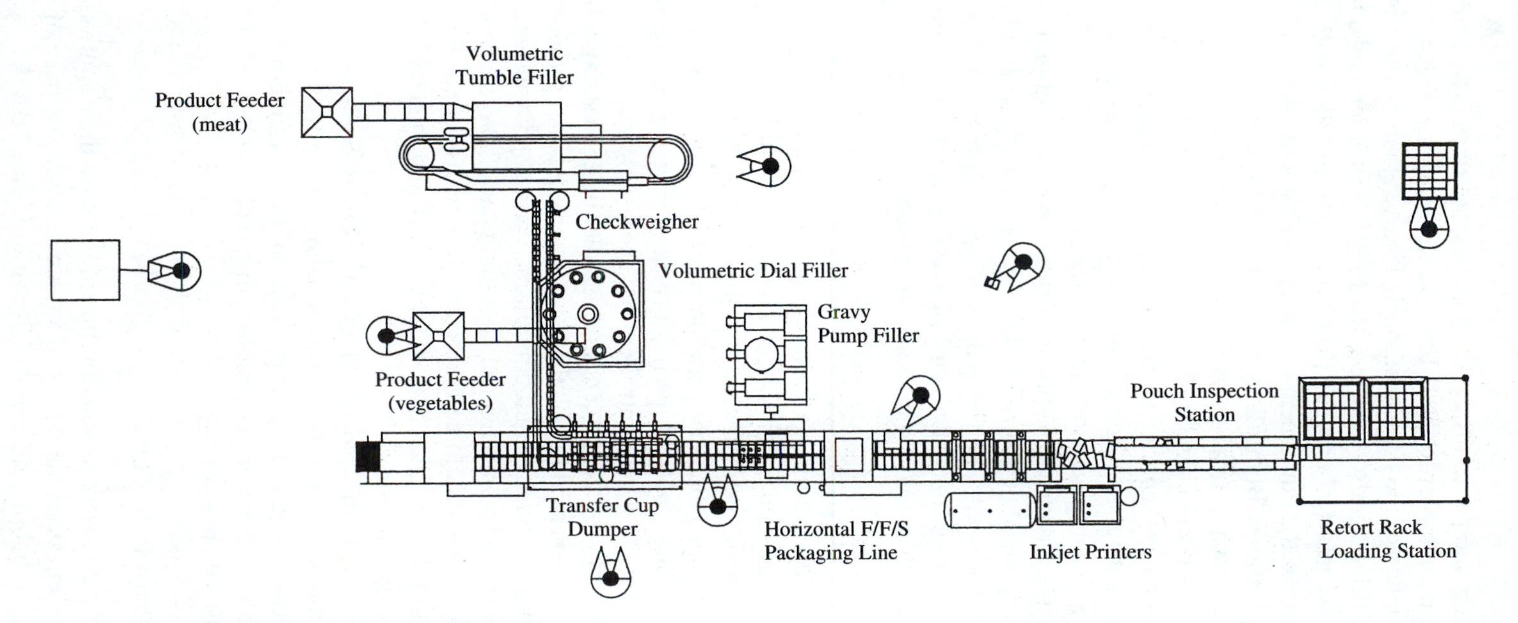

Figure 10.4. Polymer Pouch Filling and Packaging Line

turn signals the cup dumper to fill the six pouches. A return signal from the product transfer system controller to the F/F/S controller acknowledges that the fill is complete and an index of the F/F/S can begin. Start signals to the Videojet printers and the gravy filler are also sent from the F/F/S controller, and a return signal is provided from the gravy filler when the six pouches are filled with gravy.

In this system, all the equipment along the filling and packaging line is controlled as unit operations. Both fillers, product feeders, cup transfer system, and the F/F/S have their own controllers. All start and stop operations are accessible at the control panel of the individual operation. Line stoppages that can occur within any of the subsystems are reported to the relevant subsystem controller. A subsystem stoppage causes the line to stop as the appropriate handshake is not exchanged to cause the F/F/S to continue with another index cycle.

The checkweigher at the exit of the beef filler provides a digital display of the most recent package weights. However, this data is not collected and permanently logged in the factory database. The only automatic data logging occurs on the F/F/S, which keeps a permanent record of certain events occurring during the sealing operation, such as seal temperature and pressure.

Statement of the Problem

The alternative to the current operation is the addition of a production line controller and a centralized control panel incorporating all of the unit operations along the line and providing additional data logging capability.

In this new system, all the unit operation controllers report to a production line controller. It is possible to operate the line centrally from the production line controller as well as locally using controllers for each subsystem. Subsystem status information will be reported to the central controller by each subsystem controller. Information displays and readouts on the central control panel provide status of all subsystem operations including fault conditions. Fault conditions and their downtime will be kept as a permanent record, and the data will be analyzed to identify recurring conditions that should be corrected. The central controller will be able to download information to the checkweigher, beef and vegetable fillers, F/F/S machine, and Videojet printer.

The failure rates of the individual machines and controllers are given in Table 10.2. The repair rate is 0.08 repairs per hour. Investigate the effect of the proposed alternative on the overall production line availability. Recommend changes in the design of the production line that will ensure a minimum production rate of 47,000 pouches per day.

SOLUTION **THE CURRENT PRODUCTION LINE**

The equipment of the current production line along with their controllers are considered a series system since the failure of any unit causes stoppage of the line. We construct the block reliability diagram as shown in Figure 10.5.

Table 10.2. Failure Rate Data for the Food Processing Line

Equipment	*Number of Units*	*Failure Rate (Constant)*
Beef feeder	1	7.5×10^{-4} failures/hr
Tumble filler	1	8.9×10^{-4} failures/hr
Checkweigher	1	9.5×10^{-5} failures/hr
Dial filler	1	5×10^{-4} failures/hr
Vegetable feeder	1	4×10^{-4} failures/hr
Transfer cup dumper	1	12×10^{-5} failures/hr
Gravy pump filler	3	15×10^{-5} failures/hr
Horizontal F/F/S	1	20×10^{-5} failures/hr
Inkjet printers	2	8×10^{-6} failures/hr
Controller for each unit	9	5×10^{-6} failures/hr
Central controller	1	6×10^{-6} failures/hr
Central control panel	1	6×10^{-7} failures/hr

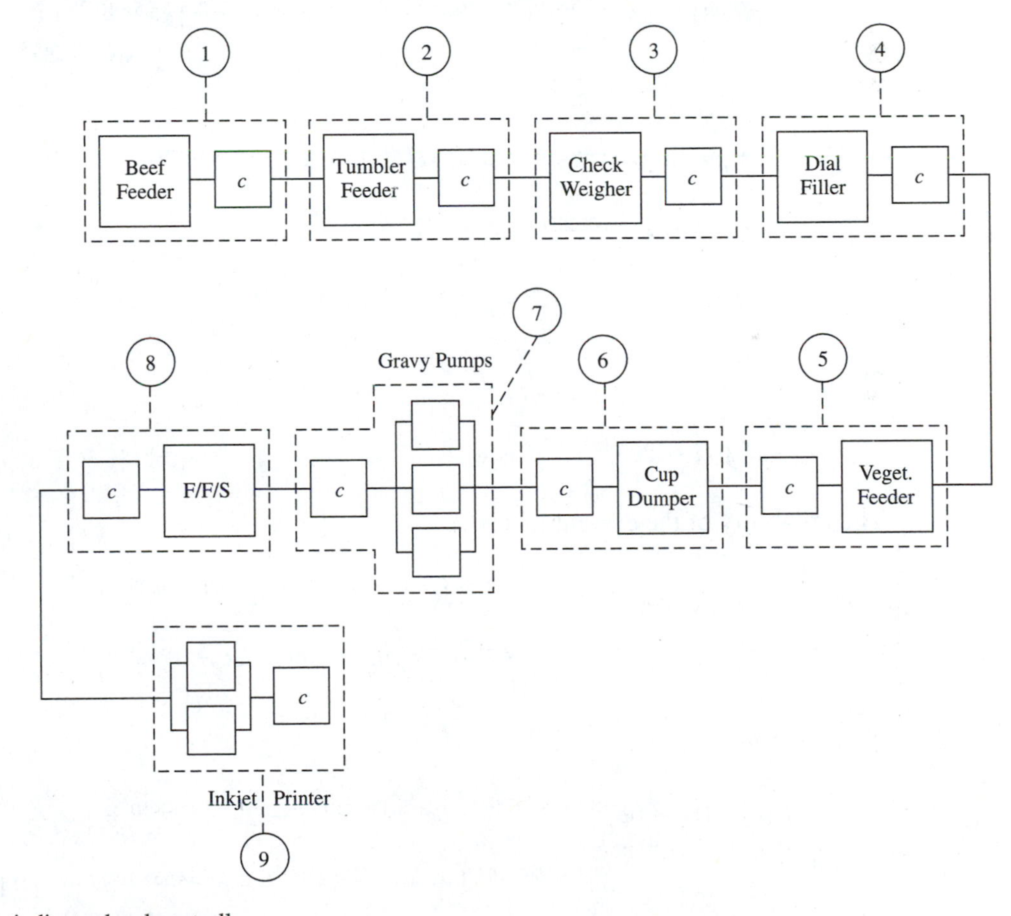

Note: *c* indicates local controller.

Figure 10.5. Block Diagram of the Current Production Line

We estimate the reliability of each block (1 through 9) as follows:

$$R_1(t) = e^{-0.000755t}$$

$$R_2(t) = e^{-0.000895t}$$

$$R_3(t) = e^{-0.0001t}$$

$$R_4(t) = e^{-0.000505t}$$

$$R_5(t) = e^{-0.000405t}$$

$$R_6(t) = e^{-0.000125t}$$

$$R_7(t) = [3e^{-1.5\times10^{-4}t} - 3e^{-3\times10^{-4}t} + e^{-4.5\times10^{-4}t}]e^{-5\times10^{-6}t}$$

or

$$R_7(t) = 3e^{-1.55\times10^{-4}t} - 3e^{-3.05\times10^{-4}t} + e^{-4.55\times10^{-4}t}$$

$$R_8(t) = e^{-0.000205t}$$

$$R_9(t) = (2e^{-8\times10^{-6}t} - e^{-16\times10^{-6}t})e^{-5\times10^{-6}t}$$

or

$$R_9(t) = 2e^{-13\times10^{-6}t} - e^{-21\times10^{-6}t}.$$

The reliability of the current line is

$$R_{\text{current}}(t) = e^{-0.00299t}(3e^{-1.55\times10^{-4}t} - 3e^{-3.05\times10^{-4}t} + e^{-4.55\times10^{-4}t})(2e^{-13\times10^{-6}t} - e^{-21\times10^{-6}t})$$

or

$$R_{\text{current}}(t) = 6e^{-3.158\times10^{-3}t} - 6e^{-3.308\times10^{-3}t} + 2e^{-3.458\times10^{-3}t} - 3e^{-3.166\times10^{-3}t} + 3e^{-3.316\times10^{-3}t} - e^{-3.466\times10^{-3}t}. \quad (10.10)$$

Figure 10.6 shows the reliability of the line over an eight-hour shift.

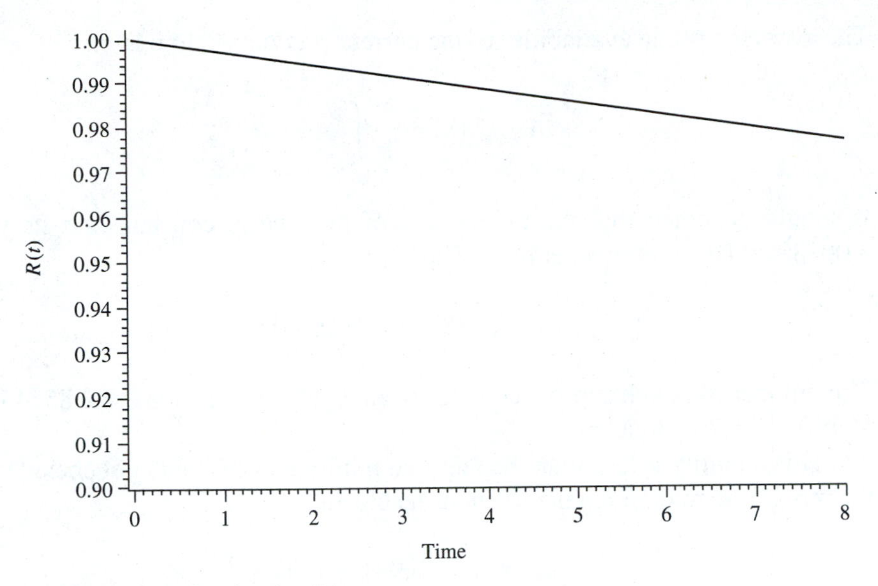

Figure 10.6. $R(t)$ of the Production Line

The p.d.f. of the current line is

$$f_{\text{current}}(t) = \frac{-dR_{\text{current}}(t)}{dt}$$

$$\begin{aligned} f_{\text{current}}(t) = {} & 18.948 \times 10^{-3}\, e^{-3.158\times10^{-3}t} - 19.848 \times 10^{-3}\, e^{-3.308\times10^{-3}t} \\ & + 6.916 \times 10^{-3}\, e^{-3.458\times10^{-3}t} - 9.498 \times 10^{-3}\, e^{-3.166\times10^{-3}t} \\ & + 9.948 \times 10^{-3}\, e^{-3.316\times10^{-3}} - 3.466 \times 10^{-3}\, e^{-3.466\times10^{-3}t}. \end{aligned} \tag{10.11}$$

The effective failure rate of the system is obtained by dividing Eq. (10.11) by Eq. (10.10):

$$h_{\text{current}}(t) = \frac{f_{\text{current}}(t)}{R_{\text{current}}(t)}. \tag{10.12}$$

The instantaneous availability of the production line, $A_{current}(t)$, is estimated by using Eq. (3.47):

$$A_{\text{current}}(t) = 1 - \frac{h_{\text{current}}(t)}{h_{\text{current}}(t) + \mu}. \tag{10.13}$$

The average uptime availability of the current production line is

$$A_{\text{current}}(8) = \frac{1}{8}\int_0^8 A(t)\,dt. \tag{10.14}$$

It should be noted that the failure rate of the line is constant and its value is 3.00780×10^{-3} failures per hour. Thus,

$$A_{\text{current}}(8) = 0.96385545.$$

The number of pouches produced during an eight-hour shift is $0.96385545 \times 480 \times 100 = 46{,}265$ pouches.

This quantity is less than the required minimum of 47,000 pouches. Doubling the repair rate to 0.16 repairs per hour results in

$$A(8) = 0.98159510.$$

The corresponding production rate is 47,116 pouches.

THE PROPOSED ALTERNATIVE

The central controller and its panel can be considered as a redundant unit for the controller of each piece of equipment. For example, if the local controller of the check weigher fails, the central controller will perform the functions of the local controller, and there will be no interruption of the production line. Similarly, if the central controller fails and the local controller is operating properly, the production line will not be interrupted. The reliability of the proposed system can be estimated by considering that the local controller ($\lambda = 5 \times 10^{-6}$) is, in effect, connected in parallel with the central controller ($\lambda = 6 \times 10^{-6}$) and its panel ($\lambda = 6 \times 10^{-7}$). Thus, the reliability of the controller system is

$$R_c(t) = (1 - e^{-6.6\times 10^{-6}t})(1 - e^{-5\times 10^{-6}t})$$

or

$$R_c(t) = e^{-6.6\times 10^{-6}t} + e^{-5\times 10^{-6}t} - e^{-11.6\times 10^{-6}t}$$

This "effective" controller is connected in series with each piece of equipment. Thus the reliability of each block becomes

$$R_1(t) = e^{-7.566\times 10^{-4}t} + e^{-7.55\times 10^{-4}t} - e^{7.616\times 10^{-6}t}$$

$$R_2(t) = e^{-8.966\times10^{-4}t} + e^{-8.95\times10^{-4}t} - e^{-9.016\times10^{-4}t}$$

$$R_3(t) = e^{-1.016\times10^{-4}t} + e^{-1\times10^{-4}t} - e^{-1.066\times10^{-4}t}$$

$$R_4(t) = e^{-5.066\times10^{-4}t} + e^{-5.05\times10^{-4}t} - e^{-5.116\times10^{-4}t}$$

$$R_5(t) = e^{-4.066\times10^{-4}t} + e^{-4.05\times10^{-4}t} - e^{-4.116\times10^{-4}t}$$

$$R_6(t) = e^{-1.266\times10^{-4}t} + e^{-1.25\times10^{-4}t} - e^{-1.316\times10^{-4}t}$$

$$R_7(t) = (3e^{-1.5\times10^{-4}t} - 3e^{-3.0\times10^{-4}t} + e^{-4.5\times10^{-4}t})R_c(t)$$

$$R_8(t) = e^{-2.066\times10^{-4}t} + e^{-2.05\times10^{-4}t} - e^{-2.116\times10^{-4}t}$$

$$R_9(t) = (2e^{-8\times10^{-6}t} - e^{-16\times10^{-6}t})R_c(t).$$

Thus,

$$R_{\text{proposed}}(t) = \prod_{i=1}^{9} R_i(t). \tag{10.15}$$

Since direct estimation of $f(t) = -dR(t)/dt$ is difficult to obtain, we calculate $R_{\text{proposed}}(t)$ numerically for the eight-hour shift. A sample of the results is shown in Table 10.3.

Table 10.3. A Partial Listing of Reliability Values

Time	*Reliability*
0.017	0.99995035
0.033	0.99990165
0.050	0.99985248
0.067	0.99980289
0.083	0.99975342
0.100	0.99970460
—	—
—	—
7.917	0.97688252
7.933	0.97683388
7.950	0.97678584
7.967	0.97673810
7.983	0.97668970
8.000	0.97664177

Examination of the results shows that the failure rate of the system is 29.54×10^{-4}. The availability of the system is

$$A(8) = 0.964384861.$$

The number of pouches produced during an eight-hour shift is $0.964384861 \times 480 \times 100 = 46{,}290$ pouches. Clearly, the proposed alternative has little effect on the system availability. Moreover, the daily production falls short, as in the current system, of the minimum required quantity of 47,000 pouches.

If the repair rate is doubled, the availability of the proposed system becomes

$$A(8) = 0.981869571.$$

The corresponding number of pouches per day is 47,129.

The availability of the system can further be improved by replacing the dial filler and the vegetable feeder by other equipment that exhibit reduced failure rates.

Case 3: An Explosive Detection System*

Introduction

Explosive detection is a major concern for law enforcement officers. Small but powerful explosive devices and material can be easily concealed in handbags, briefcases, and baggage. Methods for explosive detection have been developed over the years. This has also been paralleled with similar developments in the concealments and the mixes of the explosive material. The result is that the current methods fall short of detecting such a variety of explosives.

In an effort to detect explosives in passenger's baggage in the airport, a manufacturer of scanning systems proposes to design an X-ray system capable of classifying material in baggage as explosive or nonexplosive. The system is based on exposing the baggage to be inspected to X-rays generated at an excitation potential between 150 and 500 KeV. The X-ray beams scattered by the object being interrogated are collected and directed to the appropriate thirteen detector elements. The X-ray spectra collected by the detector elements every 30 ms are then transferred, without interfering with the spectra acquisition, to thirteen digital signal processors (DSP) that perform, using a neural network program, the discrimination between benign and suspicious spectra in real time. A summary of the technical data of this X-ray system is given in Table 10.4. The system consists of six major components as shown in Figure 10.7.

THE TOWER

The tower is manufactured from steel tubing and welded to ensure minimum flexing and movement during use. As shown in the figure, there is additional cross-bracing to further stiffen the structure. The baggage handling system is not connected in any way to the tower, so that any vibrations generated in that subsystem will not be transmitted to the tower and any of the other subsystems.

THE X-RAY GENERATOR

The chief demands on the X-ray system are the generation of X-rays at an excitation potential between 150 and 200 KeV, at the highest possible tube current, with the effective linear dimension of the X-ray source (the focal spot) not exceeding 0.8mm.

The X-ray tube is a closed, high vacuum rotating anode type tube, capable of generating X-rays continuously at 160 KeV. The anode is a circular tungsten strip mounted on an appropriately profiled molybdenum disc of 200 mm radius, rotating at 10,000 rpm on a nearly frictionless, heat-conducting liquid metal bearing. Of course, continuous cooling must take place whenever the tube is operated. In

* This system was developed by SCAN-TECH Security L.P., Northvale, New Jersey.

Table 10.4. Technical Data for the Explosive Detection System

Inspection principle	Coherent X-ray scattering (CXRS) spectroscopy
Classification principle	Neural net run on digital signal processor
Classification result	Benign versus explosive, masked or obscured
Detection probability	> 99%
False alarm rate	< 1%
Maximum bag size	L = 900mm, W = 700mm, H = 500mm
Special resolution (voxel size)	L = 50mm, W = 60mm to 90mm, H = 50mm
Minimum detectable explosive	100g (estimated)
Bag throughput	600 bags per hour
X-ray high tension	160kV
X-ray energy range	20 to 160 KeV
X-ray power	4.2kW continuous, 9.6kW pulsed (60s)
Detector energy range	20 to 120 KeV
Detector energy resolution	< 1.6 KeV at 60 KeV
Detector count rate	< 40,000/s
Spectrum acquisition time	30ms (for 10 bags per minute)
Power for X-ray subsystem	3 phase 480V/30A 50/60Hz
Power for detector subsystem	1 phase 220V/20A 50/60Hz
Power for computer subsystem	1 phase 220V/10A 50/60Hz
Operating temperature	+10°C . . . +40°C
Storage temperature	−20°C . . . +70°C

addition, active cooling must be continued for thirty minutes after X-ray generation is switched off. Because of this requirement, an uninterruptible power supply is needed to drive the circulating pumps in the event of a power failure.

THE INCIDENT BEAM COLLIMATORS

The incident beam collimator is attached at its top to the tube shield to ensure continuous radiation shielding. At the bottom of the collimator is a complex set of movable slits to define the beam shape. The housing encasing these collimators (see Figure 10.8) is of a sliding telescope design that permits vertical movement of the components during alignment. Most of the components in the housing are made from steel. But any component directly in contact with the X-ray beam is manufactured from a tungsten-10 percent copper alloy. The primary reason for choosing tungsten is that any characteristic radiation that is produced will superimpose with the lines from the X-ray tube itself. The tungsten is also an excellent X-ray absorber, and the slits need only be 5 to 10mm thick for effective shielding.

THE SCATTERED BEAM COLLIMATORS

The purpose of the scattered beam collimators is to collect and direct the X-ray beam scattered by the object being interrogated to the appropriate detector element. In order to have a high resolving power, the collimator must control the

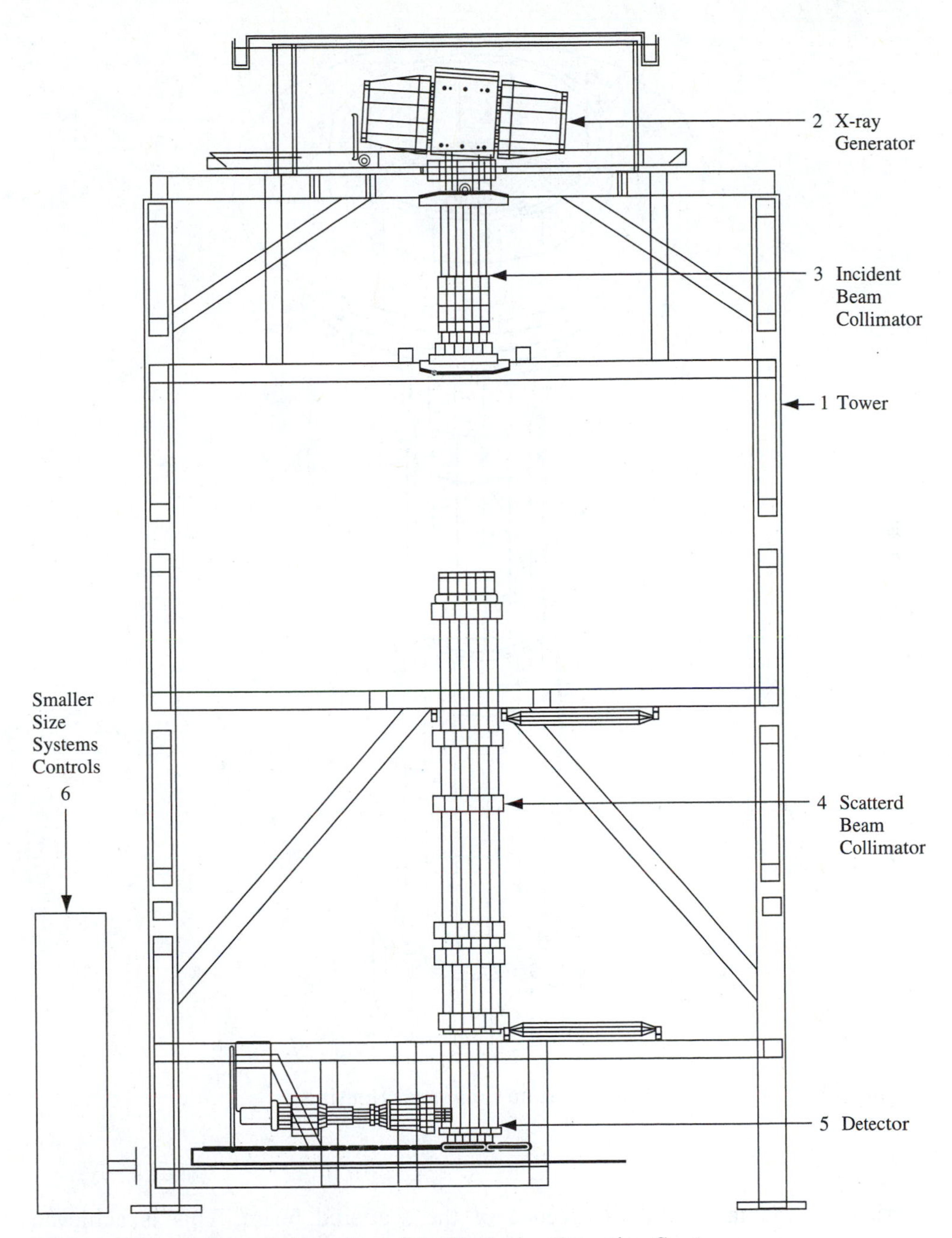

Figure 10.7. Schematic Drawing of the Explosive Detection System

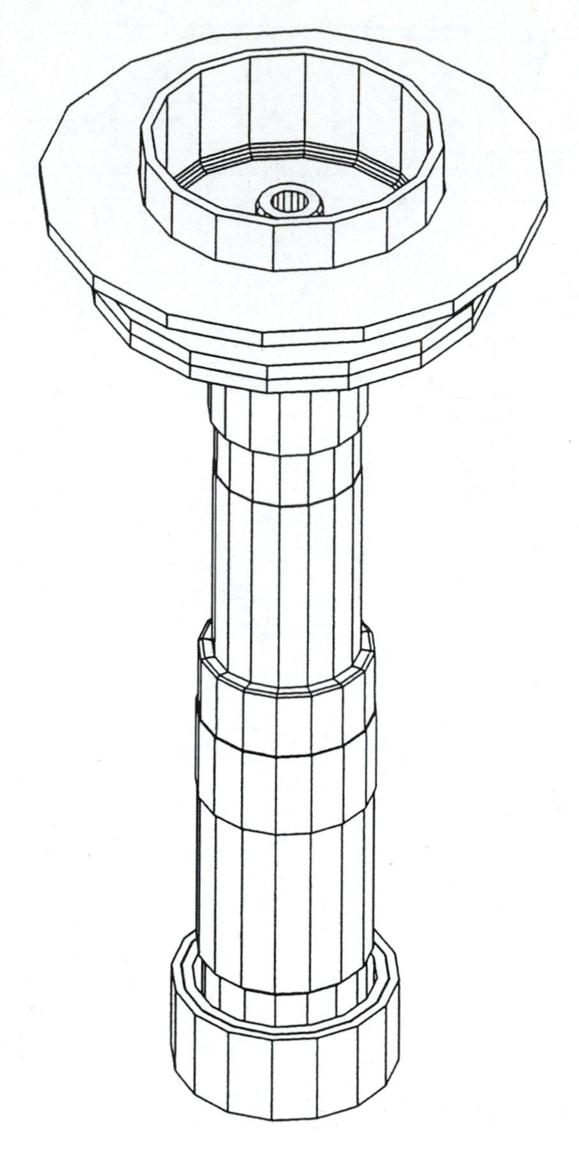

Figure 10.8. Telescoping Primary Beam Collimator Housing

horizontal and tangential divergence of the scattered beam. This is achieved through the four radial and twelve star collimators, respectively. The radial collimators limit the angular divergence as measured in the plane containing the system axis and the diffracting point. The star collimators, on the other hand, control the divergence perpendicular to this plane. If the beam diverts in either plane, the resolving power will be reduced.

THE DETECTOR SYSTEM

The main component of the detector system is a cryogenically cooled, single Ge crystal X-ray detector. When an X-ray is absorbed in this crystal, it creates a shower of electron-hole pairs, the number of which is proportional to the X-ray energy (E). A large bias voltage (1,000V) across the crystal sweeps the photo-induced charge to the electrodes on either side of the crystal, creating a current pulse. The integrated area of this pulse is proportional to the photo-induced charge and hence to the energy of the absorbed X-ray. The external electronics first amplify and shape this current pulse. A multichannel analyzer then sorts the pulses according to their net charge (X-ray energy) and increments the photon count in the appropriate energy bin.

SYSTEM CONTROLS AND ELECTRONICS

The system is controlled by a computer that acquires X-ray spectra from each of the thirteen detector segments every 30 ms. This process consists of reading the energy data from the detector and forming energy spectra in the computer RAM by classifying these energy data. These spectra must then be transferred, without interfering with the spectra acquisition, to the thirteen digital signal processors (DSP) that perform the discrimination between benign and suspicious spectra in real time.

Statement of the Problem

The explosive detection system has more than 400 components. The high-voltage excursions in the X-ray generator have a direct effect on the failure times of the system's components. In order to provide availability measures of the system, the management conducts an accelerated life test on the most critical subsystem that is closest to the X-ray generator. This subsystem is identified as the detector. Thirty detector elements are subjected to an environment of 200 KeV and their failure times, in hours, are recorded as shown in Table 10.5. The system normally operates at 160 KeV (acceleration factor is 10).

The failure times of the remaining subsystems are observed over a two-year period of operation. The failure rates of these subsystems are shown in Table 10.6. The explosive detection system is required to inspect baggage at a rate of one bag every six seconds or 600 bags per hour. When the system fails, it requires thirty

Table 10.5. Failure Times at 200 KeV

12.86	32.62	34.29	34.44	75.17	80.88	92.53	96.44	118.27	142.99
150.87	152.68	158.37	177.80	178.89	198.48	237.67	241.26	317.85	364.38
390.61	470.03	470.58	472.80	476.14	768.47[a]	768.47[a]	768.47[a]	768.47[a]	768.47[a]

[a] Indicates censoring.

Table 10.6. Failure Data of the Subsystems

Subsystem	*Constant Failure Rate*
X-ray generator	8.5×10^{-5} failures/hr
Incident beam optics	7.2×10^{-6} failures/hr
Scattered beam optics	10.2×10^{-6} failures/hr
Control	7.35×10^{-5} failures/hr

minutes to cool down before repairs begin. The average time of the actual repair is eleven hours, which is then followed by a warm-up period of thirty minutes.

A major airport receives in excess of 40,000 pieces of baggage per day for inspection. (The busy period of the airport is twelve hours per day.) The management of the airport is interested in determining the number and the configuration of several explosive detection systems that are capable of inspecting 40,000 pieces of baggage per day.

SOLUTION We use the failure-time data of the detectors to estimate the failure rate at normal operating conditions. Using Eq. (5.8) we obtain

$$\lambda_s = \frac{r}{\sum_{i=1}^{r} t_i + \sum_{i=1}^{n-r} t_i^+},$$

where

λ_s = the failure rate at stress level s,

r = the number of noncensored failure data,

t_i = the ith failure time, and

t_i^+ = the ith censored time.

$$\lambda_s = \frac{25}{5{,}178.9 + 3{,}842.35} = 27.71 \times 10^{-4} \text{ failures per hour.}$$

The failure rate of a detector element at the normal operating conditions is

$$\lambda_{\text{element}} = \frac{\lambda_s}{A_F} = 2.771 \times 10^{-4} \text{ failures per hour.}$$

The detector system is composed of thirteen detector elements connected in series. Thus, the failure rate of the detector subsystem is

$$\lambda_{\text{detector}} = 36.02 \times 10^{-4} \text{ failures per hour.}$$

All the subsystems of the explosive detection unit must operate properly for the system to function. Therefore, the subsystems are considered a series configuration with a failure rate of 0.00377 failures per hour (sum of all failure rates of the subsystems, including the detector).

If we assume that the availability of an explosive detection system is 1.0, then the number of systems needed to meet the inspection requirements is

$$\text{Number of systems} = \frac{40{,}000}{(600 \text{ bags per hour} \times 12)} \approx 6.$$

However, the failure and repair rates of the systems cause its availability to be less than 1.0. In order to estimate the availability of the six systems during the twelve hours of the airport operation, we develop the state-transition probability as shown below.

Let $P_i(t)$ be the probability that there are i systems operational at time t ($i = 0, 1, \ldots, 6$). Following Eqs. (3.99) through (3.103), we write

$$\dot{P}_0(t) = -\lambda P_0(t) + \mu P_1(t) \tag{10.16}$$

$$\dot{P}_i(t) = -(\lambda + \mu)P_i(t) + \lambda P_{i-1}(t) + \mu P_{i+1}(t) \qquad (i = 1, 2, 3, 4, 5) \tag{10.17}$$

$$\dot{P}_6(t) = -\mu P_6(t) + \lambda P_5(t), \tag{10.18}$$

where μ is the repair rate of the system. Substituting $\lambda = 37.7 \times 10^{-4}$ failures per hour and $\mu = 0.333$ repairs per hour into Eqs. (10.16) through (10.18) and solving numerically under the condition $\Sigma_{i=0}^{6} P_i(t) = 1$, we obtain the values of $P_i(t)$. A partial listing of the results is shown in Table 10.7.

Table 10.7. Partial Listing of the Solution

Time (secs)	$P_0(t)$	$P_1(t)$	$P_2(t)$	$P_3(t)$	$P_4(t)$	$P_5(t)$	$P_6(t)$
1	0.99981	0.00019	0.00000	0.00000	0.00000	0.00000	0.00000
2	0.99962	0.00038	0.00000	0.00000	0.00000	0.00000	0.00000
3	0.99944	0.00056	0.00000	0.00000	0.00000	0.00000	0.00000
4	0.99925	0.00075	0.00000	0.00000	0.00000	0.00000	0.00000
....	....	....	....	....	....	....	....
....	....	....	....	....	....	....	....
43,197	0.95476	0.04319	0.00195	0.00009	0.00000	0.00000	0.00000
43,198	0.95476	0.04319	0.00195	0.00009	0.00000	0.00000	0.00000
43,199	0.95476	0.04319	0.00195	0.00009	0.00000	0.00000	0.00000
43,200	0.95476	0.04319	0.00195	0.00009	0.00000	0.00000	0.00000

If we define the availability as $P_0(t)$, i.e., no failures of the explosive detection systems during the twelve-hour period, then

$$A(T = 12 \text{ hours}) = \frac{1}{43{,}200} \sum_{t=1}^{43{,}200} P_0(t) = \frac{41{,}238.67188}{43{,}200} = 0.954598,$$

and the number of baggage inspected per twelve hours is 41,238. This meets the minimum required baggage to be inspected.

Case 4: Reliability of Furnace Tubes*

Introduction

A major oil company produces 100 million barrels of a Sweet Oil Blend (SOB) per year. The production of the SOB requires hydrogen, which is supplied by five hydrogen plants. The production rate is proportional to the amount of hydrogen supplied—that is, more hydrogen production results in more production of oil until the maximum capacity of the plant is reached. Therefore, it is important that hydrogen plants operate without interruption or equipment failure.

Every hydrogen producing plant operates a methane reformer furnace (MRF). Each furnace has hundreds of tubes that are filled with catalyst. Methane and steam pass through these tubes at high temperature where hydrogen is produced. The tubes are fabricated from a centrifugally cast alloy steel (chrome, nickel, carbon) in order to minimize corrosion and sustain the creep stress resulting from the high temperatures and pressures within the tubes. The cost of the tubes ranges between $10 to $20 million and represents a high proportion of the total cost of the furnace.

The life of the furnace tubes is dependent on the operating conditions—namely, temperature and pressure. As mentioned earlier, increasing the hydrogen production increases the SOB production. However, increasing the hydrogen production decreases the tube's life and increases the risk of on-line tube failures.

The cost of the furnace tubes represents a high proportion of the total cost of the furnace. Therefore, the remaining life of the tubes should be accurately estimated so that the tubes are not replaced prematurely. Moreover, the tubes should be periodically inspected for possible crack propagations.

Statement of the Problem

The tubes are placed vertically in the furnace. The tubes have an internal diameter of 5.00 inches, a wall thickness of 0.4 inch, and a length of 45 feet. The design temperature of the tubes is 1710°F, and the design internal pressure is 400 psi (pounds per square inch). The tubes are flanged at the top end with a reduced diameter at the bottom end that leads into a smaller tube (common to a set of fifteen tubes), which in turn feeds into an outlet collection header.

The expected design life of the tubes when the furnace operates at the normal operating conditions is 100,000 hours. This design life is calculated based on the Larson-Miller design formula, which relates the properties of the tube material to the operating temperature and pressure. The formula is empirically developed.

* A partial description of this case was reprinted with permission. © 1995. Syncrude, Edmonton, Canada.

Increasing the oil production requires an increase in the hydrogen production, which in turn increases the furnace burner rate. As a result, the temperature in the furnace tends to increase, which causes a significant reduction in the remaining life of the tubes. Analysis of failure data collected over eight years of operation shows that operating a tube at 25°F above the design temperature of 1,710°F results in a loss of one-half of the tube remaining life. Temperature readings taken by an optical pyrometer show that approximately five out of fifteen tubes within the same set operate at temperatures of 1,724°F.

The furnace fails when four out of fifteen tubes fail or when two consecutive tubes fail. The engineers of the oil company are interested in estimating the reliability of the furnace and the remaining life of each set of tubes. The furnace has ten sets, and all are required for the proper function of the furnace. The engineers are also interested in determining the optimal preventive maintenance schedule that minimizes the downtime of the furnace.

The manufacturer of the tubes performs an accelerated life testing at 1,835°F on twenty-five tubes and records the following failure times:

1,958, 1,013, 12,416, 755, 2,901, 7,225, 511, 2,044, 191, 8,034, 6,038, 886, 1,441, 11,479, 734, 327, 1,986, 6,701, 12,822, 3,090, 3,521, 1,292, 1,245, 8,106, 8,163.

A 25°F increase in the operating temperature of the furnace results in a 10 percent increase in the number of oil barrels produced. The net profit per barrel is $2, and the cost of replacing all the tubes is $15 million and requires one year. What is the operating temperature, above the design temperature that maximizes the profit? It should be noted that the furnace can not operate beyond 1,810°F.

SOLUTION We first test the validity of using a constant failure-rate model by calculating the Bartlett value B_r as follows:

$$\sum_{i=1}^{25} \ln t_i = 194.38$$

$$T = \sum_{i=1}^{n} t_i = 104{,}879.$$

Using Eq. (5.1), we obtain

$$B_r = \frac{2 \times 25 \left[\ln\left(\frac{104{,}879}{25}\right) - \frac{1}{25} \times 194.38 \right]}{1 + (26)/(6 \times 25)} = 24.14.$$

The critical values for a two-tailed test with $\alpha = 0.10$ are

$$\chi^2_{0.95,\,24} = 13.8484 \quad \text{and} \quad \chi^2_{0.05,\,24} = 36.4151.$$

Therefore, B_{25} does not contradict the hypothesis that the failure times can be modeled by an exponential distribution.

The failure rate at the stress level of 1,835°F is

$$\lambda_{1{,}835°\mathrm{F}} = \frac{25}{104{,}879} = 2.3837 \times 10^{-4} \text{ failures per hour.}$$

Since operating the furnace at 25°F above the design temperature results in a loss of one-half the remaining life of the tubes, the acceleration factor between the design temperature (1,710°F) and the accelerated test temperature (1,835°F) is 25, and

$$\lambda_{1{,}710°\mathrm{F}} = 9.53 \times 10^{-6} \text{ failures per hour.}$$

The reliability of a single tube is

$$R(t) = e^{-9.53\times 10^{-6} t}. \tag{10.19}$$

The p.d.f. of the failure time of a single tube is

$$f(t) = \frac{-dR(t)}{dt} = 9.53 \times 10^{-6} e^{-9.53\times 10^{-6} t}. \tag{10.20}$$

The mean life at the normal conditions is $1/\lambda = 104{,}931$ hours.

If we assume that the tubes have been operating for five years (50,000 hours), then the residual life of a tube is obtained using Eq. (1.90) as follows:

$$L(t) = \frac{1}{R(t)} \int_t^\infty \tau f(\tau)\, d\tau - t$$

$$L(50{,}000) = \frac{1}{R(50{,}000)} \int_{50{,}000}^\infty 9.53 \times 10^{-6} t e^{-9.53\times 10^{-6} t}\, dt - 50{,}000$$

or

$$L(50{,}000) = \frac{1}{\lambda} = 104{,}931 \text{ hours.}$$

Since the exponential distribution has the memoryless property, then the remaining life at any time t is always $1/\lambda$.

The reliability of a set of fifteen tubes is obtained by examining the two possible failure modes: (1) four out of fifteen tubes fail or (2) consecutive-2-out-of-15 F: system

Reliability of the 4-out-of-15 system is

$$R_a(t) = \sum_{r=11}^{15} \binom{15}{r} (e^{-\lambda t})^r (1 - e^{-\lambda t})^{15-r}$$

or

$$\begin{aligned} R_a(t) = {} & 1{,}365\,(e^{-9.53\times 10^{-6}t})^{11}(1 - e^{-9.53\times 10^{-6}t})^4 \\ & + 455(e^{-9.53\times 10^{-6}t})^{12}(1 - e^{-9.53\times 10^{-6}t})^3 \\ & + 105(e^{-9.53\times 10^{-6}t})^{13}(1 - e^{-9.53\times 10^{-6}t})^2 \\ & + 15(e^{-9.53\times 10^{-6}t})^{14}(1 - e^{-9.53\times 10^{-6}t}) \\ & + (e^{-9.53\times 10^{-6}t})^{15} \end{aligned}$$

or

$$\begin{aligned} R_a(t) = {} & 1{,}365e^{-10.483\times 10^{-5}t} - 5{,}005e^{-11.436\times 10^{-5}t} + 6{,}930e^{-12.389\times 10^{-5}t} \\ & - 4{,}290e^{-13.342\times 10^{5}t} + 1{,}001e^{-14.295\times 10^{-5}t}. \end{aligned} \tag{10.21}$$

Reliability of a consecutive-2-out-of-15 F: system is obtained using Eq. (2.15) as

$$R_b(p, 2, n) = \sum_{j=0}^{\lfloor (n+1)/2 \rfloor} \binom{n-j+1}{j} (1-p)^j p^{n-j}.$$

Thus,

$$R_b(p, 2, 15) = \sum_{j=0}^{8} \binom{15-j+1}{j} (1-p)^j p^{n-j}$$

$$\begin{aligned} R_b(e^{-\lambda t}, 2, 15) = {} & e^{-6.671\times 10^{-5}t} + 28e^{-7.624\times 10^{-5}t} - 14e^{-8.577\times 10^{-5}t} \\ & - 98e^{-9.53\times 10^{-5}t} + 145e^{-10.483\times 10^{-5}t} - 70e^{-11.436\times 10^{-5}t} \\ & + 5e^{-12.389\times 10^{-5}t} + 5e^{-13.342\times 10^{-5}t} - e^{-14.295\times 10^{-5}t}. \end{aligned} \tag{10.22}$$

If we assume that the probability of a failure due to consecutive-2-out-of-15 equals the probability of a failure due to the 4-out-of-15 system, then the reliability of a set of tubes is

$$\begin{aligned} R_{\text{set}}(t) &= 0.5R_a(t) + 0.5R_b(t) \\ &= 0.5e^{-6.671\times10^{-5}t} + 14e^{-7.624\times10^{-5}t} - 7e^{-8.577\times10^{-5}t} \\ &\quad - 49e^{-9.53\times10^{-5}t} + 755e^{-10.483\times10^{-5}t} - 2{,}537.5e^{-11.436\times10^{-5}t} \\ &\quad + 3{,}467.5e^{-12.389\times10^{-5}t} - 2{,}142.5e^{-13.342\times10^{-5}t} + 500e^{-14.295\times10^{-5}t}. \end{aligned} \tag{10.23}$$

The system consists of ten sets of furnace tubes connected in series. Therefore, the reliability of the tubing system is

$$R_{\text{system}} = \prod_{l=1}^{10} R_{\text{set}}(t).$$

A plot of $R_{\text{system}}(t)$ versus time is shown in Figure 10.9.

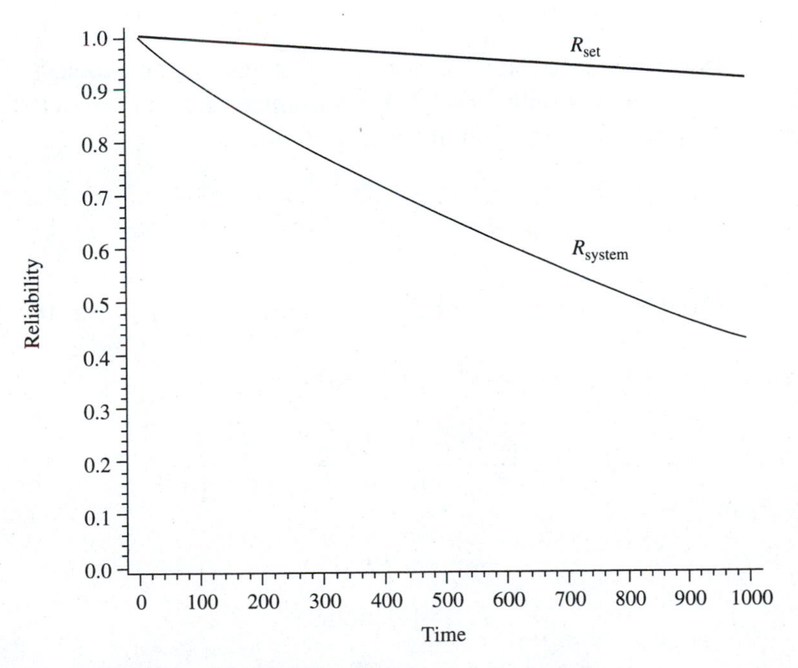

Figure 10.9. Reliability of the Furnace Tubes

MAXIMIZATION OF PROFIT

Operating the furnace at its maximum temperature of 1,810°F results in a 40 percent increase in the oil production and an increase of the tubes' failure rate to

$$\lambda_{1,810°\mathrm{F}} = \frac{2.3837 \times 10^{-4}}{2} = 1.1918 \times 10^{-4} \text{ failures per hour.}$$

The mean life of the tubes is 8,390 hours or one year of operation. The profit resulting from increasing the oil production by 40 percent per year is 40×10^6 barrels $\times\ 2 =$ \$80 million. After one year; the tubes are replaced at a cost of \$15 million; and the plant is interrupted for one year with a loss of a profit of \$200 million. Thus, it is not economical to increase the temperature to 1,810°F.

Increasing the temperature to 1,735°F results in a 10 percent increase per year in the oil production and an increase of the tubes' failure rate to

$$\lambda_{1,735°\mathrm{F}} = 1.9 \times 10^{-5} \text{ failures per hour.}$$

The mean life of the tubes at 1,735°F is 52,465 hours or five years. The profit resulting from increasing the oil production by 10 percent per year is $50 \times 10^6 \times 2 =$ \$100 million, and the cost of replacing the tubes after five years is \$15 + cost of plant interruption for one year—that is, \$115 million. Thus, it is not economically justified to operate at temperatures above 1,710°F.

PREVENTIVE MAINTENANCE

Let c_1 be the cost per inspection for a tube and c_2 the cost of an undetected failure. Then the optimum inspection interval T that minimizes the total expected cost per unit time is the value of T^* that minimizes Eq. (9.90) or

$$c(t) = \frac{c_1 + c_2 T}{1 - e^{-\lambda T}} - \frac{c_2}{\lambda}.$$

From Figure 10.9, the failure rate of a set of tubes is $\lambda = 8.45 \times 10^{-5}$ failures per hour.

Assume $c_2 =$ \$900 and $c_1 =$ \$1,200. Then

$$c(T^*) = \frac{1{,}200 + 900\,T^*}{1 - e^{-8.45 \times 10^{-5}}} - \frac{900}{8.45 \times 10^{-5}}$$

and

$$T^* = 178 \text{ hours.}$$

In other words, every set of tubes should be inspected after 178 hours of operation.

Case 5: Reliability Modeling of Telecommunication Networks for the Air Traffic Control System*

Introduction

Aircraft operating outside surveillance radar coverage areas, such as oceanic airspace, rely on high-frequency (HF) radio for reporting position information (latitude, longitude, altitude, and so on) to the air traffic control system. Figure 10.10 shows the hardware subsystems used in the current oceanic operating environment. The HF radio link suffers from congestion, electrostatic, and sun spot interference, which cause frequent losses of contact between aircraft and the air traffic controller. This has necessitated a relatively large longitudinal separation of sixty miles between aircraft.

As part of the effort to improve the current air communication, navigation, and surveillance systems in order to meet the demand created by future increases in airspace traffic, the International Civil Aviation Organization (ICAO) defines the automatic dependent surveillance function (ADSF) as "A function for use by air traffic services (ATS) in which aircraft automatically transmit via data-link, at intervals established by the ground ATS system, data derived from on-board navigation systems. As a minimum, the data include aircraft identification and three dimensional position, additional data may be provided as appropriate" (International Civil Aviation Organization, 1988).

Figure 10.11 shows one of several potential hardware subsystem configurations being considered by the Federal Aviation Administration (FAA) that could be used to carry out the ADSF in an oceanic operating environment. An ADSF equipped aircraft, or aeronautical earth station (AES), will generate position data from on-board navigation systems and automatically, without pilot involvement, transmit the information to communication satellites, such as those of the International Maritime Satellite (INMARSAT) system. In turn, the message is sent to a ground earth station (GES), such as the Communication Satellite Corporation (COMSAT) facility in Southbury, Connecticut. The message is then received by a ground communication network service, similar to the network provided by Aeronautical Radio, Inc. (ARINC), which transfers the message to its intended destination, an en-route or oceanic controller's terminal at an air route traffic control center (ARTCC), for control actions.

In addition to the main components shown in Figure 10.11, the ADSF can be considered as a collection of hardware, communication, and procedural systems.

The *hardware system* ranges in complexity from the orbital control components of a satellite to the simple telephone lines used to connect the ground communications network with the air traffic control center. The *communication system* allows the exchange of information between the various hardware subsystems.

* This case was developed in collaboration with the FAA Technical Center, Atlantic City Airport, New Jersey.

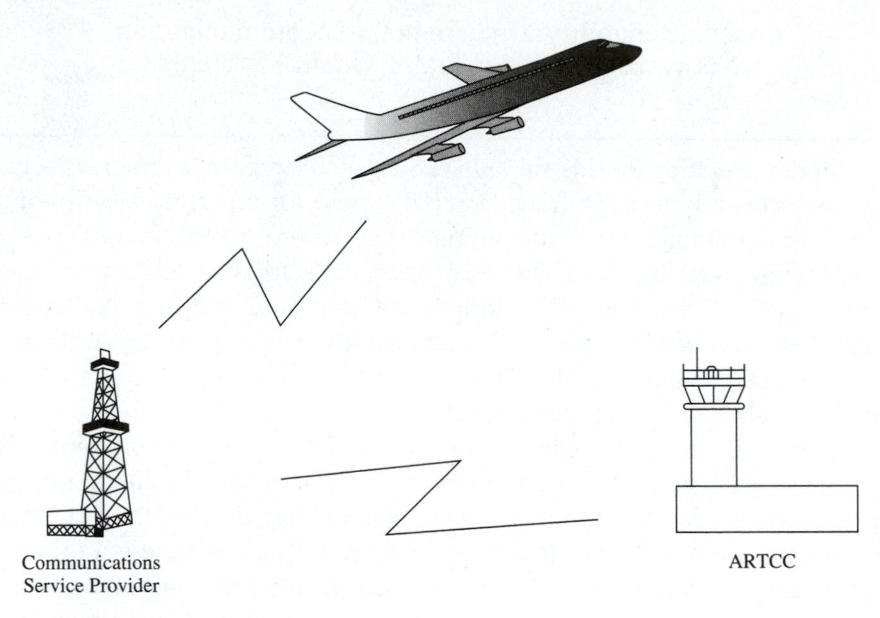

Figure 10.10. Current Oceanic Operating Environment (HF Radio)

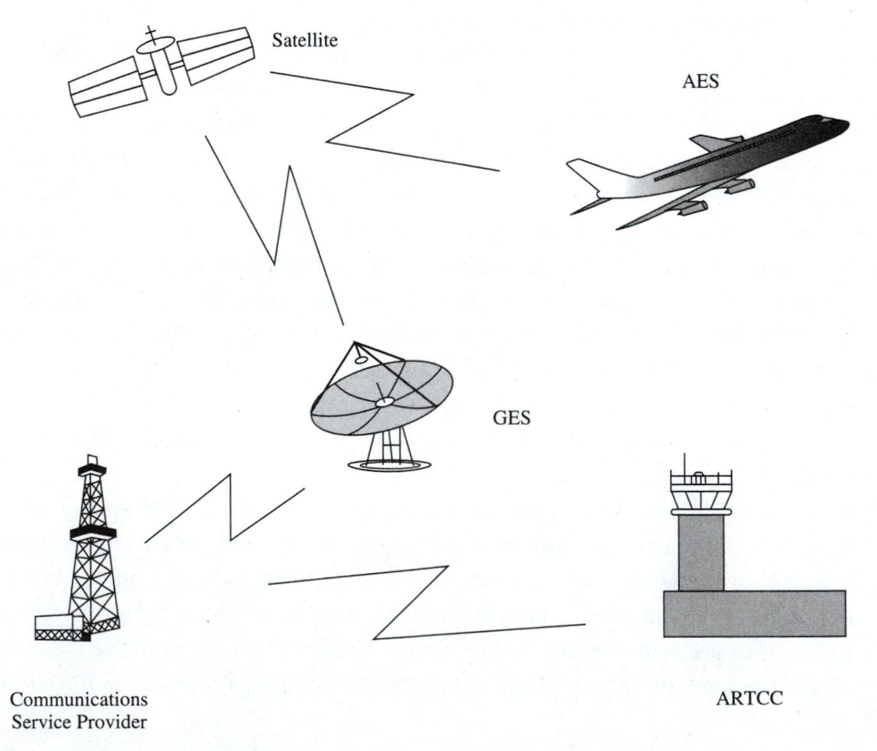

Figure 10.11. Proposed Oceanic Operating Environment (ADS)

The hardware components (HC blocks) communicate with one another via these communication components (lines). That is, a hardware component is a physical piece of equipment or transmission medium that must be operational or accessible by the ADSF to be operational. A communication component is any protocol, channel, or software code that ensures that these hardware components remain accessible and connected with one another.

The *procedural system* is necessary to coordinate the use of the hardware and communication systems under different operating conditions. For example, in emergency or catastrophic failure situations, it may be necessary to establish a link with a satellite or GES that has a higher gain (signal transmission rate) or a higher level of reliability in order to ensure that messages are received by the ARTCC controller in the required amount of time.

Statement of the Problem

The ADS system is currently under development. The FAA is interested in analyzing the performance of different configurations of the system and in specifying reliability and availability values for the manufacturers of the system's equipment. Typical availability values for critical subsystems or equipment used in the air traffic control system are 0.99999 or higher.

More important, the critical components of the system should be identified. This will enable the FAA to recommend design changes of such components to ensure that the reliability objectives of the overall system are realized.

We now describe, in detail, the hardware components of the ADS system (refer to Figure 10.11).

1. *Aeronautical earth station (AES)*: Any fixed or rotary wing aircraft is considered an aeronautical earth station. The minimum equipment required for an aircraft to be capable of operating in an ADS environment that utilizes a satellite data link is

    - *Navigation systems*: These systems are responsible for generating information describing the location of the aircraft, such as latitude, longitude, and altitude. Many oceanic aircraft use what is known as an inertial reference system (IRS).
    - *Automatic dependent surveillance unit (ADSU)*: This can exist as a standalone single rack-mounted unit or can be software implementable in a communications management unit, line replacement unit (LRU), or a flight management computer (FMC), as is the case in all Boeing 747-400's. It is the primary unit responsible for executing the ADS function on board the AES.
    - *Communications management unit (CMU)*: This acts as a "switcher," routing and forwarding messages to the desired air-ground link.

- *Satellite data unit (SDU)*: This unit determined modulation and demodulation, error correction, coding, data rates, and other signal parameters.
- *Radio frequency unit (RFU)*: The RFU, operating in full duplex mode, consists of low power amplifiers and frequency conversion electronics.
- *Antenna subsystem*: This consists of splitters, combiners, high-power amplifiers, low-noise amplifiers, low- and high-gain antennas, and other RF (radio frequency) distribution units; the combination thereof depends on the level of service required by the AES.

A possible configuration of an AES avionics system is shown in Figure 10.12.

2. *Satellite communications*: The ADSF data link service will be supported by the satellites that occupy the International Maritime Satellite (INMARSAT)

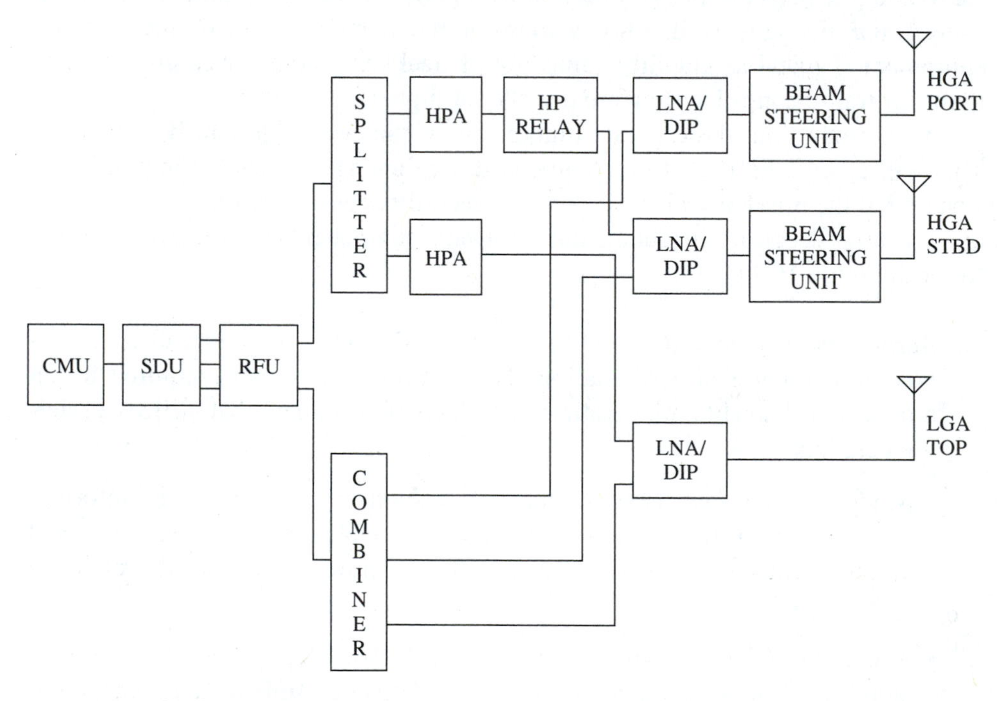

Note: HPA = high-power antenna.
HP Relay = high-power relay.
HGA = high-gain antenna.
LGA = low-gain antenna.
LNA/DIP = low-noise amplifier/diplexer.

Figure 10.12. Possible AES Avionics Configuration

constellation. The present constellation consists of four primary and seven back-up satellites. There are thirteen ground earth stations (GES) available that receive the satellite signals directly.

3. *Terrestrial subnetwork-communications service provider*: The terrestrial subnetwork connects the GES with the ARTCC. This network has a primary link and backup or secondary link. The components of each link include modems, an air ground interface system, and a data network service.
4. *Air route traffic control center*: The air route traffic control center (ARTCC) provides the control service, such as the assignments of airplanes to tracks, and ensures that separation standards between the airplanes are maintained. The major components of the ARTCC are the National Airspace Data Interchange Network, a FAA router, and modems.

There are three components for the entire telecommunications networks: hardware, communication, and procedural systems. Their interactions are complex and difficult to model within the scope of this case study. We limit this case study to the modeling of the hardware components.

RELIABILITY DATA

The failure rates of all components are constant. Since most of the equipment is under development, we utilize the reliability data projected by the manufacturer. They are shown in Table 10.8.

Table 10.8. Failure Data of the System's Components

Component/Subsystem	*Failure Rate (Failures/Hr)*
Satellite data units (SDU)	2.5×10^{-6}
Communications management unit (CMU)	1.42×10^{-6}
Radio frequency unit (RFU)	0.8×10^{-6}
Aeronautical telecommunications network (ATN)	1.75×10^{-4}
Air router traffic services (ATS)	2.85×10^{-4}
Automatic dependent surveillance unit (ADSU)	5×10^{-4}
Splitter	3×10^{-6}
Combiner	5×10^{-6}
High-power antenna (HPA)	6×10^{-5}
High-power relay (HPR)	4×10^{-6}
High-gain antenna (HGA)	4×10^{-5}
Low-gain antenna (LGA)	3.5×10^{-5}
Low-noise antenna (LNA)	2×10^{-5}
Beam steering unit (BSU)	8.7×10^{-6}

SOLUTION In order to analyze the reliability of the telecommunication networks for the air traffic control system, we first estimate the reliability of each major component separately as follows.

1. *Aeronautical earth station (AES)*: One of the proposed avionic configurations of the AES is shown in Figure 10.12, which illustrates how the components of the AES are *physically* connected to each other. It does not directly show, however, the relationship between the components in terms of the reliability of the avionics subsystem. We assume that the dual high-gain antenna (HGA) performs the same function (in terms of reliability) as the top-mounted low-gain antenna (LGA). Therefore, the components of the two antennae are connected in parallel to indicate redundancy as shown in Figure 10.13. Moreover, the AES is considered operational only if it is able to send *and* receive information. This implies that the splitter and the combiner are connected in series, not in parallel, as the physical diagram suggests.

There are two beam steering units connected in parallel, therefore the reliability of the beam steering is

$$R_{\mathrm{BSU}}(t) = 1 - (1 - e^{-8.7\times 10^{-6}t})\,(1 - e^{-8.7\times 10^{-6}t})$$

$$R_{\mathrm{BSU}}(t) = 2e^{-8.7\times 10^{-6}t} - e^{-17.4\times 10^{-6}t}.$$

Similarly, the reliability of the LNA/DIP is

$$R_{\mathrm{LNA}}(t) = 2e^{-3.5\times 10^{-5}t} - e^{-7\times 10^{-5}t}.$$

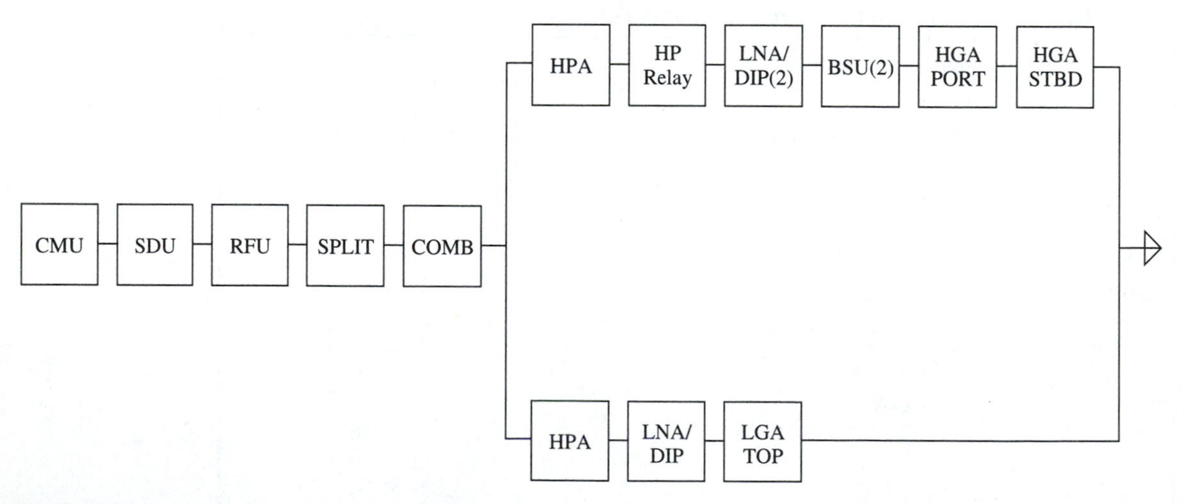

Figure 10.13. Reliability Block Diagram for the AES Avionics

The reliability of the upper path of the parallel configuration is

$$R_{\text{upper}}(t) = e^{-14.4\times10^{-5}t}(2e^{-8.7\times10^{-6}t} - e^{-17.4\times10^{-6}t})(2e^{-3.5\times10^{-5}t} - e^{-7\times10^{-5}t})$$

$$R_{\text{upper}}(t) = 4e^{-18.77\times10^{-5}t} - 2e^{-19.64\times10^{-5}t} - 2e^{-22.27\times10^{-5}t} + e^{-23.14\times10^{-5}t}.$$

The reliability of the lower path is

$$R_{\text{lower}}(t) = e^{-11.5\times10^{-5}t}$$

$$\begin{aligned} R_{\text{parallel}}(t) &= 1 - (1 - R_{\text{upper}}(t))\,(1 - R_{\text{lower}}(t)) \\ &= 4e^{-18.77\times10^{-5}t} - 2e^{-19.64\times10^{-5}t} - 2e^{-22.27\times10^{-5}t} \\ &\quad + e^{-23.14\times10^{-5}t} + e^{-11.5\times10^{-5}t} - 4e^{-30.27\times10^{-5}t} \\ &\quad + 2e^{-31.14\times10^{-5}t} + 2e^{-33.77\times10^{-5}t} - e^{-34.64\times10^{-5}t}. \end{aligned} \tag{10.24}$$

The reliability of an AES is

$$R_{\text{AES}}(t) = e^{-12.72\times10^{-6}t}R_{\text{parallel}}(t)$$

$$\begin{aligned} R_{\text{AES}}(t) &= 4e^{-2.004\times10^{-4}t} - 2e^{-2.091\times10^{-4}t} - 2e^{-2.354\times10^{-4}t} \\ &\quad + e^{-2.441\times10^{-4}t} + e^{-1.277\times10^{-4}t} - 4e^{-3.154\times10^{-4}t} \\ &\quad + 2e^{-3.214\times10^{-4}t} + 2e^{-3.5042\times10^{-4}t} - e^{-3.591\times10^{-4}t}. \end{aligned} \tag{10.25}$$

2. *Satellite communications*: The current satellite communications system includes four primary satellites and seven backup satellites. The failure rates of the primary satellites equal those of the backup satellites. We assume that the satellites are nonrepairable. The satellite subsystem can be modeled as a k-out-of-n or (4-out-of-11) system. Thus,

$$R_{\text{satellite}}(t) = \sum_{r=4}^{11}\binom{11}{r}(e^{-\lambda_s t})^r(1 - e^{-\lambda_s t})^{11-r}$$

or

$$R_{\text{satellite}}(t) = 1 - \sum_{r=0}^{3}\binom{11}{r}(e^{-\lambda_s t})^r(1 - e^{-\lambda_s t})^{11-r},$$

where λ_s is the satellite failure rate:

$$R_{\text{satellite}}(t) = 1 - [(1 - e^{-\lambda_s t})^{11} + 11e^{-\lambda_s t}(1 - e^{-\lambda_s t})^{10} + 55(e^{-\lambda_s t})^2(1 - e^{-\lambda_s t})^9 + 165(e^{-\lambda_s t})^3(1 - e^{-\lambda_s t})^8]. \quad (10.26)$$

The projected failure rate, λ_s is 9.5×10^{-5} failures per hour.

3. *Terrestrial subnetwork*: This subnetwork has two identical links in parallel. Each link consists of modems, an air ground interface system, and a data network service. The reliability of the subnetwork is

$$R_{\text{subnetwork}}(t) = 2e^{-\lambda_{\text{net}} t} - e^{-2\lambda_{\text{net}} t}, \quad (10.27)$$

where λ_{net} is the failure rate of the subnetwork. Its projected value is 3×10^{-6} failures per hour.

4. *Air route traffic control center (ARTCC)*: The major components of the ARTCC are the National Airspace Data Interchange Network, FAA router and modems, all connected in series. Therefore, the reliability of the ARTCC is

$$R_{\text{ARTCC}}(t) = e^{-\lambda_{cc} t}, \quad (10.28)$$

where λ_{cc} is the sum of the failure rates of the individual components of the center. Again, the projected failure rate of the ARTCC is 7.2×10^{-6} failures per hour.

5. *The ground earth stations (GES)*: There are thirteen GES available to receive the satellites' signals. Successful communication between the satellites and the GES requires a minimum of nine stations operating at any time. Thus, the reliability of the GES is

$$R_{\text{GES}}(t) = \sum_{r=9}^{13} \binom{13}{r} (e^{-\lambda_{\text{GES}} t})^r (1 - e^{-\lambda_{\text{GES}} t})^{13-r}$$

or

$$R_{\text{GES}}(t) = e^{-9\lambda_{\text{GES}} t}[715(1 - e^{-\lambda_{\text{GES}} t})^4 + 286e^{-\lambda_{\text{GES}} t}(1 - e^{-\lambda_{\text{GES}} t})^3 + 78e^{-2\lambda_{\text{GES}} t}(1 - e^{-\lambda_{\text{GES}} t})^2 + 13e^{-3\lambda_{\text{GES}} t}(1 - e^{-\lambda_{\text{GES}} t}) + e^{-4\lambda_{\text{GES}} t}], \quad (10.29)$$

where λ_{GES} is the failure rate of the ground earth station and its projected value is 3.75×10^{-6} failures per hour.

The reliability of the telecommunication network for the air traffic control system is obtained by considering its five major hardware components as a series system. Thus,

$$R_{\text{system}}(t) = R_{\text{AES}}(t) \cdot R_{\text{satellite}}(t) \cdot R_{\text{subnetwork}}(t) \cdot R_{\text{ARTCC}}(t) \cdot R_{\text{GES}}(t). \quad (10.30)$$

The reliability of the individual components and that of the entire system are shown in Figure 10.14.

IMPORTANCE OF THE COMPONENTS

Using Birnbaum's importance measure, as given by Eq. (2.71), we obtain

$$G(\mathbf{q}(t)) = 1 - (1 - q_{\text{AES}}(t))(1 - q_{\text{satellite}}(t))(1 - q_{\text{subnetwork}}(t))(1 - q_{\text{ARTCC}}(t)) \times (1 - q_{\text{GES}}(t))$$

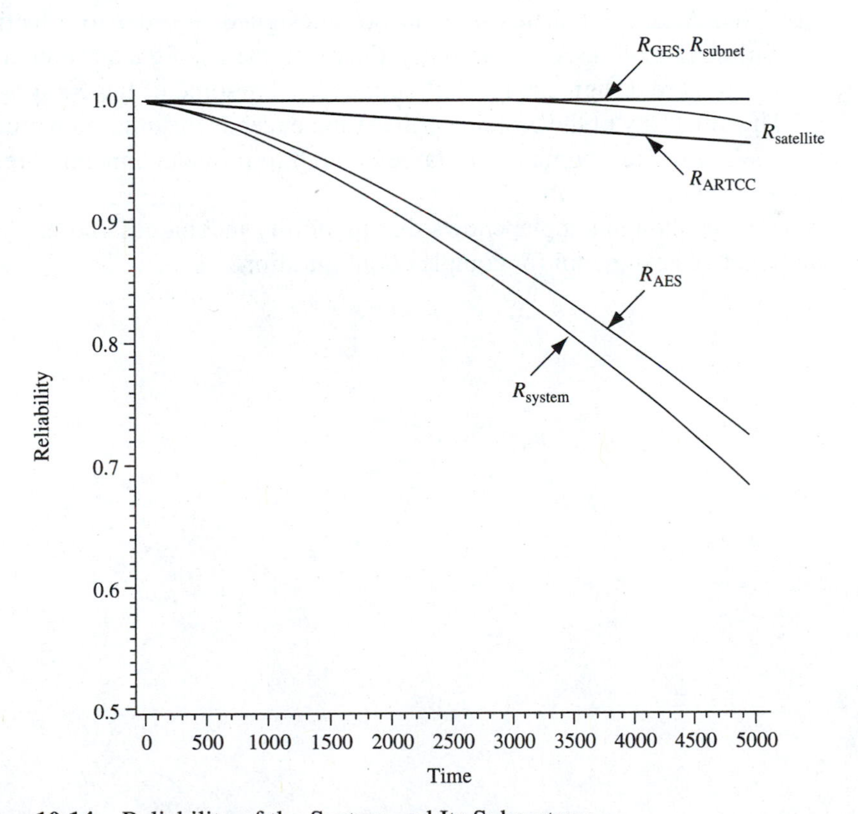

Figure 10.14. Reliability of the System and Its Subsystems

At t = 1,000 hours

$$I_B^{\text{AES}}(1{,}000) = (1 - q_{\text{satellite}}(1{,}000))(1 - q_{\text{subnetwork}}(1{,}000))$$
$$\times (1 - q_{\text{ARTCC}}(1{,}000))(1 - q_{\text{GES}}(1{,}000))$$

$$I_B^{\text{AES}}(1{,}000) = 0.992816$$

$$I_B^{\text{satellite}}(1{,}000) = 0.966502$$

$$I_B^{\text{subnetwork}}(1{,}000) = 0.966510$$

$$I_B^{\text{ARTCC}}(1{,}000) = 0.973387$$

$$I_B^{\text{GES}}(1{,}000) = 0.966502.$$

The AES has the highest importance measure. Accordingly, it has the most impact on the overall system reliability.

Most of the AES components need to be redesigned in order to effectively reduce the failure rate of the AES. Similarly, the components of the air route traffic control center require design changes or redundancy for some of the components and links. The reliability of the ground earth station exceeds the minimum requirement of the system. Their numbers are large enough to provide "inherent" redundancy in the system.

This analysis, though simple, shows that reliability techniques and modeling can be an effective design tool for complex configurations.

Case 6: System Design Using Reliability Objectives*

Introduction

Telecommunications service continuity is controlled by availability design objectives applicable to networks, network segments, and network elements. In general, these objectives are intended to control the amount of time that networks or portions of networks are unable to perform their required function. Availability objectives are typically stated as a single number equal to the long-term percentage of time that a system is expected to provide service. (*System* is a generic term used to describe an entity to which availability objectives apply. This could be a network, a network segment, or a network element.) As such, these objectives can significantly influence end-user perception of service quality.

This case discusses some of the implications of using traditional availability design objectives to control performance. This is done by first considering how current availability objectives apply to the *design of a single system* and then by examining the resulting *availability service performance across a population of systems, each of which is designed to meet the same availability design objective*. New methods for describing end-user availability performance are discussed, and how they can be used to (1) evaluate current network and service objectives, (2) aid in the development of a top-down approach to setting availability objectives for new services, and (3) provide better network performance quality control.

The topics discussed in this case are addressed because there continue to be questions about the interpretation of availability design objectives for new services and supporting technologies. Two such questions follow:

- An objective for total downtime of a common control Bell System switching system was first established in the late 1950s and evolved during the 1960s. The resulting objective required that there be "no more than two hours total downtime in forty years." The same objective is used today and is stated as a downtime objective of three minutes per year for current switching systems. If a switching system is down for more than three minutes in a year, has it failed to meet its design objective?
- If, after a period of several years, there are switching systems in a given population that have experienced (1) no downtime, (2) downtime less than the objective, (3) downtime greater than the objective, has the system design objective been met?

* This case is contributed by Norman A. Marlow and Michael Tortorella of AT&T Bell Laboratories. It is modified by the author in its present form. Copyright © 1995, AT&T. Used by permission.

As we shall discuss, the answer to the first question is "not necessarily." The answer to the second question is that we interpret the *design objective* as having been met if the *average of the population downtime distribution* does not exceed the objective value.

We first discuss traditional availability objectives, their relation to equipment reliability design objectives, and how they can be interpreted as a measure of long-term average service performance. Next we consider a population of identical systems, each designed to meet the same availability objective, and describe how downtime performance can vary across the population of system end users. Also discussed are differences in cumulative downtime performance that could be expected from simplex and duplex systems having the same availability objective and average unit restoration time. Extreme performance is considered by showing how the "longest restoration time" experienced over a given time period could vary across a population of systems. We conclude by discussing how these measures of population downtime performance can be used to assess the effects of system availability design objectives on end-user service performance and aid in the development of top-down availability objectives.

Availability Design Objectives

A typical availability objective might state that a system or service "be available at least 99.8 percent of the time." Letting *uptime* denote the time during which a system or service is performing its required function, a consistent interpretation of this objective (and an interpretation that is in common use) is that, *when measured over a sufficiently long-time interval*,

$$\frac{\text{Cumulative system uptime}}{\text{Cumulative observed time}} \geq 0.998. \tag{10.31}$$

"Cumulative observed time" in Eq. (10.31) is cumulative system uptime, plus time when the system is not providing service but is supposed to be. The latter includes downtime in general and consists of cumulative restoration times following system failures, planned maintenance downtimes, and other "out of service" times.

The definition of *instantaneous availability* given in Section 3.4 is consistent with Eq. (10.31) and, as noted above, can apply both to systems and to the services they support. In the above example, the unavailability is 0.002, or 0.2 percent. This corresponds to 1,051.2 minutes, or about 17.5 hours *expected downtime* per year.

Similarly, the steady state availability, A, of a system is defined as

$$A = \lim_{t \to \infty} P\{\text{system is operating at time } t\}. \tag{10.32}$$

From Eqs. (10.31) and (10.32), we obtain

$$E\,\{\text{cumulative system uptime in a steady state period of length } T\} = A \times T, \tag{10.33}$$

where E denotes expected value. From Chapter 3, we rewrite the steady-state availability

$$A = \frac{MTTF}{MTTF + MTTR}, \tag{10.34}$$

where MTTF and MTTR are the mean time to failure and the mean time to repair, respectively.

The corresponding steady state unavailability $\bar{A}$ in this example is then

$$\bar{A} = 1 - A$$

$$\bar{A} = \frac{MTTR}{MTTF + MTTR}. \tag{10.35}$$

Letting $\lambda = 1/MTTF$ and $\mu = 1/MTTR$ (λ is the system failure rate and μ is the system restoration rate when the time to failure and time to repair distributions are exponential, as is henceforth assumed in this case), it follows from Eqs. (10.33) and (10.35) that the expected cumulative steady-state system downtime can be expressed as

$$E\,\{\text{cumulative downtime in a steady state period of length } T\} = \frac{\lambda T}{\lambda + \mu}. \tag{10.36}$$

The performance measures given by the steady-state availability in Eq. (10.34), the steady-state unavailability in Eq. (10.35), or the expected downtime in Eq. (10.36), are determined by the system MTTF and MTTR. MTTF is a basic design characteristic, while the MTTR is characteristic of a particular maintenance or operations policy. The MTTR may also depend on design features such as system modularity, self-diagnostic capability, or other factors.

Systems are designed to meet availability objectives by adjusting their MTTF and MTTR values within a model like Eq. (10.36). For example, if the availability objective for a single simplex unit is 99.8 percent, then the corresponding *downtime objective* is 1,051.2 minutes *expected downtime* per year. Using 525,600 minutes per year as a base, Eq. (10.36) can be used to write this objective in the form

$$\frac{\lambda}{\lambda + \mu} \times 525{,}600 \leq 1{,}051.2 \text{ minutes per year.} \tag{10.37}$$

Assuming an average restoration time of at most four hours ($1/\mu \leq 240$ minutes), it follows from Eq. (10.37) that the availability objective will be met if the unit failure rate λ satisfies

$$\lambda \leq 4.38 \text{ failures per year.} \tag{10.38}$$

This is equivalent to a mean time to failure ($1/\lambda$) of at least 1,996 hours and is a system reliability design objective.

The same principles apply to more complex systems. For example, two identical simplex units operating *independently* in a load sharing parallel mode will have a system unavailability given by

$$\bar{A} = [(\lambda/(\lambda + \mu))]^2. \tag{10.39}$$

In Eq. (10.39), λ and μ are, respectively, the failure and restoration rates for each simplex unit. For example, if the parallel system downtime objective is two minutes per year (two minutes per year is the downtime objective for parallel A-link access to the SS7 Common Channel Signaling network) (Bellcore, 1993), Eq. (10.39) can be used to write the objective in the form

$$[\lambda/(\lambda + \mu)]^2 \times 525{,}600 \leq 2 \text{ minutes per year.} \tag{10.40}$$

To meet the objective specified by Eq. (10.40), each *simplex unit* in the parallel system must satisfy

$$\frac{\lambda}{\lambda + \mu} \times 525{,}600 \leq 1{,}025.28 \text{ minutes (17 hours) per year.} \tag{10.41}$$

Assuming an average restoration time of at most four hours, it follows from Eq. (10.41) that the parallel system downtime objective would be met if the failure rate λ of each simplex unit satisfies

$$\lambda \leq 4.2 \text{ failures per year.} \tag{10.42}$$

This corresponds to a MTTF of at least 2,047 hours for each simplex unit in the duplex system.

In the above example, the same downtime objective of two minutes per year could be met by using one simplex unit instead of two in parallel. However, with a mean restoration time of four hours, the MTTF of the simplex unit would have to be at least 1,051,196 hours or about 119 years.

In general, the unavailability of a complex system will be some function of its simplex network element failure and restoration rates

$$\bar{A} = \phi(\lambda_1, \mu_1, \lambda_2, \mu_2, \ldots).$$

The form of this function generally depends on the system architecture, configuration of the components, and operating procedures (Birolini, 1985) as discussed in Chapters 2 and 3. As in the above examples, system downtime objectives can be used to select design values for network element failure rates and restoration rates by applying the relation

$$\phi(\lambda_1, \mu_1, \lambda_2, \mu_2, \ldots) \times 525{,}600 \leq \text{Objective downtime (minutes) per year.} \tag{10.43}$$

In principle, Eq. (10.43) represents a performance constraint subject to which system cost could be minimized.

With this as background, in this case we explore some ideas related to the following question: Suppose a system is designed to meet an availability objective using the procedure outlined above. What then are the properties of a performance indicator, or statistic, like Eq. (10.31), when a large number of systems is put into service?

Availability Service Performance

Downtime objectives are intended to control the amount of time that networks, network segments, and network elements are unable to provide service. As discussed, these objectives are typically stated as a single number equal to the maximum expected downtime per year. As also discussed, the *same downtime objective can be met using different architectures*, and it is important to understand possible performance differences resulting from different implementations. Of course, by definition, the probability that a system is unable to perform its required function is equal to the system unavailability. The "design to availability objectives" procedure outlined above leads to the following interpretation of a downtime objective:

$$P\left\{\begin{matrix}\textit{system cannot perform}\\ \textit{its required function}\end{matrix}\right\} \leq \frac{\text{Yearly downtime objective in minutes}}{\text{525,600 minutes per year}}.$$

While this is an important *system performance measure*, analogous to a system "ineffective attempt rate," it does not adequately describe the full range of performance that can be expected in a population of such systems. In particular, if each system in a population meets the objective, all that can be said is that each system has a MTTF and MTTR meeting the specified objectives. However, individual times to system failure and corresponding restoration times will vary randomly and will be different from the average or expected values to which they have been designed. The result is that observed yearly downtimes across a population will differ from the objective value. As such, variations in cumulative downtime over a given year will occur *across a population of systems*. If these systems are designed to meet a specified objective, then the mean or *average* of the population downtime distribution in one year should not exceed the objective value.

In assessing compliance with reliability objectives, one could simply stop here by ascertaining whether the sample mean of the population annual downtime exceeds the objective value. A simple hypothesis test would provide the understanding of statistical significance needed here. However, we maintain that by using appropriately the additional information contained in the model that underlies the "design to availability objectives" procedure outlined above, additional insight into the quality control of this key service satisfaction parameter can be obtained. We explore this idea later following further discussion of distributions of downtime in the population.

CUMULATIVE DOWNTIME DISTRIBUTIONS

To illustrate how yearly downtime can vary across a population, consider the single-unit system. Assume again that the cumulative downtime objective is 17.5 hours (1,051.2 minutes) per year and that the average restoration time is 4 hours. Then, using Eq. (10.37), the parameters MTTF $= 1/\lambda$ and MTTR $= 1/\mu$ are

$$\text{MTTF} = 1{,}996 \text{ hours}$$

$$\text{MTTR} = 4.0 \text{ hours.}$$

Figure 10.15 shows how the cumulative downtime over a one-year period might appear for a single unit.

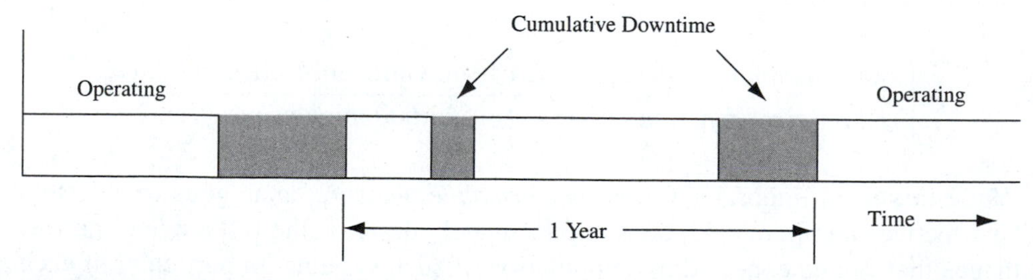

Figure 10.15. Cumulative Yearly Downtime

The cumulative downtime during a given time period depends on both the times to failure and the corresponding restoration times. As discussed above, these times will vary from failure to failure and from system to system. This implies that the downtime in a given year will have some distribution across a population of units. In particular, using a Markov model it is possible to obtain the predicted yearly cumulative downtime distribution (Barlow and Proschan, 1965; Brownlee, 1960; Puri, 1971; Takacs, 1957) as discussed in Chapter 3. This distribution is a function, $D(T)$, that depends on the simplex unit MTTF and MTTR, and is defined by

$$D(t) = P\,\{\text{cumulative system downtime in one year is} \leq T\}.$$

Using one year as a base, Figure 10.16 shows the complementary steady state cumulative downtime distribution $1 - D(T)$ when the downtime objective is an average of 17.5 hours (1,051.2 minutes) per year and the average restoration time is 4 hours.

Figure 10.16 illustrates important service consequences of using a design objective of 17.5 hours average downtime per year and an average restoration time of 4 hours. In particular, note that during one year, 43.2 percent of the units *in the population* are expected to have cumulative downtime *exceeding the objective*. For example, a particular unit in a population meeting an average cumulative down-

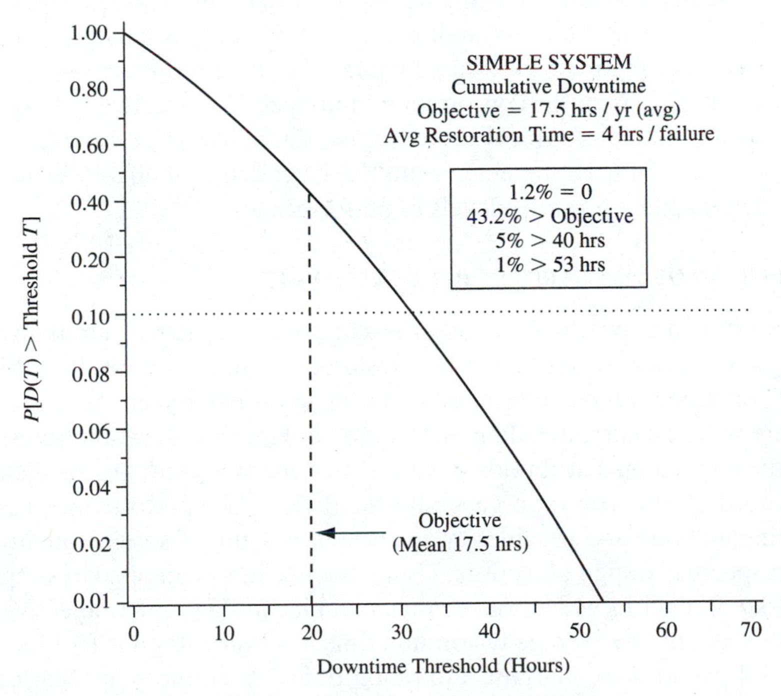

Figure 10.16. Simplex Unit Cumulative Downtime Distribution

time objective of 17.5 hours has about a 20 percent chance of being down for 25 hours or more in one year. Two other important service indicators are the probability of "zero downtime" in a given year, and the probability of exceeding a given threshold during a year. As Figure 10.16 also shows, about 1.2 percent of the unit population is expected to experience zero downtime during one year. In addition, about 5 percent of the population is expected to have cumulative downtime greater than forty hours.

Several important conclusions for a system designed to meet an unavailability objective interpreted as a long-term average follow:

- The objective value will be approached *as a time average* over a sufficiently long measurement period.
- In a large population of such systems, the *population average downtime* in a year should correspond to the objective.
- A single system may have downtime in a year exceeding objective value *while the population design objective is still met.*

In particular, *it is not currently accepted practice to interpret the design objective as a maximum value that no system may exceed in service.*

If a population of systems meets an average cumulative downtime objective of 17.5 hours per year, then normal operating conditions will result in some large cumulative yearly downtimes as Figure 10.16 shows. On the other hand, if some systems in a population have cumulative yearly downtimes exceeding the limits one deduces from Figure 10.16, then factors other than those caused by nominal statistical variation may be contributing, and special corrective action may be needed. As a measure of typical performance, distributions of cumulative yearly downtimes could be used, together with "statistical control chart" techniques, to monitor and maintain objective levels of performance.

DISTRIBUTION OF THE LARGEST RESTORATION TIME

Each system in a population meeting the same average yearly cumulative downtime objective will experience variable restoration times and will be subject to a "largest" or maximum restoration time during a particular year. Across a population, there will be a corresponding distribution of largest restoration times. In contrast to the population distribution of cumulative yearly downtime, the distribution of the largest restoration time generally highlights poor performance associated with a single failure and provides an important measure of service quality resulting from specified design objectives. Using the previous example of a simplex system Markov model in which the average cumulative downtime objective is 17.5 hours per year and the average restoration time is 4 hours, Figure 10.17 shows the predicted distribution of "largest restoration times" over one year (Marlow, n.d.).

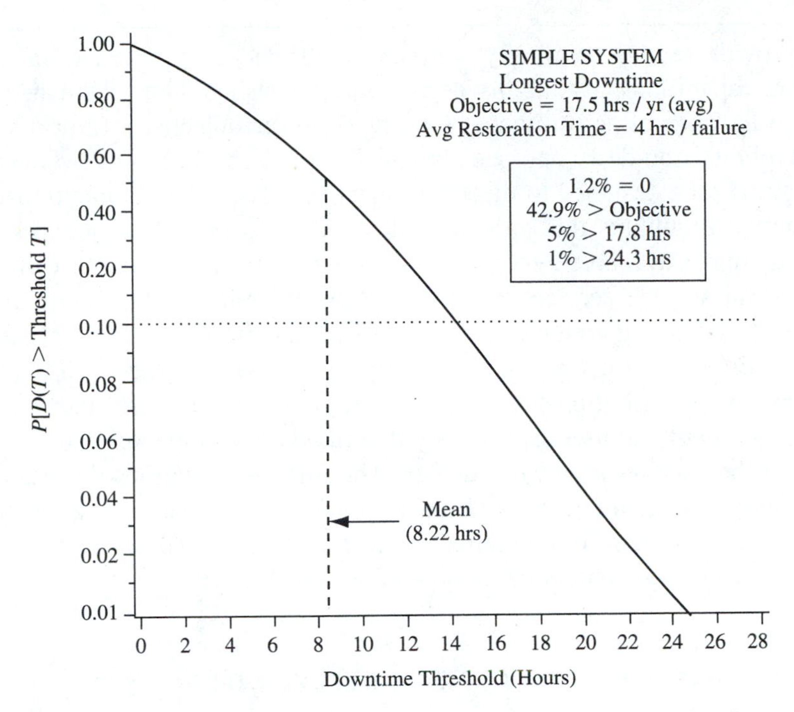

Figure 10.17. Distribution of the Largest Restoration Time for a Simplex System

This figure shows that while a MTTR of four hours is used, the largest experienced restoration time is likely to be larger than this average value. In general, as more failures occur, the potential for larger restoration times increases.

Because the cumulative downtime experienced during a year will be no smaller than the largest single restoration time, Figure 10.17 also illustrates that the cumulative system downtime experienced during a year may be larger than the average design value. In particular, this example shows that 5.3 percent of the systems are expected to have a "largest" single downtime exceeding the design value of 17.5 hours *cumulative downtime* per year.

If a population of systems meets an average cumulative downtime objective of 17.5 hours per year, then nominal operating conditions will result in some long individual restoration times as shown in Figure 10.17. On the other hand, if some systems in a population have restoration times exceeding these expected limits, then factors other than those caused by nominal statistical variation may be contributing to the increase in the restoration times, and special corrective maintenance action may be needed. As a measure of extreme performance, distributions of longest restoration times could be used, together with "statistical control chart" techniques, to monitor and maintain objective levels of performance.

Evaluation of Availability Design Objectives

To provide service continuity consistent with end-user expectations and network capabilities, availability design objectives should be established to control the full range of performance resulting from the objectives. In principle, this should follow a top-down approach in which service-level objectives are specified and networks are designed to meet the objectives. The example mentioned in the introduction highlights this approach: the goal is to provide a network service capability that is available 99.8 percent of the time. As discussed, this can be interpreted as an average service downtime of 1,051.2 minutes per year and can be achieved by selecting architectures and network elements whose combined average downtime performance meets this objective. However, the resulting population downtime distribution shows that the *range* of service performance expected across a group of end users is not fully described by a single downtime objective for either the service or network design. The following sections discuss how the selection of service and network design availability objectives can be better evaluated when the full range of downtime performance has been described using information provided by the population downtime distribution.

EVALUATION OF CURRENT NETWORK AND SERVICE OBJECTIVES

Contemporary telecommunication network standards include downtime objectives for many network elements and for corresponding services. As an example, end office switching system access to the SS7 Common Channel Signaling network is provided by A-links connected to a mated pair of signaling transfer points (Bellcore, 1993). The average CCS network access downtime objective is two minutes per year. To evaluate the full range of performance that can be expected from this objective, a parallel system Markov model was used to predict the cumulative yearly downtime distribution (Hamilton and Marlow, 1991). Figure 10.18 shows the complementary downtime distribution for an objective of two minutes per year and compares it to distributions that would result from objectives of one and eight minutes per year, respectively. Each distribution is based on an average link restoration time of four hours per failure.

Figure 10.18 highlights important consequences of using a particular average downtime objective for signaling network access. In particular, with an objective of two minutes per year, about 1 percent of the end offices each year are expected to be unable to access CCS services for at least one hour. With a tighter objective of one minute, the percentage decreases to 0.5 percent, while it increases to about 4 percent with an objective of eight minutes per year. Each downtime objective also corresponds to an ineffective call attempt rate caused by access signaling failures and resulting downtime. For example, the two-minute objective is equivalent to an ineffective call attempt rate of $2/525{,}600 = 3.8 \times 10^{-6}$. Table 10.9 summarizes these performance characteristics.

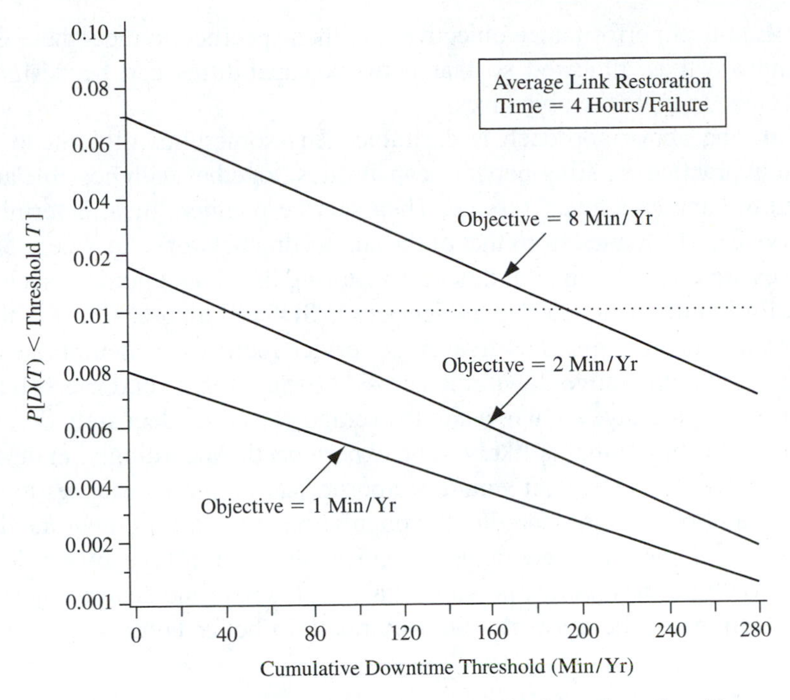

Figure 10.18. CCS Network Access Downtime Distributions Over One Year

Table 10.9. SS7 Network Access Performance

Access Downtime Objective (Min/Yr)	*Ineffective Attempt Rate*	*Offices with Access Downtime > 1 Hour/Yr*
1	1.9×10^{-6}	0.5%
2	3.8×10^{-6}	1.0%
8	15×10^{-6}	4.0%

The downtime objective for CCS network access was first published as a CCITT Recommendation in 1984 (CCITT, 1984a, 1984b). At that time, the objective was believed to be achievable and was considered adequate for call setup performance expressed as an ineffective attempt rate. As the signaling network evolves, however, more services will be supported by this network and critical reviews of all downtime objectives will be needed. Using information such as that in Table 10.9, these reviews can be made more service oriented and should result in improved guidelines for future CCS network performance planning.

ESTABLISHING TOP-DOWN AVAILABILITY OBJECTIVES

Top-down performance planning for a new service begins with a comprehensive understanding of customer needs, expectations, and perceptions of performance quality. End-user performance objectives are then specified to meet these expectations and are then allocated so that network capabilities can be designed and implemented to meet the objectives.

While the above approach is desirable, it is sometimes difficult to achieve because in practice, existing network capabilities, together with possible adjuncts, are used to support a given service. Then service planners must determine end-user service performance from that of the supporting network. Service availability objectives are often obtained in this way, resulting in a single number such as 99.8 percent for Public Switched Digital Services (PSDS) (Bellcore, 1985). This single number could be interpreted as an average service ineffective attempt rate or as an average yearly cumulative downtime across a population of end users, but as the previous examples have shown, a single average does not adequately describe the range of performance that is likely to be experienced. Accordingly, in developing end-user service objectives, it would be appropriate in many instances to use network availability models to predict the population cumulative downtime distribution and ask if the expected range is acceptable for the type of service being planned. With this as a guide, possible changes in downtime design objectives for critical portions of the network could be made to better control service performance.

REFERENCES

Barlow, R. E., and Proschan, F. (1965). *Mathematical Theory of Reliability*. New York: John Wiley.

Bellcore Special Report. (1985). SR-NPL-000224. *ISDN System Plan*, Issue 1.

Bellcore Technical Requirements. (1993). TR-TSV-000905. *Common Channel Signaling (CCS) Network Interface Specification,* Issue 2, August.

Birolini, A. (1985). *On the Use of Stochastic Processes in Modeling Reliability Problems: Lecture Notes in Economics and Mathematical Systems*. New York: Springer-Verlag.

Boucher, T. O., Gogus, O., Bruins, R., Descovich, T., and Litman, N. (1996). "Multi-criteria Evaluation of a Centralized Control Station for a Production Line: A Case Study." *Journal of Engineering Valuation and Cost Analysis* 1(1), 17–32.

Brownlee, K. A. (1960). *Statistical Theory and Methodology in Science and Engineering*. New York: John Wiley.

CCITT (International Telegraph and Telephone Consultative Committee) Recommendation G. 106. (1984a). *Terms and Definitions Related to Quality of Service, Availability, and Reliability*. Geneva: Telecommunication Union.

CCITT (International Telegraph and Telephone Consultative Committee) Recommendation Q. 706. (1984b). *Message Transfer Part Signalling Performance,* Vol. 6.7, Eighth Plenary Assembly, October. Geneva: Telecommunication Union.

Hamilton, C. M., and Marlow, N. A. (1991). "Analyzing Telecommunications Network Availability Performance Using the Downtime Probability Distribution." Paper presented at Globecom '91, Phoenix, AZ, December 2–5.

International Civil Aviation Organization, Special Committee On Future Air Navigation and Surveillance. (1988). Fourth Meeting, Report Folder, May.

Kogan, J. (1976). *Crane Design: Theory and Calculations of Reliability*. New York: John Wiley.

Marlow, N. A. (n.d.). "Distribution of the Largest Restoration Time in an Alternating Renewal Process Reliability Model." Unpublished Manuscript.

Puri, P. S. (1971). "A Method for Studying the Integral Functionals of Stochastic Processes with Applications: I. The Markov Chain Case." *Journal of Applied Probability* 8, 331–348.

Takacs, L. (1957). *On Certain Sojourn Time Problems in the Theory of Stochastic Processes*. Acta Mathematica Academiae Scientiarum Hungaricae, Budapest: Akademiai Kiado.

Appendixes

APPENDIX A

Gamma Table

Gamma Function

n	$\Gamma(n)$	n	$\Gamma(n)$	n	$\Gamma(n)$	n	$\Gamma(n)$
0.0100	99.4327	0.5100	1.7384	1.0100	0.9943	1.5100	0.8866
0.0200	49.4423	0.5200	1.7058	1.0200	0.9888	1.5200	0.8870
0.0300	32.7850	0.5300	1.6747	1.0300	0.9836	1.5300	0.8876
0.0400	24.4610	0.5400	1.6448	1.0400	0.9784	1.5400	0.8882
0.0500	19.4701	0.5500	1.6161	1.0500	0.9735	1.5500	0.8889
0.0600	16.1457	0.5600	1.5886	1.0600	0.9687	1.5600	0.8896
0.0700	13.7736	0.5700	1.5623	1.0700	0.9642	1.5700	0.8905
0.0800	11.9966	0.5800	1.5369	1.0800	0.9597	1.5800	0.8914
0.0900	10.6162	0.5900	1.5126	1.0900	0.9555	1.5900	0.8924
0.1000	9.5135	0.6000	1.4892	1.1000	0.9513	1.6000	0.8935
0.1100	8.6127	0.6100	1.4667	1.1100	0.9474	1.6100	0.8947
0.1200	7.8632	0.6200	1.4450	1.1200	0.9436	1.6200	0.8959
0.1300	7.2302	0.6300	1.4242	1.1300	0.9399	1.6300	0.8972
0.1400	6.6887	0.6400	1.4041	1.1400	0.9364	1.6400	0.8986
0.1500	6.2203	0.6500	1.3848	1.1500	0.9330	1.6500	0.9001
0.1600	5.8113	0.6600	1.3662	1.1600	0.9298	1.6600	0.9017
0.1700	5.4512	0.6700	1.3482	1.1700	0.9267	1.6700	0.9033
0.1800	5.1318	0.6800	1.3309	1.1800	0.9237	1.6800	0.9050
0.1900	4.8468	0.6900	1.3142	1.1900	0.9209	1.6900	0.9068
0.2000	4.5908	0.7000	1.2981	1.2000	0.9182	1.7000	0.9086
0.2100	4.3599	0.7100	1.2825	1.2100	0.9156	1.7100	0.9106
0.2200	4.1505	0.7200	1.2675	1.2200	0.9131	1.7200	0.9126
0.2300	3.9598	0.7300	1.2530	1.2300	0.9108	1.7300	0.9147
0.2400	3.7855	0.7400	1.2390	1.2400	0.9085	1.7400	0.9168
0.2500	3.6256	0.7500	1.2254	1.2500	0.9064	1.7500	0.9191
0.2600	3.4785	0.7600	1.2123	1.2600	0.9044	1.7600	0.9214
0.2700	3.3426	0.7700	1.1997	1.2700	0.9025	1.7700	0.9238
0.2800	3.2169	0.7800	1.1875	1.2800	0.9007	1.7800	0.9262
0.2900	3.1001	0.7900	1.1757	1.2900	0.8990	1.7900	0.9288
0.3000	2.9916	0.8000	1.1642	1.3000	0.8975	1.8000	0.9314
0.3100	2.8903	0.8100	1.1532	1.3100	0.8960	1.8100	0.9341
0.3200	2.7958	0.8200	1.1425	1.3200	0.8946	1.8200	0.9368
0.3300	2.7072	0.8300	1.1322	1.3300	0.8934	1.8300	0.9397
0.3400	2.6242	0.8400	1.1222	1.3400	0.8922	1.8400	0.9426
0.3500	2.5461	0.8500	1.1125	1.3500	0.8912	1.8500	0.9456
0.3600	2.4727	0.8600	1.1031	1.3600	0.8902	1.8600	0.9487
0.3700	2.4036	0.8700	1.0941	1.3700	0.8893	1.8700	0.9518
0.3800	2.3383	0.8800	1.0853	1.3800	0.8885	1.8800	0.9551
0.3900	2.2765	0.8900	1.0768	1.3900	0.8879	1.8900	0.9584
0.4000	2.2182	0.9000	1.0686	1.4000	0.8873	1.9000	0.9618
0.4100	2.1628	0.9100	1.0607	1.4100	0.8868	1.9100	0.9652
0.4200	2.1104	0.9200	1.0530	1.4200	0.8864	1.9200	0.9688
0.4300	2.0605	0.9300	1.0456	1.4300	0.8860	1.9300	0.9724
0.4400	2.0132	0.9400	1.0384	1.4400	0.8858	1.9400	0.9761
0.4500	1.9681	0.9500	1.0315	1.4500	0.8857	1.9500	0.9799
0.4600	1.9252	0.9600	1.0247	1.4600	0.8856	1.9600	0.9837
0.4700	1.8843	0.9700	1.0182	1.4700	0.8856	1.9700	0.9877
0.4800	1.8453	0.9800	1.0119	1.4800	0.8857	1.9800	0.9917
0.4900	1.8080	0.9900	1.0059	1.4900	0.8859	1.9900	0.9958
0.5000	1.7725	1.0000	1.0000	1.5000	0.8862	2.0000	1.0000

Gamma Function

n	$\Gamma(n)$	n	$\Gamma(n)$	n	$\Gamma(n)$	n	$\Gamma(n)$
2.0100	1.0043	2.5100	1.3388	3.0100	2.0186	3.5100	3.3603
2.0200	1.0086	2.5200	1.3483	3.0200	2.0374	3.5200	3.3977
2.0300	1.0131	2.5300	1.3580	3.0300	2.0565	3.5300	3.4357
2.0400	1.0176	2.5400	1.3678	3.0400	2.0759	3.5400	3.4742
2.0500	1.0222	2.5500	1.3777	3.0500	2.0955	3.5500	3.5132
2.0600	1.0269	2.5600	1.3878	3.0600	2.1153	3.5600	3.5529
2.0700	1.0316	2.5700	1.3981	3.0700	2.1355	3.5700	3.5930
2.0800	1.0365	2.5800	1.4084	3.0800	2.1559	3.5800	3.6338
2.0900	1.0415	2.5900	1.4190	3.0900	2.1766	3.5900	3.6751
2.1000	1.0465	2.6000	1.4296	3.1000	2.1976	3.6000	3.7170
2.1100	1.0516	2.6100	1.4404	3.1100	2.2189	3.6100	3.7595
2.1200	1.0568	2.6200	1.4514	3.1200	2.2405	3.6200	3.8027
2.1300	1.0621	2.6300	1.4625	3.1300	2.2623	3.6300	3.8464
2.1400	1.0675	2.6400	1.4738	3.1400	2.2845	3.6400	3.8908
2.1500	1.0730	2.6500	1.4852	3.1500	2.3069	3.6500	3.9358
2.1600	1.0786	2.6600	1.4968	3.1600	2.3297	3.6600	3.9814
2.1700	1.0842	2.6700	1.5085	3.1700	2.3528	3.6700	4.0277
2.1800	1.0900	2.6800	1.5204	3.1800	2.3762	3.6800	4.0747
2.1900	1.0959	2.6900	1.5325	3.1900	2.3999	3.6900	4.1223
2.2000	1.1018	2.7000	1.5447	3.2000	2.4240	3.7000	4.1707
2.2100	1.1078	2.7100	1.5571	3.2100	2.4483	3.7100	4.2197
2.2200	1.1140	2.7200	1.5696	3.2200	2.4731	3.7200	4.2694
2.2300	1.1202	2.7300	1.5824	3.2300	2.4981	3.7300	4.3199
2.2400	1.1266	2.7400	1.5953	3.2400	2.5235	3.7400	4.3711
2.2500	1.1330	2.7500	1.6084	3.2500	2.5493	3.7500	4.4230
2.2600	1.1395	2.7600	1.6216	3.2600	2.5754	3.7600	4.4757
2.2700	1.1462	2.7700	1.6351	3.2700	2.6018	3.7700	4.5291
2.2800	1.1529	2.7800	1.6487	3.2800	2.6287	3.7800	4.5833
2.2900	1.1598	2.7900	1.6625	3.2900	2.6559	3.7900	4.6384
2.3000	1.1667	2.8000	1.6765	3.3000	2.6834	3.8000	4.6942
2.3100	1.1738	2.8100	1.6907	3.3100	2.7114	3.8100	4.7508
2.3200	1.1809	2.8200	1.7051	3.3200	2.7398	3.8200	4.8083
2.3300	1.1882	2.8300	1.7196	3.3300	2.7685	3.8300	4.8666
2.3400	1.1956	2.8400	1.7344	3.3400	2.7976	3.8400	4.9257
2.3500	1.2031	2.8500	1.7494	3.3500	2.8272	3.8500	4.9857
2.3600	1.2107	2.8600	1.7646	3.3600	2.8571	3.8600	5.0466
2.3700	1.2184	2.8700	1.7799	3.3700	2.8875	3.8700	5.1084
2.3800	1.2262	2.8800	1.7955	3.3800	2.9183	3.8800	5.1711
2.3900	1.2341	2.8900	1.8113	3.3900	2.9495	3.8900	5.2348
2.4000	1.2422	2.9000	1.8274	3.4000	2.9812	3.9000	5.2993
2.4100	1.2503	2.9100	1.8436	3.4100	3.0133	3.9100	5.3648
2.4200	1.2586	2.9200	1.8600	3.4200	3.0459	3.9200	5.4313
2.4300	1.2670	2.9300	1.8767	3.4300	3.0789	3.9300	5.4988
2.4400	1.2756	2.9400	1.8936	3.4400	3.1124	3.9400	5.5673
2.4500	1.2842	2.9500	1.9108	3.4500	3.1463	3.9500	5.6368
2.4600	1.2930	2.9600	1.9281	3.4600	3.1807	3.9600	5.7073
2.4700	1.3019	2.9700	1.9457	3.4700	3.2156	3.9700	5.7789
2.4800	1.3109	2.9800	1.9636	3.4800	3.2510	3.9800	5.8515
2.4900	1.3201	2.9900	1.9817	3.4900	3.2869	3.9900	5.9252
2.5000	1.3293	3.0000	2.0000	3.5000	3.3233	4.0000	6.0000

Gamma Function

n	$\Gamma(n)$	n	$\Gamma(n)$	n	$\Gamma(n)$	n	$\Gamma(n)$
4.0100	6.0759	4.5100	11.7945	5.0100	24.3645	5.5100	53.1933
4.0200	6.1530	4.5200	11.9599	5.0200	24.7351	5.5200	54.0589
4.0300	6.2312	4.5300	12.1280	5.0300	25.1118	5.5300	54.9396
4.0400	6.3106	4.5400	12.2986	5.0400	25.4948	5.5400	55.8358
4.0500	6.3912	4.5500	12.4720	5.0500	25.8843	5.5500	56.7477
4.0600	6.4730	4.5600	12.6482	5.0600	26.2803	5.5600	57.6757
4.0700	6.5560	4.5700	12.8271	5.0700	26.6829	5.5700	58.6200
4.0800	6.6403	4.5800	13.0089	5.0800	27.0922	5.5800	59.5809
4.0900	6.7258	4.5900	13.1936	5.0900	27.5085	5.5900	60.5588
4.1000	6.8126	4.6000	13.3813	5.1000	27.9317	5.6000	61.5539
4.1100	6.9008	4.6100	13.5719	5.1100	28.3621	5.6100	62.5666
4.1200	6.9902	4.6200	13.7656	5.1200	28.7997	5.6200	63.5972
4.1300	7.0811	4.6300	13.9624	5.1300	29.2448	5.6300	64.6460
4.1400	7.1733	4.6400	14.1624	5.1400	29.6973	5.6400	65.7135
4.1500	7.2669	4.6500	14.3655	5.1500	30.1575	5.6500	66.7998
4.1600	7.3619	4.6600	14.5720	5.1600	30.6255	5.6600	67.9054
4.1700	7.4584	4.6700	14.7817	5.1700	31.1014	5.6700	69.0306
4.1800	7.5563	4.6800	14.9948	5.1800	31.5853	5.6800	70.1758
4.1900	7.6557	4.6900	15.2114	5.1900	32.0775	5.6900	71.3414
4.2000	7.7567	4.7000	15.4314	5.2000	32.5781	5.7000	72.5277
4.2100	7.8592	4.7100	15.6550	5.2100	33.0872	5.7100	73.7352
4.2200	7.9632	4.7200	15.8822	5.2200	33.6049	5.7200	74.9642
4.2300	8.0689	4.7300	16.1131	5.2300	34.1314	5.7300	76.2152
4.2400	8.1762	4.7400	16.3478	5.2400	34.6670	5.7400	77.4884
4.2500	8.2851	4.7500	16.5862	5.2500	35.2117	5.7500	78.7845
4.2600	8.3957	4.7600	16.8285	5.2600	35.7656	5.7600	80.1038
4.2700	8.5080	4.7700	17.0748	5.2700	36.3291	5.7700	81.4467
4.2800	8.6220	4.7800	17.3250	5.2800	36.9022	5.7800	82.8136
4.2900	8.7378	4.7900	17.5794	5.2900	37.4851	5.7900	84.2052
4.3000	8.8554	4.8000	17.8378	5.3000	38.0780	5.8000	85.6216
4.3100	8.9747	4.8100	18.1005	5.3100	38.6811	5.8100	87.0636
4.3200	9.0960	4.8200	18.3675	5.3200	39.2946	5.8200	88.5315
4.3300	9.2191	4.8300	18.6389	5.3300	39.9186	5.8300	90.0259
4.3400	9.3441	4.8400	18.9147	5.3400	40.5534	5.8400	91.5472
4.3500	9.4711	4.8500	19.1950	5.3500	41.1991	5.8500	93.0960
4.3600	9.6000	4.8600	19.4800	5.3600	41.8559	5.8600	94.6727
4.3700	9.7309	4.8700	19.7696	5.3700	42.5241	5.8700	96.2780
4.3800	9.8639	4.8800	20.0640	5.3800	43.2039	5.8800	97.9122
4.3900	9.9989	4.8900	20.3632	5.3900	43.8953	5.8900	99.5761
4.4000	10.1361	4.9000	20.6674	5.4000	44.5988	5.9000	101.2701
4.4100	10.2754	4.9100	20.9765	5.4100	45.3145	5.9100	102.9949
4.4200	10.4169	4.9200	21.2908	5.4200	46.0426	5.9200	104.7509
4.4300	10.5606	4.9300	21.6103	5.4300	46.7833	5.9300	106.5389
4.4400	10.7065	4.9400	21.9351	5.4400	47.5370	5.9400	108.3594
4.4500	10.8548	4.9500	22.2652	5.4500	48.3037	5.9500	110.2129
4.4600	11.0053	4.9600	22.6009	5.4600	49.0838	5.9600	112.1003
4.4700	11.1583	4.9700	22.9420	5.4700	49.8775	5.9700	114.0219
4.4800	11.3136	4.9800	23.2889	5.4800	50.6850	5.9800	115.9787
4.4900	11.4714	4.9900	23.6415	5.4900	51.5067	5.9900	117.9711
4.5000	11.6317	5.0000	24.0000	5.5000	52.3427	6.0000	120.0000

Gamma Function

n	$\Gamma(n)$	n	$\Gamma(n)$	n	$\Gamma(n)$	n	$\Gamma(n)$
6.0100	122.0661	6.5100	293.0953	7.0100	733.6171	7.5100	1908.0504
6.0200	124.1700	6.5200	298.4052	7.0200	747.5034	7.5200	1945.6019
6.0300	126.3123	6.5300	303.8161	7.0300	761.6632	7.5300	1983.9192
6.0400	128.4940	6.5400	309.3305	7.0400	776.1037	7.5400	2023.0216
6.0500	130.7156	6.5500	314.9500	7.0500	790.8292	7.5500	2062.9221
6.0600	132.9781	6.5600	320.6770	7.0600	805.8471	7.5600	2103.6414
6.0700	135.2820	6.5700	326.5134	7.0700	821.1620	7.5700	2145.1926
6.0800	137.6285	6.5800	332.4616	7.0800	836.7813	7.5800	2187.5977
6.0900	140.0181	6.5900	338.5236	7.0900	852.7099	7.5900	2230.8706
6.1000	142.4518	6.6000	344.7020	7.1000	868.9559	7.6000	2275.0332
6.1100	144.9303	6.6100	350.9986	7.1100	885.5239	7.6100	2320.1006
6.1200	147.4546	6.6200	357.4164	7.1200	902.4222	7.6200	2366.0967
6.1300	150.0255	6.6300	363.9571	7.1300	919.6564	7.6300	2413.0356
6.1400	152.6441	6.6400	370.6239	7.1400	937.2346	7.6400	2460.9426
6.1500	155.3111	6.6500	377.4185	7.1500	955.1622	7.6500	2509.8330
6.1600	158.0274	6.6600	384.3443	7.1600	973.4484	7.6600	2559.7332
6.1700	160.7941	6.6700	391.4035	7.1700	992.0996	7.6700	2610.6589
6.1800	163.6120	6.6800	398.5985	7.1800	1011.1224	7.6800	2662.6379
6.1900	166.4825	6.6900	405.9326	7.1900	1030.5265	7.6900	2715.6887
6.2000	169.4060	6.7000	413.4079	7.2000	1050.3174	7.7000	2769.8330
6.2100	172.3841	6.7100	421.0280	7.2100	1070.5054	7.7100	2825.0981
6.2200	175.4175	6.7200	428.7951	7.2200	1091.0967	7.7200	2881.5032
6.2300	178.5075	6.7300	436.7129	7.2300	1112.1016	7.7300	2939.0776
6.2400	181.6549	6.7400	444.7835	7.2400	1133.5264	7.7400	2997.8406
6.2500	184.8612	6.7500	453.0110	7.2500	1155.3823	7.7500	3057.8242
6.2600	188.1272	6.7600	461.3976	7.2600	1177.6760	7.7600	3119.0474
6.2700	191.4543	6.7700	469.9473	7.2700	1200.4185	7.7700	3181.5435
6.2800	194.8435	6.7800	478.6627	7.2800	1223.6171	7.7800	3245.3328
6.2900	198.2962	6.7900	487.5479	7.2900	1247.2832	7.7900	3310.4497
6.3000	201.8134	6.8000	496.6054	7.3000	1271.4244	7.8000	3376.9170
6.3100	205.3968	6.8100	505.8398	7.3100	1296.0537	7.8100	3444.7686
6.3200	209.0471	6.8200	515.2535	7.3200	1321.1777	7.8200	3514.0283
6.3300	212.7661	6.8300	524.8511	7.3300	1346.8097	7.8300	3584.7332
6.3400	216.5549	6.8400	534.6356	7.3400	1372.9580	7.8400	3656.9072
6.3500	220.4150	6.8500	544.6116	7.3500	1399.6354	7.8500	3730.5891
6.3600	224.3476	6.8600	554.7819	7.3600	1426.8507	7.8600	3805.8037
6.3700	228.3544	6.8700	565.1516	7.3700	1454.6176	7.8700	3882.5908
6.3800	232.4366	6.8800	575.7236	7.3800	1482.9451	7.8800	3960.9780
6.3900	236.5959	6.8900	586.5032	7.3900	1511.8477	7.8900	4041.0063
6.4000	240.8335	6.9000	597.4933	7.4000	1541.3342	7.9000	4122.7036
6.4100	245.1514	6.9100	608.6996	7.4100	1571.4203	7.9100	4206.1133
6.4200	249.5509	6.9200	620.1256	7.4200	1602.1152	7.9200	4291.2651
6.4300	254.0334	6.9300	631.7754	7.4300	1633.4349	7.9300	4378.2036
6.4400	258.6011	6.9400	643.6547	7.4400	1665.3906	7.9400	4466.9629
6.4500	263.2550	6.9500	655.7667	7.4500	1697.9950	7.9500	4557.5786
6.4600	267.9975	6.9600	668.1176	7.4600	1731.2637	7.9600	4650.0986
6.4700	272.8297	6.9700	680.7109	7.4700	1765.2081	7.9700	4744.5557
6.4800	277.7539	6.9800	693.5528	7.4800	1799.8456	7.9800	4840.9985
6.4900	282.7716	6.9900	706.6470	7.4900	1835.1874	7.9900	4939.4629
6.5000	287.8849	7.0000	720.0000	7.5000	1871.2517	8.0000	5040.0000

Gamma Function

n	$\Gamma(n)$	n	$\Gamma(n)$	n	$\Gamma(n)$	n	$\Gamma(n)$
8.0100	5142.6606	8.5100	14329.4697	9.0100	41192.7070	9.5100	121943.7969
8.0200	5247.4688	8.5200	14630.9121	9.0200	42084.6953	9.5200	124655.3359
8.0300	5354.4927	8.5300	14938.9092	9.0300	42996.5742	9.5300	127428.8906
8.0400	5463.7700	8.5400	15253.5820	9.0400	43928.7188	9.5400	130265.6094
8.0500	5575.3521	8.5500	15575.0781	9.0500	44881.5820	9.5500	133166.9062
8.0600	5689.2749	8.5600	15903.5146	9.0600	45855.5547	9.5600	136134.0625
8.0700	5805.6143	8.5700	16239.1074	9.0700	46851.3047	9.5700	139169.1562
8.0800	5924.4116	8.5800	16581.9902	9.0800	47869.2461	9.5800	142273.4688
8.0900	6045.7188	8.5900	16932.3242	9.0900	48909.8711	9.5900	145448.6562
8.1000	6169.5806	8.6000	17290.2344	9.1000	49973.5977	9.6000	148696.0156
8.1100	6296.0747	8.6100	17655.9668	9.1100	51061.1602	9.6100	152017.8594
8.1200	6425.2466	8.6200	18029.6543	9.1200	52173.0078	9.6200	155415.6406
8.1300	6557.1558	8.6300	18411.4805	9.1300	53309.6797	9.6300	158891.0625
8.1400	6691.8481	8.6400	18801.5801	9.1400	54471.6328	9.6400	162445.6406
8.1500	6829.4102	8.6500	19200.2207	9.1500	55659.6914	9.6500	166081.8906
8.1600	6969.8911	8.6600	19607.5547	9.1600	56874.3047	9.6600	169801.4062
8.1700	7113.3540	8.6700	20023.7734	9.1700	58116.1055	9.6700	173606.1250
8.1800	7259.8521	8.6800	20449.0371	9.1800	59385.5820	9.6800	177497.6250
8.1900	7409.4775	8.6900	20883.6270	9.1900	60683.6133	9.6900	181478.7188
8.2000	7562.2842	8.7000	21327.7109	9.2000	62010.7266	9.7000	185551.0938
8.2100	7718.3447	8.7100	21781.5059	9.2100	63367.6016	9.7100	189716.9219
8.2200	7877.7256	8.7200	22245.2266	9.2200	64754.9023	9.7200	193977.9688
8.2300	8040.4858	8.7300	22719.0449	9.2300	66173.1953	9.7300	198337.2812
8.2400	8206.7314	8.7400	23203.2871	9.2400	67623.4688	9.7400	202796.7500
8.2500	8376.5215	8.7500	23698.1367	9.2500	69106.3047	9.7500	207358.7031
8.2600	8549.9346	8.7600	24203.8340	9.2600	70622.4609	9.7600	212025.5625
8.2700	8727.0332	8.7700	24720.5664	9.2700	72172.5547	9.7700	216799.3438
8.2800	8907.9307	8.7800	25248.6895	9.2800	73757.6641	9.7800	221683.4844
8.2900	9092.6934	8.7900	25788.4023	9.2900	75378.4297	9.7900	226680.0938
8.3000	9281.4092	8.8000	26339.9766	9.3000	77035.6953	9.8000	231791.7969
8.3100	9474.1416	8.8100	26903.6133	9.3100	78730.1172	9.8100	237020.8281
8.3200	9671.0205	8.8200	27479.7012	9.3200	80462.8984	9.8200	242370.9844
8.3300	9872.1152	8.8300	28068.4609	9.3300	82234.7188	9.8300	247844.5156
8.3400	10077.5205	8.8400	28670.1797	9.3400	84046.5312	9.8400	253444.3906
8.3500	10287.3096	8.8500	29285.0918	9.3500	85899.0312	9.8500	259173.0625
8.3600	10501.6201	8.8600	29913.6172	9.3600	87793.5469	9.8600	265034.6250
8.3700	10720.5322	8.8700	30555.9902	9.3700	89730.8516	9.8700	271031.6250
8.3800	10944.1436	8.8800	31212.5391	9.3800	91711.9297	9.8800	277167.3125
8.3900	11172.5430	8.8900	31883.5117	9.3900	93737.6328	9.8900	283444.3438
8.4000	11405.8721	8.9000	32569.3535	9.4000	95809.3203	9.9000	289867.2188
8.4100	11644.2227	8.9100	33270.3555	9.4100	97927.9297	9.9100	296438.8438
8.4200	11887.7080	8.9200	33986.8477	9.4200	100094.4922	9.9200	303162.7500
8.4300	12136.4102	8.9300	34719.1172	9.4300	102309.9219	9.9300	310041.6562
8.4400	12390.4961	8.9400	35467.6445	9.4400	104575.7812	9.9400	317080.7500
8.4500	12650.0625	8.9500	36232.7461	9.4500	106893.0078	9.9500	324283.0938
8.4600	12915.2285	8.9600	37014.7852	9.4600	109262.8281	9.9600	331652.4688
8.4700	13186.1191	8.9700	37814.1406	9.4700	111686.1875	9.9700	339192.1875
8.4800	13462.8301	8.9800	38631.1328	9.4800	114164.8047	9.9800	346907.5625
8.4900	13745.5537	8.9900	39466.3086	9.4900	116699.7344	9.9900	354802.0625
8.5000	14034.3877	9.0000	40320.0000	9.5000	119292.2969	10.0000	362880.0000

APPENDIX B

Coefficients of b_i's for $i = 1, \ldots, n$*

* From Hassanein, K. M., Saleh, A. K., and Brown, E. (1995), "Best Linear Unbiased Estimate and Confidence Interval for Rayleigh's Scale Parameter When the Threshold Parameter Is Known for Data with Censored Observations from the Right," Report from the University of Kansas Medical School. Reproduced by permission from the authors.

The best linear unbiased estimator $\theta_2^* = \sum b_i y_{n(i)}$ when the threshold parameter θ_1 is known based on k order statistics with optimum ranks from the Rayleigh distribution for sample sizes $n = 5(1)25(5)45$ with censoring from the right for $r = 0(1)n - 2$.

i	b_i
$n=5$	$r=0$
1	0.06038
2	0.09279
3	0.12404
4	0.16048
5	0.23520
$n=5$	$r=1$
1	0.07530
2	0.11534
3	0.15324
4	0.43358
$n=5$	$r=2$
1	0.09957
2	0.15151
3	0.67477
$n=5$	$r=3$
1	0.14600
2	1.03016
$n=6$	$r=0$
1	0.04599
2	0.06969
3	0.09142
4	0.11434
5	0.14236
6	0.20229
$n=6$	$r=1$
1	0.05512
2	0.08341
3	0.10912
4	0.13579
5	0.36696
$n=6$	$r=2$
1	0.06864
2	0.10354
3	0.13477
4	0.55338
$n=6$	$r=3$
1	0.09063
2	0.13594
3	0.79308
$n=6$	$r=4$
1	0.13274
2	1.15833
$n=7$	$r=0$
1	0.03655
2	0.05485
3	0.07110
4	0.08739
5	0.10537
6	0.12799
7	0.17785
$n=7$	$r=1$
1	0.04261
2	0.06390
3	0.08272
4	0.10145
5	0.12177
6	0.31934
$n=7$	$r=2$
1	0.05103
2	0.07640
3	0.09867
4	0.12049
5	0.47254
$n=7$	$r=3$
1	0.06348
2	0.09478
3	0.12187
4	0.65522
$n=7$	$r=4$
1	0.08373
2	0.12440
3	0.89716
$n=7$	$r=5$
1	0.12255
2	1.27394
$n=8$	$r=0$
1	0.02996
2	0.04464
3	0.05739
4	0.06977
5	0.08274
6	0.09746
7	0.11634
8	0.15894
$n=8$	$r=1$
1	0.03422
2	0.05097
3	0.06549
4	0.07953
5	0.09412
6	0.11041
7	0.28344
$n=8$	$r=2$
1	0.03988
2	0.05935
3	0.07615
4	0.09228
5	0.10882
6	0.41420
$n=8$	$r=3$
1	0.04773
2	0.07092
3	0.09079
4	0.10964
5	0.56289
$n=8$	$r=4$
1	0.05933
2	0.08793
3	0.11216
4	0.74509
$n=8$	$r=5$
1	0.07820
2	0.11540
3	0.99106
$n=8$	$r=6$
1	0.11440
2	1.38007
$n=9$	$r=0$
1	0.02514
2	0.03725
3	0.04761
4	0.05745
5	0.06743
6	0.07816
7	0.09057
8	0.10672
9	0.14384
$n=9$	$r=1$
1	0.02827
2	0.04189
3	0.05351
4	0.06453
5	0.07566
6	0.08755
7	0.10106
8	0.25529
$n=9$	$r=2$
1	0.03229
2	0.04781
3	0.06103
4	0.07351
5	0.08604
6	0.09924
7	0.36983
$n=9$	$r=3$
1	0.03762
2	0.05565
3	0.07094
4	0.08529
5	0.09952
6	0.49590
$n=9$	$r=4$
1	0.04500
2	0.06647
3	0.08456
4	0.10137
5	0.64294
$n=9$	$r=5$
1	0.05590
2	0.08238
3	0.10448
4	0.82629
$n=9$	$r=6$
1	0.07364
2	0.10812
3	1.07720
$n=9$	$r=7$
1	0.10768
2	1.47869
$n=10$	$r=0$
1	0.02149
2	0.03171
3	0.04034
4	0.04841
5	0.05641
6	0.06475
7	0.07388
8	0.08457
9	0.09865
10	0.13149
$n=10$	$r=1$
1	0.02387
2	0.03522
3	0.04479
4	0.05373
5	0.06258
6	0.07176
7	0.08174
8	0.09325
9	0.23258
$n=10$	$r=2$
1	0.02685
2	0.03959
3	0.05033
4	0.06033
5	0.07019
6	0.08036
7	0.09127
8	0.33479
$n=10$	$r=3$
1	0.03066
2	0.04518
3	0.05739
4	0.06872
5	0.07982
6	0.09114
7	0.44466
$n=10$	$r=4$
1	0.03570
2	0.05256
3	0.06668
4	0.07971
5	0.09237
6	0.56855
$n=10$	$r=5$
1	0.04268
2	0.06276
3	0.07948
4	0.09477
5	0.71543
$n=10$	$r=6$
1	0.05301
2	0.07777
3	0.09821
4	0.90085
$n=10$	$r=7$
1	0.06980
2	0.10207
3	1.15721
$n=10$	$r=8$
1	0.10203
2	1.57120
$n=11$	$r=0$
1	0.01865
2	0.02742
3	0.03476
4	0.04153
5	0.04815
6	0.05488
7	0.06202
8	0.06994
9	0.07932
10	0.09177
11	0.12119
$n=11$	$r=1$
1	0.02051
2	0.03015
3	0.03821
4	0.04565
5	0.05290
6	0.06027
7	0.06805
8	0.07663
9	0.08662
10	0.21383
$n=11$	$r=2$
1	0.02279
2	0.03349
3	0.04243
4	0.05066
5	0.05867
6	0.06678
7	0.07529
8	0.08456
9	0.30633
$n=11$	$r=3$
1	0.02562
2	0.03764
3	0.04766
4	0.05687
5	0.06580
6	0.07479
7	0.08413
8	0.40400
$n=11$	$r=4$
1	0.02925
2	0.04294
3	0.05433
4	0.06476
5	0.07482
6	0.08487
7	0.51146
$n=11$	$r=5$
1	0.03405
2	0.04994
3	0.06312
4	0.07513
5	0.08662
6	0.63451
$n=11$	$r=6$
1	0.04069
2	0.05962
3	0.07522
4	0.08934
5	0.78208
$n=11$	$r=7$
1	0.05052
2	0.07386
3	0.09296
4	0.97012
$n=11$	$r=8$
1	0.06650
2	0.09693
3	1.23222
$n=11$	$r=9$
1	0.09718
2	1.65860
$n=12$	$r=0$
1	0.01639
2	0.02402
3	0.03036
4	0.03616
5	0.04175
6	0.04735
7	0.05315
8	0.05938
9	0.06636
10	0.07469
11	0.08583
12	0.11246
$n=12$	$r=1$
1	0.01788
2	0.02620
3	0.03310
4	0.03942
5	0.04550
6	0.05159
7	0.05789
8	0.06462
9	0.07212
10	0.08092
11	0.19806
$n=12$	$r=2$
1	0.01966
2	0.02881
3	0.03640
4	0.04332
5	0.04999
6	0.05665
7	0.06351
8	0.07080
9	0.07882
10	0.28269
$n=12$	$r=3$
1	0.02184
2	0.03199
3	0.04040
4	0.04807
5	0.05543
6	0.06276
7	0.07027
8	0.07817
9	0.37083
$n=12$	$r=4$
1	0.02455
2	0.03595
3	0.04537
4	0.05395
5	0.06216
6	0.07029
7	0.07855
8	0.46601
$n=12$	$r=5$
1	0.02801
2	0.04100
3	0.05171
4	0.06143
5	0.07069
6	0.07979
7	0.57226
$n=12$	$r=6$
1	0.03260
2	0.04768
3	0.06007
4	0.07126
5	0.08186
6	0.69526
$n=12$	$r=7$
1	0.03896
2	0.05690
3	0.07159

i	b_i
4	0.08476
5	0.84405
$n=12$	$r=8$
1	0.04835
2	0.07049
3	0.08848
4	1.03506
$n=12$	$r=9$
1	0.06362
2	0.09251
3	1.30303
$n=12$	$r=10$
1	0.09297
2	1.74166
$n=13$	$r=0$
1	0.01455
2	0.02127
3	0.02682
4	0.03185
5	0.03666
6	0.04142
7	0.04627
8	0.05136
9	0.05688
10	0.06311
11	0.07059
12	0.08066
13	0.10496
$n=13$	$r=1$
1	0.01576
2	0.02304
3	0.02904
4	0.03449
5	0.03969
6	0.04483
7	0.05007
8	0.05555
9	0.06147
10	0.06811
11	0.07597
12	0.18460
$n=13$	$r=2$
1	0.01719
2	0.02513
3	0.03167
4	0.03760
5	0.04326
6	0.04885
7	0.05453
8	0.06045
9	0.06681
10	0.07386
11	0.26271
$n=13$	$r=3$
1	0.01891
2	0.02763

i	b_i
3	0.03481
4	0.04132
5	0.04752
6	0.05363
7	0.05981
8	0.06623
9	0.07306
10	0.34317
$n=13$	$r=4$
1	0.02100
2	0.03068
3	0.03864
4	0.04584
5	0.05269
6	0.05941
7	0.06619
8	0.07317
9	0.42880
$n=13$	$r=5$
1	0.02360
2	0.03447
3	0.04339
4	0.05144
5	0.05908
6	0.06655
7	0.07402
8	0.52258
$n=13$	$r=6$
1	0.02692
2	0.03930
3	0.04945
4	0.05857
5	0.06719
6	0.07557
7	0.62836
$n=13$	$r=7$
1	0.03133
2	0.04570
3	0.05743
4	0.06795
5	0.07782
6	0.75181
$n=13$	$r=8$
1	0.03743
2	0.05453
3	0.06844
4	0.08083
5	0.90218
$n=13$	$r=9$
1	0.04643
2	0.06754
3	0.08460
4	1.09637
$n=13$	$r=10$
1	0.06109
2	0.08864

i	b_i
3	1.37029
$n=13$	$r=11$
1	0.08925
2	1.82095
$n=14$	$r=0$
1	0.01303
2	0.01901
3	0.02392
4	0.02834
5	0.03254
6	0.03665
7	0.04079
8	0.04506
9	0.04959
10	0.05453
11	0.06015
12	0.06693
13	0.07611
14	0.09844
$n=14$	$r=1$
1	0.01403
2	0.02047
3	0.02575
4	0.03051
5	0.03502
6	0.03944
7	0.04389
8	0.04848
9	0.05332
10	0.05859
11	0.06455
12	0.07163
13	0.17297
$n=14$	$r=2$
1	0.01520
2	0.02217
3	0.02789
4	0.03304
5	0.03792
6	0.04269
7	0.04749
8	0.05243
9	0.05762
10	0.06325
11	0.06953
12	0.24558
$n=14$	$r=3$
1	0.01658
2	0.02418
3	0.03041
4	0.03602
5	0.04133
6	0.04651
7	0.05171
8	0.05704
9	0.06263
10	0.06863
11	0.31971

i	b_i
$n=14$	$r=4$
1	0.01823
2	0.02659
3	0.03342
4	0.03958
5	0.04539
6	0.05106
7	0.05673
8	0.06251
9	0.06854
10	0.39769
$n=14$	$r=5$
1	0.02024
2	0.02951
3	0.03709
4	0.04390
5	0.05032
6	0.05656
7	0.06278
8	0.06909
9	0.48182
$n=14$	$r=6$
1	0.02275
2	0.03315
3	0.04165
4	0.04926
5	0.05642
6	0.06336
7	0.07023
8	0.57487
$n=14$	$r=7$
1	0.02595
2	0.03780
3	0.04745
4	0.05608
5	0.06417
6	0.07197
7	0.68065
$n=14$	$r=8$
1	0.03019
2	0.04394
3	0.05511
4	0.06506
5	0.07434
6	0.80489
$n=14$	$r=9$
1	0.03606
2	0.05243
3	0.06567
4	0.07741
5	0.95708
$n=14$	$r=10$
1	0.04473
2	0.06494
3	0.08118
4	1.15458

i	b_i
$n=14$	$r=11$
1	0.05884
2	0.08522
3	1.43445
$n=14$	$r=12$
1	0.08596
2	1.89695
$n=15$	$r=0$
1	0.01176
2	0.01713
3	0.02151
4	0.02544
5	0.02914
6	0.03274
7	0.03633
8	0.03999
9	0.04381
10	0.04788
11	0.05235
12	0.05745
13	0.06365
14	0.07208
15	0.09273
$n=15$	$r=1$
1	0.01260
2	0.01835
3	0.02304
4	0.02725
5	0.03121
6	0.03506
7	0.03890
$n=10$	$r=5$
9	0.04688
10	0.05122
11	0.05596
12	0.06135
13	0.06779
14	0.16280
$n=15$	$r=2$
1	0.01357
2	0.01976
3	0.02481
4	0.02933
5	0.03359
6	0.03774
7	0.04186
8	0.04605
9	0.05041
10	0.05503
11	0.06006
12	0.06571
13	0.23071
$n=15$	$r=3$
1	0.01470
2	0.02140
3	0.02686
4	0.03176
5	0.03637
6	0.04084

i	b_i
7	0.04529
8	0.04980
9	0.05447
10	0.05941
11	0.06474
12	0.29953
$n=15$	$r=4$
1	0.01603
2	0.02334
3	0.02929
4	0.03462
5	0.03963
6	0.04449
7	0.04931
8	0.05419
9	0.05922
10	0.06449
11	0.37123
$n=15$	$r=5$
1	0.01763
2	0.02565
3	0.03219
4	0.03804
5	0.04353
6	0.04884
7	0.05409
8	0.05939
9	0.06482
10	0.44766
$n=15$	$r=6$
1	0.01957
2	0.02847
3	0.03572
4	0.04219
5	0.04825
6	0.05410
7	0.05987
8	0.06566
9	0.53091
$n=15$	$r=7$
1	0.02199
2	0.03198
3	0.04010
4	0.04734
5	0.05410
6	0.06061
7	0.06699
8	0.62367
$n=15$	$r=8$
1	0.02508
2	0.03646
3	0.04568
4	0.05389
5	0.06153
6	0.06885
7	0.72977
$n=15$	$r=9$
1	0.02917

i	b_i
2	0.04238
3	0.05305
4	0.06252
5	0.07129
6	0.85505
$n=15$	$r=10$
1	0.03484
2	0.05056
3	0.06322
4	0.07439
5	1.00921
$n=15$	$r=11$
1	0.04321
2	0.06261
3	0.07816
4	1.21011
$n=15$	$r=12$
1	0.05683
2	0.08216
3	1.49592
$n=15$	$r=13$
1	0.08300
2	1.97003
$n=16$	$r=0$
1	0.01069
2	0.01554
3	0.01948
4	0.02300
5	0.02630
6	0.02949
7	0.03264
8	0.03583
9	0.03910
10	0.04254
11	0.04624
12	0.05031
13	0.05499
14	0.06069
15	0.06847
16	0.08767
$n=16$	$r=1$
1	0.01140
2	0.01657
3	0.02077
4	0.02453
5	0.02804
6	0.03144
7	0.03480
8	0.03819
9	0.04168
10	0.04533
11	0.04925
12	0.05355
13	0.05847
14	0.06436
15	0.15384

i	b_i
$n=16$	$r=2$
1	0.01221
2	0.01775
3	0.02225
4	0.02627
5	0.03004
6	0.03367
7	0.03726
8	0.04088
9	0.04460
10	0.04849
11	0.05264
12	0.05719
13	0.06232
14	0.21767
$n=16$	$r=3$
1	0.01315
2	0.01911
3	0.02396
4	0.02828
5	0.03233
6	0.03623
7	0.04009
8	0.04397
9	0.04795
10	0.05210
11	0.05651
12	0.06130
13	0.28196
$n=16$	$r=4$
1	0.01424
2	0.02070
3	0.02594
4	0.03062
5	0.03500
6	0.03921
7	0.04337
8	0.04755
9	0.05182
10	0.05626
11	0.06094
12	0.34841
$n=16$	$r=5$
1	0.01553
2	0.02257
3	0.02828
4	0.03337
5	0.03813
6	0.04271
7	0.04722
8	0.05174
9	0.05634
10	0.06110
11	0.41855
$n=16$	$r=6$
1	0.01708
2	0.02481
3	0.03108
4	0.03666
5	0.04188
6	0.04688

i	b_i
7	0.05180
8	0.05672
9	0.06170
10	0.49401

$n = 16 \quad r = 7$

i	b_i
1	0.01896
2	0.02753
3	0.03448
4	0.04066
5	0.04642
6	0.05194
7	0.05734
8	0.06272
9	0.57679

$n = 16 \quad r = 8$

i	b_i
1	0.02130
2	0.03092
3	0.03871
4	0.04562
5	0.05205
6	0.05819
7	0.06418
8	0.66957

$n = 16 \quad r = 9$

i	b_i
1	0.02429
2	0.03525
3	0.04410
4	0.05194
5	0.05920
6	0.06612
7	0.77622

$n = 16 \quad r = 10$

i	b_i
1	0.02825
2	0.04097
3	0.05121
4	0.06025
5	0.06860
6	0.90270

$n = 16 \quad r = 11$

i	b_i
1	0.03373
2	0.04887
3	0.06102
4	0.07170
5	1.05894

$n = 16 \quad r = 12$

i	b_i
1	0.04182
2	0.06052
3	0.07545
4	1.26329

$n = 16 \quad r = 13$

i	b_i
1	0.05500
2	0.07942
3	1.55500

$n = 16 \quad r = 14$

i	b_i
1	0.08032
2	2.04050

$n = 17 \quad r = 0$

i	b_i
1	0.00977
2	0.01418
3	0.01775
4	0.02093
5	0.02389
6	0.02674
7	0.02955
8	0.03235
9	0.03521
10	0.03818
11	0.04131
12	0.04468
13	0.04843
14	0.05274
15	0.05801
16	0.06523
17	0.08316

$n = 17 \quad r = 1$

i	b_i
1	0.01037
2	0.01506
3	0.01886
4	0.02223
5	0.02538
6	0.02841
7	0.03138
8	0.03436
9	0.03739
10	0.04053
11	0.04384
12	0.04741
13	0.05135
14	0.05586
15	0.06129
16	0.14587

$n = 17 \quad r = 2$

i	b_i
1	0.01107
2	0.01606
3	0.02011
4	0.02371
5	0.02706
6	0.03029
7	0.03346
8	0.03662
9	0.03985
10	0.04318
11	0.04669
12	0.05046
13	0.05459
14	0.05928
15	0.20613

$n = 17 \quad r = 3$

i	b_i
1	0.01186
2	0.01721
3	0.02154
4	0.02539
5	0.02899
6	0.03243
7	0.03582
8	0.03920
9	0.04264
10	0.04619
11	0.04991
12	0.05389
13	0.05823
14	0.26650

$n = 17 \quad r = 4$

i	b_i
1	0.01277
2	0.01853
3	0.02319
4	0.02734
5	0.03120
6	0.03490
7	0.03854
8	0.04216
9	0.04584
10	0.04963
11	0.05359
12	0.05779
13	0.32850

$n = 17 \quad r = 5$

i	b_i
1	0.01383
2	0.02006
3	0.02511
4	0.02960
5	0.03377
6	0.03777
7	0.04169
8	0.04559
9	0.04954
10	0.05360
11	0.05781
12	0.39340

$n = 17 \quad r = 6$

i	b_i
1	0.01508
2	0.02188
3	0.02737
4	0.03225
5	0.03679
6	0.04114
7	0.04539
8	0.04961
9	0.05388
10	0.05823
11	0.46251

$n = 17 \quad r = 7$

i	b_i
1	0.01657
2	0.02404
3	0.03008
4	0.03543
5	0.04040
6	0.04515
7	0.04979
8	0.05439
9	0.05901
10	0.53738

$n = 17 \quad r = 8$

i	b_i
1	0.01840
2	0.02668
3	0.03337
4	0.03929
5	0.04478
6	0.05002
7	0.05512
8	0.06016
9	0.61998

$n = 17 \quad r = 9$

i	b_i
1	0.02067
2	0.02996
3	0.03746
4	0.04408
5	0.05021
6	0.05605
7	0.06170
8	0.71300

$n = 17 \quad r = 10$

i	b_i
1	0.02357
2	0.03415
3	0.04267
4	0.05018
5	0.05712
6	0.06369
7	0.82036

$n = 17 \quad r = 11$

i	b_i
1	0.02741
2	0.03969
3	0.04955
4	0.05822
5	0.06620
6	0.94816

$n = 17 \quad r = 12$

i	b_i
1	0.03272
2	0.04734
3	0.05904
4	0.06929
5	1.10656

$n = 17 \quad r = 13$

i	b_i
1	0.04057
2	0.05862
3	0.07300
4	1.31439

$n = 17 \quad r = 14$

i	b_i
1	0.05334
2	0.07693
3	1.61195

$n = 17 \quad r = 15$

i	b_i
1	0.07790
2	2.10863

$n = 18 \quad r = 0$

i	b_i
1	0.00897
2	0.01301
3	0.01626
4	0.01915
5	0.02184
6	0.02441
7	0.02692
8	0.02942
9	0.03194
10	0.03453
11	0.03724
12	0.04010
13	0.04321
14	0.04667
15	0.05067
16	0.05557
17	0.06230
18	0.07911

$n = 18 \quad r = 1$

i	b_i
1	0.00950
2	0.01377
3	0.01722
4	0.02027
5	0.02311
6	0.02583
7	0.02849
8	0.03113
9	0.03380
10	0.03654
11	0.03940
12	0.04242
13	0.04569
14	0.04932
15	0.05348
16	0.05852
17	0.13873

$n = 18 \quad r = 2$

i	b_i
1	0.01009
2	0.01463
3	0.01829
4	0.02154
5	0.02455
6	0.02744
7	0.03026
8	0.03306
9	0.03589
10	0.03879
11	0.04181
12	0.04500
13	0.04844
14	0.05224
15	0.05655
16	0.19584

$n = 18 \quad r = 3$

i	b_i
1	0.01076
2	0.01560
3	0.01951
4	0.02297
5	0.02618
6	0.02926
7	0.03226
8	0.03524
9	0.03825
10	0.04133
11	0.04453
12	0.04790
13	0.05152
14	0.05548
15	0.25279

$n = 18 \quad r = 4$

i	b_i
1	0.01153
2	0.01671
3	0.02090
4	0.02460
5	0.02804
6	0.03133
7	0.03454
8	0.03772
9	0.04093
10	0.04421
11	0.04761
12	0.05117
13	0.05498
14	0.31094

$n = 18 \quad r = 5$

i	b_i
1	0.01241
2	0.01799
3	0.02249
4	0.02648
5	0.03018
6	0.03371
7	0.03715
8	0.04057
9	0.04400
10	0.04750
11	0.05112
12	0.05490
13	0.37140

$n = 18 \quad r = 6$

i	b_i
1	0.01344
2	0.01949
3	0.02436
4	0.02867
5	0.03266
6	0.03648
7	0.04019
8	0.04387
9	0.04756
10	0.05131
11	0.05517
12	0.43525

$n = 18 \quad r = 7$

i	b_i
1	0.01466
2	0.02124
3	0.02655
4	0.03124
5	0.03559
6	0.03973
7	0.04376
8	0.04774
9	0.05172
10	0.05576
11	0.50370

$n = 18 \quad r = 8$

i	b_i
1	0.01611
2	0.02335
3	0.02917
4	0.03432
5	0.03908
6	0.04361
7	0.04800
8	0.05234
9	0.05666
10	0.57825

$n = 18 \quad r = 9$

i	b_i
1	0.01789
2	0.02591
3	0.03236
4	0.03805
5	0.04331
6	0.04831
7	0.05315
8	0.05790
9	0.66088

$n = 18 \quad r = 10$

i	b_i
1	0.02009
2	0.02909
3	0.03632
4	0.04269
5	0.04856
6	0.05413
7	0.05950
8	0.75430

$n = 18 \quad r = 11$

i	b_i
1	0.02291
2	0.03315
3	0.04137
4	0.04860
5	0.05524
6	0.06152
7	0.86250

$n = 18 \quad r = 12$

i	b_i
1	0.02663
2	0.03852
3	0.04804
4	0.05638
5	0.06403
6	0.99171

$n = 18 \quad r = 13$

i	b_i
1	0.03180
2	0.04595
3	0.05724
4	0.06711
5	1.15232

$n = 18 \quad r = 14$

i	b_i
1	0.03942
2	0.05690
3	0.07078
4	1.36364

$n = 18 \quad r = 15$

i	b_i
1	0.05183
2	0.07466
3	1.66698

$n = 18 \quad r = 16$

i	b_i
1	0.07567
2	2.17464

$n = 19 \quad r = 0$

i	b_i
1	0.00828
2	0.01199
3	0.01497
4	0.01761
5	0.02006
6	0.02239
7	0.02466
8	0.02690
9	0.02916
10	0.03145
11	0.03382
12	0.03631
13	0.03895
14	0.04182
15	0.04504
16	0.04876
17	0.05333
18	0.05964
19	0.07545

$n = 19 \quad r = 1$

i	b_i
1	0.00874
2	0.01265
3	0.01580
4	0.01859
5	0.02117
6	0.02363
7	0.02602
8	0.02839
9	0.03076
10	0.03318
11	0.03568
12	0.03829
13	0.04107
14	0.04409
15	0.04744
16	0.05131
17	0.05600
18	0.13231

$n = 19 \quad r = 2$

i	b_i
1	0.00925
2	0.01339
3	0.01673
4	0.01968
5	0.02241
6	0.02501
7	0.02754
8	0.03005
9	0.03256
10	0.03511
11	0.03775
12	0.04050
13	0.04342
14	0.04659
15	0.05009
16	0.05408
17	0.18660

$n = 19$ $r = 3$

i	b_i
1	0.00983
2	0.01423
3	0.01777
4	0.02091
5	0.02380
6	0.02657
7	0.02925
8	0.03191
9	0.03457
10	0.03727
11	0.04006
12	0.04297
13	0.04604
14	0.04936
15	0.05300
16	0.24053

$n = 19$ $r = 4$

i	b_i
1	0.01048
2	0.01518
3	0.01896
4	0.02229
5	0.02538
6	0.02832
7	0.03118
8	0.03401
9	0.03684
10	0.03971
11	0.04266
12	0.04573
13	0.04898
14	0.05245
15	0.29534

$n = 19$ $r = 5$

i	b_i
1	0.01123
2	0.01626
3	0.02030
4	0.02388
5	0.02718
6	0.03033
7	0.03338
8	0.03640
9	0.03942
10	0.04248
11	0.04561
12	0.04887
13	0.05228
14	0.35199

$n = 19$ $r = 6$

i	b_i
1	0.01209
2	0.01750
3	0.02186
4	0.02570
5	0.02925
6	0.03263
7	0.03591
8	0.03915
9	0.04238
10	0.04564
11	0.04899
12	0.05244
13	0.41138

$n = 19$ $r = 7$

i	b_i
1	0.01309
2	0.01895
3	0.02366
4	0.02782
5	0.03166
6	0.03531
7	0.03885
8	0.04233
9	0.04580
10	0.04931
11	0.05288
12	0.47451

$n = 19$ $r = 8$

i	b_i
1	0.01428
2	0.02066
3	0.02579
4	0.03032
5	0.03449
6	0.03846
7	0.04229
8	0.04606
9	0.04982
10	0.05359
11	0.54255

$n = 19$ $r = 9$

i	b_i
1	0.01569
2	0.02271
3	0.02834
4	0.03330
5	0.03787
6	0.04221
7	0.04640
8	0.05051
9	0.05459
10	0.61699

$n = 19$ $r = 10$

i	b_i
1	0.01741
2	0.02519
3	0.03143
4	0.03692
5	0.04198
6	0.04676
7	0.05137
8	0.05588
9	0.69980

$n = 19$ $r = 11$

i	b_i
1	0.01956
2	0.02829
3	0.03528
4	0.04142
5	0.04707
6	0.05240
7	0.05752
8	0.79374

$n = 19$ $r = 12$

i	b_i
1	0.02230
2	0.03224
3	0.04018
4	0.04715
5	0.05354
6	0.05956
7	0.90288

$n = 19$ $r = 13$

i	b_i
1	0.02593
2	0.03745
3	0.04666
4	0.05471
5	0.06207
6	1.03355

$n = 19$ $r = 14$

i	b_i
1	0.03095
2	0.04467
3	0.05560
4	0.06512
5	1.19641

$n = 19$ $r = 15$

i	b_i
1	0.03836
2	0.05531
3	0.06875
4	1.41121

$n = 19$ $r = 16$

i	b_i
1	0.05043
2	0.07259
3	1.72027

$n = 19$ $r = 17$

i	b_i
1	0.07363
2	2.23870

$n = 20$ $r = 0$

i	b_i
1	0.00767
2	0.01110
3	0.01385
4	0.01627
5	0.01851
6	0.02064
7	0.02270
8	0.02473
9	0.02676
10	0.02882
11	0.03092
12	0.03310
13	0.03539
14	0.03784
15	0.04052
16	0.04351
17	0.04699
18	0.05128
19	0.05721
20	0.07214

$n = 20$ $r = 1$

i	b_i
1	0.00807
2	0.01168
3	0.01457
4	0.01713
5	0.01948
6	0.02172
7	0.02389
8	0.02603
9	0.02816
10	0.03032
11	0.03253
12	0.03482
13	0.03723
14	0.03980
15	0.04259
16	0.04571
17	0.04932
18	0.05370
19	0.12648

$n = 20$ $r = 2$

i	b_i
1	0.00852
2	0.01233
3	0.01538
4	0.01807
5	0.02056
6	0.02292
7	0.02521
8	0.02746
9	0.02971
10	0.03199
11	0.03431
12	0.03672
13	0.03925
14	0.04195
15	0.04487
16	0.04812
17	0.05183
18	0.17825

$n = 20$ $r = 3$

i	b_i
1	0.00902
2	0.01305
3	0.01628
4	0.01914
5	0.02177
6	0.02426
7	0.02669
8	0.02907
9	0.03145
10	0.03385
11	0.03630
12	0.03884
13	0.04150
14	0.04433
15	0.04738
16	0.05075
17	0.22950

$n = 20$ $r = 4$

i	b_i
1	0.00958
2	0.01386
3	0.01730
4	0.02033
5	0.02312
6	0.02577
7	0.02834
8	0.03087
9	0.03339
10	0.03593
11	0.03852
12	0.04120
13	0.04401
14	0.04697
15	0.05016
16	0.28137

$n = 20$ $r = 5$

i	b_i
1	0.01022
2	0.01479
3	0.01845
4	0.02168
5	0.02465
6	0.02748
7	0.03021
8	0.03290
9	0.03558
10	0.03828
11	0.04103
12	0.04386
13	0.04682
14	0.04993
15	0.33470

$n = 20$ $r = 6$

i	b_i
1	0.01095
2	0.01584
3	0.01976
4	0.02321
5	0.02640
6	0.02942
7	0.03234
8	0.03521
9	0.03807
10	0.04094
11	0.04387
12	0.04687
13	0.04999
14	0.39028

$n = 20$ $r = 7$

i	b_i
1	0.01179
2	0.01705
3	0.02127
4	0.02499
5	0.02841
6	0.03165
7	0.03479
8	0.03787
9	0.04093
10	0.04400
11	0.04712
12	0.05031
13	0.44893

$n = 20$ $r = 8$

i	b_i
1	0.01277
2	0.01846
3	0.02303
4	0.02704
5	0.03074
6	0.03425
7	0.03763
8	0.04095
9	0.04424
10	0.04753
11	0.05087
12	0.51159

$n = 20$ $r = 9$

i	b_i
1	0.01392
2	0.02012
3	0.02510
4	0.02947
5	0.03349
6	0.03730
7	0.04097
8	0.04456
9	0.04812
10	0.05167
11	0.57941

$n = 20$ $r = 10$

i	b_i
1	0.01530
2	0.02211
3	0.02757
4	0.03237
5	0.03678
6	0.04094
7	0.04495
8	0.04886
9	0.05273
10	0.65388

$n = 20$ $r = 11$

i	b_i
1	0.01698
2	0.02454
3	0.03058
4	0.03589
5	0.04076
6	0.04536
7	0.04977
8	0.05407
9	0.73699

$n = 20$ $r = 12$

i	b_i
1	0.01907
2	0.02755
3	0.03432
4	0.04026
5	0.04570
6	0.05083
7	0.05573
8	0.83154

$n = 20$ $r = 13$

i	b_i
1	0.02174
2	0.03139
3	0.03909
4	0.04583
5	0.05199
6	0.05778
7	0.94168

$n = 20$ $r = 14$

i	b_i
1	0.02527
2	0.03647
3	0.04539
4	0.05318
5	0.06027
6	1.07387

$n = 20$ $r = 15$

i	b_i
1	0.03016
2	0.04350
3	0.05409
4	0.06330
5	1.23900

$n = 20$ $r = 16$

i	b_i
1	0.03738
2	0.05385
3	0.06688
4	1.45726

$n = 20$ $r = 17$

i	b_i
1	0.04914
2	0.07067
3	1.77197

$n = 20$ $r = 18$

i	b_i
1	0.07175
2	2.30099

$n = 21$ $r = 0$

i	b_i
1	0.00713
2	0.01031
3	0.01285
4	0.01509
5	0.01715
6	0.01911
7	0.02099
8	0.02284
9	0.02468
10	0.02654
11	0.02842
12	0.03036
13	0.03237
14	0.03450
15	0.03679
16	0.03929
17	0.04209
18	0.04536
19	0.04939
20	0.05498
21	0.06911

$n = 21$ $r = 1$

i	b_i
1	0.00749
2	0.01082
3	0.01349
4	0.01584
5	0.01801
6	0.02006
7	0.02204
8	0.02398
9	0.02591
10	0.02785
11	0.02983
12	0.03186
13	0.03397
14	0.03620
15	0.03859
16	0.04120
17	0.04411
18	0.04749
19	0.05160
20	0.12118

$n = 21$ $r = 2$

i	b_i
1	0.00788
2	0.01139
3	0.01420
4	0.01668
5	0.01895
6	0.02111
7	0.02319
8	0.02523
9	0.02727
10	0.02931
11	0.03138
12	0.03351
13	0.03573
14	0.03807
15	0.04056
16	0.04328
17	0.04631
18	0.04977
19	0.17067

$n = 21$ $r = 3$

i	b_i
1	0.00832
2	0.01202
3	0.01499
4	0.01760
5	0.02000
6	0.02228
7	0.02447
8	0.02663
9	0.02877
10	0.03092
11	0.03310
12	0.03535
13	0.03768
14	0.04012
15	0.04274
16	0.04557
17	0.04869
18	0.21952

$n = 21$ $r = 4$

i	b_i
1	0.00881
2	0.01273
3	0.01587
4	0.01863
5	0.02118
6	0.02358
7	0.02590
8	0.02818
9	0.03044
10	0.03271
11	0.03502
12	0.03738
13	0.03983
14	0.04240
15	0.04514
16	0.04808
17	0.26877

$n = 21$ $r = 5$

i	b_i
1	0.00936
2	0.01352

i	b_i
3	0.01686
4	0.01979
5	0.02249
6	0.02505
7	0.02751
8	0.02992
9	0.03232
10	0.03472
11	0.03716
12	0.03966
13	0.04224
14	0.04494
15	0.04780
16	0.31920
$n = 21$ $r = 6$	
1	0.00998
2	0.01442
3	0.01798
4	0.02111
5	0.02398
6	0.02670
7	0.02932
8	0.03189
9	0.03444
10	0.03699
11	0.03958
12	0.04222
13	0.04495
14	0.04779
15	0.37147
$n = 21$ $r = 7$	
1	0.01069
2	0.01545
3	0.01926
4	0.02260
5	0.02568
6	0.02859
7	0.03139
8	0.03413
9	0.03685
10	0.03957
11	0.04232
12	0.04512
13	0.04801
14	0.42629
$n = 21$ $r = 8$	
1	0.01151
2	0.01663
3	0.02073
4	0.02433
5	0.02763
6	0.03076
7	0.03377
8	0.03671
9	0.03962
10	0.04253
11	0.04546
12	0.04844
13	0.48442
$n = 21$ $r = 9$	
1	0.01247
2	0.01801
3	0.02244
4	0.02633
5	0.02990
6	0.03328
7	0.03652
8	0.03969
9	0.04283
10	0.04595
11	0.04909
12	0.54678
$n = 21$ $r = 10$	
1	0.01359
2	0.01963
3	0.02446
4	0.02869
5	0.03258
6	0.03624
7	0.03977
8	0.04320
9	0.04659
10	0.04996
11	0.61453
$n = 21$ $r = 11$	
1	0.01493
2	0.02157
3	0.02687
4	0.03151
5	0.03577
6	0.03978
7	0.04363
8	0.04737
9	0.05106
10	0.68914
$n = 21$ $r = 12$	
1	0.01657
2	0.02393
3	0.02980
4	0.03494
5	0.03964
6	0.04407
7	0.04831
8	0.05243
9	0.77264
$n = 21$ $r = 13$	
1	0.01861
2	0.02686
3	0.03344
4	0.03919
5	0.04445
6	0.04939
7	0.05411
8	0.86788
$n = 21$ $r = 14$	
1	0.02122
2	0.03061
3	0.03809
4	0.04462
5	0.05057
6	0.05615
7	0.97907
$n = 21$ $r = 15$	
1	0.02466
2	0.03556
3	0.04422
4	0.05177
5	0.05863
6	1.11282
$n = 21$ $r = 16$	
1	0.02943
2	0.04241
3	0.05269
4	0.06162
5	1.28023
$n = 21$ $r = 17$	
1	0.03648
2	0.05251
3	0.06516
4	1.50194
$n = 21$ $r = 18$	
1	0.04795
2	0.06891
3	1.82222
$n = 21$ $r = 19$	
1	0.07000
2	2.36163
$n = 22$ $r = 0$	
1	0.00666
2	0.00961
3	0.01198
4	0.01405
5	0.01596
6	0.01776
7	0.01949
8	0.02119
9	0.02287
10	0.02455
11	0.02625
12	0.02799
13	0.02978
14	0.03166
15	0.03365
16	0.03578
17	0.03812
18	0.04076
19	0.04383
20	0.04764
21	0.05292
22	0.06634
$n = 22$ $r = 1$	
1	0.00697
2	0.01007
3	0.01254
4	0.01472
5	0.01671
6	0.01860
7	0.02041
8	0.02219
9	0.02395
10	0.02571
11	0.02749
12	0.02931
13	0.03118
14	0.03315
15	0.03522
16	0.03745
17	0.03989
18	0.04262
19	0.04579
20	0.04967
21	0.11633
$n = 22$ $r = 2$	
1	0.00732
2	0.01057
3	0.01317
4	0.01545
5	0.01755
6	0.01952
7	0.02143
8	0.02329
9	0.02514
10	0.02698
11	0.02885
12	0.03076
13	0.03272
14	0.03478
15	0.03695
16	0.03927
17	0.04181
18	0.04463
19	0.04788
20	0.16375
$n = 22$ $r = 3$	
1	0.00771
2	0.01113
3	0.01386
4	0.01626
5	0.01847
6	0.02055
7	0.02255
8	0.02451
9	0.02645
10	0.02839
11	0.03035
12	0.03235
13	0.03442
14	0.03657
15	0.03884
16	0.04126
17	0.04389
18	0.04681
19	0.21043
$n = 22$ $r = 4$	
1	0.00813
2	0.01174
3	0.01463
4	0.01716
5	0.01949
6	0.02169
7	0.02380
8	0.02586
9	0.02791
10	0.02995
11	0.03202
12	0.03412
13	0.03629
14	0.03855
15	0.04092
16	0.04345
17	0.04618
18	0.25735
$n = 22$ $r = 5$	
1	0.00861
2	0.01243
3	0.01549
4	0.01817
5	0.02063
6	0.02295
7	0.02519
8	0.02737
9	0.02953
10	0.03169
11	0.03387
12	0.03609
13	0.03837
14	0.04074
15	0.04322
16	0.04586
17	0.30520
$n = 22$ $r = 6$	
1	0.00915
2	0.01321
3	0.01645
4	0.01930
5	0.02191
6	0.02438
7	0.02675
8	0.02907
9	0.03135
10	0.03364
11	0.03594
12	0.03829
13	0.04069
14	0.04318
15	0.04579
16	0.35458
$n = 22$ $r = 7$	
1	0.00976
2	0.01409
3	0.01755
4	0.02058
5	0.02336
6	0.02599
7	0.02851
8	0.03098
9	0.03341
10	0.03584
11	0.03828
12	0.04076
13	0.04330
14	0.04593
15	0.40608
$n = 22$ $r = 8$	
1	0.01045
2	0.01509
3	0.01879
4	0.02204
5	0.02502
6	0.02782
7	0.03052
8	0.03315
9	0.03575
10	0.03833
11	0.04093
12	0.04357
13	0.04626
14	0.46035
$n = 22$ $r = 9$	
1	0.01125
2	0.01624
3	0.02023
4	0.02372
5	0.02692
6	0.02993
7	0.03283
8	0.03565
9	0.03843
10	0.04120
11	0.04397
12	0.04678
13	0.51813
$n = 22$ $r = 10$	
1	0.01218
2	0.01758
3	0.02190
4	0.02567
5	0.02913
6	0.03239
7	0.03551
8	0.03855
9	0.04154
10	0.04452
11	0.04749
12	0.58034
$n = 22$ $r = 11$	
1	0.01328
2	0.01917
3	0.02386
4	0.02797
5	0.03173
6	0.03527
7	0.03866
8	0.04196
9	0.04520
10	0.04840
11	0.64812
$n = 22$ $r = 12$	
1	0.01460
2	0.02106
3	0.02621
4	0.03072
5	0.03484
6	0.03871
7	0.04242
8	0.04602
9	0.04954
10	0.72296
$n = 22$ $r = 13$	
1	0.01620
2	0.02336
3	0.02907
4	0.03406
5	0.03862
6	0.04289
7	0.04698
8	0.05093
9	0.80693
$n = 22$ $r = 14$	
1	0.01819
2	0.02623
3	0.03262
4	0.03821
5	0.04330
6	0.04807
7	0.05261
8	0.90291
$n = 22$ $r = 15$	
1	0.02073
2	0.02988
3	0.03716
4	0.04349
5	0.04926
6	0.05465
7	1.01520
$n = 22$ $r = 16$	
1	0.02410
2	0.03471
3	0.04314
4	0.05046
5	0.05711
6	1.15052
$n = 22$ $r = 17$	
1	0.02876
2	0.04140
3	0.05140
4	0.06008
5	1.32022
$n = 22$ $r = 18$	
1	0.03564
2	0.05125
3	0.06357
4	1.54534
$n = 22$ $r = 19$	
1	0.04684
2	0.06726
3	1.87113
$n = 22$ $r = 20$	
1	0.06837
2	2.42076
$n = 23$ $r = 0$	
1	0.00623
2	0.00899
3	0.01119
4	0.01312
5	0.01489
6	0.01656
7	0.01816
8	0.01972
9	0.02126
10	0.02280
11	0.02435
12	0.02592
13	0.02753
14	0.02921
15	0.03096
16	0.03282
17	0.03483
18	0.03703
19	0.03951
20	0.04242
21	0.04602
22	0.05103
23	0.06380
$n = 23$ $r = 1$	
1	0.00651
2	0.00940
3	0.01170
4	0.01372
5	0.01556
6	0.01731
7	0.01898
8	0.02061
9	0.02222
10	0.02383
11	0.02544
12	0.02709
13	0.02877
14	0.03052
15	0.03235
16	0.03429
17	0.03637
18	0.03866
19	0.04124
20	0.04422
21	0.04788
22	0.11187
$n = 23$ $r = 2$	
1	0.00682
2	0.00984
3	0.01226
4	0.01437
5	0.01630
6	0.01813
7	0.01988
8	0.02159
9	0.02327
10	0.02495
11	0.02665
12	0.02836
13	0.03013
14	0.03195
15	0.03386
16	0.03588
17	0.03806

i	b_i
18	0.04043
19	0.04309
20	0.04614
21	0.15740
$n=23$ $r=3$	
1	0.00716
2	0.01034
3	0.01287
4	0.01509
5	0.01712
6	0.01903
7	0.02087
8	0.02266
9	0.02443
10	0.02619
11	0.02797
12	0.02977
13	0.03161
14	0.03352
15	0.03552
16	0.03763
17	0.03989
18	0.04235
19	0.04508
20	0.20213
$n=23$ $r=4$	
1	0.00754
2	0.01088
3	0.01354
4	0.01588
5	0.01802
6	0.02003
7	0.02196
8	0.02385
9	0.02571
10	0.02756
11	0.02942
12	0.03131
13	0.03325
14	0.03525
15	0.03734
16	0.03954
17	0.04189
18	0.04444
19	0.24695
$n=23$ $r=5$	
1	0.00796
2	0.01148
3	0.01430
4	0.01676
5	0.01901
6	0.02114
7	0.02318
8	0.02516
9	0.02712
10	0.02907
11	0.03103
12	0.03302
13	0.03506
14	0.03715
15	0.03934
16	0.04164
17	0.04409
18	0.29250
$n=23$ $r=6$	
1	0.00843
2	0.01216
3	0.01513
4	0.01774
5	0.02013
6	0.02237
7	0.02453
8	0.02663
9	0.02870
10	0.03076
11	0.03283
12	0.03493
13	0.03707
14	0.03927
15	0.04156
16	0.04396
17	0.33931
$n=23$ $r=7$	
1	0.00895
2	0.01291
3	0.01608
4	0.01884
5	0.02138
6	0.02376
7	0.02605
8	0.02828
9	0.03047
10	0.03265
11	0.03484
12	0.03705
13	0.03931
14	0.04163
15	0.04404
16	0.38791
$n=23$ $r=8$	
1	0.00955
2	0.01377
3	0.01714
4	0.02009
5	0.02279
6	0.02533
7	0.02777
8	0.03014
9	0.03247
10	0.03478
11	0.03711
12	0.03945
13	0.04184
14	0.04429
15	0.43885
$n=23$ $r=9$	
1	0.01023
2	0.01475
3	0.01836
4	0.02151
5	0.02440
6	0.02712
7	0.02972
8	0.03225
9	0.03474
10	0.03721
11	0.03968
12	0.04217
13	0.04471
14	0.49274
$n=23$ $r=10$	
1	0.01101
2	0.01588
3	0.01976
4	0.02315
5	0.02626
6	0.02917
7	0.03197
8	0.03468
9	0.03735
10	0.03999
11	0.04263
12	0.04529
13	0.55030
$n=23$ $r=11$	
1	0.01192
2	0.01719
3	0.02139
4	0.02506
5	0.02841
6	0.03156
7	0.03458
8	0.03751
9	0.04038
10	0.04322
11	0.04605
12	0.61246
$n=23$ $r=12$	
1	0.01299
2	0.01874
3	0.02331
4	0.02730
5	0.03095
6	0.03438
7	0.03765
8	0.04082
9	0.04393
10	0.04700
11	0.68035
$n=23$ $r=13$	
1	0.01428
2	0.02058
3	0.02560
4	0.02998
5	0.03398
6	0.03773
7	0.04131
8	0.04477
9	0.04815
10	0.75550
$n=23$ $r=14$	
1	0.01584
2	0.02283
3	0.02839
4	0.03324
5	0.03766
6	0.04180
7	0.04575
8	0.04955
9	0.83999
$n=23$ $r=15$	
1	0.01779
2	0.02563
3	0.03186
4	0.03729
5	0.04223
6	0.04685
7	0.05124
8	0.93675
$n=23$ $r=16$	
1	0.02028
2	0.02921
3	0.03629
4	0.04245
5	0.04805
6	0.05327
7	1.05017
$n=23$ $r=17$	
1	0.02357
2	0.03393
3	0.04213
4	0.04925
5	0.05571
6	1.18709
$n=23$ $r=18$	
1	0.02812
2	0.04046
3	0.05020
4	0.05864
5	1.35906
$n=23$ $r=19$	
1	0.03485
2	0.05009
3	0.06209
4	1.58759
$n=23$ $r=20$	
1	0.04580
2	0.06573
3	1.91881
$n=23$ $r=21$	
1	0.06686
2	2.47849
$n=24$ $r=0$	
1	0.00585
2	0.00843
3	0.01049
4	0.01229
5	0.01394
6	0.01549
7	0.01697
8	0.01842
9	0.01984
10	0.02125
11	0.02267
12	0.02410
13	0.02556
14	0.02707
15	0.02863
16	0.03028
17	0.03203
18	0.03392
19	0.03599
20	0.03834
21	0.04109
22	0.04451
23	0.04927
24	0.06145
$n=24$ $r=1$	
1	0.00610
2	0.00880
3	0.01095
4	0.01282
5	0.01454
6	0.01616
7	0.01771
8	0.01921
9	0.02070
10	0.02217
11	0.02365
12	0.02514
13	0.02666
14	0.02823
15	0.02986
16	0.03157
17	0.03339
18	0.03536
19	0.03751
20	0.03994
21	0.04277
22	0.04623
23	0.10777
$n=24$ $r=2$	
1	0.00638
2	0.00920
3	0.01144
4	0.01341
5	0.01520
6	0.01689
7	0.01851
8	0.02008
9	0.02163
10	0.02317
11	0.02471
12	0.02627
13	0.02786
14	0.02950
15	0.03120
16	0.03298
17	0.03488
18	0.03692
19	0.03915
20	0.04165
21	0.04453
22	0.15157
$n=24$ $r=3$	
1	0.00668
2	0.00964
3	0.01199
4	0.01404
5	0.01593
6	0.01769
7	0.01939
8	0.02104
9	0.02266
10	0.02427
11	0.02588
12	0.02751
13	0.02918
14	0.03089
15	0.03266
16	0.03452
17	0.03649
18	0.03861
19	0.04092
20	0.04348
21	0.19451
$n=24$ $r=4$	
1	0.00702
2	0.01012
3	0.01259
4	0.01475
5	0.01672
6	0.01858
7	0.02036
8	0.02208
9	0.02378
10	0.02547
11	0.02716
12	0.02887
13	0.03061
14	0.03240
15	0.03426
16	0.03620
17	0.03825
18	0.04045
19	0.04283
20	0.23742
$n=24$ $r=5$	
1	0.00739
2	0.01065
3	0.01325
4	0.01552
5	0.01760
6	0.01955
7	0.02142
8	0.02324
9	0.02503
10	0.02680
11	0.02858
12	0.03037
13	0.03220
14	0.03407
15	0.03601
16	0.03804
17	0.04018
18	0.04245
19	0.28091
$n=24$ $r=6$	
1	0.00780
2	0.01124
3	0.01398
4	0.01638
5	0.01857
6	0.02063
7	0.02260
8	0.02452
9	0.02640
10	0.02827
11	0.03014
12	0.03203
13	0.03395
14	0.03592
15	0.03795
16	0.04007
17	0.04229
18	0.32544
$n=24$ $r=7$	
1	0.00825
2	0.01190
3	0.01480
4	0.01734
5	0.01966
6	0.02184
7	0.02392
8	0.02595
9	0.02794
10	0.02991
11	0.03188
12	0.03387
13	0.03590
14	0.03796
15	0.04010
16	0.04231
17	0.37148
$n=24$ $r=8$	
1	0.00877
2	0.01264
3	0.01572
4	0.01842
5	0.02088
6	0.02319
7	0.02540
8	0.02755
9	0.02966
10	0.03175
11	0.03384
12	0.03594
13	0.03807
14	0.04025
15	0.04249
16	0.41951
$n=24$ $r=9$	
1	0.00935
2	0.01348
3	0.01676
4	0.01963
5	0.02226
6	0.02472
7	0.02708
8	0.02936
9	0.03160
10	0.03382
11	0.03604
12	0.03827
13	0.04053
14	0.04283
15	0.47003
$n=24$ $r=10$	
1	0.01002
2	0.01444
3	0.01795
4	0.02103
5	0.02383
6	0.02646
7	0.02898
8	0.03142
9	0.03382
10	0.03618
11	0.03854
12	0.04091
13	0.04331
14	0.52366
$n=24$ $r=11$	
1	0.01078
2	0.01554
3	0.01932
4	0.02263
5	0.02564
6	0.02847
7	0.03117
8	0.03379
9	0.03636
10	0.03889
11	0.04141
12	0.04394
13	0.58111
$n=24$ $r=12$	
1	0.01167
2	0.01682
3	0.02091
4	0.02449
5	0.02775
6	0.03080
7	0.03372
8	0.03654
9	0.03930
10	0.04203
11	0.04473
12	0.64329
$n=24$ $r=13$	
1	0.01272
2	0.01833
3	0.02279
4	0.02668
5	0.03022
6	0.03355
7	0.03671
8	0.03977
9	0.04276
10	0.04571
11	0.71137

i	b_i
$n=24$ $r=14$	
1	0.01398
2	0.02014
3	0.02503
4	0.02930
5	0.03318
6	0.03682
7	0.04028
8	0.04362
9	0.04688
10	0.78688
$n=24$ $r=15$	
1	0.01551
2	0.02234
3	0.02776
4	0.03248
5	0.03678
6	0.04079
7	0.04461
8	0.04829
9	0.87194
$n=24$ $r=16$	
1	0.01742
2	0.02508
3	0.03115
4	0.03644
5	0.04124
6	0.04572
7	0.04997
8	0.96952
$n=24$ $r=17$	
1	0.01985
2	0.02857
3	0.03548
4	0.04148
5	0.04692
6	0.05199
7	1.08408
$n=24$ $r=18$	
1	0.02307
2	0.03319
3	0.04119
4	0.04813
5	0.05441
6	1.22261
$n=24$ $r=19$	
1	0.02753
2	0.03958
3	0.04908
4	0.05730
5	1.39686
$n=24$ $r=20$	
1	0.03411
2	0.04900
3	0.06071
4	1.62875
$n=24$ $r=21$	
1	0.04483
2	0.06430
3	1.96534
$n=24$ $r=22$	
1	0.06544
2	2.53490
$n=25$ $r=0$	
1	0.00550
2	0.00793
3	0.00986
4	0.01155
5	0.01309
6	0.01453
7	0.01591
8	0.01725
9	0.01857
10	0.01987
11	0.02118
12	0.02249
13	0.02382
14	0.02519
15	0.02660
16	0.02807
17	0.02961
18	0.03126
19	0.03305
20	0.03502
21	0.03724
22	0.03985
23	0.04310
24	0.04763
25	0.05927
$n=25$ $r=1$	
1	0.00573
2	0.00826
3	0.01027
4	0.01203
5	0.01363
6	0.01513
7	0.01657
8	0.01797
9	0.01934
10	0.02070
11	0.02205
12	0.02342
13	0.02481
14	0.02623
15	0.02769
16	0.02922
17	0.03083
18	0.03254
19	0.03439
20	0.03643
21	0.03873
22	0.04141
23	0.04469
24	0.10397
$n=25$ $r=2$	
1	0.00598
2	0.00862
3	0.01072
4	0.01255
5	0.01422
6	0.01579
7	0.01729
8	0.01875
9	0.02018
10	0.02159
11	0.02301
12	0.02443
13	0.02588
14	0.02736
15	0.02888
16	0.03047
17	0.03215
18	0.03393
19	0.03584
20	0.03795
21	0.04031
22	0.04304
23	0.14618
$n=25$ $r=3$	
1	0.00625
2	0.00901
3	0.01120
4	0.01312
5	0.01487
6	0.01651
7	0.01808
8	0.01960
9	0.02109
10	0.02257
11	0.02404
12	0.02553
13	0.02704
14	0.02859
15	0.03018
16	0.03183
17	0.03357
18	0.03542
19	0.03741
20	0.03958
21	0.04201
22	0.18748
$n=25$ $r=4$	
1	0.00655
2	0.00944
3	0.01174
4	0.01374
5	0.01557
6	0.01729
7	0.01893
8	0.02053
9	0.02209
10	0.02364
11	0.02518
12	0.02674
13	0.02831
14	0.02993
15	0.03159
16	0.03332
17	0.03513
18	0.03705
19	0.03911
20	0.04135
21	0.22867
$n=25$ $r=5$	
1	0.00688
2	0.00991
3	0.01232
4	0.01443
5	0.01635
6	0.01815
7	0.01988
8	0.02155
9	0.02319
10	0.02481
11	0.02643
12	0.02806
13	0.02971
14	0.03140
15	0.03313
16	0.03494
17	0.03682
18	0.03882
19	0.04095
20	0.27028
$n=25$ $r=6$	
1	0.00724
2	0.01043
3	0.01297
4	0.01518
5	0.01721
6	0.01910
7	0.02092
8	0.02267
9	0.02440
10	0.02610
11	0.02780
12	0.02951
13	0.03125
14	0.03301
15	0.03483
16	0.03672
17	0.03869
18	0.04076
19	0.31277
$n=25$ $r=7$	
1	0.00764
2	0.01101
3	0.01369
4	0.01602
5	0.01816
6	0.02016
7	0.02207
8	0.02392
9	0.02574
10	0.02753
11	0.02932
12	0.03112
13	0.03295
14	0.03480
15	0.03671
16	0.03868
17	0.04073
18	0.35654
$n=25$ $r=8$	
1	0.00809
2	0.01166
3	0.01449
4	0.01696
5	0.01922
6	0.02134
7	0.02336
8	0.02532
9	0.02723
10	0.02913
11	0.03102
12	0.03292
13	0.03484
14	0.03679
15	0.03879
16	0.04086
17	0.40200
$n=25$ $r=9$	
1	0.00859
2	0.01238
3	0.01539
4	0.01802
5	0.02041
6	0.02266
7	0.02480
8	0.02688
9	0.02891
10	0.03092
11	0.03292
12	0.03493
13	0.03695
14	0.03901
15	0.04112
16	0.44959
$n=25$ $r=10$	
1	0.00917
2	0.01320
3	0.01641
4	0.01921
5	0.02176
6	0.02415
7	0.02644
8	0.02864
9	0.03081
10	0.03294
11	0.03506
12	0.03719
13	0.03933
14	0.04151
15	0.49982
$n=25$ $r=11$	
1	0.00982
2	0.01414
3	0.01757
4	0.02057
5	0.02330
6	0.02586
7	0.02830
8	0.03066
9	0.03296
10	0.03524
11	0.03750
12	0.03976
13	0.04204
14	0.55328
$n=25$ $r=12$	
1	0.01057
2	0.01522
3	0.01891
4	0.02213
5	0.02507
6	0.02782
7	0.03043
8	0.03297
9	0.03544
10	0.03787
11	0.04029
12	0.04271
13	0.61070
$n=25$ $r=13$	
1	0.01144
2	0.01647
3	0.02047
4	0.02395
5	0.02713
6	0.03009
7	0.03292
8	0.03565
9	0.03831
10	0.04093
11	0.04353
12	0.67298
$n=25$ $r=14$	
1	0.01247
2	0.01795
3	0.02231
4	0.02610
5	0.02955
6	0.03277
7	0.03584
8	0.03880
9	0.04169
10	0.04452
11	0.74130
$n=25$ $r=15$	
1	0.01370
2	0.01972
3	0.02450
4	0.02866
5	0.03244
6	0.03597
7	0.03933
8	0.04256
9	0.04570
10	0.81722
$n=25$ $r=16$	
1	0.01520
2	0.02188
3	0.02717
4	0.03177
5	0.03595
6	0.03985
7	0.04356
8	0.04711
9	0.90287
$n=25$ $r=17$	
1	0.01707
2	0.02456
3	0.03049
4	0.03564
5	0.04031
6	0.04467
7	0.04879
8	1.00130
$n=25$ $r=18$	
1	0.01945
2	0.02798
3	0.03472
4	0.04057
5	0.04587
6	0.05079
7	1.11703
$n=25$ $r=19$	
1	0.02261
2	0.03250
3	0.04031
4	0.04708
5	0.05319
6	1.25717
$n=25$ $r=20$	
1	0.02697
2	0.03875
3	0.04804
4	0.05605
5	1.43370
$n=25$ $r=21$	
1	0.03342
2	0.04798
3	0.05941
4	1.66892
$n=25$ $r=22$	
1	0.04392
2	0.06296
3	2.01080
$n=25$ $r=23$	
1	0.06410
2	2.59008
$n=30$ $r=0$	
1	0.00420
2	0.00603
3	0.00748
4	0.00874
5	0.00988
6	0.01094
7	0.01195
8	0.01292
9	0.01387
10	0.01479
11	0.01570
12	0.01661
13	0.01751
14	0.01843
15	0.01935
16	0.02029
17	0.02124
18	0.02223
19	0.02325
20	0.02432
21	0.02544
22	0.02663
23	0.02791
24	0.02930
25	0.03084
26	0.03260
27	0.03468
28	0.03728
29	0.04093
30	0.05044
$n=30$ $r=1$	
1	0.00434
2	0.00624
3	0.00774
4	0.00904
5	0.01022
6	0.01132
7	0.01236
8	0.01337
9	0.01434
10	0.01530
11	0.01624
12	0.01718
13	0.01812
14	0.01906
15	0.02001
16	0.02098
17	0.02197
18	0.02299
19	0.02405
20	0.02515
21	0.02630
22	0.02753
23	0.02885
24	0.03028
25	0.03187
26	0.03367
27	0.03579
28	0.03840
29	0.08856
$n=30$ $r=2$	
1	0.00450
2	0.00646
3	0.00801
4	0.00936
5	0.01058
6	0.01172
7	0.01280
8	0.01384
9	0.01485
10	0.01584
11	0.01682
12	0.01779
13	0.01876
14	0.01973
15	0.02072
16	0.02172
17	0.02275

i	b_i
18	0.02380
19	0.02489
20	0.02603
21	0.02723
22	0.02849
23	0.02985
24	0.03133
25	0.03296
26	0.03480
27	0.03694
28	0.12439
$n=30$ $r=3$	
1	0.00466
2	0.00670
3	0.00831
4	0.00971
5	0.01097
6	0.01216
7	0.01328
8	0.01435
9	0.01540
10	0.01643
11	0.01744
12	0.01844
13	0.01945
14	0.02046
15	0.02148
16	0.02252
17	0.02358
18	0.02468
19	0.02580
20	0.02698
21	0.02822
22	0.02952
23	0.03092
24	0.03244
25	0.03411
26	0.03598
27	0.15920
$n=30$ $r=4$	
1	0.00484
2	0.00696
3	0.00863
4	0.01008
5	0.01140
6	0.01262
7	0.01379
8	0.01490
9	0.01599
10	0.01706
11	0.01811
12	0.01915
13	0.02019
14	0.02124
15	0.02230
16	0.02338
17	0.02448
18	0.02561
19	0.02678
20	0.02800
21	0.02927
22	0.03062
23	0.03207
24	0.03362
25	0.03533
26	0.19362
$n=30$ $r=5$	
1	0.00504
2	0.00723
3	0.00897
4	0.01048
5	0.01185
6	0.01313
7	0.01434
8	0.01550
9	0.01663
10	0.01774
11	0.01883
12	0.01991
13	0.02100
14	0.02209
15	0.02319
16	0.02431
17	0.02545
18	0.02662
19	0.02783
20	0.02909
21	0.03041
22	0.03181
23	0.03329
24	0.03489
25	0.22806
$n=30$ $r=6$	
1	0.00524
2	0.00754
3	0.00935
4	0.01092
5	0.01234
6	0.01367
7	0.01493
8	0.01614
9	0.01732
10	0.01847
11	0.01961
12	0.02074
13	0.02186
14	0.02300
15	0.02414
16	0.02530
17	0.02649
18	0.02771
19	0.02897
20	0.03027
21	0.03164
22	0.03308
23	0.03460
24	0.26281
$n=30$ $r=7$	
1	0.00547
2	0.00786
3	0.00975
4	0.01139
5	0.01288
6	0.01427
7	0.01558
8	0.01684
9	0.01807
10	0.01927
11	0.02045
12	0.02163
13	0.02281
14	0.02399
15	0.02518
16	0.02639
17	0.02762
18	0.02889
19	0.03019
20	0.03155
21	0.03296
22	0.03444
23	0.29813
$n=30$ $r=8$	
1	0.00572
2	0.00822
3	0.01020
4	0.01191
5	0.01346
6	0.01491
7	0.01628
8	0.01760
9	0.01888
10	0.02014
11	0.02138
12	0.02260
13	0.02383
14	0.02506
15	0.02631
16	0.02757
17	0.02885
18	0.03017
19	0.03152
20	0.03293
21	0.03439
22	0.33427
$n=30$ $r=9$	
1	0.00599
2	0.00861
3	0.01068
4	0.01248
5	0.01410
6	0.01562
7	0.01706
8	0.01844
9	0.01978
10	0.02109
11	0.02238
12	0.02367
13	0.02495
14	0.02624
15	0.02754
16	0.02885
17	0.03019
18	0.03156
19	0.03297
20	0.03443
21	0.37145
$n=30$ $r=10$	
1	0.00629
2	0.00904
3	0.01121
4	0.01310
5	0.01480
6	0.01640
7	0.01790
8	0.01935
9	0.02076
10	0.02213
11	0.02349
12	0.02484
13	0.02618
14	0.02753
15	0.02888
16	0.03026
17	0.03166
18	0.03309
19	0.03456
20	0.40994
$n=30$ $r=11$	
1	0.00662
2	0.00952
3	0.01180
4	0.01378
5	0.01558
6	0.01725
7	0.01884
8	0.02036
9	0.02184
10	0.02329
11	0.02471
12	0.02612
13	0.02753
14	0.02895
15	0.03037
16	0.03181
17	0.03327
18	0.03477
19	0.44998
$n=30$ $r=12$	
1	0.00699
2	0.01004
3	0.01245
4	0.01454
5	0.01644
6	0.01820
7	0.01988
8	0.02148
9	0.02304
10	0.02456
11	0.02606
12	0.02755
13	0.02903
14	0.03052
15	0.03201
16	0.03352
17	0.03506
18	0.49190
$n=30$ $r=13$	
1	0.00740
2	0.01063
3	0.01318
4	0.01539
5	0.01740
6	0.01927
7	0.02103
8	0.02273
9	0.02438
10	0.02599
11	0.02757
12	0.02914
13	0.03070
14	0.03227
15	0.03384
16	0.03543
17	0.53602
$n=30$ $r=14$	
1	0.00786
2	0.01129
3	0.01400
4	0.01635
5	0.01848
6	0.02046
7	0.02233
8	0.02413
9	0.02588
10	0.02758
11	0.02926
12	0.03092
13	0.03257
14	0.03423
15	0.03589
16	0.58276
$n=30$ $r=15$	
1	0.00838
2	0.01204
3	0.01493
4	0.01743
5	0.01970
6	0.02181
7	0.02380
8	0.02572
9	0.02757
10	0.02939
11	0.03117
12	0.03293
13	0.03468
14	0.03643
15	0.63261
$n=30$ $r=16$	
1	0.00897
2	0.01289
3	0.01598
4	0.01866
5	0.02109
6	0.02334
7	0.02548
8	0.02752
9	0.02950
10	0.03144
11	0.03334
12	0.03522
13	0.03708
14	0.68616
$n=30$ $r=17$	
1	0.00966
2	0.01387
3	0.01720
4	0.02008
5	0.02269
6	0.02511
7	0.02740
8	0.02960
9	0.03172
10	0.03380
11	0.03583
12	0.03784
13	0.74416
$n=30$ $r=18$	
1	0.01046
2	0.01502
3	0.01861
4	0.02173
5	0.02455
6	0.02716
7	0.02964
8	0.03201
9	0.03430
10	0.03653
11	0.03872
12	0.80758
$n=30$ $r=19$	
1	0.01140
2	0.01636
3	0.02028
4	0.02367
5	0.02674
6	0.02958
7	0.03227
8	0.03484
9	0.03733
10	0.03974
11	0.87764
$n=30$ $r=20$	
1	0.01252
2	0.01797
3	0.02227
4	0.02599
5	0.02935
6	0.03247
7	0.03541
8	0.03822
9	0.04093
10	0.95603
$n=30$ $r=21$	
1	0.01389
2	0.01993
3	0.02470
4	0.02881
5	0.03253
6	0.03598
7	0.03922
8	0.04232
9	1.04503
$n=30$ $r=22$	
1	0.01559
2	0.02237
3	0.02771
4	0.03232
5	0.03648
6	0.04033
7	0.04395
8	1.14793
$n=30$ $r=23$	
1	0.01776
2	0.02548
3	0.03155
4	0.03679
5	0.04151
6	0.04587
7	1.26963
$n=30$ $r=24$	
1	0.02064
2	0.02960
3	0.03663
4	0.04269
5	0.04814
6	1.41782
$n=30$ $r=25$	
1	0.02462
2	0.03529
3	0.04365
4	0.05084
5	1.60550
$n=30$ $r=26$	
1	0.03050
2	0.04368
3	0.05399
4	1.85693
$n=30$ $r=27$	
1	0.04007
2	0.05733
3	2.22428
$n=30$ $r=28$	
1	0.05848
2	2.85005
$n=35$ $r=0$	
1	0.00334
2	0.00479
3	0.00593
4	0.00691
5	0.00780
6	0.00863
7	0.00940
8	0.01015
9	0.01087
10	0.01157
11	0.01225
12	0.01293
13	0.01360
14	0.01427
15	0.01493
16	0.01560
17	0.01628
18	0.01696
19	0.01766
20	0.01837
21	0.01910
22	0.01985
23	0.02062
24	0.02144
25	0.02229
26	0.02319
27	0.02415
28	0.02518
29	0.02632
30	0.02759
31	0.02903
32	0.03075
33	0.03291
34	0.03595
35	0.04398
$n=35$ $r=1$	
1	0.00343
2	0.00493
3	0.00610
4	0.00711
5	0.00803
6	0.00888
7	0.00968
8	0.01044
9	0.01118
10	0.01190
11	0.01261
12	0.01331
13	0.01400
14	0.01468
15	0.01537
16	0.01606
17	0.01675
18	0.01746
19	0.01817
20	0.01890
21	0.01965
22	0.02042
23	0.02122
24	0.02206
25	0.02293
26	0.02386
27	0.02485
28	0.02591
29	0.02707
30	0.02837
31	0.02984
32	0.03158
33	0.03375
34	0.07730
$n=35$ $r=2$	
1	0.00354
2	0.00507
3	0.00628
4	0.00733
5	0.00827
6	0.00915
7	0.00997
8	0.01076
9	0.01152
10	0.01226
11	0.01299
12	0.01371
13	0.01442
14	0.01513
15	0.01583
16	0.01654
17	0.01726
18	0.01798

i	b_i
19	0.01872
20	0.01947
21	0.02024
22	0.02104
23	0.02186
24	0.02272
25	0.02362
26	0.02457
27	0.02558
28	0.02667
29	0.02786
30	0.02919
31	0.03069
32	0.03244
33	0.10854

$n = 35$ $r = 3$

i	b_i
1	0.00365
2	0.00523
3	0.00648
4	0.00756
5	0.00853
6	0.00943
7	0.01028
8	0.01110
9	0.01188
10	0.01265
11	0.01340
12	0.01414
13	0.01487
14	0.01560
15	0.01633
16	0.01706
17	0.01779
18	0.01854
19	0.01930
20	0.02008
21	0.02087
22	0.02169
23	0.02254
24	0.02342
25	0.02434
26	0.02532
27	0.02636
28	0.02748
29	0.02870
30	0.03004
31	0.03156
32	0.13876

$n = 35$ $r = 4$

i	b_i
1	0.00377
2	0.00540
3	0.00669
4	0.00780
5	0.00880
6	0.00974
7	0.01061
8	0.01145
9	0.01226
10	0.01305
11	0.01383
12	0.01459
13	0.01535
14	0.01610
15	0.01685
16	0.01760
17	0.01837
18	0.01914
19	0.01992
20	0.02072
21	0.02154
22	0.02238
23	0.02325
24	0.02416
25	0.02511
26	0.02612
27	0.02718
28	0.02833
29	0.02957
30	0.03094
31	0.16847

$n = 35$ $r = 5$

i	b_i
1	0.00389
2	0.00558
3	0.00691
4	0.00806
5	0.00910
6	0.01006
7	0.01097
8	0.01183
9	0.01267
10	0.01349
11	0.01429
12	0.01508
13	0.01586
14	0.01663
15	0.01741
16	0.01819
17	0.01897
18	0.01977
19	0.02058
20	0.02140
21	0.02225
22	0.02311
23	0.02401
24	0.02495
25	0.02593
26	0.02696
27	0.02805
28	0.02923
29	0.03050
30	0.19801

$n = 35$ $r = 6$

i	b_i
1	0.00403
2	0.00577
3	0.00715
4	0.00834
5	0.00941
6	0.01041
7	0.01134
8	0.01224
9	0.01311
10	0.01395
11	0.01478
12	0.01559
13	0.01640
14	0.01720
15	0.01801
16	0.01881
17	0.01962
18	0.02044
19	0.02128
20	0.02213
21	0.02300
22	0.02390
23	0.02483
24	0.02579
25	0.02680
26	0.02786
27	0.02898
28	0.03018
29	0.22760

$n = 35$ $r = 7$

i	b_i
1	0.00417
2	0.00598
3	0.00740
4	0.00864
5	0.00975
6	0.01078
7	0.01175
8	0.01268
9	0.01358
10	0.01445
11	0.01530
12	0.01615
13	0.01698
14	0.01781
15	0.01864
16	0.01948
17	0.02032
18	0.02117
19	0.02203
20	0.02291
21	0.02381
22	0.02474
23	0.02569
24	0.02669
25	0.02772
26	0.02881
27	0.02996
28	0.25742

$n = 35$ $r = 8$

i	b_i
1	0.00432
2	0.00620
3	0.00768
4	0.00896
5	0.01011
6	0.01118
7	0.01218
8	0.01315
9	0.01408
10	0.01498
11	0.01587
12	0.01674
13	0.01761
14	0:01847
15	0.01933
16	0.02019
17	0.02106
18	0.02194
19	0.02283
20	0.02374
21	0.02467
22	0.02563
23	0.02662
24	0.02764
25	0.02871
26	0.02983
27	0.28766

$n = 35$ $r = 9$

i	b_i
1	0.00449
2	0.00644
3	0.00797
4	0.00930
5	0.01050
6	0.01160
7	0.01265
8	0.01365
9	0.01462
10	0.01556
11	0.01648
12	0.01738
13	0.01828
14	0.01917
15	0.02007
16	0.02096
17	0.02186
18	0.02277
19	0.02370
20	0.02464
21	0.02560
22	0.02659
23	0.02761
24	0.02867
25	0.02977
26	0.31845

$n = 35$ $r = 10$

i	b_i
1	0.00467
2	0.00670
3	0.00829
4	0.00967
5	0.01091
6	0.01207
7	0.01315
8	0.01419
9	0.01520
10	0.01617
11	0.01713
12	0.01807
13	0.01901
14	0.01993
15	0.02086
16	0.02179
17	0.02272
18	0.02367
19	0.02463
20	0.02561
21	0.02660
22	0.02763
23	0.02868
24	0.02977
25	0.34994

$n = 35$ $r = 11$

i	b_i
1	0.00486
2	0.00698
3	0.00864
4	0.01007
5	0.01137
6	0.01257
7	0.01370
8	0.01478
9	0.01583
10	0.01684
11	0.01784
12	0.01882
13	0.01979
14	0.02075
15	0.02172
16	0.02268
17	0.02366
18	0.02464
19	0.02563
20	0.02665
21	0.02768
22	0.02874
23	0.02983
24	0.38230

$n = 35$ $r = 12$

i	b_i
1	0.00508
2	0.00728
3	0.00901
4	0.01051
5	0.01186
6	0.01311
7	0.01429
8	0.01542
9	0.01651
10	0.01757
11	0.01861
12	0.01963
13	0.02064
14	0.02165
15	0.02265
16	0.02366
17	0.02467
18	0.02569
19	0.02672
20	0.02778
21	0.02885
22	0.02995
23	0.41567

$n = 35$ $r = 13$

i	b_i
1	0.00531
2	0.00761
3	0.00942
4	0.01098
5	0.01240
6	0.01371
7	0.01494
8	0.01612
9	0.01726
10	0.01836
11	0.01945
12	0.02051
13	0.02157
14	0.02262
15	0.02366
16	0.02471
17	0.02577
18	0.02683
19	0.02791
20	0.02900
21	0.03012
22	0.45024

$n = 35$ $r = 14$

i	b_i
1	0.00556
2	0.00797
3	0.00987
4	0.01151
5	0.01298
6	0.01435
7	0.01565
8	0.01688
9	0.01807
10	0.01923
11	0.02036
12	0.02148
13	0.02258
14	0.02368
15	0.02477
16	0.02587
17	0.02697
18	0.02808
19	0.02920
20	0.03034
21	0.48620

$n = 35$ $r = 15$

i	b_i
1	0.00583
2	0.00837
3	0.01036
4	0.01208
5	0.01363
6	0.01507
7	0.01642
8	0.01772
9	0.01897
10	0.02018
11	0.02137
12	0.02254
13	0.02369
14	0.02484
15	0.02599
16	0.02713
17	0.02828
18	0.02944
19	0.03061
20	0.52376

$n = 35$ $r = 16$

i	b_i
1	0.00614
2	0.00880
3	0.01090
4	0.01271
5	0.01434
6	0.01585
7	0.01728
8	0.01864
9	0.01995
10	0.02123
11	0.02248
12	0.02371
13	0.02492
14	0.02613
15	0.02733
16	0.02853
17	0.02973
18	0.03095
19	0.56318

$n = 35$ $r = 17$

i	b_i
1	0.00648
2	0.00929
3	0.01150
4	0.01341
5	0.01513
6	0.01673
7	0.01823
8	0.01967
9	0.02105
10	0.02239
11	0.02371
12	0.02500
13	0.02628
14	0.02755
15	0.02881
16	0.03007
17	0.03134
18	0.60476

$n = 35$ $r = 18$

i	b_i
1	0.00686
2	0.00983
3	0.01217
4	0.01419
5	0.01602
6	0.01770
7	0.01929
8	0.02081
9	0.02227
10	0.02369
11	0.02508
12	0.02645
13	0.02779
14	0.02913
15	0.03046
16	0.03179
17	0.64882

$n = 35$ $r = 19$

i	b_i
1	0.00728
2	0.01044
3	0.01293
4	0.01507
5	0.01701
6	0.01880
7	0.02048
8	0.02209
9	0.02364
10	0.02515
11	0.02662
12	0.02806
13	0.02949
14	0.03090
15	0.03231
16	0.69580

$n = 35$ $r = 20$

i	b_i
1	0.00777
2	0.01113
3	0.01378
4	0.01607
5	0.01813
6	0.02003
7	0.02183
8	0.02354
9	0.02519
10	0.02679
11	0.02836
12	0.02989
13	0.03140
14	0.03290
15	0.74620

$n = 35$ $r = 21$

i	b_i
1	0.00832
2	0.01192
3	0.01476
4	0.01720
5	0.01941
6	0.02144
7	0.02336
8	0.02519
9	0.02696
10	0.02866
11	0.03033
12	0.03197
13	0.03358
14	0.80064

$n = 35$ $r = 22$

i	b_i
1	0.00895
2	0.01283
3	0.01588
4	0.01851
5	0.02088
6	0.02307
7	0.02513
8	0.02709
9	0.02898
10	0.03082
11	0.03261
12	0.03436
13	0.85991

$n = 35$ $r = 23$

i	b_i
1	0.00969
2	0.01389
3	0.01718
4	0.02003
5	0.02259
6	0.02495
7	0.02718
8	0.02930
9	0.03134
10	0.03332
11	0.03524
12	0.92503

$n = 35$ $r = 24$

i	b_i
1	0.01056
2	0.01513
3	0.01872
4	0.02182
5	0.02460
6	0.02717
7	0.02959
8	0.03190
9	0.03411
10	0.03625
11	0.99731

$n = 35$ $r = 25$

i	b_i
1	0.01160
2	0.01662
3	0.02056
4	0.02395

i	b_i
5	0.02701
6	0.02983
7	0.03248
8	0.03500
9	0.03742
10	1.07854

$n=35$ $r=26$

i	b_i
1	0.01286
2	0.01843
3	0.02279
4	0.02655
5	0.02993
6	0.03305
7	0.03598
8	0.03876
9	1.17117

$n=35$ $r=27$

i	b_i
1	0.01444
2	0.02068
3	0.02557
4	0.02978
5	0.03356
6	0.03705
7	0.04032
8	1.27871

$n=35$ $r=28$

i	b_i
1	0.01645
2	0.02355
3	0.02912
4	0.03390
5	0.03819
6	0.04215
7	1.40639

$n=35$ $r=29$

i	b_i
1	0.01911
2	0.02735
3	0.03381
4	0.03934
5	0.04430
6	1.56247

$n=35$ $r=30$

i	b_i
1	0.02279
2	0.03261
3	0.04028
4	0.04685
5	1.76091

$n=35$ $r=31$

i	b_i
1	0.02823
2	0.04037
3	0.04983
4	2.02774

$n=35$ $r=32$

i	b_i
1	0.03708
2	0.05298
3	2.41907

$n=35$ $r=33$

i	b_i
1	0.05411
2	3.08823

$n=40$ $r=0$

i	b_i
1	0.00274
2	0.00392
3	0.00484
4	0.00564
5	0.00636
6	0.00702
7	0.00765
8	0.00824
9	0.00882
10	0.00937
11	0.00991
12	0.01044
13	0.01096
14	0.01148
15	0.01199
16	0.01250
17	0.01301
18	0.01353
19	0.01404
20	0.01456
21	0.01509
22	0.01563
23	0.01618
24	0.01674
25	0.01731
26	0.01790
27	0.01852
28	0.01916
29	0.01983
30	0.02053
31	0.02128
32	0.02208
33	0.02295
34	0.02391
35	0.02498
36	0.02620
37	0.02766
38	0.02950
39	0.03211
40	0.03903

$n=40$ $r=1$

i	b_i
1	0.00280
2	0.00402
3	0.00497
4	0.00579
5	0.00652
6	0.00720
7	0.00784
8	0.00845
9	0.00904
10	0.00961
11	0.01016
12	0.01071
13	0.01124
14	0.01177
15	0.01230
16	0.01282
17	0.01334
18	0.01387
19	0.01440
20	0.01493
21	0.01548
22	0.01603
23	0.01659
24	0.01716
25	0.01775
26	0.01836
27	0.01899
28	0.01964
29	0.02033
30	0.02105
31	0.02182
32	0.02264
33	0.02353
34	0.02450
35	0.02559
36	0.02684
37	0.02831
38	0.03015
39	0.06870

$n=40$ $r=2$

i	b_i
1	0.00288
2	0.00412
3	0.00510
4	0.00594
5	0.00669
6	0.00739
7	0.00805
8	0.00868
9	0.00928
10	0.00986
11	0.01043
12	0.01099
13	0.01154
14	0.01208
15	0.01262
16	0.01316
17	0.01369
18	0.01423
19	0.01478
20	0.01533
21	0.01588
22	0.01645
23	0.01702
24	0.01761
25	0.01821
26	0.01884
27	0.01948
28	0.02015
29	0.02086
30	0.02160
31	0.02238
32	0.02322
33	0.02413
34	0.02512
35	0.02623
36	0.02749
37	0.02897
38	0.09646

$n=40$ $r=3$

i	b_i
1	0.00296
2	0.00423
3	0.00524
4	0.00610
5	0.00688
6	0.00759
7	0.00827
8	0.00891
9	0.00953
10	0.01013
11	0.01071
12	0.01128
13	0.01185
14	0.01240
15	0.01296
16	0.01351
17	0.01406
18	0.01462
19	0.01517
20	0.01574
21	0.01631
22	0.01689
23	0.01748
24	0.01808
25	0.01870
26	0.01934
27	0.02000
28	0.02069
29	0.02141
30	0.02217
31	0.02297
32	0.02383
33	0.02476
34	0.02577
35	0.02690
36	0.02817
37	0.12323

$n=40$ $r=4$

i	b_i
1	0.00304
2	0.00435
3	0.00538
4	0.00627
5	0.00707
6	0.00780
7	0.00850
8	0.00916
9	0.00979
10	0.01041
11	0.01101
12	0.01160
13	0.01217
14	0.01275
15	0.01332
16	0.01388
17	0.01445
18	0.01502
19	0.01559
20	0.01617
21	0.01676
22	0.01735
23	0.01796
24	0.01858
25	0.01922
26	0.01987
27	0.02055
28	0.02126
29	0.02200
30	0.02277
31	0.02359
32	0.02447
33	0.02542
34	0.02645
35	0.02759
36	0.14946

$n=40$ $r=5$

i	b_i
1	0.00313
2	0.00448
3	0.00553
4	0.00645
5	0.00727
6	0.00803
7	0.00874
8	0.00942
9	0.01007
10	0.01070
11	0.01132
12	0.01193
13	0.01252
14	0.01311
15	0.01370
16	0.01428
17	0.01486
18	0.01545
19	0.01604
20	0.01663
21	0.01723
22	0.01785
23	0.01847
24	0.01911
25	0.01976
26	0.02043
27	0.02113
28	0.02185
29	0.02261
30	0.02341
31	0.02425
32	0.02514
33	0.02611
34	0.02715
35	0.17542

$n=40$ $r=6$

i	b_i
1	0.00322
2	0.00461
3	0.00570
4	0.00664
5	0.00748
6	0.00826
7	0.00900
8	0.00970
9	0.01037
10	0.01102
11	0.01165
12	0.01228
13	0.01289
14	0.01350
15	0.01410
16	0.01470
17	0.01530
18	0.01590
19	0.01651
20	0.01712
21	0.01774
22	0.01837
23	0.01901
24	0.01966
25	0.02033
26	0.02102
27	0.02174
28	0.02248
29	0.02326
30	0.02407
31	0.02493
32	0.02585
33	0.02683
34	0.20130

$n=40$ $r=7$

i	b_i
1	0.00331
2	0.00475
3	0.00587
4	0.00684
5	0.00771
6	0.00851
7	0.00927
8	0.00999
9	0.01068
10	0.01135
11	0.01201
12	0.01265
13	0.01328
14	0.01390
15	0.01452
16	0.01514
17	0.01576
18	0.01638
19	0.01700
20	0.01763
21	0.01827
22	0.01892
23	0.01958
24	0.02025
25	0.02094
26	0.02165
27	0.02239
28	0.02315
29	0.02394
30	0.02478
31	0.02566
32	0.02659
33	0.22724

$n=40$ $r=8$

i	b_i
1	0.00342
2	0.00489
3	0.00605
4	0.00705
5	0.00795
6	0.00878
7	0.00956
8	0.01030
9	0.01101
10	0.01171
11	0.01238
12	0.01304
13	0.01369
14	0.01434
15	0.01498
16	0.01561
17	0.01625
18	0.01689
19	0.01753
20	0.01818
21	0.01883
22	0.01950
23	0.02018
24	0.02087
25	0.02158
26	0.02231
27	0.02307
28	0.02385
29	0.02467
30	0.02552
31	0.02642
32	0.25336

$n=40$ $r=9$

i	b_i
1	0.00353
2	0.00505
3	0.00625
4	0.00728
5	0.00820
6	0.00906
7	0.00987
8	0.01063
9	0.01137
10	0.01208
11	0.01278
12	0.01346
13	0.01413
14	0.01480
15	0.01546
16	0.01611
17	0.01677
18	0.01743
19	0.01809
20	0.01876
21	0.01944
22	0.02012
23	0.02082
24	0.02153
25	0.02227
26	0.02302
27	0.02379
28	0.02460
29	0.02544
30	0.02631
31	0.27979

$n=40$ $r=10$

i	b_i
1	0.00365
2	0.00522
3	0.00646
4	0.00752
5	0.00848
6	0.00936
7	0.01019
8	0.01099
9	0.01175
10	0.01248
11	0.01320
12	0.01391
13	0.01460
14	0.01529
15	0.01597
16	0.01665
17	0.01732
18	0.01800
19	0.01869
20	0.01938
21	0.02008
22	0.02078
23	0.02150
24	0.02224
25	0.02299
26	0.02377
27	0.02456
28	0.02539
29	0.02625
30	0.30662

$n=40$ $r=11$

i	b_i
1	0.00377
2	0.00540
3	0.00668
4	0.00778
5	0.00877
6	0.00968
7	0.01054
8	0.01136
9	0.01215
10	0.01291
11	0.01366
12	0.01438
13	0.01510
14	0.01581
15	0.01651
16	0.01722
17	0.01792
18	0.01862
19	0.01933
20	0.02004
21	0.02076
22	0.02149
23	0.02223
24	0.02299
25	0.02377
26	0.02456
27	0.02538
28	0.02623
29	0.33395

$n=40$ $r=12$

i	b_i
1	0.00391
2	0.00559
3	0.00692
4	0.00806
5	0.00908
6	0.01003
7	0.01092
8	0.01177
9	0.01258
10	0.01337
11	0.01414
12	0.01489
13	0.01564
14	0.01637
15	0.01710
16	0.01783
17	0.01855
18	0.01928
19	0.02001
20	0.02074
21	0.02149
22	0.02224
23	0.02301
24	0.02379
25	0.02459
26	0.02541
27	0.02626
28	0.36189

$n=40$ $r=13$

i	b_i
1	0.00405
2	0.00580

i	b_i
3	0.00717
4	0.00835
5	0.00942
6	0.01040
7	0.01132
8	0.01220
9	0.01305
10	0.01386
11	0.01466
12	0.01544
13	0.01621
14	0.01697
15	0.01773
16	0.01848
17	0.01923
18	0.01998
19	0.02074
20	0.02150
21	0.02227
22	0.02305
23	0.02384
24	0.02465
25	0.02548
26	0.02632
27	0.39054
$n = 40$ $r = 14$	
1	0.00421
2	0.00602
3	0.00745
4	0.00868
5	0.00978
6	0.01080
7	0.01176
8	0.01267
9	0.01355
10	0.01439
11	0.01522
12	0.01603
13	0.01683
14	0.01762
15	0.01840
16	0.01918
17	0.01996
18	0.02074
19	0.02152
20	0.02231
21	0.02311
22	0.02392
23	0.02474
24	0.02557
25	0.02643
26	0.42001
$n = 40$ $r = 15$	
1	0.00437
2	0.00626
3	0.00775
4	0.00902
5	0.01017
6	0.01123
7	0.01222
8	0.01317
9	0.01408
10	0.01497
11	0.01583
12	0.01667
13	0.01750
14	0.01832
15	0.01913
16	0.01994
17	0.02075
18	0.02156
19	0.02237
20	0.02319
21	0.02402
22	0.02485
23	0.02570
24	0.02657
25	0.45042
$n = 40$ $r = 16$	
1	0.00456
2	0.00652
3	0.00807
4	0.00940
5	0.01059
6	0.01169
7	0.01273
8	0.01372
9	0.01467
10	0.01559
11	0.01648
12	0.01736
13	0.01822
14	0.01907
15	0.01992
16	0.02076
17	0.02160
18	0.02244
19	0.02329
20	0.02413
21	0.02499
22	0.02586
23	0.02674
24	0.48190
$n = 40$ $r = 17$	
1	0.00475
2	0.00681
3	0.00842
4	0.00980
5	0.01105
6	0.01220
7	0.01328
8	0.01431
9	0.01530
10	0.01626
11	0.01719
12	0.01811
13	0.01900
14	0.01989
15	0.02077
16	0.02165
17	0.02252
18	0.02340
19	0.02428
20	0.02516
21	0.02605
22	0.02695
23	0.51461
$n = 40$ $r = 18$	
1	0.00497
2	0.00711
3	0.00880
4	0.01025
5	0.01155
6	0.01275
7	0.01388
8	0.01496
9	0.01599
10	0.01699
11	0.01797
12	0.01892
13	0.01986
14	0.02078
15	0.02170
16	0.02262
17	0.02353
18	0.02444
19	0.02536
20	0.02628
21	0.02720
22	0.54870
$n = 40$ $r = 19$	
1	0.00520
2	0.00745
3	0.00921
4	0.01073
5	0.01210
6	0.01335
7	0.01454
8	0.01566
9	0.01675
10	0.01779
11	0.01881
12	0.01981
13	0.02079
14	0.02176
15	0.02272
16	0.02368
17	0.02463
18	0.02558
19	0.02653
20	0.02749
21	0.58436
$n = 40$ $r = 20$	
1	0.00546
2	0.00782
3	0.00967
4	0.01126
5	0.01270
6	0.01402
7	0.01526
8	0.01644
9	0.01757
10	0.01867
11	0.01974
12	0.02079
13	0.02182
14	0.02283
15	0.02384
16	0.02484
17	0.02583
18	0.02683
19	0.02783
20	0.62181
$n = 40$ $r = 21$	
1	0.00575
2	0.00823
3	0.01018
4	0.01185
5	0.01336
6	0.01475
7	0.01605
8	0.01730
9	0.01849
10	0.01964
11	0.02077
12	0.02186
13	0.02294
14	0.02401
15	0.02507
16	0.02611
17	0.02716
18	0.02820
19	0.66131
$n = 40$ $r = 22$	
1	0.00607
2	0.00869
3	0.01074
4	0.01251
5	0.01410
6	0.01556
7	0.01694
8	0.01825
9	0.01950
10	0.02072
11	0.02190
12	0.02306
13	0.02420
14	0.02532
15	0.02643
16	0.02753
17	0.02863
18	0.70316
$n = 40$ $r = 23$	
1	0.00642
2	0.00919
3	0.01137
4	0.01324
5	0.01492
6	0.01647
7	0.01792
8	0.01931
9	0.02063
10	0.02192
11	0.02317
12	0.02439
13	0.02559
14	0.02678
15	0.02795
16	0.02911
17	0.74772
$n = 40$ $r = 24$	
1	0.00682
2	0.00976
3	0.01207
4	0.01406
5	0.01584
6	0.01748
7	0.01903
8	0.02050
9	0.02190
10	0.02327
11	0.02459
12	0.02589
13	0.02716
14	0.02841
15	0.02965
16	0.79541
$n = 40$ $r = 25$	
1	0.00727
2	0.01041
3	0.01287
4	0.01498
5	0.01688
6	0.01863
7	0.02028
8	0.02184
9	0.02334
10	0.02479
11	0.02620
12	0.02757
13	0.02892
14	0.03025
15	0.84677
$n = 40$ $r = 26$	
1	0.00779
2	0.01114
3	0.01378
4	0.01604
5	0.01807
6	0.01995
7	0.02170
8	0.02337
9	0.02498
10	0.02652
11	0.02803
12	0.02949
13	0.03093
14	0.90246
$n = 40$ $r = 27$	
1	0.00838
2	0.01199
3	0.01482
4	0.01726
5	0.01944
6	0.02145
7	0.02334
8	0.02514
9	0.02686
10	0.02852
11	0.03013
12	0.03170
13	0.96330
$n = 40$ $r = 28$	
1	0.00907
2	0.01298
3	0.01604
4	0.01867
5	0.02103
6	0.02321
7	0.02525
8	0.02719
9	0.02904
10	0.03083
11	0.03257
12	1.03038
$n = 40$ $r = 29$	
1	0.00988
2	0.01414
3	0.01747
4	0.02034
5	0.02291
6	0.02527
7	0.02749
8	0.02960
9	0.03161
10	0.03355
11	1.10508
$n = 40$ $r = 30$	
1	0.01085
2	0.01553
3	0.01919
4	0.02233
5	0.02515
6	0.02774
7	0.03017
8	0.03247
9	0.03468
10	1.18929
$n = 40$ $r = 31$	
1	0.01204
2	0.01722
3	0.02127
4	0.02475
5	0.02787
6	0.03074
7	0.03343
8	0.03597
9	1.28561
$n = 40$ $r = 32$	
1	0.01351
2	0.01932
3	0.02387
4	0.02776
5	0.03125
6	0.03446
7	0.03746
8	1.39777
$n = 40$ $r = 33$	
1	0.01539
2	0.02201
3	0.02718
4	0.03160
5	0.03557
6	0.03921
7	1.53133
$n = 40$ $r = 34$	
1	0.01787
2	0.02555
3	0.03155
4	0.03667
5	0.04126
6	1.69507
$n = 40$ $r = 35$	
1	0.02132
2	0.03046
3	0.03759
4	0.04368
5	1.90384
$n = 40$ $r = 36$	
1	0.02640
2	0.03771
3	0.04650
4	2.18534
$n = 40$ $r = 37$	
1	0.03468
2	0.04949
3	2.59935
$n = 40$ $r = 38$	
1	0.05060
2	3.30932
$n = 45$ $r = 0$	
1	0.00230
2	0.00328
3	0.00406
4	0.00472
5	0.00532
6	0.00587
7	0.00638
8	0.00687
9	0.00734
10	0.00779
11	0.00823
12	0.00865
13	0.00909
14	0.00951
15	0.00988
16	0.01034
17	0.01070
18	0.01122
19	0.01142
20	0.01197
21	0.01236
22	0.01273
23	0.01318
24	0.01359
25	0.01401
26	0.01445
27	0.01490
28	0.01535
29	0.01582
30	0.01630
31	0.01679
32	0.01732
33	0.01785
34	0.01842
35	0.01904
36	0.01965
37	0.02035
38	0.02111
39	0.02191
40	0.02284
41	0.02390
42	0.02516
43	0.02676
44	0.02904
45	0.03513
$n = 45$ $r = 1$	
1	0.00235
2	0.00336
3	0.00415
4	0.00483
5	0.00544
6	0.00600
7	0.00652
8	0.00702
9	0.00751
10	0.00796
11	0.00842
12	0.00885
13	0.00929
14	0.00972
15	0.01010
16	0.01057
17	0.01094
18	0.01147
19	0.01167
20	0.01224
21	0.01264
22	0.01302
23	0.01348
24	0.01390
25	0.01433
26	0.01477
27	0.01524
28	0.01569
29	0.01617
30	0.01666
31	0.01717
32	0.01771
33	0.01825
34	0.01883
35	0.01947
36	0.02009
37	0.02080
38	0.02158
39	0.02239
40	0.02334
41	0.02441
42	0.02569
43	0.02728
44	0.06190
$n = 45$ $r = 2$	
1	0.00240
2	0.00343
3	0.00424
4	0.00494
5	0.00556
6	0.00614
7	0.00668
8	0.00719

i	b_i
9	0.00768
10	0.00815
11	0.00862
12	0.00905
13	0.00951
14	0.00995
15	0.01034
16	0.01081
17	0.01120
18	0.01173
19	0.01195
20	0.01252
21	0.01294
22	0.01332
23	0.01379
24	0.01422
25	0.01465
26	0.01512
27	0.01558
28	0.01606
29	0.01654
30	0.01705
31	0.01756
32	0.01812
33	0.01867
34	0.01926
35	0.01991
36	0.02055
37	0.02127
38	0.02207
39	0.02289
40	0.02386
41	0.02494
42	0.02621
43	0.08692

$n = 45 \quad r = 3$

i	b_i
1	0.00246
2	0.00352
3	0.00434
4	0.00506
5	0.00569
6	0.00628
7	0.00683
8	0.00736
9	0.00786
10	0.00834
11	0.00862
12	0.00927
13	0.00973
14	0.01019
15	0.01058
16	0.01108
17	0.01144
18	0.01204
19	0.01219
20	0.01287
21	0.01319
22	0.01367
23	0.01410
24	0.01457
25	0.01501
26	0.01547
27	0.01596
28	0.01643
29	0.01694
30	0.01745
31	0.01798
32	0.01854
33	0.01911
34	0.01971
35	0.02038
36	0.02102
37	0.02176
38	0.02257
39	0.02341
40	0.02439
41	0.02547
42	0.11101

$n = 45 \quad r = 4$

i	b_i
1	0.00252
2	0.00360
3	0.00445
4	0.00518
5	0.00583
6	0.00644
7	0.00700
8	0.00754
9	0.00805
10	0.00855
11	0.00904
12	0.00949
13	0.00997
14	0.01042
15	0.01086
16	0.01130
17	0.01180
18	0.01225
19	0.01259
20	0.01305
21	0.01367
22	0.01385
23	0.01457
24	0.01486
25	0.01538
26	0.01584
27	0.01635
28	0.01683
29	0.01735
30	0.01787
31	0.01841
32	0.01899
33	0.01957
34	0.02018
35	0.02086
36	0.02152
37	0.02227
38	0.02310
39	0.02395
40	0.02493
41	0.13454

$n = 45 \quad r = 5$

i	b_i
1	0.00258
2	0.00369
3	0.00456
4	0.00531
5	0.00598
6	0.00660
7	0.00718
8	0.00773
9	0.00825
10	0.00876
11	0.00926
12	0.00973
13	0.01022
14	0.01071
15	0.01107
16	0.01169
17	0.01193
18	0.01273
19	0.01270
20	0.01370
21	0.01360
22	0.01457
23	0.01465
24	0.01535
25	0.01574
26	0.01624
27	0.01675
28	0.01725
29	0.01777
30	0.01831
31	0.01887
32	0.01946
33	0.02005
34	0.02068
35	0.02137
36	0.02204
37	0.02281
38	0.02365
39	0.02451
40	0.15777

$n = 45 \quad r = 6$

i	b_i
1	0.00265
2	0.00379
3	0.00468
4	0.00544
5	0.00613
6	0.00677
7	0.00736
8	0.00792
9	0.00847
10	0.00898
11	0.00950
12	0.00998
13	0.01050
14	0.01094
15	0.01145
16	0.01179
17	0.01244
18	0.01290
19	0.01333
20	0.01344
21	0.01476
22	0.01419
23	0.01548
24	0.01558
25	0.01617
26	0.01664
27	0.01717
28	0.01769
29	0.01823
30	0.01878
31	0.01934
32	0.01995
33	0.02055
34	0.02120
35	0.02190
36	0.02259
37	0.02336
38	0.02422
39	0.18082

$n = 45 \quad r = 7$

i	b_i
1	0.00272
2	0.00389
3	0.00480
4	0.00559
5	0.00629
6	0.00694
7	0.00755
8	0.00813
9	0.00869
10	0.00922
11	0.00976
12	0.01024
13	0.01074
14	0.01128
15	0.01165
16	0.01214
17	0.01289
18	0.01333
19	0.01316
20	0.01440
21	0.01454
22	0.01527
23	0.01534
24	0.01629
25	0.01645
26	0.01712
27	0.01761
28	0.01815
29	0.01870
30	0.01926
31	0.01984
32	0.02046
33	0.02108
34	0.02173
35	0.02246
36	0.02315
37	0.02395
38	0.20386

$n = 45 \quad r = 8$

i	b_i
1	0.00279
2	0.00399
3	0.00493
4	0.00574
5	0.00646
6	0.00713
7	0.00776
8	0.00835
9	0.00892
10	0.00947
11	0.00999
12	0.01056
13	0.01096
14	0.01173
15	0.01182
16	0.01281
17	0.01245
18	0.01449
19	0.01323
20	0.01471
21	0.01497
22	0.01548
23	0.01620
24	0.01627
25	0.01716
26	0.01751
27	0.01808
28	0.01864
29	0.01920
30	0.01977
31	0.02037
32	0.02100
33	0.02164
34	0.02230
35	0.02304
36	0.02375
37	0.22695

$n = 45 \quad r = 9$

i	b_i
1	0.00287
2	0.00410
3	0.00507
4	0.00590
5	0.00664
6	0.00733
7	0.00797
8	0.00858
9	0.00917
10	0.00972
11	0.01033
12	0.01070
13	0.01155
14	0.01158
15	0.01279
16	0.01255
17	0.01372
18	0.01356
19	0.01452
20	0.01467
21	0.01576
22	0.01581
23	0.01642
24	0.01702
25	0.01743
26	0.01805
27	0.01858
28	0.01914
29	0.01972
30	0.02031
31	0.02092
32	0.02157
33	0.02221
34	0.02290
35	0.02365
36	0.25018

$n = 45 \quad r = 10$

i	b_i
1	0.00295
2	0.00422
3	0.00521
4	0.00607
5	0.00683
6	0.00754
7	0.00820
8	0.00882
9	0.00944
10	0.01001
11	0.01055
12	0.01120
13	0.01152
14	0.01261
15	0.01187
16	0.01432
17	0.01296
18	0.01490
19	0.01413
20	0.01610
21	0.01518
22	0.01679
23	0.01677
24	0.01742
25	0.01801
26	0.01853
27	0.01911
28	0.01968
29	0.02027
30	0.02088
31	0.02150
32	0.02216
33	0.02284
34	0.02352
35	0.27367

$n = 45 \quad r = 11$

i	b_i
1	0.00304
2	0.00434
3	0.00537
4	0.00625
5	0.00703
6	0.00776
7	0.00844
8	0.00909
9	0.00971
10	0.01029
11	0.01094
12	0.01134
13	0.01221
14	0.01224
15	0.01358
16	0.01306
17	0.01484
18	0.01384
19	0.01621
20	0.01502
21	0.01677
22	0.01657
23	0.01755
24	0.01789
25	0.01852
26	0.01907
27	0.01966
28	0.02025
29	0.02086
30	0.02148
31	0.02211
32	0.02280
33	0.02347
34	0.29744

$n = 45 \quad r = 12$

i	b_i
1	0.00313
2	0.00448
3	0.00553
4	0.00643
5	0.00725
6	0.00799
7	0.00870
8	0.00936
9	0.01001
10	0.01061
11	0.01121
12	0.01186
13	0.01218
14	0.01334
15	0.01286
16	0.01484
17	0.01378
18	0.01592
19	0.01507
20	0.01676
21	0.01641
22	0.01757
23	0.01782
24	0.01853
25	0.01905
26	0.01965
27	0.02024
28	0.02085
29	0.02147
30	0.02211
31	0.02277
32	0.02346
33	0.32159

$n = 45 \quad r = 13$

i	b_i
1	0.00323
2	0.00462
3	0.00570
4	0.00663
5	0.00747
6	0.00824
7	0.00897
8	0.00966
9	0.01031
10	0.01095
11	0.01158
12	0.01206
13	0.01302
14	0.01291
15	0.01449
16	0.01384
17	0.01570
18	0.01507
19	0.01660
20	0.01647
21	0.01749
22	0.01779
23	0.01852
24	0.01904
25	0.01965
26	0.02025
27	0.02086
28	0.02149
29	0.02213
30	0.02279
31	0.02345
32	0.34620

$n = 45 \quad r = 14$

i	b_i
1	0.00333
2	0.00476
3	0.00589
4	0.00685
5	0.00771
6	0.00851
7	0.00926
8	0.00996
9	0.01066
10	0.01127
11	0.01195
12	0.01259
13	0.01305
14	0.01401
15	0.01406
16	0.01527
17	0.01525
18	0.01630
19	0.01659
20	0.01737
21	0.01782
22	0.01850
23	0.01904
24	0.01967
25	0.02027
26	0.02089
27	0.02152
28	0.02216
29	0.02282
30	0.02350
31	0.37131

$n = 45 \quad r = 15$

i	b_i
1	0.00344
2	0.00492
3	0.00608
4	0.00708
5	0.00797
6	0.00879
7	0.00956
8	0.01030
9	0.01099
10	0.01169
11	0.01231
12	0.01297
13	0.01367
14	0.01411
15	0.01498
16	0.01528
17	0.01621
18	0.01655
19	0.01731
20	0.01783
21	0.01848
22	0.01906
23	0.01969
24	0.02030
25	0.02093
26	0.02157
27	0.02222
28	0.02288
29	0.02356
30	0.39702

$n = 45 \quad r = 16$

i	b_i
1	0.00356
2	0.00509
3	0.00629
4	0.00732
5	0.00824
6	0.00909
7	0.00989
8	0.01065
9	0.01138
10	0.01206
11	0.01277
12	0.01341
13	0.01406
14	0.01476
15	0.01528

i	b_i
16	0.01604
17	0.01655
18	0.01726
19	0.01782
20	0.01847
21	0.01909
22	0.01972
23	0.02035
24	0.02099
25	0.02164
26	0.02230
27	0.02297
28	0.02365
29	0.42340

$n = 45$ $r = 17$

i	b_i
1	0.00369
2	0.00527
3	0.00651
4	0.00758
5	0.00854
6	0.00942
7	0.01024
8	0.01103
9	0.01178
10	0.01251
11	0.01320
12	0.01391
13	0.01457
14	0.01523
15	0.01590
16	0.01652
17	0.01721
18	0.01782
19	0.01848
20	0.01911
21	0.01976
22	0.02041
23	0.02107
24	0.02173
25	0.02240
26	0.02307
27	0.02376
28	0.45055

$n = 45$ $r = 18$

i	b_i
1	0.00382
2	0.00547
3	0.00676
4	0.00786
5	0.00885
6	0.00977
7	0.01062
8	0.01144
9	0.01222
10	0.01296
11	0.01370
12	0.01440
13	0.01511
14	0.01580
15	0.01647
16	0.01715
17	0.01781
18	0.01849
19	0.01915
20	0.01981
21	0.02048
22	0.02115
23	0.02183
24	0.02251
25	0.02320
26	0.02391
27	0.47855

$n = 45$ $r = 19$

i	b_i
1	0.00397
2	0.00568
3	0.00701
4	0.00816
5	0.00919
6	0.01014
7	0.01103
8	0.01187
9	0.01268
10	0.01346
11	0.01422
12	0.01496
13	0.01568
14	0.01640
15	0.01710
16	0.01780
17	0.01849
18	0.01918
19	0.01987
20	0.02056
21	0.02126
22	0.02195
23	0.02265
24	0.02336
25	0.02407
26	0.50750

$n = 45$ $r = 20$

i	b_i
1	0.00413
2	0.00590
3	0.00729
4	0.00849
5	0.00956
6	0.01054
7	0.01147
8	0.01235
9	0.01319
10	0.01400
11	0.01478
12	0.01555
13	0.01631
14	0.01705
15	0.01778
16	0.01850
17	0.01922
18	0.01994
19	0.02066
20	0.02137
21	0.02209
22	0.02281
23	0.02354
24	0.02427
25	0.53753

$n = 45$ $r = 21$

i	b_i
1	0.00430
2	0.00615
3	0.00760
4	0.00884
5	0.00995
6	0.01098
7	0.01194
8	0.01286
9	0.01373
10	0.01457
11	0.01539
12	0.01619
13	0.01698
14	0.01775
15	0.01851
16	0.01926
17	0.02001
18	0.02076
19	0.02150
20	0.02225
21	0.02299
22	0.02374
23	0.02449
24	0.56877

$n = 45$ $r = 22$

i	b_i
1	0.00449
2	0.00642
3	0.00793
4	0.00922
5	0.01038
6	0.01146
7	0.01246
8	0.01341
9	0.01432
10	0.01520
11	0.01606
12	0.01689
13	0.01771
14	0.01851
15	0.01930
16	0.02009
17	0.02087
18	0.02165
19	0.02242
20	0.02319
21	0.02397
22	0.02475
23	0.60134

$n = 45$ $r = 23$

i	b_i
1	0.00469
2	0.00671
3	0.00828
4	0.00964
5	0.01085
6	0.01197
7	0.01302
8	0.01402
9	0.01497
10	0.01589
11	0.01678
12	0.01765
13	0.01850
14	0.01934
15	0.02017
16	0.02099
17	0.02180
18	0.02261
19	0.02342
20	0.02422
21	0.02503
22	0.63544

$n = 45$ $r = 24$

i	b_i
1	0.00491
2	0.00702
3	0.00868
4	0.01009
5	0.01137
6	0.01254
7	0.01364
8	0.01468
9	0.01567
10	0.01664
11	0.01757
12	0.01848
13	0.01937
14	0.02025
15	0.02112
16	0.02197
17	0.02282
18	0.02367
19	0.02451
20	0.02535
21	0.67124

$n = 45$ $r = 25$

i	b_i
1	0.00516
2	0.00737
3	0.00911
4	0.01060
5	0.01193
6	0.01316
7	0.01431
8	0.01540
9	0.01645
10	0.01746
11	0.01844
12	0.01939
13	0.02033
14	0.02125
15	0.02215
16	0.02305
17	0.02394
18	0.02482
19	0.02570
20	0.70897

$n = 45$ $r = 26$

i	b_i
1	0.00543
2	0.00776
3	0.00958
4	0.01115
5	0.01255
6	0.01385
7	0.01506
8	0.01621
9	0.01731
10	0.01837
11	0.01940
12	0.02040
13	0.02138
14	0.02235
15	0.02330
16	0.02424
17	0.02517
18	0.02610
19	0.74889

$n = 45$ $r = 27$

i	b_i
1	0.00573
2	0.00819
3	0.01011
4	0.01176
5	0.01325
6	0.01461
7	0.01589
8	0.01710
9	0.01826
10	0.01937
11	0.02046
12	0.02151
13	0.02255
14	0.02356
15	0.02457
16	0.02556
17	0.02654
18	0.79132

$n = 45$ $r = 28$

i	b_i
1	0.00606
2	0.00866
3	0.01070
4	0.01245
5	0.01402
6	0.01546
7	0.01681
8	0.01809
9	0.01931
10	0.02050
11	0.02164
12	0.02276
13	0.02385
14	0.02492
15	0.02598
16	0.02702
17	0.83664

$n = 45$ $r = 29$

i	b_i
1	0.00643
2	0.00920
3	0.01136
4	0.01322
5	0.01488
6	0.01641
7	0.01785
8	0.01920
9	0.02050
10	0.02176
11	0.02297
12	0.02415
13	0.02531
14	0.02644
15	0.02756
16	0.88528

$n = 45$ $r = 30$

i	b_i
1	0.00686
2	0.00981
3	0.01211
4	0.01409
5	0.01586
6	0.01749
7	0.01902
8	0.02046
9	0.02185
10	0.02318
11	0.02447
12	0.02573
13	0.02696
14	0.02816
15	0.93781

$n = 45$ $r = 31$

i	b_i
1	0.00734
2	0.01050
3	0.01297
4	0.01508
5	0.01698
6	0.01872
7	0.02036
8	0.02190
9	0.02338
10	0.02480
11	0.02618
12	0.02752
13	0.02883
14	0.99492

$n = 45$ $r = 32$

i	b_i
1	0.00790
2	0.01130
3	0.01395
4	0.01623
5	0.01827
6	0.02014
7	0.02189
8	0.02355
9	0.02514
10	0.02667
11	0.02815
12	0.02958
13	1.05748

$n = 45$ $r = 33$

i	b_i
1	0.00855
2	0.01223
3	0.01510
4	0.01756
5	0.01976
6	0.02179
7	0.02368
8	0.02547
9	0.02719
10	0.02883
11	0.03043
12	1.12662

$n = 45$ $r = 34$

i	b_i
1	0.00932
2	0.01332
3	0.01645
4	0.01912
5	0.02152
6	0.02373
7	0.02578
8	0.02773
9	0.02959
10	0.03138
11	1.20381

$n = 45$ $r = 35$

i	b_i
1	0.01024
2	0.01463
3	0.01806
4	0.02100
5	0.02363
6	0.02604
7	0.02830
8	0.03043
9	0.03247
10	1.29103

$n = 45$ $r = 36$

i	b_i
1	0.01135
2	0.01622
3	0.02002
4	0.02328
5	0.02619
6	0.02886
7	0.03135
8	0.03371
9	1.39103

$n = 45$ $r = 37$

i	b_i
1	0.01274
2	0.01820
3	0.02246
4	0.02611
5	0.02936
6	0.03235
7	0.03514
8	1.50774

$n = 45$ $r = 38$

i	b_i
1	0.01451
2	0.02073
3	0.02557
4	0.02972
5	0.03342
6	0.03681
7	1.64704

$n = 45$ $r = 39$

i	b_i
1	0.01685
2	0.02407
3	0.02969
4	0.03449
5	0.03877
6	1.81818

$n = 45$ $r = 40$

i	b_i
1	0.02010
2	0.02869
3	0.03538
4	0.04108
5	2.03686

$n = 45$ $r = 41$

i	b_i
1	0.02489
2	0.03551
3	0.04376
4	2.33237

$n = 45$ $r = 42$

i	b_i
1	0.03268
2	0.04661
3	2.76794

$n = 45$ $r = 43$

i	b_i
1	0.04769
2	3.51655

APPENDIX C

Variance of θ_2^*'s in Terms of θ_2^2/n and K_3/K_2^*

[*] From Hassanein, K. M., Saleh, A. K., and Brown, E. (1995), "Best Linear Unbiased Estimate and Confidence Interval for Rayleigh's Scale Parameter When the Threshold Parameter Is Known for Data with Censored Observations from the Right," Report from the University of Kansas Medical Center. Reproduced by permission from the authors.

The variance, K_3/K_2 and relative efficiency of the scale parameter θ_2^* when the threshold parameter θ_1 is known for sample sizes $n = 5(1)25(5)45$ from the Rayleigh distribution with censoring from the right for $r = 0(1)n - 2$.

n	r	Var(θ_2^*)	K_3/K_2	Eff(θ_2^*)
5	0	0.05135	0.67288	1.00000
5	1	0.06443	0.77746	0.79699
5	2	0.08657	0.92585	0.59318
5	3	0.13181	1.17616	0.38959
6	0	0.04263	0.66609	1.00000
6	1	0.05127	0.75039	0.83139
6	2	0.06439	0.86033	0.66203
6	3	0.08655	1.01965	0.49256
6	4	0.13180	1.29107	0.32344
7	0	0.03643	0.66109	1.00000
7	1	0.04257	0.73179	0.85584
7	2	0.05124	0.81914	0.71102
7	3	0.06437	0.93534	0.56602
7	4	0.08653	1.10530	0.42105
7	5	0.13179	1.39649	0.27646
8	0	0.03181	0.65724	1.00000
8	1	0.03639	0.71817	0.87410
8	2	0.04255	0.79067	0.74764
8	3	0.05123	0.88197	0.62097
8	4	0.06436	1.00450	0.49426
8	5	0.08652	1.18466	0.36764
8	6	0.13178	1.49446	0.24138
9	0	0.02823	0.65419	1.00000
9	1	0.03178	0.70776	0.88826
9	2	0.03637	0.76976	0.77603
9	3	0.04253	0.84491	0.66361
9	4	0.05122	0.94034	0.55112
9	5	0.06435	1.06906	0.43863
9	6	0.08652	1.25896	0.32624
9	7	0.13178	1.58637	0.21418
10	0	0.02537	0.65170	1.00000
10	1	0.02820	0.69952	0.89956
10	2	0.03176	0.75371	0.79868
10	3	0.03636	0.81757	0.69764
10	4	0.04252	0.89558	0.59651
10	5	0.05121	0.99513	0.49536
10	6	0.06434	1.12984	0.39423
10	7	0.08651	1.32908	0.29321
10	8	0.13178	1.67323	0.19249
11	0	0.02303	0.64964	1.00000
11	1	0.02534	0.69284	0.90879
11	2	0.02819	0.74099	0.81718
11	3	0.03175	0.79651	0.72542
11	4	0.03635	0.86243	0.63358
11	5	0.04252	0.94336	0.54171
11	6	0.05120	1.04695	0.44983
11	7	0.06434	1.18747	0.35798
11	8	0.08651	1.39565	0.26624
11	9	0.13178	1.75578	0.17479
12	0	0.02109	0.64790	1.00000
12	1	0.02301	0.68731	0.91646
12	2	0.02533	0.73064	0.83257
12	3	0.02818	0.77976	0.74853
12	4	0.03174	0.83683	0.66442
12	5	0.03635	0.90491	0.58027
12	6	0.04251	0.98873	0.49611
12	7	0.05120	1.09627	0.41195
12	8	0.06434	1.24238	0.32783
12	9	0.08651	1.45916	0.24381
12	10	0.13178	1.83462	0.16006
13	0	0.01945	0.64641	1.00000
13	1	0.02108	0.68265	0.92294
13	2	0.02301	0.72206	0.84556
13	3	0.02533	0.76611	0.76805
13	4	0.02817	0.81642	0.69048
13	5	0.03174	0.87512	0.61286
13	6	0.03634	0.94536	0.53523
13	7	0.04251	1.03203	0.45758
13	8	0.05120	1.14341	0.37995
13	9	0.06434	1.29495	0.30236
13	10	0.08651	1.52002	0.22487
13	11	0.13177	1.91020	0.14762
14	0	0.01805	0.64512	1.00000
14	1	0.01944	0.67867	0.92849
14	2	0.02107	0.71481	0.85669
14	3	0.02300	0.75476	0.78477
14	4	0.02532	0.79974	0.71278
14	5	0.02817	0.85132	0.64076
14	6	0.03174	0.91169	0.56872
14	7	0.03634	0.98408	0.49666
14	8	0.04251	1.07354	0.42460
14	9	0.05120	1.18865	0.35256
14	10	0.06433	1.34544	0.28056
14	11	0.08651	1.57851	0.20865
14	12	0.13177	1.98290	0.13697
15	0	0.01684	0.64399	1.00000
15	1	0.01804	0.67522	0.93330
15	2	0.01943	0.70862	0.86632
15	3	0.02106	0.74517	0.79923
15	4	0.02300	0.78584	0.73209
15	5	0.02532	0.83183	0.66491
15	6	0.02817	0.88473	0.59771
15	7	0.03174	0.94677	0.53049
15	8	0.03634	1.02127	0.46327
15	9	0.04251	1.11346	0.39605
15	10	0.05119	1.23221	0.32885
15	11	0.06433	1.39408	0.26169
15	12	0.08650	1.63491	0.19462
15	13	0.13177	2.05302	0.12776
16	0	0.01577	0.64299	1.00000
16	1	0.01683	0.67222	0.93749
16	2	0.01803	0.70325	0.87474
16	3	0.01943	0.73694	0.81188
16	4	0.02106	0.77407	0.74897
16	5	0.02299	0.81557	0.68602
16	6	0.02532	0.86262	0.62305
16	7	0.02816	0.91684	0.56007
16	8	0.03173	0.98053	0.49708
16	9	0.03634	1.05711	0.43408
16	10	0.04251	1.15197	0.37110
16	11	0.05119	1.27426	0.30813
16	12	0.06433	1.44108	0.24520
16	13	0.08650	1.68942	0.18235
16	14	0.13177	2.12083	0.11971
17	0	0.01484	0.64211	1.00000
17	1	0.01577	0.66957	0.94120
17	2	0.01682	0.69856	0.88216
17	3	0.01803	0.72982	0.82302
17	4	0.01943	0.76397	0.76384
17	5	0.02106	0.80177	0.70463
17	6	0.02299	0.84413	0.64539
17	7	0.02532	0.89226	0.58614
17	8	0.02816	0.94781	0.52688
17	9	0.03173	1.01313	0.46762
17	10	0.03634	1.09175	0.40835
17	11	0.04250	1.18922	0.34910
17	12	0.05119	1.31495	0.28986
17	13	0.06433	1.48658	0.23066
17	14	0.08650	1.74222	0.17154
17	15	0.13177	2.18653	0.11261
18	0	0.01401	0.64131	1.00000
18	1	0.01483	0.66722	0.94448
18	2	0.01576	0.69443	0.88875
18	3	0.01682	0.72357	0.83292
18	4	0.01803	0.75520	0.77705
18	5	0.01942	0.78992	0.72115
18	6	0.02106	0.82843	0.66523
18	7	0.02299	0.87168	0.60930
18	8	0.02531	0.92088	0.55335
18	9	0.02816	0.97774	0.49740
18	10	0.03173	1.04468	0.44145
18	11	0.03634	1.12530	0.38550
18	12	0.04250	1.22531	0.32956
18	13	0.05119	1.35441	0.27363
18	14	0.06433	1.53073	0.21775
18	15	0.08650	1.79347	0.16193
18	16	0.13177	2.25031	0.10630
19	0	0.01326	0.64060	1.00000
19	1	0.01400	0.66512	0.94742
19	2	0.01483	0.69075	0.89464
19	3	0.01576	0.71806	0.84177
19	4	0.01682	0.74752	0.78887
19	5	0.01802	0.77961	0.73593
19	6	0.01942	0.81492	0.68297
19	7	0.02106	0.85417	0.63000
19	8	0.02299	0.89831	0.57702
19	9	0.02531	0.94860	0.52404
19	10	0.02816	1.00676	0.47105
19	11	0.03173	1.07527	0.41806
19	12	0.03634	1.15785	0.36507
19	13	0.04250	1.26036	0.31209
19	14	0.05119	1.39275	0.25913
19	15	0.06433	1.57363	0.20620
19	16	0.08650	1.84328	0.15335
19	17	0.13177	2.31233	0.10067
20	0	0.01260	0.63995	1.00000
20	1	0.01326	0.66323	0.95007
20	2	0.01400	0.68746	0.89994
20	3	0.01482	0.71316	0.84973
20	4	0.01576	0.74072	0.79949
20	5	0.01681	0.77057	0.74922
20	6	0.01802	0.80316	0.69893
20	7	0.01942	0.83909	0.64862
20	8	0.02105	0.87910	0.59831
20	9	0.02299	0.92414	0.54799
20	10	0.02531	0.97549	0.49767
20	11	0.02816	1.03493	0.44734
20	12	0.03173	1.10500	0.39701
20	13	0.03633	1.18950	0.34669
20	14	0.04250	1.29445	0.29638
20	15	0.05119	1.43005	0.24608
20	16	0.06433	1.61538	0.19582
20	17	0.08650	1.89179	0.14563
20	18	0.13177	2.37273	0.09560
21	0	0.01199	0.63937	1.00000
21	1	0.01259	0.66152	0.95246
21	2	0.01326	0.68450	0.90473
21	3	0.01400	0.70877	0.85693
21	4	0.01482	0.73467	0.80909
21	5	0.01575	0.76256	0.76123
21	6	0.01681	0.79283	0.71335
21	7	0.01802	0.82596	0.66546

n	r	Var(θ_2^*)	K_3/K_2	Eff(θ_2^*)
21	8	0.01942	0.86253	0.61756
21	9	0.02105	0.90329	0.56965
21	10	0.02299	0.94922	0.52174
21	11	0.02531	1.00163	0.47382
21	12	0.02816	1.06233	0.42590
21	13	0.03173	1.13393	0.37799
21	14	0.03633	1.22032	0.33007
21	15	0.04250	1.32766	0.28217
21	16	0.05119	1.46639	0.23429
21	17	0.06433	1.65608	0.18643
21	18	0.08650	1.93907	0.13864
21	19	0.13177	2.43163	0.09101
22	0	0.01144	0.63883	1.00000
22	1	0.01199	0.65996	0.95463
22	2	0.01259	0.68182	0.90908
22	3	0.01325	0.70481	0.86347
22	4	0.01399	0.72925	0.81782
22	5	0.01482	0.75542	0.77215
22	6	0.01575	0.78368	0.72646
22	7	0.01681	0.81441	0.68076
22	8	0.01802	0.84808	0.63505
22	9	0.01942	0.88529	0.58933
22	10	0.02105	0.92681	0.54361
22	11	0.02299	0.97363	0.49788
22	12	0.02531	1.02708	0.45215
22	13	0.02816	1.08903	0.40643
22	14	0.03173	1.16213	0.36070
22	15	0.03633	1.25037	0.31498
22	16	0.04250	1.36005	0.26926
22	17	0.05119	1.50185	0.22357
22	18	0.06433	1.69580	0.17790
22	19	0.08650	1.98523	0.13230
22	20	0.13177	2.48914	0.08685
23	0	0.01094	0.63834	1.00000
23	1	0.01144	0.65854	0.95661
23	2	0.01199	0.67938	0.91306
23	3	0.01259	0.70123	0.86944
23	4	0.01325	0.72435	0.82579
23	5	0.01399	0.74902	0.78211
23	6	0.01482	0.77552	0.73842
23	7	0.01575	0.80417	0.69472
23	8	0.01681	0.83536	0.65101
23	9	0.01802	0.86959	0.60729
23	10	0.01942	0.90745	0.56357
23	11	0.02105	0.94972	0.51984
23	12	0.02299	0.99741	0.47611
23	13	0.02531	1.05190	0.43238
23	14	0.02816	1.11507	0.38865
23	15	0.03173	1.18965	0.34492
23	16	0.03633	1.27970	0.30120
23	17	0.04250	1.39167	0.25749
23	18	0.05119	1.53649	0.21379
23	19	0.06433	1.73461	0.17012
23	20	0.08650	2.03034	0.12651
23	21	0.13177	2.54534	0.08305
24	0	0.01048	0.63788	1.00000
24	1	0.01094	0.65724	0.95843
24	2	0.01144	0.67715	0.91670
24	3	0.01198	0.69796	0.87491
24	4	0.01259	0.71991	0.83308
24	5	0.01325	0.74324	0.79124
24	6	0.01399	0.76818	0.74938
24	7	0.01482	0.79502	0.70751
24	8	0.01575	0.82408	0.66563
24	9	0.01681	0.85576	0.62375
24	10	0.01802	0.89054	0.58186
24	11	0.01942	0.92905	0.53996
24	12	0.02105	0.97206	0.49806
24	13	0.02298	1.02062	0.45616
24	14	0.02531	1.07613	0.41426
24	15	0.02816	1.14051	0.37237
24	16	0.03173	1.21654	0.33047
24	17	0.03633	1.30837	0.28858
24	18	0.04250	1.42259	0.24569
24	19	0.05119	1.57036	0.20483
24	20	0.06433	1.77257	0.16299
24	21	0.08650	2.07447	0.12121
24	22	0.13177	2.60034	0.07957
25	0	0.01006	0.63746	1.00000
25	1	0.01048	0.65605	0.96010
25	2	0.01094	0.67511	0.92004
25	3	0.01144	0.69498	0.87993
25	4	0.01198	0.71587	0.83979
25	5	0.01258	0.73799	0.79963
25	6	0.01325	0.76156	0.75946
25	7	0.01399	0.78680	0.71927
25	8	0.01482	0.81399	0.67908
25	9	0.01575	0.84347	0.63888
25	10	0.01681	0.87564	0.59867
25	11	0.01802	0.91098	0.55846
25	12	0.01942	0.95013	0.51825
25	13	0.02105	0.99388	0.47804
25	14	0.02298	1.04330	0.43782
25	15	0.02531	1.09981	0.39760
25	16	0.02816	1.16538	0.35739
25	17	0.03173	1.24283	0.31718
25	18	0.03633	1.33642	0.27697
25	19	0.04250	1.45285	0.23677
25	20	0.05119	1.60351	0.19659
25	21	0.06433	1.80973	0.15643
25	22	0.08650	2.11768	0.11633
25	23	0.13177	2.65419	0.07637
30	0	0.00838	0.63577	1.00000
30	1	0.00867	0.65125	0.96677
30	2	0.00898	0.66699	0.93342
30	3	0.00931	0.68321	0.90002
30	4	0.00967	0.70006	0.86661
30	5	0.01005	0.71767	0.83317
30	6	0.01048	0.73614	0.79972
30	7	0.01093	0.75561	0.76627
30	8	0.01143	0.77620	0.73280
30	9	0.01198	0.79808	0.69934
30	10	0.01258	0.82141	0.66586
30	11	0.01325	0.84640	0.63239
30	12	0.01399	0.87329	0.59891
30	13	0.01482	0.90237	0.56543
30	14	0.01575	0.93399	0.53195
30	15	0.01681	0.96858	0.49846
30	16	0.01802	1.00667	0.46498
30	17	0.01942	1.04895	0.43149
30	18	0.02105	1.09629	0.39800
30	19	0.02298	1.14985	0.36451
30	20	0.02531	1.21117	0.33103
30	21	0.02816	1.28241	0.29755
30	22	0.03173	1.36667	0.26406
30	23	0.03633	1.46860	0.23059
30	24	0.04250	1.59551	0.19712
30	25	0.05119	1.75990	0.16367
30	26	0.06433	1.98510	0.13023
30	27	0.08650	2.32167	0.09685
30	28	0.13177	2.90852	0.06358
35	0	0.00718	0.63454	1.00000
35	1	0.00739	0.64782	0.97153
35	2	0.00761	0.66123	0.94296
35	3	0.00785	0.67495	0.91435
35	4	0.00810	0.68909	0.88573
35	5	0.00837	0.70373	0.85709
35	6	0.00866	0.71894	0.82844
35	7	0.00897	0.73480	0.79978
35	8	0.00931	0.75139	0.77112
35	9	0.00966	0.76878	0.74245
35	10	0.01005	0.78707	0.71378
35	11	0.01047	0.80638	0.68511
35	12	0.01093	0.82680	0.65644
35	13	0.01143	0.84849	0.62776
35	14	0.01198	0.87159	0.59908
35	15	0.01258	0.89629	0.57040
35	16	0.01325	0.92281	0.54171
35	17	0.01399	0.95139	0.51303
35	18	0.01481	0.98236	0.48435
35	19	0.01575	1.01608	0.45566
35	20	0.01681	1.05302	0.42697
35	21	0.01802	1.09375	0.39829
35	22	0.01941	1.13901	0.36960
35	23	0.02105	1.18973	0.34092
35	24	0.02298	1.24717	0.31223
35	25	0.02531	1.31299	0.28355
35	26	0.02815	1.38952	0.25486
35	27	0.03172	1.48011	0.22618
35	28	0.03633	1.58976	0.19751
35	29	0.04250	1.72638	0.16884
35	30	0.05119	1.90345	0.14019
35	31	0.06433	2.14617	0.11155
35	32	0.08650	2.50913	0.08296
35	33	0.13177	3.14234	0.05446
40	0	0.00628	0.63361	1.00000
40	1	0.00644	0.64524	0.97510
40	2	0.00660	0.65693	0.95011
40	3	0.00678	0.66883	0.92509
40	4	0.00697	0.68102	0.90005
40	5	0.00717	0.69356	0.87501
40	6	0.00738	0.70650	0.84995
40	7	0.00761	0.71989	0.82489
40	8	0.00785	0.73379	0.79982
40	9	0.00810	0.74823	0.77475
40	10	0.00837	0.76328	0.74968
40	11	0.00866	0.77901	0.72461
40	12	0.00897	0.79546	0.69953
40	13	0.00930	0.81273	0.67445
40	14	0.00966	0.83090	0.64937
40	15	0.01005	0.85005	0.62429
40	16	0.01047	0.87030	0.59920
40	17	0.01093	0.89177	0.57412
40	18	0.01143	0.91460	0.54903
40	19	0.01198	0.93895	0.52395
40	20	0.01258	0.96503	0.49886
40	21	0.01325	0.99305	0.47377
40	22	0.01399	1.02329	0.44868
40	23	0.01481	1.05608	0.42359
40	24	0.01575	1.09183	0.39850
40	25	0.01680	1.13101	0.37341
40	26	0.01802	1.17425	0.34832
40	27	0.01941	1.22233	0.32324
40	28	0.02105	1.27625	0.29815
40	29	0.02298	1.33735	0.27306
40	30	0.02531	1.40740	0.24797
40	31	0.02815	1.48890	0.22289
40	32	0.03172	1.58541	0.19781
40	33	0.03633	1.70228	0.17273
40	34	0.04250	1.84798	0.14766
40	35	0.05119	2.03689	0.12260
40	36	0.06433	2.29595	0.09755
40	37	0.08650	2.68351	0.07255
40	38	0.13177	3.35992	0.04762
45	0	0.00558	0.63288	1.00000
45	1	0.00570	0.64323	0.97787
45	2	0.00583	0.65360	0.95567
45	3	0.00597	0.66411	0.93343
45	4	0.00612	0.67483	0.91119
45	5	0.00627	0.68581	0.88893
45	6	0.00643	0.69707	0.86667
45	7	0.00660	0.70867	0.84440
45	8	0.00678	0.72063	0.82213
45	9	0.00697	0.73299	0.79985
45	10	0.00717	0.74578	0.77758
45	11	0.00738	0.75905	0.75530

n	r	Var(θ_2^*)	K_3/K_2	Eff(θ_2^*)
45	12	0.00761	0.77283	0.73302
45	13	0.00784	0.78718	0.71073
45	14	0.00810	0.80214	0.68845
45	15	0.00837	0.81776	0.66616
45	16	0.00866	0.83412	0.64388
45	17	0.00897	0.85127	0.62159
45	18	0.00930	0.86929	0.59930
45	19	0.00966	0.88827	0.57701
45	20	0.01005	0.90831	0.55472
45	21	0.01047	0.92952	0.53243
45	22	0.01093	0.95204	0.51014
45	23	0.01143	0.97600	0.48784
45	24	0.01198	1.00158	0.46555
45	25	0.01258	1.02899	0.44326
45	26	0.01324	1.05848	0.42096
45	27	0.01399	1.09032	0.39867
45	28	0.01481	1.12486	0.37638
45	29	0.01575	1.16254	0.35408
45	30	0.01680	1.20387	0.33179
45	31	0.01802	1.24950	0.30950
45	32	0.01941	1.30025	0.28720
45	33	0.02105	1.35721	0.26491
45	34	0.02298	1.42176	0.24262
45	35	0.02531	1.49582	0.22033
45	36	0.02815	1.58201	0.19804
45	37	0.03172	1.68411	0.17576
45	38	0.03633	1.80779	0.15347
45	39	0.04250	1.96204	0.13120
45	40	0.05119	2.16210	0.10893
45	41	0.06433	2.43654	0.08668
45	42	0.08650	2.84723	0.06446
45	43	0.13177	3.56424	0.04231

APPENDIX D

Coefficients (a_i and b_i) of the Best Estimates of the Mean (μ) and Standard Deviation (σ) in Censored Samples Up to $n = 20$ from a Normal Population*

$n = 2$

$n-r$		$t_{(1)}$	$t_{(2)}$
0	μ	.5000	.5000
	σ	−.8862	.8862

$n = 3$

$n-r$		$t_{(1)}$	$t_{(2)}$	$t_{(3)}$
0	μ	.3333	.3333	.3333
	σ	−.5908	.0000	.5908
1	μ	.0000	1.0000	
	σ	−1.1816	1.1816	

$n = 4$

$n-r$		$t_{(1)}$	$t_{(2)}$	$t_{(3)}$	$t_{(4)}$
0	μ	.2500	.2500	.2500	.2500
	σ	−.4539	−.1102	.1102	.4539
1	μ	.1161	.2408	.6431	
	σ	−.6971	−.1268	.8239	
2	μ	−.4506	1.4056		
	σ	−1.3654	1.3654		

* From A. E. Sarhan and B. G. Greenberg, "Estimation of Location and Scale Parameters by Order Statistics from Singly and Doubly Censored Samples, Parts I and II," *Annals of Mathematics Statistics* Part I, Vol. 27, No. 2 and Part II, Vol. 29, No. 1 (1956) pp. 427–451. Reproduced by permission of *Annals of Mathematical Statistics*.

$n = 5$						
$n - r$		$t_{(1)}$	$t_{(2)}$	$t_{(3)}$	$t_{(4)}$	$t_{(5)}$
0	μ	.2000	.2000	.2000	.2000	.2000
	σ	−.3724	−.1352	.0000	.1352	.3724
1	μ	.1252	.1830	.2147	.4771	
	σ	−.5117	−.1668	.0274	.6511	
2	μ	−.0638	.1498	.9139		
	σ	−.7696	−.2121	.9817		
3	μ	−.7411	1.7411			
	σ	−1.4971	−1.4971			

$n = 6$							
$n - r$		$t_{(1)}$	$t_{(2)}$	$t_{(3)}$	$t_{(4)}$	$t_{(5)}$	$t_{(6)}$
0	μ	.1667	.1667	.1667	.1667	.1667	.1667
	σ	−.3175	−.1386	−.0432	.0432	.1386	.3175
1	μ	.1183	.1510	.1680	.1828	.3799	
	σ	−.4097	−.1685	−.0406	.0740	.5448	
2	μ	.0185	.1226	.1761	.6828		
	σ	−.5528	−.2091	−.0290	.7909		
3	μ	−.2159	.0649	1.1511			
	σ	−.8244	−.2760	1.1004			
4	μ	−1.0261	2.0261				
	σ	−1.5988	1.5988				

$n = 7$								
$n - r$		$t_{(1)}$	$t_{(2)}$	$t_{(3)}$	$t_{(4)}$	$t_{(5)}$	$t_{(6)}$	$t_{(7)}$
0	μ	.1429	.1429	.1429	.1429	.1429	.1429	.1429
	σ	−.2778	−.1351	−.0625	.0000	.0625	.1351	.2778
1	μ	.1088	.1295	.1400	.1487	.1571	.3159	
	σ	−.3440	−.1610	−.0681	.0114	.0901	.4716	
2	μ	.0465	.1072	.1375	.1626	.5462		
	σ	−.4370	−.1943	−.0718	.0321	.6709		
3	μ	−.0738	.0677	.1375	.8686			
	σ	−.5848	−.2428	−.0717	.8994			
4	μ	−.3474	−.0135	1.3609				
	σ	−.8682	−.3269	1.1951				
5	μ	−1.2733	2.2733					
	σ	−1.6812	1.6812					

$n = 8$									
$n-r$		$t_{(1)}$	$t_{(2)}$	$t_{(3)}$	$t_{(4)}$	$t_{(5)}$	$t_{(6)}$	$t_{(7)}$	$t_{(8)}$
0	μ	.1250	.1250	.1250	.1250	.1250	.1250	.1250	.1250
	σ	−.2476	−.1294	−.0713	−.0230	.0230	.0713	.1294	.2476
1	μ	.0997	.1139	.1208	.1265	.1318	.1370	.2704	
	σ	−.2978	−.1515	−.0796	−.0200	.0364	.0951	.4175	
2	μ	.0569	.0962	.1153	.1309	.1451	.4555		
	σ	−.3638	−.1788	−.0881	−.0132	.0570	.5868		
3	μ	−.0167	.0677	.1084	.1413	.6993			
	σ	−.4586	−.2156	−.0970	.0002	.7709			
4	μ	−.1549	.0176	.1001	1.0372				
	σ	−.6110	−.2707	−.1061	.9878				
5	μ	−.4632	−.0855	1.5487					
	σ	−.9045	−.3690	1.2735					
6	μ	−1.4915	2.4915						
	σ	−1.7502	1.7502						

$n = 9$										
$n-r$		$t_{(1)}$	$t_{(2)}$	$t_{(3)}$	$t_{(4)}$	$t_{(5)}$	$t_{(6)}$	$t_{(7)}$	$t_{(8)}$	$t_{(9)}$
0	μ	.1111	.1111	.1111	.1111	.1111	.1111	.1111	.1111	.1111
	σ	−.2237	−.1233	−.0751	−.0360	.0000	.0360	.0751	.1233	.2237
1	μ	.0915	.1018	.1067	.1106	.1142	.1177	.1212	.2365	
	σ	−.2633	−.1421	−.0841	−.0370	.0062	.0492	.0954	.3757	
2	μ	.0602	.0876	.1006	.1110	.1204	.1294	.3909		
	σ	−.3129	−.1647	−.0938	−.0364	.0160	.0678	.5239		
3	μ	.0104	.0660	.0923	.1133	.1320	.5860			
	σ	−.3797	−.1936	−.1048	.0333	.0317	.6797			
4	μ	−.0731	.0316	.0809	.1199	.8408				
	σ	−.4766	−.2335	−.1181	−.0256	.8537				
5	μ	−.2272	−.0284	.0644	1.1912					
	σ	−.6330	−.2944	−.1348	1.0622					
6	μ	−.5664	−.1521	1.7185						
	σ	−.9355	−.4047	1.3402						
7	μ	−1.6868	2.6868							
	σ	−1.8092	1.8092							

$n = 10$											
$n-r$		$t_{(1)}$	$t_{(2)}$	$t_{(3)}$	$t_{(4)}$	$t_{(5)}$	$t_{(6)}$	$t_{(7)}$	$t_{(8)}$	$t_{(9)}$	$t_{(10)}$
0	μ	.1000	.1000	.1000	.1000	.1000	.1000	.1000	.1000	.1000	.1000
	σ	−.2044	−.1172	−.0763	−.0436	−.0142	.0142	.0436	.0763	.1172	.2044
1	μ	.0843	.0921	.0957	.0986	.1011	.1036	.1060	.1085	.2101	
	σ	−.2364	−.1334	−.0851	−.0465	−.0119	.0215	.0559	.0937	.3423	
2	μ	.0605	.0804	.0898	.0972	.1037	.1099	.1161	.3424		
	σ	−.2753	−.1523	−.0947	−.0488	−.0077	.0319	.0722	.4746		
3	μ	.0244	.0636	.0818	.0962	.1089	.1207	.5045			
	σ	−.3252	−.1758	−.1058	−.0502	−.0006	.0469	.6107			
4	μ	−.0316	.0383	.0707	.0962	.1185	.7078				
	σ	−.3930	−.2063	−.1192	−.0501	.0111	.7576				
5	μ	−.1240	−.0016	.0549	.0990	.9718					
	σ	−.4919	−.2491	−.1362	−.0472	.9243					
6	μ	−.2923	−.0709	.0305	1.3327						
	σ	−.6520	−.3150	−.1593	1.1263						
7	μ	−.6596	−.2138	1.8734							
	σ	−.9625	−.4357	1.3981							
8	μ	−1.8634	2.8634								
	σ	−1.8608	1.8608								

$n = 11$												
$n-r$		$t_{(1)}$	$t_{(2)}$	$t_{(3)}$	$t_{(4)}$	$t_{(5)}$	$t_{(6)}$	$t_{(7)}$	$t_{(8)}$	$t_{(9)}$	$t_{(10)}$	$t_{(11)}$
0	μ	.0909	.0909	.0909	.0909	.0909	.0909	.0909	.0909	.0909	.0909	.0909
	σ	−.1883	−.1115	−.0760	−.0481	−.0234	.0000	.0234	.0481	.0760	.1115	.1883
1	μ	.0781	.0841	.0869	.0891	.0910	.0928	.0945	.0963	.0982	.1891	
	σ	−.2149	−.1256	−.0843	−.0519	−.0233	.0038	.0309	.0593	.0911	.3149	
2	μ	.0592	.0744	.0814	.0869	.0917	.0962	.1005	.1049	.3047		
	σ	−.2463	−.1417	−.0934	−.0555	−.0220	.0095	.0409	.0736	.4349		
3	μ	.0320	.0609	.0741	.0845	.0935	.1020	.1101	.4430			
	σ	−.2852	−.1610	−.1038	−.0589	−.0194	.0178	.0545	.5562			
4	μ	−.0082	.0415	.0642	.0820	.0974	.1116	.6116				
	σ	−.3357	−.1854	−.1163	−.0621	−.0146	.0299	.6842				
5	μ	−.0698	.0128	.0504	.0797	.1049	.8220					
	σ	−.4045	−.2175	−.1317	−.0647	−.0061	.8246					
6	μ	−.1702	−.0323	.0303	.0786	1.0937						
	σ	−.5053	−.2627	−.1519	−.0657	.9857						
7	μ	−.3516	−.1104	−.0016	1.4636							
	σ	−.6687	−.3331	−.1807	1.1825							
8	μ	−.7445	−.2712	2.0157								
	σ	−.9862	−.4630	1.4402								
9	μ	−2.0245	3.0245									
	σ	−1.9065	1.9065									

$n = 12$

$n-r$		$t_{(1)}$	$t_{(2)}$	$t_{(3)}$	$t_{(4)}$	$t_{(5)}$	$t_{(6)}$	$t_{(7)}$	$t_{(8)}$	$t_{(9)}$	$t_{(10)}$	$t_{(11)}$	$t_{(12)}$
0	μ	.0833	.0833	.0833	.0833	.0833	.0833	.0833	.0833	.0833	.0833	.0833	.0833
	σ	−.1748	−.1061	−.0749	−.0506	−.0294	−.0097	.0097	.0294	.0506	.0749	.1061	.1748
1	μ	.0726	.0775	.0796	.0813	.0828	.0842	.0855	.0868	.0882	.0896	.1719	
	σ	−.1972	−.1185	−.0827	−.0548	−.0305	−.0079	.0142	.0367	.0608	.0881	.2919	
2	μ	.0574	.0693	.0747	.0789	.0825	.0859	.0891	.0923	.0956	.2745		
	σ	−.2232	−.1324	−.0911	−.0590	−.0310	−.0050	.0203	.0461	.0733	.4020		
3	μ	.0360	.0581	.0682	.0759	.0827	.0888	.0948	.1006	.3950			
	σ	−.2545	−.1487	−.1007	−.0633	−.0308	−.0007	.0286	.0582	.5119			
4	μ	.0057	.0428	.0595	.0724	.0836	.0938	.1036	.5386				
	σ	−.2937	−.1686	−.1119	−.0678	−.0296	.0058	.0400	.6259				
5	μ	−.0382	.0210	.0477	.0684	.0861	.1022	.7128					
	σ	−.3448	−.1939	−.1255	−.0726	−.0267	.0155	.7479					
6	μ	−.1048	−.0109	.0313	.0637	.0915	.9292						
	σ	−.4146	−.2274	−.1428	−.0774	−.0210	.8833						
7	μ	−.2125	−.0609	.0070	.0589	1.2075							
	σ	−.5171	−.2749	−.1659	−.0820	1.0399							
8	μ	−.4059	−.1472	−.0321	1.5852								
	σ	−.6836	−.3493	−.1996	1.2324								
9	μ	−.8225	−.3249	2.1474									
	σ	−1.0075	−.4874	1.4948									
10	μ	−2.1728	3.1728										
	σ	−1.9474	1.9474										

$n = 13$

$n-r$	$t_{(1)}$	$t_{(2)}$	$t_{(3)}$	$t_{(4)}$	$t_{(5)}$	$t_{(6)}$	$t_{(7)}$	$t_{(8)}$	$t_{(9)}$	$t_{(10)}$	$t_{(11)}$	$t_{(12)}$	$t_{(13)}$
0 μ	.0769	.0769	.0769	.0769	.0769	.0769	.0769	.0769	.0769	.0769	.0769	.0769	.0769
σ	−.1632	−.1013	−.0735	−.0520	−.0335	−.0164	.0000	.0164	.0335	.0520	.0735	.1013	.1632
1 μ	.0679	.0718	.0735	.0749	.0761	.0771	.0781	.0792	.0802	.0813	.0824	.1576	
σ	−.1824	−.1122	−.0806	−.0563	−.0353	−.0160	.0026	.0212	.0404	.0612	.0850	.2724	
2 μ	.0552	.0648	.0691	.0724	.0752	.0778	.0803	.0827	.0852	.0877	.2497		
σ	−.2043	−.1243	−.0884	−.0607	−.0368	−.0148	.0063	.0273	.0490	.0723	.3743		
3 μ	.0380	.0555	.0633	.0693	.0745	.0792	.0836	.0880	.0924	.3564			
σ	−.2301	−.1382	−.0970	−.0653	−.0379	−.0128	.0113	.0352	.0598	.4750			
4 μ	.0144	.0430	.0557	.0655	.0739	.0816	.0888	.0958	.4813				
σ	−.2616	−.1549	−.1071	−.0703	−.0386	−.0095	.0182	.0456	.5781				
5 μ	−.0185	.0259	.0457	.0610	.0740	.0857	.0968	.6294					
σ	−.3011	−.1754	−.1191	−.0758	−.0386	−.0046	.0278	.6867					
6 μ	−.0659	.0020	.0322	.0553	.0750	.0928	.8085						
σ	−.3528	−.2015	−.1339	−.0819	−.0374	.0032	.8042						
7 μ	−.1371	−.0330	.0132	.0484	.0784	1.0301							
σ	−.4236	−.2363	−.1528	−.0888	−.0341	.9355							
8 μ	−.2516	−.0876	−.0151	.0400	1.3143								
σ	−.5276	−.2859	−.1785	−.0964	1.0884								
9 μ	−.4561	−.1817	−.0610	1.6988									
σ	−.6969	−.3638	−.2165	1.2773									
10 μ	−.8946	−.3753	2.2699										
σ	−1.0266	−.5094	1.5360										
11 μ	−2.3101	3.3101											
σ	−1.9845	1.9845											

$n = 14$

$n-r$		$t_{(1)}$	$t_{(2)}$	$t_{(3)}$	$t_{(4)}$	$t_{(5)}$	$t_{(6)}$	$t_{(7)}$	$t_{(8)}$	$t_{(9)}$	$t_{(10)}$	$t_{(11)}$	$t_{(12)}$	$t_{(13)}$	$t_{(14)}$
0	μ	.0714	.0714	.0714	.0714	.0714	.0714	.0714	.0714	.0714	.0714	.0714	.0714	.0714	.0714
	σ	−.1532	−.0968	−.0717	−.0526	−.0362	−.0212	−.0070	.0070	.0212	.0362	.0526	.0717	.0968	.1532
1	μ	.0637	.0669	.0683	.0694	.0704	.0712	.0721	.0728	.0736	.0745	.0753	.0762	.1455	
	σ	−.1698	−.1065	−.0784	−.0568	−.0384	−.0216	−.0056	.0100	.0259	.0426	.0609	.0820	.2556	
2	μ	.0530	.0609	.0643	.0670	.0692	.0713	.0732	.0751	.0770	.0789	.0809	.2291		
	σ	−.1885	−.1171	−.0854	−.0612	−.0404	−.0215	−.0036	.0140	.0319	.0505	.0707	.3506		
3	μ	.0388	.0529	.0592	.0639	.0680	.0717	.0752	.0785	.0819	.0852	.3247			
	σ	−.2102	−.1292	−.0933	−.0658	−.0423	−.0209	−.0006	.0192	.0393	.0601	.4438			
4	μ	.0199	.0426	.0526	.0602	.0667	.0726	.0782	.0835	.0887	.4350				
	σ	−.2361	−.1434	−.1023	−.0709	−.0440	−.0196	.0035	.0260	.0487	.5382				
5	μ	−.0057	.0288	.0440	.0557	.0655	.0744	.0828	.0908	.5637					
	σ	−.2678	−.1604	−.1129	−.0765	−.0455	−.0174	.0092	.0350	.6363					
6	μ	−.0411	.0102	.0328	.0500	.0646	.0777	.0899	.7159						
	σ	−.3077	−.1815	−.1256	−.0829	−.0466	−.0137	.0172	.7407						
7	μ	−.0915	−.0158	.0175	.0429	.0643	.0835	.8992							
	σ	−.3599	−.2084	−.1414	−.0903	−.0469	−.0077	.8546							
8	μ	−.1670	−.0537	−.0040	.0338	.0655	1.1255								
	σ	−.4317	−.2444	−.1618	−.0990	−.0457	.9825								
9	μ	−.2879	−.1127	−.0360	.0218	1.4148									
	σ	−.5372	−.2959	−.1898	−.1094	1.1322									
10	μ	−.5027	−.2142	−.0886	1.8054										
	σ	−.7091	−.3771	−.2318	1.3180										
11	μ	−.9616	−.4228	2.3843											
	σ	−1.0441	−.5293	1.5734											
12	μ	−2.4378	3.4378												
	σ	−2.0182	2.0182												

$n = 15$

$n-r$	$t_{(1)}$	$t_{(2)}$	$t_{(3)}$	$t_{(4)}$	$t_{(5)}$	$t_{(6)}$	$t_{(7)}$	$t_{(8)}$	$t_{(9)}$	$t_{(10)}$	$t_{(11)}$	$t_{(12)}$	$t_{(13)}$	$t_{(14)}$	$t_{(15)}$
0 μ	.0667	.0667	.0667	.0667	.0667	.0667	.0667	.0667	.0667	.0667	.0667	.0667	.0667	.0667	.0667
σ	−.1444	−.0927	−.0699	−.0526	−.0379	−.0247	−.0122	.0000	.0122	.0247	.0379	.0526	.0699	.0927	.1444
1 μ	.0599	.0627	.0639	.0648	.0655	.0662	.0669	.0675	.0682	.0688	.0695	.0702	.0709	.1351	
σ	−.1590	−.1013	−.0760	−.0568	−.0404	−.0256	−.0116	.0019	.0154	.0293	.0440	.0602	.0791	.2409	
2 μ	.0508	.0574	.0602	.0624	.0642	.0659	.0675	.0690	.0704	.0719	.0735	.0751	.2116		
σ	−.1752	−.1108	−.0825	−.0610	−.0427	−.0262	−.0106	.0044	.0195	.0349	.0512	.0690	.3300		
3 μ	.0390	.0506	.0556	.0595	.0628	.0657	.0685	.0711	.0737	.0763	.0790	.2982			
σ	−.1937	−.1214	−.0897	−.0655	−.0450	−.0265	−.0091	.0078	.0246	.0417	.0598	.4169			
4 μ	.0234	.0418	.0498	.0560	.0611	.0658	.0701	.0743	.0784	.0824	.3969				
σ	−.2154	−.1336	−.0977	−.0705	−.0473	−.0264	−.0068	.0122	.0310	.0502	.5042				
5 μ	.0030	.0305	.0425	.0516	.0593	.0663	.0727	.0789	.0849	.5104					
σ	−.2414	−.1481	−.1071	−.0760	−.0496	−.0258	−.0035	.0180	.0393	.5940					
6 μ	−.0244	.0155	.0330	.0462	.0574	.0674	.0767	.0856	.6425						
σ	−.2733	−.1654	−.1181	−.0822	−.0518	−.0244	.0012	.0258	.6882						
7 μ	−.0621	−.0046	.0205	.0395	.0555	.0698	.0830	.7983							
σ	−.3136	−.1870	−.1315	−.0894	−.0538	−.0219	.0079	.7892							
8 μ	−.1155	−.0326	.0036	.0309	.0539	.0743	.9854								
σ	−.3664	−.2146	−.1482	−.0979	−.0555	−.0174	.9001								
9 μ	−.1950	−.0732	−.0203	.0196	.0531	1.2157									
σ	−.4390	−.2518	−.1700	−.1082	−.0562	1.0252									
10 μ	−.3217	−.1364	−.0560	.0043	1.5097										
σ	−.5459	−.3050	−.2002	−.1211	1.1722										
11 μ	−.5462	−.2448	−.1148	1.9058											
σ	−.7201	−.3892	−.2458	1.3552											
12 μ	−1.0242	−.4676	2.4918												
σ	−1.0601	−.5477	1.6077												
13 μ	−2.5574	3.5574													
σ	−2.0493	2.0493													

$n = 16$																
$n-r$	$t_{(1)}$	$t_{(2)}$	$t_{(3)}$	$t_{(4)}$	$t_{(5)}$	$t_{(6)}$	$t_{(7)}$	$t_{(8)}$	$t_{(9)}$	$t_{(10)}$	$t_{(11)}$	$t_{(12)}$	$t_{(13)}$	$t_{(14)}$	$t_{(15)}$	$t_{(16)}$
0 μ	.0625	.0625	.0625	.0625	.0625	.0625	.0625	.0625	.0625	.0625	.0625	.0625	.0625	.0625	.0625	.0625
σ	−.1366	−.0889	−.0681	−.0524	−.0391	−.0272	−.0160	−.0053	.0053	.0160	.0272	.0391	.0524	.0681	.0889	.1366
1 μ	.0566	.0589	.0599	.0607	.0613	.0619	.0625	.0630	.0635	.0640	.0645	.0651	.0657	.0663	.1261	
σ	−.1495	−.0967	−.0737	−.0563	−.0416	−.0284	−.0161	−.0042	.0075	.0193	.0316	.0447	.0593	.0763	.2279	
2 μ	.0487	.0543	.0566	.0585	.0600	.0614	.0626	.0638	.0650	.0662	.0674	.0687	.0700	.1967		
σ	−.1637	−.1051	−.0797	−.0604	−.0441	−.0294	−.0158	−.0027	.0103	.0233	.0369	.0513	.0671	.3120		
3 μ	.0386	.0483	.0525	.0557	.0584	.0608	.0630	.0652	.0673	.0693	.0714	.0736	.2757			
σ	−.1797	−.1145	−.0862	−.0647	−.0466	−.0303	−.0151	−.0006	.0138	.0282	.0432	.0590	.3935			
4 μ	.0257	.0408	.0474	.0524	.0566	.0604	.0638	.0671	.0704	.0736	.0768	.3649				
σ	−.1982	−.1252	−.0935	−.0694	−.0491	−.0310	−.0140	.0022	.0182	.0343	.0508	.4748				
5 μ	.0090	.0313	.0410	.0484	.0545	.0601	.0652	.0700	.0747	.0794	.4664					
σ	−.2200	−.1376	−.1018	−.0747	−.0518	−.0313	−.0123	.0060	.0239	.0419	.5577					
6 μ	−.0129	.0191	.0329	.0434	.0522	.0601	.0673	.0742	.0808	.5829						
σ	−.2461	−.1523	−.1114	−.0806	−.0546	−.0313	−.0097	.0110	.0312	.6439						
7 μ	−.0420	.0030	.0225	.0373	.0496	.0606	.0707	.0803	.7180							
σ	−.2782	−.1700	−.1229	−.0874	−.0575	−.0307	−.0059	.0177	.7350							
8 μ	−.0817	−.0185	.0089	.0295	.0467	.0621	.0762	.8769								
σ	−.3189	−.1920	−.1369	−.0954	−.0604	−.0292	−.0004	.8333								
9 μ	−.1379	−.0484	−.0096	.0194	.0438	.0653	1.0674									
σ	−.3723	−.2204	−.1545	−.1049	−.0632	−.0262	.9415									
10 μ	−.2211	−.0916	−.0358	.0061	.0410	1.3015										
σ	−.4457	−.2586	−.1776	−.1167	−.0657	1.0642										
11 μ	−.3534	−.1587	−.0750	−.0125	1.5996											
σ	−.5538	−.3134	−.2097	−.1319	1.2088											
12 μ	−.5869	−.2739	−.1398	2.0006												
σ	−.7303	−.4004	−.2586	1.3894												
13 μ	−1.0829	−.5101	2.5931													
σ	−1.0748	−.5645	1.6394													
14 μ	−2.6696	3.6696														
σ	−2.0779	2.0779														

$n = 17$

$n-r$	$t_{(1)}$	$t_{(2)}$	$t_{(3)}$	$t_{(4)}$	$t_{(5)}$	$t_{(6)}$	$t_{(7)}$	$t_{(8)}$	$t_{(9)}$	$t_{(10)}$	$t_{(11)}$	$t_{(12)}$	$t_{(13)}$	$t_{(14)}$	$t_{(15)}$	$t_{(16)}$	$t_{(17)}$
0 μ	.0588	.0588	.0588	.0588	.0588	.0588	.0588	.0588	.0588	.0588	.0588	.0588	.0588	.0588	.0588	.0588	.0588
σ	–.1297	–.0854	–.0663	–.0519	–.0398	–.0290	–.0189	–.0094	.0000	.0094	.0189	.0290	.0398	.0519	.0663	.0854	.1297
1 μ	.0536	.0596	.0565	.0571	.0577	.0582	.0586	.0590	.0595	.0599	.0603	.0607	.0612	.0617	.0622	.1183	
σ	–.1412	–.0925	–.0715	–.0556	–.0423	–.0304	–.0194	–.0089	.0014	.0117	.0222	.0332	.0450	.0582	.0737	.2164	
2 μ	.0468	.0515	.0535	.0550	.0563	.0574	.0585	.0595	.0605	.0615	.0624	.0634	.0645	.0656	.1837		
σ	–.1537	–.1001	–.0769	–.0595	–.0448	–.0317	–.0196	–.0080	.0033	.0146	.0261	.0381	.0510	.0653	.2960		
3 μ	.0381	.0463	.0498	.0525	.0547	.0567	.0585	.0603	.0620	.0636	.0653	.0670	.0688	.2564			
σ	–.1677	–.1085	–.0829	–.0636	–.0474	–.0329	–.0196	–.0068	.0057	.0181	.0308	.0439	.0580	.3728			
4 μ	.0271	.0398	.0453	.0494	.0528	.0559	.0588	.0615	.0641	.0666	.0692	.0718	.3378				
σ	–.1837	–.1179	–.0895	–.0681	–.0501	–.0341	–.0102	–.0051	.0087	.0225	.0364	.0509	.4491				
5 μ	.0131	.0317	.0397	.0457	.0507	.0552	.0593	.0632	.0670	.0707	.0744	.4294					
σ	–.2022	–.1287	–.0969	–.0730	–.0529	–.0351	–.0185	–.0028	.0126	.0278	.0433	.5263					
6 μ	–.0047	.0214	.0327	.0411	.0482	.0545	.0603	.0658	.0710	.0762	.5334						
σ	–.2241	–.1412	–.1055	–.0786	–.0560	–.0359	–.0173	.0004	.0176	.0346	.6059						
7 μ	–.0278	.0083	.0239	.0356	.0454	.0540	.0620	.0695	.0767	.6525							
σ	–.2504	–.1561	–.1154	–.0849	–.0592	–.0364	–.0154	.0046	.0240	.6892							
8 μ	–.0585	–.0088	.0126	.0286	.0421	.0539	.0648	.0750	.7903								
σ	–.2828	–.1742	–.1274	–.0922	–.0627	–.0365	–.0124	.0105	.7777								
9 μ	–.1002	–.0317	–.0022	.0199	.0383	.0545	.0694	.9521									
σ	–.3238	–.1967	–.1419	–.1009	–.0665	–.0359	–.0079	.8736									
10 μ	–.1589	–.0633	–.0223	.0084	.0340	.0565	1.1456										
σ	–.3777	–.2257	–.1603	–.1113	–.0704	–.0341	.9796										
11 μ	–.2457	–.1091	–.0506	–.0070	.0293	1.3831											
σ	–.4519	–.2649	–.1846	–.1245	–.0744	1.1002											
12 μ	–.3832	–.1799	–.0932	–.0287	1.6850												
σ	–.5612	–.3212	–.2185	–.1418	1.2427												
13 μ	–.6253	–.3014	–.1637	2.0904													
σ	–.7398	–.4108	–.2705	1.4211													
14 μ	–1.1383	–.5506	2.6888														
σ	–1.0885	–.5802	1.6687														
15 μ	–2.7754	3.7754															
σ	–2.1046	2.1046															

$n = 18$

$n-r$	$t_{(1)}$	$t_{(2)}$	$t_{(3)}$	$t_{(4)}$	$t_{(5)}$	$t_{(6)}$	$t_{(7)}$	$t_{(8)}$	$t_{(9)}$	$t_{(10)}$	$t_{(11)}$	$t_{(12)}$	$t_{(13)}$	$t_{(14)}$	$t_{(15)}$	$t_{(16)}$	$t_{(17)}$	$t_{(18)}$
0 μ	.0556	.0556	.0556	.0556	.0556	.0556	.0556	.0556	.0556	.0556	.0556	.0556	.0556	.0556	.0556	.0556	.0556	.0556
σ	−.1235	−.0822	−.0645	−.0512	−.0401	−.0302	−.0211	−.0125	−.0041	.0041	.0125	.0211	.0302	.0401	.0512	.0645	.0822	.1235
1 μ	.0509	.0526	.0534	.0540	.0544	.0548	.0552	.0556	.0559	.0563	.0566	.0570	.0574	.0577	.0582	.0586	.1113	
σ	−.1338	−.0887	−.0693	−.0548	−.0426	−.0318	−.0219	−.0124	−.0033	.0058	.0149	.0243	.0342	.0450	.0570	.0712	.2061	
2 μ	.0449	.0489	.0507	.0520	.0531	.0540	.0549	.0558	.0566	.0574	.0582	.0590	.0598	.0607	.0616	.1723		
σ	−.1449	−.0955	−.0743	−.0584	−.0451	−.0333	−.0224	−.0121	−.0021	.0078	.0178	.0281	.0389	.0505	.0634	.2818		
3 μ	.0373	.0443	.0474	.0496	.0515	.0532	.0547	.0562	.0576	.0589	.0603	.0617	.0631	.0646	.2396			
σ	−.1573	−.1030	−.0798	−.0623	−.0477	−.0347	−.0228	−.0115	−.0005	.0103	.0213	.0325	.0442	.0568	.3545			
4 μ	.0279	.0387	.0433	.0468	.0497	.0522	.0546	.0568	.0589	.0610	.0631	.0652	.0674	.3144				
σ	−.1713	−.1114	−.0858	−.0665	−.0504	−.0361	−.0230	−.0105	.0015	.0135	.0254	.0377	.0505	.4264				
5 μ	.0161	.0317	.0384	.0434	.0475	.0512	.0546	.0578	.0609	.0639	.0669	.0698	.3979					
σ	−.1872	−.1209	−.0925	−.0711	−.0533	−.0375	−.0230	−.0092	.0042	.0173	.0305	.0440	.4988					
6 μ	.0013	.0230	.0323	.0392	.0450	.0502	.0549	.0593	.0636	.0677	.0718	.4918						
σ	−.2058	−.1318	−.1001	−.0763	−.0564	−.0388	−.0226	−.0073	.0075	.0221	.0367	.5728						
7 μ	−.0176	.0121	.0247	.0342	.0421	.0491	.0555	.0615	.0673	.0729	.5980							
σ	−.2278	−.1445	−.1089	−.0821	−.0598	−.0400	−.0219	−.0047	.0119	.0282	.6497							
8 μ	−.0419	−.0019	.0153	.0281	.0387	.0482	.0568	.0648	.0726	.7194								
σ	−.2543	−.1597	−.1191	−.0888	−.0635	−.0411	−.0205	−.0011	.0176	.7305								
9 μ	−.0741	−.0200	.0031	.0204	.0348	.0474	.0590	.0698	.8597									
σ	−.2869	−.1781	−.1315	−.0966	−.0676	−.0418	−.0183	.0039	.8168									
10 μ	−.1176	−.0441	−.0128	.0106	.0300	.0471	.0627	1.0241										
σ	−.3283	−.2010	−.1466	−.1059	−.0720	−.0421	−.0147	.9107										
11 μ	−.1788	−.0775	−.0343	−.0022	.0245	.0479	1.2204											
σ	−.3827	−.2307	−.1657	−.1173	−.0770	−.0414	1.0148											
12 μ	−.2689	−.1257	−.0648	−.0195	.0180	1.4609												
σ	−.4576	−.2707	−.1911	−.1317	−.0824	1.1335												
13 μ	−.4113	−.2001	−.1106	−.0442	1.7662													
σ	−.5681	−.3285	−.2266	−.1509	1.2741													
14 μ	−.6615	−.3276	−.1867	2.1758														
σ	−.7486	−.4205	−.2815	1.4505														
15 μ	−1.1905	−.5891	2.7796															
σ	−1.1013	−.5948	1.6960															
16 μ	−2.8756	3.8756																
σ	−2.1294	2.1294																

$n = 19$																			
$n-r$	$t_{(1)}$	$t_{(2)}$	$t_{(3)}$	$t_{(4)}$	$t_{(5)}$	$t_{(6)}$	$t_{(7)}$	$t_{(8)}$	$t_{(9)}$	$t_{(10)}$	$t_{(11)}$	$t_{(12)}$	$t_{(13)}$	$t_{(14)}$	$t_{(15)}$	$t_{(16)}$	$t_{(17)}$	$t_{(18)}$	$t_{(19)}$
0 μ	.0526	.0526	.0526	.0526	.0526	.0526	.0526	.0526	.0526	.0526	.0526	.0526	.0526	.0526	.0526	.0526	.0526	.0526	.0526
σ	−.1178	−.0792	−.0628	−.0505	−.0402	−.0312	−.0228	−.0150	−.0074	.0000	.0074	.0150	.0228	.0312	.0402	.0505	.0628	.0792	.1178
1 μ	.0485	.0500	.0506	.0511	.0515	.0519	.0522	.0525	.0528	.0531	.0534	.0537	.0540	.0543	.0547	.0550	.0554	.1052	
σ	−.1272	−.0851	−.0672	−.0538	−.0427	−.0328	−.0238	−.0152	−.0070	.0011	.0092	.0174	.0259	.0349	.0447	.0558	.0689	.1969	
2 μ	.0431	.0467	.0482	.0493	.0502	.0511	.0518	.0525	.0532	.0539	.0546	.0552	.0559	.0566	.0574	.0581	.1623		
σ	−.1372	−.0914	−.0719	−.0573	−.0451	−.0344	−.0245	−.0152	−.0063	.0025	.0113	.0202	.0295	.0392	.0498	.0616	.2690		
3 μ	.0365	.0426	.0451	.0471	.0487	.0501	.0514	.0526	.0538	.0550	.0561	.0572	.0584	.0596	.0608	.2249			
σ	−.1482	−.0982	−.0769	−.0609	−.0477	−.0359	−.0252	−.0150	−.0053	.0043	.0139	.0236	.0336	.0442	.0556	.3381			
4 μ	.0283	.0376	.0415	.0445	.0469	.0491	.0510	.0529	.0547	.0564	.0582	.0599	.0616	.0634	.2940				
σ	−.1605	−.1057	−.0823	−.0649	−.0504	−.0375	−.0257	−.0146	−.0040	.0065	.0170	.0276	.0385	.0499	.4062				
5 μ	.0182	.0315	.0371	.0413	.0448	.0479	.0507	.0534	.0559	.0584	.0609	.0633	.0658	.3707					
σ	−.1744	−.1141	−.0884	−.0692	−.0532	−.0391	−.0262	−.0140	−.0022	.0093	.0207	.0323	.0442	.4744					
6 μ	.0057	.0240	.0318	.0376	.0424	.0467	.0506	.0542	.0577	.0611	.0644	.0678	.4561						
σ	−.1905	−.1237	−.0952	−.0740	−.0563	−.0407	−.0264	−.0129	.0000	.0127	.0253	.0380	.5437						
7 μ	−.0100	.0147	.0253	.0331	.0396	.0453	.0506	.0554	.0601	.0647	.0691	.5520							
σ	−.2091	−.1347	−.1030	−.0794	−.0597	−.0423	−.0264	−.0114	.0029	.0170	.0309	.6152							
8 μ	−.0298	−.0032	.0172	.0276	.0363	.0439	.0509	.0574	.0636	.0696	.6603								
σ	−.2312	−.1476	−.1120	−.0855	−.0634	−.0439	−.0261	−.0093	.0067	.0224	.6898								
9 μ	−.0553	−.0115	.0070	.0209	.0324	.0425	.0516	.0602	.0684	.7839									
σ	−.2578	−.1630	−.1226	−.0925	−.0675	−.0454	−.0252	−.0063	.0118	.7685									
10 μ	−.0888	−.0306	−.0059	.0125	.0277	.0411	.0532	.0646	.9264										
σ	−.2907	−.1817	−.1353	−.1007	−.0720	−.0468	−.0237	−.0021	.8530										
11 μ	−.1341	−.0560	−.0229	.0017	.0221	.0399	.0561	1.0931											
σ	−.3324	−.2051	−.1509	−.1106	−.0772	−.0478	−.0210	.9450											
12 μ	−.1976	−.0910	−.0459	−.0124	.0153	.0395	1.2920												
σ	−.3873	−.2353	−.1708	−.1228	−.0831	−.0482	1.0474												
13 μ	−.2909	−.1415	−.0783	−.0316	.0070	1.5353													
σ	−.4629	−.2762	−.1971	−.1384	−.0898	1.1645													
14 μ	−.4379	−.2194	−.1273	−.0593	1.8438														
σ	−.5744	−.3353	−.2342	−.1594	1.3034														
15 μ	−.6958	−.3527	−.2086	2.2572															
σ	−.7568	−.4295	−.2918	1.4781															
16 μ	−1.2401	−.6258	2.8659																
σ	−1.1132	−.6084	1.7216																
17 μ	−2.9705	3.9705																	
σ	−2.1527	2.1527																	

$n = 20$

$n-r$	$t_{(1)}$	$t_{(2)}$	$t_{(3)}$	$t_{(4)}$	$t_{(5)}$	$t_{(6)}$	$t_{(7)}$	$t_{(8)}$	$t_{(9)}$	$t_{(10)}$	$t_{(11)}$	$t_{(12)}$	$t_{(13)}$	$t_{(14)}$	$t_{(15)}$	$t_{(16)}$	$t_{(17)}$	$t_{(18)}$	$t_{(19)}$	$t_{(20)}$
0 μ	.0500	.0500	.0500	.0500	.0500	.0500	.0500	.0500	.0500	.0500	.0500	.0500	.0500	.0500	.0500	.0500	.0500	.0500	.0500	.0500
σ	−.1128	−.0765	−.0611	−.0497	−.0402	−.0318	−.0241	−.0169	−.0101	−.0033	.0033	.0101	.0169	.0241	.0318	.0402	.0497	.0611	.0765	.1128
1 μ	.0462	.0476	.0482	.0486	.0489	.0493	.0495	.0498	.0501	.0503	.0506	.0508	.0511	.0513	.0516	.0519	.0522	.0525	.0996	
σ	−.1212	−.0819	−.0652	−.0528	−.0425	−.0335	−.0252	−.0174	−.0099	−.0026	.0046	.0119	.0193	.0271	.0354	.0444	.0546	.0667	.1184	
2 μ	.0415	.0446	.0459	.0469	.0477	.0484	.0491	.0497	.0502	.0508	.0514	.0519	.0525	.0531	.0537	.0544	.0550	.1533		
σ	−.1303	−.0876	−.0695	−.0561	−.0449	−.0351	−.0261	−.0177	−.0096	−.0017	.0061	.0140	.0221	.0305	.0394	.0491	.0599	.2574		
3 μ	.0356	.0409	.0431	.0448	.0462	.0474	.0485	.0496	.0506	.0516	.0525	.0535	.0544	.0554	.0564	.0575	.2119			
σ	−.1401	−.0938	−.0741	−.0595	−.0474	−.0367	−.0269	−.0178	−.0090	−.0004	.0080	.0166	.0253	.0344	.0440	.0543	.3233			
4 μ	.0284	.0365	.0399	.0424	.0445	.0463	.0480	.0496	.0511	.0526	.0540	.0554	.0569	.0584	.0599	.2762				
σ	−.1511	−.1006	−.0792	−.0632	−.0500	−.0384	−.0277	−.0178	−.0082	.0011	.0103	.0196	.0291	.0389	.0492	.3880				
5 μ	.0197	.0311	.0359	.0395	.0425	.0451	.0475	.0497	.0519	.0539	.0560	.0580	.0600	.0621	.3470					
σ	−.1634	−.1081	−.0847	−.0673	−.0528	−.0401	−.0285	−.0176	−.0071	.0030	.0131	.0232	.0335	.0441	.4526					
6 μ	.0090	.0246	.0312	.0361	.0402	.0438	.0470	.0501	.0530	.0558	.0586	.0613	.0640	.4254						
σ	−.1773	−.1166	−.0908	−.0717	−.0558	−.0418	−.0291	−.0171	−.0057	.0055	.0165	.0275	.0387	.5179						
7 μ	−.0042	.0167	.0255	.0321	.0375	.0423	.0466	.0507	.0545	.0583	.0619	.0656	.5126							
σ	−.1934	−.1262	−.0978	−.0766	−.0591	−.0437	−.0296	−.0164	−.0038	.0085	.0206	.0327	.5848							
8 μ	−.0206	.0069	.0185	.0272	.0343	.0406	.0463	.0517	.0567	.0616	.0664	.6103								
σ	−.2121	−.1374	−.1057	−.0822	−.0627	−.0455	−.0299	−.0153	−.0013	.0123	.0257	.6541								
9 μ	−.0414	−.0053	.0099	.0213	.0306	.0388	.0463	.0532	.0598	.0662	.7205									
σ	−.2343	−.1504	−.1149	−.0886	−.0667	−.0475	−.0300	−.0136	.0020	.0172	.7268									
10 μ	−.0680	−.0208	−.0008	.0140	.0262	.0369	.0466	.0557	.0642	.8460										
σ	−.2612	−.1660	−.1258	−.0959	−.0712	−.0494	−.0296	−.0111	.0065	.8038										
11 μ	−.1028	−.0408	−.0146	.0048	.0209	.0349	.0476	.0594	.9905											
σ	−.2943	−.1850	−.1389	−.1046	−.0762	−.0513	−.0287	−.0076	.8866											
12 μ	−.1498	−.0673	−.0326	−.0068	.0144	.0329	.0497	1.1595												
σ	−.3363	−.2088	−.1550	−.1151	−.0820	−.0531	−.0268	.9771												
13 μ	−.2154	−.1039	−.0569	−.0222	.0064	.0313	1.3607													
σ	−.3916	−.2396	−.1755	−.1280	−.0888	−.0544	1.0779													
14 μ	−.3117	−.1565	−.0914	−.0433	−.0037	1.6066														
σ	−.4679	−.2813	−.2028	−.1447	−.0967	1.1934														
15 μ	−.4632	−.2378	−.1433	−.0738	1.9180															
σ	−.5804	−.3417	−.2414	−.1673	1.3308															
16 μ	−.7284	−.3766	−.2298	2.3348																
σ	−.7645	−.4380	−.3014	1.5039																
17 μ	−1.2872	−.6610	2.9482																	
σ	−1.1244	−.6212	1.7456																	
18 μ	−3.0609	4.0609																		
σ	−2.1745	2.1745																		

APPENDIX E

Standard Normal Distribution*

Normal Distribution and Related Functions

$$F(x) = \int_{-\infty}^{x} \frac{1}{\sqrt{2\pi}} e^{-\frac{1}{2}t^2} \, dt$$

$$f(x) = \frac{1}{\sqrt{2\pi}} e^{-\frac{1}{2}x^2}$$

* Reprinted with permission from the *Handbook of Tables for Mathematical Problems and Statistics* (2nd ed.). Copyright © 1976 CRC Press, Boca Raton, Florida.

x	$F(x)$	$1 - F(x)$	$f(x)$	x	$F(x)$	$1 - F(x)$	$f(x)$
.00	.5000	.5000	.3989	.51	.6950	.3050	.3503
.01	.5040	.4960	.3989	.52	.6985	.3015	.3485
.02	.5080	.4920	.3989	.53	.7019	.2981	.3467
.03	.5120	.4880	.3988	.54	.7054	.2946	.3448
.04	.5160	.4840	.3986	.55	.7088	.2912	.3429
.05	.5199	.4801	.3984	.56	.7123	.2877	.3410
.06	.5239	.4761	.3982	.57	.7157	.2843	.3391
.07	.5279	.4721	.3980	.58	.7190	.2810	.3372
.08	.5319	.4681	.3977	.59	.7224	.2776	.3352
.09	.5359	.4641	.3973	.60	.7257	.2743	.3332
.10	.5398	.4602	.3970	.61	.7291	.2709	.3312
.11	.5438	.4562	.3965	.62	.7324	.2676	.3292
.12	.5478	.4522	.3961	.63	.7357	.2643	.3271
.13	.5517	.4483	.3956	.64	.7389	.2611	.3251
.14	.5557	.4443	.3951	.65	.7422	.2578	.3230
.15	.5596	.4404	.3945	.66	.7454	.2546	.3209
.16	.5636	.4364	.3939	.67	.7486	.2514	.3187
.17	.5675	.4325	.3932	.68	.7517	.2483	.3166
.18	.5714	.4286	.3925	.69	.7549	.2451	.3144
.19	.5753	.4247	.3918	.70	.7580	.2420	.3123
.20	.5793	.4207	.3910	.71	.7611	.2389	.3101
.21	.5832	.4168	.3902	.72	.7642	.2358	.3079
.22	.5871	.4129	.3894	.73	.7673	.2327	.3056
.23	.5910	.4090	.3885	.74	.7704	.2296	.3034
.24	.5948	.4052	.3876	.75	.7734	.2266	.3011
.25	.5987	.4013	.3867	.76	.7764	.2236	.2989
.26	.6026	.3974	.3857	.77	.7794	.2206	.2966
.27	.6064	.3936	.3847	.78	.7823	.2177	.2943
.28	.6103	.3897	.3836	.79	.7852	.2148	.2920
.29	.6141	.3859	.3825	.80	.7881	.2119	.2897
.30	.6179	.3821	.3814	.81	.7910	.2090	.2874
.31	.6217	.3783	.3802	.82	.7939	.2061	.2850
.32	.6255	.3745	.3790	.83	.7967	.2033	.2827
.33	.6293	.3707	.3778	.84	.7995	.2005	.2803
.34	.6331	.3669	.3765	.85	.8023	.1977	.2780
.35	.6368	.3632	.3752	.86	.8051	.1949	.2756
.36	.6406	.3594	.3739	.87	.8078	.1922	.2732
.37	.6443	.3557	.3725	.88	.8106	.1894	.2709
.38	.6480	.3520	.3712	.89	.8133	.1867	.2685
.39	.6517	.3483	.3697	.90	.8159	.1841	.2661
.40	.6554	.3446	.3683	.91	.8186	.1814	.2637
.41	.6591	.3409	.3668	.92	.8212	.1788	.2613
.42	.6628	.3372	.3653	.93	.8238	.1762	.2589
.43	.6664	.3336	.3637	.94	.8264	.1736	.2565
.44	.6700	.3300	.3621	.95	.8289	.1711	.2541

x	$F(x)$	$1 - F(x)$	$f(x)$	x	$F(x)$	$1 - F(x)$	$f(x)$
.45	.6736	.3264	.3605	.96	.8315	.1685	.2516
.46	.6772	.3228	.3589	.97	.8340	.1660	.2492
.47	.6808	.3192	.3572	.98	.8365	.1635	.2468
.48	.6844	.3156	.3555	.99	.8389	.1611	.2444
.49	.6879	.3121	.3538	1.00	.8413	.1587	.2420
.50	.6915	.3085	.3521	1.01	.8438	.1562	.2396
1.02	.8461	.1539	.2371	1.51	.9345	.0655	.1276
1.03	.8485	.1515	.2347	1.52	.9357	.0643	.1257
1.04	.8508	.1492	.2323	1.53	.9370	.0630	.1238
1.05	.8531	.1469	.2299	1.54	.9382	.0618	.1219
1.06	.8554	.1446	.2275	1.55	.9394	.0606	.1200
1.07	.8577	.1423	.2251	1.56	.9406	.0594	.1182
1.08	.8599	.1401	.2227	1.57	.9418	.0582	.1163
1.09	.8621	.1379	.2203	1.58	.9429	.0571	.1145
1.10	.8643	.1357	.2179	1.59	.9441	.0559	.1127
1.11	.8665	.1335	.2155	1.60	.9452	.0548	.1109
1.12	.8686	.1314	.2131	1.61	.9463	.0537	.1092
1.13	.8708	.1292	.2107	1.62	.9474	.0526	.1074
1.14	.8729	.1271	.2083	1.63	.9484	.0516	.1057
1.15	.8749	.1251	.2059	1.64	.9495	.0505	.1040
1.16	.8770	.1230	.2036	1.65	.9505	.0495	.1023
1.17	.8790	.1210	.2012	1.66	.9515	.0485	.1006
1.18	.8810	.1190	.1989	1.67	.9525	.0475	.0989
1.19	.8830	.1170	.1965	1.68	.9535	.0465	.0973
1.20	.8849	.1151	.1942	1.69	.9545	.0455	.0957
1.21	.8869	.1131	.1919	1.70	.9554	.0446	.0940
1.22	.8888	.1112	.1895	1.71	.9564	.0436	.0925
1.23	.8907	.1093	.1872	1.72	.9573	.0427	.0909
1.24	.8925	.1075	.1849	1.73	.9582	.0418	.0893
1.25	.8944	.1056	.1826	1.74	.9591	.0409	.0878
1.26	.8962	.1038	.1804	1.75	.9599	.0401	.0863
1.27	.8980	.1020	.1781	1.76	.9608	.0392	.0848
1.28	.8997	.1003	.1758	1.77	.9616	.0384	.0833
1.29	.9015	.0985	.1736	1.78	.9625	.0375	.0818
1.30	.9032	.0968	.1714	1.79	.9633	.0367	.0804
1.31	.9049	.0951	.1691	1.80	.9641	.0359	.0790
1.32	.9066	.0934	.1669	1.81	.9649	.0351	.0775
1.33	.9082	.0918	.1647	1.82	.9656	.0344	.0761
1.34	.9099	.0901	.1626	1.83	.9664	.0336	.0748
1.35	.9115	.0885	.1604	1.84	.9671	.0329	.0734
1.36	.9131	.0869	.1582	1.85	.9678	.0322	.0721
1.37	.9147	.0853	.1561	1.86	.9686	.0314	.0707
1.38	.9162	.0838	.1539	1.87	.9693	.0307	.0694
1.39	.9177	.0823	.1518	1.88	.9699	.0301	.0681

x	$F(x)$	$1 - F(x)$	$f(x)$	x	$F(x)$	$1 - F(x)$	$f(x)$
1.40	.9192	.0808	.1497	1.89	.9706	.0294	.0669
1.41	.9207	.0793	.1476	1.90	.9713	.0287	.0656
1.42	.9222	.0778	.1456	1.91	.9719	.0281	.0644
1.43	.9236	.0764	.1435	1.92	.9726	.0274	.0632
1.44	.9251	.0749	.1415	1.93	.9732	.0268	.0620
1.45	.9265	.0735	.1394	1.94	.9738	.0262	.0608
1.46	.9279	.0721	.1374	1.95	.9744	.0256	.0596
1.47	.9292	.0708	.1354	1.96	.9750	.0250	.0584
1.48	.9306	.0694	.1334	1.97	.9756	.0244	.0573
1.49	.9319	.0681	.1315	1.98	.9761	.0239	.0562
1.50	.9332	.0668	.1295	1.99	.9767	.0233	.0551
2.00	.9772	.0228	.0540	2.51	.9940	.0060	.0717
2.01	.9778	.0222	.0529	2.52	.9941	.0059	.0167
2.02	.9783	.0217	.0519	2.53	.9943	.0057	.0163
2.03	.9788	.0212	.0508	2.54	.9945	.0055	.0158
2.04	.9793	.0207	.0498	2.55	.9946	.0054	.0155
2.05	.9798	.0202	.0488	2.56	.9948	.0052	.0151
2.06	.9803	.0197	.0478	2.57	.9949	.0051	.0147
2.07	.9808	.0192	.0468	2.58	.9951	.0049	.0143
2.08	.9812	.0188	.0459	2.59	.9952	.0048	.0139
2.09	.9817	.0183	.0449	2.60	.9953	.0047	.0136
2.10	.9821	.0179	.0440	2.61	.9955	.0045	.0132
2.11	.9826	.0174	.0431	2.62	.9956	.0044	.0129
2.12	.9830	.0170	.0422	2.63	.9957	.0043	.0126
2.13	.9834	.0166	.0413	2.64	.9959	.0041	.0122
2.14	.9838	.0162	.0404	2.65	.9960	.0040	.0119
2.15	.9842	.0158	.0396	2.66	.9961	.0039	.0116
2.16	.9846	.0154	.0387	2.67	.9962	.0038	.0113
2.17	.9850	.0150	.0379	2.68	.9963	.0037	.0110
2.18	.9854	.0146	.0371	2.69	.9964	.0036	.0107
2.19	.9857	.0143	.0363	2.70	.9965	.0035	.0104
2.20	.9861	.0139	.0355	2.71	.9966	.0034	.0101
2.21	.9864	.0136	.0347	2.72	.9967	.0033	.0099
2.22	.9868	.0132	.0339	2.73	.9968	.0032	.0096
2.23	.9871	.0129	.0332	2.74	.9969	.0031	.0093
2.24	.9875	.0125	.0325	2.75	.9970	.0030	.0091
2.25	.9878	.0122	.0317	2.76	.9971	.0029	.0088
2.26	.9881	.0119	.0310	2.77	.9972	.0028	.0086
2.27	.9884	.0116	.0303	2.78	.9973	.0027	.0084
2.28	.9887	.0113	.0297	2.79	.9974	.0026	.0081
2.29	.9890	.0110	.0290	2.80	.9974	.0026	.0079
2.30	.9893	.0107	.0283	2.81	.9975	.0025	.0077
2.31	.9896	.0104	.0277	2.82	.9976	.0024	.0075
2.32	.9898	.0102	.0270	2.83	.9977	.0023	.0073

x	$F(x)$	$1 - F(x)$	$f(x)$	x	$F(x)$	$1 - F(x)$	$f(x)$
2.33	.9901	.0099	.0264	2.84	.9977	.0023	.0071
2.34	.9904	.0096	.0258	2.85	.9978	.0022	.0069
2.35	.9906	.0094	.0252	2.86	.9979	.0021	.0067
2.36	.9909	.0091	.0246	2.87	.9979	.0021	.0065
2.37	.9911	.0089	.0241	2.88	.9980	.0020	.0063
2.38	.9913	.0087	.0235	2.89	.9981	.0019	.0061
2.39	.9916	.0084	.0229	2.90	.9981	.0019	.0060
2.40	.9918	.0082	.0224	2.91	.9982	.0018	.0058
2.41	.9920	.0080	.0219	2.92	.9982	.0018	.0056
2.42	.9922	.0078	.0213	2.93	.9983	.0017	.0055
2.43	.9925	.0075	.0208	2.94	.9984	.0016	.0053
2.44	.9927	.0073	.0203	2.95	.9984	.0016	.0051
2.45	.9929	.0071	.0198	2.96	.9985	.0015	.0050
2.46	.9931	.0069	.0194	2.97	.9985	.0015	.0048
2.47	.9932	.0068	.0189	2.98	.9986	.0014	.0047
2.48	.9934	.0066	.0184	2.99	.9986	.0014	.0046
2.49	.9936	.0064	.0180	3.00	.9987	.0013	.0044
2.50	.9938	.0062	.0175	3.01	.9987	.0013	.0043
3.02	.9987	.0013	.0042	3.51	.9998	.0002	.0008
3.03	.9988	.0012	.0040	3.52	.9998	.0002	.0008
3.04	.9988	.0012	.0039	3.53	.9998	.0002	.0008
3.05	.9989	.0011	.0038	3.54	.9998	.0002	.0008
3.06	.9989	.0011	.0037	3.55	.9998	.0002	.0007
3.07	.9989	.0011	.0036	3.56	.9998	.0002	.0007
3.08	.9990	.0010	.0035	3.57	.9998	.0002	.0007
3.09	.9990	.0010	.0034	3.58	.9998	.0002	.0007
3.10	.9990	.0010	.0033	3.59	.9998	.0002	.0006
3.11	.9991	.0009	.0032	3.60	.9998	.0002	.0006
3.12	.9991	.0009	.0031	3.61	.9998	.0002	.0006
3.13	.9991	.0009	.0030	3.62	.9999	.0001	.0006
3.14	.9992	.0008	.0029	3.63	.9999	.0001	.0005
3.15	.9992	.0008	.0028	3.64	.9999	.0001	.0005
3.16	.9992	.0008	.0027	3.65	.9999	.0001	.0005
3.17	.9992	.0008	.0026	3.66	.9999	.0001	.0005
3.18	.9993	.0007	.0025	3.67	.9999	.0001	.0005
3.19	.9993	.0007	.0025	3.68	.9999	.0001	.0005
3.20	.9993	.0007	.0024	3.69	.9999	.0001	.0004
3.21	.9993	.0007	.0023	3.70	.9999	.0001	.0004
3.22	.9994	.0006	.0022	3.71	.9999	.0001	.0004
3.23	.9994	.0006	.0022	3.72	.9999	.0001	.0004
3.24	.9994	.0006	.0021	3.73	.9999	.0001	.0004
3.25	.9994	.0006	.0020	3.74	.9999	.0001	.0004
3.26	.9994	.0006	.0020	3.75	.9999	.0001	.0004

x	$F(x)$	$1 - F(x)$	$f(x)$	x	$F(x)$	$1 - F(x)$	$f(x)$
3.27	.9995	.0005	.0019	3.76	.9999	.0001	.0003
3.28	.9995	.0005	.0018	3.77	.9999	.0001	.0003
3.29	.9995	.0005	.0018	3.78	.9999	.0001	.0003
3.30	.9995	.0005	.0017	3.79	.9999	.0001	.0003
3.31	.9995	.0005	.0017	3.80	.9999	.0001	.0003
3.32	.9995	.0005	.0016	3.81	.9999	.0001	.0003
3.33	.9996	.0004	.0016	3.82	.9999	.0001	.0003
3.34	.9996	.0004	.0015	3.83	.9999	.0001	.0003
3.35	.9996	.0004	.0015	3.84	.9999	.0001	.0003
3.36	.9996	.0004	.0014	3.85	.9999	.0001	.0002
3.37	.9996	.0004	.0014	3.86	.9999	.0001	.0002
3.38	.9996	.0004	.0013	3.87	.9999	.0001	.0002
3.39	.9997	.0003	.0013	3.88	.9999	.0001	.0002
3.40	.9997	.0003	.0012	3.89	1.0000	.0000	.0002
3.41	.9997	.0003	.0012	3.90	1.0000	.0000	.0002
3.42	.9997	.0003	.0012	3.91	1.0000	.0000	.0002
3.43	.9997	.0003	.0011	3.92	1.0000	.0000	.0002
3.44	.9997	.0003	.0011	3.93	1.0000	.0000	.0002
3.45	.9997	.0003	.0010	3.94	1.0000	.0000	.0002
3.46	.9997	.0003	.0010	3.95	1.0000	.0000	.0002
3.47	.9997	.0003	.0010	3.96	1.0000	.0000	.0002
3.48	.9997	.0003	.0009	3.97	1.0000	.0000	.0002
3.49	.9998	.0002	.0009	3.98	1.0000	.0000	.0001
3.50	.9998	.0002	.0009	3.99	1.0000	.0000	.0001
				4.00	1.0000	.0000	.0001

APPENDIX F

Computer Program to Calculate the Reliability of a Consecutive-*k*-out-of-*n*:*F* System

```
        DIMENSION Q(20),P(20),F(0:20)
10      PRINT*,"Enter Total Number of Units   n"
        READ*,N
        PRINT*,"Enter Total Number of Consecutive Failed Units k"
        READ*,K
        IF (K .LE. 1 .OR. K .GE. N) GOTO 10
20      DO 100 I =1,N
        PRINT*,"Enter the reliability of component ", I
        READ*,P(I)
        IF (P(I) .LE. 0 .OR. P(I) .GT. 1) GOTO 20
        Q(I)=1-P(I)
100     CONTINUE
C
C
        DO 200 J=1,K
        IJ=J-1
        F(J)=0.0
200     CONTINUE
C
        QP=1
        DO 300 II=1,K
        QP=QP*Q(II)
300     CONTINUE
C
        F(K)=QP
        DO 400 IK=K+1, N
        QP=QP*(Q(IK)/Q(IK-K))
        F(IK)=F(IK-1)+ (1-F(IK-K-1))*P(IK-K)*QP
```

```
400     CONTINUE
        RS=1-F(N)
        PRINT*,"Reliability of the system is ", RS
C
        STOP
        END
```

APPENDIX G

Optimum Arrangement of Components in Consecutive-2-out-of-n:F Systems

```
c
c
      integer n,nmax,i,j,permutation,temp,k
      parameter(nmax=20)
c     Components are stored in component(nmax)
      integer component(nmax),pre_order(nmax)
c     Probabilities are stored in q(nmax)
      double precision q(nmax),previous,swap,seed,reliability
      real rand
      logical flag,disregard
      common /block/ q,previous,pre_order,n,reliability


      print*,"Please input the number of components : "
      read*,n

      do i=1,n
      print*,"Enter the unreliability of component ",i
        read*,q(i)
      enddo

c     Sort them in descending order
```

```
          do i=1,n
            do j=i+1,n
              if (q(i).lt.q(j)) then
                swap=q(i)
                q(i)=q(j)
                q(j)=swap
              endif
            enddo
          enddo
c         Maximum reliability and the corresponding order of components
c         will be stored in "previous" and "pre_order" respectively.

          previous=0.0
          do i=1,n
            pre_order(i)=i
          enddo

c         Now start enumerating the components.(There will be n!
c         of them. If the ones in reversed sequence are eliminated,
c         then there will be n!/2 sequences with distinct reliabilities. )
c         Initialize component(i) to 1 2 3 ....n

          do i=1,n
            component(i)=i
          enddo

c         Now calculate the reliability of the first sequence

          call calculate_reliability(component)
          permutation = 1

c         Swap the last two elements

5         temp=component(n-1)
          component(n-1)=component(n)
          component(n)=temp

c         Check whether this sequence appeared in reversed order before.

          disregard=.false.
          do i=1,component(1)-1
            if (component(n).eq.i) disregard=.true.
          enddo
```

```
c       Calculate the reliability of the next sequence (if it didn't
c       appear in reversed order before) which is obtained
c       from the previous one by swapping the last two elements.

          if (.not.disregard) then
            permutation=permutation+1
            call calculate_reliability(component,permutation)
          endif

c       Now in so-called dynamical do-loops the next sequence is generated.
c       First (n-2)nd element is increased by one, the position of its
c       new entry in the old sequence is found and replaced by the old
c       entry. If (n-2)nd element cannot be increased any further, then
c       the (n-3)rd entry is checked upon, and so on. If a new sequence
c       is obtained by increasing the entry in the kth position, then the
c       entries to the right of k (k+1,k+2,..,n) are sorted in ascending order.
c       That way it is possible to generate n! sequences.

          k=n-2
10      continue

c       If k is zero, this signals that all n!/2 sequences have been
c       generated.

        if (k.gt.0) then
          temp = component(k)
20        component(k)=component(k)+1
          if (component(k).eq.(n+1)) then
            component(k)=temp
            k=k-1
            goto 10
          endif
          flag=.false.
          do i=1,k-1
            if (component(i).eq.component(k)) flag=.true.
          enddo
          if (flag) goto 20
          component(k)=component(k)
          do i=k+1,n
            if (component(k).eq.component(i)) then
              component(i)=temp
            endif
          enddo
```

```
c       Now sort the elements to the right of k. (Starting from
c       (k+1) until n.)

          do i=k+1,n
            do j=i+1,n
              if (component(i).gt.component(j)) then
                temp=component(i)
                component(i)=component(j)
                component(j)=temp
              endif
            enddo
          enddo

c       Check whether this sequence appeared in reversed order before.

          disregard=.false.
          do i=1,component(1)-1
            if (component(n).eq.i) disregard=.true.
          enddo

c       If a new sequence is found , then its reliability will be calculated.

          if (.not.disregard) then
            permutation=permutation+1
            call calculate_reliability(component)
          endif
          goto 5
        endif
c
c       Now the result will be printed
c
          print*,"The number of components in this run is ",n
          print*," "
          print *,"The following sequence of components has the maximum",
          1       "reliability : "
          write (6,*)(pre_order(i),i=1,n)
          print*," "
          print*,"Its Reliability is ",previous
          print*," "
          do i=1,n
          print*,pre_order(i)," has the probability of failure ",
          1       q(pre_order(i))
```

```
      enddo
      end
c
c     The following subroutine calculates the reliability
c     of a given sequence of components.
c

      subroutine calculate_reliability (component)
      integer n,nmax,i,k,step,j1,j2,j,temp
      parameter(nmax=20)

c     counter(nmax) will index the dynamical do-loops.
c     neighbour(nmax,2) keeps the two consecutive components.
c     Note that there are (n-1) of them in an n-component system.

      integer counter(nmax),component(nmax),pre_order(nmax)
     1       ,neighbour(nmax,2)
      double precision q(nmax),probability,coef,previous,product
      double precision reliability
      common/block/ q,previous,pre_order,n,reliability

c     First set the register "neighbour".

      do i=1,n-1
        neighbour(i,1)=component(i)
        neighbour(i,2)=component(i+1)
      enddo

c     Calculate the sum of the products of failure probabilities
c     for two consecutive components.

      probability=0.0
      do i=1,n-1
        probability=probability+q(neighbour(i,1))*q(neighbour(i,2))
      enddo

c     Now there will be "step" many dynamical "do-loops", which
c     will be indexed by counter(1..step). This way the rest of
c     the terms in the equation will be calculated.

      do step=2,n-1
```

```
          coef = 1.0d0
          if ((step-2*(step/2)).eq.0) coef=-1.0d0

c     Initialize the counters for the dynamical do-loops.

        do i=1,step
          counter(i)=i
        enddo
5         product=1.0

c     For the enumerated sequence the probability is calculated.

        do i=1,step
         j=counter(i)
         product=product*q(neighbour(j,1))*q(neighbour(j,2))
        enddo
        do i=1,step-1
         j1=counter(i)
         j2=counter(i+1)
         if(neighbour(j1,2).eq.neighbour(j2,1)) then
           product = product/q(neighbour(j1,2))
         endif
        enddo
        probability = probability + coef*product

c     Now increment the counter of the innermost do-loop. If the
c     upper limit is reached, then the counter of the next innermost
c     do-loop is incremented, and so on.

        k=step
10      temp=counter(k)
        counter(k)=counter(k)+1
        if(counter(k).eq.(n-step+k)) then
         counter(k)=temp
         k=k-1
         if (k.gt.0) goto 10
        else
         do i=k+1,step
           counter(i)=counter(i-1)+1
         enddo
         if (counter(step).lt.n) goto 5
         endif
        enddo
```

```
c
c       At this point the probability of failure has been calculated
c       and now the reliability will be obtained.
c
          reliability = 1.0d0-probability
c
c       If the calculated reliability is greater than the largest
c       of the previous sequences, then store the present reliability
c       as the largest one.
c
          if (reliability.gt.previous) then
            previous=reliability
            do i=1,n
              pre_order(i)=component(i)
            enddo
          endif
        end
```

APPENDIX H

Computer Program for Solving the Time-Dependent Equations Using Runge-Kutta's Method

```
      EXTERNAL VECTOR
      DIMENSION TT(1000)
      REAL*8 X(10),XDOT(10)
      COMMON ALMDA1, ALMDA2, ALMDA3, AMU
      ALMDA1=0.00001
      ALMDA2=ALMDA1/10
      ALMDA3=ALMDA2/10
      AMU=0.0001
      N=5
      H=0.05
      T=0.0
      DO 100 I=1,5
      X(I)=0.0
100   CONTINUE
      X(1)=1.0
      DO 200 I=1,1000
      TT(I)=T
      CALL RKINT(T,X,N,H,VECTOR)
      WRITE(6,44) T, X(1) , X(2),X(3),X(4)
44    FORMAT (1X,'T=',F15.3,3X,F10.7,3X,F10.7,3X,F10.7,3X,F10.7)
200   CONTINUE
      STOP
      END
C
C
      SUBROUTINE RKINT(T,X,N,H,VECTOR)
      EXTERNAL VECTOR
      REAL*8 X(10),XDOT(10),K1(10),K2(10),K3(10),K4(10),SAVEX(10)
      DO 10 J=1,N
      SAVEX(J)=X(J)
```

```
10      CONTINUE
        T=T+1
        CALL VECTOR(T,X,XDOT,N)
        DO 11 J=1,N
        K1(J)=XDOT(J)
11      X(J)=SAVEX(J)+0.5*H*K1(J)
        T=T+0.5*H
        CALL VECTOR(T,X,XDOT,N)
        DO 12 J=1,N
        K2(J)=XDOT(J)
12      X(J)=SAVEX(J)+0.5*H*K2(J)
        T=T+0.5*H
        CALL VECTOR(T,X,XDOT,N)
        DO 13 J=1,N
        K3(J)=XDOT(J)
13      X(J)=SAVEX(J)+0.5*H*K3(J)
        T=T+0.5*H
        CALL VECTOR(T,X,XDOT,N)
        DO 14 J=1,N
        K4(J)=XDOT(J)
14      X(J)=SAVEX(J)+(H/6.)*(K1(J)+2*K2(J)+2*K3(J)+K4(J))
        RETURN
        END
C
        SUBROUTINE VECTOR(T,X,XDOT,N)
        REAL*8 XDOT(10),X(10)
        COMMON ALMDA, ALMDBR,AMU
        XDOT(1)=-(ALMDA1+ALMDA2+ALMDA3)*X(1)+AMU*X(2)+AMU*X(3)+AMU*X(4)
        XDOT(2)=-AMU*X(2)+ALMDA1*X(1)
        XDOT(3)=-AMU*X(3)+ALMDA2*X(1)
        XDOT(4)=-AMU*X(4)+ALMDA3*X(1)
        RETURN
        END
```

APPENDIX I

The Newton-Raphson Method

This method is used for solving nonlinear equations iteratively. We consider first a single variable equation. Let x_0 be a point that is not a root of the function $f(x)$ but a close estimate of the root. So the function $f(x)$ can be expanded using Taylor series about x_0 as

$$f(x) = f(x_0) + (x - x_0)f'(x_0) + \frac{(x - x_0)^2}{2!}f''(x_0) + \ldots$$

If $f(x) = 0$, then x must be a root, and the right side of the above equation constitutes an equation for the root x. The equation is a polynomial of degree infinity and an approximate value of x can be obtained by setting $f(x)$ to zero and taking the first two terms of the right side to yield

$$0 = f(x_0) + (x + x_0)f'(x_0).$$

Solving for x gives

$$x = x_0 - \frac{f(x_0)}{f'(x_0)}.$$

Now x represents an improved estimate of the root and can be used in lieu of x_0 in the above equation to obtain a better estimate of the root. The process is repeated until the difference between two consecutive estimates of the root is acceptable.

These steps can be summarized as follows:

1. Determine an initial estimate of x—say, $\hat{x}_0$—such that $f(\hat{x}_0) \simeq 0$.
2. $\hat{x}_1 = \hat{x}_0 - (f(\hat{x}_0)/f'(\hat{x}_0))$. $f'(\hat{x}_0)$ first derivative of $f(x)$ at $x = \hat{x}_0$.

3. $\hat{x}_{k+1} = \hat{x}_k - \ (f(\hat{x}_k)/f'(\hat{x}_k))$.

4. Stop when $|d| = \hat{x}_k - \hat{x}_{k+1}$ is less than or equal to ε.

EXAMPLE I.1

Find the value of x that results in the following function $f(x) = 0$:

$$f(x) = x^3 - 2x^2 + 5.$$

SOLUTION

$$f'(x) = 3x^2 - 4x$$

$$\text{Let } \hat{x}_0 = -1$$

$$f(\hat{x}_0) = 2 \qquad f'(\hat{x}_0) = 7$$

$$\hat{x}_1 = -1 - \frac{2}{7} = -1.285714$$

$$f(\hat{x}_1) = 0.431484$$

$$f_1'(\hat{x}_1) = 10.102037$$

• $\hat{x}_2 = -1.285714 - \dfrac{0.431484}{10.102037} = -1.243001$

$f(\hat{x}_2) = -0.010607 \qquad f'(\hat{x}_2) = 9.607163$

• $\hat{x}_3 = -1.243001 + \dfrac{0.010607}{9.607163} = -1.241897$

$f(\hat{x}_3) = -4.0673 \times 10^{-6} \qquad f'(\hat{x}_3) = 9.594511$

• $\hat{x}_4 = -1.241897 + \dfrac{4.0673 \times 10^{-6}}{9.594511} = -1.241896.$

The value of x that minimizes the function $f(x)$ is -1.241896.

This method can be extended to solve a system of equations with more than one unknown. For example, determine $x_1, x_2, \ldots, x_p$ such that

$$f_1(x_1, x_2, \ldots, x_p) = 0$$

$$f_2(x_1, x_2, \ldots, x_p) = 0$$

$$f_p(x_1, x_2, \ldots, x_p) = 0.$$

Let a_{ij} be the partial derivative of f_i w.r.t. x_j and $a_{ij} = \partial f_i / \partial x_j$. Construct the Jacobian Matrix J as

$$J = \begin{bmatrix} a_{11} & \cdots & a_{1p} \\ a_{21} & \cdots & a_{2p} \\ \cdot & \cdots & \cdot \\ \cdot & \cdots & \cdot \\ \cdot & \cdots & \cdot \\ \cdot & \cdots & \cdot \\ a_{p1} & \cdots & a_{pp} \end{bmatrix}.$$

Let $x_1^k, x_2^k, \ldots x_p^k$ be the approximate roots at the kth iteration. Let $f_1^k, \ldots, f_p^k$ be the corresponding values of the functions $f_1, \ldots, f_p$—that is,

$$f_1^k = f_1(x_1^k, \ldots, x_p^k)$$

$$f_2^k = f_2(x_1^k, \ldots, x_p^k)$$

$$f_p^k = f_p(x_1^k, \ldots, x_p^k).$$

Let b_{ij}^k be the ijth element of J^{-1} evaluated at $x_1^k, x_2^k, \ldots, x_p^k$. The net approximation is given by

$$x_1^{k+1} = x_1^k - (b_{11}^k f_1^k + b_{12}^k f_2^k + \ldots + f_{1p}^k f_p^k)$$

$$x_2^{k+1} = x_2^k - (b_{21}^k f_1^k + b_{22}^k f_2^k + \ldots + b_{2p}^k f_p^k)$$

$$x_p^{k+1} = x_p^k - (b_{p1}^k f_1^k + b_{p2}^k f_2^k + \ldots + b_{pp}^k f_p^k).$$

Let $x_1^0, x_2^0, \ldots, x_p^0$ be the initial values of x_i. The above iteration steps are continued until either $f_1, f_2, \ldots, f_p$ are close enough to zero or when the differences in the x values between two consecutive iterations are less than a specified amount ε.

EXAMPLE I.2

Find the values of x_1 and x_2, such that

$$x_1^2 - x_1 x_2 + 2x_2 - 4 = 0$$

$$x_2^2 + x_1 x_2 - 4x_1 = 0.$$

SOLUTION

$$p = 2$$

$$f_1 = x_1^2 - x_1x_2 + 2x_2 - 4$$

$$f_2 = x_2^2 + x_1x_2 - 4x_1.$$

We obtain the partial derivatives as

$$\frac{\partial f_1}{\partial x_1} = 2x_1 - x_2$$

$$\frac{\partial f_1}{\partial x_2} = -x_1 + 2$$

$$\frac{\partial f_2}{\partial x_1} = x_2 - 4$$

$$\frac{\partial f_2}{\partial x_2} = 2x_2 + x_1.$$

The Jacobian Matrix is

$$J = \begin{bmatrix} 2x_1 - x_2 & -x_1 + 2 \\ x_2 - 4 & 2x_2 + x_1 \end{bmatrix}.$$

Let the initial estimates of $x_1^0 = 1, x_2^0 = 2, f_1^0 = -1, f_2^0 = -2$

$$J = \begin{bmatrix} 0 & 1 \\ -2 & 5 \end{bmatrix}$$

$$J^{-1} = \begin{bmatrix} 2.5 & -0.5 \\ 1 & 0 \end{bmatrix}.$$

ITERATION 1

$$x_1^1 = 1 - [(-1)(2.5) + (-2)(-0.5)] = 2.5$$

$$x_2^1 = 2 - [(-1)(1) + (0)(2)] = 3$$

$$f_1^1 = 0.75, \qquad f_2^1 = 6.5$$

$$J = \begin{bmatrix} 2 & -0.5 \\ -1 & 8.5 \end{bmatrix}$$

$$J^{-1} = \begin{bmatrix} 0.5152 & 0.0303 \\ 0.0606 & 0.1212 \end{bmatrix}.$$

ITERATION 2

$$x_1^2 = 2.5 - [(0.5152)(0.75) + (0.0303)(6.5)] = 1.91665$$

$$x_2^2 = 3 - [(0.0606)(0.75) + (0.1212)(6.5)] = 2.16675$$

Substituting in f_1 and f_2 to obtain

$$f_1^2 = -0.14585, \qquad f_2^2 = 1.1811$$

$$J = \begin{bmatrix} 1.666 & 0.0833 \\ -1.8333 & 6.250 \end{bmatrix}$$

$$J^{-1} = \begin{bmatrix} 0.5916 & 0.1735 \\ -0.0079 & 0.1577 \end{bmatrix}.$$

ITERATION 3

$$x_1^3 = 1.9166 - [(0.5916)(-0.1458) + (-0.0079)(1.1811)] = 2.0017$$

$$x_2^3 = 2.16675 - [(0.1735)(-0.1458) + (0.1577)(1.1811)] = 2.00579$$

Substituting in f_1 and f_2 to obtain

$$f_1^3 = 0.003558, \qquad f_2^3 = 0.03125$$

$$J = \begin{bmatrix} 1.9977 & -0.0017 \\ -1.9942 & 0.0133 \end{bmatrix}$$

$$J^{-1} = \begin{bmatrix} 0.5007 & 0.0001 \\ 0.1661 & 0.1663 \end{bmatrix}.$$

ITERATION 4

$$x_1^4 = 2.0017 - [(0.5007)(0.003558) + (0.0001)(0.03125)] = 1.9999$$

$$x_2^4 = 2.0057 - [(0.1661)(0.003558) + (0.1663)(0.03125)] = 2.000007.$$

Substituting in f_1 and f_2 to obtain

$$f_1^4 = -0.00002, \qquad f_2^4 = 0.00006.$$

Since the values of the functions are very close to zero, the iteration is terminated, and the solution of the equations is

$$x_1 = 2, \qquad x_2 = 2.$$

APPENDIX J

Computer Listing of the Newton-Raphson Method

```
      implicit real*8(a-h,o-z)
      dimension t(100),s(100)
      common teta
      print *, 'what is the number of observations?'
      read *,n
      print *, 'please enter failure time associated
     % with each observation,pressing CR after each one'
      do 10 i=1,n
      read *,t(i)
      print *,i,':',t(i)
10    continue
      gam=3.0
      print *, gam
11    gkap=gam
      print *,'gamma:', gkap
      gam=gkap-dif(gkap,n,t)/dp(gkap,n,t)
      ep=gkap-gam
      epp=abs(ep)
      print *,'ep',ep
      if(epp.le.0.000001) goto 20
      goto 11
20    print *,'estimated gamma',gam
      fundif=dif(gam,n,t)
      print *,'value of difference func.',fundif
      print *,'theta:',teta
      print *,' reliabilities at given times'
      do 22 i=1,n
      s(i)=exp(-t(i)**gam/teta)
      print *,'s',i,':',s(i)
```

```
22      continue
        stop
        end
c
        function dif(gg,m,tt)
        implicit real*8(a-h,o-z)
        dimension tt(m)
        common teta
c       print *,'n is',m
        s1=0.0
        s2=0.0
        s3=0.0
        do 123 i=1,m
        s1=s1+tt(i)**gg*dlog(tt(i))
        s2=s2+tt(i)**gg
123     s3=s3+dlog(tt(i))
        print *,'sums',s1,s2,s3
        teta=s2/m
        dffif=s1/s2-s3/m-1/gg
        print *,'d func.',dffif
        dif=dffif
        return
        end
c
c
        function dp(ggg,mm,tt)
        implicit real*8(a-h,o-z)
        dimension tt(mm)
        sum1=0.0
        sum2=0.0
        sum3=0.0
        do 125 i=1,mm
        sum1=sum1+tt(i)**ggg*dlog(tt(i)**2
        sum2=sum2+tt(i)**ggg
125     sum3=sum3+tt(i)**ggg*dlog(tt(i))
        dpp=(sum1*sum2-sum3**2)/sum2**2+1/ggg**2
        dp=dpp
        return
        end
```

APPENDIX K

Baker's Algorithm*

```
#include <stdio.h> /* for reading data file */
#include <math.h> /* for square root */
#include <stdlib.h>

#define MAXNUM 600 /* max. sample size */
#define MAXBIN 101 /* max array size for H array */
#define LEN 40 /* max. length of filename */
int compare(float *x, float *y)
{
   if(*x==*y)return(0);
   return *x>*y?1:-1;
}
int geth(float *,int ,float ,float *,float *,int *,
int *,float *,float *);
void main()
{

   FILE *fp;
   float a[MAXNUM], dt, h[MAXBIN],period,mu,sigsq;
   char filename[LEN]; /* array to hold filename */
   int flag, hsize=101,n,i,scanval,binsreq;
   for(i=0;i<MAXBIN;i++)h[i]=0.;
   puts("1nter input file name: ");
   scanf("%40s",filename); /* get file name (MAX 40 CHARS) */
```

---

* Reprinted with permission from Rose D. Baker, © Rose D. Baker 1993.

```
    if ((fp = fopen(filename, "r")) == 0 ) /* open file */
    {
       puts ("Can't open input file"); /* unsuccessful */
       return;
    }
    fscanf(fp,"%f",&period);
    n=0;
    while(( scanval=fscanf(fp,"%f",&a[n])) !=EOF && scanval !=NULL)
    {
       ++n;
       if(n>=MAXNUM)
       {
          printf("sample size too large. Maximum is %d0,MAXNUM);
          return;
       }
    }
    fclose(fp);
    printf("read in %d numbers0,n);
    flag=geth(a,n,period,&dt,h,&hsize,&binsreq,&mu,&sigsq);
    if(flag<0){
       printf("%d array elements requested, which is too many.0,
       binsreq);
       return;
    }
    else {
       printf("%d array elements used0,binsreq);
       printf("h array size %d, time step %f0,hsize,dt);
       for(i=1;i<hsize;i++)printf("%f %f %f0,(float)(i*dt),h[i],
    (float)(i*dt)/mu+sigsq/(2.*mu*mu)-0.5);
    }
}
int geth(float *a,int n,float period,float *dt,float *h,int *hsize,
int *binsreq,float *mu,float *sigsq)
{
    float sigma=0., *perm_factor,scale,en,
    *cumsum,*total, *b;
    int i,k,n0,n0dash,s,bin,m, ndash,ntop, *x, *dope;
    *mu=0.;
    cumsum=(float *)calloc(n,sizeof(float)); /* allocate workspace */
    dope=(int *)calloc(n,sizeof(int));
    x=(int *)calloc(n,sizeof(int));
```

```
total=(float *)calloc(n,sizeof(float));
if(!cumsum || !dope || !x || !total )return (-1);
/*return if cant allocate workspace */

qsort(a,n,sizeof(float),compare); /* sort failure times */
for(i=0;i<n;i++){
   *mu+=a[i]; /* find mean */
   sigma+=a[i]*a[i];
}
*mu/=n;
sigma=(sigma-n*(*mu)*(*mu))/(n-1.);
*sigsq=sigma;
bin-*hsize;

*dt=period/(bin-1.);
scale=(bin-1.)/period;
for(i=0;i<n;i++) x[i]=(int)(scale*a[i]+0.5);
cumsum[0]=a[0]; /* find sums of order statistics */
for(i=1,m=1;i<n;i++){
   cumsum[i]=cumsum[i-1]+a[i];
   if(cumsum[i] < period)m=i+1;
}
if(m==n)printf("Maximum value of m, i.e. %d attained0,m);
*binsreq=(m+1)*(bin+1);
b=(float *)calloc(*binsreq,sizeof(float));
if(!b) return(-1.); /* not enough workspace */
for(i=0;i<m;i++)dope[i]=i*(bin+1); /* offsets to mimic 2 dim. array */
if(x[0]<bin)++b[dope[0]+x[0]]; /* start histogram off */

n0=x[0];
for(s=1;s<n;s++)
{
   n0dash=n0+x[s]; /* translate each level by x[s] */
   for(k=(s>(m-1))?(m-1):s;k>=1;k--) /* only build up histograms as far as
         mth level, */
   {
      ntop=(bin>n0dash)?n0dash:bin;
      for(ndash=x[s];ndash<=ntop;ndash++)
         b[dope[k]+ndash]+=b[dope[k-1]+ndash-x[s]];
   }
```

```
        if(x[s]<bin)++b[dope[0]+x[s]]; /* add new term to lowest level */
        n0=n0dash;
    }
    /* convert histograms to p.d.f.s, cumulate to dist. funs, and add. */
    perm_factor=cumsum; /* re-use cumsum space for perm numbers */
    en=(float)n;
    perm_factor[0]=1./en;
    for(i=1;i<m;i++)
        perm_factor[i]=(float)(perm_factor[i-1]*(i+1)/(en-1));
    for(i=0;i<bin;i++){
        h[i]=0.;
        for(s=0;s<m;s++){
            total[s]+=b[dope[s]+i]*perm_factor[s];
            h[i]+=total[s];
        }
    }
    /* make continuity correction...only want half the mass at last point */
    for(i=bin-1;i>0;i--)
        h[i]=0.5*(h[i]+h[i-1]);
    free(b);
    free(cumsum);
    free(x);
    free(dope);
    free(total);
    return(0);
}
```

APPENDIX L

Critical Values of χ^{2*}

* Reprinted from *Biometrika*, Vol. 32, "Tables of the Percentage Points of the χ^2 Distribution," C. M. Thompson, 1941, pp. 188–9. © 1941 *Biometrika* Trustees.

Critical Values of χ^2

Degrees of Freedom	$\chi^2_{0.995}$	$\chi^2_{0.990}$	$\chi^2_{0.975}$	$\chi^2_{0.950}$	$\chi^2_{0.900}$
1	0.0000393	0.0001571	0.0009821	0.0039321	0.0157908
2	0.0100251	0.0201007	0.0506356	0.102587	0.210720
3	0.0717212	0.114832	0.215795	0.351846	0.584375
4	0.206990	0.297110	0.484419	0.710721	1.063623
5	0.411740	0.554300	0.831211	1.145476	1.61031
6	0.675727	0.872085	1.237347	1.63539	2.20413
7	0.989265	1.239043	1.68987	2.16735	2.83311
8	1.344419	1.646482	2.17973	2.73264	3.48954
9	1.734926	2.087912	2.70039	3.32511	4.16816
10	2.15585	2.55821	3.24697	3.94030	4.86518
11	2.60321	3.05347	3.81575	4.57481	5.57779
12	3.07382	3.57056	4.40379	5.22603	6.30380
13	3.56503	4.10691	5.00874	5.89186	7.04150
14	4.07468	4.66043	5.62872	6.57063	7.78953
15	4.60094	5.22935	6.26214	7.26094	8.54675
16	5.14224	5.81221	6.90766	7.96164	9.31223
17	5.69724	6.40776	7.56418	8.67176	10.0852
18	6.26481	7.01491	8.23075	9.39046	10.8649
19	6.84398	7.63273	8.90655	10.1170	11.6509
20	7.43386	8.26040	9.59083	10.8508	12.4426
21	8.03366	8.89720	10.28293	11.5913	13.2396
22	8.64272	9.54249	10.9823	12.3380	14.0415
23	9.26042	10.19567	11.6885	13.0905	14.8479
24	9.88623	10.8564	12.4011	13.8484	15.6587
25	10.5197	11.5240	13.1197	14.6114	16.4734
26	11.1603	12.1981	13.8439	15.3791	17.2919
27	11.8076	12.8786	14.5733	16.1513	18.1138
28	12.4613	13.5648	15.3079	16.9279	18.9392
29	13.1211	14.2565	16.0471	17.7083	19.7677
30	13.7867	14.9535	16.7908	18.4926	20.5992
40	20.7065	22.1643	24.4331	26.5093	29.0505
50	27.9907	29.7067	32.3574	34.7642	37.6886
60	35.5346	37.4848	40.4817	43.1879	46.4589
70	43.2752	45.4418	48.7576	51.7393	55.3290
80	51.1720	53.5400	57.1532	60.3915	64.2778
90	59.1963	61.7541	65.6466	69.1260	73.2912
100	67.3276	70.0648	74.2219	77.9295	82.3581

Critical Values of χ^2 (Continued)

Degrees of Freedom	$\chi^2_{0.100}$	$\chi^2_{0.050}$	$\chi^2_{0.025}$	$\chi^2_{0.010}$	$\chi^2_{0.005}$
1	2.70554	3.84146	5.02389	6.63490	7.87944
2	4.60517	5.99147	7.37776	9.21034	10.5966
3	6.25139	7.81473	9.34840	11.3449	12.8381
4	7.77944	9.48773	11.1433	13.2767	14.8602
5	9.23635	11.0705	12.8325	15.0863	16.7496
6	10.6446	12.5916	14.4494	16.8119	18.5476
7	12.0170	14.0671	16.0128	18.4753	20.2777
8	13.3616	15.5073	17.5346	20.0902	21.9550
9	14.6837	16.9190	19.0228	21.6660	23.5893
10	15.9871	18.3070	20.4831	23.2093	25.1882
11	17.2750	19.6751	21.9200	24.7250	26.7569
12	18.5494	21.0261	23.3367	26.2170	28.2995
13	19.8119	22.3621	24.7356	27.6883	29.8194
14	21.0642	23.6848	26.1190	29.1413	31.3193
15	22.3072	24.9958	27.4884	30.5779	32.8013
16	23.5418	26.2962	28.8454	31.9999	34.2672
17	24.7690	27.5871	30.1910	33.4087	35.7185
18	25.9894	28.8693	31.5264	34.8053	37.1564
19	27.2036	30.1435	32.8523	36.1908	38.5822
20	28.4120	31.4104	34.1696	37.5662	39.9968
21	29.6151	32.6705	35.4789	38.9321	41.4010
22	30.8133	33.9244	36.7807	40.2894	42.7956
23	32.0069	35.1725	38.0757	41.6384	44.1813
24	33.1963	36.4151	39.3641	42.9798	45.5585
25	34.3816	37.6525	40.6465	44.3141	46.9278
26	35.5631	38.8852	41.9232	45.6417	48.2899
27	36.7412	40.1133	43.1944	46.9630	49.6449
28	37.9159	41.3372	44.4607	48.2782	50.9933
29	39.0875	42.5569	45.7222	49.5879	52.3356
30	40.2560	43.7729	46.9792	50.8922	53.6720
40	51.8050	55.7585	59.3417	63.6907	66.7659
50	63.1671	67.5048	71.4202	76.1539	79.4900
60	74.3970	79.0819	83.2976	88.3794	91.9517
70	85.5271	90.5312	95.0231	100.425	104.215
80	96.5782	101.879	106.629	112.329	116.321
90	107.565	113.145	118.136	124.116	128.299
100	118.498	124.342	129.561	135.807	140.169

APPENDIX M

Solutions of Selected Problems

Chapter 1

1-1. $\mu = \dfrac{a + b}{2}$

$$\sigma^2 = \frac{(b - a)^2}{12}$$

1-3. var $(t) = e^{2\mu + \sigma^2}(e^{\sigma^2} - 1)$

$P[t \leq \text{med}] = 0.5$

then

$$0 = \frac{\ln(\text{med}) - \mu}{\sigma}$$

$\mu = \ln(\text{med})$

$\text{med} = e^{\mu}$

1-10. $R(t) = \exp\left[-(k\lambda t^c + (1 - k)b(e^{\beta t^b} - 1))\right]$

If $c = 1$, $k = 1$, then $h(t) = \text{constant} = \lambda$.

If $c > 1$, $k = 1$, then $h(t)$ is an increasing function with t.

If $c = 1$, $0 \leq k < 1$, $b > 1$, then $h(t)$ is a decreasing function with t.

1-11. $R(t) = e^{-t/290} \sum_{k=0}^{2} \frac{(t/290)^k}{k!}$

$R(100) = e^{-0.345}(1 + 0.3448 + 0.05945) = 0.9945$

$E(t) = \gamma\theta = 870$ hours

Residual life = 770 hours.

1-12. $h(t) = \dfrac{\frac{8}{7}e^{-t} - \frac{8}{7}e^{-8t}}{\frac{8}{7}e^{-t} - \frac{1}{7}e^{-8t}}$

$MTTF = \dfrac{63}{56}$

1-15. a. $R(10^5) = 0.94205$

b. MTTF − no closed form expression

c. $R(10^3) = 0.99875$, so 9 systems will survive.

$R(10^4) = 0.99168$, so 9 systems will survive.

1-22. a. 0.768 failures

b. 10^4 hours

c. 0.9824

Chapter 2

2-3.

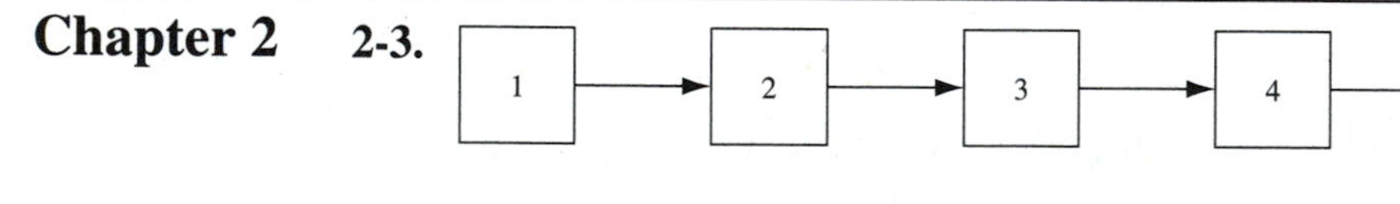

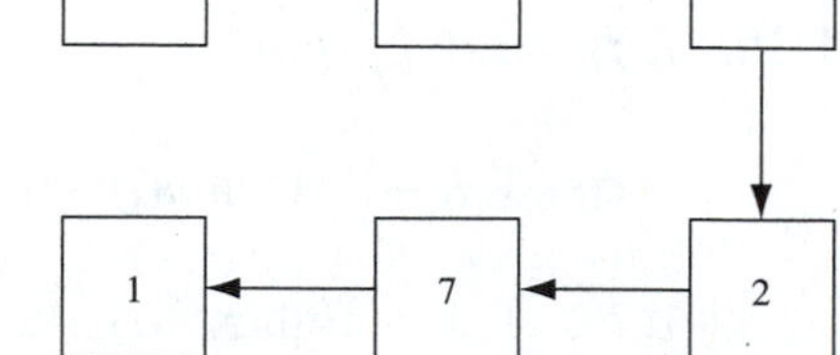

$R(t) = e^{-\sum \lambda_i t} = e^{-0.00285t}$

This is the reliability of one cassette. The reliability of recording from one cassette to another is

$R(t) = e^{-0.0057t}$

2-5.

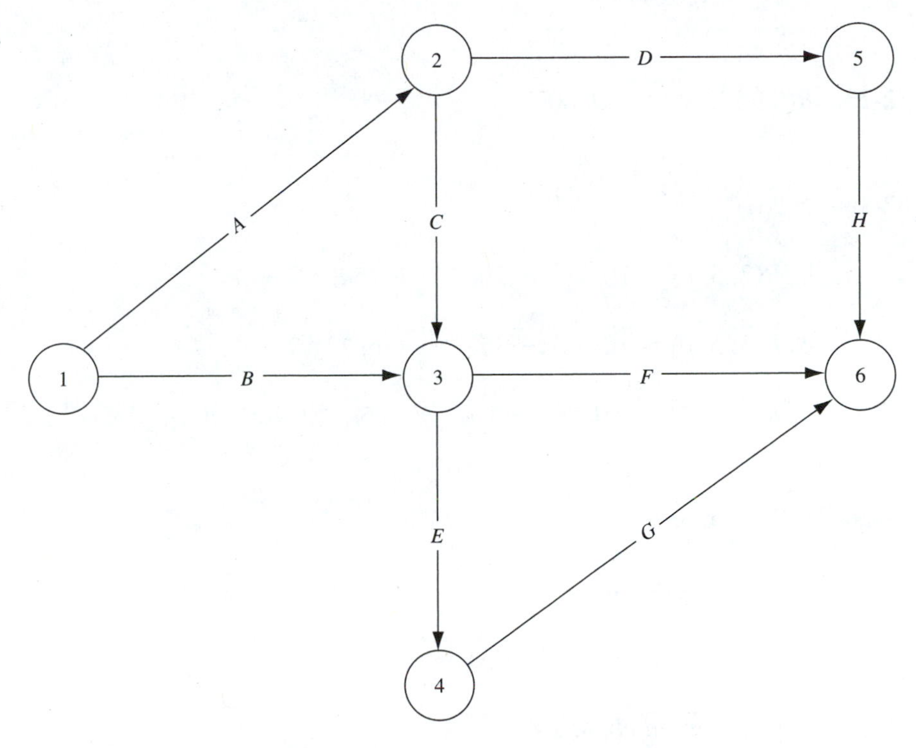

a. $R_{\text{system}} = 0.57632$

b. $R_A(t) = e^{-\int_0^t 0.2577\, e^{0.0027t} dt}$

$R_A(t) = e^{-9.544\left(e^{0.0027t} - 1\right)}$

$R_B(t), R_C(t), \ldots, R_h(t)$ can be determined as above. We then substitute the corresponding tie sets into the R_{system} equation to yield an expression that is a function of time. Integrating this expression from zero to infinity results in the MTTF. No closed-form expression will exist, therefore, use a numerical approach to estimate the MTTF.

c. To ensure a reliability of 0.98 after two years of service, the pipes connecting nodes 1 and 2 and those connecting 1 and 3, 2 and 3, and 3 and 6 must have redundant pipes in parallel with each of them.

2-9. a. $R = 0.90961$

b. $I_B^{10} = 0.00006$

$I_B^{13} = 0.02491$

2-13. $R(0.98, 2, 6) = 0.998032$

2-16. a. $R(t) = e^{-5\lambda t}$

b. $R(2; 5, p) = 10e^{-2\lambda t} - 20e^{-3\lambda t} + 15e^{-4\lambda t} - 4e^{-5\lambda t}$

c. $R(3; 5, p) = 10e^{-3\lambda t} - 15e^{-4\lambda t} + 6e^{-5\lambda t}$

d. $R = 5e^{-\lambda t} - 10e^{-2\lambda t} + 10e^{-3\lambda t} - 5e^{-4\lambda t} + e^{-5\lambda t}$

2-20. a. The cut sets are

$C_1 = \overline{ab}\ \overline{ac}\ \overline{ad},$

$C_2 = \overline{ab}\ \overline{db}\ \overline{cb},$

$C_3 = \overline{ab}\ \overline{ad}\ \overline{cb},$ and

$C_4 = \overline{ab}\ \overline{dc}\ \overline{ac}\ \overline{db}.$

The tie sets are

$T_1 = ab,$

$T_2 = ac\ cb,$

$T_3 = ad\ db,$

$T_4 = ad\ dc\ cb,$ and

$T_5 = ad\ dc\ ca\ ab.$

b. $R = P + 2P^2 - P^3 - 4P^4 + 6P^5 - 2P^6$

2-23. a. $MTTF = \int_0^{\infty} R(t)\,dt = \dfrac{1}{\lambda_1 + \lambda_2 + \lambda_3}$

b. $MTTF = \sqrt{\dfrac{\pi}{2(\lambda_1 + \lambda_2 + \lambda_3)}}$

c. No closed-form expression.

Chapter 3

3-1.
$$MTTF = \sum^{\lfloor (n+1)/2 \rfloor} \frac{n-j+1}{} \left[\int^{\infty} (e^{-kt^2/2})^{n-j}\,dt - j\int^{\infty} (e^{-kt^2/2})^{n-j+1}\,dt \right.$$
$$\left. + \frac{j(j-1)}{s} \int_0^{\infty} (-e^{-kt^2/2})^{n-j+2}\,dt + j\int_0^{\infty} (-e^{-kt^2/2})^{n-1}\,dt + \int_0^{\infty} (-e^{-kt^2/2})^{n}\,dt \right]$$

There is no closed form for this integral. For certain values of n and j, some integrals may be evaluated numerically.

3-4. $R_S(t) = 4e^{-4t^{\gamma}/\theta} - 6e^{-5t^{\gamma}/\theta} + 4e^{-6t^{\gamma}/\theta} - e^{-7t^{\gamma}/\theta}$

$$MTTF = \frac{\Gamma(1/\gamma)}{\gamma}\left(4\left(\frac{\theta}{4}\right)^{1/\gamma} - 6\left(\frac{\theta}{5}\right)^{1/\gamma} + 4\left(\frac{\theta}{6}\right)^{1/\gamma} + \left(\frac{\theta}{7}\right)^{1/\gamma}\right)$$

$$h_S(t) = \frac{f_s(t)}{R_s(t)} = \frac{\dfrac{\gamma}{\theta} t^{\gamma-1}\left[16 - 30e^{\frac{-t^{\gamma}}{\theta}} + 24e^{\frac{-2t^{\gamma}}{\theta}} - 7e^{\frac{-3t^{\gamma}}{\theta}}\right]}{\left[4 - 6e^{\frac{-t^{\gamma}}{\theta}} + 4e^{\frac{-2t^{\gamma}}{\theta}} - 7e^{\frac{-3t^{\gamma}}{\theta}}\right]}$$

3-9. a. $\lambda_1 = 5.70 \times 10^{-5}$, $\lambda_2 = 1.14 \times 10^{-4}$, $\lambda_3 = 3.42 \times 10^{-4}$ failures per hour

b. MTTF = 1,949.32 hours

c. 6.3072×10^{-4}

d. $\lambda_1 = 2.9236 \times 10^{-6}$

$\lambda_2 = 5.8473 \times 10^{-6}$

$\lambda_3 = 1.7542 \times 10^{-5}$

3-10. a. MTTF $= 2\lambda$

$$\sigma^2_{\text{MTTF}} = \int_0^\infty t^2 f(t)\, dt - (\text{MTTF})^2 = 6\lambda^2 - 4\lambda^2 = 2\lambda^2$$

b. $R(t) = 0.001497$

MTTF = 3,666 hours

3-11. $$R_{sys} = e^{-2.17\times 10^{-9} t^{2.3}} + e^{-2.5\times 10^{-8} t^2 - 1.14\times 10^{-6} t^{2.2}} + e^{-0.5\times 10^{-7} t - 1.14\times 10^{-6} t^{2.2}}$$

$$- e^{-0.5\times 10^{-7} t - 2.5\times 10^{-8} t^2 - 1.14\times 10^{-6} t^{2.2}}$$

$$- e^{-2.5\times 10^{-8} t^2 - 1.14\times 10^{-6} t^{2.2} - 2.17\times 10^{-9} t^{2.3}}$$

$$- e^{-0.5\times 10^{-7} t - 1.14\times 10^{-6} t^{2.2} - 2.17\times 10^{-9} t^{2.3}}$$

$$+ e^{-0.5\times 10^{-7} t - 2.5\times 10^{-8} t^2 - 1.14\times 10^{-6} t^{2.2} - 2.17\times 10^{-9} t^{2.3}}$$

$R_{sys}(1{,}000) = 0.9830935$

$$\text{MTTF} = \int_0^\infty R_{sys}(t) = \int_0^\infty e^{-2.17\times 10^{-9} t^{2.3}}\, dt + \int_0^\infty e^{-2.5\times 10^{-8} t^2 + 1.14\times 10^{-6} t^{2.2}}\, dt$$

$$+ \ldots + \int_0^\infty e^{-0.5\times 10^{-7} t - 2.5\times 10^{-8} t^2 - 1.14\times 10^{-6} t^{2.2} - 2.17\times 10^{-9} t^{2.3}}\, dt$$

There is no closed-form expression for the MTTF. An approximate value may be obtained through numerical analysis.

3-16. $R_s(t) = 4(R_{ps}(t))^3 - 3(R_{ps}(t))^4$

$$R_{ps}(t) = \exp\left[-0.00133t - 0.00144t^2 - \frac{t^{1.3}}{2.3\times 10^{-3}} - 3333.33e^{0.3t}\right]$$

3-20. $n = 3$

MTTF = 10,233 hours

3-21. One component

3-23. $R = 0.98597$

Chapter 4

4-1. a. $\hat{\lambda} = 0.010353$

b. $\lambda = \frac{1}{n} \sum x_i$

c. $R(49) = 0.6021$

d.

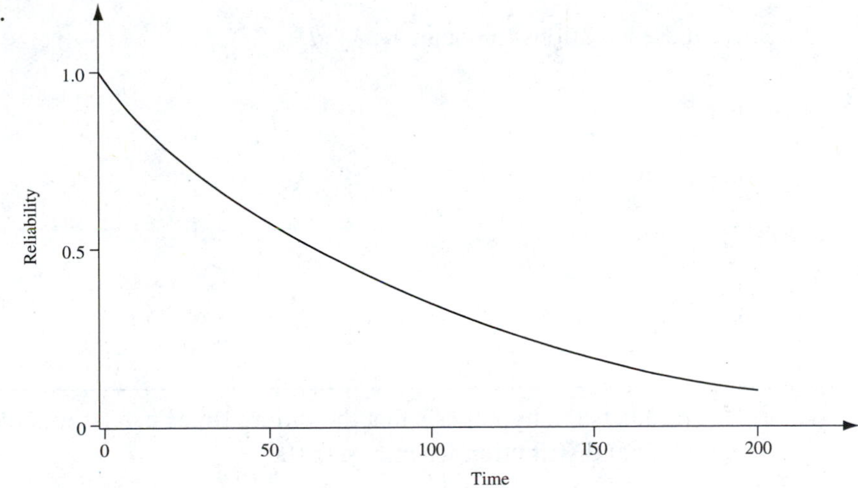

4-2. a. $\alpha = 3.83$

$\beta = 5.9766$

b. $R(t) = 1 - \left(400t^{4.830} - 178.6t^{10.806}\right)$

$$h(t) = \frac{f(t)}{R(t)} = \frac{1{,}930.17t^{4.830} - 1{,}930.17t^{10.806}}{1 - 400t^{4.830} + 178.6t^{10.806}}$$

4-9. a. From the first moment:

$$\theta = \frac{\sum t_i(\lambda_1 \lambda_2) - n\lambda_1}{n(\lambda_2 - \lambda_1)}.$$

From the second moment:

$$\theta = \frac{(\lambda_1^2 \lambda_2^2)\left(\frac{1}{n}\sum t_i^2\right) - 2\,\lambda_1^2}{2\,(\lambda_2^2 - \lambda_1^2)}.$$

From the third moment:

$$\theta = \frac{(\lambda_1^3 \lambda_2^3)\left(\frac{1}{n}\sum t_i^3\right) - 6\,\lambda_1^3}{6\,(\lambda_2^3 - \lambda_1^3)}.$$

Solve these equations to obtain $\lambda_1, \lambda_2, \theta$.

4-16. a. $M_1 = \nu$

$M_2 = 2\nu + \nu^2$

or

$\hat{\nu} = -1 \pm \sqrt{1 + M_2}$

Chapter 5

5-1. a. It contradicts the hypothesis that the failure times can be modeled by an exponential distribution when $\alpha = 0.10$.

b. Yes

c. Yes

d. MTTF = 29,038.5 hours

5-3. a. When $B_{10} > \chi^2_{0.99,\,9}$, the hypothesis that the data can be modeled by an exponential distribution cannot be rejected. t_1 is not abnormally short. t_{last} is not abnormally long.

b. $\hat{\lambda} = 3.55 \times 10^{-5}$

c. $R(20{,}000) = 0.4916$

d. MTTF = 28,169 hours

5-7. a. $\hat{\gamma} = 1.125$

$\hat{\theta} = 130{,}979.19$

b. $R(50{,}000) = 0.22850$

5-11. a. $\hat{\mu} = 2.8998$

$\hat{\sigma} = 0.7107$

$2.5497 < \mu < 3.2498$

$0.3890 < \sigma < 2.2350$

b. $\hat{\gamma} = 1.97, \hat{\theta}_1 = 22.8337$

$0.9423 < \gamma < 2.7573$

$22.8337 \exp[-U_{\alpha/2}/6.8243] \leq \theta_1 \leq 22.8337[-U_{1-\alpha/2}/6.8243]$

5-17. $\hat{\lambda} = \frac{1}{2r}\left[\sum_{i=1}^{r} t_i + \sum_{i=1}^{n-r} t_i^{+}\right]$

5-24. $R_{PLE}(8.5 \times 10^9) = 0.088506$

$R_{CHE}(8.5 \times 10^9) = 0.022823$

$MTTF_{PLE} = 4.818 \times 10^9$ cycles

$MTTF_{CHE} = 4.705 \times 10^9$ cycles

5-25. $\hat{\lambda} = 10{,}656.12$

$Var(\hat{\lambda}) = 2.2710 \times 10^8$

$5{,}985.872 < \lambda < 15{,}326.377$

$MTTF = 21{,}312.25$

Chapter 6

6-4. $\hat{\lambda}_o = 5.615 \times 10^{-5}$ failures per hour

$R(10^4) = 0.57035$

6-5. $\hat{\theta}_o = 188{,}720$

$R(50{,}000) = 0.99876$

The reliability requirements are met.

6-6. $f_o(t) = \dfrac{t}{A_F^2 \lambda_s^2} e^{-t/A_F\lambda_s}$

$$F_o(t) = 1 - \left(\frac{t}{A_F} + \lambda_s\right) / \lambda_s \, e^{-t/A_F\lambda_s}$$

$$R_o(t) = \left(\frac{t}{A_F} + \lambda_s\right) / \lambda_s \, e^{-t/A_F\lambda_s}$$

$$h_o(t) = \frac{t}{A_F\lambda_s \, (t + \lambda_s A_F)}$$

6-10. Life at 30°C and 5V is 61,206 hours.

6-12. a. Life at normal conditions is 32,764,059 hours.

b. $L_{OT} = 917.359$ hours

6-14. a. $R(10{,}000) = 0.573$

b. A_F(bet. 50 V and 5V) = 42.3

A_F(bet. 80V and 5V) = 162.1

c. $L_o = 3.14 \times 10^{11}$ hours

Chapter 7

7-1. 0.5 failures

7-2. $M(10^4) = 2.25$ failures

$A = 0.9523$

7-4. $M(20)_{\text{asymptotic}} = 0.214$

$M(40)_{\text{asymptotic}} = 0.9285$

$M(20)_{\text{exponential}} = 0.7137$

$M(40)_{\text{exponential}} = 1.427$

7-8. $M(10^4)$ brush motors $= 2.039$

$M(10^4)$ BLDC motors $= 0.861$

$x = 1.178y$

7-10. a. $M_A(200) = 41{,}371.21$

$M_B(200) = 41{,}370.89$

b. $P_A(200) = 0.68$

7-14. a. $\lambda = 0.01008$

$M(10^4) = 101.042$

$\text{Var}\,(N(10^4)) = 0.0$

b. Confidence interval

$101.042 \leq N(t) \leq 101.042$

7-16. a. $E[N(600) - N(200)] = 708.15$

b. $R(600) = 0.0$

Chapter 8

8-4. Warranty cost = \$252,530

8-5. a. Warranty cost = \$779.53

8-11. c = \$160.86

R = \$2,451,506

8-16. a. $M(12) \cong 16.66$

b. $W_0 \cong 4$ months

Chapter 9

9-1. a. $t_p^* = 0.999$

$L = 297.97$

9-2. Replace at failure; no inventory

9-7. $N^* = 1$

ϕ^* = \$95.334

9-11. Perform replacements on failure regardless of group size

Index

A

Abdel-Hameed, M., 545, 547, 548, 549, 581
Accelerated failure data models
  degradation models, 392–96
  physics-experimental-based, 388–92
  physics-statistics-based, 378–88
  statistic-based nonparametric, 368–78
  statistic-based parametric, 355–68
Accelerated failure time (AFT), 353, 354
Accelerated life testing (ALT)
  how it is used, 261–63, 353–54
  plans, 396–98
Accelerated stress, 353
Achieved availability, 184
Acoustic emission (AE), 571
Activation energy, 384
Active redundancy, 118–19, 196
Additive hazards model (AHM), 377–78
Aeronautical earth station (AES), 615, 617–18, 620–21, 624
Agrawal, A., 109, 110, 147
Ahn, J. H., 571, 583
Air route traffic control center (ARTCC), 615, 617, 619, 622
Air traffic services (ATS), 615
Allmen, C. R., 400, 409
Al-Najjar, B., 579, 581
Alternating renewal process, 170–75, 443–46
Altiok, T., 180, 216
Amato, H. N., 484, 525
Anderson, E. E., 484
Aneja, Y., 81, 148
Arnold, B. C., 349
Arrhenius model, 378–80
Ascher, H., 541, 581
Asher, J., 573, 581–82
Asymptotic relative efficiency (ARE), 222
AT&T, 4, 557, 582
AT&T Reliability Manual, 259
Atomic absorption, 572
Automatic dependent surveillance function (ADSF), 615, 618
Availability
  achieved, 184
  analysis and renewals, 421–28
  average uptime, 181–83
  importance of, 180
  inherent, 184
  instantaneous (point), 181
  mission-oriented, 181, 185
  number of spares and, 554–57
  operational, 184–85
  pointwise, 173
  repairable systems, 169–80
  steady-state, 171–72, 183–84
  time-interval versus downtime, 181
  work-mission, 187
Average uptime availability, 181–83

B

Baglee, D. A., 365, 366, 410
Bain, L. J., 296, 297, 300, 302, 303, 313, 315, 317, 319, 349
Baker, Rose D., 439, 441–43, 474, 705
Baker's algorithm, 705–8
Balakrishnan, N., 322, 327, 329, 331, 349–50
Barlow, R. E., 82, 109, 110, 120, 131, 147, 465, 474, 506, 525, 533, 541, 557, 558, 563, 582, 631, 636
Barlow-Proschan importance, 131
Bartholomew, D. J., 434, 446, 474
Bartlett's test, 265–66
Barton, R. R., 397, 409
Baruh, H., 180, 216
Baxter, L. A., 81, 148, 457, 458, 474
Bell Communications Research Reliability Manual, 259, 260, 349
Bellcore Special Report, 636
Bellcore Technical Requirements, 636
Bendell, T., 378, 411
Bergman, B., 261, 349
Best linear unbiased estimator (BLUE), 262
  coefficients, 647–60

Best linear unbiased estimator (BLUE) (*cont.*)
for Rayleigh parameters, 283–89
Beta model hazard function, 40–42
Beyne, J., 409–10
Binomial distribution, 157
likelihood function for, 230
Birnbaum's importance measure, 123–26, 134, 623
Birolini, A., 185, 187, 216, 629, 636
Bivariate distribution, 45–47
Black, J. R., 389, 405, 409
Blanks, H. S., 409, 534, 537, 582
Blischke, W. R., 477, 525
Block replacement policy, 529
BLUE. *See* Best linear unbiased estimator
Bohoris, G. A., 335, 349
Boland, P. J., 129, 148
Bollinger, R. C., 90, 148
Boltzmann's constant, 365, 381
Boolean truth table method, 104–6
Bottom-up heuristic (BUH), 82
Boucher, T. O., 247, 249, 257, 592, 636
Boulanger, M., 392, 394, 409
Brender, D. M., 563, 565, 582
Brombacher, A. C., 397, 409
Brown, E., 647, 661
Brown, J. R., 582
Brownlee, K. A., 631, 636
Bruins, R., 636
Buehler, M., 389, 409
Burn-in, 16, 261, 511
Buzacott, J. A., 65, 68, 528, 534, 582

C

CCITT (International Telegraph and Telephone Consultative Committee), 636
Camenga, R. E., 390, 410
Case, T., 106, 148
Cauchy distribution, 62
Censoring, types of, 263–65
Chan, C. K., 371, 392, 394, 409–10
Check-weigher example, 246–47
Chen, I. C., 366, 409
Chernobyl, 4
Chiang, D. T., 84, 86, 148
Christou, A., 359–60, 383, 389, 409
Cléroux, R., 541, 583
Coefficient(s)
of best estimates of mean and standard deviation, 665–77
of correlation, 249
of determination, 249
Cohen, A. C., Jr., 289, 292, 310, 349
Cold standby, 119, 196
Collins, J. A., 22, 67
Collision-avoidance system for robot manipulators example, 92–93
Combination model, 387–88
Comeford, R., 44, 67
Communications cables, example, 461–62
Complementary metal-oxide-silicon (CMOS) examples, 358, 385–87
Complex reliability systems
Boolean truth table method, 104–6
decomposition method, 96–100
event-space method, 103–4
factoring algorithm, 108–12
path-tracing method, 107–8
reduction method, 106–7
tie-set and cut-set methods, 100–102
Components
doubling, 120–21
keystone, 96–97, 144
optimal assignment in consecutive-2-out-of-n:F, 91–93
optimal assignment in system configuration, 80–83
Components, methods for measuring importance of
Barlow-Proschan importance, 131
Birnbaum's importance measure, 123–26, 134
criticality importance, 126–28
Fussell-Vesely importance, 129–30
upgrading function, 131–34
Compound events, 189, 195
Computer tomography example, 70–71
Conlon, T. W., 581–82
Confidence coefficient, 228
Confidence intervals, 228–30
for censored observations, 288–89
Gamma distribution parameters and, 319–20
for noncensored observations, 287–88
renewal process and, 457–59
Weibull distribution parameters and, 300–303
Consecutive-k-out-of-n:F systems, 83
computer program for calculating, 685, 687–93

consecutive-2-out-of-4:F systems, 85
consecutive-2-out-of-n:F systems, 84–85, 687–93
consecutive-2-out-of-7:F systems, 87–88
generalization of, 86–88
optimal assignment of components in consecutive-2-out-of-n:F, 91–93
reliability estimation, 88–91, 93–96
Consistent estimator, 221
Constandy, S. B., 582
Constant failure rate examples, 419–21
Constant hazard function
description of, 15–18
mean time to failure for, 53
mean time to failure in k-out-of-n systems with, 165–66
mean time to failure in parallel systems with, 162–63
mean time to failure in series systems with, 160
Constant interval replacement policy (CIRP), 529–33, 537–38
Continuous probability distribution, likelihood function for, 234–35
Continuous time
nonparametric renewal function estimation, 432–38
parametric renewal function estimation, 416–28
Cooper, R. B., 555, 582
Corrosion, 397
monitoring, 573
Cost minimization, 528–37
Cox, D. R., 68, 371, 372, 409, 432, 434, 444, 449, 450, 452, 474, 552, 553, 582
Crane spreader subsystem, case study, 585–91
Creep fatigue, 391
Criticality importance, 126–28
Crook, D. L., 189, 216
Croun, R., 121, 148
Cumming, A. C. D., 572, 582
Cumulative distribution function
for gamma model hazard function, 36
for normal model hazard function, 27, 28
Cumulative-hazard estimator (CHE), 334–35
Cumulative hazard function, 14
Cut-set method, 100–102
Cutters, example of end mill, 276–78

D

Dale, C. J., 310, 371, 375, 409
Dallas, D., 201, 216
Data, sources for failure, 259–60
David, F. N., 349
David, H. A., 349
Debugging region, 14
Decomposition method, 96–100
De Ceuninck, W., 409–10
Decreasing failure rates (DFR), 49, 50, 51
Defense Procurement Reform Act (1985), 475
Degradation models
hot-carrier, 395–96
laser, 394–95
resistor, 392–94
Delayed renewal process, 448–49
Deligönül, Z., 446–47, 474
Dependent failure estimates
compound events, 189, 195
joint density function, 189, 191–95
Markov model, 189–91
Derman, C., 91, 92, 148
Descovich, T., 636
DeSchepper, L., 408, 409
Detection system, case study of explosive, 601–8
Dhillon, B. S., 60, 68, 112, 148, 213, 216
Dhiman, J., 389, 409
Digital signal processors (DSPs), 605
Diodes, examples using, 112, 138, 179–80, 291–95, 438
Directed networks, 96
Discrete time
nonparametric renewal function estimation, 438–43
parametric renewal function estimation, 428–32
Dixon, W. J., 222, 256
Domangue, E., 189, 216
Doubling, 120–21
Downham, E., 571, 582
Downtime availability, time-interval versus, 181
Downtime minimization, 537–40
Downton, F., 541, 582
Dry bearings example, 335–36
Dugan, M. P., 189, 216
Durand, D., 313, 349
Dynamic random access memory device (DRAM) example, 369–71

E

Early failure region, 14
Edwards, D. G., 189, 216
Efficient estimator, 222
Eimar, B., 338, 349
Eisentraut, K. J., 573, 582
Elandt-Johnson, R. C., 245, 256
Electrical-discharge machining (EDM) example, 201–6
Electrical resistance, measuring, 124–26
Electromigration, 397
  examples, 43–45, 388
  model, 389
El-Neweihi, E., 81, 129, 148
Elsayed, E. A., 213, 216, 247, 249, 257, 371, 410, 550, 558, 559, 583
Engelhardt, M., 296, 297, 300, 302, 303, 313, 315, 317, 319, 349
Engelmaier, W., 391, 410
Equilibrium renewal process, 449
Erlang distribution, 36, 209, 452, 465, 504–5
Erlang loss formula, 554–55
Esaklul, K., 364, 410
Estimators
  consistent, 221
  efficient, 222
  point, 222
  sufficient, 222
  unbiased, 221
Event-space method, 103–4
Expected number of failures
  alternating, 444
  continuous time (nonparametric), 432–38
  continuous time (parametric), 416–28
  discrete time (nonparametric), 438–43
  discrete time (parametric), 428–32
Explosive detection system, case study, 601–8
Exponential distribution
  acceleration model, 356–58
  Bartlett's test, 265–66
  impact of Type 1 censoring on, 275–78
  impact of Type 2 censoring on, 278–80
  long failure times, testing for, 271–75
  maximum likelihood method for estimating, 237–39
  method of moments in estimating, 223–24
  parameter estimation, 265–80
  short failure times, testing for, 268–71
Exponential model hazard function, 26–27
Extreme value distribution, 26–27
  with censoring, 320–23
Eyring model, 380–83

F

Factoring algorithm, 108–12
Failure rate(s)
  instantaneous, 14
  mixture of, 49–52
Failure-time distributions, estimating parameters
  least-squares method, 222, 247–51
  likelihood method, 222, 230–47
  method of moments, 222–30
Fair, P. S., 582
Fang, P., 396, 411
Fatigue failures model, 391–92
Fatigue limit, estimating, 22–23
Feingold, H., 541, 581
Feldman, R. M., 540, 583
Feller, W., 85, 148
Fisher information matrix, 244–47
Flaherty, J. M., 391, 410
Flight data recorders example, 322–23, 329–30
Fluid monitoring, 572–73
Fostner, F., 193, 216
Freak failures, 268
Frees, E. W., 439, 440, 457, 474
Fubini's Theorem, 546
Fuh, D., 492, 494
Full rebate policy, 486–92
Fundamental renewal equation, 418
Furnace tubes reliability, case study, 609–14
Fussell-Vesely importance, 129–30

G

Gamma density, 312
Gamma distribution
  confidence intervals and, 319–20
  method of moments in estimating, 224–25
  parameter estimation, 312–20
  variance and, 317–19
  with censoring, 315–17
  without censoring, 312–14
Gamma function, 312
  table, 641–46
Gamma model hazard function, 35–39
Gandini, A., 126, 148

Gardner, J. W., 574, 582
Gas distribution system example, 109–11
Generalized Pareto model, 43
Generator regulator example, 193–95
Gnanadesikan, R., 315, 351
Gogus, O., 636
Gold-aluminum bonds example, 359–61
Gompertz distribution, 26
Gompertz-Makeham model, 43
Good-as-new repair policy, 504–7
Gossing, P., 401, 410
Government-Industry Data Exchange Program (GIDEP), 259
Gradient of likelihood method, 243
Greenberg, B. G., 306, 309, 350, 665
Greenwood, J. A., 313, 349
Gross, A. J., 306, 351
Grouchko, D., 121, 148
Ground earth station (GES), 615, 619, 622–23
Group maintenance, 557–61
Grubbs, F. E., 222, 257
Gryna, F. M., 439, 474
Gunn, J. E., 390, 410
Gurland, J., 50, 51, 67, 68

H

Hahn, G. J., 262, 350, 357, 375, 397, 410–11
Hale, P., 406, 410
Half-logistic distribution, 323–31
Hamilton, C. M., 634, 637
Harche, F., 81, 148
Harter, H. L., 289, 318, 349
Hassanein, K. M., 647, 661
Hastie, T. J., 378, 410
Hawkins, C. F., 189, 216
Hawkins, D. M., 222, 257
Hazard function(s)
  beta model, 40–42
  constant, 15–18
  cumulative, 14
  defined, 6, 14
  exponential model/extreme value distribution, 26–27
  gamma model, 35–39
  generalized Pareto model, 43
  Gompertz-Makeham model, 43
  linearly decreasing, 20
  linearly increasing, 18–20
  log-logistic model, 40
  lognormal model, 31–35
  mixed Weibull model, 24–26
  normal model, 27–31
  power series model, 43
  summary of, 57
  Weibull model, 20–24
Hazard rate(s)
  censoring and estimating, 264–65
  estimating, 7–11
  exponentially increasing, 11–14
  multivariate, 45–49
  roller-coaster, 51
Henley, E. J., 123, 126, 148, 196, 216
Hessian matrix, 244–47
Heyman, D. P., 546, 547, 582
High-frequency radio (HF), 615
Highly accelerated life testing (HALT), 262, 342
Highly accelerated stress screening (HASS), 59, 345
Highly accelerated stress testing (HAST), 358, 390
Hjorth, U., 62, 68
Holcomb, D. P., 174, 217
Homogeneous Poisson process (HPP), 462–63
Hot-carrier degradation model, 395–96
Hot standby, 119, 196
Hoyland, A., 460, 474
Hsiang, T., 550, 583
Hu, C., 366, 396, 409, 411
Humidity dependence failures model, 389–91
Hunter, L. C., 465, 474, 533, 557, 558
Huyett, M. J., 315
Hwang, C. L., 181, 183, 184, 217
Hwang, F. K., 92, 148
Hydraulic equipment example, 186–87

I

Importance measures
  Barlow-Proschan importance, 131
  Birnbaum's importance measure, 123–26, 134
  criticality importance, 126–28
  Fussell-Vesely importance, 129–30
  upgrading function, 131–34
Inactive redundancy, 118, 119–20, 196
Incident beam collimators, 602

Increasing failure rates (IFR), 49, 50, 51, 557
Infant mortality region, 14, 511
Information matrix, Fisher, 244–47
Inherent availability, 184
Inspection policy, periodic
  maintenance and, 565–70
  on-line surveillance and monitoring, 570–74
  optimum, 561–65
Instantaneous availability, 181
Instantaneous failure rate, 14
Integrated circuits (ICs) examples, 65, 210–11, 259, 357
  complementary metal-oxide-silicon (CMOS) examples, 358, 385–87
  electromigration model, 389
  fracture substrate example, 363–64
  gold-aluminum bonds example, 359–61
  humidity dependence failures model, 389–91
  metal-oxide semiconductor (MOS) failure of, 189–91, 365
  thermal fatigue crack example, 362–63
Inverse power rule model, 384–87

J

Jacks, J., 381, 387, 410
Jackson, B. S., 207, 217
Jardine, A. K. S., 65, 68, 416, 474, 528, 533, 534, 537, 538, 561, 563, 582
Jarvik heart, 4
Jensen, F., 261, 349
Johns, D., 187, 217
Johnson, N. L., 48, 68, 245, 256, 331, 349–50
Joint density function (j.d.f.), 189, 191–95
Juran, J. M., 439, 474

K

Kalbfleisch, J. D., 40, 68, 375, 410, 513, 514
Kamm, L. J., 131, 148
Kang, S-M., 395, 396, 410
Kao, J. H. K., 25, 68
Kaplan, E. L., 333, 350
Kapur, K., 269, 350
Karlin, S., 447, 474
Karmarkar, U. S., 511, 513, 525
Kaufmann, A., 121, 148
Kececioglu, D., 381, 387, 410
Keystone component, 96–97, 144
Kielpinski, T. J., 397, 410
Kinetic theory, 384
Klinger, D. J., 259, 260, 350
Ko, P. K., 396, 411
Kogan, J., 585, 637
Kolb, J., 439, 474
Kotz, S., 48, 68, 331, 349–50
k-out-of-*n* systems
  mean time to failure in, 165–67
  nonrepairable, 157–59
Kumamoto, H., 123, 126, 148, 196, 216
Kuo, W., 90, 148

L

Lam, Y., 541, 543, 544, 576, 582–83
Lamberson, L. R., 269, 350
Lambert, H. E., 131, 148
Lambiris, M., 90, 148
Laplace transform, 45, 418, 419
  of renewal density equation, 170–71
  state-transition equations and, 179
Laser degradation model, 394–95
Laser diodes (LD) example, 438
Laser printer example, 71–72
Lawless, J. F., 260, 350, 513, 514, 525
Least-squares method
  linear, 222, 247–51
  nonlinear, 247, 251
Leblebici, Y., 395, 396, 410
Lee, Elisa T., 276, 279, 290, 313, 350
Leemis, L. M., 55, 68, 322, 350
Lekens, G., 409–10
L'Hospital's rule, 198
Li, L., 457, 458, 474
Lie, C. H., 181, 183, 184, 217
Lieberman, G. L., 91, 92, 148
Likelihood method, 222, 230–47
  Fisher information matrix, 244–47
  gradient of, 243
  logarithmic values of, 230–33, 236, 238, 243
  maximum, 222, 236–42
  Newton's iterative method, 243
  variance-covariance matrix, 244–47
Lindley, D. V., 45, 48, 68
Linear least-squares method, 222, 247–51
Linearly decreasing hazard function, 20
Linearly increasing hazard function
  description of, 18–20

mean time to failure for, 53
mean time to failure in k-out-of-n systems with, 166
mean time to failure in parallel systems with, 163–64
mean time to failure in series systems with, 160
Linear models, 331–33
acceleration, 368–71
Litman, N., 636
Logarithmic values of likelihood method, 230–33, 236, 238, 243
Log-logistic model hazard function, 40
Lognormal distribution
acceleration model, 363–68
parameter estimation, 303–12
with censoring, 309–12
without censoring, 305–9
Lognormal model, 31–35
Lomax distribution, 48
Long failure times, testing for, 271–75
Lower confidence limit (LCL), 228
Lu, M-W., 400, 409
Lump-sum rebate, 481–86

M

McCollin, C., 378, 411
McPherson, J. W., 365, 366, 383, 410
Mactaggart, I., 406, 410
Mahlke, G., 401, 410
Maintenance. *See* Preventive maintenance, replacements, and inspection (PMRI)
Makino, T., 21, 68
Malik, S. K., 390, 410
Malon, D. M., 92–93, 148
Mamer, J. W., 476, 525
Mann, N. R., 384, 410
Markov models, 175
for dependent failures, 189–91
nonrepairable component, 176–78
repairable component, 178–80
semi-Markov process, 444
Marlow, N. A., 632, 634, 637
Massey, F. J., Jr., 222
Maximum likelihood method/estimators (MLE), 222, 236
for exponential distribution, 237–39
for normal distribution, 240–42
for parameter estimation, 262
for Rayleigh distribution, 239–40
Mays, L. W., 211, 217
Mean residual life (MRL), 55–56
Mean time between failure (MTBF), 52
Mean time to failure (MTTF)
defined, 52–54, 159
for k-out-of-n systems, 165–67
for other systems, 167–68
for parallel systems, 162–65
for series systems, 160–62
summary of, 169
Mean time to replace, 19
Mechanical fatigue, 397
Median time to failure (MTF), 389
Meeker, W. Q., 262, 350, 357, 397, 410–11
Meier, P., 333, 350
Membrane keyboard example, 187–88
Menendez, M. A., 259, 260, 350
Menke, W. W., 478, 483, 484, 525
Merz, R., 445, 474
Metal-oxide semiconductor (MOS), failure of, 189–91, 365
Method of moments, 222–30
Microcasting example, 445–46
MIL-HDBK-217D, 259
Miller, R. G., Jr., 369, 411
Minimal repair policy, 503, 540–45
Mission-oriented availability, 181, 185
Mixed-parallel series, 79–80
Mixed repair policy, 507–13
Mixed warranty policies, 486–92
Mixed Weibull model, 24–26
Mixture of failure rates, 49–52
Model identification, 262
Modified renewal process, 448–49
Monitoring. *See* On-line surveillance and monitoring
Monte Carlo simulation, 296–97
Moore, A. H., 289, 349
Mortenson, R. L., 207, 217
Motorettes example, 375–77
Multicensored data
cumulative-hazard estimator, 334–35
product-limit estimator, 333–34
Multistate models
parallel-series system, 116
parallel systems, 114–15
series-parallel system, 116–18
series systems, 112–14

Multivariate hazard rate, 45–49
Murthy, D. N. P., 476, 477, 502, 508, 510, 525, 550, 583

N

Nair, K. P. K., 81, 148
Nakada, Y., 259, 260, 350
Nakagawa, T., 566, 568, 583
NASA, 71, 148
Natrella, M. G., 222, 257
Natrella-Dixon test, 222
Nelson, W., 333, 350, 375, 397, 407, 410–11
Networks
  directed, 96
  undirected, 96
Newton-Raphson method
  computer listing of, 703–4
  description of, 290, 294, 295, 296, 323, 697–701
Newton's iterative method for likelihood method, 243
Nguyen, D. G., 476, 477, 502, 508, 510, 525
Niebel, B. W., 571, 572, 583
Niu, S. C., 84, 86, 148
Nondestructive testing (NDT), 570
Nondetection cost, 561, 566
Nonhomogeneous Poisson process (NHPP), 463–66
Nonparametric renewal function estimation
  continuous time, 432–38
  discrete time, 438–43
Nonrepairable products, warranties for, 477–97
Nonrepairable standby, 196
  multiunit, 198–200
  simple, 197–98
Nonrepairable systems
  examples of, 151–52
  k-out-of-*n* systems, 157–59
  parallel systems, 153–57
  series systems, 152–53
Normal distribution
  maximum likelihood method for estimating, 240–42
  method of moments in estimating, 226–28
  table for, 679–84
Normal model hazard function, 27–31
Number of spares
  availability and, 554–57
  determining, 549–57

O

O'Connor, L., 186, 217, 322, 350
Okumoto, K., 558, 559, 583
On-line surveillance and monitoring
  acoustic emission, 571
  corrosion monitoring, 573
  fluid monitoring, 572–73
  other diagnostic methods, 573–74
  sound recognition, 571
  temperature monitoring, 572
  vibration analysis, 570–71
Operational availability, 184–85
Operational life testing (OLT), 17, 260–61
Optimal replacements for items under warranty, 492–97
Ordinary free replacement warranty, 476
Ordinary renewal process, 448, 449
Outliers, 222
Ozbaykal, T., 446, 474

P

Papastavridis, S., 90, 134, 148
Parallel-series system
  description of, 77–78
  multistate components in, 116
Parallel systems
  description of, 75–77
  mean time to failure in, 162–65
  multistate components in, 114–15
  nonrepairable, 153–57
Parameter estimation, 262
  least-squares method, 222, 247–51
  likelihood method, 222, 230–47
  method of moments, 222–30
Parametric reliability models
  censoring, types of, 263–65
  exponential distribution, 265–80
  extreme value distribution, 320–23
  gamma distribution, 312–20
  half-logistic distribution, 323–31
  linear models, 331–33
  lognormal distribution, 303–12
  multicensored data, 333–36
  Rayleigh distribution, 280–89
  Weibull distribution, 289–303
Parametric reliability models, approaches to
  accelerated life testing, 261–63
  burn-in testing, 261
  failure data, use of, 259–60
  operational life testing, 260–61

Parametric renewal function estimation
continuous time, 416–28
discrete time, 428–32
Pareto distribution of the second kind, 48
Pareto model, generalized, 43
Park, K. S., 497, 498, 525–26
Partial-fraction-expansion formula, 433
Partial redundancy, 214
Parzen, E., 465, 474
Path-tracing method, 107–8
Pearson, K., 36, 68
Pearson Type V, 54, 63
Pearson Type VI, 48
Pease, R. W., 85, 148
Pelliccia, A., 137, 148
Periodic
See also Inspection policy, periodic replacement, 546–49
Permanent magnet synchronous motor (PMSM) example, 426–28
Petersen, N. E., 261, 349
Pham, H., 90, 148, 212, 217
Physics-experimental-based models, 388
electromigration model, 389
fatigue failures model, 391–92
humidity dependence failures model, 389–91
Physics-statistics-based models
Arrhenius model, 378–80
combination model, 387–88
Eyring model, 380–83
inverse power rule model, 384–87
Point availability, 181
Point estimator, 222
Pointwise availability, 173
Poisson distribution, likelihood function for, 232
Poisson processes
homogeneous, 462–63
nonhomogeneous, 463–66
Power series model hazard function, 43
Prasad, V. R., 81, 148
Prentice, R. L., 40, 375, 410
Preventive maintenance, replacements, and inspection (PMRI)
constant interval replacement policy, 529–33, 537–38
cost minimization, 528–37
downtime minimization, 537–40
function of, 527–28
group, 557–61
inspection policy, 561–70
minimal repair, 540–45
number of spares, determining, 549–57
on-line surveillance and monitoring, 570–74
optimum for systems subject to shocks, 545–49
replacement at predetermined age, 533–37, 538–40
Printed circuit boards (PCBs), 62, 250
Prinz, F. B., 445, 474
Probability density function
exponential, 16
of gamma distribution, 35
of log-logistic model, 40
of lognormal distribution, 31
Rayleigh distribution, 18–19
for standard normal distribution, 28
Production line design, case study, 592–600
Product-limit estimator (PLE), 333–34
Proportional hazards model (PHM), 371–78
Pro-rata warranty, 476
Proschan, F., 49, 68, 81, 131, 147, 148, 465, 474, 506, 525, 541, 557, 558, 563, 631, 636
Proschan, R., 82, 120
Prot method, 22–23
Puri, P. S., 631, 637

Q

Quader, K. N., 396, 411

R

Ramamurthy, K. G., 123, 148
Random censoring, 263
Rausand, M., 460, 474
Rayleigh distribution
acceleration model, 361–63
best linear unbiased estimator for parameters, 283–89
description of, 18–19, 21
maximum likelihood method for estimating, 239–40
parameter estimation, 280–89
variance, 281
with censored observations, 282–83
without censored observations, 280–82
Reduction method, 106–7, 112

Redundancy
active, 118–19, 196
allocation for a series system, 120–22
cold standby, 119
defined, 118
difference between active and inactive, 119–20
hot standby, 119
inactive, 118, 119–20, 196
partial, 214
system, 118–22, 196
warm standby, 119
Redundancy and standby
cold standby, 196
examples of, 196
hot standby, 196
nonrepairable, 196–97
nonrepairable multiunit, 198–200
nonrepairable simple, 197–98
repairable, 197, 201–6
warm standby, 196
Relative efficiency, 222
Reliability
block diagrams, 70–73
defining, 3, 4–5
estimating, 5–15
graph, 70, 71, 72
importance of, 3–4
Reliability function, 16
for exponential model hazard function, 26
for gamma model hazard function, 36
for linearly increasing hazard function, 18
for log-logistic model hazard function, 40
for the mixture of two increasing failure rates (IFR), 50
for normal model hazard function, 31
for power series model hazard function, 43
Renewal
availability analysis and, 421–28
confidence intervals, 457–59
density equation, 419
estimating, 446–48
function, 417
fundamental renewal equation, 418
remaining life at time, 459–62
variance of, 449–57
Renewal function estimation, nonparametric
continuous time, 432–38
discrete time, 438–43
Renewal function estimation, parametric
continuous time, 416–28
discrete time, 428–32
Renewal processes
alternating, 170–75, 443–46
equilibrium, 449
modified/delayed, 448–49
ordinary, 448, 449
Renewal theory approach, 416
Renner, K. M., 214, 217
Repair, minimal, 503, 540–45
Repairable products, warranties for, 497–513
Repairable standby, 197, 201–6
Repairable systems
alternating renewal process, 170–75
examples of, 169
Markov models, 175–80
Poisson processes, 462–66
Repeaters, examples using, 89–91, 555–57
bipolar transistors for, examples, 367–68, 387–88
Replacement
block, 529
constant interval replacement policy (CIRP), 529–33, 537–38
for items subject to shocks, 545–49
for items under warranty, optimal, 492–97
periodic (time-dependent cost), 548–49
periodic (time-independent cost), 546–48
at predetermined age, 533–37, 538–40
under minimal repair, 541–45
Resistance, measuring electrical, 124–26
Resistor degradation model, 392–94
Reverse-biased second breakdown (RBSB), 404
Rhine, W. E., 582
Ritchken, P. H., 486, 488, 491, 492, 494, 526
Rivera, R., 189, 216
Robinson, J. A., 513, 514, 525
Roggen, J., 409–10
Root mean square (RMS), 262
Ross, S. M., 91, 92, 148, 439, 494, 503, 526
Ross, S. S., 474
Runge-Kutta method, 180, 182, 695–96

S

Saleh, A. K., 647, 661
Sarhan, A. E., 306, 309, 350, 665
Satellite communications, 615, 621–22
Scattered beam collimators, 602, 604

Schafer, R. E., 384, 410
Schendel, U., 180, 217
Sears, R. W., 197, 217
Semi-Markov process, 444
Senju, S., 533, 583
Sensors, examples using, 284–87
Series-parallel system
  description of, 78–79
  multistate components in, 116–18
Series systems
  description of, 73–75
  mean time to failure in, 160–62
  multistate components in, 112–14
  nonrepairable, 152–53
  redundancy allocation for, 120–22
Seth, A., 123, 148
Sethuraman, J., 50, 51, 67, 68, 81
Shake-down region, 14
Shanthikumar, J. G., 90, 148
Shapiro, S., 207, 217, 306, 351
Shaw, M., 406, 410
Shepard, C., 189, 216
Shooman, M. L., 96, 103, 148, 156, 176, 192, 217
Short failure times, testing for, 268–71
Shortwave radios example, 480–81, 483–86
Siemens Aktiengesellschaft, 72, 148
Silicon controlled rectifier (SCR), 64
Singh, C., 112, 148
Singpurwalla, N. D., 45, 48, 68, 384, 410
Sirocky, W. F., 207, 217
Sobel, M. J., 546, 547, 582
Soden, J. M., 189, 216
Sös, V. T., 92, 148
Sound recognition, 571
Spectrographic emission, 572, 573
Stadje, W., 541, 542, 583
Stals, L., 409–10
Standard normal distribution, 28
Standby. *See* Redundancy and standby
Statistic-based nonparametric models
  linear model, 368–71
  proportional hazards model, 371–78
Statistic-based parametric models, 355
  exponential distribution acceleration model, 356–58
  lognormal distribution acceleration model, 363–68
  Rayleigh distribution acceleration model, 361–63
  Weibull distribution acceleration model, 359–61
Steady-state availability, 171–72, 183–84
Strain-gauge technique, 564–65
Stress testing, 59, 261–63
Structure function, 122–23
Sufficient estimator, 222
Surface mount technology (SMT) example, 250–51, 391
Suzuki, N., 394, 411
Swartz, G. A., 189, 217, 442, 474
System
  defined, 69
  impact of components on, 69
  redundancy, 118–22, 196
  reliability block diagrams used to evaluate, 70–73
  structure function, 122–23
System configurations
  See also under type of
  complex reliability, 96–12
  consecutive-k-out-of-n:F, 83–96
  k-out-of-n, 157–59, 165–67
  mixed-parallel, 79–80
  multistate models, 112–18
  optimal assignment of components, 80–83
  optimal assignment of components in consecutive-2-out-of-n:F, 91–93
  parallel, 75–77, 114–15
  parallel-series, 77–78, 116
  series, 73–75, 112–14
  series-parallel, 78–79, 116–18
  types of, 69
System design using reliability objectives, case study, 625–36

T

Taguchi, G., 550, 583
Takacs, L., 631, 637
Takata, S., 571, 583
Takeda, E., 394, 411
Taylor, H. M., 447, 474
Taylor's expansion, 243, 432
Telecommunication networks reliability, case study, 615–24
Telecommunication system example, 94–96
Temperature acceleration testing examples, 373–77
  Arrhenius model, 378–80
  Eyring model, 380–83

Temperature monitoring, 572
Test censoring time, 263
Thermal fatigue crack example, 362–63, 397
Thermocouple example, 198–200
Thin layer activation (TLA), 573
Thomas, M. U., 488, 526
Thompson, W. A., 541, 583
Thornton, T. J., 582
Thresher, U.S.S., 3
Tibshirani, R. J., 378, 410
Tie-set method, 100–102
Tielemans, L., 409–10
Tillman, F. A., 181, 183, 184, 217
Tilquin, C., 541, 583
Time-dependent dielectric breakdown (TDDB), 189–91, 365, 442
Time-dependent equations, computer program for solving, 695–96
Time-dependent reliability estimates
  See also Mean time between failure (MTBF)
  alternating renewal process, 170–75
  Markov models, 175–80
  nonrepairable systems, 151–59
  repairable systems, 169–80
Time-interval versus downtime availability, 181
Time to failure (TTF), 159, 388–92
Tobias, P. A., 355, 392, 411
Tofield, B. C., 581–82
Tomography example, computer, 70–71
Top down heuristic (TDH), 81–82
Tordan, M. J., 534, 537, 582
Tortorella, M., 392, 394, 409
Transistors, example of testing, 112, 266–68
Trindade, D., 355, 392, 411
Trivedi, K. S., 422, 474
Truth table method, Boolean, 104–6
Turbine example, 544–45
Type 1 censoring, 263
  impact on estimation, 275–78
Type 2 censoring, 263
  impact on estimation, 278–80

U

Unbiased estimator, 221
  Weibull distribution parameters and, 296, 297–300
Unbiasing factor, 297
Undirected networks, 96
Unlimited free replacement warranty, 476
Upgıading function, 131–34
Upper confidence limit (UCL), 228

V

Valdez-Flores, C., 540, 583
Vanhecke, B., 409–10
Varadan, J., 322, 349
Variance-covariance matrix, 244–47
Variance of number of renewals, 449–57
Variance of parameters, 661–64
Very large-scale integrated (VLSI) circuits, 395–96
Vibrations
  analysis of, 570–71
  excessive, 27
Video display terminals (VDTs) example, 307–9
Von Alven, W. H., 115, 148

W

Walski, T. M., 137, 148
Warm standby, 119, 196
Warranties
  estimating warranty cost, 477–81
  for fixed lot size (arbitrary failure-time distribution), 501–3
  for fixed lot size (good-as-new repair policy), 504–7
  for fixed lot size (minimal repair policy), 503
  for fixed lot size (mixed repair policy), 507–13
  full rebate policy, 486–92
  importance of, 475
  lump-sum rebate, 481–86
  mixed policies, 486–92
  for nonrepairable products, 477–97
  optimal replacements for items under, 492–97
  ordinary free replacement, 476
  pro-rata, 476
  for repairable products, 497–513
  time constraints on, 475
  unlimited free replacement, 476
Warranty claims, 513
  for grouped data, 519–20
  with lag times, 514–19
Wear-out region, 15
Wei, V. K., 92, 148

Weibull distribution
acceleration model, 359–61
confidence interval and, 300–303
parameter estimation, 289–303
unbiased estimates for parameters, 296, 297–300
variance of maximum likelihood estimates, 296–97
with censoring, 295
without censoring, 290–95
Weibull hazard
mean time to failure in *k*-out-of-*n* systems with, 167
mean time to failure in parallel systems with, 164–65
mean time to failure in series systems with, 161–62
Weibull model, 20–24
mean time to failure for, 53
mixed, 24–26
Weiss, L. E., 445, 474
Wetherill, G. B., 234, 235, 243, 251, 257
White, G. L., 404, 411
Wightman, D., 378, 411
Wilk, M. B., 315, 351
Wilkins, N. J. M., 581–82
Wong, K. H. T., 327, 349
Wong, K. L., 51, 68
Work-mission availability, 187

X

X-ray generator, 601–2
x^2, critical values of, 709–11

Y

Yadigaroglu, G., 131, 148
Yee, S. R., 497, 526
Yue, J. T., 396, 411
Yule process, 547

Z

Zmani, N., 389, 409
Zuckerman, D., 541, 542, 583
Zuo, M., 90, 148

ERAU - Prescott Library

Addison Wesley Longman, Inc. warrants the enclosed disk to be free of defects in materials and faulty workmanship under normal use for a period of ninety days after purchase. If a defect is discovered in the disk during this warranty period, a replacement disk can be obtained at no charge by sending the defective disk, postage prepaid, with proof of purchase to:

Addison Wesley Longman, Inc.
Editorial Department
Corporate & Professional Publishing Group
One Jacob Way
Reading, Massachusetts 01867

After the 90-day period, a replacement will be sent upon receipt of the defective disk and a check or money order for $10.00, payable to Addison Wesley Longman, Inc.

Addison Wesley Longman, Inc. makes no warranty or representation, either express or implied, with respect to this software, its quality, performance, merchantability, or fitness for a particular purpose. In no event will Addison Wesley Longman, Inc., its distributors, or dealers be liable for direct, indirect, special, incidental, or consequential damages arising out of the use or inability to use the software. The exclusion of implied warranties is not permitted in some states. Therefore, the above exclusion may not apply to you. This warranty provides you with specific legal rights. There may be other rights that you may have that vary from state to state.

The Reliability Analysis Software™ was designed to run on IBM PCs or compatibles with Windows 3.1. It requires a 386 or greater CPU, 3MB of free hard disk and at least 4MB of RAM. Questions, comments, or technical support regarding the Reliability Analysis Software™, please contact:

Industrial Engineering Department
Rutgers University
Phone: (908) 445-3654
Fax: (908) 445-5467
Email: elsayed@gandalf.rutgers.edu